SILICON PROCESSING
FOR
THE VLSI ERA

VOLUME 1:

PROCESS TECHNOLOGY

SILICON PROCESSING

FOR

THE VLSI ERA

VOLUME 1:

PROCESS TECHNOLOGY

STANLEY WOLF Ph.D.
Professor, Department of Electrical Engineering
California State University, Long Beach
Long Beach, California
and
Instructor, Engineering Extension, University of California, Irvine

RICHARD N. TAUBER Ph.D.
Manager of VLSI Fabrication
TRW - Microelectronics Center
Redondo Beach, California
and
Instructor, Engineering Extension, University of California, Irvine

LATTICE
PRESS
Sunset Beach, California

DISCLAIMER

This publication is based on sources and information believed to be reliable, but the authors and Lattice Press disclaim any warranty or liability based on or relating to the contents of this publication.

Published by:

Lattice Press,
Post Office Box 340
Sunset Beach, California 90742, U.S.A.

Cover design by Roy Montibon and Donald Strout, Visionary Art Resources, Inc., Santa Ana, CA.

Library of Congress Cataloging in Publication Data
Wolf, Stanley   and   Tauber, Richard N.

Silicon Processing for the VLSI Era
  Volume 1 : Process Technology

Includes Index
1. Integrated circuits-Very large scale
 integration. 2. Silicon. I. Title

86-081923

ISBN 0-961672-3-7

Reprinted with Corrections June, 1987

9 8 7 6 5 4 3 2

PRINTED IN THE UNITED STATES OF AMERICA

To my wife, Carrol Ann,
and my children, Jennifer Laura and Stanley Charles Ross

Stanley Wolf

———————

To my wife, Barbara

Richard N. Tauber

———————

PREFACE

SILICON PROCESSING FOR THE VLSI ERA is a text designed to provide a comprehensive and up-to-date treatment of this important and rapidly changing field. The text will consist of two volumes of which this book is the first, subtitled, *Process Technology*. Volume 2, subtitled, *Manufacturing Technology* is scheduled for publication in 1988. In this first volume, the individual processes utilized in the fabrication of silicon VLSI circuits are covered in depth (e.g. epitaxial growth, chemical vapor and physical vapor deposition of amorphous and polycrystalline films, thermal oxidation of silicon, diffusion, ion implantation, microlithography, and etching processes). In addition, chapters are also provided on technical subjects that are common to many of the individual processes, such as vacuum technology, properties of thin films, material characterization for VLSI, and the structured design of experiments for process optimization. The topics covered in the book are listed in more detail in the Table of Contents. In Volume 2, *Manufacturing Technology*, other issues related to VLSI fabrication such as process integration, process simulation, manufacturing yield, VLSI manufacturing facilities, assembly, packaging, and testing will be covered.

The purpose of writing this text was to provide professionals involved in the microelectronics industry with a single source that offers a complete overview of the technology associated with the manufacture of silicon integrated circuits. Other texts on the subject are available only in the form of specialized books (i.e. that treat just a small subset of all of the processes), or in the form of edited volumes (i.e. books in which a group of authors each contributes a small portion of the contents). Such edited volumes typically suffer from a lack of unity in the presented material from chapter-to-chapter, as well as an unevenness in writing style and level of presentation. In addition, in multi-disciplinary fields, such as microelectronic fabrication, it is difficult for most readers to follow technical arguments in such books, especially if the information is presented without defining each technical "buzzword" as it is first introduced. In our book we hope to overcome such drawbacks by treating the subject of VLSI fabrication from a unified and more pedagogical viewpoint, and by being careful to define technical terms when they are first used. The result is intended to be a *user friendly* book for workers who have come to the semiconductor industry after having been trained in but one of the many traditional technical disciplines.

An important technical breakthrough has occurred in publishing, that the authors also felt could be exploited in creating a unique book on silicon processing. That is, revolutionary electronic publishing techniques have recently become available, which can cut the time required to produce a published book from a finished manuscript. This task traditionally took 15-18 months, but can be now reduced to less than 4 months. If *traditional* techniques are used to produce books in such fast-breaking fields as VLSI fabrication, these books automatically possess a built-in obsolescence, even upon being first published. The authors took advantage of the *rapid production techniques,* and were able to successfully meet the reduced production-time schedule. As a result, they were able to include information contained in technical journals and conferences which was available within four months of the book's publication date. Earlier books written on silicon processing, unfortunately, suffer from having had to undergo an 15-18 month production cycle. *This is the first book on the subject in which such time-delay effects have been eliminated from the production process!*

Written for the professional, the book belongs on the bookshelf of workers in several microelectronic disciplines. *Microelectronic fabrication engineers* who seek to develop a more complete perspective of the subject, or who are new to the field, will find it invaluable. *Integrated circuit designers, test engineers,* and *integrated circuit equipment designers,* who must understand VLSI processing issues to effectively interface with the fabrication environment, will also find it a uniquely useful reference. The book should also be very suitable as a text for graduate-level courses on silicon processing techniques, offered to students of electrical engineering, applied physics, and materials science. It is assumed that such students already possess a basic familiarity with semiconductor device physics. Problems are included at the end of each chapter to assist readers in gauging how well they have assimilated the material in the text.

The book is an outgrowth of an intensive seminar conducted by the authors through the Engineering Extension of the University of California, Irvine. In the first three years that it was offered, over three hundred engineers and managers from more than 75 companies and government agencies, enrolled in the course. Its fine reputation is attested to by the fact that many firms have sent participants each time the course has been offered, presumably based on the recommendations of earlier enrollees.

In setting out to create a comprehensive text on VLSI fabrication, the authors each contributed a set of unique and complementary skills to the project. Professor Wolf's proficiency as a teacher and writer were utilized to produce a clearly written and logically organized book. Some of this expertise was gained in authoring an earlier best-selling text *Electronic Measurements and Laboratory Practice,* Prentice-Hall, 1983. Dr. Tauber brought a technical expertise acquired from his long involvement in the semiconductor industry. He used this background to insure that the most important topics of VLSI fabrication were addressed, and that the information was up-to-date and presented in a technically correct fashion. Note that for over twenty years, Dr. Tauber has held positions at Bell Telephone Laboratories, Xerox, And Hughes Aircraft Company. Currently he directs advanced VLSI Fabrication efforts at the Microelectronics Center of TRW. The labor of the writing effort was divided between the authors in the following manner: Professor Wolf was responsible for writing Chapters 1, 2, 3, 9, 10, 12, 13, 15, 16, 17, and 18, and Dr. Tauber undertook the writing of Chapters 4, 5, 7, 8, 11, and 14. Material for Chap. 6 was jointly contributed by Andrew R. Coulson and Dr. Tauber.

A book of this length and diversity would not have been possible without the indirect and direct assistance of many other workers. To begin with, virtually all of the information presented in this text is based on the research efforts of a countless number of scientists and engineers. Their contributions are recognized to a small degree by citing some of their articles in the references given at the end of each chapter. The direct help came in a variety of forms, and was generously provided by many people. The text is a much better work as a result of this aid, and the authors express heartfelt thanks to those who gave of their time, energy, and intellect.

Each of the chapters was reviewed after the writing was completed. The engineers and scientists who participated in this review were numerous. Special thanks are given to Leonard Braun and Ethelyn Motley, who provided extensive and incisive editing services. We also thank Warren Flack, Stephen Franz, Kenneth Tokunaga, Dean Denison, Simon Prussin, and Vitus Matare for their critical reviews. Simon Prussin also provided clarification of many concepts during the course of numerous technical discussions with the authors, and can be rightly considered as being the technical underpining of Chapters 1, 2, 8, and 9. Extra thanks are also extended to Mark Miscione for bringing valuable technical input on the subject of the physics of microlithography, to Susan Curry for donating SEM photographs, and to Andrew Coulson for creating some of the drawings. Ada Mae Hardeman, of the Engineering Extension of the University of California, Irvine is owed special thanks for helping to make a success of the seminar from which this book grew. Otto Gruneberg, of QBI, Inc. was also a benefactor of the project. He graciously agreed to share his exhibition space with Lattice Press at Semicon-West, 1986, where the book made its debut.

Superlative computer support and access to computer resources was generously made available by Donald E. Carlile, Harry T. Hayes, and Dale Lambertson of the Personal Computer Support Section of the Electronic Systems Group of TRW. Henry Nicholas was a computer expert and friend who lit the fire of inspiration that led to the undertaking of the project. The management of the Electronics Systems Group and the Microelectronics Center of TRW, including most notably Dr. Barry Dunbridge and Phillip Reid, are warmly thanked for providing a supportive environment, conducive to producing such an intensive technical project. They made available technical literature and other resources to the authors, especially S. Wolf, who was able to avail himself of this generosity while writing during a Sabbatical leave from his teaching duties at California State University, Long Beach. Roy Montibon and Donald Strout of Visionary Art Resources, Inc., Santa Ana, CA designed the cover. Finally, we wish to thank Shirley Rome, Carrol Ann Wolf, and Barbara Tauber for typing the manuscript.

Stanley Wolf and Richard N. Tauber

P.S. Additional copies of the book can be obtained from:

Lattice Press,
P.O. Box 340-V
Sunset Beach, CA, 90742

An order form, for your convenience, is provided on the final leaf of the book.

CONTENTS

PROLOGUE

Since the creation of the first integrated circuit in 1960, there has been an ever increasing density of devices manufacturable on semiconductor substrates. Silicon technology has remained the dominant force in integrated circuit fabrication and is likely to retain this position for the forseeable future. The number of devices manufactured on a chip exceeded the generally accepted definition of *very large scale integration,* or *VLSI* (i.e. more than 100,000 devices per chip), somewhere in the mid-70's (Fig. 1a). By 1986, this number had, in fact, grown to over 1 million devices per chip. The increasing device count was accompanied by a shrinking minimum feature size (Fig. 1b), which is expected to be smaller than 1 μm before 1990. Progress in VLSI manufacturing technology seems likely to continue to proceed in this manner, and even further reductions in the unit cost per function, and in the power-delay product of VLSI devices, are projected. The entire adventure of VLSI represents a remarkable application of scientific knowledge to the requirements of technology, and this book represents an enthusiastic report on the state-of-the-art of VLSI silicon processing, as practiced at the time of publication.

Figure 2 illustrates the sequence of steps that occurs in the course of manufacturing an integrated circuit. These steps can be grouped into two phases: 1) the design phase; and b) the fabrication phase. This book is concerned only with the fabrication phase of this undertaking, but it is useful to briefly outline the steps of the design phase. This provides a context which allows the reader to perceive the role of silicon processing, within the totality of integrated circuit manufacturing. Readers wishing to explore steps of the design phase in further detail, are referred to other technical literature, including the texts listed in Refs. 1-3 at the end of this prologue.

In the *design phase* of this sequence, the desired functions and necessary operating specifications of the circuit are initially decided upon. The chip is then designed from the "top

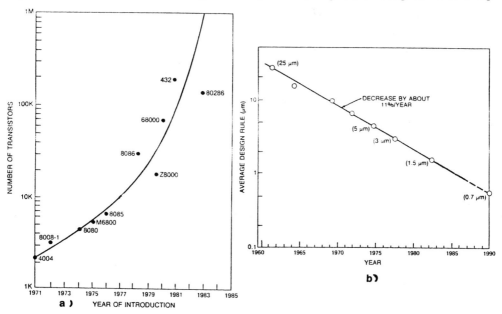

Fig. 1 (a) Increase in the number of transistors per microprocessor chip versus year of introduction, for a variety of 8-bit and 16 bit microprocessors, and (b) The decrease in minimum device feature size versus time on integrated circuits.

down". That is, the required large *functional blocks* are first identified. Next, their *sub-blocks* are selected, and then the *logic gates* needed to implement the sub-blocks are chosen. Each logic gate is designed by appropriately connecting devices (transistors, and in some cases resistors as well) that are ultimately slated for fabrication on the Si wafers. Upon completing these various levels of design, each is checked to insure that correct functionality has been achieved. When the designers are satisfied that each level of the design is correct, *test vectors*, that will be used to test the manufactured circuits, are generated from the schematic of the logic gates.

The circuit is then layed out. The *layout* consists of sets of patterns that will be transferred to the silicon wafer. The patterns correspond to device regions, or interconnect structures, and such patterns are sequentially transferred to the wafers through the use of photolithographic processes and a set of masks, as part of the wafer fabrication sequence (Fig. 4a). The result of each pattern transfer step is a set of features created on the wafer surface. These features are

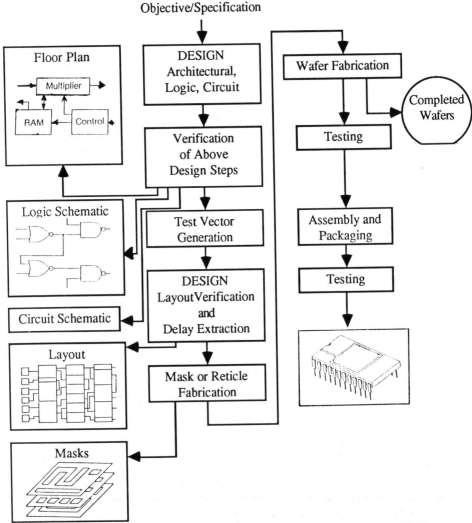

Fig. 2 Steps required for the manufacture of very large scale integrated circuits (VLSI).

generally either in the form of: a) an etched opening in a film (or region of the substrate); or b) a patterned feature of a film present on the surface (e.g. an interconnect line or pad).   After the openings (or *windows*) are created by the pattern transfer step, either: a) controlled quantities of dopant are added to the silicon substrate through the openings, or b) another layer is deposited that makes contact to the underlying layer through the opening. In any case, device regions, and structures that interconnect device regions, are produced by the patterning processes and associated fabrication steps. A cross-section of a completed device, resulting from having carried out a sequence of such fabrication steps is shown in Fig. 4b.

The creation of the layout proceeds from the "bottom up". That is, a variety of typical devices are first layed out. Then, a set of cells representing the required primitive logic gates are created by interconnecting appropriate devices. Next, sub-blocks are generated by connecting these logic gates, and finally the functional blocks are layed out by connecting the sub-blocks. Power busses, clock-lines, input-output pads required by the circuit design are also incorporated during the layout process. The completed layout is then subjected to a set of *design rule checks* and *propagation delay simulations* to verify that correct implementation of the circuit has been

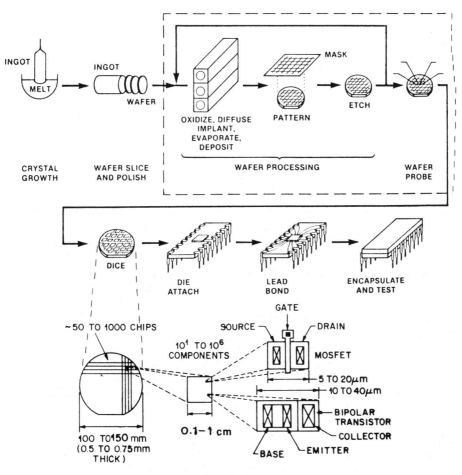

Fig. 3 The fabrication process sequence of integrated circuits.

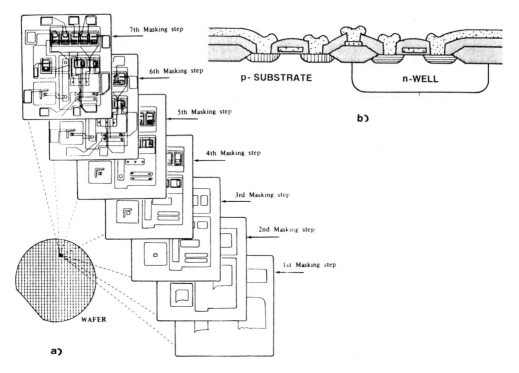

Fig. 4 (a) Example of the patterns transferred to a wafer during a seven-mask process sequence, and (b) Cross section of completed devices in a basic CMOS process.

achieved in layout form. After this checking procedure is completed, the layout information is ready to be used to generate a set of masks that will serve as tools for specifying the circuit patterns on silicon wafers. For VLSI circuits, layout information is stored on a computer.

In this two volume text, we describe the integrated circuit manufacturing steps that occur from the point at which the layout information becomes available (Fig. 3). That is, the first step of the manufacturing process considered in this book, is the set of the procedures utilized to convert the layout information, as stored on the computer, into a set of masks or reticles. All of the individual fabrication processes associated with creating patterns, introducing dopants, and depositing films on silicon substrates (to form integrated circuit features) are also subjects of this volume of our text. In addition, information about the manufacture of Si wafers (i.e. the *starting material* on which integrated circuits are fabricated), and direct supporting technologies (e.g. vacuum technology and material characterization) is also presented. In Volume 2 of this text, the remaining aspects silicon processing are covered, including process simulation, process integration, assembly, packaging, test, and it is thus aptly subtitled *Manufacturing Technology*.

REFERENCES

1. C. Mead & L. Conway, *Introduction to VLSI Systems.* Addison-Wesley, Reading, MA. 1980.
2. S. Muroga, *VLSI Systems Design,* John Wiley & Sons, New York, 1983.
3. A. Mukherjee, *Introduction to nMOS and CMOS VLSI Systems Design,* Prentice-Hall, 1986.

1

SILICON :

SINGLE CRYSTAL GROWTH

and WAFER PREPARATION

Silicon (Si) is presently the most important semiconductor for the electronics industry, with VLSI technology being based almost entirely on silicon. The name *silicon* comes from the Latin *silex* or *silicis* meaning flint. Second in abundance only to oxygen, silicon comprises 25.7% of the earths crust. The parameter values of some of the most useful properties of silicon[1] are given in Appendix 2.

Solid-state electronics was launched with the invention of the bipolar transistor by Bardeen, Brattain, and Shockley in 1947[2]. Germanium (Ge) was the original semiconductor material used to fabricate diodes and transistors. The narrow bandgap of Ge (0.66 eV), however, causes reverse-biased p-n junctions in Ge to exhibit relatively large leakage currents. This limits Ge device operation to temperatures below about 100°C. In addition, integrated circuit planar processing requires the capability of fabricating a passivation layer on the semiconductor surface. Germanium oxide (GeO_2) could act as such a layer but it is difficult to form, is water soluble, and dissociates at 800°C. These limitations make Ge an inferior material for the fabrication of integrated circuits, compared to silicon. The larger bandgap of silicon (1.1 eV) results in smaller leakage currents and thereby allows silicon devices to be built with maximum operating temperature of about 150°C. The oxide of silicon, SiO_2, is also easy to form and is chemically very stable. Finally electronic grade silicon is about one tenth as costly as germanium. As a result, Si has almost completely replaced Ge for fabricating microelectronic components.

In this chapter we consider the subject of *single-crystal silicon starting material*. We first show how raw material is processed to obtain electronic grade polysilicon. Then we examine the Czochralski and float-zone methods of growing single crystal silicon from such polysilicon. Next, the technology for forming the *silicon wafers* from single crystal ingots is covered. Finally, a discussion of the material properties that such silicon wafers must possess in order to be suitable for VLSI (and ULSI) fabrication is presented. In Chap. 2, we cover the crystalline defects that occur in silicon, and techniques for suppressing their formation.

TERMINOLOGY OF CRYSTAL STRUCTURE

Solid matter exists in crystalline and amorphous forms. In crystalline solids, the atoms which make up the solid are spatially arranged in a periodic fashion. If the periodic arrangement exists throughout the entire solid, the substance is defined as being formed of a *single crystal*. If

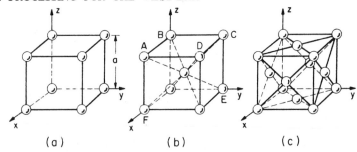

Fig. 1 Cubic-crystal unit cells. (a) Simple cubic. (b) Body-centered cubic. (c) Face-centered cubic.

the solid is composed of a myriad of small single crystal regions the solid is referred to as *polycrystalline* material. Nevertheless, within any single crystal region a crystal appears exactly the same at one point as it does at a series of equivalent points, once the basic periodicity of the crystal is recognized. *Amorphous* materials are devoid of long-range periodic structure. The starting-silicon material (wafers) on which VLSI circuits are fabricated, have single crystal form.

The periodic arrangement of atoms in a crystal is called the *lattice*. The crystal lattice always contains a volume which is representative of the entire lattice (referred to as a *unit cell*), and it is regularly repeated throughout the crystal. As shown in Fig. 1, for each unit cell, vectors can be defined (*basis vectors*) such that if the unit cell is translated by integral multiples of these vectors, a new unit cell identical to the original is found. The importance of the unit cell is that the crystal as a whole can be studied by analyzing a representative volume. For example, distances of nearest neighbor atoms can be calculated, as can the periodic properties of the crystal which define the allowed energy levels for conduction electrons.

The simplest lattices are the three cubic lattices seen in Fig. 1 (simple cubic, body-centered cubic, and face-centered cubic). The dimension **a** for a cubic cell is called the lattice constant, but the distance of nearest neighbor atoms may be smaller than **a**. For example, in the face-centered cubic (fcc) lattice the nearest neighbor distance is one half the diagonal of a cube face, or $(a\sqrt{2})/2$.

The directions in a lattice are expressed as a set of three integers with the same relationship as the components of a vector in that direction. The three vector components are given in multiples of the basis vectors. For example in cubic lattices, the body diagonal has the components of 1a, 1b, and 1c. Therefore, this diagonal exists along the [111] direction. (The [ ] brackets are used to denote a specific direction.) From a crystallographic point of view, however, many directions in a crystal are equivalent, depending only on the arbitrary choice of orientation for the axes. Such equivalent directions are expressed with angular brackets < >. For example the crystal directions in the cubic lattice [100], [010], and [001] are all crystallographically equivalent (Fig. 1), and are referred to as <100> directions.

It is also useful to describe *planes* in a crystal. A set of three integers h, k, and l, called the *Miller indices* [3], are used to define a set of parallel planes. These are derived as follows (Fig. 2):

a) The intercepts of the plane with the crystal axes are found, and these intercepts are expressed as integral multiples of the basis vectors.

b) The reciprocal of the three integers found in step a) are taken.

c) The smallest set of integers h, k, and l are found that have the same relationship to each other as the three reciprocals (i.e., if 1/4, 1/3, and 1/2 are the three reciprocals, then 3, 4, and 6 are the three smallest integers whose relative values are the same as the relative values of the three reciprocals). These integers are the *Miller indices* of the plane, and the plane is labeled (h k l). Note that ( ) parentheses are used to denote a specific plane.

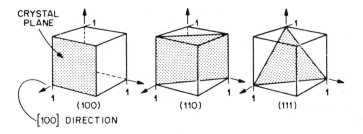

Fig. 2   Miller indices of some important planes in a cubic crystal.

One advantage of taking reciprocals of intercepts is that infinities are avoided in the notation that identifies a plane. The reciprocal of the intercept of a plane that is parallel to an axis (and thus would intercept the axis at an infinite distance) is zero. From a crystallographic point of view, as in the case of directions, many planes in a lattice are equivalent. The indices of such equivalent planes are collectively indicated by one set of integers, placed within { } parentheses. It should also be noted that for *cubic lattices*, direction [h k l] is perpendicular to a plane with the identical three integers (h k l). This fact makes it convenient to analyze cubic cells. That is, if either a direction or a plane is known, its perpendicular counterpart can be quickly determined without further calculation.

Silicon has a *diamond cubic lattice*, a structure which can be represented as two interpenetrating face-centered cubic lattices (Fig. 3a). Thus the simplicity of analyzing and visualiz-

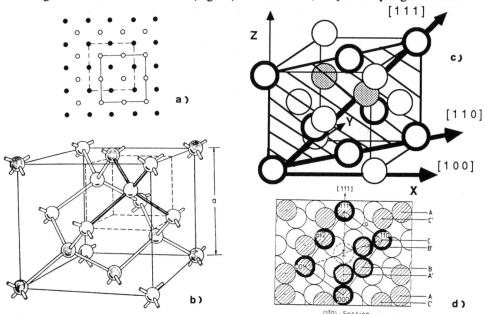

Fig. 3   (a) Top view (along any <100> direction) of an extended diamond lattice cell. The open circles indicate one fcc sublattice and the solid circles indicate the interpenetrating fcc. (b) Schematic representation of the diamond lattice unit cell. (c) A (110) section through the diamond lattice unit cell[8]. (d) (110) section showing atoms in the unit cell of (c) and neighboring atoms in the same (110) plane. Reprinted with permission of Springer-Verlag.

ing cubic lattices can be extended to characterization of silicon crystals. The three dimensional representation of a diamond-lattice unit cell is shown in Fig. 3b. In studying the diamond structure unit cell, it is evident that each atom has four nearest neighbors. The importance of this fact becomes evident when the electronic structure of silicon is considered. That is, since Si is a column IV element and has four valence electrons, each of these electrons is shared with one of its four nearest neighbor Si atoms. (This nearest neighbor atom likewise contributes one of its electrons to the bond.) The two valence electrons shared between nearest neighbors form a covalent bond, and such interactions of electrons in neighboring atoms of a solid serve the function of holding the crystal together. Since all the valence electrons in a perfect Si crystal are involved in shared bonding pairs, none are available as free electrons for conducting electrical current (as long as they remain localized in the bonding region). Such covalently bonded electrons can, however, be thermally or optically excited, and thereby become free to participate in conduction. (Note that in Si, $a = 5.43$ Å, and the nearest neighbor distance is 2.43 Å).

On the other hand, when dopant atoms (mostly from Group III or V) are substituted for Si atoms in the crystal lattice, they are said to occupy *substitutional* lattice sites. Such dopants can also become the source of additional conduction electrons or holes. For example, phosphorus is a substitutional *donor*, because four of its valence electrons become covalently bonded with other Si atoms. However, the fifth valence electron is easily separated, and once free, becomes available to conduct current. Group III elements (e.g. boron, aluminum) are substitutional *acceptors*, and become sites leading to *hole* conduction. Atoms that do not occupy lattice sites are described as occupying *interstitial* (e.g. between lattice sites) sites.

It is useful to consider the diamond lattice in further detail. For example, instead of describing it as two interpenetrating face-centered cubic lattices, it is crystallographically more precise to consider it as a *face-centered cubic lattice with two basis atoms at each lattice point*. Thus, to construct a diamond lattice unit cell, one would erect a face-centered cube, and place two atoms at each lattice site (one directly on the site, and the other offset from the first at a distance $a/4$, along all axes). The result would again be unit cell shown in Fig. 3b. It should be noted that the specific structure of crystal silicon arises from having to simultaneously satisfy two different conditions: 1) that each atom in the crystal has 4 nearest neighbors; and 2) that the atoms also be as closely packed in space as condition 1) will allow. The packing density of the *diamond lattice* is much lower than that of the *face-centered cubic* lattice (34% versus 74%).

In the study of several processing technologies, reference to various planes and directions in silicon crystals will be made, especially to the (100), (110), and (111) planes and directions. Direct examination of the diamond lattice unit cell does not allow easy conceptualization of the {110} and {111}-planes and directions. Thus, we shall pursue a simple procedure to assist the reader in visualizing them. In order to determine the {111}-planes most easily, we will make use of the concept that in *cubic lattices* a crystal direction [h k l] is perpendicular to the plane (h k l).

To begin, we draw a section of a {110}-plane of the diamond lattice unit cell as in Fig. 3c. Note the $a\sqrt{2}$ dimension is the length of a cube face diagonal [e.g. along the (110)-direction]. All atoms shown in Fig. 3c exist along the {110}-plane thus selected. Next, we observe that a [111]-direction exists along the diagonal of the {110}-plane (Fig. 3c). Finally planes perpendicular to <111> directions can be identified as {111}-planes.

In Fig. 3d, we add the additional neighboring atoms in the crystal that are in the {110}-plane, but are outside the unit cell (shaded circles). The other atoms in this figure (unshaded circles) lie in {110}-planes below the {110}-surface plane which we have cut as our section. From this diagram it can be seen that there are actually three double {111}-planes in the diamond lattice (A, A'; B, B'; C, C') arranged in a regularly repeating *stacking order*. (If one were to examine a set of hard spheres arranged in a close packed fcc structure, the third plane of three stacked{111}-planes would lie such that the sphere centers would not be directly above the centers of the spheres in either of the two planes below, and thus this structure has an ...ABCABCABC... stacking order.   On the other hand, it is also possible to place the third layer in a close packed structure such that the centers of the third layer spheres *do lie* directly over the A (or first) level spheres. This structure has an ..ABABAB...stacking order, but this is not an fcc structure.)   In addition to illustrating the stacking order of the {111} planes, the diagrams of Fig. 3 graphically depict the open structure of the silicon lattice.   They also show the unobstructed channels that exist when traversing the silicon crystal along <110>-directions (perpendicular to the {110}-plane shown in Fig. 3d).

Various structural properties of silicon depend on crystal orientation, as do many VLSI fabrication steps.   For example the tensile strength of silicon is highest for <111>- directions. In addition, {111}-planes have the highest density of atoms.   Therefore {111}-planes oxidize more rapidly than {100}-planes, as more atoms per unit surface area are available for the oxidation reaction.   MOS devices are fabricated on {100}-wafers since the smallest surface state densities are observed on such orientations.

Actual solid materials in the form of single crystals differ to some extent from the ideal crystal structure discussed in this section.   To begin with, even ideal crystals possess *surfaces* at their boundaries, and at a surface, the atoms are incompletely bonded.   In addition, real crystals have a variety of imperfections (referred to as *crystalline defects*) which can significantly alter the properties of the crystal.   A detailed discussion of such defects in silicon and their impact on crystal and device properties is given in Chap. 2.

It is also useful to mention the units used for specifying dopant and contaminant concentrations in silicon.   Most often they are given in atoms /cm^3, but at other times, in *parts per million atoms* (ppma).   Since there are about 5×10^{22} Si atoms /cm^3 in single crystal silicon, an impurity concentration of 5×10^{16} /cm^3 equals 1 ppma, and other concentration levels can likewise be converted.

MANUFACTURE OF SINGLE CRYSTAL SILICON

The fabrication of VLSI takes place on silicon substrates possessing very high crystalline perfection.   G.K.Teal originally recognized the critical importance of utilizing single crystal material for the transistor regions of microelectronic circuits[4].   He reasoned that polycrystalline material would exhibit inadequately short minority carrier lifetimes, due to the defects occurring at the grain boundaries of the polycrstalline grains.   Teal also pointed out the need to define the degree of chemical purity, degree of crystal perfection, uniformity of structure, and chemical composition of single crystal material when specifying starting material on which to fabricate semiconductor products.   The method of obtaining such single crystal Si for VLSI fabrication, involves several steps (Fig. 4):

1) Raw material (e.g. quartzite, a type of sand),is refined by a complex, multi-stage process which produces *electronic grade polysilicon* (EGS),

2) This polysilicon is used to grow single crystal silicon by *Czochralski (CZ) crystal* growth or *float zone (FZ)* growth.   Single crystal silicon is commercially available in either {100} or {111}-orientations (although other orientations, such as {110} can be obtained on special order from some suppliers).

In CZ growth, single crystal ingots are pulled from molten silicon contained in a crucible. Czochralski silicon is preferred for VLSI applications since it can withstand thermal stresses better than FZ material[40], and it is able to offer an internal gettering mechanism that can remove

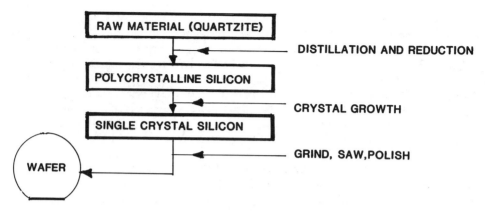

Fig. 4 Process sequence from starting material to polished wafer.

unwanted impurities from the device structures on wafer surfaces[8]. Float zone crystals, because they are grown without making contact to any container or crucible, can attain higher purity (and thereby higher resistivity) than CZ silicon. Devices and circuits calling for high purity starting material (e.g. high voltage, high power devices) are typically fabricated from FZ silicon.

From Raw Material To Electronic Grade Polysilicon

Single crystal silicon is grown from melts of *electronic grade polycrystalline silicon* (EGS). Since the CZ process (which is used to grow most single crystal silicon) adds impurities to the resultant ingot, the EGS must be extraordinarily pure. In fact, in order to achieve controlled doping during subsequent single crystal growth, EGS must have impurity levels in the parts per billion atoms (ppba) range (or ~10^{13} /cm^3), and is, in fact the purest material routinely available on earth. The raw material from which EGS is refined (quarzite), contains high levels of impurities (e.g. aluminum levels of ~3×10^{20} /cm^3). Thus the refining process must reduce impurities by approximately *eight orders of magnitude!* Such a radical refinement procedure involves four major stages[5]:

1) Reduction of quartzite to metallurgical grade silicon (MGS) with a purity of approximately 98%,
2) Conversion of MGS to trichlorosilane (SiHCl$_3$),
3) Purification of SiHCl$_3$ by distillation,
4) Chemical vapor deposition (CVD) of Si from the purified SiHCl$_3$, as EGS.

In *Step 1*, SiO$_2$ in the form of quartzite is reacted with carbon to yield silicon and carbon monoxide:

$$SiO_2 + 2C \rightarrow Si + 2CO \qquad (1)$$

Quartzite is a relatively pure form of SiO$_2$ that occurs in nature. In the presence of carbon (in the form of coal, coke, or wood chips), it is reduced to MGS. The name MGS is derived from the fact that silicon with this purity (~98%) is sufficiently refined to be used as an alloy material in the manufacture of aluminum or for producing silicone polymers. Most silicon manufactured every year is consumed by such applications, and only a small fraction is further refined into EGS for electronic applications. The quartzite, coal (or coke), and wood chips (for

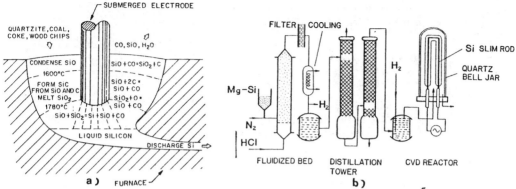

Fig. 5 (a) Schematic of submerged-electrode arc furnace for production of MGS[5]. Reprinted by permission of the publisher the Electrochemical Society. (b) Schematic of fluidized bed, distillation tower, and CVD reactor developed by Siemens[9]. Reprinted with permission of Academic Press.

porosity) are loaded into a submerged electrode arc furnace (Fig. 5a). The temperatures within the electric arcs that are generated between the furnace electrodes can exceed 2000°C, and in those regions SiC is formed. The SiC reacts with SiO_2 to form Si, SiO, and CO. The silicon is drawn off as shown in Fig. 5a, while the SiO and CO gases escape through the spaces created by the presence of the wood chips. Large quantities of electrical power are consumed in producing the reaction (12-14 kWh /kg). The MGS from this process has Al and Fe as the main impurities.

In *Step 2*, trichlorosilane ($SiHCl_3$) is formed by the reaction of anhydrous hydrogen chloride and MGS. In this step, the MGS is first ground to a fine powder. The powdered Si is then treated with HCl to form $SiHCl_3$. The reaction of the solid Si and the gaseous HCl occurs at 300°C in the presence of a catalyst. Both the formation of the $SiHCl_3$ and the chlorides of the impurities (e.g. $AlCl_3$, BCl_3) takes place in this step. The resultant $SiHCl_3$ is a liquid at room temperatures (boiling point 31.8°C), and hence can be purified by distillation.

In *Step 3*, the $SiHCl_3$ is separated from its impurities by fractional distillation. Upon the conclusion of this step, the $SiHCl_3$ is so highly refined that it is no longer possible to directly determine its impurity levels. That is, the $SiHCl_3$ must be deposited as semiconductor grade polysilicon, then converted to single crystal silicon, and finally measured electrically, in order to determine the impurity levels.

In *Step 4*, the highly purified $SiHCl_3$ is converted back into polycrystalline silicon (EGS) by CVD in the presence of hydrogen. The process takes place in a reactor (Fig. 5b) first proposed by Siemens GmbH in the late 1950s, and it is still referred to as the *Siemens process*.

$$2\ SiHCl_3\ (gas)\ +\ 2\ H_2\ (gas)\ \rightarrow\ 2\ Si\ (solid)\ +\ 6\ HCl\ (gas) \qquad (2)$$

The starting surface is a thin silicon rod (called a *slim rod*) which serves as a nucleation surface for the depositing silicon. For large rods (200 mm in diameter and several meters long), the deposition process takes several hundred hours.

After deposition, the EGS is processed into three products: a) one-piece crucible charges; b) nuggets (random sized pieces); and c) poly-rods (Fig. 6). The first two are used as charges in the CZ growth of single crystal silicon, while the rods are used for float-zoning single crystal ingots. In 1985 the world-wide consumption of EGS was approximately 3 million kilograms.

To evaluate the purity of EGS (as well as the $SiHCl_3$ as mentioned earlier) conversion to single crystal silicon must first be accomplished. This conversion is done by the FZ technique,

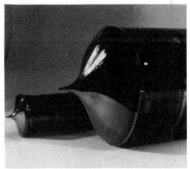

a)

Fig. 6   (a) EGS in polysilicon form[5].   Reprinted with permission of the publisher, the Electrochemical Society.   (b) 150 mm single-crystal CZ silicon ingot.   Reprinted with permission of the Monsanto Eletronics Materials Company.

since as we shall discuss later, FZ contributes negligible levels of impurities to the sample. The measurement of n-type impurities is done after a one-pass float zone. The measurement of residual boron, on the other hand, requires multiple passes of the float zone (which sweeps out other impurities but leaves behind the boron, because of its 0.8 segregation coefficient). Thus 90% of the boron is still left in the silicon after several passes. The purity of EGS is on the order of 0.3 ppba, boron. This represents approximately 1.5×10^{13} boron atoms /cm^3, and a resistivity of ~1000 Ω-cm. ASTM Std F-754 describes the evaluation of EGS by float zoning[6].

The EGS rods can also be machined to form diffusion tubes and wafer boats. Such tubes can withstand several years of temperature cycling between 900 and 1250°C without exhibiting the sagging or deformation of fused silica tubes.

CZOCHRALSKI (CZ) CRYSTAL GROWTH

Czochralski (CZ) growth, named for its inventor, involves the crystalline solidification of atoms from a liquid phase at an interface. The basic production process for producing CZ Si has undergone remarkably little change since it was pioneered by Teal[4] and others in the early 1950s. (For a more detailed description of CZ growth than is presented here, see Refs. 3, 8, and 9.)

The simplicity of the process and the relatively high degree of crystal purity helped establish the CZ process as the dominant crystal growing technology in the early years of the semiconductor industry. The development of dislocation-free ingot growth and automatic diameter control in the late 1960s, lead to a rapid growth in the sizes of ingot diameters and crystal melts. Further increases in charge size and diameters was driven by the economic advantages of even larger ingots. As shown in Fig. 7, by 1985 150 mm diameter crystals and 60 kg charges had become available.   Typical commercial ingots are grown 0.5-2.5 m in length, and a 60 kg charge can grow a 150 mm diameter ingot roughly 1 m long.)

In this section we introduce the most important aspects of CZ growth, focusing on the details that impact the material properties of the silicon wafers for VLSI. These include: a) the sequence of steps in the growth of the ingot; b) the first-order theory of how impurities are incorporated into the ingot (Normal freezing); c) modifications to the normal freezing model that must be incorporated as a result of non-ideal effects encountered in real CZ growth environments (and the impact that this causes on the material properties of CZ wafers); d) the components of CZ crystal growing equipment; and e) the evaluation of CZ silicon crystals in ingot form.

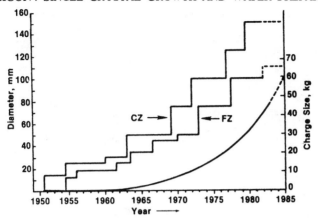

Fig. 7 The increase with time of CZ silicon crystal diameter and charge sizes[47]. Reprinted with permission of Semiconductor International.

Czochralski Crystal Growth Sequence

The steps used in growing a CZ Si crystal are shown in the sequence of photographs of Fig. 8a and a schematic diagram of a CZ crystal grower in which this process is performed is given in Fig. 14a. A fused silica crucible is first loaded with a charge of undoped EGS together with a precise amount of diluted silicon alloy. The growth chamber is then pumped out, backfilled with an inert gas (to limit incorporation of atmospheric gases into the melt during growth), and the charge is melted (melting point of silicon is 1421°C). Next, a slim seed crystal of silicon ($\approx$5 mm in diameter and 100-300 mm long) with precise orientation tolerances is lowered into the molten silicon. The seed crystal is then withdrawn at a controlled rate. Both the seed crystal and the crucible are rotated during the pulling process, but in opposite directions.

The initial pull rate is relatively rapid so that a thin neck is produced. As was first discovered by Dash[10], by forming this thin neck, the growing ingot can achieve a macroscopically dislocation-free crystalline state. At that point the melt temperature is reduced and stabilized so that the desired ingot diameter can be formed (shouldering out step). This diameter is subsequently maintained by a precise monitoring the pull rate. The pulling continues until the charge is nearly exhausted, at which point the ingot is withdrawn to form the *tang* (tail off). A shutdown procedure of the furnace completes the process.

The initial immersion of the cool seed crystal into the melt sets up very high stresses between the heated crystal exterior and the still cool interior. These stresses are high enough to cause plastic deformation, and result in macroscopic dislocations in the crystal. By rapidly pulling a thin necked region from the melt, these dislocations are driven to the cylindrical surface of the neck, and at that point they cease to propagate into the subsequently solidifying crystal regions (Fig. 8b). Thus the growing crystal becomes free of macroscopic dislocations and will continue to grow in this way unless a significant perturbation disturbs the solidification process (e.g. a particulate in the melt contacts the growing crystal, or the crystal is jolted during growth). This tendency for a dislocation-free crystal to continue to grow without dislocations is due to the very high stresses that must be imparted to a perfect crystal in order to create dislocations. Such high stresses are not incurred under well-controlled growth conditions. Further discussion of the Dash process for growing dislocation-free crystals can be found in Refs. 8 and 10.

As the crystal solidifies and is pulled from the melt, the latent heat of fusion L, is

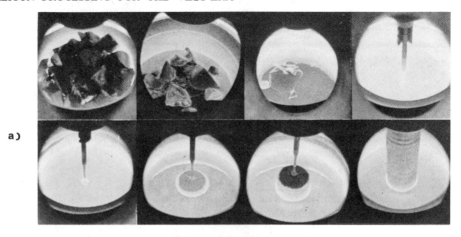

a)

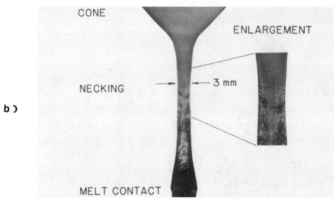

b)

Fig. 8 (a) Illustration of several process steps during CZ crystal growth. Courtesy of Dynamit-Noble-Grace.   (b) X-ray topograph of seed necking and conical part of crystal. Dislocations generated at the end of the seed crystal that contacted the molten zone grow out to the side surface of the neck and do not propagate into the main crystal[9]. Reprinted with permission of Academic Press.

transferred to the crystal. This heat is transported from the solid-liquid interface along the growing crystal. Heat is lost from the crystal surface due to radiation and convection. The upper-limit to the pull rate V_{max}, is reached when the maximum heat that can lost by the ingot is being transferred to the ingot by the solidifying silicon. The value of V_{max} can be calculated from the following relation under the condition of zero thermal gradient in the melt[11]:

$$V_{max} = k_s (dT/dx) / (L d) \qquad (3)$$

where: d is the density of silicon; k_s is the thermal conductivity of silicon; and (dT/dx) is the temperature gradient along the axis of the ingot.

As the diameter of the ingot is increased, the maximum pull-rate must decrease, because the heat loss is proportional to the surface area of the ingot (which increases only linearly with the diameter), but the heat gain is proportional to the volume being solidified (which increases as the square of the ingot radius).

The actual pull rate is typically 30% to 50% less than the maximum rate[11] calculated from Eq. 3. Such slower rates result from the effects of temperature fluctuations in the melt which occur near the interface. (The origin of these fluctuations is discussed in a subsequent section). When such a fluctuation causes a transient increase in the melt temperature, *remelt* of the crystal starts to occur and the nominal pull-rate must be reduced until solidification restarts. When a subsequent temperature reduction takes place, the solidification rate increases, leading to an increase in the ingot diameter. To maintain the desired diameter, the pull-rate at that instant must be increased. Such temperature oscillations thus lead to pull-rate variations and an overall lower average pull-rate. Automatic diameter control (ADC) mechanisms slave the pull-rate changes to optical diameter measurements.

Several ingot properties are impacted by such temperature fluctuations. First, dimensional nonuniformities occur in the diameter of the ingot. Second, these temperature instabilities cause microscopic resistivity variations in both radial and axial directions (e.g. local resistivity variations of ±8% for boron-doped, and ±10-20% for phosphorus-doped wafers from this effect are typically observed). This problem and techniques proposed to suppress it are discussed in subsequent sections.

The dislocation-free attribute of CZ crystals refers to macroscopic (edge and screw) dislocations. Small dislocation loops, however, can be formed during the growth of the main crystal by the condensation of excess point defects (Chap. 2). This phenomenon occurs as the crystal cools from the solidification temperature (1421°C) down to about 950°C. During this cooldown, the solubility of the point defects created at temperatures near meltdown is reduced, until the crystal is supersaturated with point defects. If these point defects can diffuse sufficiently, they will tend to arrive at sites at which they can agglomerate and thereby be removed from solution. If the pull-rate is high enough, the solidified region cools with sufficient rapidity that the excess point defects are immobilized (or *quenched*) in the lattice. They thereby have insufficient opportunity to diffuse and agglomerate into clusters large enough to cause dislocation loops. Both interstitial silicon and oxygen have been suggested as the point defect species responsible for forming such agglomerates. For crystals in which the pull-rate has not been sufficiently high, the presence of the resultant dislocation loops will cause the crystal to exhibit *swirl* patterns when wafers are subjected to defect-delineation chemical etching. Swirl has also been linked to the growth-induced defects stemming from fluctuations occurring at the

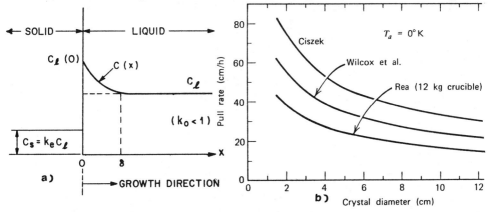

Fig. 9 (a) Doping distribution near the melt-solid interface. (b) Crystal pull rate as a function of diameter[21]. From K.V. Ravi, *Impurities in Semiconductor Silicon*, 1981, Copyright © John Wiley and Sons, Reprinted with permission of John Wiley and Sons.

liquid /solid interface, but there is no well accepted theory on their nucleation and growth[12,13].

When CZ silicon crystals were first grown, the process required highly skilled operators to control all the important growth parameters including: melt temperature; ingot diameter, pull-rate; and rotation. Today such parameters are monitored by microcomputers, resulting in more reproducible control of the physical and chemical aspects of the pulling process. A variety of growth process sequences can be carried out by modern crystal growth systems, with "one button" operations that automatically take the system from meltdown to shutdown.

Incorporation of Impurities into the Crystal (Normal Freezing)

Impurities are incorporated into the CZ crystal as it solidifies, together with the silicon. As pointed out earlier, the impurities are included in the melt by adding precise amounts of silicon alloy (heavily doped with the impurity of choice) into the crucible with the EGS. In the model used to predict how the impurity is incorporated into the crystal (Normal freezing model)[14], three assumptions are made: 1) that the impurity will become uniformly distributed throughout the melt; 2) that once solidification occurs, the impurity does not diffuse in the solidified crystal during the remainder of the ingot growth process; and 3) that the impurity arrives at the solid-liquid interface in the same concentration as exists in the melt.

In the course of solidification, however, the following phenomenon is encountered: the solubilities of the impurity in the melt C_l and in the solid C_s at a given temperature, are generally not equal. For the dilute solutions that typically occur in silicon, the ratio of these solubilities may be defined as the equilibrium segregation coefficient k_o:

$$k_o = C_s / C_l \qquad (4)$$

Let us consider how this phenomenon impacts the impurity concentration in the ingot when k_o is <1 (since this is the case for most impurities in Si, see Table 1). As the silicon solidifies, only a fraction of the impurity concentration that was present in the molten volume that solidified (as specified by k_o), is incorporated into the crystal. The remaining fraction remains in the melt (Fig. 9). As the ingot grows the melt volume is consumed, and therefore decreases. However, a fraction of the impurity arriving at the solidifying interface between the melt and solid is continuously rejected by the crystal and stays in the melt. Thereby, the melt becomes progressively more concentrated with the impurity. As a result, as the ingot grows, the molten material arriving at melt-crystal interface contains an ever larger proportion of impurity atoms. Since the same *fraction* of total impurities present in the arriving materials gets incorporated into the crystal, the ingot also becomes more heavily doped along its length.

If a crystal solidifies and incorporates the impurity from the melt in the manner described, a mathematical relation, known as the *Normal freezing relation* can be used to predict the impurity

Table 1. SEGREGATION COEFFICIENTS FOR COMMON IMPURITIES IN SILICON

IMPURITY	Al	As	B	C	Cu	Fe	O	P	Sb
k_o	0.002	0.3	0.8	0.007	4x10-4	8x10-6	1.25	0.35	0.023

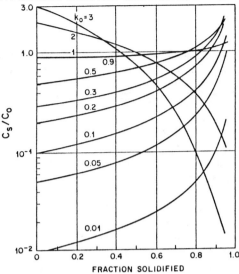

FRACTION SOLIDIFIED

Fig. 10 Curves fron growth from the melt, showing the doping concentration in a solid as a function of ther fraction solidified. From W.G. Pfann, *Zone Melting*, 2nd Ed. 1966. Copyright © John Wiley and Sons. Reprinted with permission of John Wiley and Sons.

concentration along the length of the pulled ingot:

$$C_s = k_o C_o (1 - X)^{k_o - 1} \qquad (5)$$

where: X is the melt fraction solidified; and C_o is the initial impurity concentration in the melt. Figure 10 shows how the predicted impurity concentration in an ingot will vary for various k_o values (including those of B and P) if Eq. 5 is obeyed. Although impurity incorporation does not exactly follow Eq. 5, the information in Fig. 10 does support the following observations in CZ ingots: 1) when k_o is close to 1 (e.g. for B, $k_o = 0.8$), it is less difficult to achieve relatively level *axial* resistivity values (and, as we shall see, adequate *radial* resistivity uniformity as well); 2) when $k_o \ll 1$ (e.g. for P, $k_o = 0.3$, or for Sb, $k_o = 0.023$), these desired uniformities are more difficult to obtain and control.

> **Example :** It is desired to grow a CZ silicon ingot containing 1×10^{15} boron (B) atoms $/cm^3$. What concentration of B atoms must be present in the melt to yield this result ? For a 60 kg silicon charge, how many grams of B (atomic weight 10.8) are needed ?

> **Solution :** Since for B, k_o is 0.8, the initial concentration required in the melt should be: $1 \times 10^{15} / 0.8 = 1.25 \times 10^{15}$ B atoms $/cm^3$.

> Next calculate the volume of the 60 kg silicon charge. The density of the molten silicon is 2.53 g $/cm^3$, so the volume of 60 kg of Si is:

> $$\frac{60 \times 10^3 \text{ grams}}{2.53 \text{ grams} / cm^3} = 2.37 \times 10^4 \ cm^3$$

The number of B atoms in the melt is

$$1.25 \times 10^{15} \text{ atoms /cm}^3 \times 2.37 \times 10^4 \text{ cm}^3 = 2.96 \times 10^{19} \text{ B atoms, and}$$

$$\frac{2.96 \times 10^{19} \text{ atoms} \times 10.8 \text{ g /mole}}{6.023 \times 10^{23} \text{ atoms / mole}} = 5.31 \times 10^{-4} \text{ g of B.}$$

Modifications Encountered to Normal Freezing in CZ Growth

In actual CZ growth of silicon, some of the assumptions on which Normal freezing behavior is based, are not entirely correct. The most important deviation is that the impurities are not homogeneously distributed throughout the melt at all times. In fact, at the solid-melt interface (shown in Fig. 11), a *boundary* layer exists. In this layer, the material is largely unmoved by fluid flow (i.e. does not move by macroscopic flow, but only by diffusion of individual melt atoms), and the impurities rejected by the growing ingot cause the boundary layer impurity concentration to rise above that of the melt. (If such boundary layer did not exist, and if macroscopic motion of the melt caused mixing to establish complete homogeniety of impurities, such local concentration increases of impurities near the interface would not arise.)

Convective flow of the melt is always present and Fig. 11 shows the basic flow patterns. In general, *thermal convection* occurs as a result of hot silicon rising along the crucible walls and descending towards the center of the crucible[45]. Temperature gradients and solute concentrations are the driving forces of thermal convection. When such flows become irregular and unstable, the melt exhibits intermittent release of "thermals". (This phenomenon is also known as *pluming*.) Such pluming causes variations in the temperature and boundary layer thickness near the melt-crystal interface. These temperature and boundary layer fluctuations degrade the uniformity of several electrical properties of the ingot (e.g. macroscopic and microscopic radial resistivity).

It is common practice in CZ growth to use *forced convection* in an attempt to suppress thermal convection (and thereby reduce the unwanted effects arising from the flow instabilities).

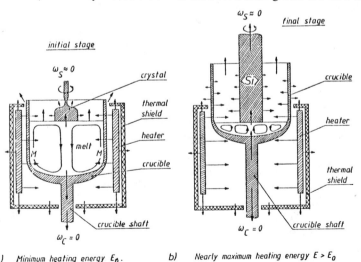

a) *Minimum heating energy E_0.*   b) *Nearly maximum heating energy $E > E_0$*

Fig. 11 Heat flow conditions during crystal growth. Arrows indicate approximate direction of heat flow[8]. Reprinted with permission of Springer-Verlag.

Forced convection is introduced by rotating the crystal during growth. For small melts this technique is relatively effective, but for larger melts only a very limited reduction of thermal convection is achieved through crystal rotation. The crucible is also rotated (in the opposite direction to crystal rotation), to homogenize the melt and achieve thermal symmetry.

It has been determined experimentally that the Normal freezing model must be modified such that the effects of boundary layers are taken into account to give more accurate predictions of the impurity incorporation. The modifications lead to the definition of an *effective segregation coefficient*, k_e, which produces closer correlations between observed and calculated axial dopant incorporation in an ingot[15,16]:

$$k_e = \frac{k_o}{k_o + (1 - k_o)\, e^{(\frac{-VB}{D})}}\qquad (6)$$

where: B is the boundary layer thickness; V is the growth velocity (pull-rate); and D is the diffusion coefficient of the impurity in the melt. Equation 6 predicts that as the boundary layer thickness increases, k_e increases (thus more impurity is incorporated into the growing crystal).

The concepts used to arrive at the effective segregation coefficient and the thermal convection behavior of the melt described above, can help to explain the radial resistivity variation observed in CZ growth wafers (and the cause of microscopic resistivity variations in the ingot). The *macroscopic radial resistivity variation* (Fig. 12) arises because the motion of the molten silicon (due to the dominant thermal convection flow in large melts) [17], causes a thinner boundary layer to occur near the periphery of the growing ingot (and a thicker boundary layer at the ingot center). Thus, if $k_o < 1$, more impurities will be incorporated near the center of the ingot, and progressively fewer towards the edge. The *microscopic resistivity variations* are caused by transient temperature and boundary layer fluctuations brought about by thermal convection flow instabilities.

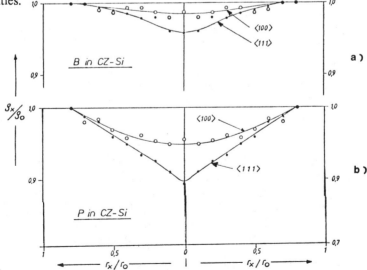

Fig. 12 Radial resistivity variation in (a) boron doped, and (b) phosphorous doped CZ single crystals showing the influence of the crystal orientation on the dopant concentration[8]. Reprinted with permission of Springer-Verlag.

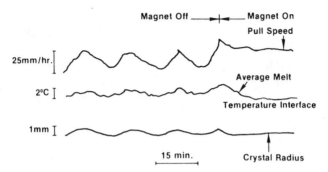

Fig. 13 Correlation between temperature, pull-speed, crystal diameter[35]. Reprinted by permission of Solid State Technology, published by Technical Publishing, a company of Dun & Bradstreet.

The values of k_e are larger than those of k_o, and can approach 1 if the parameter (VB /D) is large. This phenomenon (when $k_e \rightarrow 1$) may be used to improve *axial* doping uniformity. By increasing pull-rate and decreasing seed rotation, B is increased (since B is inversely proportional to rotation speed). On the other hand, Abe argues that to increase *radial* uniformity, B should be thin (and therefore more uniform) across the liquid /solid interface[9]. This calls for a large value for the rotation rate. As discussed in the chapters' closing section, a variety of other techniques have been investigated to suppress thermal convection in large melts. If such thermal convection can be reduced, more uniform impurity incorporation during CZ growth should be possible.

The incorporation of precisely controlled amounts of oxygen into the crystal[8,18,19,25] is also becoming an important requirement for creating wafers suitable for fabricating VLSI with high yields (Chap. 2). Ordinarily, incorporation of oxygen is unintentional, and arises from crucible dissolution during growth, resulting in oxygen concentrations from 5×10^{17}-1×10^{18} atoms /cm^3. Variables that effect the distribution and concentration of oxygen in the growth crystal include thermal convection currents rising along the crucible walls, oxygen evaporation from the melt surface, incorporation of oxygen into the crystal, and dependence of crucible dissolution on melt level. The number of variables and their potential interactions with one another make it difficult to develop models which accurately predict oxygen incorporation in CZ crystals.

The techniques proposed to suppress thermal convection, however, are also predicted to reduce the impact of crucible dissolution effects on oxygen incorporation. In fact, significantly lower (2×10^{17} /cm^3) concentrations and more homogeneous incorporation of oxygen have been observed when thermal convection currents are eliminated and the solid-liquid interface is maintained at a nearly constant temperature (Fig. 13). With oxygen concentrations as low as 2×10^{17} atoms /cm^3, oxygen-related defects such as swirl, stacking faults, thermal donor generation, and oxide precipitation (Chap. 2) are reported to be eliminated, but such concentrations are also too low to allow intrinsic gettering techniques to be implemented (Chap. 2).

Commercially available, state-of-the-art wafers in 1985 offer 10% axial and 5% radial uniformities of interstitial oxygen (O_i) concentrations (in ranges of 6×10^{17}-1×10^{18} atoms /cm^3). As noted by Zuhlehner and Huber in Ref. 8, however, controlled oxygen incorporation into CZ Si is still being successfully accomplished only on the basis of semi-empirical relationships[8].

Czochralski Silicon Growing Equipment

The systems used to grow Czochralski Si crystals[47] (also known as *pullers*), can be considered to consist of four subsystems as shown in Fig. 14:

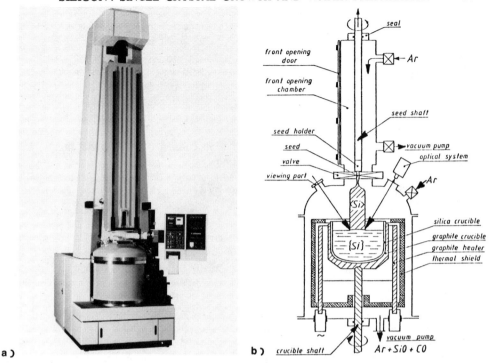

a)   **b)**

Fig. 14 (a) Modern computer controlled CZ puller. Courtesy of Cybeq Systems, a business unit of Siltec Corp. (b) Schematic drawing of CZ puller[8]. Reprinted permission of Springer-Verlag.

1) *Furnace* : crucible, susceptor, rotation mechanism, heater and power supply, and growth chamber.
2) *Crystal-pulling mechanism* : seed cable or chain, rotation mechanism, seed chuck, and crystal handling device.
3) *Ambient control* : chamber gas source, flow controller, and vacuum /exhaust system.
4) *Control system* : computer-based controller, and sensors.

Furnace - The material chosen for CZ silicon crucibles is fused silica. It has the properties of high melting point, thermal stability, and of being relatively non-reactive with molten silicon. The molten silicon to some extent does corrode the fused silica crucible, especially during the higher temperature melt-in period. Thus, substantial quantities of oxygen and silicon become dissolved into the melt. Some of this oxygen becomes incorporated into the crystal, initially taking the form of an interstitial impurity. As discussed earlier and also in more detail in Chap. 2, the presence of this oxygen can have potential beneficial as well as detrimental effects on the crystal properties. The purity of the crucible silica is also important, since the silica impurities can limit the upper values of resistivity possessed by ingots grown in the crucible (the levels of B and P in silicon are typically 8-10 ppma). In addition, silica has no distinct melting point, but gradually softens with increasing temperature.

Near the melting point of silicon, the silica material of the crucible becomes so soft that it requires the support of a heat resistant and rigid outer crucible (susceptor). Such susceptors are fabricated from high purity graphite, which exhibits adequate temperature and cleanliness properties. The other components in the chamber that are subjected to high temperatures are also fashioned from graphite, including the *pedestal* (on which the susceptor sits, and which can rotate as well as raise or lower the susceptor [and thereby also the crucible and melt]), *heater elements*, *insulation package,* and *heat shield.* Carbon contamination of the melt occurs primarily from graphite dust particles, reaction of the hot graphite parts with oxygen introduced from oxygen or water leaking into the chamber, and reaction of the susceptor /pedestal with the silica crucible to form CO. Keeping carbon pieces free of carbon dust, maintaining leak-tight growth systems, and using inert purge gases to sweep out any CO gas that may form from reactions, are effective techniques used to keep carbon incorporation in modern CZ wafers to acceptably low levels[23]. The silicon charge is melted by graphite resistance heaters connected to a dc power supply. Large pullers require tens of kilowatts of power for operation.

The use of CVD-deposited silicon nitride as an alternative crucible material has also been demonstrated. Nitride crucibles also exhibit some dissolution. This causes the crystal to be doped with nitrogen, which behaves as a weak donor in silicon.

Crystal Pulling Mechanism - The mechanism which pulls the crystal from the melt (and simultaneously rotates it), must perform this function with minimal vibration and sufficient precision. Most new systems use a cable or chain to pull the crystal, but their major problem is a tendency for the crystal to swing under certain operating conditions. The process engineer must design the pulling process to avoid these critical conditions. In newer pullers, the hot crystals have become too large to be manually handled. Thus, crystal unloading fixtures and carts are used to transport the ingots from the furnace to the cropping and evaluation areas (Fig. 24b).

Ambient Control - As mentioned, the environment in the chamber must be kept free of reactive gases (O, CO, etc.). This is accomplished by filling the chamber with an inert gas (e.g. Ar). During the process, this gas is swept out of the chamber by the vacuum system (together with any CO, SiO, etc. that may be evolving in the chamber), and replaced with fresh inert gas through the ambient gas inlet. Growth at a reduced pressure in the chamber can be used to alter evaporation from the melt. This has been suggested as a method for controlling the concentration of Sb-doped melts and to reduce the oxygen content incorporated into the ingot[39]. A typical consumption rate of inert gas for this purpose is ~1500 liters /kg of silicon grown.

Control System - The process parameters, such as pull-rate, crucible rotation, seed rotation, melt temperature, gas flow, and crystal diameter, are controlled in modern pullers with digital computer systems. As mentioned earlier, completely automated cycles (one-button operation) require manual intervention only for loading the crucible with EGS. Recipes that result in various crystal properties are stored in the computer memory, allowing the system to produce nearly identical crystals from run-to-run. Other control systems features may include: central computer interfacing capability, data logging and printing, crt terminals, and tape or magnetic disk program storage.

Ingot diameters during growth are controlled by focusing an infrared temperature sensor on the melt-crystal interface, and monitoring changes of the meniscus temperature. The pull-rate mechanism and chamber heater is slaved to the output of the sensor, and the diameter is adjusted by changes in the pull-rate. The level of the melt is detected by laser beam reflection. Some aspects of implementing digital control of ingot diameter during growth are discussed in Ref. 46.

Analysis of Czochralski Silicon in Ingot Form

Upon the completed growth of a silicon ingot, a variety of analysis steps are conducted before the ingot is sawn into wafer form. The ingot is weighed and the diameter is measured along its length. The ingot is visually inspected for evidence of any twinning or other gross crystalline imperfections which may have occurred during growth. The entire crystal is next etched in an acid bath designed to delineate any dislocations that might have been triggered and then propagated through to the tang end of the crystal. Any such dislocated segments must be removed by sawing (see ASTM Std F-47[6] for a discussion of this procedure).

After grinding the ingot to a precise diameter, the resistivity is measured at various points along its growth axis, using either a two-point or four-point probe method (see Chap. 4). This identifies the various resistivity ranges along the ingot length. Radial resistivity measurements are also commonly performed on polished ingot slices at this time for quality or process control.

Oxygen and Carbon Measurements in Silicon Using Infrared Absorbance Spectroscopy

The levels of interstitial oxygen and substitutional carbon present in single crystal Si are most commonly determined by using infrared absorbance spectroscopy (but the technique is not valid for polycrystalline silicon). A spectrum of percent transmittance is obtained from a double-sided polished sample of known thickness (typically 0.2-1.0 cm thick, where the technique exhibits its maximum detection capability). The concentration of oxygen and carbon can then be calculated from the IR absorbance due to their presence. Either a dual-beam IR spectrometer or a Fourier transform infrared (FTIR) spectrometer may be employed for this purpose, but FTIR has become the method of choice (due its increased sensitivity and ease of

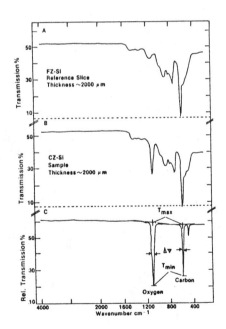

Fig. 15 Transmission spectra: a) Silicon slice free of oxygen and carbon; b) Slice containing oxygen and carbon; c) the difference between spectra hown in a) and b)[20]. Reprinted by permission of Solid State Technology, published by Technical Publishing, a company of Dun & Bradstreet.

use). The principles of FTIR analysis are described in more detail in Chap. 5. FTIR analysis can also be used to infer oxygen precipitation rates (Chap. 2). Since FTIR detects interstitial oxygen, a reduction in the oxygen concentration after thermal cycling can be interpreted as a measure of precipitation. Therefore, FTIR measurements can be included among the procedures used to characterize incoming silicon wafers and for process control.

The presence of interstitial oxygen in the Si lattice causes absorption of IR radiation at 1106 cm^{-1} (wavenumber) as a result of the antisymmetric vibration of the Si_2O complex in the host Si crystal (and substitutional carbon causes absorption at 607 cm^{-1} due to the local vibration mode of the carbon). These bands are superimposed on multiple phonon excitations of the Si, and must therefore always be subtracted from the spectrum of an oxygen- and carbon-free sample (Fig. 15). Quantitative analysis is based on measuring the percent transmission of wavelengths characteristic of an element, and comparing this to standards. The detection limits at room temperature are $2.5x10^{15}$ /cm^3 for oxygen, $5x10^{15}$ /cm^3 for carbon, and $2x10^{14}$ /cm^3 for nitrogen with FTIR. Samples may be prepared by cutting and polishing segments of a single crystal ingot, or whole wafers may be analyzed.

Two methods can be utilized to measure the absorbance due to interstitial oxygen; a) the air reference method; and b) the difference method. In the *air reference method*, the percent transmittance relative to air is measured and an appropriate baseline (I_o) drawn at the peak location. The absorbance α, is then calculated according to the relationship:

$$\alpha = (1 / x) (\ln [I_o / I]) \qquad (7)$$

where x is the sample thickness (in cm), I is the percent transmitted at peak absorption, and I_o is the baseline intensity at peak absorption. The value is then adjusted by 0.4 cm^{-1} for 300°K measurements (and 0.2 cm^{-1} for 77°K measurements) to account for the shape of the background (baseline). In the *difference method*, a reference sample is placed in the reference beam of a dual-beam spectrometer. The reference sample is usually a polished slug of float-zone Si with negligible oxygen or carbon content. Following standard procedures, the absorbance of the test sample relative to that of the reference is measured, and the value of α is calculated as in the air reference method (except that no adjustment is made to α, as the baseline is flat and well-characterized).

The value of the interstitial oxygen content in silicon $[O_i]$, is obtained by multiplying α by a *calibration constant*. The literature contains reports of several values for this constant (ranging from 4.9 to 9.63). The discrepancy is due to the difficulty in conducting the experiments from which this value is extracted. The ASTM Std F-121 for measuring $[O_i]$ specified the value of 9.63 for the calibration constant up until the F 121-77 (1977) version. In 1980, ASTM approved a new standard, ASTM F-121-80, in which the accepted value of the constant was changed to 4.9. Thus, it is now customary in literature that discusses $[O_i]$ values to report the calibration constant value that is being used. If 9.63 is utilized, the $[O_i]$ values are referred to as "old ASTM", while use of 4.9 implies "new ASTM" values. In pre-1980 literature the range of $[O_i]$ in CZ wafers was reported as 26-38 ppma, while $[O_i]$ in CZ wafers using "new ASTM" is given as 13-19 ppma. In this text we use "new ASTM" values for $[O_i]$ values. Even more recently (1983), a cooperative effort sponsored by JEIDA in Japan yielded a value of the calibration constant of 6.06[46].

Substitutional carbon present in single crystal Si can be determined in the same manner as oxygen. The diiference method must be employed when analyzing for the presence of carbon, since strong interference by phonon bands makes the air reference method ineffective. The calibration constant for carbon is 2.0 at 300°K.

The analysis of silicon crystals for oxygen and carbon can be carried out using a modified form of the ASTM Stds[6] F-120, F-121, and F-123 and FTIR. Reference 20 discusses an extension of these standard measurements in that it covers techniques for oxygen and carbon measurements with IR on samples with either one or both sides polished (whereas ASTM F-121 and F-123 only discuss the analysis of double-side-polished slabs).

FLOAT ZONE SINGLE CRYSTAL SILICON

The *float zone* (FZ) process[21,22] is the second major method used to grow single crystal Si. In FZ, a molten zone is passed through a poly-Si rod of approximately the same dimensions as the final ingot. The Si does not come into contact with any substance other than ambient gas in the growth chamber. Thus, FZ crystals have higher inherent purity than CZ ingots (which in-corporate significant amounts of oxygen, carbon, boron, and other metallic impurities from the dissolution of the crucible walls). For applications requiring high purity starting Si (e.g. high power thyristors and rectifiers, and infrared radiation detectors), FZ material becomes the material of choice. The largest quantities of FZ silicon are supplied in the resistivity range of 10-200 Ω-cm, although material up to 30,000 Ω-cm has been produced. Some economical and technical disadvantages, however, make FZ silicon less suitable for VLSI applications than CZ silicon.

From an economic perspective, FZ growth is more costly than CZ growth. In addition, CZ wafers are available in larger diameter form than FZ wafers (Fig. 7), and it appears that the maximum FZ size is ~100 mm because of stability problems of the molten zone under gravity conditions. From a technical viewpoint, as-grown FZ crystals generally exhibit more severe microscopic resistivity variations than do CZ crystals, and the oxygen impurity content of CZ

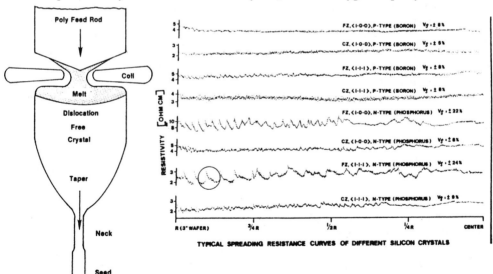

Fig. 16 Zone-refining with a needle-eye coil[22]. Reprinted with permission of Solid State Technology, published by Technical Publishing, a company of Dun & Bradstreet. (b) Comparison of spreading-resistance measurements of CZ and float-zone crystals having the same resistivity specification. The dopant striations in float-zone crystals are greater due to the pronounced remelt phenomena at each rotation of the crystal[8]. Reprinted with permission of Springer-Verlag.

wafers offer added strength and intrinsic gettering potential. The microscopic resistivity variations of FZ silicon can be essentially eliminated with neutron transmutation doping, but this technique is not as attractive for lower resistivity materials. Techniques for incorporating oxygen into FZ silicon have also proved to be difficult to implement. Typical oxygen values for FZ silicon are 0.1-0.3 ppma, and carbon is found at 0.1-1.0 ppma levels. Thus, in the forseeable future, CZ and FZ methods are likely to continue serving their respective application areas.

Figure 16 shows a schematic of the FZ process. A polysilicon rod is mounted vertically inside a chamber under vacuum or an inert atmosphere (e.g. argon). A needle-eye coil provides rf power to the rod, causing it to melt and establish a 2 cm long molten zone. The rf field maintains a stable liquid zone. In fact, because of its levitation effect, the rf field allows a much larger molten region to be maintained than would be possible by surface tension alone. The molten zone is moved along the rod to effect purification and to establish the crystal. Modern zone-refiners use stationary coils and provide for vertical motion of feed rod and crystal.

In bottom-seeded FZ, the seed crystal is brought up from below to make contact with the drop of melt formed at the lower end of the polycrystalline feed rod. As in the CZ process, a neck is then formed to establish a dislocation-free crystal region. At that point, a taper is induced to allow crystal formation at the desired diameter.

A widely used technique for conventionally doping FZ crystals is to add the dopant to the inert gas in the chamber, in the form of a dilute concentration of phosphine (PH_3), or diborane (B_2H_6). This allows the impurity of be continuously incorporated into the molten zone at a fixed ambient concentration, and results in a very level axial doping profile. On the other hand, if the refining passes are made in a vacuum, additional purification is achieved through the evaporation of volatile impurities present in the EGS. As mentioned earlier, however, the FZ process also causes microscopic radial resistivity *striations* whose amplitudes are even larger than those found in CZ ingots. These occur as a result of remelt phenomena and because FZ systems possess thermal environments that are inherently less symmetrical than those of CZ pullers. The melt in CZ processes also provides a large thermal mass that helps to stabilize some of the temperature fluctuations. In many applications the microscopic resistivity variations of as-grown FZ silicon cannot be tolerated.

To overcome the resistivity striation limitation, a technique known as *neutron transmutation doping* (NTD) has been developed[21,22]. In this method, FZ crystals with very low impurity levels are placed in a nuclear reactor and exposed to a stream of thermal neutrons. The neutrons react with the stable silicon isotope ^{30}Si, to form an unstable isotope ^{31}Si, that decays into a stable phosphorus isotope, ^{31}P with a half-life of 2.6 hours,

$$^{30}Si \ (n, \gamma) \quad \rightarrow \quad ^{31}Si \quad \rightarrow \quad ^{31}P \ + \ \beta^- \qquad (\ 8 \ )$$
$$(\ 2.6 \ h \ )$$

The formation of ^{31}Si occurs uniformly throughout the ingot, and by controlling the neutron flux, a very uniform level of phosphorus at the desired level can be obtained. Typically, the average resistivity in the starting EGS rod is about an order of magnitude higher than that of the desired NTD FZ material. Although radiation damage is caused in the crystal, annealing can be performed to restore lattice regularity and resistivity. Microscopic resistivity variations in NTD FZ ingots can be reduced to ± 5% as compared to ± 30% for conventionally doped FZ ingots (Fig. 17).

As the desired resistivity decreases, irradiation cost and crystal damage increases, and the silicon stays radioactively "hot" for excessively long times. As a result, NTD is most attractive for higher resistivity material. In addition, only n-type Si can be produced by this technique.

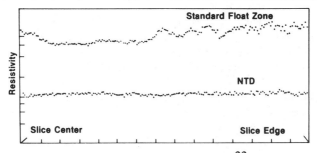

Fig. 17 Spreading resistance profiles of non-NTD Si and NTD Si[22]. Reprinted with permission of Solid State Technology, published by Technical Publishing, a company of Dun & Bradstreet.

NTD Si is commercially available with target resistivities between 5-500 Ω-cm.

Swirls can be exhibited by FZ as well as CZ crystals. Work has been done that indicates that such swirls can be suppressed in non-NTD silicon by maintaining growth rates above 5 mm /min. (Ref. 11). In NTD silicon, oxygen present at low levels (0.15-0.5 ppma) appears to precipitate on radiation damage sites during the annealing step. Further reduction of oxygen has been suggested to eliminate swirls in NTD silicon.

Since oxygen in FZ crystals is difficult to introduce to the sufficient levels needed to improve warpage resistance, work on the adding nitrogen to create the same improvement has been reported. The nitrogen was added to the argon FZ ambient and was then incorporated into the growing crystal at less than the solubility limit at the melting point (without reaction with the silicon). The incorporated nitrogen was reported to increase warpage resistance by pinning dislocations, and did not substantially impact the electrical properties.

FROM INGOT TO FINISHED WAFER : SLICING, ETCHING, and POLISHING

After a single crystal ingot has been grown, a complex sequence of shaping and polishing

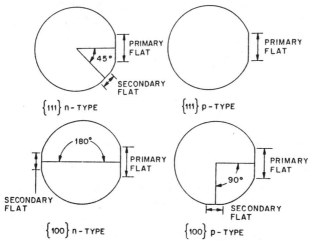

Fig. 18 Identifying flats on a silicon wafer. Reprinted with permission, from the Semiconductor Equipment and Materials Institute, Inc. Copyright the Semiconductor Equipment and Materials Institute, Inc., 625 Ellis St., Suite 212, Mountain View, CA 94043.

steps must be performed on it to produce starting material suitable for fabricating semiconductor devices (i.e. wafers) [26,51]. The procedure (known as *wafering*) includes the following steps:

a) The grown single crystal ingot undergoes routine evaluation of resistivity, impurity content, crystal perfection, size, and weight. Sections of the ingot which are crystallographically defective, irregularly shaped, or undersized, are cut off and discarded (as are the seed and butt [or *tang*] ends). Loss of the ingot material may approach 50% at this point.

b) Since ingots do not grow perfectly round, nor with sufficiently uniform diameters, they must shaped to the desired form and dimension. Thus, ingots are deliberately grown slightly larger than the final desired wafer diameter, and a grinding operation removes the excess material as it reduces as-grown crystals to a cylindrical shape of precise diameter. (Exact diameter dimensions are required for making wafers compatible with automated process and wafer transport equipment.) Since silicon is a hard, brittle material (registering 72.6 on the Rockwell "A" hardness scale), grinding machines with industrial grade diamond wheels are used to perform this cylindrical shaping process. The grinding step is followed by an etch step that removes the grinding work-damage. The ingot diameter is reduced 0.25-1.0 cm by these grinding and etch steps.

c) One or more *flats* are next ground along the length of the ingot. The largest, called the *primary flat*, is usually positioned relative to a specific crystal direction. (The primary flat orientation is found by using x-rays, to produce a Laue photograph.) The primary flat is used for several purposes. Automated wafer handling equipment utilizes the primary flat to obtain correct alignment, and in addition, devices on the wafer can be oriented to specific crystal directions with this flat as a reference. Smaller flats are called *secondary flats*, and they are utilized to identify the orientation and conductivity type of the wafer (see Fig. 18). Since automated equipment relies on the flats for correct operation, the flat dimensions must be precisely machined.

d) The sawing operation[26,27] that produces silicon slices from the shaped ingot also defines the surface orientation, thickness, taper, and bow of the slice. The ingot is rigidly mounted to maintain exact crystallographic orientation during the sawing process. Wafers of <100> orientation are normally cut "on orientation" (e.g. within ±0.5°), while <111> wafers are generally cut "off orientation" (3° ±0.5°) for epitaxial processing applications (see Chap. 5).

Wafer thickness is primarily set by the sawing operation, although some material is also removed by subsequent operations (e.g. for a 150 mm wafer of final 675 μm thickness, a total of ~1200 μm of ingot silicon is needed: 675 μm final wafer thickness; 400 μm for kerf loss during sawing; 50 μm lapping loss; 50 μm etching loss; 25 μm polishing loss. Thus, about 8, 150 mm wafers per linear cm of ingot can be produced). Larger wafers must be thicker (Table 3) to allow them to withstand thermal processes (epitaxy, oxidation, and diffusion) and handling during VLSI fabrication, and thus fewer wafers per unit length of ingot are possible than for smaller wafers. The first cut of crystal is checked for orientation using an x-ray goniometer.

The most common mode of slicing is *inner diameter slicing* (Fig. 19) using stainless steel blades with diamond particles bonded to their edges[26]. Continuous monitoring with blade deflection sensors is required to assure that slices are sawn adequately free of bow, taper, work damage, and saw marks. Cutting speeds are in the neighborhood of 0.05 cm /sec. Since one slice is cut at a time by the saw, this is a relatively slow process.

e) A technique using a *laser* to create an alphanumeric dot-matrix character identification code on the front of the wafer near the primary flat (Fig. 21) can then be employed. *Laser marking* according to SEMI [29] Std. M 1.8, specifies an 18 character field to identify the wafer manufacturer, conductivity type, resistivity, flatness, wafer number, and device type. This allows each wafer to be identified. Wafer traceability, correlation of device characteristics produced on a wafer with the wafer properties, and the ability to prevent in-process rejected wafers from being re-inserted into the fab line, are all benefits of utilizing this option. Automatic code readers at

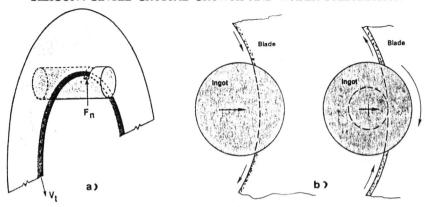

Fig. 19   (a) ID saw geometry[27]. (b) ID slicing[28].   Reprinted with permission of Solid State Technology, published by Technical Publishing, a company of Dun & Bradstreet.

various stages of processing can also track wafers and provide data on wafer movement[30,31].

f) The thickness of the as-sawn slices is sufficiently variable that a *lapping* and *grinding* step is next employed to bring all slices to within the specified thickness tolerance. This step also serves to reduce bow and warp, and to increase slice flatness. Both sides of the slice are lapped using a mixture of Al_2O_3 and glycerine, and a succession of increasingly finer polishing grits is used in multiple lapping steps. This process can achieve flatness uniformities of 2 μm.

g) The edges of the silicon slices are then shaped (rounded) as shown in Fig. 20. Edge rounding substantially reduces the incidence of chipping during normal wafer handling, and helps minimize film buildup on the edge of the wafer during photoresist and epitaxial processes. (Edge chip defect sites are known to nucleate plastic deformation [*slip* - see Chap. 2] during subsequent thermal processing step)[32].

h) The work damage and contamination caused by the previous shaping steps is next removed. This damaged and contaminated layer is chemically etched away[43]. A relatively non-porous and clean wafer backside is also produced by this step. The wet etch procedure for this step typically utilizes an etchant solution of hydrofluoric, nitric, and acetic acids to attack the silicon. Chapter 15 describes this silicon etching reaction in more detail.

i) A chemical-mechanical polishing[43] step is used to produce the highly reflective, scratch, and damage free surface on one side of the wafer. This is accomplished by mounting unpolished slices onto a carrier, and then putting them on a polishing machine. There, a powered platen drives an appropriate polishing pad material across the wafer surface. A colloidal silica slurry of sodium hydroxide and fine (~100Å) SiO_2 particles, is dripped onto the table. The frictional heat of the sliding mounted wafers causes the sodium hydroxide to oxidize the silicon (i.e. the chemical part of the process). The oxide is then abraded away by the silica particles (i.e. the mechanical part). Following the polishing cycle (in which a layer of about 25 μm is removed) the wafers are subjected to series of chemical dips and rinses to remove the polish slurry. A cleaning process concludes this step. An inspection and packing in a particle-free package finish the sequence.

j) Epitaxial layers are required for some VLSI process technologies. Wafer manufacturers are now supplying silicon wafers with a variety of epitaxial layers for CMOS VLSI technologies. Another option being offered is a layer of polysilicon or mechanical damage on the backside of the wafers for extrinsic gettering purposes (see Chap. 2).

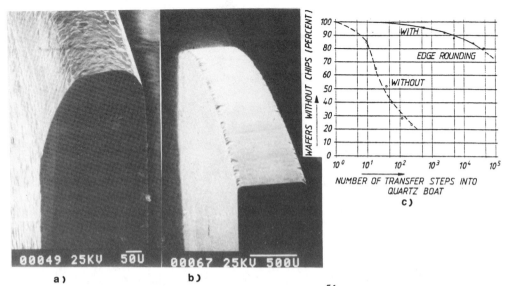

Fig. 20 (a) Edge-rounded, and (b) non-edge rounded wafers[51]. Reprinted with permission of Academic Press. (c) Mechanical strength of edge-rounded vs. non-edge-rounded wafers[8]. Reprinted with permission of Springer-Verlag.

SPECIFICATIONS OF SILICON WAFERS FOR VLSI

There are many properties that a wafer must exhibit (Table 2) in order to be an appropriate starting substrate for the fabrication of VLSI (and ULSI). In this section the terms used to specify these properties are defined and measured. Readers are also directed to recommended test procedures that have been developed to verify that received silicon wafers conform to purchase specifications. Many of these tests have been standardized by the American Society for Testing and Materials (ASTM)[6] and the Semiconductor Equipment and Materials Institute (SEMI)[29]. The standard test procedures are available (updated when appropriate) in publications annually

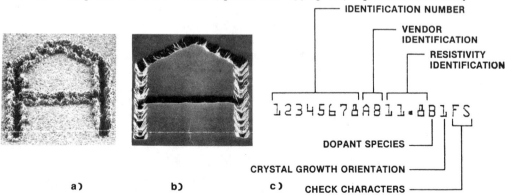

Fig. 21 (a) SEM of letter "A" marked on wafer, as marked on lapped wafer (a), and after polishing (b). Courtesy of Wacker Chemitronic. (c) Marking codes on wafers. Reprinted with permission, from the Semiconductor Equipment and Materials Institute, Inc. Copyright the Semiconductor Equipment and Materials Institute, Inc., 625 Ellis St., Suite 212, Mountain View, CA 94043.

Table 2.COMPARISON of MATERIAL PROPERTIES and REQUIREMENTS for VLSI

Property	Czochralski	Float zone	Requirements for VLSI
Resistivity (phosphorus) n-type (ohm-cm)	1–50	1–300 and up	5–50 and up
Resistivity (antimony) n-type (ohm-cm)	0.005–10	–	0.001–0.02
Resistivity (boron) p-type (ohm-cm)	0.005–50	1–300	5–50 and up
Resistivity gradient (four-point probe) (%)	5–10	20	< 1
Minority carrier lifetime (μs)	30–300	50–500	300–1000
Oxygen (ppma)	5–25	Not detected	Uniform and controlled
Carbon (ppma)	1–5	0.1–1	< 0.1
Dislocation (before processing)(per cm^2)	$\leqslant$ 500	$\leqslant$ 500	$\leqslant$ 1
Diameter (mm)	Up to 200	Up to 100	Up to 150
Slice bow (μm)	$\leqslant$ 25	$\leqslant$ 25	< 5
Slice taper (μm)	$\leqslant$ 15	$\leqslant$ 15	< 5
Surface flatness (μm)	$\leqslant$ 5	$\leqslant$ 5	< 1
Heavy-metal impurities (ppba)	$\leqslant$ 1	$\leqslant$ 0.01	< 0.001

issued by these organizations. Table 3 provides some typical specification ranges for 100, 125, and 150 mm diameter CZ wafers. Note that the specifications have become tighter with increasing wafer size as a result of automatic wafer handling and photolithographic requirements.

Electrical Specifications

Conductivity Type - Specifies whether wafers are n or p type. Information about which element was used to dope the wafer should also be provided.

Resistivity or Resistivity Ranges (Ω-cm) - specifies the average (or range of) resistivities of the wafers. The resistivity is related to the doping density[49] as shown in Fig. 22. The wafer resistivity can be verified using ASTM Std F-43 with a four-point probe.

Radial Resistivity Gradient (% Variation) - gives a measure of the variation of the resistivity between the center and the selected outer regions of a wafer. The radial resistivity variations of wafers are a function of the crystal growth process and dopant type. The ability to predictably control a variety of device characteristics depends on having uniformly doped substrate material (an example is the threshold voltage of MOS devices). Radial resistivity gradient is measured using ASTM Std F-81 with a four-point probe.

Resistivity Variations - are local   variations of the resistivity that occur primarily due to

Table 3. SPECIFICATIONS FOR POLISHED SINGLE-CRYSTAL SILICON WAFERS

Parameter	100 mm	125 mm	150 mm
Diameter (mm)	100 $\pm$ 1	125 $\pm$ 1	150 $\pm$ 1
Thickness (mm)	0.5–0.55	0.6–0.65	0.65–0.7
Primary flat length (mm)	30–35	40–45	55–60
Secondary flat length (mm)	16–20	25–30	35–40
Bow (μm)	60	70	60
Total thickness variation (μm)	50	65	50
Surface orientation	(100) $\pm$ 1°	Same	Same
	(111) $\pm$ 1°	Same	Same

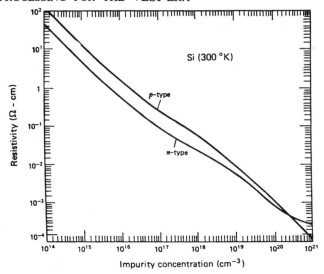

Fig. 22 Resistivity versus impurity concentration for Si. From W.F. Beadle, R.D. Plummer, and J.C. Tsai, *Quick Reference Manual for Silicon Integrated Circuit Technology*, 1985, Copyright© John Wiley & Sons. Reprinted with permission of John Wiley & Sons.

remelt phenomena. The effect is more pronounced in FZ than CZ material, although the effect can be substantially eliminated in FZ ingots by subjecting them to neutron-transmutation doping. Resistivity variations are measured with a four-point probe (ASTM Stds F-81 and F-525).

Mechanical /Dimensional Specifications

Crystal Growth and Doping Method - describes the type of crystal growing procedure (CZ or FZ) and doping method (e.g. ion implantation, NAA).

Diameter - specifies the linear dimension across the surface of the wafer, passing through the center of the wafer, excluding flats. This specification has been continuously tightened to improve prealignment placement of wafers onto stepper equipment, and to facilitate automated wafer handling. For 150 mm wafers the diameter specification is 150 ± 0.2mm. It is measured according to ASTM Std F-613 using a ring template.

Thickness - must be controlled because if wafers are too thin they may break or warp during normal wafer processing, or if too thick, they may not work in all processing equipment and fixtures. Thickness is measured at the wafer center according to ASTM Standard F-533 with a thickness gauge[51]. See SEMI specifications of wafer thickness according to their diameter.

Total Thickness Variation (TTV) - Excessive thickness variations may also cause problems in mechanical handling and photolithographic processes especially if stepper processing employs back-surface referencing. TTV is the difference between the maximum and minimum values of thickness on a wafer (Fig. 23), and is primarily a function of the sawing process. It is measured according to ASTM Standard F-533 using a non-contact type thickness gauge[43,50,51].

Bow - is the concavity due to sawing, or the deformation from thermal processing of the wafer centerline (Fig. 23). Bow is a *bulk* property of the wafer, not of the wafer surface, and units of bow are μm. Bow is measured according to ASTM Std F-534-84. Parameters such as bow and warp are measured on free wafers (rather than on wafers flattened by a vacuum chuck)[43,50,51].

Warp - is the deviation (difference between maximum and minimum distance) exhibited by the centerline of a wafer from a planar condition, when such a deviation includes both concave

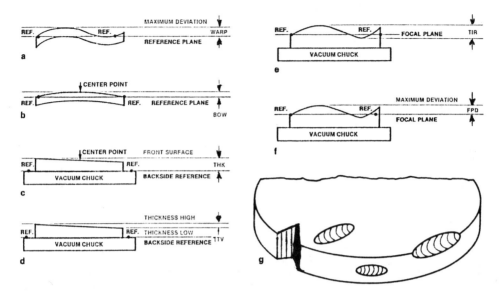

Fig. 23 Typical wafer flatness parameters (a) warp (b) bow (c) thickness (d) total thickness variation, TTV (e) total indicator reading, TIR (f) focal plane deviation, FPD. Chips and indents in wafer shown in (g). Reprinted with permission of Microelectronics Manufacturing and Testing.

and convex regions (Fig. 23). Warp is also a *bulk* property (expressed in μm) and should not be confused with flatness. Warp can occur during wafer processing, as discussed in Chap. 2. It can be measured according to ASTM Std F-657-80[43,50,51].

Flat dimensions - must be machined to a precise length and orientation. Edge rounding of the wafer can degrade the ends of the flatted area, and can thus change the effective flat length[44]. Flat length and any such flat edge ambiguity is measured using an optical comparator according to ASTM F-671.

Chips - are regions of silicon missing from the wafer surface or edge which do not go completely through the slice (Fig. 23). Chips are detected visually, or by automatic surface scanners.

Indents - are edge defects that go all the way through the slice (Fig. 23). They are detected in the same manner as chips.

Edge contour - the wafer edges are contoured (rounded) to minimize occurrence of chipping and to help reduce film buildup at the wafer edge during photoresist and epitaxial deposition. The contour shape should meet the SEMI specification M1.5. The contour quality is measured by comparison to a template.

Laser Marking - as described earlier, a laser can be used to create a dot-matrix alphanumeric code on the wafer surface[30,31]. See SEMI specification M1.8 for character generation information and interpretation.

Chemical /Structural Specifications

Bulk Structural Defects - Rapid inspection of large diameter wafers by x-ray topography (Chap. 17) should reveal no structural defects in starting wafers.

Surface Orientation - specifies the orientation of the surface of the semiconductor wafer and the allowed degree of misorientation. Normally <100> material is cut *on orientation* ±0.5°, and <111> material is cut *off orientation* 3° ±0.5°. Surface orientation of wafers is verified according to ASTM Std F-26.

Oxygen Content - is important because it affects wafer strength, impurity gettering processes, thermal donors, etc., as discussed in Chap. 2. Oxygen content is measured using ASTM F-121 and FTIR. The interstitial oxygen content should be tailored to the customers specification and a tolerance of 2 ppma may be acceptable.

Carbon Content - Measured according to ASTM F-123 and FTIR, although measurement is difficult due to low carbon levels (~1ppma) in modern wafers.

Swirl - before processing, swirl should not be present, but after an 1100°C step, can be high. Measured using ASTM F-416.

Oxidation Induced Defects - are defects induced or enhanced by oxidation steps that are a part of the processing cycle. (Strictly speaking, these are not a part of the specificaion of unprocessed wafers, but such defects may have their origin in pre-existing stacking faults, dislocations, or shallow etch pits). These defects are detected according to ASTM Std F-416.

Surface /Near Surface Specifications

Flatness - is the maximum peak-to-valley deviation of a wafer surface as measured from a reference plane (Fig. 23). Flatness is a *surface* property, expressed in µm and can be measured according to ASTM Std F-775-83. Flatness parameters (global and local flatness), wafer thickness, and TTV are usually measured with the wafer pulled flat on a vacuum chuck. Flatness of approximately 3-4 µm may be required for small feature size (≤2 µm) projection printer applications. A local site flatness of ±1 µm over a 20 x 20 mm field of view will ensure that each step and repeat field is sufficiently flat, and thus remains within the depth of focus of lenses designed to resolve 1 µm dimensions[33,43,50,51].

Particles - are detected visually by the unaided eye and a collimated light beam, optical microscopy, or automatic surface scanners (ASTM Std F-154 and F-24). Their presence is quantified as the number of particles /unit area.

Haze - is light scattering caused by microscopic surface irregularities, whose origin can be crystalline defects or externally caused surface damage. Detected using a narrow-beam tungsten light source according to ASTM Std F-154.

Saucer Pits (S-Pits) - are a class of microdefects associated with metallic impurities that manifest themselves after oxidation, oxide strip, and preferential chemical etching (see Chap. 2). Densities of <100 cm^{-2} may be tolerable.

Saw Marks - are caused in the slicing operation by the saw blade. If not removed by lapping, these marks may still be visible after polishing (see ASTM Std. F-154).

Cracks /Crows Feet - are cleavages of fracture that extend to the surface of the wafer, resulting from wafer surface impact with sharp objects (e.g. burrs on a vacuum chuck that holds the wafers during edge rounding). Crows feet are cleavages or fractures characterized by a *crows foot* pattern in <111> material and a *cross* pattern in <100> material . (See ASTM Std F-154).

Stains, Streaks, Residue, Smudges, Dimples, Orangepeel, Mounds, and Scratches - surface defects due to a variety of causes (see ASTM Std F-154).

TRENDS IN SILICON CRYSTAL GROWTH AND VLSI WAFERS

The increasing size of silicon wafers is the most obvious trend in silicon material technology. In 1984, 150 mm wafers were introduced and 200 mm appears to be the next incremental step[48]. The use of larger diameter wafers for maintaining productivity as die sizes increase, presents several major challenges to semiconductor manufacturers including: a) retooling equipment for larger wafers requires large capital expenditures; b) patterning techniques must be changed to print the smaller feature sizes over the larger area; c) the need to maintain flatness requires the development of low-temperature or RTP processing; d) larger wafers must also be thicker to increase their resistance to warpage. (As noted by Takasu[32], the projected thickness may make the wafers too stiff to be adequately held down by vacuum chucking, and hence new wafer hold-down techniques may be required); and e) larger sizes will substantially increase the weight of the wafers and their cassettes, forcing changes in wafer transport mechanisms.

Problems associated with impurity incorporation into the crystal are being addressed in a number of ways. To increase axial impurity uniformity several procedures have been investigated, including use of double crucibles[34], continuous liquid feed (CLF) systems[35], magnetic Czochralski growth (MCZ)[35,36], and controlled evaporation from the melt[37].

The *double crucible* (Fig. 24) uses a large outer crucible and a smaller, low aspect-ratio (wide and shallow) inner crucible. A connection between them is maintained by a capillary tube. By feeding the inner crucible from a larger reservoir in which a uniform doping level is maintained, the doping level in the inner crucible can be held more constant. In this manner, an improved uniformity in the axial doping level can be achieved. In addition, thermal convection in the melt of the inner crucible is reduced as a result of its low-aspect ratio dimensions. Thermal fluctuation effects are thus also less pronounced.

The *CLF method* is similar in that additional molten silicon is supplied to the growth crucible from a second crucible (Fig. 24). The feed mechanism utilizes a siphon technique, and the doping uniformity improvements stem from the same phenomena as in the double-crucible method. Since the second crucible is out of the growth chamber, larger total charges may be possible than with the double-crucible.

Magnetic fields surrounding the crucible have been found to suppress the thermal convection in the melt. Lenz's law states that when a conductor crosses magnetic field lines, a current is induced. The magnetic field generated by the induced current opposes the conductor motion. This principle also applies to the silicon melt in a magnetic field. The resultant suppression of the thermal convection brings about several benefits including: a) smaller temperature fluctuations occur at the crystal-melt interface, and thus ingot diameter and microscopic resistivity variations due to remelt are smaller; and b) the transport of oxygen from the crucible

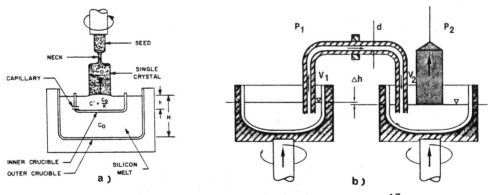

Fig. 24 (a) Schematic representation of double-crucible arrangement[17]. Reprinted with permission of the publisher, the Electrochemical Society. (b) Concept of continuous liquid-feed furnace[35]. Reprinted with permission of Solid State Technology, published by Technical Publishing, a company of Dun & Bradstreet.

Fig. 25 (a) A large solenoid magnet encircles the lower chamber in this axial-field MCZ version of a puller. (b) 4" crystal in removal cart[51]. Courtesy, Microelectronics Manufacturing & Testing.

walls to the solidification interface is reduced, and less oxygen is incorporated into the ingot (7 ppma for MCZ, versus 13-19 ppma for conventional CZ silicon). With lower oxygen content, significantly less oxygen thermal donor formation is observed, and oxygen-associated defects (precipitates and stacking faults) are reported to be suppressed. MCZ is being investigated by several silicon manufacturers (Fig. 25)[35].

Controlled evaporation of the dopant from the melt has reportedly been successfully utilized to grow an antimony-doped CZ silicon ingot with uniform axial doping. Antimony (Sb) shows a much higher evaporation rate than P, B, or As, and if it can be evaporated in a controlled manner to keep the melt concentration of Sb constant (in spite of its small segregation coefficient $k_o = 0.023$) a more uniformly doped ingot will result[9,39].

The *incorporation of nitrogen* in CZ and FZ silicon[38] is being studied as a technique to increase the warpage resistance of wafers. Oxygen is known to increase wafer mechanical strength, and nitrogen appears to play a similar role. Unlike oxygen, which can outdiffuse during thermal processing (and thereby cause the wafer surface region strength to be reduced), the surface

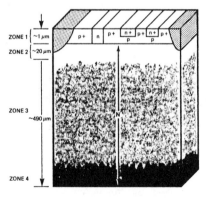

Fig. 26 Diagram of principle zones in a wafer design for ULSI circuits[33]. Reprinted with permission of Academic Press.

concentration of nitrogen is much more constant due to its small diffusion coefficient in Si. The intrinsic gettering properties of nitrogen (Chap. 2) are as yet unknown.

As memory devices on wafers become smaller, *radiation from uranium and thorium* can become a significant source of soft errors. Silicon material currently contains ppba levels of these radioactive elements, and this level must be reduced to allow future memory devices to be successfully fabricated. New analytical techniques, such as solid-state detection (after concentration) will have to be used to detect such low levels, and chemical extraction methods will be needed to reduce their concentration[39].

Wafer specifications will also become more stringent. Increased flatness will be needed to successfully perform submicron lithography with flatness of <1 μm required by some technologies. Controlled wafer curvature[32] (e.g. concave or convex) may need to be specified depending on the specific types of chucks used to hold down the wafers during lithography steps. Chipless edge contours will be mandatory as will tighter diameter specifications (for required pre-alignment accuracy). A technique for determining backside particle existence will be needed, since it has been shown that particles on the backside of a wafer can impact wafer flatness. The concentration of C and O of every wafer will be specified more precisely and a maximum 2 ppma variation in O may be needed to control intrinsic gettering procedures. Wafer cleanliness and front-surface particle control will become even more important.

As a result, the *design of wafers* is likely to become a standard practice[41,44]. As an example, before wafers for a process will be specified, the process will have to be studied. Based on the projected thermal cycles, lithography processes, etc., a specific wafer structure will be specified (Fig. 26). Each region of the wafer will be called out to have a set of properties. For example, device regions might need a denuded zone (with or without an epitaxial layer) possessing certain defect, structural, and chemical properties. The bulk region below the device region may be specified to contain required concentrations of O, C, or N. The wafer backside may be specified as having a polysilicon layer or mechanical damage, for extrinsic gettering behavior.

REFERENCES

1. C.L. Yaws, R. Lutwack, L. Dickens, and G. Hsiu, "Semiconductor Industry Silicon: Physical and Thermodynamic Properties", *Solid State Technol.*, **24**, 87, (1981).
2. W. Shockley, "The Theory of p-n Junctions in Semiconductors and p-n Junction Transistors," *Bell Syst. Tech. J.*, **28**, 435, (1949).
3. R.A. Laudise, *The Growth of Single Crystals*, Prentice-Hall, Englewood Cliffs, N.J., 1970.
4. G.K. Teal, *IEEE Trans. El. Dev.* **ED-23**, 621, (1976).
5. L.D. Crossman and J.A. Baker, "Polysilicon Technology", *Semiconductor Silicon 1977*, Electrochem. Soc., Pennington, New Jersey, 1977, p. 18.
6. The American Society for Testing and Materials (ASTM), Committee F-1 on Electronics, Philadelphia, Pennsylvania.
7. J.R. Carruthers, *et al*, "Czochralski Growth of Large Diameter Silicon Crystal-Convection and Segregation", *Semiconductor Silicon, 1977*, Electrochem. So., Pennington, N.J., 1977, p. 61.
8. W. Zuhlehner and D. Huber, "Czochralski Grown Silicon", in *Crystals 8*, Springer-Verlag, Berlin (1982).
9. T. Abe, "Crystal Fabrication", in [N.G. Einspruch and H. Huff, Eds.], *VLSI Electronics-Microstructure Science*, Vol. 12, Academic Press, Orlando, Fla. (1985), Chap. 1, p. 3.
10. W.C. Dash, *J. Appl. Physics*, **29**, 705, 736, (1958), *ibid.*, **30**, (1959), p. 459.
11. S.N. Rea, "Czochralski Silicon Pull Rates", *J. Cryst. Growth*, **54**, 267, (1981).

12. H. Kolker, "Behavior of Nonrotational Striations in Si", *J. Cryst. Growth*, **50**, 852, (1980).

13. I. Chikawa and S. Yoshikawa, "Swirl Defects in Silicon Single Crystals", *Solid State Technol.*, 23, 65, (1980).

14. W.R. Runyan, *Silicon Semiconductor Technology*, McGraw-Hill, New York, 1965.

15. J.A. Burton, R.C. Prim, and P. Slichter, *J. Chem. Phys.*, **21**, 1977, (1953).

16. M.H. Liepold, T.P. O'Donnell, and M.A. Hagan, "Materials of Construction for Silicon Crystal Growth", *J. Cryst. Growth*, **40**, 366, (1980).

17. K.E. Benson, W. Lin, and E.P. Martin, "Fundamental Aspects of Czochralski Silicon-Crystal Growth for VLSI", *Semiconductor Silicon, 1981*, Electrochem. Soc., Pennington, NJ, 1981.

18. J.R. Patel, "Current Problems on Oxygen in Silicon", *Semiconductor Silicon, 1977*, Electrochemical Society, 1977, p. 189.

19. L. Jastzerbski, "Origin and Control of Material Defects in Silicon VLSI Technologies: An Overview", *IEEE Trans. on Electron Devices*, **ED-29**, No. 4, Apr. '82, p. 475.

20. P. Stallhofer and D. Huber, "Oxygen and Carbon Measurements on Silicon Slices by the IR Method", *Solid State Technol.*, Aug. '83, p. 233.

21. K.V. Ravi, *Imperfections and Impurities in Semiconductor Silicon*, Wiley, New York, 1981.

22. H.G. Kramer, "Float-Zoning of Semiconductor Silicon: A Perspective", *Solid State Technol.*, Jan. '83, p. 137.

23. J.W. Moody and R.A. Frederick, "Developments in Czochralski Silicon Crystal Growth", *Solid State Technol.*, Aug. '83, p. 221.

24. M. Watabe, *et al*, "Oxygen-Free Silicon Single Crystal from Silicon-Nitride Crucible", *Semiconductor Silicon*, Electrochem. Soc., Pennington, N.J., (1981), p. 126.

25. R.S. Swaroop, "Advances in Silicon Technology for the Semiconductor Industry, Part II", *Solid State Technology*, July '83, p. 97.

26. A. Bonora, "Flex-Mount Polishing of Si Wafers", *Solid State Technol.*, **20**, 55 1977.

27. N. Jackson, "Materials and Technology of Wafering", *Solid State Technol.*, July '85, p. 107.

28. R.L. Lane, "ID Slicing Technol. for Large Diameters", *Solid State Technol.*, July '85, p. 119.

29. Semiconductor Equipment and Materials Institute, (SEMI), Mountain View, California.

30. I.A. Morris and J. Cobb, "Wafer Identification Increases Device Yields", *Microelectronics Manuf. and Test*, Sept. '85, p. 50.

31. J. Scaroni, "Wafer Identification Making via Laser", *Microelectronic Manuf. and Testing*, Nov. '85, p. 29.

32. S. Takasu, "Silicon Wafer for VLSI Technology", *VLSI Technology 1983*, Electrochem. Soc., Pennington, N.J. (1983), p. 490.

33. J.E. Lawrence and H.R. Huff, "Silicon Material Properties for VLSI Circuitry", in *VLSI Electronics*, Vol. 5 (N.G. Einspruch, Ed.) Academic Press, Orlando, Fla. (1982), p. 51.

34. ibid, Ref. 17.

35. G. Feigl, "Recent Advances and Future Directions in CZ-Silicon Crystal Growth Technology", *Solid State Technol.*, Aug. '83, p. 121.

36. T. Suzuki, N. Isawa, Y. Okubo, and K. Hoshi, "CZ Silicon Growth in a Transverse Magnetic Field", *Semiconductor Silicon 1981*, Electrochem. Soc., Pennington, N.J., 1981, p. 90.

37. *ibid*, Ref. 39, p. 404.

38. T. Abe, *et al*, "The Characteristics of Nitrogen in Si Crystals", *VLSI Science and Technology*, Electrochem. Soc., Pennington, N.J, (1985), p. 543.

39. S. Kishino, "Material Characterization for VLSI Applications", *VLSI Science and Technology*, Spring, 1985, Electrochem. Soc., 1985, p. 399.

40. J. Doerschel and F.G. Kirscht, "Differences in Plastic Deformation Behavior of CZ and FZ Grown Si Crystals, *Phys. Status Solid: A*, 64, K85 (1981).

41. H. Huff, "Silicon Wafers Engineered for VLSI Circuits, *Semiconductor Intntl.*, July '85, p. 82.

42. H. Herzer, "Neutron Transmutation Doping", *Semiconductor Silicon, 1977*, Electrochem. Soc., Pennington, N.J., 1977, p. 106.

43. D. Biddle, "Characterizing Semiconductor Wafer Topography", *Microelectroncs Manuf. and Testing*, Mar. '85, p. 15.

44. H. Huff and R.F. Holt, "Computer Automated IC Manufacturing Demands on VLSI Wafers", *Solid State Technol.*, Sept. '85, p. 193.

45. *ibid*, Ref. 7.

46. K.M. Kim, *et al.*, "Digital Control of CZ Silicon Crystal Growth, *Solid State Technol.*, Jan. '85, p. 165.

47. R. Iscoff, "Crystal Growing & Fabrication Equipment", *Semiconductor Intntl*, Nov. '83, p. 68.

48. P. Burggraaf, "Silicon Crystal Growth Trends", *Semiconductor International*, Oct. '84, p. 56.

49. W.R. Thurber, R.L. Mattis, and Y.M. Liu, "Resistivity Dopant Density Relationships for Si", *Semicon. Characterization Techniques*, Electrochem. Soc., Pennington, N.J., 1978, p. 81.

50. W.A. Baylies, "A Review of Flatness Effects in Microlithographic Technology," *Solid State Technol.*, 24, 132 (1981).

51. W. Helgeland, *et al.*, "CZ Crystal Growth into the 90's", *Microel. Mfg. & Test.*, May '86. p.1

PROBLEMS

1. (a) The lattice constant of Si is 5.43Å.   Find the nearest neighbor distance in a diamond lattice which has this lattice constant. (b) Calculate the density of Si from the lattice constant, the atomic weight, and Avogadros number. Compare the results with those given in Table 1.

2. Sketch a view down a <110> directon of a diamond lattice.

3. Show that in fcc lattice, each atoms has 12 nearest neighbor atoms.

4. Show that the maximum fraction of the unit cell which can be filled by hard spheres in the simple cubic, fcc, and diamond lattices are 0.52,   0.74, and 0.34, respectively.

5. Label the planes illustrated in Fig. P1.

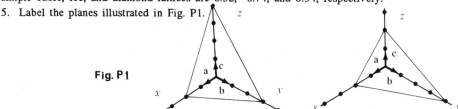

Fig. P1

6. Determine the number of atoms in the (100), (110), and (111) planes of a Si crystal.

7. A Si crystal doped with boron is to have a reisitivty of 4.0 Ω-cm   when one-half of the ingot is grown. If the Si charge weighs 30 kg, how many grams of boron-doped Si with a reistivity of 0.02 Ω-cm should be introduced into the melt to achieve the desired ingot resistivity.

8. Plot the doping distribution of (a) arsenic, and (b) phosphorus at distances of 20, 40, 60, 80, and 90 cm from the seed in 100 cm long ingot that has been pulled from a melt with an initial doping concentration of 10^{16} cm^{-3}.

9. The seed crystal used in CZ growth is "necked down" to ~3 mm (Fig. 8), to help produce a dislocation-free state in the pulled ingot. Using the yield strength of silicon ($2x10^6$ g /cm^2), calculate the maximum weight of an ingot that can be supported by such a neck. For ingots of 150 and 200 mm in diameter, calculate their length for this weight.

10. Using the maximum growth rates shown in Fig. 9b, calculate the time required to pull a 100 mm diameter, and a125 mm diameter ingot of 1 m in length.

11. Explain the difference between *bow* and*TLV*.

2

CRYSTALLINE DEFECTS,

THERMAL PROCESSING,

and GETTERING

In this chapter we cover three separate but interrelated subjects: defects, thermal processing, and gettering. The discussion on defects will be limited to the important class of *structural defects in single crystal silicon lattices*. Thermal processes involve exposing wafers to elevated temperatures (sometimes in the presence of specific gas species), in order to effect a variety of changes in material properties of VLSI wafers. They are performed as a part of many procedures and can be carried out before, during, or after various steps, depending on the application. The decision to include a discussion of thermal processing at this point in the book is based on a number of considerations. First, thermal cycles are an inherent part of many gettering methods. Second, thermal steps are utilized in so many other processes that an early consideration of the topic is appropriate. Finally *rapid thermal processing* (RTP) is a relatively new technique for which new applications are still being discovered. Thus, it deserves to be addressed separately, rather than just as part of a discussion of a particular processing technology, such as ion implantation or diffusion. *Gettering* is a term used to describe a variety of processing techniques that remove harmful defects or impurities from the regions on a wafer in which devices are fabricated.

As only a description of *defects in single-crystal silicon* will be undertaken in this chapter, the term *defect* should be briefly considered from a broader perspective. It is useful to remember the most general meaning of the term, even though it is used in many specific contexts. That is, *a defect can be any material, process, or design flaw that may interfere with the successful fabrication of microelectronic components*. A partial list of the types of defects encountered in manufacture of VLSI is given in Table 1. It also lists chapters of this text in which additional information concerning various non-crystalline defects are discussed.

Table 1. TYPES OF DEFECTS ENCOUNTERED IN FABRICATING VLSI DEVICES

Crystalline Defects	Chap. 2	Oxidation-Induced Stacking Faults	Chaps. 2, 5
Particulates	Chap. 15	Pinholes, Scratches, and Chips	Chap. 1
Contaminants	Chap. 15	Hillocks in Metallization Films	Chap. 10
Mask Errors and Flaws	Chap. 13	Epitaxial Film Spikes, Distortion,	
Photoresist Bubbles /Striations	Chap. 12	and Stacking Faults	Chap. 5

Table 2. **CRYSTALLINE DEFECTS IN SILICON**

Type	Dimension	Examples
Point	0	Vacancy, Interstitial, Frenkel defects (Intrinsic - silicon self-interstitial) (Extrinsic - dopants, oxygen, carbon, metals)
Line	1	Straight dislocations (edge or screw) Dislocation loops
Area	2	Stacking faults Twins Grain boundaries
Volume	3	Precipitates, voids (Oxygen precipitates, metal precipitates)

CRYSTALLINE DEFECTS IN SILICON

As discussed in Chap. 1, real crystals differ from the ideal in that they possess imperfections or defects[1]. Some, due to impurity dopant atoms, are absolutely necessary for creating devices in the crystal. Other crystalline defects may be helpful if present in moderate density. Most however, are undesirable, regardless of the density in which they may be found in the crystal. Table 2 lists crystalline defects according to their geometry. There are zero-dimensional (or *point* defects), one-dimensional (or *line* defects), two-dimensional (or *area* defects), and three-dimensional (or *volume* defects). The following sections expand the information in the table. Descriptions of defect structures, mechanisms of their formation and motion, and definitions of terminology used to describe them are also included. Data on the impact of defects on device operation, and methods for characterizing the presence and properties of defects is then provided.

Point Defects

The various forms of zero-dimensional or point defects[2] are shown in Fig. 1B. In crystal regions where no point defects exist, each Si atom occupies its normal position on the lattice. If an atom is missing from one of these sites a *vacancy* exists in the lattice. In the event that an atom leaves a lattice site (creating a vacancy) and *moves to the external surface* of the crystal, a *Schottky* defect is created (Fig. 1Ba). If an atom is found at some non-lattice site in the crystal, it is said to lie at an *interstitial* site. Silicon atoms at interstitial sites in a Si crystal are called *self-interstitials*, and self-interstitial atoms may originate from a lattice site in the crystal, or from the crystal surface. Vacancies and self-interstitials are classified as *intrinsic point defects*.

In any solid, lattice atoms are in continuous vibration about their equilibrium positions. The amplitude of these vibrations increases with increasing temperature, and may become large enough for an occasional atom to leave its equilibrium position entirely, creating a vacancy. The energy of the solid is thereby increased, and as in any physical process, equilibrium is reached when the energy is a minimum. Energy is minimized when the solid atoms have rearranged themselves to accommodate the presence of the vacancies. In equilibrium, there will be an equal

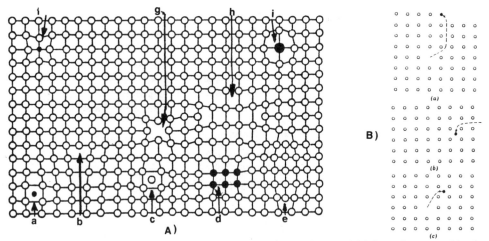

Fig. 1 (A) Various crystal defects in a simple cubic lattice a) interstitial impurity atom, b) edge dislocation, c) self-interstitial, d) coherent precipitate of substitutional atoms, e) small dislocation loop formed by agglomeration of self-interstitials, f) substitutional atom widening the lattice, g) vacancy, h) small dislocation loop formed by agglomeration of vacancies, i) substitutional impurity atom compressing the lattice[10]. Reprinted by permission of Springer-Verlag. (B) Point defects. a) Schottky defect, b) interstitial arriving from surface, c) Frenkel defect.

number of vacancies being produced and refilled, and the vacancy concentration at equilibrium, N_V, is found from:

$$N_V = N_o e^{(-E_{av} / kT)} \qquad (1)$$

where N_o is the number of lattice sites /cm^3, E_{av} is the activation energy required to create a Schottky defect, k is the Boltzmanns constant, and T is the temperature (°K). The value of E_{av} for silicon is ~2.6 eV. At room temperature (300°K), this would predict a vacancy concentration of ~1 in 10^{34} atoms, while at 1000°K the vacancy concentration is increased to ~1 in 10^{12}. Since processing involves the use of high temperatures, it is possible to form high vacancy levels which may not return to equilibrium values during the cool-down to room temperature. At 300°K it is estimated that each vacancy will succeed in migrating only once in 10^4 sec. At 1000°K, however, vacancies make ~10^8 successful hops per second, and are therefore in a state of continuous migration, creation, and recombination. If a hot crystal is rapidly cooled (quenched) the number of vacancy migration events per unit time is substantially decreased, with the result that the high temperature vacancy concentration remains (i.e. is quenched in) at the lower temperature. Thus, Eq. 1 may not describe the actual concentration of vacancies in a crystal.

The equilibrium concentration of self-interstitials, N_I, is similarly given by:

$$N_I = N_o e^{(-E_{ai} / kT)} \qquad (2)$$

where E_{ai} is the activation energy required to create a self-interstitial (~4.5 eV for Si).

Another method which also leads to the formation of vacancies consists of the removal of an atom from a lattice site, followed by its transfer to an interstitial site (Frenkel defect). This event (Fig. 1Bc) involves the formation of *two* point defects (a vacancy and an interstitial), while the Schottky defect involves only *one* (i.e. the formation of a vacancy, and the movement of the departing atom to the surface, thereby leaving only one point defect within the crystal lattice).

What is important about the fact that there are three events that can create vacancies and self-interstitials (Schottky, Frenkel, and movement of surface silicon atoms to interstitial sites), is that the formation mechanisms may act independently of one another. As a result, concentrations of vacancies and self-interstitials in a crystal cannot, in general, be expected to be equal.

Intrinsic point defects are also important in the kinetics of diffusion. The diffusion of many impurities depends on the vacancy concentration. The low packing-density of the diamond lattice (34% versus 74% for a fcc lattice) implies the existence of large spaces between atoms (interstices) into which atoms of the same size as the crystal atoms can be placed without shifting of the neighboring atoms (see Fig. 3 in Chap. 1). Thus, this structural condition of the silicon lattice favors the incorporation of interstitials, and is probably the main reason why the Si lattice has far fewer vacancies than typical metal lattices in equilibrium[27].

Even more critical than intrinsic point defects are *extrinsic point defects*, involving foreign atoms. When non-Si atoms occupy lattice sites, the defects are referred to as *substitutional impurities*. These constitute *bona fide* point defects, as such atoms are generally larger or smaller than the Si atoms that normally occupy lattice sites. Thus, crystalline regularity at that point is perturbed. Impurity atoms may also occupy non-lattice sites, in which case they are called *interstitial impurities*. To be electrically active, atoms usually must be located on lattice sites.

There is an inherent *solubility* of an impurity in a silicon crystal. That is, there is a maximum specific concentration that the Si lattice can accept in a solid solution of itself and the impurity, and this depends on the element as well as the temperature (Fig. 2). In this figure we observe that the solubility of most impurities in silicon increases up to some temperature, and then decreases as silicon approaches its melting point. This is known as *retrograde solubility*. If an impurity is introduced at temperature T_2 at the maximum concentration allowed by its solubility, the crystal will be saturated with the impurity at that temperature. If the crystal is

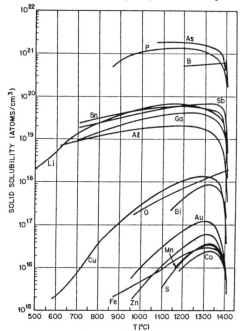

Fig. 2 Solid solubility of impurity elements in Si[60]. Reprinted with permission from the *Bell Systems Technical Journal*. Copyright 1960 AT&T.

then cooled to a lower temperature T_1, without removing the impurities from the solution, a *supersaturated* condition in the crystal is created. The crystal can only return to equilibrium (i.e. saturated condition) by precipitating impurities present above the solubility limit at the equilibrium temperature. The parameters affecting precipitation are the temperature of the crystal, the degree of supersaturation, time, the diffusion coefficients of intrinsic and extrinsic point defects, and the nucleating sites upon which precipitates can form. Extrinsic point defect precipitation creates volume defects, and these have been observed for B, O, and metallic impurities in Si. The important phenomenon of oxygen precipitation in Si is covered in detail later in the chapter.

Intrinsic point defects can also exist in supersaturated (or undersaturated) concentrations, and thus are subject to precipitation as well. In fact as we shall see, the precipitation of point defects is apparently involved in the formation of a variety of other crystalline defects.

One Dimensional Defects (Dislocations)

One dimensional (or line) defects in crystals take the form of *dislocations* [3,4]. They may be *edge* dislocations, *screw* dislocations, or *mixed* dislocations (which contain both screw and edge components). Dislocations are an important class of defect in silicon crystals, and a full treatment of their properties, formation, growth, and movement is a complex subject. Readers are directed to Ref. 4 for a comprehensive treatise on the subject of dislocations.

An *edge dislocation* is shown in Fig. 3a. Note that the upper half of the crystal contains an *extra half-plane* of atoms squeezed into the same space normally occupied by the atoms of an

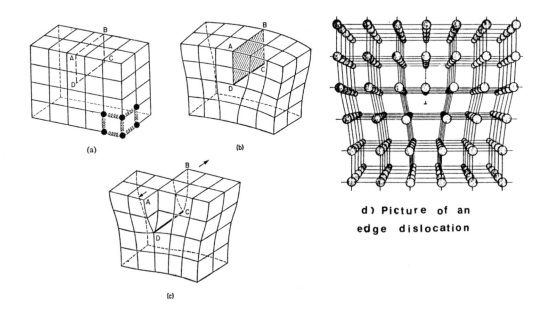

(a)

(b)

d) Picture of an edge dislocation

(c)

Fig. 3 (a) Model of an edge dislocation. (b) Positive edge dislocation formed by inserting an extra half-plane of atoms in ABCD. (c) Left-handed screw dislocation formed by displacing the faces of ABCD relative to one another in direction AB. From D. Hull, *Introduction to Dislocations*[3]. Reprinted with permission of Pergamon Press.

ideal crystal. The *dislocation* occurs at the termination of the sheet of extra atoms. An edge dislocation could be theoretically created by slicing the lattice and inserting an extra half-plane of atoms into the slice as shown in Figs. 3b. Stresses exist in the crystal as a result of the dislocation. The lattice planes in the upper half of the lattice near the dislocation are in *compressive* stress, while those in the lower half are in *tensile* stress.

In a *screw* dislocation, a displacement of half of the crystal occurs so that a ramp is formed (Fig. 3c). If one were to traverse a complete counterclockwise path around the periphery of a crystal containing a screw dislocation, as shown in Fig. 3c, the starting point would not be reached, but instead the end point would be one lattice spacing above the starting point.

The *dislocation line* in an edge dislocation is the line connecting all of the end atoms on the extra half-plane, while in a screw dislocation it is the axis of the screw. A vector notation developed by Burgers[3] characterizes dislocations in crystals, as well as dislocation interactions.

By definition, the *dislocation line* must either terminate on an external surface of the crystal, grain boundary (see Fig. 3b, for example), or form a closed curve if the dislocation lies entirely within the crystal (e.g. *dislocation loops*). That is, in a dislocation loop, an extra plane of atoms (or a missing plane of atoms) that exists entirely within the interior of the crystal must have a dislocation along the outer edge of the plane (Fig. 4a). The formation of such loops can occur in many ways, but one important mechanism has been postulated by Seitz[1]. That is, as point-defect clusters begin to form, they normally first exhibit a non-ordered (or semi-ordered) structure. As the clusters grow, ordered structures become more energetically favored, and a thin sheet, or platelet, structure is one possible type of ordered structure. That is, when enough point-defects (vacancies or extra atoms) agglomerate, they may collapse into a sheet along a plane (Fig. 4b). The edge of such a sheet (or platelet) forms the *bounding dislocation line*, or *dislocation loop*. A circular shape of such platelets is favored, as a circular shape has the shortest dislocation length (and thus has the lowest total dislocation energy). Nevertheless, platelets with other shapes (e.g. hexagonal) are also observed.

Single-crystal silicon starting wafers are free of macroscopic dislocations, although small dislocation loops can be formed in the manner described above. Macroscopic dislocations, however, can subsequently be introduced into silicon wafers by high-temperature processing.

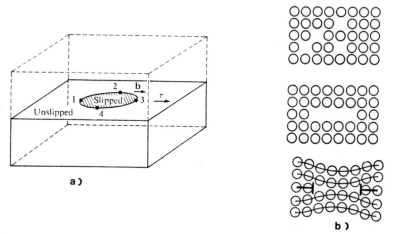

Fig. 4 (a) A dislocation line entirely within a lattice, forming a closed loop. (b) Formation of a prismatic dislocation loop by agglomeration and collapse of vacancies. (see also Fig. 8a).

These dislocations form by the growth and multiplication of dislocation loops in the bulk, or from dislocations generated at the surface in response to stresses created in the crystal. Stresses can arise in a number of ways including: a) differences in the expansion of different regions of the lattice, due to temperature variations in the crystal[5,6]; b) introduction of substitutional impurities in local regions of a crystal, which cause stresses between doped and undoped crystal regions (For example, impurity atoms smaller than silicon, such as B or P, will cause regions into which they are introduced to shrink, while larger atoms than Si will cause the region to expand. Dislocations that occur from these stresses are called *misfit dislocations*.); c) compressive stresses from volume mismatches that arise during some precipitation events (to be discussed in the section on *Bulk Defect* formation)[1]; and d) coefficient of thermal expansion induced stresses caused by layers present on the surface of the crystal[7] (e.g. stresses caused during the formation of so-called LOCOS isolation structures on VLSI wafers. LOCOS isolation [described further in Vol. 2] involves the growth of thermal oxide on local regions of the wafer surface, and suppression of the oxidation on other areas by the presence of nitride layers. Dislocation causing stresses result primarily from the tensile stress intrinsic to the silicon nitride, and from the volume expansion of the oxide formed in the walls.)

Dislocations arising from the growth and multiplication of dislocation sources as a result of thermal expansion stresses, need to be elaborated upon. These stresses can arise in a variety of ways in the thermal environments encountered by silicon crystals (both in ingot and wafer form).

For example, seed crystals undergo high thermal stress when first immersed into the Si melt during single crystal *ingot* growth. The exterior of the seed crystal is brought to the melt temperature, while the interior is initially much cooler. Stresses resulting from the expansion differences at the surface and interior of the seed crystal induce dislocations that then lead to plastic deformation of the crystal. (*Plastic deformation* is the permanent deformation of a material that remains after the stress on the material is released. *Elastic deformation*, on the other hand, is deformation that is lost upon the release of the stress. Plastic deformation occurs when the elastic limit [or *yield strength*] of a material is exceeded.)

Dislocations in *wafers* can also be induced by thermal stress during furnace operations[5]. Upon removing heated wafers from a furnace, the edges of the wafers cool faster than their centers (especially when the wafers are held vertically and spaced closely in wafer boats. The wafer edges can radiate heat to the cooler surroundings while the wafer centers are surrounded by the heated adjacent wafers). Thus, the resultant temperature gradient from edge to center causes the edges to undergo relatively more contraction than the centers. The thermal stress S from this uneven cooling can be estimated from :

$$S = \alpha \ E_f \ \Delta T \qquad\qquad (3)$$

where α is the coefficient of thermal expansion of silicon $\cong 4 \times 10^{-6}$ cm /cm°K, E_f is Youngs modulus ($\cong 1.5 \times 10^{12}$ dyn /cm^2), and ΔT is the temperature difference between the edge and the center of the wafer (temperature gradients in cooling wafers can reach >150°C). If the stress exceeds the *yield strength* of the material, dislocations will form. The stress from a 150°C gradient is $\cong 0.9 \times 10^9$ dyn /cm^2, and is in fact larger than the yield strength of silicon (0.45×10^9 dyn /cm^2 at 850°C, Fig. 16), and can thus introduce dislocations. We shall see that the yield strength of CZ Si wafers is impacted by the presence of impurities, such as oxygen.

Dislocations can move by two principal mechanisms: climb and glide. *Climb* occurs when point defects are absorbed by the dislocation line. Thus, if a self-interstitial is captured, an edge dislocation moves as shown in Fig. 5a, while the absorption of a vacancy causes the line to climb in the opposite direction. Dislocation loops also change size by climb-type events (absorption of point defects). Movement of the dislocation in the surface defined by its line and

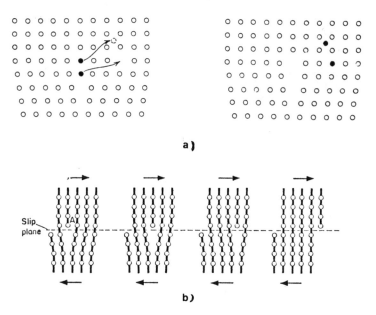

a)

b)

Fig. 5 Movement of a dislocation by (a) climb, and (b) glide. From D. Hull, *Introduction to Dislocations*[3]. Reprinted with permission of Pergamon Press.

Burgers vector is known as *glide*, and can occur when the lattice is subjected to shear stress. As shown in Fig. 5b, a lattice subjected to shear stress will cause the dislocation to move to the right by breaking the bond at A and attaching the bond at the right. The extra half-plane (and the dislocation line) is thereby moved to a position one lattice spacing to the right. The dislocation line is said to glide along the slip plane, and when an edge dislocation has glided throughout the entire crystal, the crystal is said to have suffered slip.   *Slip* is therefore seen to be a manifestation of plastic deformation. The glide plane of the silicon lattice is the {111}-plane. That is, dislocations glide along {111}-planes and in <110> directions. Some dislocation loops cannot move by glide (e.g. Frank-type partial dislocation loops) and are called *sessile*. Others, including perfect dislocations and Shockley partial dislocations, can move by glide under moderate stress, and are referred to as *glissile*.

The presence of dislocations allows slip in crystals to occur in response to much lower shear stresses than in dislocation-free crystals. In perfect crystals, slip would occur only by *simultaneous* displacement of all the atoms along a slip plane. This would require the application of much larger shear stress than that necessary to break and move just a single bond (which is all that must be done to move an edge dislocation as shown in Fig. 5b). In fact dislocations were first hypothesized to help account for the fact that the theoretical shear stress for slip was many times greater than experimentally observed stress.

The thermally induced dislocations as a result of wafer cooling upon being withdrawn from a furnace are generated at the wafer edge, where surface irregularities and lattice damage act preferentially as the dislocation sources. The dislocations propagate from the edge toward the center of the wafer, relieving some of the mechanical stress, and producing slip. Figure 6a shows an x-ray topograph of slip patterns in a <111> wafer.   These dislocations are composed

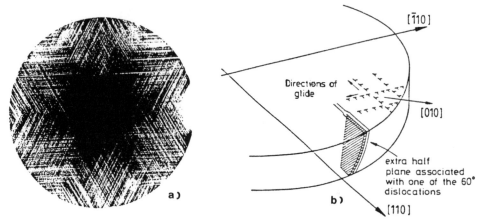

Fig. 6 (a) A (111)-oriented Si wafer in which the severe slip has been delineated by etching to show the emergence of the dislocations. (b) Schematic of the slippage distribution of (a)[5]. Reprinted with permission of the American Physical Society.

predominantly of 60° dislocations with surface-parallel Burgers vectors as illustrated in Fig. 6b. These dislocations and their attendant slip and warpage must be prevented. Various procedures for this purpose are discussed in the later section entitled *Wafer Resistance To Warpage*.

When a material contains dislocations and is subject to sufficient stress, the dislocations can also multiply. One important mechanism that causes dislocation multiplication is the *Frank-Read* source. It consists of a segment of a dislocation (Fig. 7) which is *anchored* at two points (e.g. by impurity atoms or precipitates). Such anchoring, often also called *pinning* or *locking*, is due to the interaction of the impurity atom and the dislocation. If a sufficiently large stress is applied, the dislocation is caused to bow, just as a bubble of soap film bows under applied pressure. The bowing goes through successive stages as illustrated in Fig. 7a, until the two parts of the dislocation loop meet behind the pinning points. They then pinch off to form a loop, which may subsequently glide over the entire slip plane. The source is restored to its original state, and another loop can be formed if the stress is still present. Thus, the multiplication process can be repeated many times, and a large number of new dislocation loops can be produced. Figure 7b is a photograph of an actual Frank-Read source in silicon.

The distortion of the lattice causes the dislocation to become a site that is energetically favored for occupancy by impurity atoms. That is, diffusing impurity atoms are likely to be *captured* at dislocation sites, because they distort the lattice less at such sites than elsewhere in the perfect lattice. This results in a lower free energy of the crystal. Impurity atoms larger than silicon, occupy the regions nearby the dislocation that are in compression, while those smaller than silicon locate in the tensile-stressed portions. This phenomenon is very important in the processing of silicon, as dislocations trap fast-diffusing impurities in silicon. The effect can be harmful to device operation, or if correctly harnessed, can be used to advantage (e.g. through the intentional application of gettering techniques). When impurities have segregated themselves at dislocations sites, the dislocation is said to be *decorated* with them, while *pure* dislocations are free of any decorated impurities.

Area Defects

Area defects in single crystals include stacking faults, grain boundaries, and twin

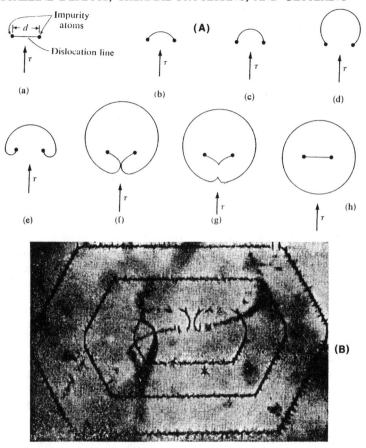

Fig. 7 (A) A Frank-Read source is capable of producing dislocations under the action of an applied shear stress. Various stages of operation of this source are illustrated. **(B)** Micrograph of a Frank-Read source in a Si crystal. (Dash, 1957 *Dislocation and Mechanical Properties of Crystals*, Wiley)

boundaries[1]. Stacking faults are the most important area defects that degrade the device performance of integrated circuit components, and thus our discussion shall focus on them.

In Chap. 1, it was shown that the diamond lattice consists of three double {111}-planes (i.e. A, B, and C) that are stacked in a regular order in a perfect crystal (...ABCABCABC...). If an extra double {111}-plane is inserted into (or removed from) a diamond lattice, the correct stacking order of the {111}-planes is disturbed, and a *stacking fault* is said to be created. For example, if a {111}-A plane is inserted between the B and C planes, an ...ABCABÁCA... structure is created. As illustrated in Fig. 8a, the insertion of an extra plane creates an *extrinsic* stacking fault, while the removal of a plane creates an *intrinsic* stacking fault.

In bulk silicon, all stacking faults occur on the {111}-lattice planes and extrinsic faults (Frank loops, bounded by Frank partial dislocations[10], with a Burgers vector of $\mathbf{b} = (a/3)<111>$) have predominantly been observed (Fig. 8b). Those stacking faults that intersect the surface (*surface stacking faults*) are bounded by such partial dislocations along the interior stacking fault boundary (Fig. 8c).

The formation of extrinsic stacking faults (both bulk and surface) involves the local accumulation and condensation of excess atoms, in a manner similar to the phenomena that cause dislocations. Stacking faults can form in IC processing during: a) epitaxy (see Chap. 5); b) ion-implantation (see Chap. 9); and c) oxidation. The sources of these excess atoms include supersaturated self-interstitials, self-interstitials created during the growth of oxygen precipitates, or silicon atoms generated by the formation of SiO_2 at the wafer surface. Dislocations formed during oxidation are known as *oxidation induced stacking faults* (OISF). Booker and Tunstall[43], using TEM, identified OISF as extrinsic stacking faults bounded by Frank partial dislocations. Their extrinsic nature suggests that they are formed by the injection of Si self-interstitials into the lattice, which are generated in the course of oxidation. It appears that there are a variety of surface sites at which OISF can nucleate, including: a) swirl-defects (point defect clusters present in the as-grown crystal); b) surface mechanical damage from the wafering process; c) chemical contamination; d) s-pits (whose presence is revealed in the form of *haze* after a defect etch), due to metal precipitates; and e) oxygen precipitates.

It is also important to note that since stacking faults are bounded by dislocations, fast-diffusing impurities from the surrounding lattice can be absorbed at these dislocations. Various device-degrading mechanisms are manifested when such stacking faults become electrically active, as the result of being *decorated* by impurities. For example, excess reverse-bias currents in p-n junctions, and storage-time degradation in MOS circuits, both of which are described in more detail in a subsequent section, can be caused by electrically active stacking faults.

OISF not only grow during oxidation, but can be caused to shrink as well (Fig. 9). At the retrogrowth temperatures (>1200°C), stacking fault formation is suppressed, and existing faults grow smaller[12]. Apparently at these temperatures, the higher solubility limits allow the clustered point defects (which are one type of OISF nucleation site) to dissolve back into solution. The presence of HCl during oxidation can also produce stacking fault shrinkage[13]. The model proposed to explain this mechanism is as follows: If wafers are heated in a non-chlorine oxidizing ambient, vacant sites on the wafer surface become occupied by oxygen atoms. This has the effect of reducing the level of vacancies below the *surface* thermal equilibrium value. One manner by which the surface vacancy level is restored, is through the growth of OISF (i.e. OISF expand by emitting vacancies). When chlorine is added to the oxidizing ambient, a possible new source of vacancies is introduced. It is suggested that the high-temperature chlorine atoms reaching the Si-SiO$_2$ interface remove silicon atoms from the silicon region. This generates an

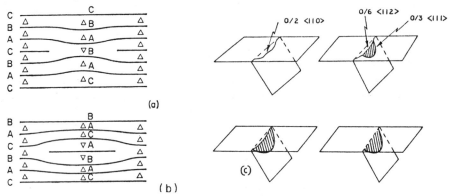

Fig. 8 (a) Intrinsic, and (b) extrinsic stacking faults in a fcc lattice. (c) Generation of an extrinsic fault from a dislocation at a surface. From Ravi, *Imperfections and Impurities in Semiconductor Silicon*, Copyright © 1981 John Wiley & Sons. Reprinted by Permission of John Wiley & Sons.

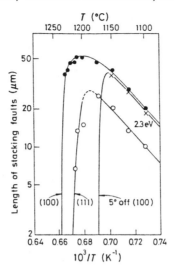

Fig. 9 Temperature dependence of the length of oxidation-induced stacking faults after 3 hr of dry oxidation for different wafer orientations[12]. Reprinted with permission of American Physical Soc.

increased number of surface vacancies to interact with the excess Si interstitials that exist in the OISF. The OISF thereby shrink, possibly to annihilation. A last technique for reducing the incidence of OISF is to eliminate the sources that produce the other types of nucleation sites[17].

A *twin boundary* is a defect that occurs during the growth of the silicon ingot and can be detected with the unaided eye upon inspecting as grown crystals. *Microtwins* also form during epitaxial film growth (Chap. 5). During crystal growth, a perturbation may cause the mirror image of the regular lattice to be formed, with the twin boundary acting as the mirror plane.

The boundaries between two grains which have only a small angle of misorientation (*low-angle grain boundaries*) are viewed as consisting of an array of edge dislocations (Fig. 10). Thus, *grain boundaries* are considered to be area defects in the crystal in which certain planes of atoms terminate and fill-in the wedge-shaped region between the two grains. The atoms in the region of the boundary are displaced from their normal position, but this displacement is confined to a region very close to the grain boundary.

Bulk Defects and Precipitation

Bulk (or volume) defects in crystals can include *voids* and *local amorphous regions*, but the most important volume defects are *precipitates* of extrinsic or intrinsic point defects. The formation and growth of precipitates is a very complex subject. It is believed that precipitates first form (or *nucleate*) an intermediate crystal structure that is later transformed into the final structure. The initially formed particles are called *nuclei*. The nature of nuclei structures is difficult to determine because of their small size at the time of their formation. The nucleation process can occur in two basic ways: a) *heterogeneous nucleation*, or nucleation at a crystalline defect (e.g. a dislocation, grain boundary, or impurity); and b) *homogeneous nucleation*, or formation of nuclei by a random composition fluctuation of the solute in an otherwise perfect crystal (i.e. in which the solute atoms coincidentally cluster in the crystalline lattice, and thereby form a nucleus). Homogeneous nucleation can take place if there is a large enough driving force (e.g. a departure from equilibrium, such as supersaturation of solute atoms in the crystal).

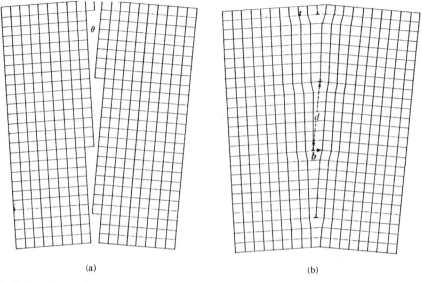

(a)   (b)

Fig. 10 Dislocation model of grain boundary. Two grains with an angle of misorientation θ are shown in (a). As the two grains form a wedge-shaped intersection, certain rows of atoms cannot extend their full lengths and the dislocation of (b) results.

In general, the more rapid the diffusion of the solute atoms in the crystal, the more quickly precipitation occurs. In dislocation-free Si ingots and starting wafers, homogeneous nucleation is the dominant nucleation mechanism of bulk defects.

The nuclei formed in homogeneous nucleation must become large enough to exceed a certain critical radius in order for the precipitate to grow. The reason for this can be illustrated by considering the combination of the surface energy and volume energy that is associated with an object as it grows. (This argument is based on the classical *Capillarity Model for Homogeneous Nucleation from the Vapor Phase*, developed by Volmer and Weber[55].) Figure 11 shows how the free energy vs. the radius of a growing precipitate particle varies. As a particle grows due to precipitation, new surface is created. To create this surface, energy must be added, and thus the *surface free energy* is always positive. The *volume free energy* is negative, since as the precipitates grow, solute is removed from solution and the overall material system returns closer to equilibrium. When a particle is very small, it has a very large surface-to-volume ratio. Therefore, as such a small particle grows, the total free energy must increase. That is, the two opposing free energies are added together, and the magnitude of the incremental surface free energy is larger than the magnitude of the volume free energy. As the radius of the particle grows in size, the sum of the free energies goes through a maximum at a radius r_0 (called the critical radius). The condition of maximum free energy corresponds to the state of minimum stability of the precipitate particle. Therefore, if the radius is less than r_0 a particle can lower its free energy by *decreasing* its size, and thus the particle will tend to dissolve. Particles with radius larger than r_0 can decrease their overall free energy by *growing*. Therefore, in order to create permanent precipitates, nuclei of critical size (or larger) must first be formed. Note that the size of r_0 increases for increasing temperature of the solid solute.

The formation of such nuclei for homogeneous precipitation (Fig. 11b & c) has a low probability, since the coincidental simultaneous conglomeration of a large number of point defects is an unlikely event, and agglomerates of only a few atoms are not stable. For example, it has been reported[3] that r_o is about 10 Å for oxygen (with a volume of ~4×10^3 Å, which contains ~200 oxygen atoms) in silicon at 1150°C[14]. This unlikely formation of point-defect clusters composed of oxygen atoms must nevertheless occur from time-to-time in CZ Si, since such clusters are ultimately responsible for the precipitation and gettering of point defects, and also contribute to the mechanical strength of wafers during processing. (See *Oxygen in Silicon* section.)

The rate of formation of precipitates depends on the temperature, degree of supersaturation, and time of cooling (or annealing). Each point defect requires a certain time, depending on its diffusivity, to reach the precipitate. Thus, faster diffusing point defects may form the first precipitates, even though slower diffusing point defects may also be present in a crystal at higher levels of supersaturation. Once precipitation occurs, and nuclei of one point-defect type are formed, further precipitation by condensation at the first nuclei can take place more easily for all types of point defects[14].

The mechanism of precipitation that occurs in a freshly-grown CZ crystal as it cools to room temperature provides a useful example of the general mechanism of precipitation. In such a cooling crystal, there are more point defects (self-interstitials, vacancies, and impurities) than are soluble at the lowered temperature. This creates a departure from equilibrium as the crystal cools, and establishes a driving force that invites precipitation. Specifically, while the interior of the ingot is still hot (~400-1200°C), any fast-diffusing supersaturated point defects (e.g. oxygen or self-interstitials) can still move over significant distances in the crystal. Precipitates can nucleate on extended crystal defects, such as dislocations and stacking faults. In the days prior to availability of dislocation-free silicon, point-defect precipitation and excess point defect annihilation occurred primarily on such grown-in dislocations. However, if there are no extended crystal defects that can act as nucleation centers, as in dislocation-free CZ silicon crystals, the point-defect type which first exceeds the solubility limit during cooling is then able to form precipitation nuclei with its own species (*homogeneous nucleation*). These types of nuclei (Fig. 11b) are sometimes referred to as *point-defect clusters* [10].

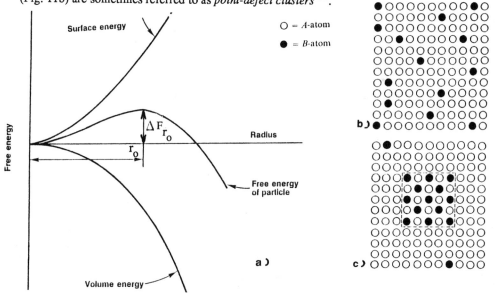

Fig. 11 (a) Free energy of a precipitate particle as a function of its radius. (b) and (c) Formation of nuclei. A supersaturated solid solution of B in A contains excess of B atoms. By holding at a temperature where atoms have sufficient mobility, rearrangement will occur and B-rich nuclei form.

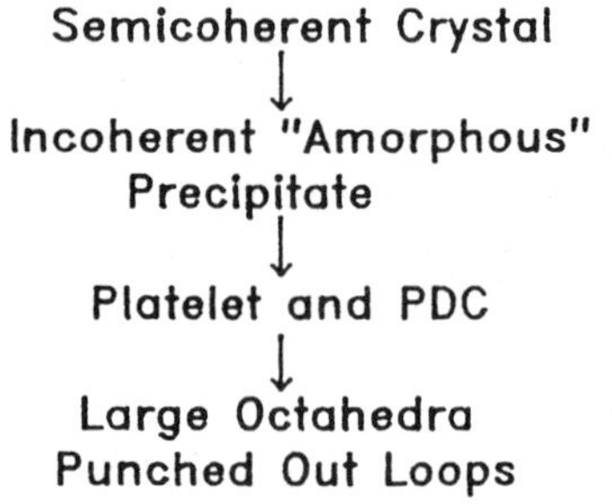

Semicoherent Crystal

$\downarrow$

Incoherent "Amorphous"
Precipitate

$\downarrow$

Platelet and PDC

$\downarrow$

Large Octahedra
Punched Out Loops

Fig. 12 Steps associated with the precipitation of oxygen in Si[51].

The *nucleation and growth of oxygen precipitates* in Si appears to follow the path of events shown in Fig. 12. The first stage in the formation of *oxygen point-defect clusters* in Si is believed to consist of a *semi-coherent crystallite* which is metastable below 900°C[51](Fig. 13a). The second stage involves a phase transition to the poorly crystallized SiO_x structure, which is often seen as a *platelet* after additional oxygen atoms are added to the crystallite, and it has reached a critical size. The phase transition is triggered by the size of the precipitate, since addition of oxygen atoms changes the relative surface-to-volume energy of the crystallite. The poorly crystallized precipitates grow by taking up oxygen atoms along <110> edges. At higher temperatures (>1100°C), the platelets appear to grow into large octahdera. These large symmetric precipitates are often associated with prismatic punching of perfect dislocation loops.

The bulk defects (platelets and octahedra) themselves can be responsible for creating other crystalline defects as they grow. For example, the growth of oxygen precipitates in silicon occurs by the reaction of oxygen with silicon at the precipitate surface to form SiO_2. The volume change from Si to SiO_2 is quite large (Volume $_{SiO_2}$/Volume$_{Si}$ ~2). Consequently the lattice surrounding such a precipitate is subject to large compressive stresses. This compression can be relieved in a number of ways, all involving the generation of other crystal defects, including: a) emission of Si self-interstitials; b) absorption of vacancies; or c) punching out of dislocation loops[10] (Fig. 13b).

The production of self-interstitials to accommodate volume expansion is apparently predominant, and thus such oxygen precipitates are sources of Si self-interstitials. These emitted self-interstitials also have a propensity to form dislocation loops (containing an extrinsic stacking fault between two {111}-planes), since a loop represents an energetically more favorable state than self-interstitials distributed randomly in the lattice. These stacking faults also grow by absorbing the self-interstitials (at the bounding dislocation loop) that are continually emitted by the growing precipitate[16]. It is observed that the oxygen precipitates themselves do not severely degrade device performance and yield, but they indirectly cause such effects by generating dislocation loops and stacking faults during their growth.

Another important class of point-defect clusters involves metallic impurities that interact with other defects. This class of defects is macroscopically observed as *haze*, and after preferential chemical etching, is revealed in the form of microscopic saucer or *s-pits* [17,18]. Haze and s-pits cannot be ignored since they have been identified as nuclei for stacking faults during epitaxial deposition, and as degenerating into stacking faults during subsequent oxidation. Metallic precipitates, Fe, Co, Ni, and Cu are most often found to be associated with haze, and such particles are found to form during furnace operations with contamination native to the wafer, or introduced by a process. One important source of such metal impurities has been found to be the heating coils that surround furnace tubes. At high temperatures the metal atoms from these coils can diffuse through the fused silica tubes and then deposit on the wafer surface. Back-surface gettering, wafer clean-up procedures, use of HCl in the oxidation ambient[19], and double-walled furnace tubes have all been suggested as techniques to suppress haze and its effects on device performance (see *Gettering* section for more information).

INFLUENCE OF CRYSTAL DEFECTS ON DEVICE PROPERTIES

The information gained from the study of defects in crystalline silicon of most interest to process engineers is the manner in which the defects influence integrated circuit device behavior. That is the subject of this section[1,20,57]. Some defects which can negatively impact device performance are present in the starting material, but other defects which adversely alter device properties are induced by subsequent processing. Among the most important device /material properties that are influenced by crystalline defects include: a) leakage currents in p-n junctions; b) collector-emitter leakage currents in bipolar transistors; c) minority carrier lifetimes; d) gate-oxide quality; e) threshold voltage uniformity in MOS devices; and f) resistance to warpage by wafers during thermal process steps.

Leakage Currents in P-N Junctions

Reverse-biased p-n junctions must exhibit low leakage current levels for VLSI circuits (especially for CMOS circuits) in order to maintain low power-delay products. Excess current flow will clearly increase the required power dissipated by a circuit. Large leakage currents also degrade refresh times in dynamic RAMs and adversely impacts bipolar IC performance. Both

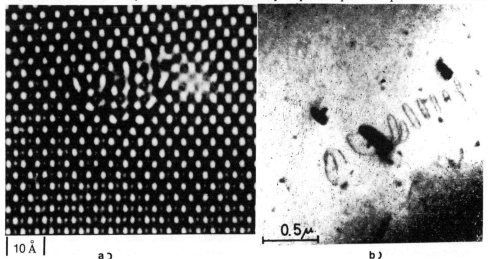

10 Å   a)   b)

Fig. 13 (a) Image of the core of a dislocation in Si obtained by high resolution TEM depicts an extra plane of atoms that terminates at the core of the dislocation. A precipitate is at the core[58]. (© 1985 IEEE) (b) Dislocation loops generated in Si due to prismatic punching by SiO_2 particles[59].

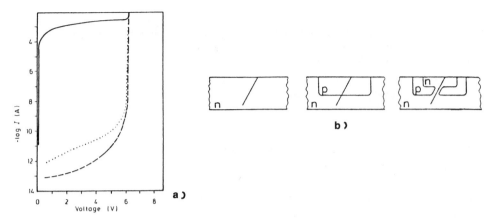

Fig. 14 (a) Typical characteristics of single-emitter test transistor (emitter area 3 x 8 μm) with emitter collector short showing up in I_{CEO} /V_{CE} (---), I_{CB} /V_{CB} (- - -), and I_{EB} /V_{EB} (...) are "hard"[57]. Reprinted with permission of Academic Press. (b) Schematic of enhanced emitter diffusion model to explain collector-emitter pipe formation[57]. Reprinted with permission of the publisher, the Electrochemical Society.

precipitates and dislocations can increase p-n junction leakage.

The presence of transition metal precipitates in the p-n junction produces leakage due to mid-gap energy levels at low voltages, and a "soft" leakage component at higher voltages. Both phenomena adversely alter the ideal p-n junction reverse-bias I-V characteristic[20].

Dislocations, and thereby extrinsic stacking faults which are bounded by dislocations that cross p-n junctions, can also degrade the reverse-bias I-V characteristics of p-n junctions. The formation of generation-recombination (g-r) centers at defect sites in the p-n junctions, and the decoration of dislocations (including the bounding dislocations of stacking faults that pass through the junction) with electrically active (metal) impurities, lead to excess leakage currents. The g-r centers arise both from strain fields of the pure dislocation in the lattice, and from the strain generated by the impurities that such dislocations accumulate.

Collector-Emitter Leakage in Bipolar Transistors

The largest yield-reducing failures in bipolar transistors are due to various collector-to-emitter leakage currents. In devices exhibiting such failures, only the collector-emitter (C-E) reverse characteristics exhibit high leakage current, while the E-B and C-B reverse characteristics usually continue to have low-leakage (Fig. 14a). Such collector-emitter failures can arise from a number of defects, including dislocations, stacking faults, or precipitates.

Collector-to-emitter leakages have been correlated with dislocations that pass through the transistor from emitter to collector. If the dislocation is decorated with metallic impurities, it can be conductive enough to permit significant current flow between collector and emitter even when the base terminal is open (I_{CEO}). The dislocation passing through the same region can also play a role in enhancing diffusion along the dislocation during emitter formation (Fig. 14b). This can lead to emitter-collector *pipes*, which allow large collector-to-emitter current flow when there is no base current. Since stacking faults are bounded by dislocations, their role is simi-lar[20]. Most experimental data regarding pipes supports the enhanced diffusion model.

It has also been suggested that the presence of precipitates contributes to locally *retarded diffusion* of dopant atoms in shallow double-diffused structures. The dopant appears not to diffuse as rapidly in the region of the precipitate, forming a localized spike pointing upward toward the wafer surface[20]. The precipitate apparently dissolves during the first diffusion. The second diffusion is thus not effected by the precipitate, and its diffusion front is uniform. The locally narrow separation of emitter and collector at the spike causes excessively large reverse currents. However, this failure is not expected to occur in ion-implanted devices.

Minority Carrier Lifetimes

Minority carrier lifetime is defined as the mean time spent by excess charge carriers before they recombine to re-establish the thermal equilibrium carrier concentration. In general, long minority carrier lifetimes are beneficial to device operation. For example, long lifetimes (as well as low leakage currents) in MOS dynamic RAMs permit longer retention of charge, thereby allowing an increase in the time between refresh cycles. Lifetimes, however, are reduced by various defects through the introduction of localized intermediate energy levels within the silicon bandgap. These localized energy levels can be introduced by point defects, point-defect clusters, dislocations, and crystal strain[20]. Figure 15 shows a Frank partial dislocation, decorated with impurities, that terminates within the space charge region of an MOS capacitor. This defect caused the minority-carrier surface generation lifetime to be degraded from 500 µs to 2 µs.

On the other hand, reduced lifetimes are an advantage in regions of some devices away from where the active devices reside. For example, in CMOS circuits, short lifetimes in the bulk can help suppress *latch-up* by reducing the current gain of the parasitic transistor.

Gate Oxide Quality

Defects in the silicon substrate apparently impact two important parameters of the thin gate oxides used in MOS devices: 1) the oxide leakage current; and 2) the oxide breakdown voltage. The degradation of both of these properties correspond to stacking faults at the silicon substrate that are generated by metallic contamination during oxidation[21]. In addition, low breakdown voltages in thin oxides has been found to correlate with high defect density on the wafer surface. A higher defect density in gate oxide layers was also found on wafers with oxygen precipitates at the silicon surface[25].

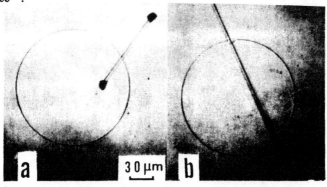

Fig. 15 (a) Minority carrier surface generation lifetime τ_g = 2 µs when a Frank partial dislocation is present in the gate area of an MOS capacitor. (b) When the stacking fault completely crosses the gate area τ_g = 500 µs (i.e. no dislocation line intersects surface and junction)[56]. Copyright ASTM. Reprinted with permission.

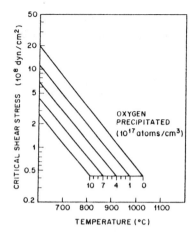

Fig. 16 Yield strength of Si showing the influence of oxygen precipitates[22]. Reprinted with permission of the publisher, the Electrochemical Society.

Threshold Voltage Control

One of the factors that determines the threshold voltage V_T of MOS transistors, is the resistivity of the substrate. Due to the limitations of the CZ growth process discussed in Chap. 1, the resistivity from wafer-to-wafer in a lot can vary, as can the macroscopic and microscopic radial resistivity. In addition, thermal donors from oxygen precipitates can significantly alter the carrier concentrations in low resistivity material (see the *Oxygen in Silicon* section). It has been calculated that for an MOS process with a substrate concentration of $N_A = 4 \times 10^{15}$ cm^{-3}, a 250Å gate oxide and with a maximum allowed variation in V_T between bits of 0.03V, a maximum 10 percent variation in the resistivity could be tolerated[21].

Wafer Resistance to Warpage and Slippage

The flatness of a wafer must be tightly maintained throughout the entire fabrication process in order to allow fine geometrical features to be precisely printed. It has been shown that plastic deformation of silicon wafers can contribute to *lateral* wafer deformation up to 0.5 μm, which can have significant impact on layer-to-layer registration. Note that use of steppers (Chap. 13) alleviates this problem, as registration is performed over smaller fields. Warpage causing stresses, however, arise in thermal processing steps. When the stress in a wafer exceeds the yield strength of Si (Fig. 16), dislocations grow and multiply, and plastic deformation occurs to partially relieve the stress. When wafers are cooled to room temperature, a reversed stress distribution, which cannot be relieved by plastic deformation, is induced. This stress causes wafers to buckle (warp), destroying their planar form. Since larger gradients occur at higher temperatures during cooling, and since yield strength decreases significantly at higher temperatures, slip primarily occurs during cooling portions of thermal steps.

The thermal stress that brings about warpage is a cumulative effect. Therefore, warpage gets progressively worse throughout a process, and will be most pronounced at contact and metal patterning. As the diameter of wafers increases, the magnitude of such forces also increases, necessitating the use of thicker wafers. On the other hand, edge rounding, improved polishing techniques, and suppression of microdefects during growth are recent advances that have lowered the density of potential dislocation sources in wafers. These measures aid in decreasing slippage and warpage. Process optimization steps to reduce thermal stresses, such as slow push-pull of wafers into furnaces and ramping furnace temperatures up and back down, are also recognized as being necessary. Finally, the strength-increasing properties of oxygen[22] (or possibly nitrogen[23]) in silicon must be utilized. That is, a concentration level of oxygen that produces optimum yield strength needs to be present in the wafer in interstitial (or small cluster) form. If the oxygen is allowed to precipitate such that either the density of precipitates exceeds 25×10^9 defects /cm^{-3}, or the size of the precipitates becomes too large[25], excessive warpage can occur (see Fig. 23a).

CHARACTERIZATION OF CRYSTAL DEFECTS

The detection of crystalline defects in silicon is performed with a variety of measurement techniques, each utilized for an inherent measurement advantage. Table 3 summarizes these techniques, and where appropriate, lists other chapters of the book that contain further details on them. Chemical etching to delineate defects is discussed in this section. The most useful application of these techniques is to detect impurities, defects, and other properties that are significant to device operation, while ignoring insignificant properties. It should be noted that most of the available defect characterization techniques are *inferential* rather that direct. That is, the measurements produce data about the manifestation of an interaction between the object of the measurement and an externally imposed stimulus. This requires good models on which to base the interpretation and suggests that it is wise to compare results from several different methods to insure that the desired properties are in fact measured[26].

Chemical etching techniques can be used to rapidly and inexpensively evaluate the degree of crystal perfection, as well as to determine the presence and type of defects present in a crystal. Normally, dislocations, stacking faults, swirls, and saucer pit defects are observable with chemical etching and inspection by optical microscopy. The method employs strong chemically oxidizing etchants to delineate crystalline defects, by preferentially etching the points on a wafer where crystal defects have intersected the surfaces. Excellent definition of the density and

Table 3. SOME IMPORTANT TECHNIQUES FOR ANALYZING DEFECTS

Class of Analysis	Characterization	Where Discussed in Text
Electronic	1. Electrical measurements	Chaps. 4, 5, 7, 8, 9
	2. Scanning electron microscope in EBIC mode.	Chap. 17
Structural	1. X-Ray topography	Chap. 17
	2. Preferential etching	This section, and Chap. 15
	3. Transmission electron microscopy	Chap. 17
Chemical	1. Neutron activation analysis	Chap. 17
	2. Auger electron spectroscopy	Chap. 17
	3. Secondary ion mass spectroscopy	Chap. 17
	4. Deep level transient spectroscopy	

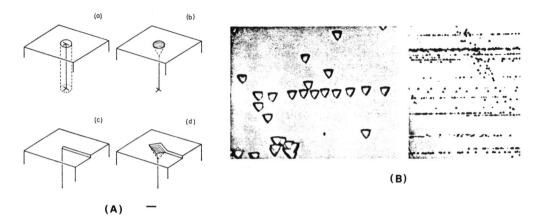

Fig. 17 (A) Example of how etch pits are formed where dislocations intersect a surface. The shape of the pit depends on the type of etchant and the crystal orientation. (B) Micrograph of etch pits due to dislocations.

distribution of structural defects across the surface is obtained. The type of defect is inferred from the type of structure observed after the preferential etch.

Depending on the type of defect, etchant, and crystal orientation, the etched features may exhibit a specific structure. The region near *dislocation lines* usually etches more rapidly than the rest of the crystal, and thus develops etch pits as indicated in Fig. 17. Dislocation pits, however, usually continue to etch rapidly and thus maintain pointed bottoms, whereas the pits resulting from surface damage tend to become flat bottomed. *Stacking faults,* when appropriately etched, produce grooves where the fault planes intersect the wafer surface, and thus give the outlines of various shapes (see for example Fig. 17). Three etchants (*Sirtl, Secco, and Wright*) are commonly used to delineate surface defects. Their formulations and comments about conditions under which each gives the best results are given in Chap. 15.

There are however, several limitations to this technique. First, the detailed structure of the defect cannot be determined, and impurities associated with the defect cannot be identified. Second, the method is destructive, and the object of interest (the defect) is consumed before the optical analysis is performed. Finally, the etchants normally contain (and hence heavily contaminate the silicon slice with) elements such as Cr and Cu. Thus, such etched wafers must be kept out of furnace tubes, and the etching must be performed in containers that are used only for that express purpose.

THERMAL PROCESSING

Silicon wafers are subjected to many elevated temperature steps during the fabrication of VLSI. In general, VLSI devices are fabricated by creating precisely controlled regions of dopants in silicon wafers. The thermal treatments however, can cause the dopants to diffuse and therby reduce the control of their concentration and locations. For devices fabricated with feature sizes in the 1-2 μm range, dopant diffusion (both vertically and horizontally) must be reduced to

Table. 4 · THERMAL PROCESSES IN VLSI FABRICATION

1. Defect Removal -
   - Thermal donor annihilation: Chap. 2
   - Formation of denuded zone on wafer surface: 2
   - Oxygen precipitate formation for intrinsic gettering: 2
   - Thermal cycle for extrinsic gettering: 2
   - Thermal reduction of oxidation-induced stacking faults: 2
   - Removal of crystal damage from ion-implantation: 2
   - Removal of radiation damage from E-gun evaporation or reactive-ion-etching: Chaps 10 & 16.

2. Active Device Fabrication -
   - Dopant incorporation by diffusion: Chap. 8
   - CVD of poly-Si, dielectrics, conductors: Chaps. 6 & 11
   - Dopant activation following ion-implantation: Chap. 9

3. Interconnect Formation -
   - Silicide formation: Chap. 11
   - Contact sintering: Chap. 10
   - CVD glass densification: Chap. 6
   - CVD glass reflow: Chap. 6

   - Epitaxial deposition: Chap. 5
   - Thermal Oxidation: Chap. 7
   - Interface state density reduction: Chap. 7

maintain shallow junctions, controlled gate lengths, etc. Extensive research has been conducted on reducing the magnitudes of the thermal treatments to which wafers must be subjected. In this section we shall provide an overview of the role of thermal processing and describe techniques for minimizing thermal cycles. VLSI fabrication processes that entail thermal steps are first listed, although a detailed discussion is reserved for later chapters. The techniques and equipment used to perform thermal processing will be discussed, emphasizing the important topic of rapid thermal processing (RTP). Thermal treatments during VLSI fabrication can be divided according to their applications into three categories, defect removal, active device fabrication, interconnect formation. Table 4 shows the thermal cycles involved in each of these process categories.

Rapid Thermal Processing (RTP)

Many of the thermal steps in VLSI fabrication are performed at temperatures high enough (900°C) to cause unwanted dopant diffusion. Two approaches are utilized to minimize such diffusion: 1) *low-temperature processing*, such as high pressure oxidation and lower reflow temperature of CVD glasses; and 2) *short-time, high temperature processing*. Some processes are better suited to one or the other of these approaches, and thus both find useful application in VLSI fabrication. Low-temperature thermal processes are discussed in subsequent chapters.

The durations of short-time, high-temperature techniques range from seconds to a few minutes. Thus, wafers are subjected to high temperatures only long enough to achieve the desired process effect, and dopant diffusion is minimized. Short-time thermal processes, however, must be performed in specially designed systems. In conventional furnaces or reactors, the large thermal mass of the susceptors, wafer boats, and reactor walls, rules them out as possible systems for performing short thermal cycles. In addition, if large diameter wafers are heated or cooled too rapidly in a furnace, wafer warpage or slip can result.

Initial work on the rapid heating of wafers was conducted using lasers as the energy sources. Lasers allowed high heating to occur within fractions of a microsecond without significant thermal diffusion. However, the wafer surfaces had to be scanned by the small spot-size laser beams, and this caused lateral thermal gradients and resultant wafer warpage.

Subsequently, large-area incoherent energy sources were developed to overcome these limitations. These sources emit radiant light, which then heats the wafers. This allows very rapid and uniform heating and cooling. Systems are available in which such rapid thermal processing (RTP) is performed[30,31] (Fig. 18). Wafers in RTP systems are thermally isolated, so that radiant (not conductive) heating and cooling is dominant. Temperature uniformity is an important design consideration in these systems so that thermal gradients, which can cause slip or warpage, are minimized. Various heat sources are utilized, including arc lamps, tungsten-halogen lamps, and resistively-heated slotted graphite sheets. The heating chamber provides a controlled environment for the wafer, and for coupling energy from the radiant energy sources to the wafers. Most heating is done in inert atmospheres (Ar or N_2) or vacuum, although oxygen or ammonia for growth of SiO_2 and Si_3N_4 are introduced into the chamber in RTP systems.

Precise control of time and temperature is necessary to obtain reproducible RTP results. Optical pyrometers and closed-loop temperature control mechanisms are used to ensure set-point and cycle repeatability. A wide range of temperatures is also a desirable option in RTP systems since the thermal processes that can be carried out by RTP extend from 420-1150°C. A flexible, easy-to-program machine interface is also useful to allow a variety of processes to be performed in the system. The throughput of RTP systems can be quite high. Typically 60-200 125 mm wafers can be processed per hour by a single RTP system.

RTP has been successfully implemented in many thermal fabrication processes, and new applications for the technique are still being discovered. It was first widely adopted in the VLSI production environment for annealing implantation monitors (see Chap. 9). It has also has been used to grow films of SiO_2 and Si_3N_4, to form junctions in bipolar transistors through controlled diffusion, to annihilate thermal donors[29] (see following section on *Oxygen in Silicon*), and to activate dopants after ion-implantation in both single-crystal silicon and poly-Si. In interconnect formation, RTP has found use in contact sintering (with apparent reduction of hillock formation), CVD glass reflow, and the formation of Ti, Ta, Mo, and W silicides[28].

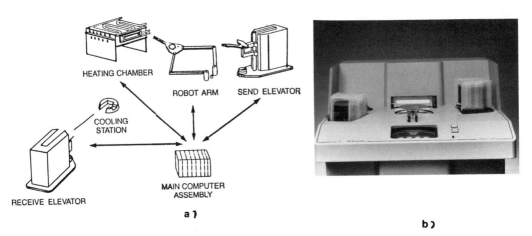

Fig. 18 (a) HEATPULSE 2106 system components. (b) Photograph of a rapid thermal processing system. Courtesy of AG Associates.

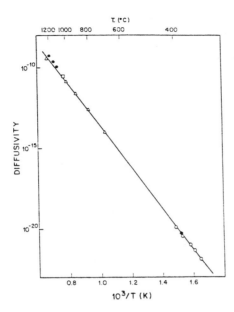

Fig. 19 Diffusion of oxygen in Si[55].   Reprinted with permission of Academic Press.

OXYGEN IN SILICON

Other than in heavily doped crystals, the most abundant impurity in Czochralski (CZ) Si crystal is oxygen, with concentrations typically ranging from $5x10^{17}$-$1x10^{18}$ atoms /cm^3 (or 10-20 ppma). The majority of the oxygen (~95%) in the as-grown crystal is atomically dissolved, and occupies interstitial sites. The diffusivity of oxygen is moderately high (Fig. 19), whereas its solubility decreases strongly at lower temperatures (Fig. 2). These two properties make oxygen the most important precipitate-forming element in CZ-silicon. The precipitation behavior affects all of the important effects influenced by the presence of oxygen in silicon including: a) donor formation; b) surface oxygen precipitates; c) bulk oxygen precipitates (that provide the basis of intrinsic gettering); and d) wafer warpage resistance[32].

The mechanism by which oxygen is incorporated into CZ silicon was described in Chap. 1, and the kinetics of precipitate formation was introduced in the *Bulk Defects* section. The characterization methods to measure oxygen concentration levels in silicon was presented sections dealing with *FTIR* in Chaps. 1 and 5. In this section, and the one following on *Gettering*, we complete the precipitation formation discussion as it applies to oxygen in silicon, and thereby also cover thermal donor formation.

Oxygen is incorporated as a result of crucible erosion during CZ growth. In general, higher concentrations of oxygen are found at the seed end of the ingot because more crucible wall area is in contact with the melt when the crucible is full. In addition, the convection currents which contribute to the crucible erosion rate, are much stronger in a full crucible than at later stages of growth when the crucible is emptier.

Oxygen in silicon is able to form up to $3x10^{16}$ donors /cm^3 in the temperature range of 300-500°C (highest formation rate is at ~450°C), and this is an undesirable property from the viewpoint of manufacturing devices. Apparently, oxygen moves from interstitial locations to substitutional sites and thus forms a donor effect. More recent investigations show that a second type of donor can also be formed at temperatures of 550-800°C, although these tend to form more slowly. *Low-temperature oxygen donors* arise from the formation of oxygen-silicon complexes, SiO_x, and these can be annihilated by 650°C heat treatment (Fig. 20). The growth of *high-temperature oxygen donors* seems to involve carbon, since carbon at levels of ~1 ppma is found in the crystals in which they form. The higher-temperature donors are destroyed by heat treatments at 900°C. None of the existing donor formation models however have been able to closely confirm the experimentally observed relationships. A review of the proposed models of the structure of oxygen donors, and the diffusivity of oxygen in Si at temperatures between 300-400°C can be found in Ref. 56. Whatever the exact model for these donors, they are significant because considerable "back end" processing of MOS devices occurs near 450°C (e.g. during passivation, wire bonding, and packaging processes). In some cases, degradation of device parametric performance has been attributed to low-temperature donor formation.

The standard 650°C heat treatment to eliminate low-temperature donors is limited by the inability to quickly quench large wafers, and the possibility that high-temperature donors may be forming during the anneal. In addition, the normal 1-2 hour 600-700°C anneal can cause the formation of very small oxygen precipitates that allow more rapid oxygen precipitation at higher

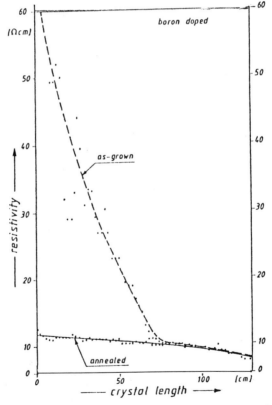

Fig. 20 Oxygen donors in as-grown CZ Si. The boron acceptors $(1.2 \times 10^{15}$ cm$^{-3})$ were totally compensated by oxygen donors in the seed-end of the as-grown crystal. In the second half-of the crystal only few oxygen donors had formed. After annealing the crystal for 2 h at 650°C, followed by rapid cooling, oxygen donors were eliminated[10]. Reprinted with permission of Springer-Verlag.

temperatures. Short RTP cycles, on the order of seconds at 650-750°C, have been suggested as an effective alternative annealing step[28].

In silicon which contains oxygen at levels <6 ppma, so few oxygen donors are formed that their presence is not measurable. Thus, another way to effectively eliminate the effects of oxygen donors is to form a denuded zone at the wafer surface, and to create oxygen precipitation in the wafer bulk. This will reduce the interstitial oxygen content in both the surface and the bulk to levels below the minimum required to form oxygen donors.

As described in Chap. 1, the time /temperature conditions endured by different parts of the ingot can vary substantially. For example, the seed-end of the ingot undergoes a long exposure to 400°C, while at the tang end this temperature exposure is short. Thus, wafers that indicate identical measurable levels of interstitial oxygen, may in fact contain different actual oxygen concentrations, since some of the oxygen may be present in precipitate (rather than in interstitial) form. As a result, the subsequent precipitation behavior of the two wafers may be substantially different. Therefore, starting wafers will need to be supplied in a form that will exhibit predictable and reproducible precipitation kinetics at specific oxygen concentrations and temperature cycles. This is especially true if intrinsic gettering is to be used. It has been demonstrated that only reproducible precipitation will lead to reliable intrinsic gettering[33]. The precipitation behavior can be inferred from measurements of the interstitial oxygen content before and after a high heat treatment[44] (e.g. 1050°C for 24 h).

The presence of surface oxygen precipitates adversely impacts the quality of thin gate oxides and nucleates oxygen-induced stacking faults, which can cause device failures by excessive leakage currents, etc[25]. It would be useful if these were not present at the surface. In addition, if the precipitates are to be used to getter unwanted impurities from the active device regions, it is necessary that the precipitates be prohibited from forming in such regions. The procedure utilized to prevent oxygen precipitates from forming near the wafer surface involves creation of a *denuded* zone, or zone that contains less interstitial oxygen than required to form oxygen precipitates. The thermal steps used to create denuded zones are outlined in the section that describes gettering.

It was mentioned in Chap. 1, and again earlier in this chapter, that the presence of oxygen enhances the warpage resistance of silicon. However, it has been reported that if high oxygen content and high temperature cycles occur simultaneously, very large (~2000Å) precipitates can be formed. These are surrounded by large dislocations, which can grow and move through the lattice under the presence of strain, and result in warpage[10,20,25]. On the other hand, silicon wafers in which the oxygen is interstitially dispersed, or forms small clusters (~500Å), appear to strengthen and cause wafers to exhibit a higher elastic limit. Figure 16 shows the comparative values of the critical shear stress in silicon at different temperatures and varying precipitate concentrations[22]. The critical shear stress increases with both reduced wafer temperature and reduced oxygen precipitate formation. However, such results also appear to be dependent on detailed material and experimental process conditions.

GETTERING

As has been discussed in this chapter, unwanted crystalline defects and impurities can be introduced during silicon crystal growth or subsequent wafer fabrication processes. The degradation of VLSI device properties by *extended crystalline defects* (dislocations, stacking faults, and precipitates) was described in an earlier section. The damaging effects by *impurities* is discussed here, and can be grouped into two classes: a) impurities (e.g. oxygen and carbon) that participate in the process of nucleating extended crystalline defects; and b) other impurities that decorate pre-existing extended crystalline defects, such as Cu, Ni, Au, Fe. The metallic impurities are highly mobile and diffuse long distances in the crystal at moderate process temperatures. For example, Cu can diffuse ~600 μm in 1 minute at 900°C, and Fe ~3000 μm in 30 min at 1000°C[10]. These impurities thus have a high probability of encountering extended defect sites, and being captured by them. The electrical activity of such decorated extended defects degrades device operation by causing larger leakage currents and lower breakdown voltages.

Efforts undertaken to eliminate the effects of impurities and defects have involved three approaches[16]: 1) suppression of extended defect nucleation; 2) annihilation of existing extended defects; and 3) removal of device degrading impurities (and *point defects*) from the active device regions. This third technique is referred to as *gettering*, which is term originally used to describe the process in which an appropriate material (such as cesium), is used to getter (remove) the last traces of gas in vacuum tubes, so that the required high vacuums can be attained.

By combining the three approaches, product yield is enhanced. That is, proper application of the first two approaches minimizes the number of extended defects that exist in the active regions. Gettering is then used to remove impurities introduced during crystal growth or processing. Thus, the probability that any of the remaining extended defects will become decorated is significantly reduced.

Methods used to suppress extended defect nucleation include: a) control of the thermal process conditions during wafer fabrication that lead to excess thermal-stress and resultant dislocations; and b) creation of oxygen denuded zones at the wafer surface regions in which active devices are fabricated. This procedure eliminates oxygen precipitates in these device regions.

Annihilation of pre-existing or process-induced extended defects, such as stacking faults, is achieved by using various techniques, including elevated temperatures (>1200°C)[12], and the addition of HCl to the growth ambient[13] during the growth of SiO_2. These techniques are discussed elsewhere in this chapter. The subject of healing crystalline damage due to ion-implantation is discussed in Chap. 9.

It is also useful to mention at this point, that in addition to gettering, the control of unwanted impurities (especially metals), should involve the removal of the sources of contamination whenever possible. Such an approach involves the thorough characterization of the material and processing equipment used to fabricate devices. Some examples of such contaminant sources that have been identified in silicon processing include: a) stainless steel fixtures in ion-implant machines that ended up being sputtered during the ion-implant process[35]; b) impurities originating from the sputtering of components of reactive ion etching systems[34]; and c) the diffusion of heavy metals from the heater coils through the quartz walls of diffusion furnaces; d) release of transition metals (e.g. Fe and Ti) by the graphite susceptors of epitaxial reactors; and e) handling with metal tweezers.

A variety of procedures have been implemented to reduce the level of metal contamination due to these sources. Stainless steel parts which may release metal atoms as a result of sputtering in ion implanters, plasma and reactive ion etchers, have been replaced with parts made from aluminum, silicon, and carbon. Suggested procedures to contamination by sources of heavy metals have included the following: 1) reducing process temperatures, since quartz tubes become increasingly permeable to the metal atoms of the furnace heating coils as the temperature increases; 2) subjecting the quartz furnace tubes to periodic high-temperature HCl cleaning; and 3) use of double-wall furnace tubes. Contamination from epitaxial susceptors can be reduced by performing the epitaxy process at lower temperatures and reduced pressures, and by *in situ* cleaning of the wafers with HCl prior to initiating epitaxial growth. The increasingly widespread use of cassette-to-cassette operations in handling of wafers has reduced the contamination due to handling with metal tweezers.

Basic Gettering Principles

The mechanism of gettering removes impurities from device regions by three steps: 1) the impurities to be gettered are released into solid solution from any stable precipitate form that they have assumed; 2) they then diffuse through the silicon; and 3) they are "captured" by extended defects (dislocations or precipitates) at a position away from the device regions, and are prevented from being re-released into the wafer during subsequent heat treatments. In general, the transition metal impurities are initially considered to rapidly disperse themselves uniformly throughout the wafer. This assumption is based on their high mobility and long diffusion lengths at typical processing temperatures (as described earlier).

Gettering processes are divided into two categories: extrinsic and intrinsic. *Extrinsic gettering* involves the use of external means to create the damage or stress in a silicon lattice that leads to the creation of the extended defects or chemically reactive sites at which the mobile impurities are captured. Normally such capturing sites are generated on the backside of the wafer, although front side extrinsic gettering has also been explored. Extrinsic gettering can be performed prior to the first oxidation step (*pre-oxidation gettering*), or concurrently with the remainder of the fabrication steps (*gettering during,* or *subsequent to, junction formation*). *Intrinsic gettering* involves the localization of impurities at extended defects which exist within the bulk material of the silicon wafer, and whose origin is due to an "intrinsic" property of the starting wafer, such as its oxygen content acquired during CZ crystal growth.

The gettering of gold atoms in silicon by phosphorus diffusion is described, as this process serves to illustrate the basic mechanism of gettering. A known concentration of gold atoms (1×10^{14} /cm^3) is first diffused uniformly throughout a group of silicon wafers. Phosphorus diffusions at 900°C, and with varying P surface concentrations (10^{19}-10^{21} P atoms /cm^3), are then carried out in an attempt to getter the gold. After the P diffusions are completed, the gettering efficiency of the procedure is measured. It is observed that with increasing surface concentration of P, a strongly decreasing concentration of gold in the silicon bulk occurs. At P concentrations of 10^{21} atoms /cm^3, the gold concentration is decreased below the detection limit. Further investigations show that the gold has diffused into the phosphorus layer.

A model believed by some investigators to occur describes this gettering process according to the following steps[10]: a) P diffusion is accompanied by the generation of phosphorus-vacancy complexes and /or misfit dislocations which become possible capture sites (sinks) for the gold atoms; b) during the P-diffusion step the highly mobile gold atoms diffuse long distances in the wafer; c) the gold is captured at the phosphorus-vacancy or dislocation sites, and is thus removed from solution (this maintains a favorable concentration gradient of gold in solution toward the phosphorus-rich zone); and d) once a gold atom is captured, the electrostatic interaction energy with the capture site is so high that subsequent heat treatment does not release the impurity back into the wafer. Sufficient phosphorus must be deposited to create enough capture sites to getter the gold present in the wafer.

Diffused phosphorus layers (and other backside gettering methods) are also effective in relieving supersaturated concentrations of point defects, by providing a source of vacancies. Oxidation induced stacking faults (OISF), are regions of such excess point defects (i.e. Si self-interstitials). As the excess self-interstitials are absorbed by the vacancies of the gettering region, the OISFs undergo shrinkage.

Fig. 21 Cross-sectional TEM micrographs of (a) as-deposited poly-Si, then wet oxidized at 1150°C for (b) 100 min, (c) 765 min, and (d) 1300 min[37]. Reprinted with permission of the publisher, the Electrochemical Society.

Extrinsic Gettering

Historically, gettering of mobile metallic impurities has been accomplished by *extrinsic gettering*, in which the wafer backside is subjected to a process that creates damage or stress, and subsequently traps mobile impurities.

Mechanical damage by abrasion, grooving, or sandblasting have all been used to create stress fields at the backside of wafers. During subsequent annealing steps, dislocations which relieve these stresses, are generated. The dislocations then serve as gettering sites. Unfortunately, microcracks and dislocations in wafers may lower their mechanical strength, and thus make them more prone to warpage during heat cycles[10]. In addition, the silicon dust created is difficult to remove, and the degree of damage is difficult to control and /or reproduce.

Phosphorus diffusion to produce gettering including a proposed model of the gettering process was presented earlier[10]. Work on Ni gettering in P-diffused Si[38] indicates that SiP precipitates in the diffused P regions apparently nucleate $NiSi_2$ particles. The important observation is that the $NiSi_2$ particles do not contain any surrounding dislocations. This indicates that the removal of the Ni impurities occurs through interaction between Si self-interstitials and Ni atoms, which results in the formation of $NiSi_2$ particles. Thus, the gettering mechanism by P-diffusion may in fact be similar to the intrinsic gettering mechanism by oxygen precipitates. A disadvantage of the phosphorus diffusion procedure, however, is that the large amounts of P on the wafer backside can cause problems in later process steps by contaminating or auto-doping epitaxial layers on the active side of the wafer.

The gettering action by *laser induced damage*, created by a focused laser beam scanned across the surface, is in principle similar to mechanically induced damage. However, the damage is introduced by a cleaner and more controllable procedure. A high power (10-1000 J /cm^2) laser beam is employed to cause enough thermal shock to create dislocation nests in the irradiated region. These then act as gettering sites[36].

The use of *ion-implantation* to introduce backside damage has also been investigated. Various ions (e.g. Ar, P, As, and B) have been used in this method[39]. The number and energy of the ions can be controlled to create the desired stress in the lattice. A concern with this technique is that the damage may anneal out at high temperatures, and that the step would thus have to be repeated several times during the process. (Chap. 9 describes such damage annealing.)

A *layer of polysilicon* deposited on the wafer backside (1.2-1.5 μm thick) is another extrinsic gettering technique that has been developed[40]. Starting wafers can be purchased with such a layer already deposited. The grain boundaries, and high degree of lattice disorder in polysilicon, act as sinks for mobile impurities. However, as shown in Fig. 21, even such a stable gettering procedure cannot fully survive after processing at 1150°C under oxidizing ambient[41]. During such heat treatment, the number of grain boundaries decreases and the lattice disorder recovers as the grains grow. More importantly, a part of the polysilicon is consumed to form SiO_2 at the surface. Fortunately, when the growing oxide front encounters the oxide precipitates present at the original polysilicon /single crystal silicon interface, new defects (dislocation and stacking faults) are generated (Fig. 21). These new defects then continue to getter impurities (but, it is presumed, with much less efficiency than the original polysilicon).

Extrinsic gettering procedures, as discussed above, can be applied prior to the initial oxidation of the starting wafer (preoxidation gettering), during, or subsequent to the formation of pn junctions. Rozgonyi, *et al*, suggested that a heavily P-diffused layer be created on the wafer backside prior to device processing (i.e. the so-called *pre-oxidation gettering at other side*, or POGO process). This technique effectively prevents the formation of stacking faults at the wafer front surface during subsequent oxidation. Initially S-pits (haze), which are due to metallic precipitates at the wafer surface, are suppressed.

Extrinsic gettering techniques, however, suffer from a variety of limitations. Some have been mentioned when the individual methods were described. The primary general limitation of extrinsic gettering procedures is their inherent lack of stability, especially at high temperatures up to 1250°C for bipolar structures. Such instability manifests itself as the dissolution of gettered metal back into the wafer, and the annealing out of dislocations as the wafer undergoes a number of high temperature steps. Upon being annealed out, dislocations can no longer trap

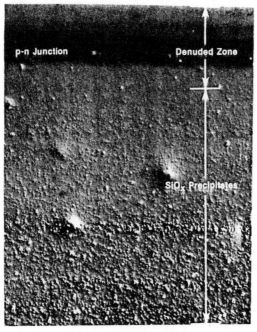

Fig. 22   SEM micrograph of p-n junction created by 450°C anneal for 70 h in denuded-zoned Si with SiO_x precipitates formed by thermal treatment[54]. Reprinted by permission of Solid State Technology, published by Technical Publishing, a company of Dun & Bradstreet.

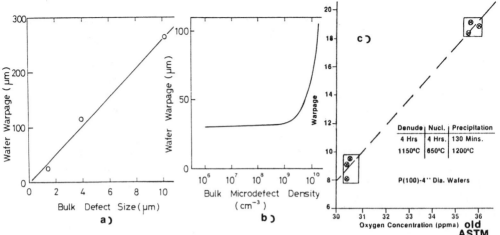

Fig. 23 (a) Dependence of wafer warpage on bulk microdefect size. (b) Wafer warpage as a function of bulk microdefect density[25]. Reprinted with permission of the publisher, the Electrochemical Society. (c) Effect of initial oxygen content of the wafer on warpage[33]. Reprinted by perm. of Solid State Technology, publ. by Technical Publishing, company of Dun & Bradstreet.

newly introduced impurities, nor recapture the impurities released back into the wafer from getter sites. This limitation is one reason why extensive investigation of intrinsic gettering as a complementary gettering procedure has been undertaken, and why it is being increasingly implemented into VLSI fabrication sequences.

Intrinsic Gettering

Intrinsic gettering is based on the principle that under proper conditions, supersaturated oxygen in silicon wafers will precipitate out of solution, and form clusters within the wafer during thermal processing. The stresses that result as these clusters grow into larger precipitates can be relieved by punching out dislocation loops (or by emitting Si-self interstitials, that agglomerate to form stacking faults and dislocation loops). These dislocations become sites at which unwanted impurities can be trapped and localized[10]. In an effectively designed intrinsic gettering process, these precipitates are only allowed to form in the bulk regions of the wafer. They are prevented from forming in the active device regions by reducing the oxygen concentrations to levels below the threshold required for precipitation (denuded zone formation). In this manner, unwanted impurities are localized (gettered) only in regions not containing active devices (Fig. 22).

The advantages of intrinsic over extrinsic gettering include the following: a) the technique may be employed without any external treatment of the wafer other than heating; b) the volume of the wafer provides a sink during intrinsic gettering that is roughly two orders of magnitude larger than the volume of the gettering region on the wafer backside; and c) the gettering region in the bulk is much closer to the device regions than the backside of the wafer[36], so that the distance required for impurities to diffuse from the circuit regions to the sinks is 25-50 times shorter. Therefore, intrinsic gettering allows diffusion cycles which are significantly shorter than those required for extrinsic gettering to remove the same contamination

The first requirement for successfully implementing this technique is that the starting wafers must have appropriate oxygen concentration levels (i.e. 15-19 ppma). The lower limit of this range is somewhat above the threshold concentration of oxygen required for precipitation

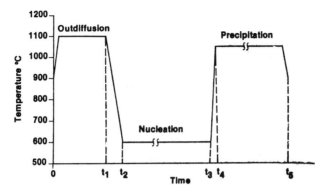

Fig. 24 Example of a thermal cycle to form the denuded zone[54]. Reprinted by permission of Solid State Technology, published by Technical Publishing, a company of Dun & Bradstreet.

($\sim 6 \times 10^{17}$ atoms /cm^3 = 12 ppma). Precipitation apparently does not occur for concentrations below this level[14]. Wafers from CZ-grown ingots, however, normally exceed this lower limit, the oxygen being incorporated during crystal growth by erosion of the SiO_2 crucible. On the other hand, if oxygen concentrations exceed the upper bound of the high concentration range (>19 ppma), too high a density of precipitates is formed, and this may contribute to both wafer warpage and dislocation generation in the vicinity of the active devices[20,25] (Fig. 23).

Assume that the wafers contain appropriate concentrations of oxygen, and that the oxygen is uniformly distributed across the wafer diameter. The wafers can then be subjected to a series of temperature cycles to achieve intrinsic gettering (Fig. 24). A *high-low-high* sequence of such steps is then:

Step 1) Denuded Zone Formation - a *high* temperature step which causes
reduced oxygen concentration in the region near the wafer surface;
Step 2) Nucleation of SiO_x Clusters - a *lower* temperature step whose purpose is
to homogeneously nucleate SiO_x precipitation sites;
Step 3) Precipitate Growth and Gettering - a *high* temperature step for growing
the SiO_x nuclei formed in Step 2. The result of their growth is the formation
of dislocation loops, which function as the desired gettering sites.

In *Step 1*, a region near the surface, containing less oxygen than is required to form precipitates (*denuded zone*), is created. Thus, such precipitates or their punched out dislocation loops, which could intersect active device junctions or could become decorated with impurities near device surfaces, are prevented from forming. Denuded zones are formed by subjecting wafers to a high temperature step[14,45,46]. This allows the oxygen near the top and bottom wafer surfaces to diffuse out, reducing its concentration near the surface to about the solubility level at that temperature. The denuded zone required for a VLSI process must have a depth that exceeds the deepest junction, plus its surrounding space-charge region.   Since the depletion region thickness depends on the wafer resistivity and the maximum operating voltage of the device, different devices made on the same substrate resistivity may need different denuded zone depths. Therefore an important question that must be answered is: How shallow a denuded zone can the design and process tolerate?   There is no real shortcut to determining this information. Functional circuits must be fabricated on wafers containing a variety of initial oxygen levels. These circuits must be tested, and then cleaved and etched to determine the denuded zone depth and precipitate densities   Temperature cycles of 1100-1200°C for 30-240 minutes reportedly create denuded zones 10-40 μm deep. Figure 25 shows the outdiffusion depth for various temperatures and initial oxygen concentrations. Note that while nitrogen and argon are typically suggested as appropriate ambient gases for the denuding cycle, oxygen has been recommended for high denuding temperature cycles (≥ 1200°C) to avoid pitting[45].

In *Step 2*, a lower temperature cycle (600-800°C) is next used to cause interstitial oxygen in the wafer bulk to form the nuclei required for the subsequent precipitation and gettering events of *Step 3*. The supersaturated (interstitial) oxygen can diffuse and homogeneously nucleate into small clusters (see section on *Bulk Defects*) if wafers are subjected to temperatures between 650-800°C for prolonged periods (e.g. 4-64 hrs). Note that above 950°C, homogeneous nucleation appears to be negligible. Thus, the first temperature step on starting wafers should be >950°C, to prevent formation of clusters near wafer surfaces during the initial heat treatment.

A large number of relatively small oxygen clusters are formed by this step. This number, for a fixed temperature and time, depends on the oxygen content and increases with initial oxygen concentration. Since the oxygen content of the starting wafers is one important parameter that determines the density of oxygen clusters, it is crucial that this concentration be controlled in CZ silicon. Figure 26 shows the nucleation rate and final radius of the clusters that are created at 800°C for an initial oxygen content of 15 ppma[14]. If nucleation cycles are carried out at the lower end of the temperature range (~650°C), significantly longer annealing times are required to achieve a sufficient concentration of adequately sized clusters (16-64 hr). To enable an initial 650°C temperature to be used for this step, and yet allow a reduction in the overall nucleation cycle time, a *ramping up* of the temperature during *Step 2* has been suggested[25,50]. The rate of cluster formation can be enhanced by this procedure if the rate of change of the critical size, r_o, with temperature, is smaller than the rate of growth of the clusters during annealing. That is, by the time a growing cluster reaches a higher temperature, it will have grown fast enough that its size exceeds the required r_o at that temperature.

The time to grow the clusters into precipitates that would be large enough to form the desired dislocations in the bulk, is prohibitively long over the range of temperatures of *Step 2*. Thus, once the clusters have grown to a size where they exceed the r_o of a higher temperature step, a third thermal cycle at that temperature can be used to grow the clusters into larger precipitates in a shorter time. If a temperature ramping technique during the nucleation cycle is

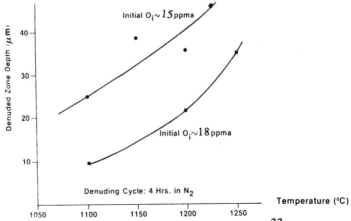

Fig. 25 Effect of initial oxygen concentration on the denuded zone depth[33]. Reprinted by permission Solid State Technology, published by Technical Publishing, company of Dun & Bradstreet.

used, the final temperature can be that called for by *Step 3*.

In *Step 3*, wafers are heated to high temperatures (900-1250°C) for extended periods (4-16 hrs). At these temperatures, oxygen clusters of a sufficient size (i.e. $> r_0$) which were formed during *Step 2*, will grow (e.g. from 30-50 Å to ~500-1000 Å in radius). Such growth is caused by the diffusion of interstitial oxygen in the bulk to the SiO_x clusters, and reaction with silicon on the precipitate surface. For each SiO_2 molecule thus formed, a Si self-interstitial is produced to accommodate the volume expansion. The large volume change associated with the precipitation is not completely relieved by the emission of self-interstitials, and the precipitate builds up a strain field which can punch out prismatic dislocation loops.

The self-interstitials emitted by the precipitate can also form dislocation loops as discussed in the section on *Bulk Defects*. The dislocation loops contain an extrinsic stacking fault which grows by absorbing the self-interstitials emitted by the growing precipitate. The dislocations formed from both mechanisms act as the gettering sites which can trap fast diffusing impurities. i

The ultimate size of the precipitates apparently also plays a role in the warpage resistance of the wafers. Kishino[25] shows that wafer warpage increases with increasing precipitate size (Fig. 23a). This data, and that of others[20,45], suggests that *Step 3* heat cycles which lead to precipitates >2000 Å can negatively impact production yield. Several reports have, in fact, indicated that the best wafers are those in which the precipitation occurs slowly, uniformly throughout the wafer, (especially with good radial uniformity), and results in small precipitates with attached precipitation defect complexes[46,51]. The concern for wafer strength has lead most users of internal gettering away from materials with high oxygen levels (e.g. >19 ppma), to material containing medium levels of oxygen (e.g. 15-19 ppma).

Summary of Gettering

Effective gettering is required to produce high yielding VLSI. We have described a variety of gettering techniques, both extrinsic and intrinsic. In practice, more than one gettering mechanism is likely to be active in a wafer during fabrication. In fact, a conservative approach to gettering is to deliberately implement more than one technique in a process. An extrinsic, backside technique and an intrinsic gettering process can be employed concurrently. Since typical CZ wafers contain oxygen in the appropriate range of 15-19 ppma, it is likely that many processes have been inadvertently benefiting from intrinsic gettering. As low temperature

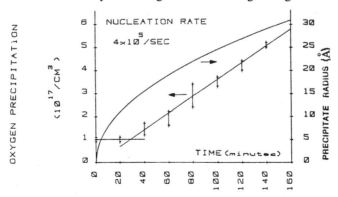

Fig. 26 Nucleation rate for oxygen precipitates. Curve is expected growth rate for precipitate at 880°C[14]. This figure was originally presented at the Spring, 1981 meeting of the Electrochemical Society. Reprinted with permission of the publisher, the Electrochemical Society.

processes, or RTP, become more extensively utilized however, it is likely that the use of intrinsic gettering will no longer be left to chance. The reduced thermal cycles of these processes will demand use of a gettering technique which allows the placement of gettering sinks closer to active device regions.

The intrinsic gettering method, however, does require the strict control of a number of oxygen-related processes, and this can be a complex task. Thus, some process engineers might be tempted to forego intrinsic gettering completely, by ordering CZ wafers with an interstitial oxygen content below that needed for precipitation. Such an approach might be a viable if there is sufficient confidence that an alternative gettering technique will remain effective throughout an entire process. If this has not been conclusively determined, the process will be operating without adequate "disaster insurance", should contamination occur. It is also important to remember that crystal defects present in the active device regions also getter impurities, and at about the same efficiency as the getter sites in the bulk or backside of the wafers. Therefore, it is as important to prevent the formation of extended crystal defects in the active regions, as it is to implement effective gettering procedures.

REFERENCES

1. K.V. Ravi, *Imperfections and Impurities in Semiconductor Silicon*, Wiley, New York, 1981.
2. R.K. Watts, *Point Defects in Crystals*, Wiley and Sons, New York, 1977.
3. D. Hull, *Introduction to Dislocations*, Pergamon Press, New York, 1965.
4. J.P. Hirth and J. Lothe, *Theory of Dislocations*, McGraw-Hill, New York, 1968.
5. J.M. Hu, "Temperature Distribution and Stresses in Circular Wafers in a Row During Radiative Cooling", *J. Appl. Phys.*, **40**, 4413, (1969).
6. K.G. Moerschel, C.W. Pearce, and R.E. Reusser, "A Study of the Effects of Oxygen Content, Initial Bow and Furnace Processing on Warpage of 3-inch Diameter Si Wafers", *Semiconductor Silicon, 1977*, Electrochem. Soc., Pennington, N.J., 1977, p. 170.
7. G. Franz, B.O. Kolbesen and R. Lemme, "Crystal Defects Induced by Local Oxidation", in *Semiconductor Silicon, 1981*, Electrochem. Soc., Pennington, N.J., 1981, p. 821.
8. A.J.R. deKock and W.M. vander Wijgert, "The Effect of Doping on the Formation of Swirl Defects in Dislocation-Free Czochralski Silicon", *J. Cryst. Growth*, 49, 718, (1980).
9. *ibid*, Ref. 1, p. 62-82.
10. W. Zuhlehner and D. Huber, " Czochralski Growth Silicon", in *Crystals 8 (Silicon Chemical Etching)*, Springer-Verlag, Berlin, (1982).
11. A. Lin, *et al.*, "The Growth of Oxidation Stacking Faults and the Point Defect Generation at Si-SiO$_2$ Interface during Thermal Oxidation of Si", *J. Electrochem. Soc.*, **128**, 1221, (1981).
12. S.M. Hu, "Anomalous Temperature Effect of Oxidation Stacking Faults in Silicon", *Appl. Phys. Lett.*, **17**, 165, (1975).
13. H. Shiraki, "Stacking Fault Generation Suppression and Grown-In Defect Elimination in Dislocation-Free Silicon Wafers by HCl Oxidation", *Jpn. J. Appl. Phys.*, **15**, 1 (1976).
14. R.A. Craven, "Oxygen Precipitation Czochralski Silicon", in *Semiconductor Silicon, 1981*, Electrochemical Society, Pennington, N.J., 1981, p. 224.
15. T. Abe and H. Harada, "Microdefects and Impurities in Dislocation-Free Si Crystals, *Materials Research Socieiy Symposium Proceedings*, Vol. 14, (1983), p. 1, Elsevier Science Publ. Co.
16. G.A. Rozgonyi, "Trends in Defect and Impurity Control for Fine Line VLSI Processing", *Defects in Silicon, 1983*, Electrochem. Soc., Pennington, N.J., 1983, p. 29.
17. C.W. Pearce and V.C. Kannan, "Saucer Pits in Si", in *Defects in Silicon*, '83, Electrochemical Soc., Pennington, N.J., (1983), p. 396.

18. W.T. Stacy, D. Allison, and T.Wu, "The Role of Metallic Impurities in the Formation of Haze Defects", *Semiconductor Silicon, 1981*, Electrochem. Soc., Pennington, N.J., 1981, p. 344.

19. A. Wang, J.D. Jensen, H.R. Huff, A. Robbins, *J. Electrochem. Soc.*, **128**, 237C, (1981).

20. J.E. Lawrence and H.R. Huff, "Si Material Properties for VLSI Circuitry", in *VLSI Electronics, Microstructure Science*, Vol. 5 (N.G. Einspruch, Ed.), Academic Press, Orlando, (1982), p. 51.

21. H. Huff and F. Shimura, "Si Material Criteria for VLSI", *Solid State Technol.*, Mar.'83, p. 103.

22. B. Leroy & C. Plougonven, "Warpage of Si Wafers", *J. Electrochem. Soc.*, **127**, 961, (1980).

23. K. Sumino, "Dislocation Behavior and Mechnical Strengths of Float-Zone Si Crystals and Czochralski Si Crystals", *Semiconductor Silicon 1981*, Electrochem. Soc., Pennington, N.J., 1981, p. 208.

24. T. Abe, *et al.*, "The Characteristics of Nitrogen in Si Crsytals", *VLSI Science and Technology*, Electrochem. Soc., Pennington, N.J. (1985), p. 543.

25. S. Kishino, "Material Characterization for VLSI Applications," *VLSI Science and Technology*, 1985, Electrochem. Soc., Pennington, N.J., 1985, p. 399.

26. W.M. Bullis and F.G. Vieweg-Gutberlet, "Current Trends in Si Characterization Techniques", *Semiconductor Silicon*, 1977, Electrochem. Soc., 1977, Pennington, N.J., p. 360.

27. U. Gosele and F. Morehead, "The Predominant Intrinsic Point Defects in Si: Vacancies or Self-Interstitials?", *Semiconductor Silicon, 1977*, Electrochem. Soc., 1977, p. 766.

28. S.R. Wilson, W.M. Paulson, and R.B. Gregory, "Rapid Annealing Technology for Future VLSI", *Solid State Technology*, June 1985, p. 185.

29. W.C. O'Mara, *et al.*, "Rapid Thermal Annealing for Oxygen Donor Annihilation", *VLSI Science and Technology, 1985*, Electrochem. Soc., Pennington, N.J., 1985, 456.

30. W. Thurston and R. Seaman, "Rapid Thermal Processing Systems Performance", *Microelectronic Manufacturing and Testing*, July '85, p. 1.

31. R. Sheets, "Rapid Thermal Processing Systems", *Microelectronic Mfg. & Test*, July '85, p.16.

32. J.R. Patel, "Current Problems on Oxygen in Silicon", *Semiconductor Silicon, 1981*, Electrochem. Soc., Pennington, N.J., 1981, p. 189.

33. R.B. Swaroop, "Advances in Silicon Technology for the Semiconductor Industry, *Solid State Technol.*, July '83, p. 97.

34. L.M. Eprath and R.S. Bennett, *Extended Abstracts, Electrochem. Soc. Fall Meeting*, Abs. 302, Vol. 80-2, Hollywood, Fla., 1980.

35. E.W. Haas, *et. al.*, *J. of Elec. Matls*, 7, 525, (1978).

36. C.W. Pearce, L.E. Katz, and T.E. Seidel, "Considerations Regarding Gettering in Integrated Circuits", *Semiconductor Silicon, 1981*, Electrochem. Soc., Pennington, N.J., 1981, p. 705.

37. W.T. Stacy, *et al, Defects in Silicon, Proc. Electrochem. Soc.* 85-9, p. 423, 1983.

38. A. Ourmazd and W. Schrotes, *Appl. Phys. Letters*, 45, p. 781, 1984.

39. A.G. Cullis, T.E. Seidel, and R.L. Meek, "Comparative Study of Annealed Neon, Argon, and Krypton Ion Implantation Damage in Silicon", *J. Appl. Phys.*, **49**, 5188, (1978).

40. M.C. Chen and V.J. Silvestri, *J. Electrochem. Soc.*, **129**, p. 1294, 1982.

41. D.K. Sadana, "Gettering in Processed Silicon", *Proceedings Material Research Society*, Nov. '84, and in *Semiconductor Int'l*, May '85, p. 362.

42. A. Goetzberger and W. Shockley, *J. Appl. Phys.*, **32**, 1821, (1960).

43. G.R. Booker and W.J. Tunstall, *Phil. Mag.*, **13**, 71, (1966).

44. H.D. Chiou and L.W. Shive, "Test Methods for Oxygen Precipitation in Silicon", *VLSI Technology*, Electrochem. Soc., Spring, 1985, p. 429.

45. *ibid.*, Ref. 33, p. 101.

46. D. Huber and J. Reffle, "Precipitation Process Design for Denuded Zone Formation in CZ Silicon Wafers", *Solid State Technol.*, Aug., 1983, p. 137.

47. J. Matlock, "Current Methods for Si Wafer Characterization", *Solid State Tech.* Nov 83, p.11.

48. D.C. Miller and G.A. Rozgonyi, "Defect Characterization by Etching, Optical Microscopy and X-Ray Topography", in *Handbook on Semiconductors*, North-Holland, 1980, p. 217.

49. R.A. Craven, *et. al.*, "Characterization Techniques for VLSI Si", *VLSI Science and Technology, 1985*, Eds. W. Bullis and S. Broydo, Electrochem. Soc., Pennington, N.J.,1985, p. 20.

50. R.F. Pinizzotto, *et. al.*, "Temperature Ramping for Nucleation of Oxygen Precipitates in Si", *Materials Research Soc. Symp. Proc.* Vol. 36, Materials Res. Soc., 1985, p. 275.

51. R. Craven "Internal Gettering in CZ Silicon", *Matls. Res. Soc. Symp. Proc.* Vol. 36, Materials Res. Soc., 1985, p. 159.

52. *ibid*, Ref. 1, p. 262.

53. *ibid.*, Ref. 1, p. 232.

54. D.C. Gupta and R.B. Swaroop, *Solid State Technology*, Aug., 1984, p. 113.

55. L.C. Kimmerling and J.R. Patel, "Silicon Defects: Structures, Chemistry, and Electrical Properties", in *VLSI Electronics* Vol. 12, Academic Press, Orlando, FA., Chap. 5, p. 223.

56. Y. Ichida, *et al.*, in *Lifetime Factors in Silicon*, ASTM, 1980. p. 107.

58. B.O. Kolbesen and H.P. Strunk, "Analysis, Electrical Effects, and Prevention of Defects", in *VLSI Electronics-Microstructure Science* Vol. 12, Academic Press, 1985, Chap. 4, p. 157.

57. J.E. Lawrence, in *Semiconductor Silicon /1973*, p.17, Electrochem. Soc., Pennington, N.J.

58. J. M. Gibson, "Tools for Probing 'Atomic' Action", *IEEE Spectrum*, Dec. 1985, p. 38.

59. T.Y. Tan and T.K. Tice, *Phil. Mag.*, **30**, 615, (1976).

60. F.A. Trumbore, *Bell Syst. Tech. Jnl.*, **39**, 210, (1960).

PROBLEMS

1. Calculate the equilibrium density of vacancies in a silicon lattice at room temperature, 800°C, 1000°C, and 1200°C.

2. Show that in a closely packed structure of hard spheres, the stacking order of planes of spheres can be either ...ABABA..., or ...ABCABCABC..., where A represents the first plane of spheres, and B represents the second plane of spheres, whose centers lie above points at which the spheres of plane do not touch one another. Therefore show that the third plane of atoms can lie such that the centers of the atoms in the third plane lie either above the centers of the atoms in the A plane (and hence, can be considered another A plane) or above the remaining points in the A plane at which the spheres in that plane do not touch.

3. Show that an fcc lattice has a stacking sequence of (111) planes that is of the...ABCABC...type.

4. Show that the diamond lattice has a stacing sequence of the ...AA'BB'CC'AA'BB'CC'...sequence. That is start with the unit cell of the diamond lattice and identify the (111) planes in the cell (and adjacent cells, to get more atoms into the picture), and show that the planes are stacked in the sequence stated. Sketch an extrinsic stacking fault in a diamond lattice.

5. Explain why a dislocation line must terminate at the surface of the crystal.

6. Describe how stacking faults are induced during oxidation, and the methods found which prevent their formation during oxidation.

7. A silicon wafer is described in a 1977 publication as having 1.5×10^{18} oxygen atoms per cm^3. This value is based on the "old ASTM" calibration constant. If the wafer were described in terms of the "new ASTM" calibration constant, what would be the reported oxygen concentration, in ppma.

8. Describe the difference between *extrinsic* and *intrinsic* gettering.

9. Explain the reason why *temperature ramping* in Step 2 of intrinsic gettering has been proposed.

3

VACUUM TECHNOLOGY for VLSI

APPLICATIONS

A discussion on vacuum technology and concepts is presented early in this volume for two principal reasons: 1) many microelectronic fabrication processes occur in reduced pressure regimes and /or gas flow environments; and 2) much of the equipment used to perform these processes therefore incorporates vacuum systems. An understanding of the terminology and concepts of vacuum technology (as well as the fundamental concepts of gases and their interaction with condensed matter) thus leads to a deeper understanding of the processes themselves.

In this chapter we describe those aspects of vacuum systems that are the most important for microelectronic applications including basic definitions, vacuum pump technology, and residual gas analysis. We are unable to cover other interesting, but less broadly applicable areas such as the theory of gas flow at low pressures, the design of vacuum systems, vacuum materials, vacuum sealing techniques, and leak detection. However, substantial information on these subjects is readily available and references are cited for interested readers[1,2,3].

FUNDAMENTAL CONCEPTS OF GASES AND VACUUMS

Molecular matter can occur in the solid, liquid, or gaseous state. The state in which molecules are most independent of one another is gas. The gas state has been modeled using the concept of an *ideal* gas and this model works well to describe the behavior of gas in vacuums. An ideal gas is treated as if the molecules are minute spheres that do not exert forces upon each other; that travel along rectilinear paths in a completely random fashion, and make perfectly elastic collisions.

Many interesting properties of ideal gases have been derived using these assumptions, including the *average velocity*, c, of a group of such particles:

$$c = \left[\frac{8 \, k \, T}{\pi \, m} \right]^{1/2}$$

(1)

where k is Boltzmanns constant, T is the absolute temperature, and m is the mass of the molecules. Note that as T increases, c becomes larger, and the molecules collide with each other or a wall with greater momentum and frequency. The concept of *average* velocity, however, also implies that gas particles travel both slower and faster than c, and the velocity distribution of such particles is given by a Maxwell-Boltzmann distribution as shown in Fig. 1a. It is pertinent to note that, while some atoms travel very much faster, and others much slower than the average, nearly 90% at any time have speeds between half and twice c.

The *pressure*, P, on a surface is defined as the force (or, rate of momentum) imparted to the

73

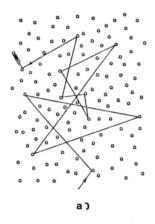

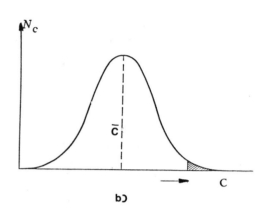

Fig. 1 (a) Molecule traveling in a gas, colliding with other molecules in its path. All other molecules are moving in similar fashion. (b) Boltzmann distribution of molecular velocities in gas.

surface. The pressure on a wall due to ideal gas molecules with mass m can be calculated from:

$$P = \frac{n\,m\,c^2}{3} \qquad (2)$$

where n is the *concentration* of gas particles (that is, the number of particles per unit volume). In combining Eqs. 1 and 2, it is evident that higher temperatures lead to higher pressures. If there is more than one gas in a system, the total pressure is the sum of the *partial pressures* that each of the gases alone would exert in the system.

A *gram molecule* is the quantity of material corresponding to the molecular weight in grams, and *one gram molecule* of material contains 6.023×10^{23} molecules (Avogadros number). A gram molecule of any gas at standard temperature ($273°K$) and pressure (1 atmosphere) contains 6.023×10^{23} molecules, and occupies a fixed volume which turns out to be 22.4 liters. From this information it can be calculated that one cubic centimeter of volume, at standard temperature and pressure (STP), contains 2.7×10^{19} molecules $/cm^3$.

Pressure Units

The System Internationale (SI) pressure unit is the *pascal* (Pa), which is 1 newton $/m^2$, and the pressure at 1 atmosphere is 1.013×10^5 Pa. The *bar* is also an accepted SI unit and corresponds to 10^5 N $/m^2 = 10^5$ Pa.

A set of pressure units that was not derived from MKS units, but from measurements of the height of a column of liquid, were the earliest pressure units used. In spite of having been officially declared obsolete, they still find wide use and therefore need to be defined here. That is, it was observed that the pressure of one atmosphere (at sea level) at 0° C was able to balance a column of mercury (Hg) 760 mm in height. The unit of 1 mm (or 1 *torr*) of pressure is understood as that gas pressure which is able to balance a column of Hg which is 1 mm high. The unit of the *micron of Hg* (10^{-3} torr or 1 mtorr) is also frequently encountered.

The following relationships are used to convert pressure values from Pa to torr units, and back:

$$1 \text{ pascal} = 7.5 \times 10^{-3} \text{ torr} = 7.5 \text{ microns of Hg} \qquad (3a)$$
$$1 \text{ torr} = 133.3 \text{ pascal} \qquad (3b)$$
$$1 \text{ bar} = 1 \times 10^{5} \text{ Pa} = 750 \text{ torr} \qquad (3c)$$
$$1 \text{ atm} = 1.013 \times 10^{5} \text{ Pa} = 760 \text{ torr} \qquad (3d)$$

Pressure Ranges

As the concentration of gas particles in a volume is reduced, it can be seen from Eq. 2 that the pressure is also decreased. If the pressure in a region of space is below that of atmospheric pressure, a *vacuum* is said to exist. It has been customary to divide the pressure scale below atmospheric into several ranges as follows :

Low	- 10^5 Pa (750 torr)	to 3.3×10^3 Pa (25 torr)	
Medium	- 3.3×10^3 Pa (25 torr)	to 10^{-1} Pa	(7.5×10^{-4} torr)
High	- 10^{-1} Pa (7.5×10^{-4} torr)	to 10^{-4} Pa	(7.5×10^{-7} torr)
Very High	- 10^{-4} Pa (7.5×10^{-7} torr)	to 10^{-7} Pa	(7.5×10^{-10} torr)
Ultra High	- 10^{-7} Pa (7.5×10^{-10} torr)	to 10^{-10} Pa	(7.5×10^{-13} torr)

Mean Free Path and Gas Flow Regimes

The fact that each molecule in a gas is randomly located in the volume and is moving at a different velocity, implies that each molecule will move in a different straight line distance before colliding with another molecule (Fig. 1b). This distance, called the *free path*, will vary because of the randomness of the particle locations and velocities, but an average, or *mean* of the free paths, λ, can be calculated from:

$$\lambda = \frac{1}{\sqrt{2} \pi d_o^2 n} \qquad (4)$$

where d_o is the molecular diameter, and n is the gas concentration. For air at 300°K a simpler and more convenient relation for λ (in millimeters) is given by:

$$\lambda = \frac{6.6}{P \text{ (Pa)}} = \frac{0.05}{P \text{ (torr)}} \qquad (5)$$

Note that this model indicates that 63% of molecules in a system undergo a collision in a distance less than λ, and only about 0.6% travel more than 5λ without undergoing a collision.

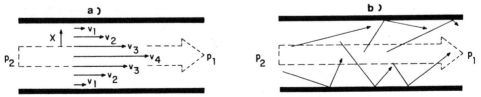

Fig. 2 (a) Schematic representation of particle velocities for viscous flow through a narrow tube. (b) Schematic representation of particle velocities for free molecular flow through a narrow tube. From L. Maissel and R. Glang, Eds., *Handbook of Thin Film Technology*, 1970[5]. Reprinted with permission of McGraw-Hill Book Company.

Example 1 : Given a vacuum chamber that contains Ar gas (atomic weight 39.94 amu) at 300°K and at pressures of: a) 200 Pa (1.5 torr); and b) 0.2 Pa (1.5 mtorr). Find, 1) the average velocity of the gas molecules, and 2) their mean free path.

Solution : a) The average velocity, c is found from Eq. 1, where $k = 1.38 \times 10^{-23}$ J /°K, and $m_{Ar} = 39.94 \times 1.66 \times 10^{-27}$ kg.

$$c = (8 \, k \, T \, / \, \pi \, m)^{1/2}$$
$$= (8 \times 300 \times 1.38 \times 10^{-23} \, / \, \pi \times 39.94 \times 1.66 \times 10^{-27})^{1/2}$$
$$\cong 4 \times 10^2 \text{ m /sec}$$

b) The approximate mean free path, λ, is calculated using Eq. 2,

$$\lambda \text{ (mm)} = 6.6 \, / \, P \text{ (Pa)} \quad \text{or}$$
$$\lambda_{200 \text{ Pa}} = 3.3 \times 10^{-2} \text{ mm}, \quad \lambda_{0.2 \text{ Pa}} = 333 \text{ mm}.$$

Up to this point our discussion has been concerned with gases in contained volumes. However, in many processes there is a steady-state population of atoms, being continuously fed by gas flow, and simultaneously pumped away by a vacuum system. Thus, a brief mention of the behavior of flowing gases must be made. An important aspect of gas flow is that its nature can vary considerably, depending upon the pressure and the dimensions of the vessel in which it is moving. At high pressures, the mean free path is short, and therefore the behavior of the particles is completely governed by collisions with other gas particles. In gas flows of this type, the particles move along as a stream, and such motion is called *viscous flow* (Fig. 2a). Such flow can be either *laminar* (wherein the gas streams move parallel to one another), or *turbulent* .

At low pressures on the other hand, the mean free path can be quite long, and molecules will encounter collisions with the walls of the container much more frequently than with each other. The particles arrive at a wall, stick, and are re-emitted in a direction independent of their arrival velocity. This form of flow is called *molecular* flow. Because gas molecules scarcely ever collide with each other in molecular flow, two gases can diffuse in opposite directions in the system, with neither being affected by the presence of the other (Fig. 2b).

In general, when the characteristic dimension, d, of a vacuum chamber or pipe in which the gas is flowing is much larger than the mean free path of a gas ($\lambda << $ d), *viscous flow* occurs. When the mean free path is much larger than d ($\lambda >> $ d), the flow regime is *molecular*. In the region where 0.01d $ < \lambda < $ d, the flow is governed by both viscous and molecular phenomena, and is referred to as *intermediate, transition,* or *Knudsen* flow (which is less well understood and characterized than viscous or molecular flow).

The *conductance*, C, of an orifice or a length of pipe is a measure of its ability to transmit material flux in the form of gas flow. The throughput, Q_a, of a gas through any conducting element (e.g. the orifice or length of pipe) is given by :

$$Q_a = C (P_1 - P_2) \qquad (6)$$

where $(P_1 - P_2)$ is the difference between the pressures at the entrance and exit of the conducting element. The conductance, C is determined by the geometry of the conducting element (e.g. C becomes larger for increasing orifice or pipe diameter, but smaller for increasing pipe length) and can be calculated. However, the calculations are usually complex. In vacuum systems of microelectronic process equipment (dry-etching, sputter deposition, etc.), vacuum components (valves, tubes) that intentionally limit conductance are sometimes used to reduce gas flow from the process chamber into a pump. This can allow high-vacuum pumps to pump gases from a process chamber, even when the pressure is above the maximum inlet pressure of the high-vacuum pump.

LANGUAGE OF GAS /SOLID INTERACTIONS

When two states of matter, a condensed phase (that is, a solid or liquid) and its vapor exist at the same temperature, and remain in contact with each other without undergoing net changes, they are said to be in *thermodynamic equilibrium*. Under these conditions the same amount of material is being evolved from the surface of the condensed phase as is condensing upon it. The pressure of the vapor in such a system is called its *vapor pressure*. (Note that by *vapor* we refer to a material which is in the gaseous state, but that is condensable at the ambient temperature. In contrast, a gas refers to gaseous matter that is not condensable at the operating temperature.) The vapor pressure of a substance increases as the temperature is increased.

When a vapor (or gas) and a condensed phase interact, gas or vapor molecules which strike the condensed phase may become bound to it. (Note that the binding process may or may not involve condensation of the non-solid phase material as it interacts with the solid surface). If the binding occurs only at the *surface*, the process is called *adsorption* and the layer of adsorbed gas (or vapor) on the surface is called the *adsorbate*. On the other hand, if the molecules penetrate the surface and become bound or captured in the bulk of a solid or liquid, the process is known as *absorption*. The weakest form of adsorption (also called *physisorption*) is usually the result of a physical attraction (van der Waals force) between the adsorbate and surface. A stronger form of adsorption involves a transfer of electrons between surface and adsorbed molecules and results in chemical binding (*chemisorption*). The adsorbed molecules may stay bound to the surface but may move about along the surface by a process called *surface migration*. On the other hand, if the adsorbed molecules are released from the surface, the process is called *desorption*. Any number of energy sources can stimulate desorption, including thermal energy, photoenergy, or impact by atoms, ions, or electrons. *Vaporization* is the process in which a condensed vapor is thermally transformed into a vapor. *Evaporation* is the conversion of a substance from the liquid state to a gas, while *sublimation* is the process of transition directly from a solid to a vapor phase (without passing through the intermediate liquid phase).

The release of gas (and or vapor) by a solid is important in several aspects of vacuum processing, including the phenomenon of *outgassing*. This effect is observed as the slow evolution of gases and vapors from interior surfaces of a vacuum container and its contents (Fig. 4). Such outgassing is the result of several processes, including vaporization, thermal desorption, diffusion (transport of the gas from the bulk of the solid to the surface, followed by desorption), permeation (i.e. penetration [by gases external to the chamber] of the outer chamber wall, diffusion through the chamber wall material to the inner surface, and subsequent desorption), and electron- and ion-stimulated desorption. Techniques are described in a later section that can be used to reduce outgassing after a chamber has been exposed to atmosphere.

The interactions of gases and condensed matter can be treated from an atomistic point of view, as well as from the macroscopic view as discussed above. For example, a useful concept is the *impingement rate*, Z_A, which describes the number of molecules that strike a unit area of surface per unit time. The impingement rate can be expressed in several useful ways including:

Table 1. WORKING PRESSURE RANGES OF VACUUM PUMPS

Ultra-high vacuum $<10^{-7}$ mbar $<10^{-5}$ Pa	High vacuum $10^{-7}...10^{-3}$ mbar $10^{-5}...10^{-1}$ Pa	Medium vacuum $10^{-3}...1$ mbar $10^{-1}...10^1$ Pa	Rough vacuum $1...ca. 10^3$ mbar $10^2...ca. 10^5$ Pa
			Piston pump
			Diaphragm pump
			Liquid-ring pump
			Sliding-vane rotary pump
			Multiple-vane rotary pump
			Rotary piston pump
			Rotary plunger pump
			Roots pump
			Turbine pump
			Gaseous-ring pump
	Turbomolecular pump		
			Liquid jet pump
			Vapour jet pump
	Diffusion pump		
		Diffusion ejector pump	
			Adsorption pump
	Sublimation pump		
	Sputter-ion pump		
	Cryopump		

Intake pressure → (10^{-11} 10^{-10} 10^{-9} 10^{-8} 10^{-7} 10^{-6} 10^{-5} 10^{-4} 10^{-3} 10^{-2} 10^{-1} 10^0 10^1 10^2 10^3 mbar)

$$Z_A = \frac{n\,c}{4} \quad \text{or} \quad Z_A = 3.51 \times 10^{22} (MT)^{-1/2} P \qquad (7)$$

where n is the concentration of gas molecules, c is their average velocity, M is the molar mass in grams, and P is the pressure (expressed in torr).

Of the number of molecules impinging, some fraction of them may become bound (stuck to) the surface for a finite time. The ratio of the number that stick, to the impinging number, is defined as the sticking coefficient, s_c. The time to form a single atomic or molecular layer (*monolayer*) of gas atoms on the surface (*monolayer time*, τ_m) can be estimated from information about Z_A and s_c.

TERMINOLOGY OF VACUUM PRODUCTION AND PUMPS

In order to reduce the gas density, and thereby the gas pressure in a gas-filled volume, gas particles must be removed. Vacuum pumps are used to perform this task. Vacuum pumps can be fundamentally divided into two classes:

a) Those which remove the gas particles from the pumped volume and convey them to the outside, by subjecting them to one or more stages of compression. These are known as *compression, discharge,* or *gas transfer* vacuum pumps.

b) Those which condense (or in some other manner bind) the particles to be removed on a solid surface. These are called *entrapment* vacuum pumps.

Compression pumps have an *inlet*, which is at the entrance of the pump (at the low pressure side), and an *outlet* to the high pressure outside. The total pressure on the outlet side of a compression pump is called the *backing, exhaust, discharge,* or *forepressure*. Entrapment pumps have an *inlet*, but no outlet.

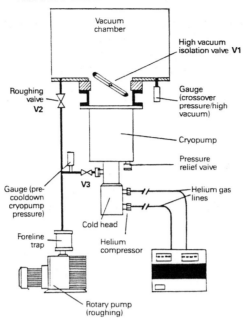

Fig. 3 Schematic of a vacuum pump system. Courtesy of Edwards High Vacuum Company.

The vacuum environment is divided into several ranges (as defined earlier), and no single pump exists which is able to pump chambers from atmosphere to the high vacuum (or ultrahigh vacuum range). As a result, many pumps have been invented, each of which can operate effectively between specific pressure levels. Table 1 shows the operating pressure ranges of a variety of pumps and pump combinations.

For VLSI applications it is useful to identify two broad vacuum ranges: a) the low and medium vacuum range (10^5-0.1 Pa); and b) the high vacuum range (0.1-10^{-4} Pa). Most microelectronic processes are operated in one or the other of these ranges. Pumps that produce low-medium vacuums are called *roughing pumps*, and the others, *high-vacuum pumps*. Figure 3 shows a schematic of a vacuum pump system, which has a high-vacuum cryogenic pump.

In order to pump from atmospheric pressure to high vacuum, a roughing pump is first used to evacuate the chamber to a medium vacuum range, then the high-vacuum pump takes over. The pressure at which the high vacuum pump is connected into the vacuum system following the initial roughing, is called the *crossover pressure*. The roughing pumps most widely used in microelectronic applications are rotary vane pumps, rotary piston pumps, and Roots blowers. The common high-vacuum pumps are diffusion pumps, cryopumps, and turbomolecular pumps.

The *pressure range* of a single pump is that range over which the pumping speed of the pump is considered useful. The *pumping speed*, S, of a pump is defined as the volume of gas which the pumping device removes from the system per unit time, and is typically expressed in liters /sec. Pumping speed is not constant, but is dependent on the pressure at the inlet valve of the pump and on the type of gas being pumped. Most pumps however, maintain a nearly constant pumping speed over several decades of pressure. Above and below this region, however, their speed drops and the pump becomes quite inefficient. The pumping speed for various gases also typically varies for each pump type, even though it is normally specified just for N_2. Light gases (H and He) often have significantly lower pumping speeds than N_2. The pumping speed is the basic parameter used to determine the ultimate pressure that a system can achieve. In addition, the faster the pumping speed, the less time it will take to reach a desired vacuum level.

The ultimate (or lowest) pressure, P_{ult}, achievable by a pump at its inlet is determined by the system gas load, Q_a (due to the leakage of the pump itself, the vapors given off by the walls of the chamber, and the vapor pressure of the pump fluid), and the pumping speed, according to $P_{ult} = Q_a / S$. The time required for a system to reach P_{ult} depends not only on S, but also on the extent of the outgassing in the vacuum chamber. That is, if all the gas to be removed were located only in the volume of the chamber, it could be removed in a short time. In practice the slow evolution of gases from interior surfaces substantially lengthens the actual pumpdown time (Fig. 4). The dominant gas load is water vapor, which is rapidly adsorbed whenever the chamber is exposed to atmosphere. This effect can be demonstrated by venting a vacuum system to room air and recording the time required to go from some crossover pressure (e.g. 15 Pa) to an easily achieved P_{ult}. If the system is then vented with dry N_2, without being opened, the pumpdown from crossover to the same P_{ult} as before will take less than one tenth the time.

Two simple methods which accelerate the desorption of gases (and thus minimize the effects of outgassing on pumpdown time), are: a) heating the chamber to drive off sorbed moisture; and b) maintaining a flow of argon during pumpdown. In the latter technique, a calibrated flow of argon that maintains a chamber pressure of between 0.01- 0.1 Pa, causes each solid surface atom in the chamber to be struck by approximately 30-300 Ar atoms /sec during pumpdown. Some of these impinging Ar atoms are energetic enough to cause the desorption of adsorbed surface atoms.

In high vacuum applications the inlet port of the high vacuum pump is normally open at all times when the chamber is at high vacuum ($<10^{-4}$ Pa). Since the gas molecules are subject to molecular flow at these pressures, they are not actually "pumped", but randomly arrive at the inlet of the pump. The task of the pump is to snare and remove them from the chamber. This shows that pumping speed is limited by the number of molecules that enter the inlet at a given pressure, which is dependent on the area of the pump inlet and molecular impingement rate. If the pump were 100% efficient at removing all molecules that happen to wander into its inlet, the maximum theoretical pumping speed would exist. Thus, to a great degree, the larger the pumping port of the chamber and the pump inlet port, the larger the theoretical maximum pumping speed of a high vacuum pump. However, a pumping system is never 100% efficient. For example, the pumping port of a chamber is rarely connected directly to the inlet port of a pump. Instead, valves, cold traps, baffles, and interconnecting tubing exist between the chamber pumping port and the high vacuum pump inlet. These components impede the motion of gas molecules from the chamber to pump, and thus lower the total effective pumping speed (Fig. 5).

The *throughput* of a pump, Q_a, specifies the *quantity* of a gas (as opposed to the specific volume of gas as with the pumping speed) which is removed from the chamber at the inlet of the pump. (Remember that a larger quantity of gas is contained in the same volume as the pressure

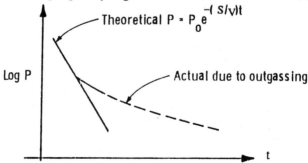

Fig. 4 Pressure in a vacuum chamber as a function of pumping time, showing outgassing effects.

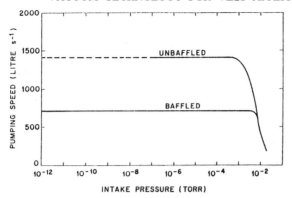

Fig. 5 Speed of a 6-inch diameter diffusion pump as a function of pressure. A speed reduction of fifty percent has been assumed for a baffled curve. From L. Maissel and R. Glang, Eds., *Handbook of Thin Film Technology*, 1970[3]. Reprinted with permission of McGraw-Hill Book Co.

increases.) The units of throughput are Pa-liters /sec (or torr-liters /sec), and like pumping speed, throughput is pressure dependent. However, it is important to note that pumping speed and throughput, although related, should not be confused. Throughput, in fact impacts several operating specifications of a pump other than P_{ult}. First, the quantity of gas that a high-vacuum compression pump can remove from a chamber is important in determining the size of the backing pump (to be defined) that it requires. Second, the steady state throughput of a pump, which specifies the quantity of gas that can be pumped on a continuous basis, is related to its *impulsive gas handling capability* (i.e. the ability to absorb, or *gulp*, an abnormally large burst of gas for a short time). Finally the throughput plays a role in the average time a gas molecule stays in a vacuum chamber. For the problem of continuous outgassing, we would like to remove desorbed gas molecules from a chamber as quickly as possible. This dictates a short residence time, and hence requires a pump with a high throughput capability.

Example 2: Given a vacuum system with a pump that has a throughput, Q_A, of 200 Pa-1 /sec. The system first pumps the chamber out to 2×10^{-4} Pa, and then during the subsequent process, maintains a pressure of 0.2 Pa (while pumping Ar gas). The chamber volume V is 100 l, and the ambient temperature is 300°K. Find the following quantities: a) the impingement rate for both pressures; b) the monolayer time, assuming a sticking coefficient of 0.1; and c) the residence time at the 0.2 Pa process pressure.

Solution : a) The impingement rate, Z_A, is found from Eq. 7 :

$$Z_A = \frac{n\ c}{4} \qquad \text{or}$$

$$Z_A(2 \times 10^{-4}\ \text{Pa}) = (\ 5 \times 10^{10}\ \text{cm}^{-3} \times 4 \times 10^{4}\ \text{cm /sec}\ ) /4 = 5 \times 10^{14}\ /\text{cm}^2\text{sec}$$

$$Z_A\ (0.2\ \text{Pa}) = 5 \times 10^{17}\ /\text{cm}^2\text{sec}$$

b) If a monolayer contains 1×10^{15} atoms/cm^2, and the sticking coefficient is 0.1, then only 10% of the molecules stick to the surface.

Thus:

$$\tau_m \text{ (sec)} = 1\times10^{15} \text{ cm}^2 \,/\, (Z_A \times s_c) \qquad \text{or}$$

$$\tau_m(\text{sec})_{(2\times10^{-4}\text{ Pa})} = 1\times10^{15} \text{ cm}^2 \,/[\, (5\times10^{14} \text{ cm}^2/\text{sec}) \times 0.1] = 20 \text{ sec}$$

$$\tau_m(\text{sec})_{(0.2\text{ Pa})} = 0.02 \text{ sec}$$

c) The residence time τ, is found from:

$$\tau \text{ (sec)} = \frac{V}{S} = \frac{P\,V}{P\,S} = \frac{P\,V}{Q_A} \qquad \text{or}$$

$$\tau\,(\,0.2 \text{ Pa}\,) = (0.2 \text{ Pa} \times 100 \text{ l}) \,/\, [200 \text{ (Pa-l /sec)}] = 0.1 \text{ sec.}$$

The pressure at the inlet of the pump is called the *inlet pressure*, and at its outlet, the *outlet (or discharge, backing, or fore) pressure*. The ratio of the outlet pressure to inlet pressure is called the *compression ratio* of the pump, and is an important specification of compression pumps. The *critical inlet pressure* specifies the inlet pressure of a pump above which an abrupt decrease in pumping speed occurs. The *critical outlet pressure* (or *critical backing pressure*, or *limiting forepressure*) is the pressure at the discharge side above which the pumping action of the pump rapidly deteriorates (e.g. as evidenced by a sudden increase of inlet pressure). If a compression-type pump (e.g. a high vacuum compression type pump) is discharging its pumped gas into a volume whose pressure is lower than atmosphere it is the function of the *forepump* (or *backing pump*) to produce this lowered pressure region, and keep the forepressure lower than the critical outlet pressure.

As mentioned, the ultimate pressure of a vacuum system, P_{ult}, may also be limited by the

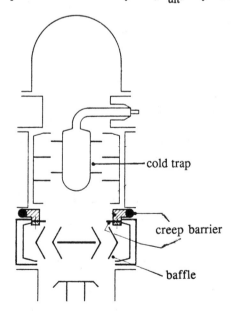

Fig. 6 Schematic arrangement of baffle, anti-creep barrier and cold trap above an oil diffusion pump. Courtesy of Leybold-Heraeus Vacuum Products, Inc.

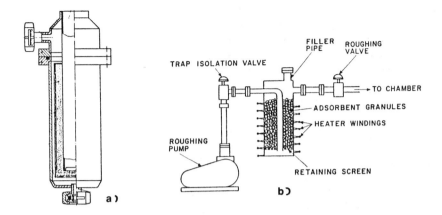

Fig. 7 (a) Section of a foreline sorption trap. Courtesy of Leybold-Heraeus Vacuum Products, Inc. (b) Bakeable foreline trap with pump and isolation valves. From L. Maissel and R. Glang, Eds., *Handbook of Thin Film Technology*, 1970[3]. Reprinted with permission of McGraw-Hill Book Co.

pump oil of the pumps used to evacuate the chamber. Pump oils are employed to seal rotary pumps, to lubricate bearings in turbomolecular and Roots pumps, and to provide pumping action in diffusion pumps. The P_{ult} limitation due to pump oils stems from two phenomena, *creep* (which is the spreading of pump fluid on surfaces by surface migration), and *backstreaming* (in which oil vapor from the pumps moves counter to the flow of gases being pumped). Such movement of pump oil into the chamber may not only limit P_{ult} but can be a source of serious contamination in vacuum processes (i.e. it can become incorporated into and degrade the properties of deposited films and /or cause adhesion problems).

Techniques to reduce backstreaming and creep include: a) selecting pumps that do not utilize oil (e.g. cryopumps); b) use of low vapor pressure pump oils; c) following careful system operating procedures designed to minimize backstreaming; and d) employing creep barriers, baffles, and traps. We will now discuss item (d), while the items (a), (b), and (c) will be covered later in the chapter in more detail.

The *creep barrier* is a structure of non-wettable material (e.g. Teflon[®]) that is placed along the path of surface migration to the vacuum chamber, and thereby prevents creep beyond that point. A *baffle* is a device that is placed in the path of gas or vapor flow to prevent straight line flow (Fig. 6a). Gas molecules colliding with a baffle have a higher probabiltiy of bouncing back into the pump than into the vacuum chamber. *Traps* are pumps for condensable vapors. However, the terms are not precise because baffles are often cooled and use the effect of condensing the returning vapor to the pump by multiple collisions. Baffles are normally placed near or at the inlet of diffusion pumps (Fig. 6b). In cryopumps they serve as radiation shields and not oil condensing structures, and may have a chevron or shell configuration.

Traps typically use either cryocondensation or nonrefrigerated adsorption to reduce backstreaming. The former use surfaces cooled to 77°K by liquid nitrogen to condense backstreaming pump fluids. Such cold traps can be placed on the pump side of the high vacuum valve (typically re-entrant type units are used for this purpose) or directly in the vacuum chamber. Those placed directly in a chamber are known as *Meissner traps*, and can be built like the reentrant types, or can consist of copper coils through which liquid N_2 is passed.

The purpose of placing a Meissner trap directly into a chamber is primarily to increase the pumping speed of water vapor. That is, if the Meissner trap has a cold surface area (77°K) equal to the area of the chamber pumping port, the Meissner trap should double the system pumping speed for water vapor, because the area for pumping water vapor is doubled. It was shown that the pumping speed of high vacuum pump is proportional to the area of the inlet port /pumping port. In fact, a Meissner trap often increases the system pumping speed of water vapor by nearly four times, because the pumping port does not remove 100% of the molecules that enter it from the vacuum chamber.

Non-refrigerated adsorption traps, also known as *molecular sieves*, are filled with materials having a very large surface to volume ratio (e.g. zeolite or activated alumina) which allows them to effectively adsorb large quantities of vapor (Fig. 7). Water vapor, however, will quickly saturate zeolite traps, and even when used to trap oil vapor, they will saturate in ~3 months under normal use, and must then be regenerated by baking. Molecular sieves can also be sources of particles, which drift into valve seats and the interior of mechanical pumps, hastening their wear.

Valves in vacuum systems are used to separate different parts of the system from one another. Vacuum valves that merely separate two spaces at different pressures, include: *roughing and foreline valves*, which serve to connect or disconnect forepumps with the chamber and the vacuum pump, respectively; *high-vacuum valves*, located between the chamber and high vacuum pump; and *relief valves*, which will automatically open when the pressure on the seat side rises above a pre-set value. The most widely used high-vacuum valve is the *gate* or *sliding* type.

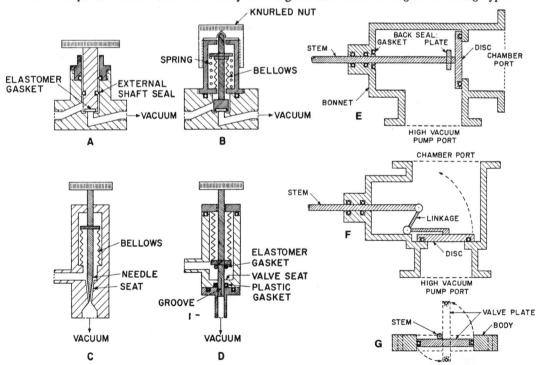

Fig. 8 Gas-admittance valves: (a) packed disk valve (b) packless disk valve (c) packless needle valve (d) packless variable-leak valve. High-vacuum valves: (e) disk valve, (f) flap valve, (g) butterfly valve. From L. Maissel and R. Glang, Eds., *Handbook of Thin Film Technology*, 1970[5]. Reprinted with permission of McGraw-Hill Book Company.

Their basic operating principle is illustrated in Fig. 8e, f, & g.   If installed as shown in Fig. 8e, the interior of the valve, with the exception of the disk surface, remains under vacuum while the chamber is opened to air.   An alternative high-vacuum valve for throttling cryopumps during sputter deposition processes is the *louvre-type*, which also doubles as a radiation shield.

Valves are also used to vent a vacuum chamber or to adjust and maintain specific gas flow rates into the vacuum chamber (e.g. needle and leak valves, Figs. 8a-d).   These are normally small valves of simple construction.   Figure 8c, shows one design of a variable leak valve (i.e. a needle valve), in which flow control is achieved by varying the space between a hard stainless-steel pointed shaft (needle) and a copper brass seat of matching shape.   See Ref. 20 for more information on variable leak valves.

ROUGHING PUMPS

Oil-Sealed Rotary Mechanical Pumps

Oil-sealed rotary mechanical pumps[1,3,4] are compression pumps widely used to perform two important functions in vacuum systems: a) they are used to pump vacuum chambers from atmosphere (10^5 Pa) to the medium vacuum range (15-0.1 Pa); and b) they serve as forepumps for diffusion and turbomolecular high vacuum pumps.   There are two common types in widest service, rotary piston pumps (Fig. 9a) and rotary vane types (see Fig. 9b, which shows a two stage version).   In both types, the gas first enters the suction chamber, is compressed by the rotor and piston (or vane), and then is ejected to the atmosphere through the discharge chamber.

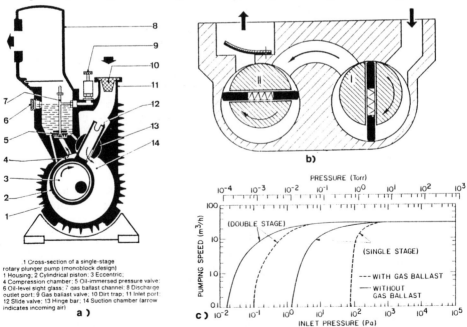

.1 Cross-section of a single-stage
rotary plunger pump (monoblock design)
1 Housing; 2 Cylindrical piston; 3 Eccentric;
4 Compression chamber; 5 Oil-immersed pressure valve;
6 Oil-level sight glass; 7 gas ballast channel; 8 Discharge
outlet port; 9 Gas ballast valve; 10 Dirt trap; 11 Inlet port;
12 Slide valve; 13 Hinge bar; 14 Suction chamber (arrow
indicates incoming air)   **a)**

b)

c)

Fig. 9   (a) Cross-section of a single stage rotary-piston pump (b) Cross-section of a two-stage sliding-vane rotary pump. Courtesy of Leybold-Heraeus Vacuum Products, Inc. (c) Pumping speed curves for a single stage and a double stage rotary vane pump.

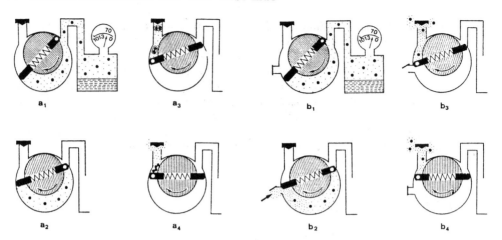

Fig. 10 Illustration of the pumping process in a rotary pump without a gas ballast device (left), and with a gas ballast device (right). Courtesy of Leybold-Heraeus Vacuum Products, Inc.

The exhaust seal is made by the discharge valve and the oil above the valve.

In the *rotary vane pump*, the vanes are spring loaded and press against the inner surfaces of the stator. An oil film serves to lubricate all the parts of the pump and also seals the space between the vanes (and rotor) and the housing. The whole rotor-stator assembly in rotary vane pumps is also submerged in pump oil. Because of the tight fit between the vane and stator wall, rotary vane pumps are able to attain lower ultimate pressures than rotary piston pumps, and are thus most widely used for non-corrosive gas pumping applications.

In *rotary piston pumps*, the piston is rotated on an eccentric rotor and slides along the wall of the stator, although it is constrained from actually contacting the stator. The oil is used to seal the space between the fixed and moving components. The clearances of the piston pump are greater than those of the vane pumps, and this makes the pump more tolerant of particle contamination. In addition, its simplicity helps increase its ruggedness. Since there are no vanes to stick or rub against the stator housing, it will pump as long as the rotor can be turned. For these reasons (and since the piston is more corrosion resistant than the vanes), rotary piston pumps have an advantage for applications that require the pumping of reactive gases.

The lowest pressure that single stage rotary pumps can attain is about 1 Pa, but substantially lower pressures than this ($\sim 3 \times 10^{-3}$ Pa) can be achieved by adding a second stage (Fig. 9c). In such configurations the exhaust of the first is connected to the inlet of the second.

Even low vapor pressure oils can cause contamination through backstreaming into the vacuum chamber. As a result, when rotary pumps are used to rough a vacuum chamber, it is best to follow a procedure that will prevent this occurrence. That is, the rough pumpdown must be restricted to pressures no lower than 15 Pa. At 15 Pa or above, the roughing line will be in viscous flow, and no oil molecules will be able to backstream into the chamber. When a roughing pressure of 15 Pa is reached, the roughing valve must be closed, the high vacuum valve opened, and the chamber pumped as quickly as possible to high vacuum.

The gas that the roughing pumps expel to atmosphere picks up large quantities of pump oil in the form of mist or oil droplets. This will lead to rapid depletion of oil from the oil reservoir. To avoid this problem, rough pumps are equipped with *mist separators*. They remove the oil from the gas and return it to the reservoir.

The last item of this section concerns the use of rotary pumps for pumping large amounts of condensable vapors (e.g. water, acetone). Such vapors will condense on the inner walls of the pump and eventually contaminate the pump oil (causing gummy deposits that may finally even cause the pump to seize). To avoid this condensation, a valve is provided which admits another gas (usually N_2) into the compression chamber during the compression stage. This added gas (or *ballast*) raises the pressure in the chamber and causes the exhaust valve to open before condensation occurs (Fig. 10). The price that is paid is a higher ultimate pressure when the gas ballast valve is operated.

Vacuum Pump Oils for Semiconductor Processing

The choice of the proper pump oil is important in obtaining the best performance from oil-using pumps[12]. The lowest ultimate pressure of oil-sealed rotary pumps is limited by the vapor pressure of the pump oil, and leakage of air past the seal between the exhaust and inlet ports. Therefore, the oil that makes the seal must have a low vapor pressure and be viscuous enough to fill the gap between the moving surfaces (without being so viscuous that the pump is hard to start). As a result, only limited types of materials will serve as good pump oils. There are other requirements for the oil as well. They include: a) the capability of the oil to resist reaction with the gases being pumped, as corrosive gases cause pump oils to polymerize, turning them into a tar-like sludge [Fig. 11]. Thus, if the oil reacts with the process gases, frequent oil changes and maintenance of the pump are likely to be necessary; and b) non-flammability. There are a broad range of pump oils available, each best suited for particular applications of semiconductor fabrication.

Hydrocarbon fluids are one type that find wide use. The hydrocarbon fluids obtained directly from the petroleum refinery are first distilled to remove volatile components. *Singly distilled hydrocarbon oils* are suitable for rotary vane pumps in non-corrosive applications. Doubly-distilled fluids have even lower vapor pressures than singly distilled varieties, and consequently exhibit lower backstreaming rates and longer life in corrosive environments.

a)   b)
Fig. 11 (a) Sludge generated on inside of rotary pump from hydrocarbon-based pump fluid. (b) Same pump insides when operated with perfluorinated polyether fluid. Reprinted with permission of Semiconductor International.

The *technical white* (or TW) series of hydrocarbon fluids are distilled from petroleum stocks that are processed to remove a variety of naturally occurring impurities. As a result, TW fluids have service lives that are two to three times longer than the doubly-distilled fluids. Hydrcarbon fluids, howver, are also subject to reaction with oxygen.

Inert mechanical fluids are utilized in applications that call for stability in the presence of halogens, hydrogen halides, and oxygen. Two types of inert fluids are currently available: a) perfluoropolyethers (PFPE); and b) polychlorotrifluoroethylene (CTFE).

The *PFPE fluids* (e.g. Fomblin®) are not flammable and are inert to most gases and process reaction by-products. They can, however, be decomposed by ammonia (used in nitriding) and aluminum chloride (a by-product of Al dry-etching). The decompostion products are normally volatile, and are exhausted into the scrubber. PFPE fluids are most useful in highly corrosive environments, and in processes in which the pump oil is exposed to large quantities of O_2. Although the PFPE fluids do not react with most process gases, many such gases are nevertheless highly soluble in PFPE. Thus, some means of filtering and neutralization must be employed. Otherwise the fluid can become so acidic that it will attack the internal components of the pump. Unfiltered PFPE can also cause skin burns or eye irritation (from the fumes), and hence safe procedures must be followed when handling spent PFPE. The highly inert behavior of PFPE, however, allows it to be completely recycled many times by removing the contaminants (e.g. by filtration of any solids, neutralization of any acids, and distillation to return it to its original low vapor pressure state). The PFPE fluids are also very costly.

The *CTFE fluids* share many of the properties of PFPE, but are considerably less expensive. That is, they are stable in the presence of O_2, halogens, and hydrogen halides, and are decomposed by ammonia and aluminum chloride. They do, however, exhibit significantly higher vapor pressures that PFPE.

Three example processes can be used to demonstrate the advantages of the various pump fluids: a) a nitride deposition process; b) a dry etch process using fluorine-based etch gases; and c) an aluminum dry etch process. In the *nitride deposition process,* silicon nitride is deposited everywhere in the chamber, and the process gases also react to form ammonium chloride and SiO_2. As all of these products are solids, use of an inert pump fluid offers no advantage. An optimum choice is to use a less expensive doubly-distilled or TW fluid instead, together with a large capacity filter to remove the solid particulates. In the *fluorine-based etch process,* some elemental fluorine and HF will be removed by the pump. These gases will attack the pump and hydrocarbon based fluids. If the gas flow is small, a TW hydrocarbon fluid may still be suitable, but for high gas flow rates, an inert fluid must be used. The choice of PFPE or CTFE is dependent on the required initial pumpdown pressure (i.e. the higher vapor pressure CTFE may not allow the pump to reach the required low pressure). In all cases, an external fluid filtration system with acid neutralization to protect the pump should be employed. When *Al is being dry etched,* the aluminum-chloride by-product will decompose PFPE and CTFE. Thus these very costly fluids will slowly dissappear from the pump, and will require frequent (and thereby expensive), topping-off. A more ecnomical solution is to use a chemical series hydrocarbon fluid (i.e. a *TW oil*) together with an external filtration unit having a neutralizing medium (such as activated alumina) that will remove the chlorine and hydrogen chloride (HCl).

Roots Pumps

Roots pumps (also often referred to as *Roots blowers*) are mechanical pumps typically used in series with rotary pumps[1,3,4]. They serve to increase the pumping speed and to achieve lower operating pressures than those obtainable with rotary pumps alone (as low as 5×10^{-3}Pa,

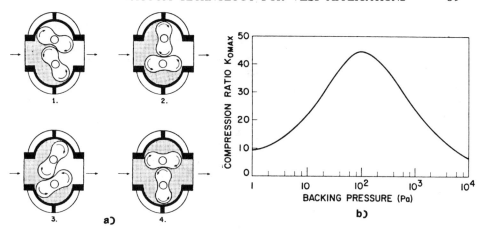

Fig. 12 (a) Operating principle of the Roots pump. From L. Maissel and R. Glang, Eds., *Handbook of Thin Film Technology*, 1970[3]. Reprinted with permission of McGraw-Hill Book Company. (b) Dependence of the compression ratio of a Root pump on vacuum pressure. Courtesy of Leybold-Heraeus Vacuum Products, Inc.

Fig. 12). Roots pumps contain two counter-rotating lobes mounted on parallel shafts which rotate synchronously in opposite directions at speeds of 3000-3500 rpm. The lobes do not touch each other or the housing, with clearances being about 0.1 mm. Since no oil is used to seal the gaps, such pumps are termed *dry* pumps. However, oil *is* used to lubricate pump bearings.

The compression ratio of the Roots pump is pressure dependent, generally having a maximum (between 10 and 100) near 100 Pa (Fig. 12). At higher pressures, the compression ratio is lower because the conductance of the gaps increases with pressure. As a result, pumping against atmosphere is possible, but causes considerable pump heating. If the heat produces excessive expansion of the rotors, damage to the pump may occur. As a result, Roots pumps are commonly operated in series with a rotary pump that has a rated speed about 1/10 that of the Roots pump. During the initial pumpdown from atmosphere, a bypass line around the Roots pump is opened. All pumping is done by the rotary pump, until a backing pressure of ~2000 Pa is reached (15 torr), at which time the Roots pump is switched on and the bypass valve closed.

Roots pumps provide two key advantages: a) they offer high capacity, low cost pumping in a range not efficiently handled by either rotary or high vacuum pumps; and b) they substantially reduce oil contamination due to rotary pumps by keeping the inlet to these pumps at a higher pressure, and by interposing the dry Roots pump between the rotary pump and vacuum chamber. On the other hand, at low enough chamber pressures, the pressure in the line connected to the outlet of the Roots pump may become low enough to allow molecular flow. In such cases oil can backstream to the Roots pump, then creep around its interior surfaces, and finally enter the process chamber. This limits the use of Roots plus rotary pump combinations to conditions at which the Roots outlet port is maintained at a pressure of 15-20 Pa.

HIGH VACUUM PUMPS I: DIFFUSION PUMPS

Diffusion pumps[1,4,17] were the most widely used high vacuum pumps in the early years of the semiconductor industry. They enjoyed this position because of their high pumping capacity, simplicity, and relatively low initial cost (although their operating costs can be higher than other high vacuum pumps if they are operated with liquid N_2 cold traps). The basic design is still similar to that invented by Gaede in 1913, although significant advances have been made that allow the pump to remain appropriate for some applications (i.e. by reducing backstreaming through the development of low vapor pressure pumping fluids and pump design enhancements). One application in which diffusion pumps are still widely used is for pumping ion sources in ion implanters, since pure ion beams are extracted from the source by the mass analyzer (Chap 9).

The operating principle of diffusion pumps is described with the aid of Fig. 13. A stable, low vapor pressure oil is heated in a boiler to a temperature on the order of 200°C (at which the oil's vapor pressure is about 200 Pa). The hot vapors rise up a chimney above the boiler. At the top of the chimney the vapor direction is reversed by a jet cap, and the vapor emerges from a ring shaped nozzle (Fig. 13a, points A-D) moving in a direction away from the high vacuum side of the system. In passing from a relatively high pressure region (inside the boiler) to one of lower pressure, the vapor expands and thereby acquires a supersonic velocity. This effect is critical, since molecules emerging with a normal velocity distribution would spread omnidirectionally and would yield no pumping action. The supersonic vapor then streams downward and toward the cooled outer walls of the pump, where it is condensed and returned to the boiler. Cooling coils are seen in Fig. 13b. Gases from the high vacuum side of the pump inlet (vacuum chamber) diffuse into the supersonic vapor stream, and on average receive velocity components from

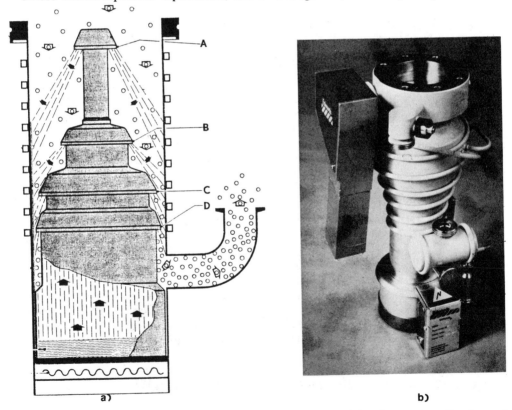

Fig. 13 (a) Cross section of a diffusion pump showing principle of operation. Courtesy of Leybold-Heraeus Vacuum Products, Inc. (b) Photograph of high-vacuum diffusion pump. Courtesy of Edwards High Vacuum Company.

collisions with molecules of the pump fluid, which carry them toward the pump outlet, and at the same time cause them to undergo compression.

Typically, a vacuum pump has three compression stages working in series (each compressing the gas to a successively higher pressure than the preceding stage as it is transported toward the outlet). A fourth stage built into the foreline, and referred to as the *ejector* stage, is used to compress the gas at the highest pressure in the pump and eject it into the foreline. The use of the ejector helps produce a pump with a higher forepressure tolerance than is possible with vapor jets alone.

Since the pressure in the boiler is about 200 Pa, the pressure at the outlet of the diffusion pump cannot be higher (it must in fact be lower than this value, or all pumping action will cease). This *critical forepressure* value (about 50-75 Pa) must be maintained by a forepump of adequate pumping capacity connected to the diffusion pump outlet. The pressure at the *inlet* of the diffusion pump also must not exceed a specified maximum value (*critical inlet pressure*). This would alter the course of the vapor from the top jet and cause pumping speed (and pressure control) instability, and increased oil backstreaming. Critical inlet pressures of diffusion pumps are typically around 0.1 Pa.

At normal operating pressures, diffusion pumps are constant-pumping speed devices (Fig. 13b), and this range extends from 10^{-1}-10^{-4} Pa for most gases. In practice, at the lowest operating pressures, the pump is limited by outgassing and vapor pressure of the pump fluids condensed on the baffles, although few applications in the semiconductor industry require vacuums below this limit. The throughput is the product of the pumping speed and the pressure at the pump inlet flange, and hence rises linearly with pressure (which corresponds to the maximum usable throughput).

Although the ultimate vacuum pressure of diffusion pumps is low enough for most all semiconductor manufacturing processes, diffusion pumps have another serious limitation. That is, oil from the diffusion pump is always transported from the pump toward the the vacuum chamber where it becomes a contaminant that degrades fabrication capabilities. This unwanted effect, defined earlier as *backstreaming*, can arise from several causes such as: a) evaporation of fluid condensed on the upper pump wall; b) overdivergence of the vapor in the top jet; and c) evaporation of oil from heated lip of the top jet. Sources (b) and (c) are substantially suppressed by use of low vapor pressure pump fluids and additional baffles, and /or traps over the pump. Transport of oil by *creep*, can also occur and is suppressed by use of *creep barriers*.

To operate a vacuum system containing a high-vacuum diffusion pump, the following approaches should be taken to minimize backstreaming: a) the diffusion pump and mechanical roughing pump oils should be low vapor pressure and stable fluids; b) the diffusion pump should possess adequate traps and baffles to intercept backstreaming oil molecules; and c) system operation should involve pumping to a rough vacuum of no lower than 15 Pa, closing the roughing valve, opening the high vacuum valve, and pumping to high vacuum as quickly as possible.

Figure 14 shows that the roughing pump will cause significant backstreaming into the chamber through the roughing line at pressures lower than about 15 Pa. By the same token, when the high vacuum valve is opened at 15 Pa, the critical inlet pressure of the diffusion pump is momentarily exceeded, and this will allow some backstreaming until the vacuum pressure is reduced to 0.1 Pa. To minimize this occurrence, conductance-limited high vacuum valves (which operate in a controlled opening manner) have been designed into some diffusion pumps to allow them to be operated without overloading during pumpdown.

If the system employs a roughing line trap, it should also be periodically cleaned. When diffusion pumps are operated with cold traps, the trap must be supplied with liquid N_2. This makes the diffusion pump considerably more expensive to operate than the other high vacuum

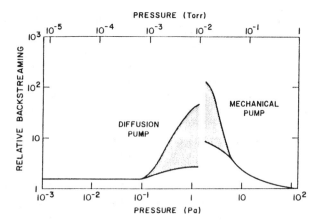

Fig. 14   Pumping and lubrication fluid backstreaming in the transition zone between the diffusion and mechanical pump operation regions.   Reprinted with permission from Japan J. Appl Phys., Suppl. 2, Pt. 1, p. 25, M.H. Hablanian. Copyright 1974, Japanese Jnl. of Applied Physics.

pump types. In some applications such traps are located on the pump side of the high vacuum valve, while in others they are located in the processing chamber (Meissner traps). Meissner traps must be vented each time the chamber is returned to atmosphere, and then heated above the ambient temperature to insure against condensation of H_2O onto the trap from the air. Consumption rates of LN_2 are therefore up to ten times higher than those in pumps whose traps are not located in the chamber.

The maintenance of diffusion pumps is a relatively simple task, and normally involves only an oil change. The frequency of oil changes varies with the process, but usually is done only when pump performance deteriorates or on a long term schedule of once or twice a year. Such maintenance is generally performed in-house. Occasionally the pump must be removed from the system, to be thoroughly cleaned, and perhaps to re-adjust the jet gaps or replace the heaters.

HIGH VACUUM PUMPS   II : CRYOGENIC PUMPS

High-vacuum cryogenic pumps for microelectronic applications[1,7,8,9,16] are closed cycle refrigerator pumps that remove gases from the vacuum chamber by capturing them on a cold surface (entrapment pumps). It is worth noting at the outset that since cryopumps are entrapment pumps they do not use (or need) backing pumps, and that since their operation is based on an oil-free pumping mechanism, no oil backstreaming occurs from such pumps.

The process of gas capture by high-vacuum cryopumps involves both cryocondensation and cryosorption. *Cryocondensation* refers to the condensation on a surface whose temperature is cold enough that the vapor pressure of the condensed substance is so low that the vapor is effectively removed from the system. Cryocondensation is thus utilized to pump condensable gases. *Cryosorption* involves the *adsorption without condensation* of a gas on a cold surface. Any solid surface has a weak attraction (van der Waals type) for at least a few monolayers of gas or vapor. After a few monolayers have been deposited (a typical monolayer is $\sim 10^{15}$ atoms $/cm^2$), it ceases to act as an effective pumping mechanism. Cryosorption, however, is an important phenomenon because it can be used to pump substances to pressures far below their saturated vapor pressures. For instance, the pumping of such gases as H, He, and Ne, is still achieved by cryosorption, which effectively immobilizes these gases onto cold surfaces. That is, such gases still have high vapor pressures even at 20°K, and would therefore quickly desorb from 20°K surfaces after colliding and briefly sticking. Thus, cryocondensation could not be used to effectively pump them.

The closed-cycle refrigerators of modern high-vacuum cryopumps operate much like household refrigerators, in that the refrigerant is compressed, cooled, and allowed to expand. The refrigerant however in this case is helium, always maintained in the gaseous phase. The closed-loop refrigeration system is usually split into two parts: 1) an expander (or *cold head*) mounted on the pump itself; and 2) a compressor (Fig. 15). The expander is constructed to give two stages of expansion, resulting in two cold zones (usually at 77°K and 15°K) in the pump.

The high-vacuum cryopumps use three pumping zones to entrap gas molecules: a) a 77°K surface that mainly condenses water vapor; b) a 15°K surface that cryocondenses N, O, and Ar; and c) a 15°K surface whose function is to cryosorb H, He, and Ne. The 77°K surface is said to belong to the *first stage* of the cryopump, and the two 15°K zones to the *second stage* (Fig. 15). The first stage cylinder also serves as a radiation shield that protects the colder second stage from exposure to ambient heat radiation. The *cryosorption* zone of the second stage is shielded as much as possible from the inlet port to lengthen and randomize the path that a molecule must traverse before striking the cryosorption surface. This increases the probability that the condensable gases will cryocondense on other surfaces before they encounter the cryosorption surface, thereby preserving it as a much as possible for the function of pumping H, He, and Ne.

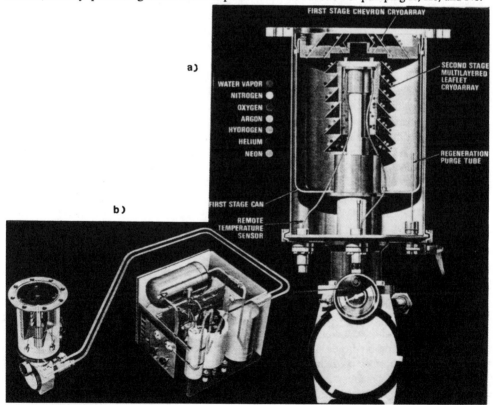

Fig. 15 (a) Cutaway view of a cryopump. (b) Cryopump and refrigerator unit. Courtesy of Varian Associates.

Activated charcoal is the most commonly used cryosorption material because it is a porous material that also has a very large surface /volume ratio. The latter property gives it a high sorption capacity. The porosity is important because even if a captured molecule desorbs from the charcoal material, it is likely to encounter another charcoal surface before escaping back into the pumping chamber.

The cryotemperature of each stage must be maintained in the face the total heat flux to the pumping surfaces. The heat arrives through radiant transfer from warmer surfaces, gas condensation, and gas conduction. Below about 0.1 Pa, the radiant transfer is dominant, but under high gas-flow conditions, heat from other sources may become greater (e.g. during pumpdown, or Ar pumping in sputter applications). For fabrication processes in which the chamber temperatures rise over 100-150°C, radiation baffles over the inlet must be used to avoid excessive radiant transfer heat loads. The refrigeration capacity of the pump must also be considered when evaluating heat loads. The first stage refrigeration capacities are typically 5-80 W, while the second stage capacities are generally between 2-15 W.

The pumping speeds of cryopumps cannot be specified as simply as those of diffusion or turbomolecular pumps because the speed depends on several factors including: the gas species; the size of the input port; the refrigeration capacity; the relative sizes of the warm and cold stages; and the pumping history. Cryopumps, however, are generally able to achieve higher pumping speeds with higher throughput than other high vacuum pumps, especially for water vapor. The pumping speeds tend to be constant below about 0.1 Pa.

The maximum steady-state throughput will depend upon the refrigeration capacity (especially of the second stage), when large quantities of gases such as Ar in sputter deposition are being pumped. For large gas flow pumping applications, the temperature of the second stage should be carefully monitored to insure that the throughput of the pump is not being exceeded. If the temperature starts to rise, this indicates that the heat load of the process is exceeding the pumps refrigeration capacity.

On the other hand, the cryopump has an important advantage over other high vaccum pumps. That is, it is able to exceed *the maximum steady-state throughput* for short periods of time, with no adverse vacuum effects. The capability to handle impulsive gas loads (or to *gulp* gas) is due to the heat capacity of the elements. As long as the gas impulse can be pumped before the elements are warmed too much, the pump will recover with negligible effects on the pumping cycle, or the pump. One reason that this capability is so important is that it allows the crossover pressure in cryopump systems to remain high enough (e.g. between 30-75 Pa) to keep the roughing line in viscous flow and prevent backstreaming into the chamber during initial pumpdown. Remember that when a diffusion pump performs the same task, it may subject the chamber to significant backstreaming while pumping from the crossover pressure down to a high vacuum pressure. The ability to implement a high crossover pressure also shortens the overall time of the process cycle, since pumpdown time is reduced.

Momentary loss of power to a cryopump, however, may cause it to release He from the 15°K stage into the pump /chamber volume. Such He can then conduct large quantities of heat from the chamber walls to the pumping surfaces, and cause the pump temperature to irreversibly rise, even if refrigeration power is restored. Pump regeneration is required at that point.

Even under normal operation the surfaces of the cryopump will eventually become less effective at capturing gases. For example, the saturation of a 6-inch cryopump by Ar would occur after pumping Ar at a speed of 1000 l /s at a pressure of 0.3 Pa (2.3mtorr) at ~100 hours of continuous running. In most cases however, hydrogen is the first gas to saturate the system, its main source being water vapor. (For example, if hydrogen is being pumped as a carrier gas in an ion implanter, at a pressure of 10^{-4} Pa and with a hydrogen pumping speed of 2000 l /s, the pump would become saturated in about two months.) When the pump becomes saturated, either the minimum pump pressure increases, or the temperature of the cold array surfaces starts to rise. At that point, the accumulated gases must be removed from the pump, a process that is referred to as *regeneration*.

The three phases of regeneration involve: a) warmup; b) pumpout; and c) chilldown. During *warmup* the elements of the pump are brought back to room temperature. This process can be hastened by purging the pump with nitrogen. A rough pump is used to evacuate the pump after it has been brought back to room temperature (*pumpout*), since the cryopump refrigerator is not powerful enough to cool the elements except when they are in an insulating vacuum. To prevent backstreaming of mechanical pump oil into the cryopump during pumpout, which would permanently contaminate the charcoal adsorbate and prevent subsequent pumping of light gases, a molecular sieve trap must be used on the vacuum line from the rough pump to the cryopump. This trap must have a valve on either side so that the pump can be isolated from the trap during trap bakeout. If the cryosorption charcoal bed ever becomes contaminated, the only fix is to replace it. Upon completing the pumpout phase, all valves to the cryopump should be closed, and the pump refrigerator started (*chilldown*). The first stage is allowed to cool before the second stage during chilldown, so that any water will be captured by it. Next, the residual O, N, and Ar are pumped on the outer surfaces of the second stage. Finally the *adsorbent stage* is cooled to permit light gas pumping. When all the elements of the pump have reached their ultimate temperatures, regeneration is complete and the pump is ready for full operation again. Several fully-automated, single-push-button regeneration controllers are available, designed to minimize regeneration time and reduce labor costs. Regeneration should be done on a regular basis (such as once every week or month), scheduled to coincide with other routine system maintenance.

All cryopumps have a large capacity, and therefore require a relief valve to prevent dangerous pressure buildup in case of unexpected pump warming. Since the condensed gases may be explosive or toxic, an enclosed exhaust should be provided to pump off any such gases that may be released during accidental warmup, as well as during regeneration.

Some questions have also been raised about the effects of corrosive gases on cryopump elements during regeneration warmup. Little information is available on the subject and hence caution is probably advisable when considering cryopumps for corrosive gas pumping.

HIGH VACUUM PUMPS III: TURBOMOLECULAR PUMPS

The turbomolecular high-vacuum pump[1,3,4] is a compression-type pump which functions in a manner analogous to the diffusion pump. Instead of imparting momentum by collisions with streaming high-speed vapor molecules, however, turbomolecular pumps cause momentum transfer through impacts with high-speed rotating blades. A series of such blades (whose form is like the blades used in turbines) are assembled on a single shaft (Fig. 16). This assembly is designed to rotate at speeds ranging from 24,000-36,000 rpm. The spaces along the axis of the shaft are alternated between rotating and stationary (stator) discs, with clearances between them on the order of millimeters. As the pumped gas moves from the inlet further into the pump, it is continually compressed by impact from the rotor blades. Since each disc represents a pumping stage (but with a relatively small compression ratio), a large number of stages is cascaded to achieve adequate pumping capability. For a series of stages, the total pump compression ratio is approximately the product of the compression ratios of each stage. The compression ratios of light gases (H and He) are somewhat smaller than those of heavier gases, although some turbopump models pump light gases significantly better than others.

Fig. 16   Turbomolecular pump.   Courtesy of Leybold-Heraeus Vacuum Products, Inc.

If the pressure in the pump is kept within the range such that the gas flow occurs in the molecular regime, essentially all collisions by gas molecules will occur with the pump blades, thereby efficiently transferring momentum from the pump to the gas. On the other hand, if viscous flow prevails (higher pressures), intermolecular collisions will dominate and the pump does not transfer momentum efficiently to the pumped gas. In this range the pump operates poorly. As a result, turbomolecular pumps, like diffusion pumps, require a backing pump to keep the forepressure low enough to sustain molecular flow everywhere in the turbopump. If the forepressure is allowed to rise to the point of viscous flow (~1 Pa), the pump experiences a sudden decrease in rotor speed, and thereby also pumping speed. Turbo-pumps will operate at constant speed over their normal operating ranges, and their ultimate pressure (~10^{-8}-10^{-9} Pa) is be limited by their compression ratio for light gases, and the effects of outgassing.

Turbomolecular pumps offer several attractive feature for high-vacuum pumping applications including: a) the ability to pump at high-vacuum ranges without contamination from oil backstreaming; b) high reliability; c) low maintenance (e.g. no regeneration required); and d) compatibilty with other pumps where they provide continuous pumping of inert gases.

The potential limitations of the pump include: a) poor compression ratios for pumping light gases; b) relatively high initial cost (i.e. more costly than comparably sized cryopumps); c) special design of the rotors and bearing assembly to resist corrosion if reactive gases must be pumped; d) potential damage or destruction of the rotor by particles falling into the pump (this is minimized by the placement of fine screens over the pump inlet); e) bearing maintenance (depends on the model); f) produces vibration (depends on the model); g) lower maximum pumping speed than the largest available diffusion and cryopumps; and h) the possibility of backstreaming of mechanical pump oil during roughing or turbopump shutdown (can be prevented by proper venting procedures that are commenced in the event of pump shutdown or power outage).

The maintenance of turbopumps involves the periodic lubrication and replacement of the pump bearings. The turbo design and lubrication system varies widely from brand to brand and among models, as do the the lubrication intervals and difficulty of bearing change. Pump maintenance should be performed by skilled workers (preferably trained by the pump manufacturer), and under clean room conditions. Typical bearing lifetimes are in the two to five year, and 15,000-50,000 hour ranges, depending on the process and other factors.

SPECIFICATIONS OF VACUUM PUMPS FOR VLSI

Users of vacuum systems are likely to find the need to evaluate the capabilities of vacuum pumps in many circumstances. These may include: a) during the design of new equipment, or the modification of existing systems that utilize vacuum pumps; b) when purchasing systems which contain vacuum pumps; and c) when specifying a new process on available equipment that contains vacuum pumps. By being able to identify the performance characteristics of a pump and then comparing them to the application requirements, the suitability of the pump for the task can be established. Listed below are some key vacuum pump specifications of interest to microelectronic workers:

1. Vacuum range of the pump. (Also, does it require a backing pump, and of what pumping specifications?)
2. Pumping speed (as a function of pressure and the species being pumped).
3. Throughput (as a function of pressure and the species being pumped).
4. *Gulp* capability (that is, the abilty to handle impulsive gas loads).
5. Low atomic weight gas pumping capability (H and He).
6. Lowest absolute pressure of the system to be pumped.
7. Critical inlet pressure (used to determine cross-over pressure).
8. Critical outlet pressure, if pump is used with a backing pump.
9. Capabilty of pumping corrosive gases.
10. Variable pumping speed capability.
11. Degree of potential contamination of the process by the pump.
12. Ability of the pump to tolerate radiative heat loads.
13. Initial cost.
14. Cost to operate, maintain, and repair.
15. Reliability.
16. Estimated down-time for routinely scheduled maintenance, regeneration, or most likely repairs.
17. Ease of operation and maintenance. (e.g. Does it need a liquid N_2 supply?)
18. Necessary conditions for safe operation and maintenance.

TOTAL PRESSURE MEASUREMENT

The total gas pressure is the quantity most often measured to characterize the degree of vacuum attained in a system. The pressures measured in vacuum technology cover the range from atmospheric to 10^{-13} torr - or over 15 orders of magnitude! Instruments used to measure

Table 2. WORKING PRESSURE RANGES OF VACUUM GAUGES

total pressure are known as vacuum gauges.   Since it is not possible to build a vacuum guage which can give quantitative measurements over the entire vacuum region, a great variety of vacuum gauges have been developed. Each type has a characteristic measuring range, which mostly extends over a few orders of magnitude (Table 2).

Since the factors involved in selecting and using vacuum gauges are not as critical as other components of vacuum systems, our discussion will be limited to a brief discussion of the following gauges (which are most commonly encountered when measuring total pressure in microelectronic vacuum environments):

Low vacuum gauges   (10 Pa - 1 atmosphere) -- Capacitance manometers
Medium vacuum gauges (0.1 - 100 Pa)   -- Thermocouple / Pirani gauges
High vacuum gauges   (10^{-6} - 1 Pa)   -- Ionization gauges (various types)

References at the end of the chapter[1,2,3,4] are given for readers that desire additional information on vacuum gauges.

Capacitance Manometers
Capacitance manonmeters provide an accurate measure (e.g. to within ±0.1%) of pressure from atmosphere down to about 10 Pa (Fig. 17b). Changes in pressure result in the displacement of a sensitive diaphragm's position relative to a fixed plate (Fig. 17a). The capacitance between the diaphragm and the fixed plate changes with displacement and is directly related to the pressure. Changes in the ambient temperature can affect the gauge readings, and some units are provided with temperature controlled sensors. Zero drift problems may occur if the gauge is cycled between atmosphere and vacuum. To avoid this problem, a mechanical or solenoid valve between the capacitance manometer and the process chamber may be used.

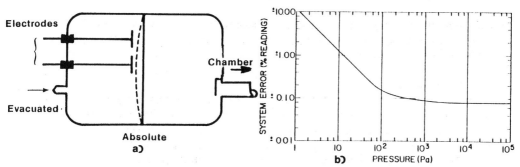

Fig. 17 (a) Capacitance manometer. (b) Performance curve for a capacitance manometer.

Thermocouple Gauges

Thermocouple (and Pirani) gauges are widely used to measure total pressure in the range from 0.1 to 100 Pa ($10-10^{-3}$ torr). For example, they are employed to measure: a) the backing pressures of turbo or diffusion pumps; b) the pressure inside a cryopump during chilldown; and c) chamber pressures as they are being rough pumped during pumpdown. Within the gauge, a filament is heated (Fig. 18b), and an attached thermocouple is used to measure the filament temperature. This temperature increases with decreasing pressure over the pressure range of the gauge (for constant input power to the filament). This occurs because the thermal conductivity of a gas decreases with decreasing pressure over the pressure range where the mean free path of molecules is larger than the spacing between opposite surfaces. As a result, the cooling to the walls of the gauge tube is reduced. Thus, the conduction of heat from a hot object to a cold one becomes linearly proportional to the gas pressure, and the temperature of the filament (as measured by the thermocouple), provides an indication of the pressure. Eventually, at low enough pressures (e.g. ~0.1 Pa), the radiative heat loss from the filament becomes dominant, and since this mechanism is pressure independent, the gauge ceases to be effective. Even over valid pressure ranges, the readings of thermocouple gauges, however, are only accurate for the gas species for which they have been calibrated. The thermal corrections for thermocouple gauges for various gases are as follows: a) nitrogen = 1.0; b) Freon-12 = 0.65; c) helium = 1.25; d) acetone = 0.5; and e) argon = 1.6. As an example, if a thermocouple gauge that is calibrated for nitrogen

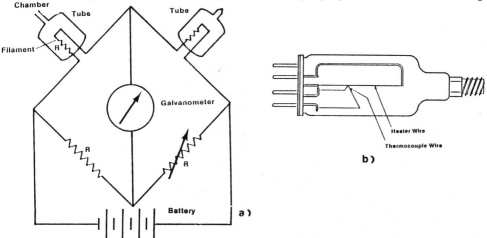

Fig. 18 (a) Pirani gauge. (b) Thermocouple gauge.

reads 1 Pa in an argon atmosphere, the actual pressure would be 1 x 1.6, or 1.6 Pa. It should be noted that high accuracy from thermocouple and Pirani gauges is rarely needed, since it is not called for in the above applications.

Pirani Gauges

Pirani gauges are used to measure pressure over the same general range as thermocouple gauges, although their upper pressure limit is somewhat lower (1-10^{-3} torr). They utilize a resistance wire (enclosed in a glass envelope) which is exposed to the pressure in the vacuum system while being part of a Wheatstone bridge circuit (Fig. 18a). A fixed current is passed through the wire resistor, causing it to be heated. Again, the temperature of the sensing wire decreases as the thermal conductivity of the gas rises. Consequently the resistance of the wire decreases, and the current in the unbalanced bridge circuit indicates is a measure of the pressure.

Ionization (High-Vacuum) Gauges

In ionization gauges, a hot filament is used to provide electrons that are attracted by positively charged grid. The electrons strike gas atoms and cause them to become ionized. These ions flow to a negatively charged cathode, and the resulting current is related to the ion density, or pressure. Two types of ionization gauges are commonly utilized: a) the Bayard-Alpert gauge; and b) the Schultz Phelps gauge. The *Bayard-Alpert gauge* is used for vacuum levels from 10^{-8}-10^{-3} torr. Such gauges have a tube containing either a double tungsten filament (one is a spare), or a single thoriated tungsten filament. The accuracy of the double tungsten filament design is more accurate, but is subject to being burned-out if exposed to air at high pressure. The thoriated tungsten filament can tolerate many exposures to air.

The *Schultz-Phelps* type gauges are used for reading higher pressures (e.g. in the ranges in which sputtering is carried out, up to ~500 mtorr) because they can tolerate these higher pressures and still operate at pressures as low as 10^{-5} torr.

Ionization gauges are also usually calibrated for nitrogen, and must be recalibrated if a different gas atmosphere is to be measured, as the ionization probability is dependent on the gas

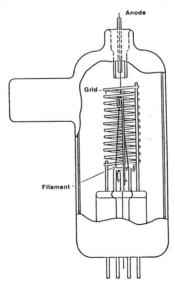

Fig. 19 Ionization gauge.

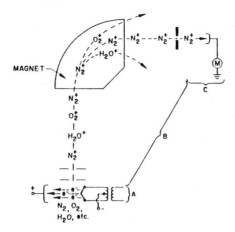

Fig. 20 Three stages of partial pressure analysis (a) Ionization, hot filament shown (b) mass separation, magnetic sector shown. (c) detection, Faraday cup shown. From J. O'Hanlon, *Users Guide to Vacuum Technology*, Copyright © 1980. Reprinted by permission of John Wiley & Sons, Inc.

species. For example, the relative ionization gauge sensitivities for various gases are: a) nitrogen = 1.0; b) argon = 1.6; c) air = 1.0; d) oxygen = 0.9; e) hydrogen = 0.5; and f) helium = 0.2, thus, a gauge calibrated for N_2, reading 1×10^{-5} torr in an O_2 atmosphere, would have an actual pressure of 9×10^{-6} torr.

MEASUREMENT OF PARTIAL PRESSURE : RESIDUAL GAS ANALYZERS

In various vacuum processes it is as important to know the composition of a gas or vapor mixture as it is to know the total pressure. For example, knowledge of partial pressures can provide information about gas purity and the gas contributions from pumps, can help identify the sources of outgassing in processes, can relate film properties to the presence specific gases, and can evaluate the effects of experimental procedures such as baking or glow-discharge cleaning. The measurements of partial pressures in high-vacuum systems is typically performed by mass spectrometers, specifically designed for high-vacuum applications. These are referred to as *residual-gas analyzers.*

Operation of Residual Gas Analyzers

Residual gas analyzers (RGAs) identify the gases present in vacuum environments (Fig. 20) by producing a beam of ions from samples of the gas, separating the resulting mixture of ions according to their charge-to-mass ratios, and providing an output signal which is a measure of the relative species present[1,3,4]. RGAs are also distinguished from other mass spectrometers by their high sensitivity and their ability to withstand baking so that gases from the RGA can be desorbed. This allows gases of low partial pressure to be identified without being obscured by contributions from the analyzer itself.

The gas molecules being analyzed are subjected to electron impact ionization by an *ionizer*. The ionizing electron beam is extracted from an emission filament by means of an electric field.

Since this filament can be rapidly destroyed by exposure to such reactive gases as O_2, all RGAs require operating pressures of less than 10^{-4} torr.

The mass-separation is done by a mass analyzer. Although almost a dozen techniques have been developed for mass separation, the *rf quadropole* has become the most popular design for use in RGAs. The quadropole consists of four cylindrical rods, to which is applied a combination of dc and ac potentials. For a given applied frequency, only ions of a particular mass-to-charge ratio pass through to the collector. Since they do not require a magnetic field, quadropoles are much less bulky than magnetic mass analyzers, and can be used directly in vacuum systems. Their acceptance has become widespread with the development of stable, high-power quadropole power supplies.

The detection of the ions is generally performed at less sensitive ranges with a Faraday cup, but electron multipliers are necessary for higher sensitivity. An oscilloscope and chart recorder can be used for data taking.

The operating parts of the RGA (the ionizer, mass analyzer, and detector) are mounted in a vacuum tight enclosure known as the *measuring head*, and this assembly is bakeable, usually up to 450°C. The control unit is separate. The RGA head is mounted on a work chamber port, and a valve is placed between the head and chamber to keep the electron multiplier clean when the chamber is vented to air.

RGAs and Non-High Vacuum Applications: Differential Pumping

It is sometimes desired to analyze the composition of gases at pressures greater than the pressures at which the mass separators and electron sources of an RGA can tolerate ($>10^{-5}$ torr). Applications like reactive ion etching or sputter deposition frequently generate such requirements. A pressure-reducing gas-inlet device (that does not alter the gas composition in the process) is required in order to allow RGAs to be used in such cases. The technique is termed *differential pumping*, and an auxiliary pump is required to pump the spectrometer chamber (Fig. 21).

The RGA may be connected to the vacuum chamber with a valve that will permit monitoring the background when the system is evacuated to the high-vacuum range and a parallel leak valve for use when the chamber is being operated at higher (e.g. sputter) pressures. In this fashion, the RGA can roughly monitor the gas composition during the process. The maximum pressure in the vacuum chamber being monitored by differential-pumped RGAs should not be greater than ~10 torr.

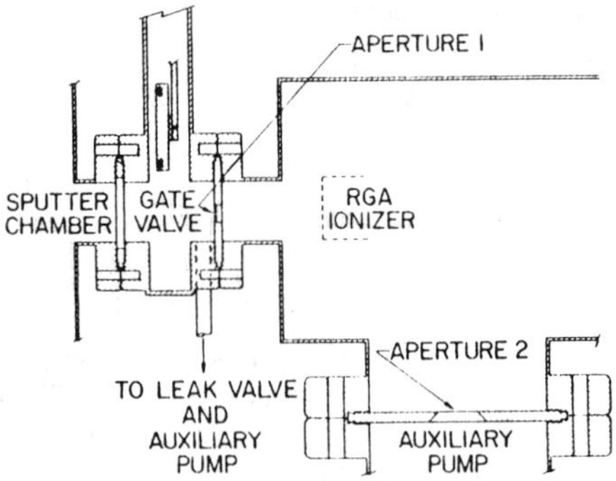

Fig. 21 Installation of a differentially pumped RGA on a sputtering chamber. From J. O'Hanlon, *A Users Guide to Vacuum Technology,* Copyright © 1980. Reprinted with permission of John Wiley and Sons, Inc.

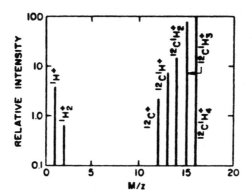

Fig. 22 Mass-spectrum obtained from residual gas analysis.

Interpretation of RGA Spectra

The output data from an RGA is called a *mass scan*, or *mass spectrum*. It is represented on a chart with the mass-to-charge ratio on the horizontal axis, and the relative intensity on the vertical. Signals from the chamber gas being analyzed (which correspond to specific mass-to-charge ratios, and their relative intensities) are plotted on the chart. The resulting mass spectra need to be interpreted so that the recorded peaks can be correlated to the gas species from which they originated. Ambiguity about the origin of peaks can arise when different molecules have the same mass. This difficulty can be resolved by considering the dissociation and double ionization behavior of molecules, processes which are specific for different species. That is, when molecules of a gas or vapor are struck by electrons whose energy can cause ionization, fragments of several mass-to-charge ratios are created. This pattern of fragments, called a *cracking pattern*, forms a fingerprint that may be used for absolute identification of a gas or vapor (Fig. 22).

A common example of gases with equal masses is that of N and CO. These gases become distinguishable by analyzing their cracking patterns.   The molecule N_2 is mostly detected as N_2^+ (m = 28), but smaller percentages dissociate into N_2^{++} and N^+, which appear as a mass-to-charge ratio of 14. The molecule of CO, on the other hand breaks up into several dissociation products, such as C^+, O^+, CO^{++} which produce peaks at mass-to-charge ratios of 12, 16, and 14. Cracking patterns of hydrocarbon fragment ions characteristic of common pump oils, have also been compiled to allow identification of backstreaming sources and to discriminate against other sources of system contamination.   In fact, one method for determining the nature of organics in systems is to become familiar with the cracking patterns of commonly used solvents, pump fluids, and elastomers (Ref. 1).

After becoming familiar with a particular RGA, *qualitative* analysis of many major constituents in the residual gas becomes rather straightforward, whereas precise *quantitative* analysis requires careful calibration and complex analysis techniques. Reference 1 provides an introduction to the methods used to obtain quantitative data from RGAs.

Qualitative analysis of the types of gas and vapor in a vacuum system are generally of most use to those employing an RGA. In many instances, the detection of a particular species points the way to fixing a leak or correcting an errant process step. The quantitative value or partial pressure of the molecular species in question usually does not need to be known because process control is frequently done empirically. The level of contaminant (for example, water vapor) that will cause the process to fail, is determined by monitoring the quality of the product. An inexpensive RGA tuned to the mass of the offending vapor is then used to indicate when the vapor has exceeded a predetermined partial pressure. With experience many gases, vapors, residues of cleaning solvents, and traces of pumping fluids, will be recognizable without much difficulty.

RGA Specification List

RGAs are essentially specified by the following characteristic properties:

1. *Peak Width* - measured in atomic mass units (amu) is specified for at least two positions of the peak-for 50% and 10% of the peak height. The peak width is characteristic of the mass resolution of the RGA.
2. *Mass Range* - specifies the lightest and heaviest singly charged ions which can be detected.
3. *Smallest Detectable Partial Pressure* - is the partial pressure that causes a collected current greater than the system noise amplitude.
4. *Smallest Detectable Partial Pressure Ratio* - is the smallest detectable partial pressure of a given gas to the partial pressure of the reference gas.
5. *Linear Range* - is the pressure region over which the sensitivity between the given limits (e.g. ±10%) remains constant.
6. *Sensitivity* - is the quotient of the ion current at the collector and the partial pressure of the gas present in the ion source.
7. *Scan Rate* - is the speed at which the RGA sweeps the ion beams of all masses in a given mass range across the collector and records the resulting spectrum.

Simple RGAs have a mass range of 1-50 amu and a resolving power of 1-3 mass units. Such instruments are capable of performing routine monitoring of background gases (fixed gases up to mass 44 and hydrocarbons at mass numbers 39, 41, and 43). More elaborate instruments are needed for detailed RGA applications. For example a mass range of 1-200 amu permits identification of most pump oils and many heavy solvents.

HIGH GAS FLOW VACUUM ENVIRONMENTS IN VLSI PROCESSING

Many of the VLSI fabrication processes take place in low and medium vacuum ranges in which a high gas flow also exists. Such applications can be divided into two groups (Fig. 23), depending on the vacuum pump types that are used to establish the required processes conditions: a) low-medium vacuum applications including CVD, plasma CVD, reduced pressure epitaxy, and some dry etch processes; and b) throttled high-vacuum applications, such as sputter deposition and some dry etch processes.

Various processes carried out in high gas flow systems often require the use of hazardous gases (i.e. gases which are potentially dangerous to personnel [toxic], or harmful to equipment [corrosive, particulate generating, flammable, and /or explosive]). Therefore, special precautions

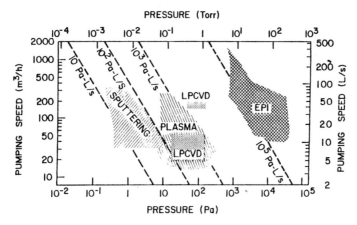

Fig. 23 Pressure-speed ranges for some thin film growth, deposition, and etching processes that require medium to low-vacuum gas flow. Courtesy of Applied Materials, Inc.

must be taken in the design, operation, and maintenance of these systems to insure operator safety and equipment/process protection. See Refs.1 and 15 for a more detailed discussion of the problems associated with hazardous gas pumping.

Medium and Low Vacuum Systems

A variety of CVD and dry etching processes are carried out at low-to-medium vacuum ranges, and may also involve the flow of large quantities of hazardous gases. Oil sealed mechanical rotary and Roots pumps are well suited to dynamically maintain the $10-10^4$ Pa pressures of such processes (Fig. 23). The flow of gas through the reactor is primarily required to replenish the reactive gas species consumed in these processes.

Since the task of pumping hazardous gases is common to most of such processes, safe and efficient operation is possible only if the following issues are adequately addressed: a) appropriate system design and pump selection; b) choice of suitable pump oil; c) use of gas source and feed plumbing interlocks; d) proper maintenance procedures; and e) adequate removal of residual gases in the process chamber before the onset of the process. Finally, operating and maintenance personnel must be made aware of the dangers and of proper safety and emergency procedures.

System design considerations include provisions for adequate gas flushing and venting, especially for the purging of gas spaces in the piping, pump filter, and pump oil reservoir. Non-reactive gas flushing at all required system points, including the building, must be used to prevent backdiffusion of the lines by air. Gas sources and associated piping should be constructed and interlocked according to established practices. As far as pump selection is concerned, rotary piston pumps for pumping corrosive gases are recommended over rotary vane pumps (for the reasons given in the section on *Rotary Pumps*. Roots pumps backed by a rotary pump can be used in high gas flows or to operate at somewhat lower pressures than rotary pumps alone.

Oil selected for the pump should not react with the process gas, nor incorporate residual gases that are subject to such reaction. Fluorochloro- and chlorocarbon pump oils (PFPE and CTFE) are the only fluids which are completely safe for reactive gases and have been highly recommended. The proper use of these oils in mechanical pumps is discussed in Ref. 1. Roots pumps, like rotary pumps require a non-reactive lubricating fluid when pumping reactive gases.

Maintenance needs to be performed with great care because pump fluids can accumulate toxic chemicals, such as arsenic, boron, and phosphorus. Changing and disposing of pump oil must be performed according to specific safe handling procedures.

Residual gases that remain in a chamber after initial pumpdown following the loading of a chamber with wafers can react with process gases and lead to system contamination and particulates. Use of an auxiliary high-vacuum pump to remove such residual gases is one suggested solution.

Throttled High Vacuum Systems

The vacuum ranges required for sputter deposition, as well as some low-pressure dry etching processes, cannot be adequately attained by rotary and Roots pumps (0.5-10 Pa), because in these ranges the roughing line from the pump to the vacuum chamber is in molecular flow. Thus, significant contamination from oil backstreaming will occur. As a result, high vacuum pumps must be used to create such vacuums[1,7,17,18,19]. However, the 0.5-10 Pa range at which these processes are carried out is above the pressure at which high pressure pumps can normally operate. Therefore, to allow these pumps to perform this pumping function, a throttle valve is used to restrict the inlet of the pump. The throttle valve maintains the pressure at the pump entrance below its critical inlet pressure. The pumping speed of the pump is also thereby reduced. The pumping speed at the chamber, S_c, under throttled conditions can be calculated from:

$$S_c = P_p S_p / P_c = Q_p / P_c \qquad (8)$$

where P_p and P_c are the pressure in the pump and chamber, S_p is the pumping speed of the pump, and Q_p is the throughput of the pump. For typical sputtering chambers, 6 inch high-vacuum pumps with a maximum throughput of 100-200 (Pa-1 /sec) are used.

The primary functions of these pumps are to establish initial cleanliness of the chamber and to flush any outgassing contaminants that desorb from chamber walls during deposition or etch. The first task could be performed by a high-vacuum pump operated in a conventional mode, but the second requires a throttled high-vacuum pump as explained above. As also pointed out earlier in the chapter, the residual outgassing species (e.g. water vapor, nitrogen) can seriously degrade the characteristics of deposited films or etching operations. The three high-vacuum pumps can all be used in throttled high-vacuum systems, although each has its limitations.

Fig. 24 Cryopump with integrated throttle valve. Courtesy of Varian Associates, Inc.

Diffusion pumps find best application in processes that require high pumping speed, high gas throughput capability, low initial capital costs, simple maintenance, and that can tolerate some degree of oil backstreaming. They are also generally resistant to corrosive attack, although it is best to consult with the manufacturer before they are use to pump corrosive gases, due to the sensitivities of some pump components such as aluminum jet assemblies. Diffusion pumps have a maximum throughput found from the product of the inlet pumping speed and the critical inlet pressure (which is about 0.1 Pa). Since all pumping will cease if the critical inlet pressure is ever exceeded, oversized forepumps are often used to back diffusion pumps in high gas-flow applications. Diffusion pumps also need to be operated in a manner which will minimize backstreaming and maximize water vapor pumping speed. A liquid nitrogen trap may be placed above the throttle valve, or in the process chamber, to satisfy the latter requirement.

Turbomolecular pumps can be well suited to the task of high-gas flow, oil-free, medium vacuum pumping. They must, however, be kept properly throttled to allow the blades to run at full velocity, and be backed by adequately large rotary pumps. They find useful application in processes that need freedom from valves and baffles, quick start-up, and a broad operating range. Occasional dust particles of less than 0.5 mm in size can usually pass through a turbo, but constant bombardment will hurt the pump. Larger particles, such as glass fragments or screws, can instantly destroy the turbine blades. Fine inlet port screens can be used to protect the pump, but they reduce the pumping speeds by about 15 percent. Turbopumps also require a relatively high initial investment and infrequent, but skilled maintenance.

Cryopumps find best use in applications that call for high pumping speeds under clean pumping conditions. They also provide low operating costs with infrequent (but skilled) maintenance. Cryopumps need throttle valves to keep the pressure lower than 0.2-0.4 Pa, and must be designed to withstand any heat loads from the process chamber. However, at high gas flows, the major heat flow to the cryogenic surfaces is carried by the gas molecules. Therefore the upper throughput limit for any gas is determined by the maximum heat flux that can be removed by the stage on which the gas or vapor condenses.

The gas handling capacity of cryopumps is also an important factor that must be considered for these applications. They have a finite capacity for different gases, and must be relieved of the stored gas by a regeneration process. The small hydrogen capacity of cryopumps implies that hydrogen-rich processes will require frequent regeneration. Furthermore, processes with a substantial steady throughput of gas, such as sputtering, may dictate frequent regeneration as well. Cryogenic pumping of toxic or explosive gases can also create unacceptable safety risks. In such applications, if the cryopump was to experience sudden warming, a large quantity of gas could be fed to the exhaust system, possibly overloading its safe venting capability. The effects of corrosive gases on the pump components as the pump is warmed during regeneration are also deleterious. Thus, cryopumps are not recommended for use in such applications.

REFERENCES

1. J.F. O'Hanlon, *A Users Guide to Vacuum Technology*, Wiley and Sons, New York, 1980. Probably the single best source on using vacuum systems available.
2. A. Roth, *Vacuum Technology*, North-Holland Publishing Company, Amsterdam, 1982. Good for information on how to design vacuum systems. Less information on their use.
3. R. Glang, R. Holmwood, J. Kurtz, "High Vacuum Technology", in L.J. Maissel and R. Glang, Eds., *Handbook of Thin Film Technology*, McGraw-Hill, New York, 1970, Chap. 2, p. 1-142.
4. *Product &Vacuum Technology Reference Book*, Leybold-Hereaus Vacuum Products, Export, PA.
5. J. Peterson and H. Steinherz, "Vacuum Pump Technology", Pt I, *Solid State Technol*, Dec. '81.

6. J. Peterson and H. Steinherz, "Vacuum Pump Technology", Pt II, *Solid State Tech.*, Jan. '82.

7. R. Scholl,"Cryopumping in Semiconductor Applications",*Solid State Technol*, Dec.'83 p. 187.

8. J. Ballingall,"Cryopump Applications & Operation", *Microel. Mfg & Testing*, Oct.'84, p. 24.

9. G. Connell, "How to Select & Use Cryopumps", *Microel. Mfg & Testing*, Nov. '85, p. 25.

10. R. Iscoff, "Vacuum Pump Update", *Semiconductor International*, Oct. '81, p. 39.

11. "An Expensive Vacuum Pump Fluid Can Save Money for Semiconductors Processors", *Semiconductor International*, Dec. '80, p. 96.

12. M. Mastroianni, M.C. Tarplee, and L. Gilbert, "Vacuum Pump Fluids for Semiconductor Processing", *Semiconductor International*, Nov. '85, p. 62.

13. D.G. Gotz and H.H. Henning, "Modern, State-of-the-Art Turbomolecular Pumps", *Microelectronic Manufacturing and Testing*, Oct. '84, p. 22.

14. L. Beavis, V. Harwood, M. Thomas,*Vacuum Hazards Manual*, American Vacuum Soc., 1979.

15. *Pumping Hazardous Gases*, American Vacuum Society Monograph.

16. S.R. Wilder and R.G. Johanson, "High Vacuum Systems for Sputtering Applications", *Solid State Technology*, Nov. '85, p. 25.

17. J.F. O'Hanlon, "Vacuum Systems for Microelectronics", *Solid State Technol*, Nov.'85, p. 103.

18. J. Freeeman, "How to Select High Vacuum Pumps", *Microel. Mfg & Testing*, Oct.'85, p. 11.

19. R. Iscoff, "Vacuum Pumps, the Critical Component", *Semiconductor Interntl*, Nov. '85, p. 47.

20. A. Roth, "Vacuum Sealing Techniques", Pergamon Press, New York, 1966, p. 616.

PROBLEMS

1. Convert the following pressure values from torr units to Pa units, (a) 1 torr, (b) 3 mtorr, (c) 5×10^{-6} torr. Convert the following pressure values from Pa to torr, (d) 150 Pa, (e) 2×10^{-2} Pa.

2. Calculate the average molcular velocity of Ar molecules at 300°K. Calculate their mean free path at atmospheric pressure.

3. Determine the number of molecules present in a gas per cubic centimeter at 300°K at the following pressures, 1 torr, 5 mtorr, 5×10^{-6} torr. Repeat the calculation for a volume of 1 μm^3.

4. Explain why backstreaming is most pronounced in a vacuum system containing an oil-sealed mechanical pump and a diffusion pump, between the pressure ranges of 10 Pa - 10^{-1} Pa.

5. Explain why cryocondensation is ineffective for pumping the gases of He and H_2 at 15-22 °K.

6. What is the removal rate of molecules (e.g. number per minute) from a chamber by a pump with a throughput of 1 torr-liter /sec.

7. Explain why increasing the pumping speed of a pump will allow both a faster pumpdown to a specified base pressure, and a lower ultimate base pressure.

8. Eleborate upon why a baffle or screen at the pump inlet reduces the effective pumping speed.

9. Explain in more detail how a *ballast* is used in the operation of an oil-sealed rotary pump.

10. Show how a throttle valve allows a high vacuum pump to maintain a higher pressure in a chamber which exceeds the maximum allowable inlet pressure of the high vacuum pump.

11. Cite at least 2 specific applications in which each of the high vacuum pump types would be best suited for the required task. Cite two disadventeges of each of the high vacuum pump types.

12. Determine the number of atoms per unit area required to form a monolayer if each atom has six nearest nieighbors in the surface plane, and the atomic diameter is 2.30 Å.

4

BASICS OF THIN FILMS

A large variety of "thin films" are used in the fabrication of VLSI devices. These films may be thermally grown or deposited from the vapor phase. They can be metals, semiconductors, or insulators. Figure 1 shows a cross-section of an advanced CMOS device illustrating some of the instances in which thin films are used. Since thin films are so important to VLSI devices, an understanding of their chemical and physical properties leads to a better understanding of their behavior in devices.

Thin films for use in VLSI fabrication must satisfy a large set of rigorous chemical, structural, and electrical requirements. Film composition and thickness must be strictly controlled to facilitate etching of submicron features. Very low densities of both particulate defects and film imperfections, such as pinholes, become critical for the small linewidths, high densities, and large areas necessary for VLSI. These small geometries also create highly rugged topography for overlying films to cover. Therefore, excellent adhesion, low stress, and conformal step coverage are required of a VLSI thin film, and its own surface topography should reduce or even planarize the underlying steps if possible. Finally, non-conducting thin films must have low dielectric constants to reduce parasitic capacitances exacerbated by the scaled down film thicknesses.

Although the properties of a bulk material may be well characterized, the same material in its thin film form may have properties substantially different from those of the bulk. One reason is that thin film properties are strongly influenced by surface properties, while in bulk materials this is not the case. The thin film, by its very definition has a substantially higher surface-to-volume ratio than does a bulk material. The structure of thin films, and their method of preparation also play a vital role in determining the film properties.

The formation of such films is accomplished by a large variety of techniques (which are the subject of subsequent chapters), but which can conceptually be divided into two groups: 1) film growth by interaction of a vapor-deposited species with the substrate; and 2) film formation by

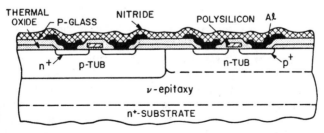

Fig. 1 Examples of types of thin films used in VLSI fabrication[14]. (© 1980 IEEE.)

deposition without causing changes to the substrate material. The first category includes thermal oxidation and nitridation of single crystal silicon (Chap. 7) and polysilicon (Chap. 6), and the formation of silicides by direct reaction of a deposited metal and the substrate (Chap. 11).

The second group includes another three subclasses of deposition: a) chemical vapor deposition, or CVD, in which solid films are formed on a substrate by the *chemical* reaction of vapor phase chemicals that contain the required constituents; b) physical vapor deposition, or PVD, in which the species of the thin film are *physically* dislodged from a source (to form a vapor), which is transported across a reduced pressure region to the substrate, where it condenses to form the thin film; and c) coating of the substrate with a liquid, which is then dried to form the solid thin film. When CVD process are used to form single-crystal thin films, the process is termed *epitaxy* and is the subject of Chap. 5. The formation of amorphous and polycrystalline thin films by CVD is discussed in Chaps 6 and 11. The formation of thin films by PVD includes the processes of sputtering and evaporation (Chap. 10), and molecular beam epitaxy (Chap. 5). The deposition of a liquid on a substrate to form thin films, most commonly by spin-coating, is discussed in Chap. 12 as part of the photoresist process presentation.

The purpose of this chapter is to discuss the terminology and properties common to thin films in general. That is, basic background information on the growth mechanism, structure, mechanical properties, and electrical properties of thin films is provided for readers not well acquainted with materials science concepts. By defining these properties in a stand-alone chapter, it is not be necessary to repeat this information each time a new thin film deposition method is introduced. The specific formation and associated processing of particular films is described in detail in later chapters.

THIN FILM GROWTH

Thin films are most commonly prepared by condensation of atoms or molecules from the vapor phase. This implies that during formation, a change of phase from the vapor to the solid occurs (also termed *condensation*). Deposition on a substrate different from the film requires the consideration of a third phase (i.e. that of the adsorbed atoms which have not yet combined with other adsorbed atoms). Condensation is initiated by the formation of small *clusters* (or *nuclei*) through the random agglomeration of several adsorbed atoms. The enlargement of these nuclei to form a coherent film is termed *growth*. Often nucleation and growth can occur simultaneously during film formation. The growth of thin films is discussed in detail in many standard reference works (e.g. Maissel and Glang[1], and Chopra[2]).

Thin Film Nucleation

The condensation of vapor atoms requires that attraction occurs as the vapor atoms impinge upon the surface. Either dipole or quadropole attractive forces can result in an atom becoming attached to the surface (which after becoming attached to the surface, is then referred to as an *adatom*). In order to undergo attachment, the atom must transfer the normal component of its velocity to the surface. If the attachment occurs without a transfer of electrons, the adsorption mechanism is known *physisorption*, while if electrons *are* transferred, it is termed *chemisorption*. The binding energies associated with physisorption are typically much smaller than those of chemisorption (i.e. ~0.25 eV vs. 8-10 eV). Adsorbed atoms may continue to move along the surface by jumping into other positions (i.e. by moving from potential well-to-potential well), as a result of the kinetic energy associated with their initial lateral velocity, or by thermal

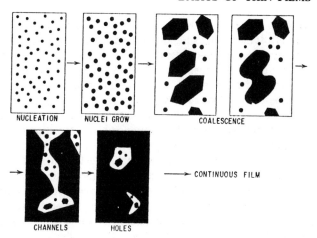

Fig. 2 Schematic of the stages of thin film growth. From Maissel and Glang, Eds. *Handbook of Thin Film Technology*, 1970[1]. Reprinted with permission of McGraw-Hill Book Co.

activation from the surface.   On the other hand, in some cases the adatoms may re-evaporate from the surface into the vapor phase. As the adatoms migrate along the surface they can interact with other adatoms to form stable clusters, called *nuclei*, and the mechanism of their formation is known as *nucleation*. The onset of *condensation* is also marked by the initial formation of the nuclei. The theory of nucleation (see Chap. 2) was developed by Gibbs for the case of liquid droplet condensation from a vapor. Basically, the theory shows that for molecular clusters greater than a critical size, the total free energy of the system decreases with increasing size, so that continued growth is energetically favorable.   This model is also applicable to cases of condensation on surfaces.   If the critical radius is not reached, atom clusters are more likely to shrink and re-evaporate. The tendency for nuclei to form and remain stable results from the lowering of the total system free energy as the nuclei grow above the critical radius. The higher the supersaturation in the vapor phase the smaller the critical radius that results.

After the formation of critical sized nuclei (which also tend to grow to supercritical dimensions by depleting the adatoms in the "capture zones" around them), the film *growth stage* begins. Film growth has been studied with the aid of electron microscopy and several stages of growth have been observed (Fig. 2). The *island stage* occurs when nuclei grow in three dimensions, but growth parallel to the substrate exceeds growth normal to the substrate. This effect arises as a result of diffusion of adatoms along the surface. In the island stage, nuclei may have well-defined crystallographic shapes. This stage is followed by *coalescence*, where nuclei contact each other and form new larger and rounded shapes. The total area projected on the substrate is reduced at this step (Fig. 2), but the height of the deposit is increased. Eventually during coalescence, larger islands may again take on a crystallographic appearance, in the form of hexagonal shapes (Fig. 2). Since the surface area covered by these structures decreases, fresh substrate surface is exposed, allowing for additional nucleation in these areas (secondary nucleation). The large islands continue to grow, leaving behind channels or holes of exposed substrate (*channel stage*). The channels are then filled by secondary nucleation, to result in a continuous film.

STRUCTURE OF THIN FILMS

Thin film generally have smaller grain size than do bulk materials. *Grain size* is a function of the deposition conditions and annealing temperatures. Larger grains are observed for increased film thickness, with the effect increasing with increasing substrate (deposition) temperature. It is also noted that the grain size saturates under these conditions, indicating that new grains nucleate upon existing ones. Larger grains are expected for increased substrate and annealing temperatures, as a result of the increased surface mobility. Annealing a film at a temperature equal to that at which it was deposited does not produce as large grains. The initial deposition temperature plays a more important role in determining the final grain size. The dependence on deposition rate is due to the fact that even if clusters have high mobility, at high deposition rate they are quickly buried under subsequent layers. For very high deposition rates, the heat of condensation can raise the substrate temperature (thereby producing increased grain size from thermal effects).

The *surface roughness* of films occurs as a result of the randomness of the deposition process. Real films almost always show surface roughness, even though this represents a higher energy state than that of a perfectly flat film. Depositions at high temperatures tend to show less surface roughness. This is because increased surface mobility from the higher substrate temperatures can lead to filling of the peaks and valleys. On the other hand, higher temperatures can also lead to the development of crystal facets, which may continue to grow in favored directions, leading to *increased* surface roughness. At low temperatures the surface roughness (as measured by surface area) tends to increase with increased film thickness. Oblique deposition which results in shadowing, also increases surface roughness. Epitaxial and amorphous deposits have shown measured surface area nearly equal to the geometrical area, implying the existence of very flat films. This has been confirmed by SEM examination of these films. Surface roughness has also been observed to increase as a result of the presence of contamination on the substrate (or in the vapor phase) during deposition.

The *density* of a thin film provides information about its physical structure. Density is usually determined by weighing the film and measuring its volume (area x thickness). If a film is porous from the deposition process, it generally has a lower density than bulk material.

The *crystallographic structure* of thin films depends on the adatom mobility, and can vary from a highly disordered (or amorphous-like) state, to a well-ordered state (e.g. epitaxial growth on a single crystal substrate). The amorphous structures are frequently observed for the deposited dielectrics such as SiO_2, SiO, and Si_3N_4, while most metals result in polycrystalline structure. Silicon can be either amorphous, polycrystalline, or single crystal, depending on the deposition parameters and the substrate material.

Some polycrystalline films deposit with a *fiber texture* or *preferred orientation*. This texture can be described as having a preponderance of grains with the same orientation in a direction normal to the substrate. Deposition of polycrystalline Si on SiO_2 can occur in this manner. The observed fiber texture in that case has been reported as {110} for poly-Si deposited at 600-650°C.

Epitaxy describes the phenomenon of an oriented overgrowth of one substance on the crystal surface of another substance. If the film and the substrate are composed of the same material, this growth is termed *homoepitaxy* (e.g. Si on Si, GaAs on GaAs), while if the film and substrate are different materials, such growth is termed *heteroepitaxy* (e.g. Si on Al_2O_3).

The single crystal substrate is the dominant factor in epitaxy. Some crystal symmetry must exist between the deposit and the substrate, and the lattice parameters of each material must be close to the same value. The lattice misfit, which is the percentage difference between the lattice constant of the deposit and that of the substrate plane in contact, must be small to achieve epitaxy. The deposition temperature also plays a role in epitaxy. Elevated temperatures increase the propensity for epitaxial growth. This is engendered by several mechanisms, including: cleaner surfaces; decreased supersaturation; and increased adatom energy.

The deposition rate also plays a role in epitaxy. Lower rates tend toward epitaxy, while higher deposition rates lead to polycrystalline or amorphous deposition. Three growth regimes are observed for the deposition of Si on a clean Si substrate: a) at low temperature and high deposition rates the deposits are amorphous; b) at high substrate temperatures and low rate deposition they tend to be single crystal; and c) at intermediate conditions polycrystalline films tend to form. Chapter 5 discusses the growth of epitaxial silicon films in much more detail.

MECHANICAL PROPERTIES OF THIN FILMS

Adhesion

The adhesion of grown and deposited films used in VLSI processing must be excellent (both *as deposited*, and after subsequent processing). If films lift from the substrate device failure can result, and thus poor adhesion represents a potential reliability problem.

Early attempts at measuring adhesion used: a) the *tape test*; or b) a method of *abrasion testing*. The tape method consists of pressing a piece of adhesive tape to the film. When the tape is pulled off, the film is either removed (in whole or in part), or remains on the substrate. This method is qualitative, and if the film remains on the substrate, it provides no data as to the magnitude of the adhesion forces. If the film is removed, a quantitative value of the adhesion can be obtained by measuring the force to remove it. For most films used in VLSI processing, failure of the tape test implies that the film is unsuitable for device fabrication.

The *abrasion* or *scratch method* gives results which depend both on the film hardness, as well as on its adhesion. A vertical load is applied to a chrome-steel point which is drawn across the film surface. The load is increased until the point strips the film completely from the substrate leaving a deep channel (i.e. critical load is reached). The stripping is generally observed with a microscope. The critical load is taken as a measure of the adhesion. Critical loads ranging from several grams to hundreds of grams are observed, depending on the film and the substrate. It has been found that the critical load depends strongly on the film thickness. The stylus scratches the film so the mechanical strength and hardness of the film also play a role in this test. Therefore the method is least valid for films greater than 2000Å thick.

A relatively simple quantitative pull test can also measure adhesion. In this test, a small metal pin is attached to the film surface (after the surrounding film material has been removed), with an epoxy that does not react with the film material (Fig. 3). An increasing force is applied normal to the substrate surface, until either the film-substrate or film-epoxy bond fails.

Adhesion energy values between film and substrate have been measured, and range from several tenths of an electron volt to 10 eV or more. In order to explain this wide range of values,

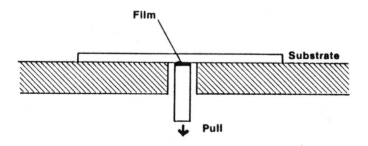

Fig. 3 Apparatus for measuring the force required to detach a film from its substrate.

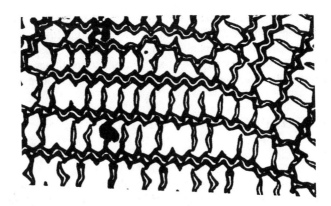

Fig. 4 W on mica at 300°C, showing buckle waves.   From K.I. Chopra, *Thin Film Phenomena*, 1969[2]. Reprinted with permission of McGraw-Hill Book Co.

two adhesion mechanisms have been postulated. For low values of adhesion, it is surmised that the electron shells of the adsorbed atoms remain intact, and these atoms are held to the surface by *Van der Waals forces*. These forces hold to approximately 0.4 eV, and the atoms are said to be physisorbed on the substrate. Above 0.4 eV, sharing of electrons between the film and substrate occurs, and the atoms are chemisorbed.   Generally adhesion is greater the higher the adsorption energy of the deposit and /or the higher the number of nucleation centers in the early growth stage of the film. Chemisorption due to an intermediate-layer formation that allows a continuous transition from one lattice to the other results in excellent adhesion. This is particularly true for the adhesion of strong oxide formers, such as Ti and Al on silicon dioxide.   If intermetallic metal alloys form, adhesion is also improved.

Adhesion is also strongly effected by the cleanliness of the substrate. Contamination generally results in reduced adhesion, as does an adsorbed gas layer. Cleaning the substrate prior to deposition is therefore important to insure good film adhesion.

Substrate surface roughness can also affect adhesion. For example, increased roughness may promote adhesion because: a) the substrate exhibits more surface area than a flat surface; and b) mechanical interlocking between the film and substrate may also occur. Excessive roughness, on the other hand, may result in coating defects, which may promote adhesion failure.

It is highly advantageous to include a layer of a strong oxide-forming element between an oxide substrate and the metallization. This is particularly true for gold-based metallization, where a chromium layer can be used to serve as an intermediate "adhesion layer". Intrinsic stress in a thin film is generally not sufficient to result in delamination, unless the film is extremely thick. More often, high stress results in the cracking of films.

Stress in Thin Films

Nearly all films are found to be in a state of internal stress, regardless of the means by which they have been produced. The stress may be *compressive* or *tensile*. Compressively stressed films would like to *expand* parallel to the substrate surface, and in the extreme, films in compressive stress will buckle up on the substrate, as shown in Fig. 4). Films in *tensile stress*, on the other hand, would like to *contract* parallel to the substrate, and may crack if their elastic limits are exceeded.   In general, the stresses in thin films are in the range of 10^8-5 x 10^{10} dynes /cm². Highly stressed films are generally undesirable for VLSI applications for several reasons, including: a) they are more likely to exhibit poor adhesion; b) they are more susceptible to corrosion; c) brittle films, such as inorganic dielectrics, may undergo cracking in tensile stress; and d) the resistivity of stressed metallic films is higher than that of their annealed counterparts. The *total stress* , σ, in a film is the sum of: a) any *external stress*, σ_{ext}, on the film, perhaps from another film; b) the *thermal stress*, σ_{th}; and c) the *intrinsic stress*, σ_{int}. The total stress is written as:

$$\sigma = \sigma_{ext} + \sigma_{th} + \sigma_{int} \tag{1}$$

Thermal stress results from the difference in the coefficients of thermal expansion between the film and the substrate. The intrinsic stress results from the structure of the growing films. The *thermal stress* is due to the constraint imposed by the film-substrate bonding and is given by:

$$\sigma_{th} = (\alpha_f - \alpha_s) (\Delta T) E_f \tag{2}$$

where: α_f and α_s are the average coefficients of thermal expansion for the film and the substrate, respectively; ΔT is the temperature of film growth (or deposition), minus the temperature of measurement; and E_f is the Young's modulus of the film.   The σ_{th} can be of either sign (positive is tensile, negative is compressive), based on the relative values of α_f and α_s. For example, SiO_2 grown or deposited on silicon at elevated temperatures will have a compressive component as part of its total stress, since $\alpha_{Si} > \alpha_{SiO2}$.

The *intrinsic stress* reflects the film structure in ways not yet completely understood. It has been observed that the intrinsic stress in a film depends on thickness, deposition rate, deposition temperature, ambient pressure, method of film preparation, and type of substrate used, among other parameters.

At low substrate temperatures, metal films tend to exhibit tensile stress. This decreases with increasing substrate temperature (often in a linear manner), finally going through zero (or even becoming compressive). The changeover to compressive stress occurs at lower temperatures for lower melting point metals.

The stress in a thin film is known to vary with depth.   This may be a manifestation of a

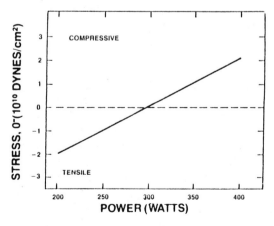

Fig. 5 Variation of stress in sputter-deposited thin film tungsten, versus sputter deposition power. Courtesy of Materials Research Corp (MRC).

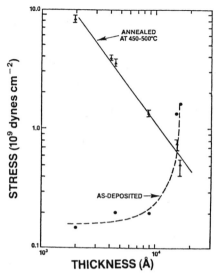

Fig. 6   Stress *versus* thickness for annealed aluminum films.   Courtesy of Materials Research Corp (MRC).

change in film structure with depth.   Figure 5 shows film stress as a function of thickness for aluminum films, both as-deposited and annealed. In the range of interest of films for VLSI the stress is relatively low and constant. The as-deposited films show an increase in stress with thickness. With no stress relief during cooling from the annealing temperature, a stress of 10^{10} dyne /cm^2 would be expected. Clearly stress relief has occurred as evidenced by a lower stress value.   Crystalline slip (see Chap. 2) is a mechanism of stress relief.

The effect of deposition rate on film stress is ambiguous.   Figure 5 shows data for stress as a function of power (deposition rate) for sputtered tungsten. The film stress starts out tensile, decreases as the power increases, and finally becomes compressive with further power increase.

The origin of thermal stress is easily understood. Intrinsic stresses however are more difficult to explain as they may arise from several causes. Several models relating to the origins of intrinsic stress have been proposed. These invoke the following stress-causing mechanisms: a) lattice mismatch between the substrate and the film; b) rapid film growth (which locks in defects); and c) incorporation of impurities into the film.

Measurement of stress is a relatively straight-forward procedure based upon beam bending. It can be performed on commercial laser wafer-flatness measuring equipment.   This technique also allows the stress to be directly determined on the substrate that the film is deposited upon. The procedure is based upon the phenomenon that the stresses in a film on a thin substrate will result in the bending of the substrate.

A tensile stress in the film bends the substrate so that it becomes concave (Fig. 7a), while a compressive stress makes the substrate convex (Fig. 7b). In the wafer or disc method, the film is deposited on one side, and stress is determined by measuring the deflection of the wafer center. Use of laser flatness equipment allows direct measurement of the disc deflection.   Previously, use of an optical flat was required, so that deflection could be determined from the interference fringes (Newton's rings) in an interferometer. Stress in films using this method is found from:

$$\sigma \; = \; \frac{E_s \, D}{6 \, r \, T} \tag{6}$$

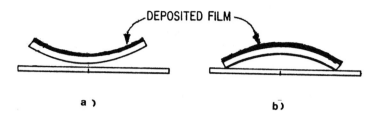

a)   b)

Fig. 7 (a) Tensile stress causes *concave* bending, and (b) compressive stress causes *convex* bending of substrate.

where: r is the measured radius of curvature of the bent substrate; E_s is Young's modulus of the substrate; D is the thickness of the substrate; and t is the thickness of the film. For films used in silicon processes, the assumption that t<<D, is generally valid. Films are typically less than 1 μm (1×10^{-4} cm) thick, while wafers are 5×10^{-2} cm thick, and D ~5×10^{-2} cm. Any curvature that is inherent in the substrate must be measured before the film is deposited and algebraically added to the final measured radius of curvature. In reality this method measures the product $\sigma t = F$, called the force per unit width, from which the average stress can be calculated. The instantaneous stress is given by dF /dt. For very thin films (i.e. t→ 0), the measurement is dubious, unless it is known that the film stress is independent of thickness.

Another method of measuring the radius of curvature or deflection of the substrate involves measuring a capacitance change. This is achieved by using the film as one plate of the capacitor. Stress may also be measured by use of x-ray or electron diffraction of the stressed thin film. In such techniques, the changes in the lattice parameter and diffraction line broadening are excellent measures of the films' strain (which in turn can be used to calculate the stress).

The stress in the film is determined from the lattice parameter change and is given by:

$$\sigma = \frac{E_f}{2\mu_f} \left[\frac{a_o - a}{a_o} \right] \qquad (4)$$

where: a_o and a are the unstressed and stressed lattice constants, respectively; and μ_f is the Poisson ratio; and E_f Young's modulus of the film.

Equation 4 is valid for the case of the lattice constant perpendicular to the film plane. For the case of the lattice constant *in* the plane of the film, the stress is given by:

$$\sigma = \frac{E_f}{1 - \mu_f} \left[\frac{a - a_o}{a_o} \right] \qquad (5)$$

The use of diffraction line broadening to determine the strain in the film requires the apportioning of the measured value between small crystallite effects, and the strain in the film.

Other Mechanical Properties

Stress-strain curves of thin films can often be obtained by several techniques. Values of the *elastic modulus* can be measured from the slope of the curve in the linear region. The total elongation at fracture in a metal film may be only 1-2%, and may exhibit a large elastic component. The *tensile strength* is the maximum value of the breaking force divided by the original cross-sectional area. The tensile strength of thin metal films has been measured to be up to 200 times larger than annealed bulk samples.

The *microhardness* of a film shows a strong correlation with tensile strength. For example, the value of both properties decreases as the deposition temperature is increased. Palatnik[7] studied the effect on the microhardness of aluminum thin film due to the presence of aluminum oxide and alloying additions of Cu, as well as other materials. Both the aluminum oxides and the alloying additions were found to increase the microhardness of the film. Knowledge of the hardness of thin films is important in developing such processes as attaching wires from integrated circuit packages to the chip bonding pads. The hardness of the film is one parameter that establishes the bonding-tool pressure required to form a strong wire-bond.

Microhardness can be measured by pressing a hard, specially shaped point into the surface, and observing the indentation. Read[8] has reviewed the available methods for hardness measurements. The Vicker microhardness measuring procedure uses a diamond pyramid of four faces with edges of equal length, while the Knoop system uses a pyramid in which the length of two edges is seven times that of the other two. The Knoop's penetration into the film is one-third less than the Vickers. The minimum limiting film thickness for valid measurements is about 1.5×10^4Å for the Knoop test, since the film thickness must be at least ten times the penetration depth. That is, for deeper indentations, as the indenter penetrates toward the interface, the substrate begins to interfere with the measurement).

ELECTRICAL PROPERTIES OF METALLIC THIN FILMS

The electrical conductivity of a material is due to the motion of charge carriers (electrons in metals, electrons or holes in semiconductors) through the lattice under the influence of applied electric fields[9]. These electrons are affected by non-periodicity in the lattice. Such non-periodicity results from the lattice vibrations (phonons) that exist at all temperatures above absolute zero (0°K), and lattice defects, such as impurities, vacancies, interstitials, dislocations, grain boundaries, and alloying additions. The total resistivity is given by Mathiessen's rule, which states the resistivity of a material is the arithmetic sum of the individual contributions made by all sources of resistance. It is convenient to combine all contributions other than the thermal component of resistivity into a term referred to as *residual resistivity*. That is:

$$\rho = \rho_{temp} + \rho_{residual} \qquad (6)$$

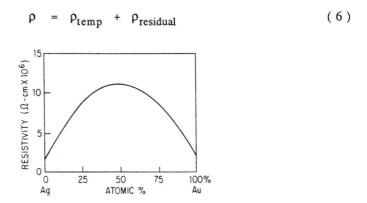

Fig. 8 Resistivity vs. composition for the gold-silver system. From Maissel and Glang, Eds., *Handbook of Thin Film Technology*, 1970[1]. Reprinted by permission of McGraw-Hill Book Co.

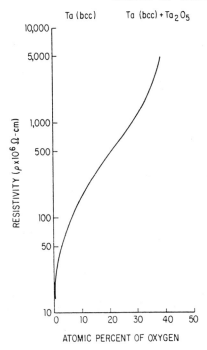

Fig. 9 Resistivity of Ta as function O impurity concentration. From Maissel and Glang, Eds., *Handbook of Thin Film Technology,* 1970[1]. Reprinted with permission of McGraw-Hill Book Co.

The resistivity of metal thin films is generally higher than that of the bulk counterpart. This results from the fact that films generally contain more defects (e.g. dislocations ~10^{10} cm^{-2}), and grain boundaries, than do bulk materials.

The composition of the thin film may also play a major role in the resistivtiy that it exhibits. Figure 8 shows the effect of alloy additions on the resistivity of the Ag-Au system (which shows complete solid solubility). The resistivity increases with the addition of solute. The addition of small amounts of impurities can cause very large changes in resistivity. For example, the addition of oxygen in tantalum (Fig. 9) results in resistivity increases several orders of magnitude, as Ta_2O_5 (an insulator) is formed. The contribution of *vacancies* and *interstitials* to the resistivity, on the other hand, is on the order of 1 $\mu\Omega$-cm per atomic percent per vacancy or interstitial, and can therefore generally be neglected. The scattering power of dislocations is also small. The effect of grain boundaries on the resistivity of bulk materials is relatively small, since grain boundaries can be treated arrays of dislocations.

Measurement of the Electrical Properties of Thin Films

As shown in Fig. 10 the resistance of a rectangular shaped section of film of length, l, width, b, and thickness, t (measured in the direction parallel to the film), is given by:

$$R = \frac{\rho \, l}{t \, b} \qquad (7)$$

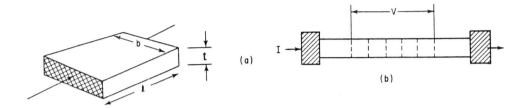

Fig. 10 (a) Definition of sheet resistance. (b) Direct measurement of sheet resistance.   From Maissel and Glang, Eds., *Handbook of Thin Film Technology,* 1970[1]. Reprinted with permission of McGraw-Hill Book Co.

where R is the resistance of the film, ρ is the resistivity, and t is the film thickness. If $1 = b$, Eq. 7 then becomes:

$$R = \rho / t = R_s \tag{8}$$

where R_s is the sheet resistance in Ω /square of the film, and is independent of the size of the square (but depends only on the film resistivity, and thickness of the film). If the thickness and sheet resistance are known, the film resistivity is given by:

$$\rho = R_s\, t \tag{9}$$

The sheet resistance, R_s, can be measured by several techniques.   In the direct measurement, a conductor strip is fabricated (Fig. 10b). The voltage drop across the length of the strip is measured, as a current is driven through the conductor. The resistance of the conductor is given by $R = V /I$. The number of squares are counted between the voltage terminals, and $R_s = R /N$, when N is the number of squares ($N = L /b$).

A second method which does not require the use of a strip is the *4-point probe method,* described in detail by Valdes[12] (Fig. 11).   If the sample is semi-infinite with regards to the probe spacings, and the probe spacings s are equal, then the resistivity is given by:

$$\rho = \frac{V}{I}\, 2\pi s \tag{10}$$

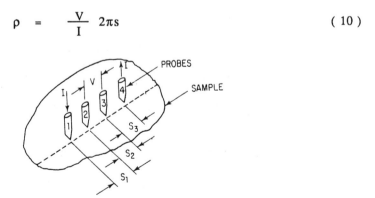

Fig. 11   In-line four point probe.   From Maissel and Glang, Eds., *Handbook of Thin Film Technology,* 1970[1]. Reprinted with permission of McGraw-Hill Book Co.

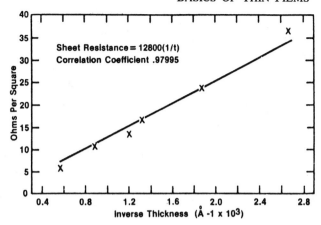

Fig. 12 TiW sheet resistance as a function of 1 /t. From S. Lim and D. Ridley, *Solid State Tech.,* Feb. 83, p. 99. Reprinted with permission of Solid State Technology, Published by Technical Publishing, A Company of Dun & Bradstreet.

If the material on which the probe is placed is an infinitely thin slice on an insulating substrate, then Eq. 10 reduces to:

$$\rho = \frac{V \ t \ \pi}{I \ \ln 2} \qquad (11)$$

Depending on the probe spacing relative to the boundaries, and thickness of the sample, there are various geometric correction factors that apply to Eq. 1. The reader is referred to Valdes paper or page 268 of this text for these corrections.   Commercial equipment is used for measuring R_s in thin metal films, and semiconductor layer is available.

The sheet resistivity measurement technique is also used to indirectly measure the thickness of conductive films deposited on insulating substrates (e.g. metal, silicide, or semiconductor films), especially if the films are less than 1000Å thick (since stylus profilometer instruments are difficult to use with such thin films). The thickness, t, and the sheet resistance are related by Eq. 9. Thus, the sheet resistance can be utilized to determine thickness only if the resistivity of the film is known.   Since the resistivity of thin films is generally different from that of the bulk resistivity, the thin film resistivity is usually measured independently.   The preferred method utilizes the fact that the sheet resistance is linear with the inverse of the thickness, and that the slope is equal to the resistivity. Thus, by depositing films of varying thickness, and measuring R_s and t (an independent technique, utilizing a surface profilometer, or SEM, is used to measure t), a plot of R_s vs 1 /t can be generated. A least square fit of the data will yield a curve from which ρ can be determined.   With this information, measurements of R_s on sample films can yield the thickness by using Eq. 9.   Figure 12 illustrates such a curve for TiW films. Implicit in this method is the assumption that resistivity is independent of thickness. This assumption is no longer valid when the thickness becomes less than the electron mean free path, since the resistivity will then exhibit a contribution from the surface scattering of the electrons.

An improvement over the 4-point probe is the use of eddy currents, as this is a contactless method.   High frequency current through a coil induces current in conducting films placed under the coil, which unbalances the oscillator circuit.   The change in the circuit is related to the resistance of the conducting film, which is dependent on the resistivity and the film thickness. Two examples of this type of instrument are the Eddy Co. Model E-1, and the Tencor M-Gage.

The 4-point probe technique also offers one great advantage for showing variations in sheet resistance or thickness by stepping the probes over a test wafer to provide contour maps. This technique is described in additional detail in Chap. 9, in the section dealing with the monitoring of ion implant doses.

The *temperature coefficient of resistance*, α_T (TCR) at a temperature T, is determined from:

$$\alpha_T = \frac{(R_1 - R_2)}{R_T(T_1 - T_2)} \tag{13}$$

where $T_1 > T > T_2$, and the R's are the resistance values at T_1, T_2, and T. A minimum of three resistance measurements must by made (the third being required to see if changes in temperature have resulted in any permanent changes in the resistance of the film). When the TCR is quoted without specifying temperature, T is generally assumed to be 20°C.

Electrical Transport in Thin Metal Films

Although most metal films used in IC fabrication are sufficiently thick to be continuous, and not have surface scattering as a major factor, it is still informative to briefly review the electrical properties exhibited by such thinner films.

As previously discussed, the resistivity of a metal arises from scattering of electrons. Electrons will be scattered when they reach the surface of the film, and the resistivity will increase when the film becomes thin enough so that collisions with surface become a significant fraction of the total number of collisions. This occurs when the film thickness approaches the electron mean free path (mfp) at a given temperature. A mathematical analysis of the problem was first treated by Thomson, who concluded that the conductivity σ at any film thickness is given by:

$$\sigma = \sigma_0 [1 - \frac{3(1 - p)}{8k}] \qquad (k >> 1) \tag{13a}$$

$$\sigma = \sigma_0 \frac{3k}{4}(1 + 2p)[\ln\frac{1}{k} + 0.423] \quad (k << 1) \tag{13b}$$

where σ_0 is the conductivity of a thick film, k is the ratio of the film thickness to the mean free path, and p is the fraction of electrons elastically scattered.

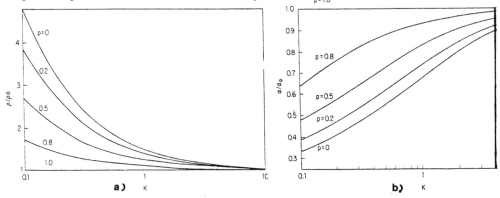

Fig. 13 (a) Effect of film thickness on resistivity. (b) Effect of film thickness on temperature coefficient of expansion. From Maissel and Glang, Eds., *Handbook of Thin Film Technology*, 1970[1]. Reprinted with permission of McGraw-Hill Book Co.

Figure 13a is a plot of ρ/ρ_0 (reciprocal of σ) as a function of the ratio k, for different values of p. Figure 13b shows the dependence of α_T on the k factor. Values of the mean free path of the electrons in metals λ depend on temperature, but are typically several hundred Å at 300°K.

REFERENCES

1. L.I. Maissel and R. Glang, *Handbook of Thin Films*, McGraw-Hill, New York, 1970.
2. K.I. Chopra, *Thin Film Phenomena*, McGraw-Hill, New York, 1969.
3. C. Neugebauer, "Condensation Nucleation and Growth of Thin Films", Ch. 8 in Ref. 1.
4. *ibid.*, Ref. 2, p. 312.
5. J. Smith, "Basics of Thin Films", in *The Book of Basics*, MRC, Orangeburg, N.Y. Chap. III,
6. *ibid.*, Ref. 5, Fig. 16.
7. L.S. Palatnik, *et al.*, *Phys. Metals Metalog.*, **182**, 296 (1965) {Russian}.
8. H. J. Read, *Proc. Am. Elctroplaters Soc.*, **50**, 37 (1963).
9. L. Maissel, "Electrical Properties of Metallic Thin Films", Chap. 13, in Ref. 1.
10. *ibid.*, Ref. 9, Fig. 13-1.
11. *ibid.*, Ref. 9, Fig. 13-2.
12. L.B. Valdes, *Proc. IRE*, **42**, 420 (1954).
13. D.S. Campbell, *The Use of Thin Films in Physical Investigations*, p. 229, Academic Press, New York, 1966.
14. L.C. Parrillo, *et al.*, "Twin-Tub CMOS", Tech. Dig. IEDM, 1980, p. 752.

PROBLEMS

1. Calculate the resistance of a Ti:W interconnect line that is 2μm wide, 1500 μm long and (a) 1000 Å thick, and (b) 3000 Å thick (see Fig. 12).

2. The linear thermal coefficient of expansion of Al is on the order of 25×10^{-6} cm /cm /°C, and that for Si is 2.6×10^{-6} cm / cm /°C. Using this information offer an explaination of why thin films of Al on Si substrates exhibit hillocks (See Fig. 36, p. 371, Chap. 10) after being subjected to elevated temperatures (e.g. > 250°C).

3. Explain why compressive stresses in thin films can cause convex bending, and tensile stresses concave bending, of the substrates on which they are deposited.

4. The correction factor for R_s when thick materials are being measured with a four-point probe is shown in Fig. P1. Equation 11 in Chap. 4 must be multiplied by this factor to obtain accurate R_s values from V /I measurements. If a uniformly doped silicon layer with a thickness equal to the probe spacings is measured and V /I = 45, calculate its R_s.

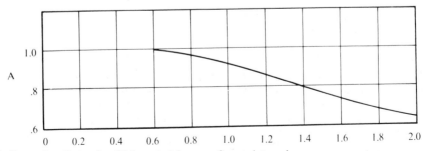

Fig. P1 Correction factor for thick materials on a four-point probe.

5

SILICON EPITAXIAL

FILM GROWTH

The epitaxial growth process is a means of depositing a thin layer (0.5 to 20 μm) of single crystal material upon the surface of a single crystal substrate. If the film is the same material as the substrate the process is called *homoepitaxy*, but is often referred to as *epitaxy* or simply *epi*. Silicon deposition on a silicon substrate is the most important technological use of homoepitaxy, and is the primary subject of this chapter. If, on the other hand, the deposit is made on a substrate that is chemically different, the process is termed *heteroepitaxy*. This process has found application in the deposition of *silicon on sapphire* (Al$_2$O$_3$) which is termed *SOS*. The term *epitaxy* is derived from two Greek words meaning "arranged upon".

Epitaxial growth can be achieved from the vapor phase (VPE), liquid phase (LPE), or solid phase (SPE). For silicon processing, VPE has met with the widest acceptance, since excellent control of the impurity concentration and crystalline perfection can be achieved. Liquid phase deposition has found its widest use in producing epitaxial layers of III-V compounds (e.g. GaAs, InP). Solid phase epitaxy is observed in the crystalline regrowth of ion implanted amorphous layers, as described in the *Annealing Amorphous Layer Damage* section of Chap. 9.

The major impetus for developing silicon epitaxy was to improve the performance of bipolar transistors and later bipolar integrated circuits. By growing a lightly doped epitaxial layer over a heavily doped silicon substrate, the bipolar device could be optimized for high breakdown voltage of the collector-substrate junction, while still maintaining low collector resistance. The low collector resistance provided high device operating speeds at moderate currents. More recently, epitaxial processes have been used in fabricating advanced complementary metal-oxide-silicon (CMOS) VLSI circuits. In these circuits the device is built in a thin (3-7 μm), lightly doped epitaxial layer over a heavily doped substrate. This structure minimizes latch-up effects that a CMOS circuit may undergo when powered-up, or when subjected to a radiation burst. Other advantages of fabricating devices (both bipolar and MOS) in an epitaxial layer are that the doping concentration of the device can be accurately controlled, and that the layer can be made oxygen and carbon free. The epitaxial process however is not without its disadvantages. These include: a) increased processing complexity and wafer costs; b) defect generation in the epitaxial layer; c) autodoping; and d) pattern shift and washout.

In this chapter we present: a) the fundamentals of epitaxial deposition; b) doping in epitaxial films; c) defects in epitaxial films; d) process considerations for VLSI epitaxial depositions; e) epitaxial deposition equipment; f) the characterization of epitaxial films; g) selective deposition of epitaxial silicon; and h) molecular beam epitaxy of silicon.

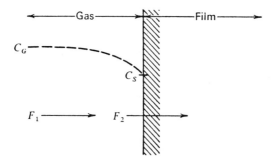

Fig. 1 Basic model of the epitaxial growth process.

FUNDAMENTALS OF EPITAXIAL DEPOSITION

In this section the basics of the chemical vapor deposition (CVD) process that are applicable to VPE of silicon are discussed. The deposition involves the transport of reactants and products across a boundary layer, as well as their chemical reaction at the substrate.

The Grove Epitaxial Film Growth Model

Epitaxial deposition is a chemical vapor deposition process. The following five steps are fundamental to all *chemical vapor deposition* (CVD) processes: 1) the reactants are transported to the substrate surface; 2) the reactants are adsorbed on the substrate surface; 3) a chemical reaction takes place on the surface leading to the formation of the film and reaction products; 4) the reaction products are desorbed from the surface; and 5) the products are transported away from the surface. In some cases, a gas phase reaction leading to the formation of film precursors (reactants) occurs prior to step 1.

Developing a mathematical relationship for all five steps has proven difficult even up to this time. Grove[1] developed a model based on the assumption that only steps 1 and 3 (i.e. transport of reactants and chemical reaction) are operative. Despite these limiting assumptions, the model explains many phenomena observed in the epitaxial deposition process.

Figure 1 is a schematic depicting the essentials of the model. It shows the concentration distribution of the reactant gas and the *flux* (number of atoms or molecules crossing a unit area in a unit time, atoms /cm^2 sec) from the bulk of the gas to the surface of the growing film, F_1. A second flux which corresponds to the consumption of the reactant gas is also shown, and is termed F_2. The reactant gas for silicon epitaxy may be any one of the following: SiH_4; $SiHCl_3$; SiH_2Cl_2; SiH_3Cl; or $SiCl_4$. The choice of gas species which serves as the silicon bearing compound does not affect the basic principles described by the model. The flux F_1 is approximated by assuming it is linearly proportional to the concentration difference between the reactant in the bulk of the gas C_g and that at the surface of the substrate, C_s. The constant of proportionality is termed the *gas-phase mass-transfer coefficient*, h_g. The relationship for F_1 is:

$$F_1 = h_g (C_g - C_s) \qquad (1)$$

The flux F_2 is assumed to be linearly proportional to the surface concentration of the reactant, and the constant of proportionality is the *chemical surface reaction rate constant*, k_s. The linear dependence implies the reaction follows first order kinetics. The expression F_2 is given by:

$$F_2 = k_s C_s \qquad (2)$$

It should be noted that although there is also a flux of products away from the substrate, it is neglected in this model.

At *steady state* conditions (no build-up or depletion of material) the two fluxes must be equal: $F_1 = F_2 = F$. By setting Eqs. 1 and 2 equal, we obtain an expression for the surface concentration of reactants, C_s:

$$C_s = \frac{C_g}{1 + \dfrac{k_s}{h_g}} \qquad (3)$$

There are two limiting cases of Eq. 3. If $h_g >> k_s$, the value of surface concentration, C_s, approaches C_g. This is termed the *surface-reaction control* case. If $h_g << k_s$ the value of the surface concentration approaches zero. This condition is known as the *mass-transfer control* case.

The growth rate of a silicon epitaxial film, V (cm /sec), is given by:

$$V = \frac{F}{N_1} \qquad (4)$$

where N_1 is the number of silicon atoms incorporated into a unit volume of the film. For silicon the value of N_1 is 5×10^{22} atoms $/cm^3$. Substituting for F_2 ($F_2 = F_1 = F$) from Eq. 2 and for C_s from Eq. 3 into Eq. 4 we obtain:

$$V = \frac{F}{N_1} = \frac{k_s h_g}{k_s + h_g} \frac{C_g}{N_1} \qquad (5)$$

The concentration of the reactant in the gas phase can be defined as:

$$C_g = C_T Y \qquad (6)$$

where Y is the mole fraction of the reaction species and C_T is the total number of molecules per cm^3 in the gas. Substituting Eq. 6 in Eq. 5 we obtain the expression for film growth rate:

$$V = \frac{k_s h_g}{k_s + h_g} \frac{C_T}{N_1} Y \qquad (7)$$

Two important effects are predicted by Eq. 7. First, it predicts that the *growth rate should be proportional to the mole fraction of reacting species in the gas phase*. This is in agreement with experimental observations at low values of Y. Figure 2 shows the growth rate dependence for silicon (produced by $SiCl_4$ reduction in H_2) on the $SiCl_4$ concentration in the vapor phase. For this reaction the growth rate rapidly departs from linearity, and at a $SiCl_4$ concentration as low as 0.1 mole fraction (10 mole %), the growth rate even starts to decrease. At concentrations greater than ~0.27 mole fraction, negative growth (i.e. *etching*) of the substrate occurs. This comes about since one of the products of the reaction is HCl. The HCl in turn reacts with the silicon, causing etching of the depositing epitaxial layer (or even the substrate). Second, Eq. 7 indicates that the *growth rate at constant Y is controlled (in the limits) by the smaller value of k_s and h_g*. In the limiting cases the growth rates are given by:

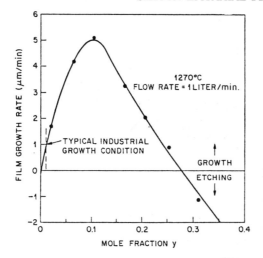

Fig. 2 Silicon growth rate as a function of $SiCl_4$ concentration[52]. Reprinted with permission of the publisher, the Electrochemical Society.

$$V = C_T k_s Y /N_1 \quad \text{(surface-reaction control)} \quad (8)$$

or by;

$$V = C_T h_g Y /N_1 \quad \text{(mass transfer control)} \quad (9)$$

The chemical surface reaction rate constant, k_s, describes the kinetics of the chemical reaction at the substrate surface. Chemical reactions are often thermally activated, and if this is the case, can be represented by an *Arrhenius type* equation (see Appendix 3). Assuming that the reactions at the surface exhibit such behavior, k_s can be written as:

$$k_s = k_o e^{[-E_a /kT]} \quad (10)$$

where k_o is a temperature independent frequency factor, and E_a is the activation energy of the reaction. The mass-transfer coefficient, h_g, on the other hand, is relatively temperature insensitive, and depends primarily on the gas flow (fluid dynamics) in the reactor.

Figure 3 shows a plot of the growth rate data (points) as a function of reciprocal temperature for the deposition of silicon by the H_2 reduction of $SiCl_4$. At *low temperatures* the growth rate follows the exponential law with $E_a = 1.9$ eV, and $k_o = 1\text{x}10^7$ cm /sec. At higher temperature the growth rate tends to become temperature insensitive. This growth rate data can be explained with the use of Eqs. 7-10. At *low temperatures* $h_g >> k_s$, and the growth rate is controlled by k_s. Equation 8 then dominates with the value of k_s given by Eq. 10. Since k_s increases rapidly with temperature, the reactant gas supply reaching the surface (which is controlled by h_g), eventually cannot keep up with the demand of the reaction, and the reaction rate tends to level off. At *high temperatures* $h_g << k_s$, and mass-transfer (which is given by Eq. 9) dominates. The value of h_g in Eq. 9 is relatively temperature insensitive. At intermediate temperatures both h_g and k_s contribute, and the growth rate does not increase as rapidly as it does at lower temperatures. Also shown in Fig. 3 are calculated plots of Eq. 7 for several values of h_g, ranging from very small (1 cm /sec), to very large. The experimental data best fit Eq. 7 if the values, $h_g \cong 5\text{-}10$ cm /sec and $k_s = 10^7$ exp (-1.9 eV /kT) cm /sec, are used.

The Grove model is a simplified approach to the deposition problem since it neglects the

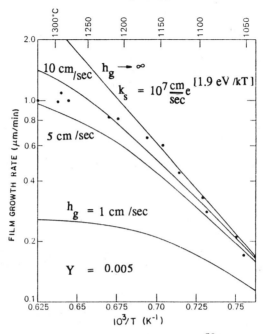

Fig. 3 Silicon growth as a function of reciprocal temperature[53]. Points represent experimental data. Reprinted with permission of the publisher, the Electrochemical Society.

flux of reaction products, and assumes the reaction rate flux depends linearly on the surface concentration. The latter is true only for low values of Y. The model also neglects the effect of temperature gradient on the gas phase mass transfer. Despite these limitations, the model predicts the two regions of the growth process, the *mass transfer* and the *surface controlled* regions, and provides a reasonable estimate to the values of k_s and h_g from the growth rate data.

Gas Phase Mass Transfer

Equation 1 approximated the flux of reactant from the gas bulk to the surface of the substrate by $F_1 = h_g (C_g - C_s)$. In this section we discuss the factors that affect the value of h_g.

The Stagnant Film Model

A simple technique used to arrive at estimates of the value of h_g is to assume that the gas is divided into two regions. In one region (away from the substrate) the gas is well mixed, and is moving with a *constant velocity*, U, parallel to the substrate. The other region is the *stagnant film* region in which the gas velocity is zero. The thickness of the stagnant layer is assumed to be δ_s. Figure 4 shows a schematic representation of the model. In this model, mass transfer of the reactant species across the stagnant layer is assumed to proceed solely by diffusion. Under these conditions, flux F_1 of Eq. 1 can be rewritten (according to Fick's First Law, Chap. 8), as:

$$F_1 = D_g (C_g - C_s) / \delta_s \qquad (11)$$

where D_g is the diffusion coefficient of the active species and $(C_g - C_s) / \delta_s$ is the concentration gradient. Setting F_1 in Eq. 1 equal to F_1 in Eq. 11 we obtain:

$$h_g = D_g / \delta_s \qquad (12)$$

This model is arbitrary since it assumes an abrupt transition from a region of finite gas flow to zero flow, and such abrupt stagnant layers in fact do not actually exist. It is nevertheless a useful first order approximation that has been applied to the solution of many gas phase mass-transfer problems.

Boundary Layer Theory

Fluid mechanics is capable of providing a more realistic, and therefore more accurate estimate of the mass-transfer coefficient, h_g. An approach to this problem was developed by Prandtl in 1904 and is termed *boundary layer theory*[2]. The *boundary layer* (which represents the events at a surface more realistically than a stagnant layer) is a transition region between the solid surface (substrate or susceptor) where the gas velocity is zero, and the free gas stream, where the gas velocity (parallel to the substrate) is that of the flow field, U. It is a region of rapidly changing gas velocity. The boundary layer thickness, $\delta (x)$, is defined by the point at which the velocity is 99% of the free stream velocity (0.99U). Figure 5 shows the formation of a boundary layer resulting from gas flow along a flat plate. This is similar to the gas flow observed along the length of the susceptor in a horizontal reactor.

It can be shown that the velocity of the gas next to the plate must be zero in order to avoid the existence of infinitely large velocity gradients. This condition of zero velocity disturbs the velocity distribution. As the fluid moves along the plate beyond the leading edge which is initially encountered by the flowing gas, the disturbance spreads further and further into the bulk of the gas, resulting in a boundary layer that increases in thickness with distance along the plate. From first principles, the boundary layer thickness as a function of distance along the plate, x, can be calculated from:

$$\delta (x) = (\mu x / d U)^{1/2} \qquad (13)$$

where μ is the gas viscosity and d the gas density. The dotted line in Fig. 5 is the locus of points whose value reaches 99% of the free stream velocity, U. More precise calculations of $\delta (x)$ differ from Eq. 13 only by the numerical coefficient of the equation, which is found to range from 0.67 to 5, depending on the precise definition of δ.

The average boundary layer thickness over the whole plate (or length of tube) is given as:

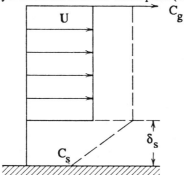

Fig. 4 The stagnant film model of gas-phase mass transfer.

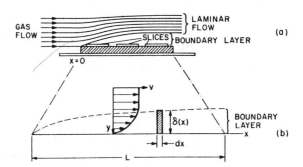

Fig. 5 (a) Development of a boundary layer in gas flowing over a flat plate, (b) expanded view of the boundary layer. From S.M. Sze, *Semiconductor Devices, Physics and Technology*, Copyright© 1985 John Wiley & Sons. Reprinted with permission of John Wiley & Sons.

$$\delta = \frac{1}{L}\int_0^L \delta\,(x)\,dx = \frac{2}{3}L\,(\frac{\mu}{d\,U\,L})^{1/2} \qquad (14)$$

or;

$$\delta = (2\,L\,)\,/\,(\,3\sqrt{Re_L}\,) \qquad (15)$$

where Re_L given by;

$$Re_L = d\ U\,L\,/\,\mu \qquad (16)$$

is the *Reynolds number* for the gas. Re_L is a dimensionless number used in fluid dynamics which represents the ratio of the magnitude of inertial effects to viscous effects in fluid motion. For low values ($Re_L < 2000$) the gas flow regime is called *laminar*, while for larger values the flow regime is *turbulent*.

The average boundary layer thickness, δ, can be substituted for δ_s in Eq. 12 of the stagnant layer model, resulting in an expression of the mass transfer coefficient:

$$h_g = \frac{D_g}{\delta_s} = \frac{3\,D_g}{2\,L}\,\sqrt{Re_L} \qquad (17)$$

Equation 17 can be rearranged into a dimensionless equation:

$$h_g\,L\,/D_g = 3\,\sqrt{Re_L}\,/\,2 \qquad (18)$$

and compared with a more exact solution by Pohlhausen[3]:

$$\frac{h_g\,L}{D_g} = (\frac{2\sqrt{Re_L}}{3})\,(\frac{\mu}{d\,D_g})^{1/3} \qquad (19)$$

where $\mu\,/dD_g$ is a dimensionless parameter called the *Schmidt number*. The value of the Schmidt number varies between 0.5 to 0.8 for most gases and is independent of temperature.

In the mass flow controlled regime, the film growth rate should depend on the square root of the gas velocity, U (since h_g is proportional to $U^{1/2}$). Figure 6 shows a plot of the silicon film growth rate as a function of the gas flow rate (which is proportional to the gas velocity in a fixed volume). The predicted square root dependence is observed. At high flow rates, however, the growth rate reaches a maximum and then becomes independent of the flow. In this regime the deposition is controlled by the reaction rate ($h_g \gg k_s$), as is evidenced by the exponential dependence of the growth rate on temperature observed at those flow rates.

Ban studied the boundary layer formation in a horizontal reactor[4]. Figure 7 shows a representation of the boundary layer that forms in the horizontal reactor. Here we observe that the boundary layer increases until the flow is fully established. It is across this layer that the reactants and the reaction by-products must diffuse for the deposition to occur. The fluxes of the various species across this layer depend in a complex way on the process variables, including, the reactant concentration in the gas stream, the deposition temperature, the system pressure, and the thickness of the boundary layer.

Using Eq. 19 we can estimate values of h_g for vertical and horizontal reactor configurations. Typical values of U are 20-30 cm /sec, leading to values of $Re_L \cong 20$. This in turn results in a value of $h_g \cong 5$ cm /sec, which is in good agreement with the data shown in Fig. 3 for the deposition of silicon in a vertical reactor.

To insure uniformity of deposition across all wafers when designing and optimizing processes in a epitaxial reactor, the effect of both the boundary layer and the temperature gradients must be taken into account. Manke and Donaghey studied the effect of Re_L on the layer growth rate along the length of a horizontal susceptor[5]. The values of Re_L used in this study were changed by varying the gas flow rate (velocity). Figure 8 shows the results of their study. It was found that the thickness uniformity along the susceptor could be optimized at a particular flow rate (Re_L) in the reactor. Ban calculated the temperature profile above the susceptor in a horizontal reactor[4]. The results are shown in Fig. 9. These calculations show that a large temperature gradient exists which affects the fluid flow, since such a gradient can create turbulence in the gas stream. In addition, the parameters of Re_L and h_g depend on temperature, and thus temperature variations above the susceptor should be taken into account for accurate modeling of the epitaxial process.

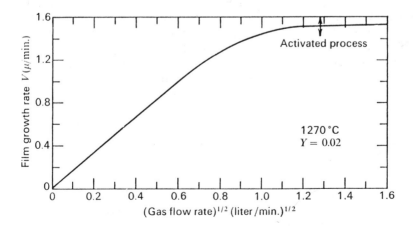

Fig. 6   Silicon growth rate as a function of the gas flow rate in a vertical reactor[52]. Reprinted with permission of the publisher, the Electrochemical Society.

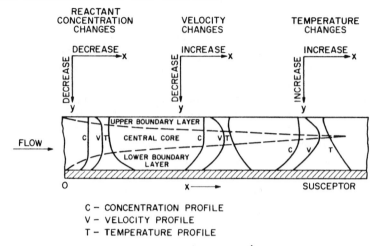

C — CONCENTRATION PROFILE
V — VELOCITY PROFILE
T — TEMPERATURE PROFILE

Fig. 7 Boundary layer formation in a horizontal reactor[4]. Reprinted with permission of the publisher, the Electrochemical Society.

Atomistic Model of the Growth of Epitaxial Layers

The growth of epitaxial layers is believed to occur via the migration of *adatoms* (A in Fig. 10) to steps on the silicon surface (B in Fig. 10)[6]. The *kink position* along a step is the most energetically favorable for growth (C in Fig. 10), since one-half of the Si-Si bonds have already been established with the crystal. The growth of the epitaxial layer then proceeds by lateral movement of the steps.

The existence of a *maximum growth rate* of epitaxial layers has been experimentally determined as a function of temperature for the atmospheric pressure deposition of silicon [7]. At a given temperature, polycrystalline growth occurs if the maximum growth rate is exceeded. The value of the maximum growth rate is found to increase exponentially with temperature, with an activation energy of ~5 eV. This behavior is illustrated in Fig. 11.

These results may be explained as follows. At high growth rates there is not sufficient time available for the adatoms to migrate to the kink sites, thereby polycrystalline growth results. As temperatures increase, the surface migration rate increases, allowing sufficient time for the adatoms to reach the kink sites. The value of the activation energy, ~5 eV for this process, is comparable to the activation energy of self-diffusion in silicon, which is believed to be the controlling mechanism for the growth of single crystal films (epitaxy).

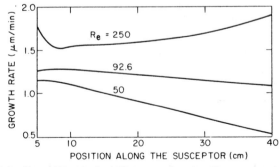

Fig. 8 The effect of the Reynold's number (Re_L) on the growth rate showing uniformity along the susceptor[5]. Reprinted with permission of the publisher, the Electrochemical Society.

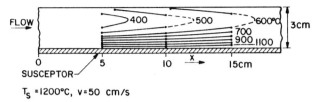

T_S =1200°C, v=50 cm/s

Fig. 9 Calculated temperature profiles in a horizontal reactor[4]. Reprinted with permission of the publisher, the Electrochemical Society.

CHEMICAL REACTIONS USED IN SILICON EPITAXY

There are four major chemical sources of silicon used commercially for epitaxial deposition. They are: *silicon tetrachloride* ($SiCl_4$), *trichlorosilane* ($SiHCl_3$), *dichlorosilane* (SiH_2Cl_2), and *silane* (SiH_4). Each of these sources has properties that make them attractive under particular deposition conditions and film requirements. The hydrogen reduction of $SiCl_4$ has been the most widely used source, and therefore the most widely studied. Demands for thinner epitaxial layers and lower temperature depositions, however, have lead to increased use of SiH_2Cl_2 and SiH_4.

Overall reactions can be written for each of the silicon sources. For example, the overall reaction for the hydrogen reduction of $SiCl_4$ can be considered to proceed as follows:

$$SiCl_4 + 2H_2 \rightarrow Si + 4HCl \qquad (20)$$

Such overall reactions, however, do not provide a complete picture of the reaction sequence, the species present in the gas phase, or the species adsorbed on the substrate. Ban and coworkers[8-11] studied the thermodynamics of the Cl-H-Si system. In these studies the starting mixture consisted of H_2 with SiH_2Cl_2, $SiHCl_3$, or $SiCl_4$. Using a mass spectrometer to determine the active species present in the reaction atmosphere, additional constituents besides those initially introduced into the chamber were observed. These additional species included HCl, and $SiCl_2$. Their presence implies that a large number of chemical reactions are taking place simultaneously in the systems. Ban postulated a number of possible reactions, and they are summarized in Table 1. The reactions are separated into two groups: surface reactions and gas phase reactions. The obvious conclusion of this work is that the chemistry of the Si deposition is not simple, and is not well represented by the overall single reaction that is often given.

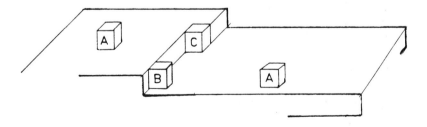

Fig. 10 Epitaxial growth model showing step-wise growth with adatoms (A), atoms on steps (B), and atoms at stable kink positions (C).

Table 1. SOME REACTIONS OF THE Si-H-Cl SYSTEM[4]

Surface reactions

Si(s) + 4 HCl (ads) ⟶ SiCl$_4$ (ads) + 2 H$_2$ (ads)	(1)
Si(s) + 3 HCl (ads) ⟶ SiHCl$_3$ (ads) + H$_2$ (ads)	(2)
Si(s) + 2 HCl (ads) ⟶ SiCl$_2$H$_2$ (ads)	(3)
Si(s) + HCl (ads) + H$_2$ (ads) ⟶ SiClH$_3$ (ads)	(4)
Si(s) + 2 H$_2$ (ads) ⟶ SiH$_4$ (ads)	(5)
Si(s) + 2 HCl (ads) ⟶ SiCl$_2$ (ads) + H$_2$ (ads)	(6)
SiCl$_2$H$_2$ (ads) ⟶ Si(s) + 2 HCl (ads)	(7)
SiCl$_4$ (ads) + Si(s) ⟶ 2 SiCl$_2$ (ads)	(8)
SiCl$_4$ (ads) + 2 H$_2$ (ads) ⟶ Si(s) + HCl (ads)	(9)

Gas phase reactions

SiCl$_2$H$_2$ (g) ⟶ SiCl$_2$ (g) + H$_2$ (g)	(10)
SiCl$_3$H (g) ⟶ SiCl$_2$ (g) + HCl (g)	(11)
SiCl$_4$ (g) + H$_2$ (g) ⟶ SiCl$_3$H (g) + HCl (g)	(12)

Figure 12 shows the concentration of the chemical species found in the vapor during epitaxial deposition of silicon from the H$_2$ reduction of SiCl$_4$, as a function of position along a horizontal reactor[12]. SiCl$_4$ concentration is found to decrease while the concentrations of the other three constituents increase with distance. Based on these results the reaction is postulated to proceed as:

$$SiCl_4 + H_2 \longleftrightarrow SiHCl_3 + HCl \qquad (21)$$

$$SiHCl_3 + H_2 \longleftrightarrow SiH_2Cl_2 + HCl \qquad (22)$$

$$SiH_2Cl_2 \longleftrightarrow SiCl_2 + H_2 \qquad (23)$$

$$SiHCl_3 \longleftrightarrow SiCl_2 + HCl \qquad (24)$$

$$SiCl_2 + H_2 \longleftrightarrow Si + 2HCl \qquad (25)$$

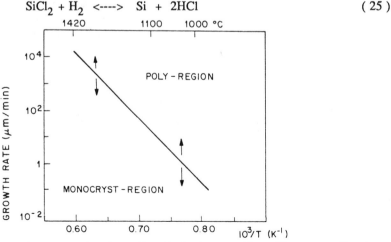

Fig. 11 Maximum growth rate for formation of single crystalline silicon as a function of temperature[7]. Reprinted with permission of North Holland Publishing Co.

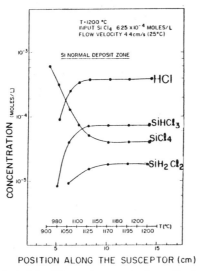

Fig. 12 Species present (detected by IR spectroscopy) along the length of a horizontal reactor with $SiCl_4 + H_2$ at 1200°C[12]. Reprinted by permission of publisher, the Electrochemical Society.

The sequence of the $SiCl_4$ reduction shows the presence of $SiHCl_3$ and SiH_2Cl_2 as intermediate species. Growth of silicon from these reactants would start with Eq. 22 or Eq. 23, repectively. It should be noted that the reactions are reversible, meaning that silicon may be etched by HCl (Eq. 25) as well as be deposited. The equilibrium of the reactions favor the etching direction at both low and elevated temperatures. The effects of temperature on the net growth rate is shown in Fig. 13. It should be noted that the etching reaction (negative growth rate) occurs at temperatures below 900°C and above 1400°C .

The growth rate of the epitaxial film depends on several parameters: a) chemical source; b) deposition temperature; and c) mole fraction of reactants. Figure 14 shows a comparison of the silicon growth rate vs. temperature (Arrhenius plots) for $SiCl_4$, $SiHCl_3$, SiH_2Cl_2, and SiH_4[13]. Two growth regimes are observed for all four reactants. In the elevated temperature regime, the gas phase mass transport control is dominant, and the growth rate is only weakly dependent on temperature (region B of Fig. 14). At low temperatures, growth rate is dependent on the surface

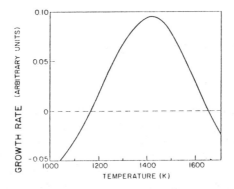

Fig. 13 The growth rate of variation of silicon films as a function of deposition temperatures for $SiCl_4 + H_2$[38]. Reprinted with permission of the publisher, the Electrochemical Society.

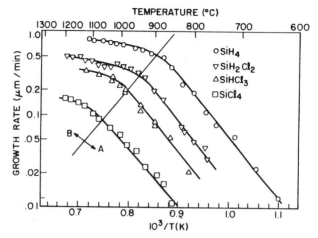

Fig. 14 Growth rate of silicon films as a function of temperature for various silicon sources[13]. Reprinted with permission of Philips Research Reports.

controlled reaction, which is strongly (exponentially) dependent on temperature (region A of Fig. 14). The transition points (shown by the line in Fig. 14) depend on the mole fraction of reactant, reactor type, gas flow rate, and particular active gas species. Production depositions are usually performed in region B, since temperature variation effects are minimized in that regime.

The silane (SiH_4) reaction proceeds by the overall reaction:

$$SiH_4 \text{ (vapor)} \rightarrow Si \text{ (solid)} + 2H_2 \text{ (vapor)} \qquad (26)$$

and is not reversible at normal processing temperatures. Since no HCl is formed as a product of the reaction, the reverse etching reaction cannot take place. The major advantage of the SiH_4 reaction is that the deposition can occur at lower temperatures than any of the chlorosilane decompositions, and as a result is more easily controlled in the mass transfer regime.   Several potential disadvantages of using the SiH_4 reaction are: a) no HCl is present (the presence of HCl can be beneficial since Cl is known to remove metallic impurities, such as Fe or Cu, from Si); and b) homogeneous gas phase reactions can occur, resulting in powdery deposits on the reactor walls, or even on wafers.

DOPING OF EPITAXIAL FILMS

Intentional Doping

In order to control the conductivity type and carrier concentration (electrical resistivity) of epitaxial layers, dopants (impurities) are added to the reaction gases (silicon source and hydrogen) during the entire growth process. Typically dopants are introduced using their hydrides. Diborane [B_2H_6] is used to incorporate boron, phosphine [PH_3] to incorporate phosphorus, and arsine [AsH_3] to incorporate arsenic. The hydrides are typically diluted in hydrogen to low concentrations (10-1000 ppm). Caution must be exercised when using these gases as they are extremely toxic. Although hydrides are relatively unstable above room temperaure, they are prevented from dissociating in the gas phase by the presence of large quantities of H_2 in the reactor.

Unfortunately, there is no simple rule to relate the incorporation of dopant atoms from the gas phase into the Si film, since the incorporation depends upon many factors, including substrate temperature, deposition rate, dopant mole volume relative to the source gas mole volume, and geometry of the reactor. *Thus, the dopant /Si ratio incorporated into the solid phase is different from that of the gas phase. The ratio of the dopant concentration in the gas phase to that in the solid must be empirically determined for each process.* Nevertheless, epitaxial films with well controlled doping concentrations in the range of 10^{14}-10^{20} atoms /cm^3 can be routinely achieved.

Autodoping and Solid-State Diffusion (Outdiffusion)

Most VLSI applications require an epitaxial deposition of a lightly doped layer (10^{14}-10^{17} atoms /cm^3) upon a heavily doped substrate (10^{19}-10^{21} atoms /cm^3). The substrate may be uniformly doped or consist of isolated regions of heavy doping (called *buried layers*) diffused in a lightly doped substrate. For most CMOS device structures, uniformly doped substrates are used, while in most bipolar applications buried layers are used.

During the epitaxial deposition, dopants from the heavily doped regions on the substrate enter the growing film by two distinct means: a) *solid state diffusion;* and b) *vapor phase autodoping*. The solid state diffusion effect simply represents the diffusion of impurity atoms from the substrate into the growing film at elevated temperatures (and is also known as *outdiffusion*). The concentration of dopant transported to the film can be calculated from the complementary error function (see Chap. 8) by:

$$C_e(x) = \frac{C_s}{2} [1 - \text{erf} \frac{x}{2\sqrt{D_s t}}] \qquad (27)$$

where C_s is the initial concentration of dopant in the substrate, $C_e(x)$ is the concentration of dopant in the epitaxial layer, D_s is the diffusion coefficient of the substrate impurity, and t is the deposition time. Use of this expression assumes that the film growth rate far exceeds the velocity of the diffusion front, which is generally the case for practical depositions.

The second type of dopant transfer (*vapor phase autodoping*) arises from the evaporation of dopants from the substrate that are subsequently reincorporated into the growing film. There are two types of autodoping, vertical and lateral (both of which are depicted schematically in

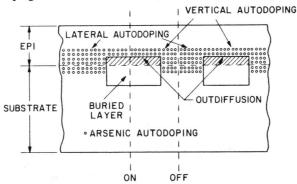

Fig. 15 Cross section of an epitaxial wafer (with an arsenic buried layer) showing the effects of lateral autodoping, vertical autodoping, and outdiffusion[14]. Reprinted with permission of the publisher, the Electrochemical Society.

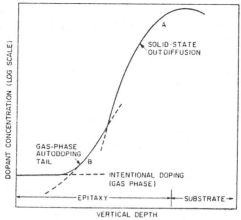

Fig. 16 Plot of the expected doping profile of an epitaxial layer grown over a heavily doped substrate showing the several regions of autodoping[15]. Reprinted with permission of the publisher, the Electrochemical Society.

Fig. 15)[14]. Vapor phase autodoping can also result from atoms evaporated from the edges and backsides of wafers, as well as from susceptors coated with heavily doped material. The latter is responsible for the so-called *memory effect* of epitaxial reactors (i.e. it is observed that high resistivity layers are difficult to grow immediately after growing a low resistivity layer). Vapor phase autodoping requires that evaporated dopants be trapped within the gas-phase boundary layer.

The outdiffusion and autodoping effects manifest themselves as an increased thickness of the transition layer from the heavily doped substrate to the uniformlly doped layer. The transition region is shown schematically in Fig. 16[15]. Close to the substrate, solid state diffusion from the substrate is the predominant effect, causing a broader transition profile (Region A in Fig. 16). The rapid growth of the film relative to the motion of diffusion limits this effect. Thereafter, the transition zone becomes controlled by the incorporation of dopants from the vapor phase. When the atoms evaporated from the substrate exceed those intentionally introduced, the autodoping tail (Region B in Fig. 16) develops. Since the high concentration buried layers quickly become covered with more lightly doped material, the autodoping effect ceases, and the desired concentration is reached. If the whole substrate is heavily doped, evaporation from the wafer edges and backside can continue after the front side is sealed. However, even this effect decreases with time, since solid state diffusion from the wafer interior must supply dopant to the surface for evaporation. The extent of the autodoping tail depends on deposition temperature, silicon source, growth rate, reactor geometry, and pressure (related to the boundary layer thickness).

Autodoping imposes a limit on the minimum thickness and doping level for an epitaxial layer and therefore must be minimized for VLSI applications. Various techniques are available to the process engineer to reduce these effects, including:

1. To prevent boron autodoping, the epitaxial deposition process should be performed at the lowest possible deposition temperature, commensurate with good quality films and reasonable deposition rates (e.g. SiH_2Cl_2 can be deposited at lower temperatures than $SiCl_4$). Note that this technique, however, is only effective for minimizing boron autodoping, since arsenic autodoping actually increases at lower temperatures;
2. Substrate and buried layer dopants that have low vapor pressure *and* low diffusivities should be used (e.g. select Sb instead of As[high vapor pressure] or P (high diffusivity]);

3. If necessary seal the backside of substrates and /or susceptors with lightly doped silicon;
4. Operate the epitaxial system at reduced pressures (sub-atmospheric) to minimize lateral autodoping. Reduction in lateral autodoping is achieved since the diffusivity of gas molecules at reduced pressures is increased, which allows the dopant atoms to rapidly reach the main gas flow and be swept out of the reactor. This method works well for phosphorus and arsenic, but not for boron;
5. Use a buried layer that is ion-implanted to reduce the substrate surface concentration;
6. A *cap-purge-grow* sequence can be utilized, in which a very thin layer of undoped film is first grown over the buried layer or substrate, to seal in the dopant. This is followed by a gas purge, and introduction of the dopant as the remainder of the epi layer is grown;
7. Use a low temperature purge after the HCl etch to insure that the dopant evolved during the etching is swept out of the system.

DEFECTS IN EPITAXIAL FILMS

During the epitaxial deposition, process-induced crystalline defects can occur in the film. These defects result from such diverse factors as: a) reactor contamination; b) substrate preparation; and c) substrate imperfection. (The nature and types of crystalline defects in silicon, their effect on device properties, and methods for their detection are thoroughly covered in Chap. 2 of this book). As noted, there are four types of crystalline defects: 1) point defects (vacancies, interstitials); 2) line defects (dislocations); 3) area defects (stacking faults); and 4) volume defects (voids, precipitates). The methods used to detect the presence of defects include chemical etching and examination by optical and electron microscopy, deep level transient spectroscopy (DLTS), and x-ray topography (see Chaps. 2 and 17). In this section we describe the mechanisms by which epitaxial defects are nucleated, and then discuss how such defect formation is minimized.

The first noteworthy concept is that the crystalline quality of the layer cannot surpass that of the substrate. Therefore, great care must be exercised in substrate preparation. A pre-epitaxial wafer cleaning sequence similar to a pre-furnace cleaning sequence (but including a wafer scrub), is required to remove particulate, organic, and inorganic (heavy metal, Na) contamination from the front and rear surfaces of the wafer. Often such cleaning procedures are not sufficient, and an additional *in-situ* anhydrous HCl (1-5% HCl in H_2) etch at temperatures in excess of 1100°C is required to remove any remaining contaminants from the surface. Care must be taken to prevent removal of significant thicknesses of silicon substrate, as the etch rate is rapid at elevated temperatures. If lower temperatures (1050°C) are required, *in-situ* etching with sulfur hexafluoride (SF_6) has been found to be a suitable alternative to HCl.

At one time it was believed that good bipolar device quality epitaxial films could not be prepared without an *in-situ* HCl etch, or the presence of a Cl-bearing species during the deposition. A recent study[16], however, has shown that excellent device quality epitaxial material can be prepared on wafers that have received only a wet pre-epitaxial clean. The films were deposited from SiH_4 and H_2 at temperatures as low as 900°C. This result also indicates that it is possible to grow high quality, thin epitaxial layers (<1 μm) with little or no outdiffusion and /or autodoping.

A potential source of impurities within the reactor is the Si-C coated graphite susceptor, which may release metallic impurities[17] if the SiC coating becomes cracked. A high temperature HCl treatment of the susceptor substantially reduces these impurities.

Defects Induced During Epitaxial Deposition and Their Nucleation Mechanisms

The most common types of defects that occur as a result of epitaxial deposition are dislocations and stacking faults. Dislocations are generated in epitaxial films by several mechanisms. These include: a) the propagation into the growing film of a dislocation line in the substrate that reaches the substrate surface. (Heavily dislocated substrates result in heavily dislocated films. Therefore, dislocation-free substrate material should be used); b) the existence of a large difference in lattice parameter between the film and the substrate (e.g. when the substrate is very heavily doped and the epi layer lightly doped), resulting in *misfit dislocations*. The mismatch is relieved by forming misfit dislocations in the epitaxial film near the substrate; and c) by thermally generated stresses that exceed the yield strength of the silicon, resulting in slip.

Thermal stresses are set up by temperature gradients that exist between the back and front of the wafers during the deposition process. These gradients cause a sequence of events that ultimately lead to slip generation. This sequence of events is as follows (and is worse for rf than for IR heating of the susceptors used for epitaxial depositions): As the rf heats the susceptor (which in turn heats the back of the wafers by radiation and conduction from the susceptor), the front sides of the wafers radiate heat to the environment. This results in a temperature difference of several degrees between the back and front of the wafers, and to different amounts of thermal expansion between the front and back. Since the back surface of the wafer expands more than the front surface, the wafer assumes a bowl-like shape to accommodate this effect (wafer bow). This causes the wafer edges to lift from the susceptor. The wafer edges now receive less heat from the susceptor, and cool to a lower temperaure than the remainder of the wafer, causing even further wafer bow. The bowing eventually results in stress levels that exceed the yield strength of the silicon at the elevated temperature, resulting in crystal slip. This effect is depicted in Fig. 17. The temperature difference tends to increase for larger diameter wafers. The problem can substantially be reduced in rf heated reactors by shaping the sample holders into pockets, so the wafers make contact with the susceptor only on the edges (see Fig. 18). By placing heat reflectors above the wafers, the temperature difference between the back and front surface can also be reduced. A recent solution to the slip problem includes the use of a radiant heater focused on the front side of the wafers to reduce the net heat flux through the wafers. In IR heated

THERMALLY INDUCED STRESS

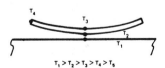

$T_1 > T_2 > T_3 > T_4 > T_5$

Fig. 17 Effect of temperature difference between front and backside of wafer on wafer bow. Courtesy of Gemini Research.

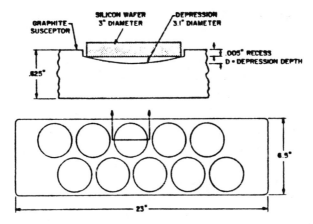

Fig. 18 Wafer pockets cut in susceptor to reduce wafer bow effect. Courtesy of Gemini Research.

reactors, on the other hand, heating from the front side is already provided, which tends to minimize the problem.

The nucleation of *stacking faults* in the epitaxial layer is believed to result from two distinct causes: a) microscopic surface steps on the substrate; and b) impurities either on the substrate or within the reactor itself. As presented earlier, an atomistic model for the growth of an epitaxial film has been developed[7].   In this model, the film tends to propagate by two-dimensional steps lying along the <110> directions.   Ravi[18] has shown that these side-moving steps may encounter an obstacle on the substrate, resulting in a fault in the stacking sequence (see Chap. 2). The presence of slip bands in the substrate, or particles on the substrate can also nucleate stacking faults. Local regions of silicon dioxide, nitride, or carbide on the surface also provide sites for nucleation of such faults. Sources of these compounds include: a) residual amounts of moisture or carbon dioxide in the reactor (which may grow SiO_2 on the substrate) during the deposition; and b) hydrocarbon contamination, which can result in pockets of SiC on the substrate.   Metallic impurities (Fe, Cu, Ni) can become incorporated in the vapor phase and subsequently be deposited with the silicon film as a solid solution. As a result of the rapid cooling from the deposition temperature, these impurities may remain in solid solution even at room temperature (quenching).   Further heat treatment, such as a subsequent oxidation, can cause these impurities to precipitate, resulting in oxidation-induced stacking faults.

Particulates within the reactor that land on the substrate can also serve as sites for stacking fault nucleation. Radiantly heated barrel reactors have an effective boundary layer (flow is parallel to the wafer surface), which protects the substrates from particle injection from the gas phase. Particle transport across this boundary layer can be further reduced by using low lamp power, which eliminates electrostatic migration of particles[19].

Techniques for Reducing Defects in Epitaxial Films

In addition to the cleaning processes discussed above, defects in epitaxial layers can be reduced by the use of denuded zones and oxygen intrinsic gettering (IG) of the substrates prior to the deposition. (These techniques are introduced in Chap. 2). Borland, *et al.*[20], investigated the effect of IG processes performed on epitaxial substrates prior to deposition. They found that improved silicon epitaxial layers were obtained by applying pre-epitaxial intrinsic gettering to the substrate wafer. These techniques were investigated on various epitaxial device structures including n on p⁻ (bipolar), as well as p on p⁺ and n on n⁺ (CMOS). It was found that heavily doped n⁺ wafers were resistant to oxygen precipitation, requiring anneals in excess of 40 hours to induce bulk micro-defects. Dyson and co-workers[21], have shown that the use of a starting substrate with a polysilicon gettered backside, along with an IG process (utilizing a *high-low-high* annealing sequence, see Chap. 2) resulted in significant reduction in the time to micro-defect formation in n⁺ substrates. These gettering treatments resulted in significantly improved carrier lifetimes, which is a measure of metallic impurity levels in the epitaxial layer.

A recent technique for achieving low defect epitaxial layers is the use of *strain layers*[22]. The strain layer is formed between the substrate and the active part of the epitaxial layer by intentionally alloying the initially deposited silicon with germanium. The resulting Si-Ge alloy has a substantially different lattice parameter than the silicon substrate, and the mismatch is accommodated by the formation of misfit dislocations. The epitaxial device layer is deposited on the Si-Ge strain layer. The misfit dislocations, in turn, serve as sites for heavy metal precipitation, thereby trapping these impurities away from the active device region. Bean[23] reviewed the use of *multilayer misfit dislocation extrinsic gettering strain fields*. Here, epitaxial Si-Ge layers are sequentially deposited with each layer having an increasing concentration of Ge, (and the density of misfit dislocations increases with Ge concentration). Each layer is interposed with a strain-free Si layer and a final Si device layer is deposited. The more highly doped layers (having greater gettering efficiency), are closer to device regions. Observation show that misfit dislocations do not reach the device layer even after high temperature (>1000°C) heat cycling. Results show this method is effective in gettering Au, Cu, Fe, and Ni from epitaxial layers[24].

PROCESS CONSIDERATIONS FOR VLSI EPITAXIAL DEPOSITION

As discussed in *Chemical Reactions Used in Silicon Epitaxy*, there are basically four chemical sources of silicon epitaxy ($SiCl_4$, $SiHCl_3$, SiH_2Cl_2, and SiH_4) available to the process engineer. Each source is used over different temperature regimes and for different applications. Table 2 lists the sources, typical deposition temperatures, and deposition rates.

The use of $SiCl_4$, requires high deposition temperatures (1150°-1250°C), which results in significant autodoping and outdiffusion (as discussed in a previous section). The major advantage of this source is that very little deposition occurs on the reactor walls, thereby reducing the

Table 2. SILICON SOURCE DEPOSITION CHARACTERISTICS

Silicon Source	Deposition Temp (°C)	Growth Rate (μm /min)	Remarks
Silicon tetrachloride, $SiCl_4$	1150-1250°	0.4–1.5	High temperature required, Good selectivity
Trichlorosilane, $SiHCl_3$	1100-1200°	0.4–2.0	Easily reduced, high deposition temperature
Dichlorosilane, SiH_2Cl_2	1050-1150°	0.4–3.0	Low temperature, good quality epitaxy
Silane, SiH_4	950-1050°	0.2–0.3	Low temperature, gas phase deposition problems, used for SOS

frequency of cleaning. This in turn also results in lower particulate contamination, since silicon flaking off the reactor is minimized. The $SiCl_4$ is used for thick deposits (>3 μm) on devices that can tolerate the elevated temperature (and its attendant unintentional doping effects).

$SiHCl_3$ has similar properties to those of $SiCl_4$, except that it can be deposited at somewhat lower temperatures (1100-1200°C). Its primary use is also in depositing thicker epitaxial layers, since it can have deposition rates in excess of 1 μm /min. Because it does not offer significant advantages over $SiCl_4$, however, this material has found little application in VLSI.

SiH_2Cl_2 is widely used to deposit thinner layers at lower temperatures (1000-1100°C). The lower temperature reduces outdiffusion and autodoping, mechanisms which must be controlled in order to achieve the target epitaxial layer resistivity.   High quality layers have been prepared at temperatures as low as 1000°C using SiH_2Cl_2. It has been reported that films prepared from SiH_2Cl_2 have lower defect densities than those prepared either from SiH_4 or $SiCl_4$[25]. Device yields have also been reported to increase when using SiH_2Cl_2, compared to films deposited at high temperatures from $SiCl_4$[26].

SiH_4 is used to deposit very thin epitaxial layers at temperatures less than 1000°C. This results in substantially less autodoping and outdiffusion when compared to films deposited at 1175°C using $SiCl_4$.   Another significant advantage of using SiH_4 is that it does not produce pattern shift (discussed in a subsequent section). The most important disadvantage of using SiH_4 is that it can pyrolyze (decompose) at low temperatures (causing heavy deposits on reactor walls). This necessitates frequent cleaning of the reactor, reducing its effective throughput.

Pattern Shift, Distortion, and Washout

When the image of the step in an epitaxial layer deposited over a buried layer is examined, it

Table 3.   AN EPITAXIAL PROCESS SEQUENCE

Step	Estimated Time
1. Load clean wafers into reactor	2 - 4 min
2. Purge with inert gas and fill with H_2 carrier. (Also pumpdown if reduced pressure epitaxy process is being conducted).	3 - 5 min
3. Heat to hydrogen bake temperature (1200°C) to remove oxide layer. (Note: this step can remove 50-100Å of an SiO_2 layer.)	10 - 12 min
4. a) Heat to HCl etch temperature; b) Introduce anhydrous HCl (or SF_6) to etch surface Si layer. c) Purge to clear HCl and dopants from system.	5 - 7 min
5. a) Cool to deposition temperature; b) Introduce silicon source and dopants to deposit desired layer; c) Clear out silicon source and dopants by purging with H_2.	10 - 12 min
6. Cool to room temperature.	10 - 12 min
7. Purge out H_2 and backfill with N_2	3 - 5 min
8. Unload wafer	2 - 4 min
Total Process Time	45 + min
Total Heated Process Time	25 + min

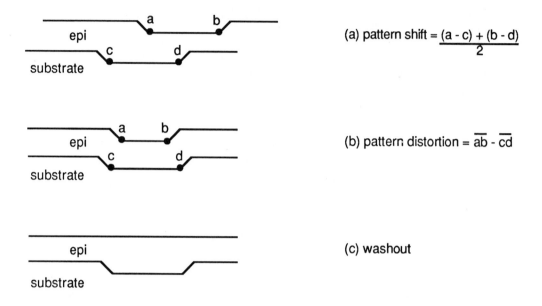

(a) pattern shift = $\dfrac{(a-c)+(b-d)}{2}$

(b) pattern distortion = $\overline{ab} - \overline{cd}$

(c) washout

Fig. 19 Schematic representation of (a) pattern shift, (b) pattern distortion, (c) pattern washout.

is sometimes found that its position is shifted relative to its position in the buried layer (or has become distorted, or even washed out). These effects are termed *pattern shift, pattern distortion*, and *pattern washout*, respectively. Figure 19 shows a schematic representation of these effects. The VLSI circuit designer must adjust the position of features on subsequent mask levels in order to compensate for the amount of shift that these effects introduce. Selecting the correct amount of adjustment, however, is complicated by the fact that the effects are dependent on substrate orientation, deposition rate, deposition temperature, and silicon source.

The *crystal orientation* has a significant effect on *pattern shift*. For the case of <100> oriented wafers, pattern shift is a minimum when oriented on-axis, while <111> wafers show a minimum effect when misoriented by 2° to 5° from the <111> direction to the nearest <110> direction. Substrates to be used for growing epitaxial layers of <111> orientation are therefore normally supplied 3° off-orientation.

Lee, *et al*[27] studied the pattern shift as a function of temperature and deposition rate, and found that the shift decreased with increasing temperature, and increased with increasing deposition rate for both <111> and <100> orientations. Reduced pressure deposition was found to substantially reduce the pattern shift. Although early studies showed no effect of reactor geometry, it was more recently observed that under similar deposition conditions, inductively heated vertical reactors produce less pattern shift than radiantly-heated cylinder reactors (Fig. 20) under similar deposition conditions. Finally, pattern shift can be eliminated by using SiH_4 as the source gas. It also appears the presence of Cl_2 or HCl is required to induce pattern shift.

Weeks[28] studied *pattern distortion* during epitaxial deposition. The results of his study showed that while distortion occurs in both <100> and <111> oriented substrates, it is exacerbated on <100> substrates. Although reduced pressure deposition was found to control distortion in <111> material, it was ineffective for reducing the effect on <100> substrates. Unlike the situation with pattern shift, it was found that increased temperature and decreased

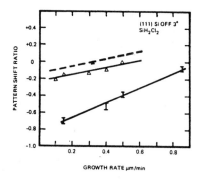

PATTERN SHIFT RATIO

(111) Si OFF 3°
SiH_2Cl_2

GROWTH RATE μm/min

△, ▲ RADIANTLY HEATED CYLINDER - 1080°C THERMOCOUPLE (∼ 1110°C TRUE),
△ = 80 TORR, ▲ = 100 TORR.

Ⅰ   INDUCTION VERTICAL - 1050°C OPTICAL, (1110°C TRUE),
110 TORR.

Fig. 20 Pattern shift as a function of growth rate. Courtesy of Gemini Research.

growth rate tended to increase pattern distortion. Furthermore, the use of SiH_4 as the silicon source actually resulted in more pattern distortion than when chlorine-based sources were utilized. Finally, increased layer thickness minimized the distortion. Note that the opposite dependencies of pattern shift and pattern distortion create a dilemma for the process engineer. As a result, trade-offs and empirical studies must be utilized to fine-tune an epitaxial process so that distortion and shift are each reduced as much as possible.

EPITAXIAL DEPOSITION EQUIPMENT

Epitaxial reactors are high temperature, batch, chemical vapor deposition (CVD) systems. The demands that VLSI devices place on epitaxial systems such as: a) tight thickness and resistivity control; b) low number of process induced defects; c) minimized pattern shift and washout; d) reduced autodoping; e) reduced temperature operation; f) the capability of handling 150 mm wafers; and g) requirements of high productivity, have been the driving functions for advances in equipment technology. Today, there are two major types of equipment that are commercially viable. These are the multiwafer rf inductively heated *pancake reactor*, and the multiwafer IR radiatively heated *barrel reactor*. The terms *pancake* and *barrel* arise from the configuration of the susceptor, which is used to hold the wafers during the deposition cycle. A third configuration (rarely used in VLSI production) is the rf inductively-heated horizontal reactor.

At a minimum, epitaxial deposition equipment consists of: a) a quartz bell-jar or reactor tube, to isolate the epitaxial growth environment; b) a gas distribution system to provide for controlled introduction of the various chemical species (the gas flow and process sequencing are generally microprocessor controlled); c) a wafer heating source (which can be either IR or rf); and d) an exhaust system to scrub the effluent gases. Temperature monitoring and control is achieved by using an IR sensitive optical pyrometer, focused upon the wafer surface. The indicated temperature of such a measurement is often 50° to 100°C below the actual temperature, due to changes in the emissivity of the silicon at the elevated temperatures. In the radiantly heated systems, thermocouples may also be used to measure temperature.

The *induction heated vertical pancake reactor* is shown schematically in Fig. 21. The wafers are placed on the silicon carbide coated graphite susceptor which is heated by the underlying rf

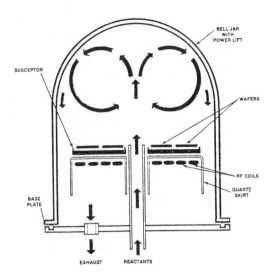

Fig. 21 Schematic of an induction heated vertical pancake reactor. Courtesy of Microelectronic Manufacturing and Testing.

coils. As discussed earlier, the susceptor is machined to hold the wafers at the edge in order to minimize thermally induced slip. The reactants enter from the center of the susceptor and are subsequently exhausted from the periphery. The gases are distributed symetrically across the wafers, resulting in a constant growth rate. The vertical gas flow minimizes autodoping effects, since a stable boundary layer is difficult to form under those conditions (Fig. 21). The pancake configuration, however, is susceptible to particulates falling on the wafer surface and becoming embedded in the growing film. The system is capable of processing wafers up to 200 mm in diameter, but the load size decreases as the wafer size increases. The system is also capable of operating at both reduced pressures (100-200 torr), or at atmospheric pressure. Reduced pressure deposition further reduces the autodoping effect, as well as the pattern shift.

The *radiantly heated barrel reactor* is shown schematically in Fig. 22. The susceptor (silicon carbide coated graphite) has a hexagonal cross-section, where each face can accommodate several wafers, the number depending on the wafer diameter. The wafers are held at 2.5° to the vertical, and the gases flow parallel to the wafer surfaces. The gases are injected at the top of the chamber and exhausted from the bottom. This gas flow results in the formation of a boundary layer, which tends to increase interwafer autodoping and source-gas depletion effects. The depletion effect tends to decrease the deposition rate downstream from the gas entry port. Heating of the wafers is accomplished by water-cooled quartz lamps, which line the inside of the reactor. This type of heating minimizes thermally induced slip in the wafers. Such systems are also capable of operating in the pressure range of 80-100 torr, for reduced autodoping, as well as at atmospheric pressure. Since the wafers are held in near-vertical positions, they are less susceptible to particulate contamination.

Both types of reactors are capable of providing resistivity and thickness uniformity better than ±5% across the wafer, wafer-to-wafer, and run-to-run. For values of resistivity less than 0.1 Ω-cm, or greater than 1 Ω-cm (and for thicknesses less than 1 μm), the uniformity degrades below the ±5% value. Both types of reactors are limited in productivity, based on maximum load sizes for large diameter wafers. A typical load sizes is 12 - 125 mm wafers, and the duration of a complete deposition process can exceed 60 minutes, resulting in low wafer throughput.

CHARACTERIZATION OF EPITAXIAL FILMS

Characterization of the grown epitaxial film is performed to estimate the epi-dependent properties of the product devices early in the fabrication process sequence. Epitaxial layers are characterized for: a) surface quality; b) crystallographic defects; c) electrical properties (including resistivity, carrier lifetime, and dopant profile); and d) film thickness. Many of the characterization techniques are common to bulk silicon and diffused silicon layers, and are therefore discussed in more detail in other chapters of the book.

Optical Inspection of Epitaxial Film Surfaces

Much can be learned about the quality of the epitaxial layer by inspection (under bright light illumination), using an optical microscope, or a surface scanning laser defect counter. These techniques elucidate the presence of haze, pits, particles, scratches, and spikes. *Bright light illumination* employs the examination of the reflection of a bright near-UV source from the wafer surface of the epitaxial wafers. Any deviation from a smooth, highly reflective surface is clearly observed in the reflected light. *Microscopic evaluation* at low magnifications (75-200X) reveals stacking faults and pyramidal defects. The use of Normarski phase contrast microscopy (see Chap. 17) reveals these defects more readily than standard microscopy, and is thus the preferred technique. *Automated laser scanners* which detect light scattering centers are now commercially available (see Chap. 15) and can be used to identify and quantify defects in epitaxial layers.

Electrical Characterization

The two electrical properties that have the most profound influence on the operation of a device fabricated in an epitaxial layer are the *film resistivity*, and the *minority carrier or MOS*

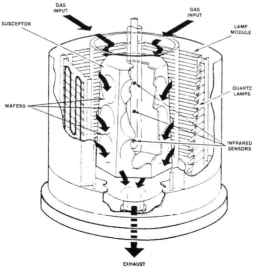

Fig. 22   Schematic of a radiantly heated barrel reactor.   Courtesy of Applied Materials Inc.

lifetimes. The resistivity and doping are related by $\rho = 1/q\,n\,\mu$, where q is the electronic charge, n is the doping concentration (N_D or N_A), and μ is the carrier mobility. The desired film resistivity is achieved by controlling the ratio of dopant gas to active silicon-bearing gas during deposition. Variations in dopant incorporation, autodoping, and depletion effects, however, are sufficient to result in variations from the prescribed resistivity. The doping concentration in the layer is generally determined using a 4-point probe sheet resistivity measurement (Chap. 4), while the doping profile is obtained from spreading resistance probe, capacitance-voltage (C-V), neutron activation analysis, and secondary ion mass spectroscopy, SIMS, techniques. See Chaps. 4, 8, and 17, respectively for more details about these measurement techniques.

When using the 4-point probe method it is desirable that the substrates have an opposite conductivity type than that of the epi layer. Under those conditions the resistivity is given by:

$$\rho = R_s\,h\,F_1 \qquad\qquad (28)$$

where R_s is the measured sheet resistivity, h is the epitaxial thickness, and F_1 is the geometrical correction factor. For thin epitaxial layers, care must be exercised to prevent the probe tips from penetrating the layer and reaching the substrate. The accuracy of this technique is only $\pm 10\%$.

The C-V techniques can be performed by measuring the capacitance as a function of voltage of either a p-n junction or a Schottky barrier. Both techniques are sensitive in the doping range of 10^{14}-10^{16} atoms /cm^3, however, they suffer from limited depth profiling capabilities, and take excessive time to prepare samples. Using a liquid Hg probe to form the Schottky diode with the silicon does offer the advantage of rapid turnaround for specimen preparation.

The spreading resistance technique has applicability over a broad range of dopant concentration (10^{14}-10^{20} atoms /cm^3), but suffers from poor accuracy ($\pm 15\%$), and the requirements of significant data reduction. The technique, however, is quite useful for elucidating the outdiffusion and autodoping that occurs with some types of epitaxial depositions.

The sensitivity of the SIMS technique is relatively low for the common dopants (e.g. 10^{15}, 10^{16}, and 2×10^{17} atoms /cm^3 for B, Sb, and P, respectively), and therefore has limited use for epitaxial layers. It can, however, be useful for studying outdiffusion near the substrates /film interface. It should also be noted that SIMS results give both the active and inactive impurity concentration, and an additional analysis technique must therefore be employed to provide a value for the active dopant concentration.

The *minority carrier lifetime* in the epitaxial layer provides a measure of the heavy metal impurity concentration. The Zerbst[29] method is widely used to measure lifetime. In order to make the Zerbst measurement, the capacitance of an MOS capacitor (which has been fabricated in the epitaxial material), is monitored as a function of time after the device is driven into deep inversion. A plot of [-d $(C_o/C)^2$ /dt] vs. $C_F/(C-1)$ is constructed, and the slope and intercept of its straight line portion is measured. The *slope* is proportional to *lifetime*, t_g, and the *intercept* is proportional to the *surface recombination velocity*, S. The measured lifetime consists of two terms, a *bulk lifetime*, t_b, and a *surface lifetime*, t_s. These terms are related through:

$$(1/t_g) = (1/t_b) + (1/t_s) \qquad\qquad (29)$$

The major failings of this technique are that: a) the measurement is confined to a thin layer near surface; and b) the lifetimes that are measured are far from the equilibrium carrier concentration, since the measurement is made in a space charge region. Photoelectric methods[30], on the other hand, could provide an average value of lifetime over a layer on the order of the carrier diffusion length. This however creates a problem in thin epitaxial layers (since their thickness dimension

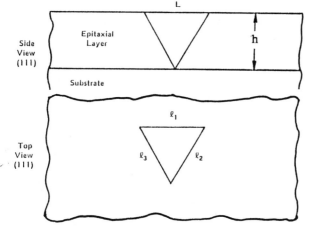

Fig. 23   Geometrical basis of the stacking fault method for determining the thickness of an epitaxial layer for the <111> case.

is less than the diffusion length of a free carrier).   Thus, photoelectric technqiues are not applicable to epitaxial films used in VLSI.

Epitaxial Film Thickness Measurements

The *epitaxial layer thickness* is a critical parameter, and therefore must be accurately measured and controlled. For example, this control is necessary to insure that outdiffusion and autodoping from the buried layer do not consume the epitaxial film. In addition, several of the bipolar transistor device parameters, such as breakdown voltage, junction capacitance, transistor gain, and ac performance, depend on the layer thickness.

Epitaxial thickness measurements fall into two categories: 1) destructive; and 2) nondestructive. Although we shall briefly describe a variety of such techniques, it should be noted that the most widely used of them for VLSI applications are the non-destructive methods based on the optical properties of the epitaxial structure (infra-red reflectance). Angle-lap and stain, groove and stain, and tapered groove and profilometry, are all excellent (albiet destructive) methods for directly measuring the epitaxial thickness. Nevertheless they are directly applicable to epitaxial layers. These techniques are discussed further in Chap. 8.

In order to achieve staining differentation for the destructive methods, it may be necessary to deposit the epitaxial layer on a monitor wafer of the opposite conductivity type. A direct measurement of the layer thickness that has been developed employs the cross-sectional examination of the p /n or n /n$^+$ junctions which have been differentiated by chemical etching and are then observed and measured using scanning electron microscopy (SEM) . Schumann[30] reviewed the applicability and pitfalls of the epitaxial thickness measurement techniques previously described. Each results in a thickness measurement that is peculiar to that technique.

An example of a non-destructive technique is the measurement of the weight gain of the silicon wafer after the epitaxial deposition. Since 1.0 μm of silicon on a 150 mm wafer represents a 30 mg weight gain, this measurement is easily made using modern analytical balances. It must be remembered, however, that this technique provides only an average thickness across the wafer, and also includes the weight of any silicon deposited on the backside.

Capacitance-voltage (C-V) techniques can be used to define an epitaxial thickness based on an arbitrarily defined dopant concentration level. The film thickness measured by C-V relates more closely to an electrical performance thickness rather than a physical thickness.   The technique is slow and not economical, and is therefore rarely used for VLSI applications.

Another nondestructive technique for measurements of the epitaxial layer thickness is based on measuring the length of the side of a stacking fault on the film surface that was nucleated at the film /substrate interface.   In some cases, it may be necessary, but undesirable, to delineate the stacking fault more clearly by etching techniques.   This measurement technique is based on the premise that the stacking fault edge length is related to the epitaxial thickness by the relationship:

$$h = K L \qquad\qquad (30)$$

where L is the length of the fault side (the side of a square for <100>, and the side of an equilateral triangle for <111>), and K is a geometric factor that depends on the crystal orientation. The values of K are 0.707 for <100> and 0.616 for <111> orientations.   Figure 23 shows the geometrical relationship expressed in Eq. 30.

Infrared Reflectance Measurement Techniques

When electromagnetic radiation is incident upon an epitaxial film on a silicon substrate the following sequence of events occur: 1) the radiation is partially reflected from the air /film surface and partially transmitted through the film where it strikes the substrate; 2) the radiation at the substrate is also reflected (or may be partially transmitted for lightly doped or sapphire substrates); 3) the radiation reflected from the substrate combines with the radiation reflected from the air /film surface.   The result of this combination is *inteference*, which amounts to intensity variations of the reflected radiation.   The inteference and the optical patterns that are formed, as a function of the frequency (wavelength), depend on the optical properties of the film and the substrate, as well as the thickness of the film.   As a result of this interference phenomenon, the patterns can be used to measure the film thickness. The lightly doped epitaxial layers typically grown for VLSI applications are relatively transparent to the infrared spectrum in the range of 2.5 to 50 μm.   The heavily doped substrate, however, behaves as a reflective surface for radiation in this range, and thus the IR reflectance technique is applicable for determining epi layer thickness.

ASTM Technique of Epi Layer Thickness Measurement with IR Reflectance
The American Soceiety for Testing Materials (ASTM)[31] has developed a standard technique for determining the epitaxial film thickness which is based on the measurement of IR reflection from a specimen.   The thickness measured by the IR reflectance technique gives a relatively good indication of the added physical thickness, if substrate outdiffusion can be neglected.   This is generally the case when the deposition is performed at temperatures below 1050°C.   For deposition temperatures where outdiffusion is significant (1150°C), an error of 2-3% with respect to the physical thickness (compared against weight gain) is observed.

Fourier-Transform Infrared (FTIR) Spectroscopy
The use of *Fourier-transform infrared (FTIR) spectroscopy* has led to more accurate and rapid measurements of epitaxial layer thickness.   A detailed discussion of the technique is beyond the scope of this book but is well documented in other publications (e.g. a recent book authored by Smith[32]).   Briefly, a Michelson interferometer, which consists of a fixed and a movable mirror, is used to collect information about all wavelengths (frequencies) used in the IR source (Fig. 24).   The incident IR beam is split and one path travels to the substrate, where it is reflected from the substrate and detected by the detector, while the second path is reflected

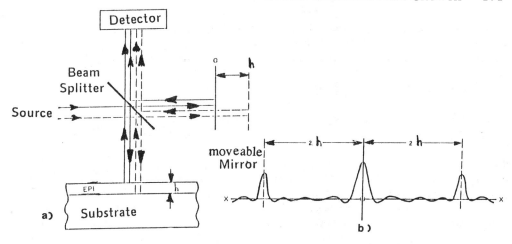

Fig. 24 Schematic representation of interferometric measurement of epitaxial layer thickness. (a) interferometer showing path lengths, (b) interferogram showing a coherence peak and side-bursts.

from the movable mirror and is similarly detected. When the two path lengths of the split beam are equal (solid line in Fig. 24), the beam is nearly coherent, and a large signal is detected. When the movable mirror is at a distance equal to the optical path length of the IR beam in the epitaxial layer, a second coherence is achieved resulting in what is termed *side-bursts*. The resulting spectrum is called an *interferogram* and an example is shown in Fig. 25. Performing a Fourier transform on the interferogram yields a spectrum of the reflected light.

The measurements from commercially available FTIR equipment, however, are empirical in nature, since corrections for phase shift or index of refraction must be entered into the system (dependent on substrate and layer doping).   A new commercially available system correlates the reflection spectrum from the epitaxial specimen with theoretically generated spectra. The maximum correlation coefficient between the theoretical and experimental spectra result in a unique value of epitaxial thickness. This technique is highly accurate and works well in the submicron thickness regime. The FTIR method of measuring epitaxial thickness is rapid (10-30 sec), accurate (± 0.05 μm), and precise (± 0.005 μm), and as such, enjoys wide applicability.

SILICON ON INSULATORS

VLSI devices require reduced parasitic capacitance (a parameter which impacts device performance and interconnect delay), and increased radiation hardness for space applications (particularly single-event and burst-upset resistance). These requirements have renewed interest in the use of *silicon on insulators* (SOI).   In SOI technology, small islands of silicon which contain the individual device are fabricated on an insulating substrate. These islands are then interconnected in the normal way. A significant amount of investigatory work, going back to the 1960's, has been performed on silicon on sapphire (SOS). There are several excellent review articles on this subject and the reader is referred to them for additional details[33,34]. Some of the recent advances in the technology will be discussed here.

In addition to SOS, attempts have been made to grow single-crystal (or large grain polycrystalline) Si on amorphous insulators.   Several techniques have been studied, but have not proven successful for VLSI device fabrication. Reference 35 discusses some of the details of these techniques. On the other hand, recent interest has been focused on a method to achieve device quality Si on an insulating layer. In this technique, either oxygen or nitrogen is deeply implanted (~0.2-0.3 µm) into Si at concentrations sufficient to form buried SiO_2 or Si_3N_4 layers.

Silicon on Sapphire (SOS)

The deposition of single crystal silicon on sapphire (Al_2O_3) is an example of *heteroepitaxy*. As such, the silicon layer must form a crystallographic relationship with the sapphire substrate. Since this technology is limited to fabrication of MOS devices, the silicon <100> orientation is desirable. In order to grow <100> silicon, several sapphire orientations have been successfully used, such as <0112>, <1012>, and <1102>. The lattice parameter mismatch between the layer and the substrate is sufficiently large to result in misfit dislocations, as well as stacking fault and edge dislocation generation, in the layer near the substrate. The defect concentration is found to decrease as the film continues to grow, which allows fabrication of devices in thick films. For VLSI devices, however, epitaxial thicknesses on the order of 0.5 µm are required, resulting in highly faulted silicon in which the device must be fabricated. Another difficulty with the SOS technology is that upon cooling from the deposition temperature, the differing thermal coefficients of expansion (between the silicon and the sapphire) result in stress in the silicon layer. The stress levels are sufficient to cause a strain-induced change in the band structure of the silicon, which results in a hole mobility that is only 80% that of the bulk hole mobility. This reduction in mobility is compounded by the high defect density in the layer, which also tends to decrease both the hole and electron mobilities. Thus MOS devices fabricated in SOS wafers tend to have lower inherent performance than those fabricated in bulk silicon. The minority carrier lifetime in this material is also too low (1-10 ns) for fabrication of bipolar devices, and SOS is thus not used for that purpose. Figure 26 shows the stacking fault density as a function of epitaxial film thickness[36].

To grow an epitaxial film on sapphire, Si is typically deposited by the pyrolysis of silane in a hydrogen carrier, according to:

$$SiH_4 \rightarrow Si + 2H_2 \tag{31}$$

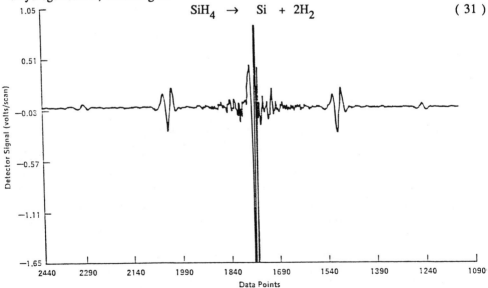

Fig. 25 Typical interferogram obtained for an epitaxial thickness measurement.

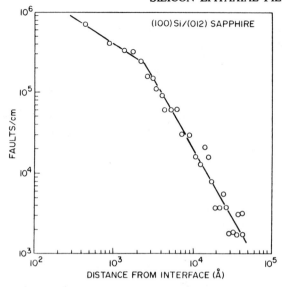

Fig. 26 Stacking fault density as a function of the distance from the silicon /sapphire interface in an MOS structure[36]. Reprinted with the permission of the American Physical Society.

The deposition temperature is generally below 1050°C to minimize autodoping of aluminum from the substrate to the growing film. Since a SiH_4 source is used, the moisture and oxygen levels must be maintained below 1 ppm to produce device quality films.

Recent studies have shown that the *solid phase epitaxial* (SPE) regrowth of defected silicon on sapphire results in much improved material[37]. To achieve SPE, the defected Si near the Si /sapphire interface is amorphized by a high dose silicon ion implantation. The undamaged single crystal Si near the surface, which has a low defect density, is used as seed material for the epitaxial regrowth of the amorphous layer (see Chap. 9). Since the regrowth is primarily along the <100> direction, stacking faults and twins do not propagate, leaving dislocations as the major type of defect in the film. A double SPE process has been described[37], in which the Si region near the interface is first amorphized and the surface regions used as a seed for regrowth. This is then followed by an implantation that amorphizes the surface. The previously recrystallized bottom layer is utilized as the seed to regrow the surface amorphous layer. Devices fabricated in such material show higher surface mobilities and lower junction leakage than either untreated SOS, or material obtained from a single SPE process.

Heteroepitaxial films require additional characterization over that required for homoepitaxial films[40], since there is a large sensitivity of the Si properties to the deposition conditions. As a result of this sensitivity, the ability to achieve reproducibility in SOS films has been difficult. Characterization techniques have been developed specifically for determining the quality of heteroepitaxial Si. These include: a) IR reflectance for determination of crystalline quality of the sapphire substrate; b) UV reflectivity for examination of the Si near the surface; c) x-ray pole figure analysis to characterize the bulk Si quality; and d) photo-voltage techniques for determining electrical properties. These methods are comprehensively reviewed by Cullen, *et al*[40].

Silicon on Other Insulators

Over the past several years, attempts have been made to fabricate single crystal silicon layers on amorphous substrates (such as SiO_2 on Si). In those studies, either amorphous or poly-crystalline Si was chemically deposited on the amorphous substrate, and then recrystallized by various techniques. The intent was to produce recrystallized films of device quality. The advantage of this approach over SOS is that relatively inexpensive and non-brittle silicon wafers serve as substrates, and that automated processing equipment may be used without the modifications that are required for sapphire substrates. The recrystallization can be achieved by melting the deposited silicon using either a laser beam, electron beam, or radiant heat. Although this technique did hold promise, the achievement of high quality devices has not been reported, and the work has been substantially abandoned.

The current approach to producing device quality SOI material employs the buried oxide technology[41-44]. This technology is based on the use of very high dose oxygen ion implantations to produce subsurface layers of silicon dioxide.

Figure 27 shows an example of the formation of a buried oxide substrate. The sequence of events in preparing such a substrate are: a) O_2^+ is implanted at a high dose ($\sim 2 \times 10^{18}$ cm^{-2}) and at sufficient energy (150-300 keV) so that the peak of the implant is deep within the silicon (0.3-0.5 μm); b) a post-implant anneal is conducted in a neutral ambient (N_2) for sufficient time (3-5 hours), and at high enough temperature (1100-1175°C) to restore the crystallinity of the top residual silicon, and to form the buried layer of silicon dioxide; and c) a layer of epitaxial silicon is then deposited to provide ~ 0.5 μm of silicon into which devices are subsequently fabricated. The drawbacks of this technique are: a) a high dose of oxygen is required to form a sharp Si /SiO_2 interface; b) high energy of implantation is required to achieve the buried layer; and c) an epitaxial deposition is necessary to thicken the silicon device layer. Recently, nitrogen implantation[45] has been used to form a buried silicon nitride layer (Si_3N_4) on which devices have been fabricated. Nitrogen layers can be formed at lower doses (40-50% of that

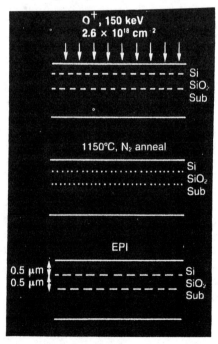

Fig. 27 A typical example of the formation of a buried oxide in a silicon substrate.

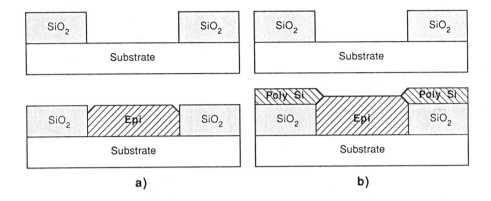

Fig. 28   (a) Type 1, selective epitaxial silicon deposition.   (b) Type 2, selective epitaxial silicon deposition.

required for SiO_2), and at sufficient depth to preclude the need for an epitaxial deposition.

Both MOS and bipolar devices have been fabricated in buried oxide and nitride material, with resulting devices exhibiting excellent properties[41]. Significant efforts continue in this area since encouraging results towards achieving the long time goal (of making silicon devices on insulating substrates) have been achieved.

SELECTIVE DEPOSITION OF EPITAXIAL SILICON

There has recently been a renewed interest (after a 20 year hiatus) in *selective epitaxial growth* of silicon.   This rebirth has been partly due to the development of the capability to etch narrow features of SiO_2 films on the Si substrates (Chap. 16).   If selective epitaxy can be successfully performed, this then allows the formation of closely spaced silicon islands isolated by SiO_2.   The advantages of this technology over other means of isolation include increased density, and reduced processing complexity.

There are basically two types of selective epitaxy.   In *Type 1*, epitaxial growth occurs only on the bare silicon substrates within the openings in the oxide film, and no growth whatsoever occurs on the oxide.   Type 1 selective epitaxy is depicted in Fig. 28a.   In *Type 2*, simultaneous deposition of epitaxial silicon on the single crystal silicon in the oxide openings, and polysilicon deposition is produced on the oxide surface.   Type 2 selective deposition is depicted in Fig. 28b.

Type 1 selective epitaxial deposition is achieved when silicon atoms possessing high surface mobility are deposited from the silicon source.   These atoms are then able to migrate to sites on the single crystal material where nucleation is favored.   It appears that the silicon mobility is enhanced by the presence of halides.   It is found that the higher the number of chlorine atoms in the silicon source, the better the degree of selectively.   Hence, the heirarchy for the degree of selectivity achieved, as a function of silicon source is as follows: $SiCl_4$; $SiHCl_3$; SiH_2Cl_2; and SiH_4.   In fact, SiH_4 shows very little Type 1 selectivity.

Based on the presence of chlorine atoms alone, $SiCl_4$ would be the silicon source of choice for Type 1 selective deposition. The high reduction temperature, however, required for $SiCl_4$ may result in the degradation of the SiO_2 masks. Using Si_3N_4 masks in place of SiO_2 can circumvent this problem, but Si_3N_4 tends to act as a better nucleation surface for silicon deposition than does SiO_2. A more optimum silicon source for Type 1 selective epitaxy is $SiHCl_3$, based on its high chlorine content, and its reduced deposition temperature. Both SiH_2Cl_2 and SiH_4 can be used if hydrogen chloride (HCl) or chlorine (Cl_2) is added to the reaction. The addition of these gases reduces the deposition rate, but has no adverse effects on the electrical or structural properties of the deposit. Bromine based silicon compounds such as $SiHBr_3$ and $SiBr_4$ have shown improved selectivity over their chlorinated counterparts. Selective Si epitaxy from a $SiBr_4$ source, has been accomplished at temperatures as low as 600°C[46].

Several factors are known to enhance the selective nature of the silicon deposition. They include[46]: a) reduced pressure; b) increased temperature; and c) decreased mole fraction of silicon source in the gas stream. It has also been found that large islands of exposed silicon separated by narrow bands of oxide are preferred to other geometrical arrangements for improved selectivity.

The presence of crystallographic defects at the SiO_2/epitaxial interface is common to both Type 1 and Type 2 selective depositions. These defects can affect the performance of devices built into the epitaxial islands. Transmission electron microscopy (TEM) studies of the interface show a band of lamellar defects resembling microtwins. Typically, the band of defects is confined to within 0.1 to 0.3 µm of the interface. Epitaxial stacking faults have also been observed in the interface region. Coating the wall of the oxide step with ~1000Å of polysilicon greatly reduces the crystalline defects in the interface region.

Type 2 selective deposition is best achieved by using SiH_4 as the Si source, since selectivity of deposition is non-critical. When this type of deposition is made, an interface between the epitaxial and polysilicon is formed. The angle of this interface relative to the film growth direction depends upon the crystallographic orientation of the substrate. The angle is observed to be 90° for <110> orientations, 72° toward the polysilicon for <100> orientations, or tapered towards the single crystal silicon at 70° for the <111> orientation. For Type 2 depositions, defected structures are observed at the epitaxial Si /SiO_2 and the epitaxial Si /poly-Si interfaces. Work continues on eliminating interfacial defects for both types of selective depositions.

MOLECULAR BEAM EPITAXY (MBE) OF SILICON

The silicon molecular beam epitaxy (MBE) process involves the evaporation of silicon and the desired dopants under ultrahigh vacuum (UHV) conditions (10^{-11} torr). The evaporant atoms are directed at high velocity to heated substrates, where they condense and grow epitaxially. The epitaxial condition is achieved since the evaporants at the substrate are in a supersaturated state, and as such, readily nucleate and grow. Under conditions of high supersaturation, neither diffusion nor chemical reactions are required to achieve epitaxy. The process is typically carried out in the temperature range of 600–900°C. These temperatures are sufficiently low to eliminate outdiffusion and autodoping. In fact, it has been shown that equally abrupt impurity profiles can be obtained using MBE, whether deposited on heavily-doped arsenic layers or undoped material.

The prototype Si-MBE system was first described by Unvala[47] and Abbink and co-workers[48] in the 1960's, in which they investigated silicon deposition under UHV conditions. The technique was not exploited at that time, since CVD epitaxy could meet production requirements.

Molecular beam epitaxy of silicon, however, has several potential advantages over the chemical deposition techniques, including: a) low growth temperatures which reduces outdiffusion

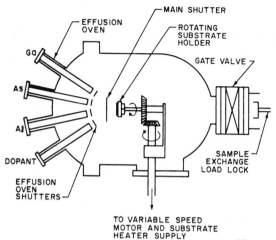

Fig. 29   Schematic of a molecular beam epitaxy growth chamber[39].

and autodoping effects; b) precise control of the layer thickness and doping profile on an atomic layer level; and c) the ability to produce novel structures such as silicon /insulator /metal sandwiches and superlattices.   Excellent reviews on silicon MBE are available[49,50] and the reader is referred to them for additional details on the subject.

Growth of MBE Silicon

In order to achieve MBE growth of silicon, electron beam evaporation from an ultrapure silicon source is required[51]. To achieve this, the center of a silicon source is heated using an electron beam. To minimize radiation damage to the substrates, they must be shielded from ions and backscattered electrons produced by the electron beam. Dopant species are introduced into the MBE film by producing dopant atom beams by evaporation from an effusion cell or by low energy ion implantation. In the evaporation mode, the dopant materials used are gallium (Ga) for p-type doping and antimony (Sb) for n-type doping, because the common dopants (P, As, and B) do not lend themselves to deposition by evaporation.   Silicon can be doped with Sb up to its saturation level of 2×10^{19} atoms /cm^3 at 750°C, while Ga can be incorporated up to values of 2×10^{18} atoms /cm^3. On the other hand, P, As, or B can be incorporated into the film by simultaneously ion implanting, at low energy and current, during MBE growth.

Very abrupt changes in doping profile can be achieved by passing or interrupting the dopant fluxes with shutters positioned above the effusion cells. The sticking coefficient of the dopants, however, are low (~10^{-4}) at the high end of the deposition temperature range, requiring a higher dopant flux than expected. The sticking coefficients do increase with decreasing temperature.

During film growth, atoms from residual gases impinge on the substrate and may be incorporated as contamination. Therefore, background partial pressures of O_2, H_2, and N_2 should be less than 10^{-14} torr. The substrate itself may be a source of contamination, with its chemisorbed oxygen and carbon. The chemisorbed contaminants may be eliminated by an *in-situ* clean using: a) thermal cleaning at elevated temperatures; b) thermal etching with a silicon or gallium beam; c) ion sputtering; or d) laser annealing.

MBE Equipment
The typical, commercially available MBE system is configured with three separate chambers

connected through load locking systems. The load lock is required to insure vacuum integrity, which dominates all other considerations in building such a system. The use of three chambers increases throughput by allowing sample preparation to occur in one chamber, while deposition is taking place in another. The ultra-high vacuum requirements of 10^{-11} torr are achieved through pumping systems that utilize a combination of the following vacuum pumps: a) cryogenic; b) turbomolecular; c) titanium sublimation; and d) ion pumps. The interior surface of the chambers are fabricated from materials such as copper and stainless steel for cold parts (cryogenically cooled), and molybdenum, titanium, boron nitride, and silicon itself for hot parts.

The next most important consideration in designing an MBE system is the accurate control of both the evaporation sources (e-beam evaporator and effusion cells) and the substrate temperature. Finally, the system requires: a) a wafer handling mechanism to move wafers through the three chambers; b) associated analytical equipment; and c) a wafer cleaning and annealing capability.

In a typical MBE processing sequence, multiple wafers (e.g. 10, 3-inch wafers) are loaded into a substrate carrier and placed in the loading chamber. This chamber is evacuated to ~10^{-9} torr. The wafers are then transported to the analysis chamber where their surfaces are cleaned. They are next transported to the growth chamber. (Both the analysis and growth chamber must be maintained at better than 10^{-11} torr.) When the cassette is in the growth chamber, a substrate is unloaded, positioned, and the film grown. After all the wafers have received a deposition, they are brought back to the analysis chamber for evaluation. Figure 29 shows a schematic representation of an MBE growth chamber.

REFERENCES

1. A.S. Grove, "Mass Transfer in Semiconductor Technol.", *Ind. & Eng. Chem.*, **58**, 48 (1966).
2. H. Schlichting, *Boundary Layer Theory*, 4th Ed., McGraw-Hill, NY, 1960, Chap. 7.
3. H. Schlichting, *Boundary Layer Theory*, 4th Ed., McGraw-Hill, NY, 1960, Chap. 14.
4. V.S. Ban, "Mass Spectrometric Studies of Chemical Reactions and Transport Phenomena in Si Epitaxy", *Proc. of the VI International Conf. on Chemical Vapor Deposition*, 1977, L.F. Donaghey, P. Rai-Chaudary and R.Tauber, Eds., Electrochemical Soc., Princeton, 1977, p. 66.
5. C.W. Manke and L.F. Donaghey, "Numerical Simulation of Transport Processes in Vertical Cylinder Epitaxy Reactors", ibid, p. 151.
6. J. Bloem and L.J. Gilling, "Epitaxial Growth by Chemical Vapor Deposition", *VLSI Electronics*, N.G. Einspruch and H. Huff, Eds., Vol. 12, (1985), p. 89, Academic Press, Orlando.
7. J. Bloem, "Nucleation and Growth of Silicon by CVD", *J. Cryst. Growth*, **50**, 581 (1980).
8. V.S. Ban and S.L. Gilbert, *J. Cryst. Growth*, **32**, 284 (1975).
9. V.S. Ban and S.L. Gilbert, *J. Electrochem. Soc.*, **122**, 1382 (1975).
10. V.S. Ban, *J. Electrochem. Soc.*, **122**, 1389 (1975).
11. V.S. Ban, *J. Electrochem. Soc.*, **125**, 317 (1978).
12. J. Nishizawa and M. Saito, "Growth Mechanism of Chemical Vapor Deposition of Silicon", *Proc. of the VIIIth International Conf. on Chemical Vapor Deposition*, 1981, Electrochemical Society, NJ, 1981, p. 317.
13. F.C. Eversteyn, "Chemical Reaction Engineering in the Semiconductor Industry", *Philips Research Rept.*, **19**, 45 (1974).
14. G.R. Srinivasen, "Autodoping Effects in Si Epitaxy", *J. Electrochem. Soc.*, **127**, 1334 (1980).
15. G.R. Srinivasen, "Kinetics of Lateral Autodoping in Silicon Epitaxy", *J. Electrochem. Soc.*, **125**, 146 (1978).
16. V. Silvestri, G.R. Srinivasen, and B. Ginsburg, "Submicron Epitaxial Films", *J. Electrochem. Soc.*, **131**, 877 (1984).

17. R.C. Rossi and K.K. Schuegraf, "Glassy Carbon-Coated Susceptors for Semiconductor CVD Processes", *Semiconductor Int'l.*, **4**, 99 (1981).

18. K.V. Ravi, *Imperfections and Impurities in Semiconductor Silicon*, Wiley, New York, 1981.

19. R.E. Logar and J.O. Borland, "Silicon Epitaxial Processing Techniques for Ultra-Low Defect Densities", *Solid State Tech.*, p. 133, June '85.

20. J.O. Borland, *et al.*, "Influence of Epi-Substrate Point Defect Properties on Getter Enhanced Si Epitaxial Processing for Advanced CMOS and Bipolar Technologies", *VLSI Science and Technology/ 1984*, K.E. Bean and G. Rozgonyi, Eds., Electrochem. Society, NJ (1984), p. 93.

21. W. Dyson, *et al.*, "N$^+$ and P$^+$ Substrate Effects on Epitaxial Silicon Properties", ibid, p. 107.

22. S.M. Ali, *et al.*, "Extrinsic Gettering with Epitaxial Misfit Dislocations", *Semiconductor Processing*, ASTM, STP 850, Gupta (1984).

23. K.E. Bean, W.R. Runyan, and R.G. Massey, "Silicon Epitaxial Growth: A Critical Technology", *Semiconductor Int'l.*, p. 136 (May 1985).

24. S.M. Ali, *et al.*, "Extrinsic Gettering Via the Controlled Introduction of Misfit Dislocations", *Appl. Phys. Lett.*, Feb. '85.

25. H.R. Chang, "Defect Control for Silicon Epitaxial Process Using Silane, Dichlorosilane, and Silicon Tetrachloride", *Proc. of the Symposium on Defects in Silicon*, W.M. Bullis and L.C. Kimerling, Eds., Electrochem. Soc., NJ, pp. 549-557 (1983).

26. S.B. Kulkarni, "Defect Reduction by Dichlorosilane Epitaxial Growth", *Proc. of the Symposium on Defects in Si*, W.M. Bullis, and L.C. Kimerling, Eds, Electrochem. Soc., p. 558-567 (1983).

27. P.H. Lee, M.T. Wauk, R. Rosler, and W. Benzing, "Epitaxial Pattern Shift Comparisons in Vertical, Horizontal, & Cylindrical Reactor Geometries", *J. Electrochem. Soc.*, **124**, 1824 (1977).

28. S.P. Weeks, "Pattern Shift and Pattern Distortion during CVD Epitaxy on <111> and <100> Silicon", *Solid State Technol.*, **24**, 111 (1981).

29. D.K. Schroder and H.C. Nathanson, *Solid State Electronics*, **13**, 577 (1970).

30. W.E. Phillips, *Solid State Elecronics*, **15**, 1097 (1972).

31. "Thickness of Epitaxial Layers of Silicon on Substrates of Same Type by Infrared Reflectance", 1984 Annual Book of ASTM Standards, Vol. 10.05, p. 213, ASTM, Philadelphia (1984).

32. A.L. Smith, "Applied Infrared Spectroscopy", J. Wiley, New York, 1979.

33. G.W. Cullen and C.C. Wang, Eds., *Heteroepitaxial Semiconductors for Electronic Devices,* Springer-Verlag, 1978.

34. G.W. Cullen, *et al, J. Crystal Growth*, **56**, 287 (1982).

35. H.W. Lam, A.F. Tasch, Jr., and R.F. Pinzzotto, "Silicon-on-Insulator for VLSI and VHSIC", in *VLSI Electronics*, Ed., N.G. Einspruch, Vol. 4, p. 1, Academic Press, NY (1982).

36. M.S. Abrahams and C.J. Buiocchi, "Cross Sectional Electron Microscopy of Silicon on Sapphire", *Appl. Phys. Lett.*, **27**, 325 (1975).

37. A. Gupta and P.K. Vasudev, "Recent Advances in Hetero-Epitaxial Silicon-on-Insulator Technology, Part II", *Solid State Technology*, **26**, 129 (1983).

38. E. Sirtl, L.P. Hunt, and D.H. Sawyer, "High Temperature Reactions in the Silicon-Hydrogen-Chlorine System", *J. Electrochem. Soc.*, **121**, 919, (1974).

39. A.Y. Cho, *Thin Solid Films*, **100**, 291, (1983).

40. G.W. Cullen, M.T. Duffy, and R.K. Smeltzer, "Recent Advances in Heteroepitaxial Si on Sapphire Technology", *VLSI Science and Technology/1984*, Electrochem. Society, NJ, 1984, p. 230.

41. C.E. Chen, *et al., 1984 IEDM Tech. Digest*, 701 (1984).

42. D.J. Forster, A.L. Bulter, and P.H. Bolbot, *1984 IEDM Tech. Digest*, 704 (1984).

43. C.G. Tuppen, M. Taylor, P.L. Hemment, and R. Arrowsmith, *Appl. Phys. Lett.*, **45**, 57 (1984).

44. R.F. Pinizzoto, *J. Cryst. Growth*, **63**, 559 (1983).

45. L. Nesbit, S. Stiffler, G. Slusser, and H. Vinton, "Formation of Silicon-on-Insulator Structure

by Implanted Nitrogen", *J. Electrochem. Soc.*, **132**, 2713 (1985).

46. J. Bosch, "Epitaxial Process is Highly Selective in Depositing Silicon", *Electronics International*, pp. 59-60, Jan. 31, 1980.

47. B.A. Unvala, *Vide*, **104**, 109 (1963).

48. H.C. Abbink, R.M. Broudy, and G.P. McCarthey, *J. Appl. Phys.*, **39**, 4673 (1968).

49. Y. Ota, *Thin Solid Films*, **106**, 1 (1983).

50. J.C. Bean and S.R. McAfee, *J. De Physique*, **43**, C5-153 (1982).

51. E. Kasper and K. Worner, "Application of Si-MBE for Integrated Circuits", in *VLSI Science and Technology/1984*, K.E. Bean and G.A. Rozgonyi, Eds, Electrochem. Society, NJ,1984, p. 429.

52. H.C Theuerer, *J. Electrochem. Soc.*, **108**, 649, (1961).

53. W.H. Shepherd, *J..Electrochemical Society*, **112**, 988 (1965).

PROBLEMS

1. Why is epitaxial deposition done in cold-wall reactors (see p. 167 for definition of this term)?

2. What is the minimum percentage of HCl that could be used at 1250°C without causing etching of the silicon substrate?

3. Find the mole fraction of $SiCl_4$ that will result in the maximum film growth rate. Why is epitaxial deposition not carried under these conditions?

4. Determine the time necessary to grow a monolayer of Si epitaxially under the conditions shown in Fig. 3 of this chapter, at 1200°C.

5. Calculate the activation energies from the Arrhenius plots of growth rate versus temperature.

6. Explain why a highly abrupt junction is difficult to achieve using epitaxial deposition.

7. Explain how mask compensation is implemented when laying out a pattern to be transferred to an epitaxial wafer.

8. Sketch a stacking fault in an epitaxial layer on a {100} surface. The fault propgates from layer to layer, retaining its shape and increasing in size with each successive atomic plane. Show the planes along which this fault will propagate. Find the relationship between the height of the stacking fault and its sides.

9. Determine the growth rate of an epitaxial film grown from a $SiCl_4$ source at 1200°C. The gas phase mass transfer coefficient of the reactor is $h_g = 5$ cm /sec, the surface reaction coefficient is $k_s = 10^7$ exp (-1.9 eV /kT) cm /sec, and $C_g = 5 \times 10^{16}$ cm^{-3}. By how much will the growth rate change if the reaction temperature is decreased by 2°C?

10. Calculate the thickness for <111> silicon of an epitaxial layer in which the etch pits have a dimension of 1.838 µm.

6

CHEMICAL VAPOR DEPOSITION of

AMORPHOUS and POLYCRYSTALLINE

THIN FILMS

Chemical vapor deposition (CVD) is defined as the formation of a non-volatile solid film on a substrate by the reaction of vapor phase chemicals (reactants) that contain the required constituents. The reactant gases are introduced into a reaction chamber and are decomposed and reacted at a heated surface to form the thin film. A wide variety of thin films utilized in VLSI fabrication is prepared by CVD[1,2]. In Chap. 5, the growth of single-crystal silicon films by CVD epitaxial techniques is presented. The deposition of amorphous and polycrystalline thin films by CVD is the subject of this chapter (and to a lesser degree, of Chap. 11).

The deposition technology and equipment used to prepare such films by CVD will first be considered. Next a discussion of the properties and deposition conditions of some of the most widely used films deposited by CVD will be undertaken (including polycrystalline silicon, silicon dioxide, and silicon nitride). Information about the properties and deposition of refractory metals and their silicides by CVD is given in Chap. 11. Also note that the measurement of many of the properties of CVD films is virtually identical to the measurement of the same properties in other thin films. Thus, only the measurement of those properties that are unique to the specific thin films under discussion will be mentioned in this chapter. Readers should consult the index for information on various generic thin film measurement methods.

As discussed in Chap. 4, thin films are used in a host of different applications in VLSI fabrication, and can be prepared using a variety of techniques. Regardless of the method by which they are formed, however, the process must be economical (high throughput), and the resultant films must exhibit the following characteristics: a) good thickness uniformity; b) high purity and density; c) controlled composition and stoichiometries; d) high degree of structural perfection; e) good electrical properties; f) excellent adhesion; and g) good step coverage.

Specific deposition methods are developed to form such thin films, based on their potential capabilities for satisfying these demanding criteria. CVD processes are often selected over competing deposition techniques because they offer the following advantages: a) high purity deposits can be achieved; b) a great variety of chemical compositions can be deposited; c) some films cannot be deposited with adequate film properties by any other method; and d) good economy and process control are possible for many films.

This chapter was written by Andrew R. Coulson *and* Richard N. Tauber   **161**

BASIC ASPECTS OF CHEMICAL VAPOR DEPOSITION

A CVD process can be summarized as consisting of the following sequence of steps: a) a given composition (and flow rate) of reactant gases and diluent inert gases is introduced into a reaction chamber; b) the gas species move to the substrate; c) the reactants are adsorbed on the substrate; d) the adatoms undergo migration and film-forming chemical reactions, and e) the gaseous by-products of the reaction are desorbed and removed from the reaction chamber. Energy to drive the reactions can be supplied by several methods (e.g. thermal, photons, or electrons), with thermal energy being the most commonly used.

In practice, the chemical reactions of the reactant gases leading to the formation of a solid material may take place not only on (or very close to) the wafer surface (*heterogeneous reaction*), but also in the gas phase (*homogeneous reaction*). Heterogeneous reactions are much more desirable, as such reactions occur selectively only on heated surfaces, and produce good quality films. Homogeneous reactions, on the other hand are undesirable, as they form *gas phase clusters* of the depositing material, which can result in poorly adhering, low density films, or defects in the depositing film. In addition, such reactions also consume reactants and can cause decreases in deposition rates. Thus, one important characteristic of a chemical reaction for CVD application is the degree to which heterogeneous reactions are favored over gas phase reactions.

Since the aforementioned steps of a CVD process are sequential, the one which occurs at the slowest rate will determine the rate of deposition. The steps can be grouped into 1) gas-phase processes, and 2) surface processes. As discussed in Chap. 5, the *gas phase phenomenon of interest* is the rate at which gases impinge on the substrate. This is modeled by the rate at which gases cross the boundary layer that separates the bulk regions of flowing gas and substrate surface. Such transport processes occur by gas-phase diffusion, which is proportional to the diffusivity of the gas, D, and concentration gradient across the boundary layer, dC /dx, see Chap. 5). The rate of mass transport is only relatively weakly influenced by temperature (D α $T^{1.5-2.0}$.

Several *surface processes* can be important once the gases arrive at the hot substrate surface, but the *surface reaction,* in general, can be modeled by a thermally activated phenomenon which proceeds at a rate, R, given by:

$$R = R_o \, e^{\,[\,-E_a\,/kT\,]}$$

(1)

where R_o is the frequency factor, E_a is the activation energy in eV, and T is temperature in °K (see Appendix 3). According to Eq. 1, the surface reaction rate increases with increasing temperature. For a given surface reaction, the temperature may rise high enough so that the reaction rate exceeds the rate at which reactant species arrive at the surface. In such cases, the reaction cannot proceed any more rapidly than the rate at which reactant gases are supplied to the substrate by mass transport, no matter how high the temperature is increased. This situation is referred to as a *mass-transport limited* deposition process.

On the other hand, at lower temperatures, the surface reaction rate is reduced, and eventually the arrival rate of reactants exceeds the rate at which they are consumed by the surface reaction process. Under such conditions the deposition rate is *reaction rate limited*. Thus, at high temperatures, the deposition is usually mass-transport limited, while at lower temperatures it is surface-reaction rate-limited (Fig. 1). In actual processes the temperature at which the deposition condition moves from one of these growth regimes to the other is dependent on the activation energy of the reaction, and the gas flow conditions in the reactor.

In processes that are run under *reaction rate-limited conditions*, the temperature of the

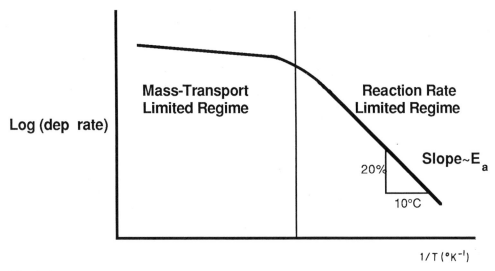

Fig. 1 Temperature dependence of growth rate for CVD films.

process is an important parameter. That is, uniform deposition rates throughout a reactor require conditions that maintain a constant reaction rate. This in turn implies that a constant temperature must also exist everywhere at all wafer surfaces. On the other hand, under such conditions the rate at which reactant species arrive at the surface is not as important, since their concentration does not limit the growth rate. Thus, it is not as critical that a reactor be designed to supply an equal flux of reactants to all locations of a wafer surface. It will be seen that in low-pressure CVD (LPCVD) reactors, wafers can be stacked vertically and at very close spacing because such systems operate in a reaction rate limited mode. The reason for this is as follows:

Under the low pressure of an LPCVD reactor (~1 torr) the diffusivity of the gas species is increased by a factor of 1000 over that at atmospheric pressure, and this is only partially offset by the fact that the boundary layer (the distance across which the reactants must diffuse) increases by less than the square root of the pressure. The net effect is that there is more than an order of magnitude increase in the transport of reactants to (and by-products away from) the substrate surface, and the rate-limiting step is thus the surface reaction[4]. Note, however, that since the surface reaction rate also generally increases with increasing surface concentration, non-uniform gas phase concentrations produced by local depletion of reactants within the reactor can result in deposition non-uniformities. An example of a such an effect is the depletion of reactants from a gas by their deposition on wafers located at the inlet of an end-feed reactor tube. Wafers near the outlet end are consequently exposed to gases containing lower concentrations of reactants than those at the inlet end of the tube.

In deposition processes that are *mass-transport limited*, however, the temperature control is not nearly as critical. As mentioned earlier, the mass transport process which limits the growth rate is only weakly dependent on temperature. On the other hand, it is very important that the same concentration of reactants be present in the bulk gas regions adjacent to all locations of a wafer, as the arrival rate is directly proportional to the concentration in the bulk gas. Thus, to insure films of uniform thickness across a wafer, reactors which are operated in the mass transport limited regime must be designed so that all locations of wafer surfaces and all wafers in a run are supplied with an equal flux of reactant species. Atmospheric pressure reactors that deposit SiO_2 at ~400°C operate in the mass-transport limited regime. The most widely used

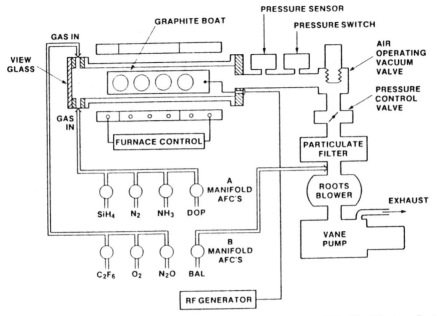

Fig. 2 Diagram of a typical commercial PECVD system. Courtesy of Pacific Western Systems.

APCVD reactor designs provide a uniform supply of reactants by horizontally positioning the wafers and moving them under a gas stream.

CHEMICAL VAPOR DEPOSITION SYSTEMS

Components of CVD Systems

A schematic diagram of a CVD system is shown in Fig. 2.   Actually, this figure shows a plasma-enhanced [PECVD] system.   Non-plasma enhanced CVD systems can be identically represented, minus the rf generator and matching network, and pump [in atmospheric systems]. CVD reactors are generally *open-flow* systems, in which gases continuously flow into the reaction chamber (where the deposition occurs), and gaseous by-products are exhausted together with unused reactant and diluent gases. Reactant gases are typically carried by diluent (*carrier*) gases such as H_2, N_2, or Ar.   Corrosive and hazardous gases pumped from the chamber are re-moved from the exhaust gas flow by a *scrubber*, and the remainder are vented to the atmosphere.

CVD systems usually contain the following components: a) gas sources; b) gas feed lines; c) mass-flow controllers (for metering the gases into the system); d) a reaction chamber (or reactor); e) a method for heating the wafers onto which the film is to be deposited (and in some types of systems, for adding additional energy by other means); and f) temperature sensors. LPCVD and PECVD systems (to be defined) also contain pumps for establishing the reduced pressure and exhausting the gases from the chamber.   With the exception of the reaction chamber configurations and mass-flow controllers, a discussion of the operating procedures of all of the components of CVD systems is undertaken in other chapters of the book since they find use in various wafer processing systems besides CVD equipment . As a result, only mass-flow controller operation is described in this section, and reactor configurations are covered in later sections of this chapter.

Mass-flow Controllers

CVD reactors and other process systems (e.g. dry-etchers, diffusion furnaces), require that the rates of introduction of process gases into the process chambers be controlled. In some applications, this is achieved by adjusting the gas influx to maintain a constant chamber pressure. More commonly, the process gas flow is directly controlled. In the latter case, an instrument known as a *mass-flow controller* performs this function. Mass flow controllers consist of a mass-flowmeter, a controller, and a valve. They are located between the gas source and the chamber, where they can monitor and dispense the gases at predetermined rates.

Gas flow is measured in units of volume /unit-time, where the volume measurement assumes standard temperature and pressure. A flow of one *standard cubic centimeter per minute* (sccm) is thus defined as a flux of one cm^3 of gas per minute at 273 °K and 760 torr. Since one mole of gas at STP occupies 22.414 liters, one sccm = (1 /22,414) moles /min.

The heart of the mass-flow controller is its mass flowmeter, of which there are two types, the *differential pressure type*, and the *thermal type*. The differential type relates a pressure drop at a flow restriction to mass rate of flow. In semiconductor applications, however, the thermal type is more widely used, as a result of to its relatively rapid response time (e.g. 2-3 sec), and its low cost. We describe its operation in more detail.

The operation of thermal mass flowmeters relies on the ability of a flowing gas to transfer heat. A schematic of a mass-flow controller is shown in Fig. 3. The mass flowmeter consists of a small sensor tube in parallel with the larger main gas flow tube. A heating coil is wrapped around the sensor tube midway along its length, and temperature sensors (typically resistance thermometers) are located both upstream and downstream of the heated point. When gas is not flowing, and the heat input is constant, the temperatures at both sensors are equal. Flowing gas causes the temperature distribution in the sensor tube to change. The temperature downstream of the heater becomes higher than the temperature upstream since the flowing gas conducts heat away from the heated point. It can be shown that the mass flow, m_f, is given by:

$$m_f = (\kappa\, W_h\, \Delta T)^{1.25} \qquad (1)$$

where W_h is the heater power, ΔT is the temperature difference between the points where the sensors are located ($\cong 1°C$), and κ is a constant that depends on heat transfer coefficients, the specific heat of the gas, the density of the gas, and the thermal conductivity of the gas. It is assumed that these characteristics remain constant over the flow range measured, and hence mass rate can obtained by measuring the temperature difference. Each of the resistance thermometers are connected to one arm of an unbalanced Wheatstone bridge, and the temperature difference is converted to a voltage signal. Calibration factors are used so that the voltage output derived from the sensor tube can express the flow (in units of sccm) through the entire flow controller. As the flow rate is determined, its value is compared by the controller to the setpoint value. If necessary, the controller adjusts the valve setting so that the flow returns to the specified value.

During the startup of a process from a condition of zero flow, the feedback and control loop operation is hindered by the slow response of the sensor (i.e. it takes time for the temperature sensors to heat). As a result, the initial gas flow may initially overshoot the setpoint value. Two techniques have been devised to reduce this problem. They are: 1) *setpoint ramping*, by which the setpoint value is slowly ramped up from zero to the final value during startup; and 2) use of *mass flow sensors* with much faster response times. Microscopic mass flow sensors micromachined on silicon substrates have been recently been developed for this application[3].

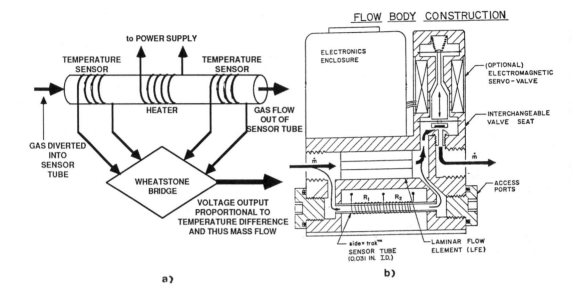

Fig. 3 (a) Operating principle, and (b) cutaway drawing of a mass flow controller. Courtesy of Sierra Instruments.

Due to their extremely small thermal mass, such sensors are capable of responding to heat flux differences (and hence to gas flow rate changes) in times on the order of milliseconds.

Some other requirements of mass flow controllers include: a) that they be immune to stray rf fields (in plasma equipment, for example); b) that they be extremely leak-tight, to prevent leakage of hazardous process gases; c) that they be resistant to damage by corrosive process gases; and d) that they be easy to clean and maintain.

Terminology of CVD Reactor Design

The design and operation of CVD reactors depends on a variety of factors, and hence they can be categorized in several ways. Figure 4 illustrates one way of grouping CVD reactor types. The first distinction between reactor types is whether they are *hot-wall* or *cold-wall* reactors, and this is dependent on the method used to heat the wafers. The next criterion used to distinguish reactor types is their pressure regime of operation (*atmospheric pressure* or *reduced pressure* reactors). Finally, the reduced pressure group is split into: a) low-pressure reactors (the so called *low-pressure CVD*, or *LPCVD* reactors, in which the energy input is entirely thermal); and b) those in which energy is partially supplied by a plasma as well (known as *plasma-enhanced CVD*, or *PECVD reactors*). Each of the reactor types in the two pressure regimes are further divided into sub-groups, defined by reactor configuration and method of heating.

Four methods of heating the wafers and their holder (e.g. a susceptor) have been adopted. They are: a) resistance heating; b) rf induction heating; c) energy from a glow discharge (plasma), and d) photons. Energy may be transferred either to the reactant gases themselves or to the substrate. When radiant heating from resistance-heated coils surrounding the reaction tube is utilized, not only the wafers but also the reaction chamber walls become hot, and hence such designs are known as *hot-wall* reactors. In these systems, film forming reactions (and as a result, film deposition) occurs on reaction chamber walls as well as on substrates. This implies that such systems will require frequent cleaning.

On the other hand, energy input via infrared lamps mounted within the reactor or rf induction, primarily heats the wafers and susceptor, and does not cause appreciable heating of the reaction chamber walls. Consequently systems heated by these methods are termed *cold-wall* reactors. In some cold-wall systems, however, significant chamber wall heating can still take place, and provisions for cooling the walls must be implemented (e.g. by water cooling) to prevent reactions or depositions from occurring there.

Reactor geometry is constrained by the pressure regime and energy source used, and is an important factor in throughput. Since atmospheric pressure reactors by and large operate in the mass-transport limited regime, they must be designed so that an equal flux of reactants is delivered to each wafer. As a result wafers are never stacked vertically at close spacing, but are rather laid flat on a horizontal surface. An undesirable consequence of this design is the high susceptibility of the wafers to incorporate falling particles. LPCVD reactors are not constrained by the mass-transfer rate limitation, allowing designs that accomodate a large number of wafers per run. That is, wafers can be stacked side by side, only a few mm apart in a quartz reaction tube. Quartz wafer holders (boats) can hold up to 200 wafers. Since LPCVD reactors operate in the surface reaction limited mode, they must, however, be capable of precise temperature control. Table 1 summarizes the characteristics and applications of the various CVD reactor designs.

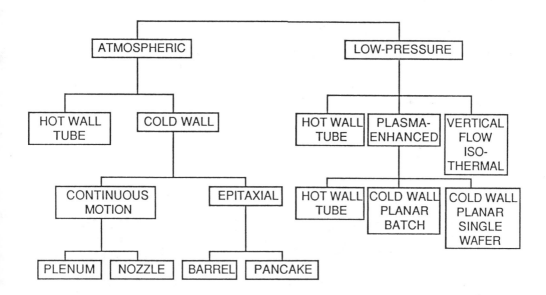

Fig. 4 CVD reactor types.

Table 1. CHARACTERISTICS and APPLICATIONS OF CVD REACTORS

PROCESS	ADVANTAGES	DISADVANTAGES	APPLICATIONS
APCVD (Low Temperature)	Simple Reactor, Fast Deposition, Low Temperature	Poor Step Coverage, Particle Contamination	Low Temperature Oxides, both doped and undoped
LPCVD	Excellent Purity and Uniformity, Conformal Step Coverage, Large Wafer Capacity	High Temperature Low Deposition Rate	High Temperature Oxides, both doped and undoped, Silicon Nitride, Poly-Si, W, WSi_2
PECVD	Low Temperature, Fast Deposition, Good Step Coverage	Chemical (e.g. H_2) and Particulate Contamination	Low Temperature Insulators over Metals, Passivation (Nitride)

Atmospheric Pressure CVD Reactors

Atmospheric pressure CVD reactors (APCVD) were the first to be used by the microelectronics industry[4,5]. Operation at atmospheric pressures kept reactor design simple and allowed high film-deposition rates. APCVD, however, is susceptible to gas phase reactions, and the films typically exhibit poor step coverage. Since APCVD processes are also generally conducted in the mass-transport limited regime, the reactant flux to all parts of every substrate in the reactor must be precisely controlled. This places constraints on reactor geometry and gas flow patterns. Nevertheless, APCVD processes for deposition of oxides, nitrides and polysilicon were developed, only to be supplanted by superior LPCVD processes (see below). Currently, atmospheric reactors are primarily used for low-temperature oxide (LTO) deposition and epitaxy.

Figure 5 shows schematics of the two types of atmospheric systems. The first is the *horizontal tube type* (Fig. 5a). Such systems consist of a horizontal quartz tube, with the wafers

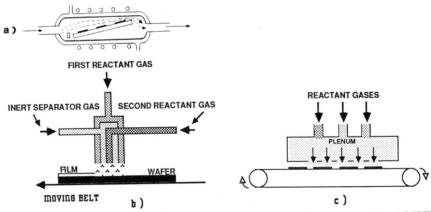

Fig. 5 (a) Horizontal tube APCVD reactor. (b) Gas injector type continuous processing APCVD reactors. (c) Plenum-type continuous processing APCVD reactor.

lying flat on a fixed horizontal plate, while gas flows parallel to the wafer surface. Reactant gases are metered into one end of the tube, and unused or by-product gases exhausted out through the other. Energy is supplied by radiant heating from resistance-heated coils that surround the tube (hot wall). It is possible to deposit polysilicon and SiO_2 in these systems, but as the systems suffer from low throughput, poor uniformity and particulate contamination, they find little use in VLSI fabrication.

The second AP type is the *continuous processing APCVD reactor* (Fig. 5b and c). This configuration is the the most widely used design for depositing low temperature CVD SiO_2 films in production applications. Continuous processing APCVD reactors move the wafers at constant speed through a heated reactor(s), either on a moving plate or continuous conveyor belt. The deposition region is carefully isolated from outside air by curtains of flowing inert gas. Reactant gases may either: a) be mixed in a *plenum* (i.e. from the Latin, full), which is a confined volume filled with flowing process gases at a positive pressure (i.e. slightly greater than atmospheric pressure); or b) be injected from cooled, nitrogen shrouded nozzles, so that mixing occurs only millimeters above the wafer surface (thus reducing gas phase reactions).

Film deposition also takes place on the plate or belt, as well as on the wafers, so frequent cleaning of the system is necessary. At least one commercial system offers an *in situ* HF vapor clean, which allows the continuous belt to be cleaned without disassembly. Modern continuous belt reactors are capable of the thickness uniformity, low contamination, and high throughputs required of VLSI CVD SiO_2 films.

Low-Pressure Chemical Vapor Deposition Reactors

Low pressure chemical vapor deposition (LPCVD) in some cases is able to overcome the uniformity, step coverage, and particulate contamination limitations of early APCVD systems[6,7,8]. By operating at medium vacuum (30-250 Pa or 0.25-2.0 torr), and higher temperatures (550-600°C), LPCVD reactors typically deposit films in the reaction rate limited regime. As described earlier, at reduced pressure the diffusivity of the reactant gas molecules is sufficiently increased so that mass-transfer to the substrate no longer limits the growth rate. The surface reaction rate is very sensitive to temperature (Eq. 1), but precise temperature control is relatively easy to achieve. The elimination of mass-transfer constraints on reactor design allows the reactor to be optimized for high wafer capacity. Low pressure operation also decreases gas phase reactions, making LPCVD films less subject to particulate contamination. LPCVD is used for depositing many types of films, including poly-Si, Si_3N_4, SiO_2, PSG, BPSG, and W.

The two main disadvantages of LPCVD are the relatively low deposition rates, and the relatively high operating temperatures. Attempting to increase deposition rates by increasing the reactant partial pressures, tends to initiate gas phase reactions. Attempting to operate at lower temperatures, results in unacceptably slow film deposition. LPCVD reactors are designed in two primary configurations: a) horizontal tube reactors; and b) vertical flow isothermal reactors.

Horizontal Tube LPCVD Reactors (Hot Wall)

Horizontal tube, hot wall reactors are the most widely used LPCVD reactors in VLSI processing. They are employed for depositing poly-Si, silicon nitride, and undoped and doped SiO_2 films. They find such broad applicability primarily because of their superior economy, throughput, uniformity, and ability to accommodate large diameter (e.g. 150 mm) wafers. Their main disadvantages are susceptibility to particulate contamination[13] and low deposition rates.

Conventional LPCVD tube reactors are very similar to APCVD tube reactors, in that the wafers are radiantly heated by resistive heating coils surrounding the tube. Reactant gases are

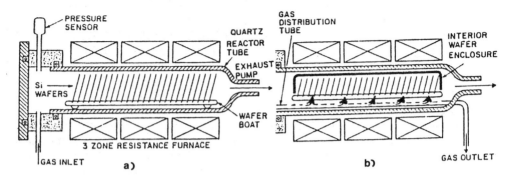

Fig. 6 (a) End feed LPCVD reactor. (b) Distributed-feed LPCVD reactor[2]. Reprinted with permission of Semiconductor International.

also metered into one end of a horizontal quartz tube (using mass flow controllers), and reaction by products are pumped out the other. Vacuum pumps are used to establish the required reduced pressure, typically 0.25-2.0 torr, in the tube. An oil-sealed rotary mechanical pump is generally utilized, often augmented with a Roots pump. A pressure controller[9], with feedback control of the gas flow or the pump inlet valve may be also be included.

The fact that LPCVD processes operate in a reaction rate limited regime allows wafers to be arranged differently inside the tube than in APCVD tube reactors (since equal mass transport to all parts of every wafer is no longer critical). That is, wafers are positioned vertically (Fig. 6) at very close spacing (e.g. 5-6 mm), perpendicular to the gas flow, in quartz *boats* (in a manner similar to that used in diffusion furnaces, as described in Chap. 7). Up to 200 wafers can be processed per run. The large number of wafers processed in a single process cycle allows greater throughput than with horizontal-tube APCVD reactors, in spite of the relatively low deposition rates of LPCVD processes (100-500Å /min).

Depletion effects which reduce gas phase concentrations as reactants are consumed by reactions on wafer surfaces, are still operative in such end-fed reactors. That is, wafers near the inlet are exposed to higher concentrations of reactant gases. As a result, deposition rates are greater on wafers placed near the inlet end of the tube. Since the reaction rates increase with increasing temperature, these deposition rate variations can be minimized by linearly increasing temperature along the length of the tube (e.g. by 25-40°C). Such temperature ramping compensates for the reactant gas depletion, and better thickness uniformity is obtained (Fig. 6a). Nonuniform temperatures along the tube length, however, will affect deposition temperature-dependent film properties. This is a particular problem for poly-Si and low-temperature (~400°C) deposition of doped SiO_2. Hence, conventional end-feed LPCVD reactors are used for high temperature deposition (800-850°C) of silicon nitrides and oxynitrides, and for undoped poly-Si films (~600°C), in which the temperature differential does not significantly impact film properties.

Another technique to alleviate the depletion effect, known as *distributed feed,* has been developed for hot-wall tube reactors. In this approach, special injection systems supply fresh reactants at a number of points along the length of the tube (Fig. 6b). Although this requires the use of specially designed quartzware, it allows the temperature along the tube to remain constant. Distributed feed LPCVD hot wall reactors are used for low-temperature (<600°C) deposition of SiO_2, PSG, and BPSG films, and *in situ* doped poly-Si films.

Vertical Flow Isothermal LPCVD Reactors (Cold-Wall)

The vertical flow isothermal LPCVD reactor further extends the distributed gas feed technique, so that each wafer receives an identical supply of fresh reactants[10]. Wafers are again stacked side by side, but are placed in perforated-quartz cages. The cages are positioned beneath long, perforated, quartz reaction-gas injector tubes (one tube for each reactant gas). Gas flows vertically from the injector tubes, through the cage perforations, past the wafers (parallel to the wafer surface, Fig. 7) and into exhaust slots below the cage. The size, number, and location of cage perforations are used to control the flow of reactant gases to the wafer surfaces. By properly optimizing cage perforation design, each wafer can be supplied with identical quantities of fresh reactants from the vertically adjacent injector tubes. Thus, this design can avoid the wafer-to-wafer reactant depletion effects of the end-feed tube reactors, requires no temperature ramping, produces highly uniform depositions, and reportedly achieves low particulate contamination.

Plasma Enhanced CVD: Physics, Chemistry, & Reactor Designs

The third and last of the major CVD deposition methods is categorized not only by pressure regime, but also by its method of energy input. Rather than relying solely on thermal energy to initiate and sustain chemical reactions, plasma enhanced CVD (or PECVD) uses an rf-induced glow discharge to transfer energy into the reactant gases, allowing the substrate to remain at a lower temperature than in APCVD or LPCVD processes. Lower substrate temperature is the major advantage of PECVD, and in fact, PECVD provides a method of depositing films on substrates that do not have the thermal stability to accept coating by other methods (the most important being the formation of silicon nitride and SiO_2 over metals). In addition, PECVD can enhance the deposition rate when compared to thermal reactions alone, and produce films of unique compositions and properties. Desirable properties such as good adhesion, low pinhole density, good step coverage, adequate electrical properties, and compatibility with fine-line pattern transfer processes, have led to application of these films in VLSI.

The plasma (more correctly a *glow discharge*) is generated by the application an rf field to a low pressure gas, thereby creating free electrons within the discharge region. There is a thorough discussion of glow discharge physics in Chap. 10. The electrons gain sufficient energy from

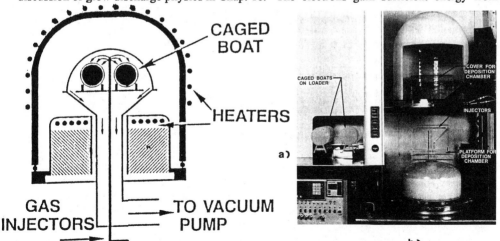

Fig. 7 Vertical isothermal LPCVD reactor. (a) Schematic drawing. (b) Photograph of system. Courtesy of Anicon, Inc.

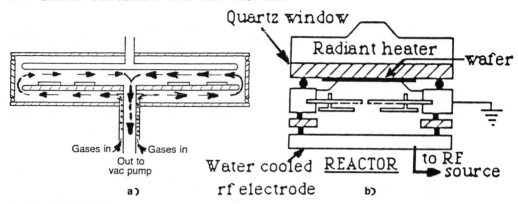

Fig. 8 (a) Radial flow reactor's inward flow (Reinberg design). Reprinted with permission of Semiconductor International[13] (b) Single wafer PECVD reactor. Courtesy of CVD Spectrum, Inc.

the electric field so that when they collide with gas molecules, gas-phase dissociation and ionization of the reactant gases (e.g. silane and nitrogen or oxygen-containing species) then occurs.

The energetic species (predominantly radicals, as described in Chap. 16) are then adsorbed on the film surface. (Note that the radicals tend to have high sticking coefficients, and also appear to migrate easily along the surface after adsorption. These two factors can lead to excellent film conformality). Upon being adsorbed on the substrate, they are subjected to ion and electron bombardment, rearrangements, reactions with other adsorbed species, new bond formations and film formation and growth. Adatom rearrangement includes the diffusion of the adsorbed atoms onto stable sites and concurrent desorption of reaction products. Desorption rates are dependent on substrate temperature, and higher temperatures produce films with fewer entrapped by-products. Note that gas-phase nucleation should be avoided to reduce particulate contamination.

The fact that the radicals formed in the plasma discharge are highly reactive, presents some options as well as some problems, to the process engineer. PECVD films, in general, are not stoichiometric because the deposition reactions are so varied and complicated. Moreover, by-products and incidental species are incorporated into the resultant films (especially hydrogen, nitrogen, and oxygen), in addition to the desired products. Excessive incorporation of these contaminants may lead to outgassing and concomittant bubbling, cracking, or peeling during later thermal cycling, and to threshold shifts in MOS circuits.

A plasma process requires control and optimization of several deposition parameters besides those of an LPCVD process, including rf power density, frequency, and duty cycle. The deposition process is dependent in a very complex and interdependent way on these parameters, as well as on the the usual parameters of gas composition, flow rates, temperature, and pressure. Furthermore, as with LPCVD, the PECVD method is surface reaction limited, and adequate substrate temperature control is thus necessary to ensure film thickness uniformity.

Three types of PECVD reactors are available: 1) the *parallel plate* type; 2) the *horizontal tube* type; and 3) the *single wafer* type. It should be noted that with one major exception, the discussion on glow discharges and rf diode sputtering in Chap. 10, and rf generated plasmas for dry etching in Chap. 16, applies to the production of rf generated plasma in PECVD reactors. Thus it is recommended that readers study these sections for information dealing with the generation of rf plasmas.

The one key difference in designing the reactor and electrode configuration in PECVD systems (compared to those in rf sputtering systems), is that in PECVD systems it is desired that the potential of both the powered and the grounded electrode(s), relative to the plasma, be approximately equal. Rf tuning networks of PECVD systems therefore usually employ an inductor which shunts the powered electrode to ground to establish this condition. Grounding prevents the powered electrode from developing a larger self-bias, thereby maintaining approximately equal average potentials between the plasma and the powered and grounded electrodes. The walls of parallel plate reactors are made of either quartz, ceramic, or aluminum oxide coated steel, in order to place them at a floating potential with respect to the plasma. This minimizes wall bombardment and sputtering, which reduces contamination of the growing films. (Note that the quartz tube of tube PECVD reactors is an insulator, and thus does not require such coating).

Parallel Plate PECVD Reactors

The radial parallel plate reactor was developed by Reinberg[52] in the early 1970's, and the first commercial model was offered in 1976. A schematic of this reactor type is shown in Fig. 8a. The reaction chamber is a short, vertically oriented cylinder, typically constructed of stainless steel. The rf power (at frequencies of 450 kHz-13.5 MHz) which establishes the plasma is applied to the upper electrode, and the wafers reside on the bottom grounded electrode, which can be rotated for improved uniformity, and be heated up to 400°C. The electrode spacing is typically 5 to 10 cm, and such systems operate in the pressure range of 0.1-5 torr. The reactants are introduced either through the center and removed from the periphery (Reinberg design), or from the periphery and removed from the center. Despite depletion effects, uniform deposition can be achieved by correctly balancing the plasma density and gas flows. Parallel plate systems, however, suffer from low throughput for large diameter wafers. In addition, particulates flaking off walls or the upper electrode, can fall onto the horizontally positioned wafers.

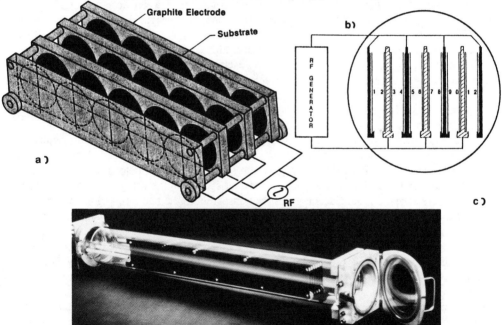

Fig. 9 (a) Long, multiple plate reactor generates plasma between the wafers facing each other on graphite electrodes[14]. (b) Cross section of electrode assembly and wafers shown in (a). Reprinted with permission of Solid State Technology, published by Technical Publishing, a company of Dun & Bradstreet. (c) Photograph of tubular PECVD reactor. Courtesy of Pacific Western Systems.

Horizontal Tube PECVD Reactors

In 1979 a new PECVD reactor design, the *horizontal tube* reactor, was introduced. This design allows PECVD reactor throughputs to be greatly increased. The reactor resembles a "hot wall" LPCVD system. It consists of a long horizontal quartz tube that is radiantly heated. Gas is fed into one end, and flows linearly to the other. Special long rectangular graphite plates (Fig. 9) serve both as the electrodes that are needed to establish a plasma, and as holders of the wafers (the electrode configuration is also designed to provide a uniform plasma environment for each wafer, to ensure uniformity of deposition). These vertically-oriented graphite electrodes are stacked parallel to one another, side-by-side, with alternating plates serving as power and ground for the rf voltage. A plasma is formed in the space between each pair of plates.

The use of several long slabs allows an increased number of wafers (up to 120 100mm wafers) to be loaded into the reactor at one time. Most systems consist of two stacked deposition tubes. In addition, since the wafers are held vertically, most particulates do not fall on the wafer surfaces. The entire graphite assembly is withdrawn from the reactor for wafer loading, and then reinserted for processing. Care must be exercised, however, to prevent particulate contamination when loading and unloading wafers from these systems.

Tubular PECVD reactors, like horizontal tube LPCVD reactors, suffer from a depletion effect, but resultant deposition nonuniformities can be minimized by temperature ramping. Another technique used to combat depletion effect nonuniformities is to pulse the plasma, so that during the off phase of the duty cycle, fresh reactant gases fill the tube and replace depleted gases.

Single Wafer PECVD Reactors

The most recently introduced PECVD reactor is the single-wafer design (Fig. 8b). The reactor is load locked, offers cassette-to-cassette operation, provides rapid radiant heating of each wafer, and allows *in situ* monitoring of the film deposition. Wafers $\geq$ 200 mm can be accomodated. Little direct information on the operational characteristics of such reactors was available at the time of publication, but such reactors are clearly being designed to allow fabrication of large wafers in low-particulate, automated environments.

Photon-Induced Chemical Vapor Deposition

A final CVD method, which may fill the need for an extremely low temperature deposition process without the film composition problems of PECVD, is PHoton-induced Chemical Vapor Deposition (PHCVD). PHCVD uses high-energy, high-intensity photons to either heat the substrate surface, or to dissociate and excite reactant species in the gas phase. In the case of substrate surface heating, the reactant gases are transparent to the photons, and the potential for gas phase reactions is completely eliminated. In the case of reactant gas excitation, the energy of the photons can be chosen for efficient transfer of energy to either the reactant molecules themselves, or to a catalytic intermediary, such as mercury vapor. This technique enables deposition at extremely low (i.e. room) substrate temperatures. PHCVD films show good step coverage, but may suffer from low density and molecular contamination as a result of the low deposition temperature.

There are two classes of PHCVD reactors, depending on energy source: a) UV lamp; and b) laser. *UV lamp reactors*[15] generally use mercury vapor for energy transfer between photons and the reactant gases. UV radiation at 2537Å is efficiently absorbed by mercury atoms, which then transfer energy to the reactant species. Deposition rates for UV PHCVD reactors are typically much slower than in other low temperature techniques. *Laser PHCVD reactors*[16] offer the advantages of *frequency tunability* and a high intensity light source. Tunability is useful in that specific photon energies targeted for particular dissociation reactions can be dialed in, enabling greater control of the deposition reaction. The energy from the high intensity laser also increases reaction rates. Laser PHCVD opens the possibility of CVD *writing* (i.e. patterned deposition by spatial control of the laser). Nevertheless, the deposition rates of current PHCVD processes are still too low to allow them to be adopted for production microelectronic applications.

POLYCRYSTALLINE SILICON: PROPERTIES AND CVD DEPOSITION METHODS

Polycrystalline silicon (also called *polysilicon, poly-Si* or *poly*) in thin film form has many important applications in integrated circuit technology. Heavily doped poly-Si films have been widely used as gate electrodes and interconnections in MOS circuits. Poly-Si is utilized in these roles because of its compatibility with subsequent high temperature processing, its excellent interface with thermal SiO_2 (low concentration of interface states), its higher reliability than Al gate materials[51], and its ability to be deposited conformally over steep topography. In some such applications more than one layer of poly-Si is utilized, and a layer of SiO_2 must be thermally grown or deposited on the first layer to electrically isolate it from subsequent layers. More recently, VLSI MOS devices have been fabricated with high conductivity films deposited onto the poly-Si, to create interconnect structures with lower sheet resistances than are possible with poly-Si alone (see Chap. 11 for more information on this subject). Heavily doped poly-Si films are also utilized to form emitter structures in advanced bipolar technologies. Selective growth of poly-Si is being studied as a planarization technique for filling contact holes[17]. Lightly doped poly-Si films are used as high-value load resistors in static memories, and to refill trenches employed in dielectric isolation technologies. In this section the properties of poly-Si films and CVD methods for preparing such films are presented.

Properties of Polysilicon Thin Films

Physical Structure and Mechanical Properties of Poly-Si

Thin films of polycrystalline silicon are made up of small (on the order of 1000Å) single crystal regions (*crystallites* or *grains*) separated by grain boundaries. It is of interest to note that *as-deposited* poly-Si films can be amorphous or polycrystalline, but subsequently exhibit a polycrystalline structure if subjected to elevated temperatures after deposition. Polycrystalline silicon films exhibit many mechanical material properties close to those of bulk single-crystal Si (e.g. the density of poly-Si is 2.3 g /cm^3, the coefficient of thermal expansion is 2×10^{-6} /°C, and the temperature coefficient of resistance is 1×10^{-3} /°C). The intrinsic stress in thin polysilicon films is compressive (e.g. the stress in 2000-5000Å films is $1-5 \times 10^9$ dynes /cm^2 *compressive* over the range of doping from undoped material to 10^{20} atoms /cm^3, and annealing temperatures from 250–1100°C)[18]. The effect of deposition conditions and post-deposition annealing on poly-Si film structure is discussed in more detail in a later section.

The grain boundaries are composed of disordered atoms, and contain large numbers of defects due to incomplete bonding; while inside each grain, atoms are arranged in a periodic structure. The silicon material in the interior (bulk) of the grains consequently behaves similarly to that of bulk single crystal silicon (e.g. diffusion constants and substitutional impurity properties are assumed to be roughly identical). The defects and departure from periodicity at the grain boundaries, however, substantially alter the behavior of two important material properties of polysilicon, its diffusion characteristics, and its dopant distribution. That is, diffusion

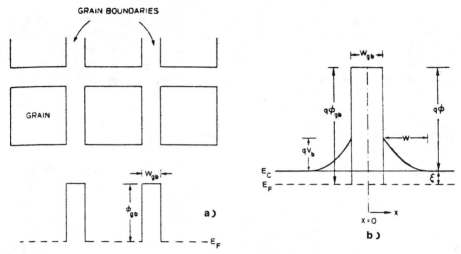

Fig. 10 (a) Schematic representation of potential barriers generated by the grain boundaries. (b) Energy band diagram near a grain boundary under zero applied voltage[19]. (© 1981, IEEE).

constants in grain boundary regions are significantly higher than those in single crystal regions. As a result, dopants are capable of diffusing much more rapidly along grain boundaries than through crystallites. Likewise, dopant redistribution is also impacted, as some dopants (e.g. As and P, but not B) exhibit segregation coefficients which favor their moving from single crystal regions into grain boundary regions at low temperatures, and back again at high temperatures[19]. See Chap. 4 for an introduction to thin film properties, and Chap. 8 for diffusion in poly-Si.

Electrical Properties of Polysilicon

The electrical properties of poly-Si films are functions of both their semiconductor nature and their polycrystalline structure and doping. That is, single crystal regions of poly-Si are assumed to behave electrically in a manner similar to bulk single crystal silicon. As in single crystal silicon, low resistivity is obtained through heavy doping with impurity atoms.

At a given dopant concentration, polysilicon exhibits significantly higher resistivity than single crystal silicon (except at very high dopant concentrations, where the resistivity is only slightly greater). This is primarily due to two mechanisms: 1) under heat treatment some of the dopant atoms (e.g. As and P, but not B) segregate to the grain boundaries, where they do not effectively produce free carriers. Thus, the concentration of dopant atoms remaining in single crystal regions, and thus still able to contribute charge carriers, is decreased; and 2) the grain boundaries are rich with incomplete bonds, which traps some free carriers at the grain boundaries. Not only does such trapping decrease the overall free carrier concentration, but the trapped charge in the grain boundaries causes local regions of neighboring single crystal material to become depleted of carriers (and creates local potential variations in the material). These changes in potential at grain boundaries (Fig. 10) are detrimental to carrier movement, and also lead to an increase in apparent resistivity. The grain boundary region itself, approximately 5-10 angstroms wide, can be modeled as a separate, amorphous silicon region with increased bandgap. This larger bandgap represents another potential barrier for charge carriers to overcome, although it is so narrow that quantum tunneling effects prevail. Finally, the defects in the grain boundaries decrease the carrier mobility, also causing the resistivity to increase.

Based on the above model for resistivity degradation effects, the behavior of poly-Si resistivity as a function of dopant concentration and grain size, can be qualitatively understood. First, films with larger grain sizes (at the same dopant concentration) exhibit lower resistivity. That is, since surface area does not increase with grain size as quickly as volume, films with larger grains have proportionally smaller grain boundary densities, and thus less deviation from bulk-silicon resistivity values is observed. Second, grain size and dopant concentration effects interact to determine the degree to which each crystallite is depleted of free charge. Very small grains will become fully depleted more easily than larger grains, and high dopant concentrations result in narrower depleted regions, thus making full depletion of the grains more difficult. If the grains do become completely depleted, the resistivity increases dramatically. As a result, there is a sharp transition region between *very high resistivity*, where grain size is small and /or doping is low and the grains are completely depleted, and *lower resistivity*, where grain size is large and /or doping is high.

In practice, sheet resistances as low as 10-30 Ω /square may be obtained by several techniques (see below) in poly-Si films used in integrated circuits. Note that the work function of poly-Si, important in MOS device threshold voltage values, is also impacted by the dopng level.

Chemical Vapor Deposition of Polysilicon

Poly-Si is generally deposited by the pyrolysis (i.e. thermal decomposition) of silane (SiH_4) in the temperature range 580-650°C. The main technique used to deposit poly-Si is LPCVD because of its uniformity, purity, and economy [20,21]. The deposition reaction sequence is:

$$SiH_4 + \text{surface site} = SiH_4 \text{ (adsorbed)} \tag{3a}$$

$$SiH_4 \text{ (adsorbed)} = SiH_2 \text{ (adsorbed)} + H_2 \text{ (gas)} \tag{3b}$$

$$SiH_2 \text{ (adatom)} = Si \text{ (solid)} + H_2 \text{ (gas)} \tag{3c}$$

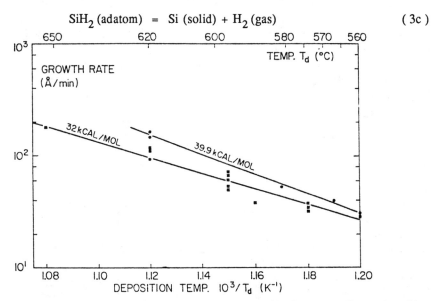

Fig. 11   Growth rate as a function of deposition temperature T_d for two different deposition conditions[20]. ($\bullet$ 350 mtorr, $SiH_4 = 200$ cm^3 /min   and $\blacksquare$ 120 mtorr, $SiH_4 = 50$ cm^3 /min). Reprinted by permission of the publisher, the Electrochemical Society, Inc.

where adsorption of the SiH_4 is followed by decomposition to an intermediate compound, SiH_2. Then, upon evolution of the remaining hydrogen, the solid film forms. The overall reaction is generally given as:

$$SiH_4 \text{ (vapor)} = Si \text{ (solid)} + 2H_2 \text{ (gas)} \qquad (4)$$

Three processes are commonly used in conventional LPCVD systems. The first uses 100% SiH_4 at total pressures from 0.3-1 torr, while the second uses approximately 25% SiH_4 in a nitrogen carrier at approximately the same pressures. A third technique, performed in vertical flow isothermal reactor configurations, uses 25% SiH_4 diluted in hydrogen, also at ~1 torr.

Deposition Parameters

The deposition rate shows an exponential dependence on temperature. Figure 11 is a plot of deposition rate vs. reciprocal temperature. Apparent activation energies are found to depend on the silane pressure and range from 1.36 to 1.7 eV. Depositions are limited to the 580-650°C range, since at higher temperatures gas phase reactions occur (leading to rough and loosely adhering films), and below 580°C the rate is too slow for practical use (<50Å /min).

The polysilicon deposition rate depends on the silane (SiH_4) pressure as shown in Fig. 12. The dependence is not linear except for low values of pressure. Such behavior may be due to homogeneous reactions, adsorption of hydrogen on the surface sites, or transport phenomena. The total pressure in the reactor can be varied by: a) changing the flow rate of silane at constant pump speed; b) changing the pump speed at constant flow rate; and c) in processes which use diluted silane, by changing the nitrogen flow to adjust the silane concentration. It appears that changing pumping speed to control total pressure provides the best control of deposition rates.

In conventional LPCVD systems it is necessary to ramp the temperature by as much as 30°C from the gas inlet tube end to the outlet. This is done to overcome the effects of the depletion of SiH_4, and the attendant decrease in deposition rate down the length of the tube. The increased deposition rate resulting from increased temperatures is intended to just compensate for the decrease due to silane depletion. A difficulty with this process is that poly-Si properties depend very strongly on deposition temperature, and will thus vary with wafer position along the tube. A newly designed vertical flow reactor (see section on LPCVD) circumvents this problem.

Polysilicon can be doped during deposition (*in-situ*), or after deposition by diffusion or ion implantation. The addition of diborane to the silane during deposition (boron doping) causes

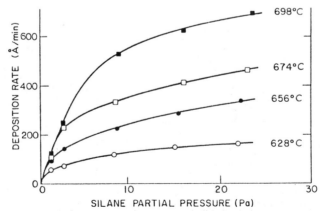

Fig. 12 The effect of silane concentration on the polysilicon deposition rate[1]. Copyright, 1983, Bell Telephone Laboratories, Incorporated, reprinted with permission.

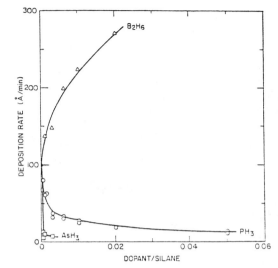

Fig. 13 The effect of dopants on the polysilicon deposition rate at 610°C[1]. Copyright, 1983, Bell Telephone Laboratories, Incorporated, reprinted with permission.

rapid increases in the deposition rates, while addition of phosphine or arsine to the silane results in significant growth rate reductions. Figure 13 shows a plot of the deposition rate of polysilicon at 610°C as a function of the volume fraction of dopant added to the silane. *In-situ* doping of polysilicon has not been popular even though it is considerably less complex than post-deposition doping, because of the difficulty of maintaining thickness and doping uniformity from wafer to wafer and across the wafer. Meyerson and Olbricht have shown that the factor of 20 reduction in deposition rate, which is observed when a PH_3/SiH_4 ratio of 1:40 is used, can be attributed to the adsorption of a phosphine layer on the substrate[22]. This inhibits subsequent silane adsorption and heterogeneous silane decomposition.

Structure of Polysilicon: Deposition Condition Dependence

The structure and properties of poly-Si depend very much on the deposition temperature, the dopant type and concentration, and subsequent thermal cycling. At temperatures below 580°C, the as-deposited film is essentially amorphous[23]. At temperatures above 580°C, the films deposit as polycrystalline, with preferred orientation. The {110} fiber axis with columnar grains is dominant in the 625°C range while the {100} predominates in the 675°C range. The {110}-preferred orientation also dominates at higher temperatures[24]. Harbeke, *et al* found a {311}-preferred orientation for films deposited between 580-600°C[20]. Films that are recrystallized at 900-1000°C from the amorphous state tend towards a strong {111} fiber texture (Fig. 14). It was also observed that the texture and grain size are exceptionally reproducible in films crystallized from the amorphous phase, and the average grain size is somewhat larger than in the as-deposited polycrystalline film. Figure 15 shows the average grain size of as deposited and recrystallized films (undoped and P-doped). In addition, it was found that as-deposited amorphous films tend to have a smoother surface than do films grown at 600°C (which occasionally show rough surfaces), and films grown at 620°C (which always show surface roughness). The smooth surfaces of the as-deposited amorphous films are maintained even after annealing at 900-1000°C. The penalty paid for achieving smoother surfaces is slower deposition rates at 580° C (~50 Å /min versus 100Å /min at 600°C). The average grain size

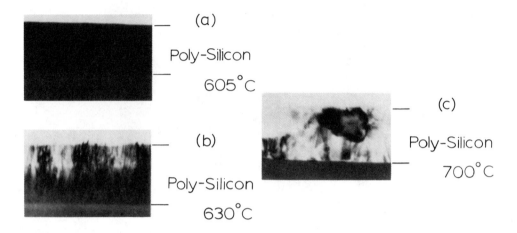

Fig. 14 TEM cross sections (60,000X) of polysilicon. (a) Amorphous structure deposited at 605°C. (b) Columnar structure deposited at 630°C. (c) Crystalline grains formed by annealing an amorphous sample at 700°C[1]. Copyright, 1983, Bell Telephone Laboratories, Incorporated, reprinted with permission.

has been observed to increase with film thickness[26], and it was also found that the resistivity in boron doped films increases exponentially as thickness decreases.

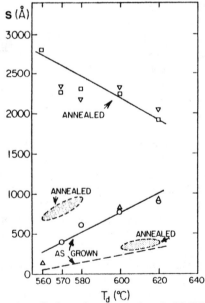

Fig. 15 Average crystallite size **S** for phosphorus-doped LPCVD silicon layers as-grown (o = interface, Δ = surface), and annealed at 1000°C (◊ = interface, ∇ = surface) as a function of deposition temperature, T_d. Undoped layers (dashed lines) are shown for comparison. From G. Harbeke, *et al.*, RCA Review, **44**, 287 (June, 1983).

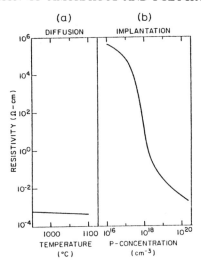

Fig. 16   Resistivity of P-doped polysilicon. (a) Diffusion, 1 h at the indicated temperature[25]. Reprinted by permission of the publisher, the Electrochemical Society, Inc. (b) Implantation, 1 h anneal at 1100°C[19]. (© 1981, IEEE).

Doping Techniques for Polysilicon

Three techniques are used to dope polysilicon: 1) diffusion; 2) ion implantation; and 3) *in situ* doping. In most applications, poly-Si is deposited undoped, and is subsequently doped by diffusion or implantation.

Diffusion Doping of Polysilicon

Diffusion doping (see Chap. 8) is a relatively high-temperature process (900-1000°C) wherein a highly doped glass is grown or deposited on undoped polysilicon, and serves as a solid source for dopant diffusion. The advantage of this method is its ability to introduce very high concentrations of dopants into the poly-Si film (above solid-solubility limits, because of grain-boundary segregation effects), resulting in low resistivity. A high temperature diffusion doping step can also serve double duty as the polysilicon anneal step. Lower final resistivities have been reported for diffusion doping of as-deposited amorphous (low deposition temperature) poly-Si compared to as-deposited polycrystalline poly-Si (Fig. 16a). The disadvantages of diffusion doping are its high temperature and the possibility of increasing film surface roughness.

Ion Implantation Doping of Polysilicon

The second doping technique is ion implantation and subsequent anneal. See Chap. 9 for more on Ion Implantation. This method has the advantage of precise control of dopant dose, and is well suited for applications that do not require a maximally doped polysilicon film (e.g. high value load resistors used in static memories, or doping studies). The resistivity of heavily doped implanted poly-Si is about ten times higher than that of diffusion doped poly-Si (Fig. 16b). The implantation energy is generally selected so that the impurity peak is produced at the center of the film. A subsequent anneal step (e.g. ~900°C, 30 min) redistributes and activates the implanted dopant. Rapid thermal processing (see Chaps. 2 and 9) is also being investigated as an annealing and activating technique[28]. With the increased diffusion rates of dopants along grain boundaries, dopant redistribution and activation of implanted polysilicon can be achieved by RTP in less than 30 seconds at 1150°C. The advantage of using RTP is its short duration, which avoids redistributing dopants in the single crystal silicon substrate. Note that ion implantation of poly-Si with As, followed by a rapid-thermal processing step, has also been investigated as a method for forming shallow emitters in advanced bipolar technologies.

In Situ Doping of Polysilicon

In situ doping involves adding doping gases such as diborane and phosphine to the CVD reactant gases[50]. Although combining doping and deposition in one step may appear simple, the control of film thickness, dopant uniformity, and deposition rate is greatly complicated by the addition of the dopant gases. Moreover, the physical properties of the film are affected. Adding phosphine can change the temperature dependence of the polycrystalline film structure, grain size, and grain orientation. In undoped films, it is reported that depositing an initially amorphous film results in superior structural perfection after anneal[21]. Note that for *in-situ* arsenic or phosphorus doped poly-Si, an oxide capping layer must either be deposited or thermally grown before or during the anneal cycle, in order to avoid outdiffusion of the dopant through the top surface of the polysilicon during the anneal. If the deposition temperature is high enough (>600°C) to result in an as-deposited polycrystalline film with sufficiently low resistivity, then the high temperature anneal step may be skipped altogether.

Oxidation of Polysilicon

Polysilicon, like single crystal silicon, can be thermally oxidized. Its oxidation rate depends on grain orientation, dopant type, and dopant concentration. In general, lightly doped poly-Si oxidizes more rapidly in wet O_2 than {111} or {100} single crystal silicon[29]. Polysilicon heavily-doped with phosphorus oxidizes more rapidly than undoped poly-Si, but not as rapidly as heavily-doped single crystal Si[28]. The ratio of poly-Si consumed during oxidation, to the thickness of the oxide, is about the same as in single-crystal Si (1:1.56).

Of particular importance in some applications (especially in which a double-poly, or even triple-poly structure is used) is that SiO_2 thermally grown on the poly-Si exhibits adequate breakdown strength. This strength is strongly influenced by the smoothness of the polysilicon surface prior to oxidation. That is, a rough surface leads to high local electric fields and lower oxide dielectric strength. Again, low temperature poly-Si depositions (<600°C) have been reported to produce smoother surfaces and grown oxides with higher breakdown voltages than obtained under higher temperature depositions. Another technique reported to increase the smoothness of the poly-Si /SiO_2 interface is to grow the oxide at higher temperatures. It is believed that the viscous flow of the oxide can moderate the surface roughness produced by oxidation. A method for reducing leakage and increasing the breakdown strength of oxides grown on phosphorus-doped poly-Si is to dope it to an optimum level (~6×10^{20} cm^{-3}), with a high temperature anneal in an inert ambient (~1100°C, 10 min, N_2) prior to a 950°C oxidation[30]. This is thought to improve interface flatness by phosphorus-enhanced grain growth during the anneal.

PROPERTIES AND DEPOSITION OF CVD SiO$_2$

Chemical vapor deposited (CVD) SiO_2 films, and their binary and ternary silicates, find wide use in VLSI processing. These materials are used as insulation between polysilicon and metal layers, between metal layers in multilevel metal systems, as getters, as diffusion sources, as diffusion and implantation masks, as capping layers to prevent outdiffusion, and as final

Table 2. PROPERTIES OF CVD AND THERMAL SILICON DIOXIDE[1]

FILM TYPE:	THERMAL	PECVD	APCVD	$SiCl_2H_2+N_2O$	TEOS
Deposition Temp. (°C):	800-1200	200	450	900	700
Step Coverage:	conformal	good	poor	conformal	conformal
Stress ($x10^9$ dynes/cm^2):	3C	3C-3T	3T	3T	1C
Dielectric Strength (10^6 V/cm):	3 - 6	8	10	10	
Etch Rate (Å /min): (100:1, H_2O:HF)		400	60	30	30

passivation layers.   In general, the deposited oxide films must exhibit uniform thickness and composition, low particulate and chemical contamination, good adhesion to the substrate, low stress to prevent cracking, good integrity for high dielectric breakdown, conformal step coverage for multilayer systems, low pinhole density, and high throughput for manufacturing.

CVD silicon dioxide is an amorphous structure of SiO_4 tetrahedra with an empirical formula SiO_2. (See Chap. 7, section on *Properties of Silica Glass*.) Depending on the deposition conditions, as summarized in Table 2, CVD silicon dioxide may have lower density and slightly different stoichiometry from thermal silicon dioxide, causing changes in mechanical and electrical film properties (such as index of refraction, etch rate, stress, dielectric constant and high electric-field breakdown strength). Deposition at high temperatures, or use of a separate high temperature post-deposition anneal step (referred to as *densification*) can make the properties of CVD films approach those of thermal oxide.

Deviation of the CVD silicon dioxide film's refractive index, n, from that of the thermal SiO_2 value of 1.46 is an often used as an indicator of film quality.   A value of n greater than 1.46 indicates a silicon rich film, while smaller values indicate a low density, porous film. CVD SiO_2 is deposited with or without dopants, and each type has unique properties and applications.

Chemical Reactions for CVD SiO_2 Formation

There are various reactions that can be used to prepare CVD SiO_2. The choice of reaction is dependent on the temperature requirements of the system, as well as the equipment available for the process.   The deposition variables that are important for CVD SiO_2 include: temperature, pressure, reactant concentrations and their ratios, presence of dopant gases, system configuration, total gas flow, and wafer spacing.   There are three temperature ranges in which SiO_2 is formed by CVD, each with its own chemical reactions and reactor configurations.   These are: 1) low temperature deposition (300-450°C); 2) medium temperature deposition (650-750°C); and 3) high temperature deposition (~900°C).

The *low temperature deposition* of SiO_2 utilizes a reaction of silane and oxygen to form undoped SiO_2 films.   The depositions are carried out in APCVD reactors (primarily of the continuous belt type), in distributed feed LPCVD reactors, or in PECVD reactors. The depletion effect precludes the use of conventional LPCVD for the $SiH_4 + O_2$ reaction.   The addition of PH_3 to the gas flow forms P_2O_5, which is incorporated into the SiO_2 film to produce a phosphosilicate glass (PSG)[31]. The reactions are given by:

$$SiH_4 + O_2 \ \text{------------>} \ SiO_2 + 2H_2 \quad\quad\quad (5)$$

$$4PH_3 + 5O_2 \ \text{------------>} \ 2P_2O_5 + 6H_2 \quad\quad\quad (6)$$

The reaction between silane and excess oxygen forms SiO_2 by heterogeneous surface reaction. Homogeneous gas-phase nucleation also occurs, leading to small SiO_2 particles that form a white powder on the reaction chamber walls (and which may potentially cause particulate contamination in the deposited films).

The deposition rate increases slowly with increased temperature between 310 and 450°C. An apparent activation energy of less than 0.4 eV has been measured which is indicative of a surface adsorption or gas phase diffusion deposition process.   The deposition rate can be increased at constant temperature (up to a limit) by increasing the O_2: SiH_4 ratio. Continued increase in the ratio eventually results in a decrease in deposition rate, as a result of O_2 being adsorbed on the substrate, thus inhibiting the SiH_4 decomposition.

Silicon dioxide films deposited at low temperatures exhibit lower densities than thermal SiO_2, and have an index of refraction of ~1.44 . They also exhibit substantially higher etch rates in buffered hydrofluoric acid (HF) solutions than thermal SiO_2. Subsequent heating of such films to temperatures between 700-1000°C causes *densification.* That is, this step causes the density of the material to increase from 2.1 g /cm^3 to 2.2 g /cm^3, the film thickness to decrease, and the etch rate in HF to decrease.

SiO_2 can also be deposited by a plasma enhanced reaction between SiH_4 and N_2O (nitrous oxide) or O_2 at temperatures between 200-400°C[32]:

$$SiH_4 + 2N_2O \xrightarrow{\text{200-400°C, rf}} SiO_2 + 2N_2 + 2H_2 \qquad (7)$$

Nitrogen and /or hydrogen is often incorporated in PECVD SiO_2. A low ratio of N_2O /SiH_4 will increase the index of refraction, due to large amounts of nitrogen incorporated in the film and the formation of silicon rich films. Nearly stoichiometric (n = 1.46) plasma oxide films can be achieved by reacting SiH_4 and O_2 mixtures.   The buffered HF etch rate is a sensitive measure of the film's stoichiometry and density. Lower deposition temperatures and higher N_2O /SiH_4 ratios lead to less dense films and faster etch rates. As with nitride films, PECVD oxides also contain 2-10 at% H_2 in the form of Si-H, Si-O-H, and H-O-H[32]. The hydrogen concentration is a strong function of the deposition parameters. Low deposition temperatures, high rf power, and high carrier-gas flow rates are required to prevent gas phase nucleation and its attendant particulate problems.

Plasma oxide films are generally deposited in compressive stress, with values ranging between $1x10^8$-$4x10^9$ dynes /cm^2, depending on deposition temperature and rate. Dielectric strengths of 4-$8x10^6$ V /cm, and dielectric constants ranging from 4-5 have been obtained. Low pinhole counts have been obtained with PECVD oxides, as have very conformal coatings. Adhesion to metal is also reported to be excellent[34].

In the *medium temperature range,* SiO_2 is deposited in LPCVD reactors by decomposing tetraethosiloxane, $Si(OC_2H_5)_4$, also known as tetraethyl orthosilicate, or *TEOS.* The deposition rate for TEOS shows an exponential increase with temperature in the range of 650-800°C with an apparent activation energy of 1.9 eV[33]. This pronounced temperature dependence can lead to thickness control problems. The deposition rate is also dependent on the TEOS partial pressure. It is linearly dependent at low partial pressures, and tends to level off as the adsorbed TEOS saturates the surface. TEOS films generally show excellent conformality.

At *high temperatures* (near 900°C) SiO_2 is formed by an LPCVD process in which dichlorosilane and nitrous oxide are reacted[8]. The reaction is given by:

$$SiH_2Cl_2 + 2N_2O \longrightarrow SiO_2 + 2N_2 + 2HCl. \qquad (8)$$

Such depositions produce films having excellent uniformity, and with properties close to those of thermal SiO_2. High-temperature LPCVD is sometimes used to deposit SiO_2 over poly-Si.

Step Coverage of As-Deposited CVD SiO_2 Films

The manner in which a thin film covers (or conforms to) the underlying features on a substrate is an important characteristic of a deposited film. Film conformality of CVD films is a function of film species and reactor type, and deposition conditions. *Conformal coverage* is coverage in which equal film thickness exists over all substrate topography regardless of its slope (i.e. vertical and horizontal surfaces are coated with equal film thickness, Fig. 17a). The thickness of deposited films at any given point in the mass-transport limited case is dependent on the supply of reactants to that point, which itself is set by process pressure and adatom migration. The following model has been proposed to explain step coverage in terms of these parameters[1].

If adsorbed reactants are able to migrate rapidly across a substrate surface, they will be found with equal probability on any part of the substrate, regardless of topography. This situation results in conformal step coverage, as shown in Fig. 17a. Adatom mobility is a function of adatom species and energy. Both higher substrate temperature and ion bombardment of adatoms enhance adatom migration. Note that under the relatively high temperature conditions used to deposit LPCVD poly-Si and Si_3N_4 films, excellent (i.e. highly conformal) step coverage is usually achieved. Thus, only SiO_2 step coverage need be further considered in this discussion.

If on the other hand, reactant surface migration is slow, then the arrival flux of gaseous reactants at a given point becomes more important in determining the local film thickness. In such cases, it is useful to introduce the concept of the *arrival angle* to help model step coverage. That is, the flux of reactant molecules arriving from an angle between θ and $\theta + d\theta$ can be expressed as $P(\theta)\, d\theta$ (in a 2-dimensional analysis). Then the total flux arriving at any

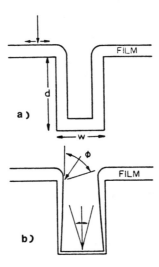

Fig. 17 Schematic diagrams showing types of step coverage. (a) Conformal coverage resulting from rapid surface migration; (b) Non-conformal coverage caused by no surface migration[33]. Reprinted with permission of Solid State Technology, published by Technical Publishing, a company of Dun & Bradstreet.

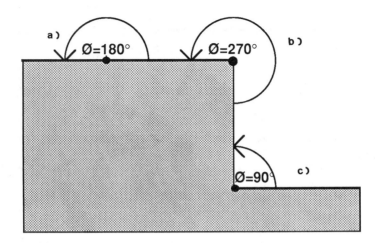

Fig. 18 The concept of *arrival angle* and its effect on step coverage. (a) 180° (b) 270° (c) 90°.

point is $\int_0^{2\pi} P(\theta)\, d\theta$.   For an atmospheric pressure reactor, the mean free path for reactant gas molecules is short, and the frequent collisions of gas molecules with each other completely randomize their velocity vectors. Under such conditions, $P(\theta)$ is a constant, independent of the value of θ. However, $P(\theta)$ is zero for values of θ which are blocked by the substrate. For example, in Fig. 18a, $P(\theta)$ is zero for any angle from 180-360°, and constant for $0° < \theta < 180°$.

In short, the value of the flux integral (and thus the eventual film thickness), is directly proportional to the range of angles for which $P(\theta)$ is not zero. This range is called the *arrival angle*. As seen in Fig. 18b, at the top corner of a step, $P(\theta)$ is non-zero over a range of 270°. The resultant film thickness is 270 /180, or 1.5 times greater than for the planar case of Fig. 18a. Similarly in Fig. 18c (i.e. at the bottom corner of a step, or trench) the arrival angle is only 90°, and the film thickness is 90 /180, or one-half that of the planar case. This explains the thickness cross-section that is observed when APCVD SiO_2 is deposited over a sharp step (Fig. 19a). The film is thickest at the upper corner and thinnest at the bottom of the step. This causes the slope in SiO_2 films at the bottom corner to be >90° (i.e. a so-called *re-entrant angle*) making deposition and anisotropic etching of subsequently deposited films extremely difficult.

On the other hand, if the mean free path of the reactant gases is high (e.g. due to the low operating pressure of LPCVD and PECVD reactors), a *shadowing* effect can occur. That is, no longer do frequent collisions over very small distances randomize the gas molecule velocity vectors. Instead, reactant molecules experience few collisions, and follow straight line trajectories over distances comparable to substrate topography dimensions.   Substrate surface features near to points being impinged upon can block the straight paths of reactant molecules. Thus such points can be *shadowed* from the reactant flux, and will experience less deposition, and thus less resulting film thickness, than those points which are not shadowed. The arrival angle for the given point now depends on the range of unobstructed *lines of sight* from that point to the reaction chamber (see Fig. 17b). In such long mean free path cases the arrival angle, and thus film thickness, decreases with depth into a trench (as shown in Fig. 17c).

The reactor type and deposition conditions impact step coverage insofar as reactant gas mean free path and adatom migration is affected. The reactor operating pressure determines mean free path lengths, while adatom migration is affected by substrate temperature and also by energy transfer method. For example, due to rapid surface migration, SiO_2 films prepared by TEOS

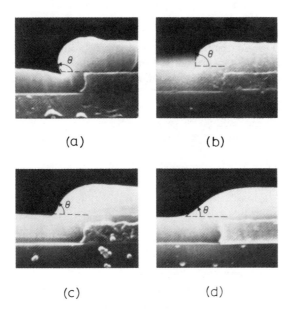

(a)   (b)

(c)   (d)

Fig. 19 SEM cross-sections (10,000X) of samples annealed in steam at 1100°C for 20 min for the following weight of phosphorus: (a) 0.0 wt.% P; (b) 2.2 wt.% P; (c) 4.6 wt.% P; 7.2 wt.% P.[33] Reprinted by permission of the publisher, the Electrochemical Society, Inc.

decomposition, or by the high temperature reaction of $SiCl_2H_2$-N_2O under LPCVD conditions, exhibit nearly conformal coverage. Conversely, during low temperature deposition of SiO_2 by APCVD, almost no surface migration takes place, and step coverage is dependent only on the arrival angle (and is quite poor, as noted earlier). On the other hand, PECVD produces reactants which arrive at the substrate with more energy (obtained from the glow-discharge) than in APCVD. This results in increased adatom migration. Thus, improved step coverage by PECVD SiO_2 films can be obtained compared to APCVD depositions made at the same temperature. Note that the mechanisms of this model can also be applied to analyzing the step coverage effects seen in evaporated and sputtered films.

CVD and Applications of Undoped and Doped SiO_2 Films

Both undoped and doped films of SiO_2 are easily deposited at low temperatures. Doping of TEOS and high temperature CVD SiO_2 films is more difficult. As we shall see, doping of the SiO_2 can produce a variety of desirable film properties for some applications.

Undoped CVD SiO_2

Undoped CVD SiO_2 is similar to its thermally grown counterpart in many respects, with some exceptions being its as-deposited density and etch rate, and a lower quality SiO_2/Si interface (higher density of interface states). Due to its poor interface qualities, CVD SiO_2 is therefore only used as a temporary structure if it is used in contact with single crystal silicon (e.g. as a capping layer over doped regions to prevent outdiffusion during thermal processes or as an ion-implantation mask). Its chief use, however, is as a permanent structure, whose function is to increase the thickness of field oxides, or to provide isolation between conductors.

When the underlying conductor is able to withstand high temperatures (e.g. polysilicon or refractory metal silicides) one of the LPCVD methods may be utilized, because of their excellent uniformity, good step coverage, low particulate contamination, and high purity.

Often the film must be deposited over aluminum, and therefore the process must take place at less than 450°C. At these temperatures LPCVD reactions exhibit extremely low deposition rates (Fig. 20b). On the other hand the reaction of silane and oxygen mixtures at atmospheric pressure and temperatures from 300-500°C provides adequately high deposition rates. Therefore such APCVD processes are widely used, despite their particulate and step coverage problems. Undoped oxide can also be deposited by PECVD techniques, with generally improved step coverage as compared to APCVD.

Phosphosilicate Glass

Adding phosphorus dopant during the deposition, typically in the form of phosphine, PH_3, forms phosphosilicate glass, or PSG. Since PSG consists of two compounds, P_2O_5 and SiO_2, it is a *binary glass* (or binary silicate), and some of the its properties are considerably different from those of undoped CVD SiO_2.   That is, APCVD PSG shows reduced stress and somewhat improved, though still relatively poor, step coverage as compared to undoped CVD SiO_2. Moreover PSG is a diffusion barrier to moisture, and getters alkali ions. Finally PSG can be *flowed* at high temperatures to create a smoother surface topography , and thereby facilitate the step coverage of subsequently deposited films. The flow step is performed at 1000-1100°C, at pressures from 1-25 atm, and in gas ambients of H_2O, N_2 and O_2. The glass becomes viscous and responds to surface tension forces, rounding sharp corners. The extent of flow is often measured as the reduction in slope angle of the flowed film surface over an underlying step (Fig. 19), and is a function of flow temperature and phosphorus concentration. Increasing temperature, duration of the flow step, or phosphorus concentration in the oxide will increase film flow. Note that for concentrations less than 6 wt% phosphorus, PSG does not readily flow.

It is relatively easy to incorporate the P_2O_5 into the PSG, as the SiH_4:PH_3 ratio in the gas flow controls the SiO_2: P_2O_5 ratio in the deposited film. PSG, however, becomes increasingly hygroscopic at high phosphorus levels. Thus, it is recommended that its concentration in the oxide film be limited to 6-8 wt% P to minimize phosphoric acid formation, and consequent Al corrosion. Rapid thermal processes have also been reported to successfully flow PSG. It should be noted, however, that if PSG is to be used as a final passivation layer (in which no glass flow

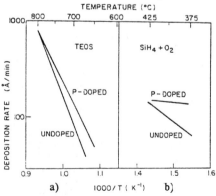

Fig. 20 Arrhenius plots for the deposition of (a) LPCVD TEOS, and (b) LPCVD SiO_2 (P-doped and undoped)[33]. Reprinted with permission of the publisher, the Electrochemical Society, Inc.

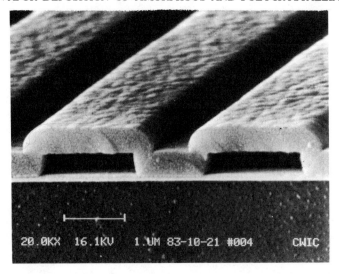

Fig. 21 SEM of PECVD oxide over polysilicon lines. Poly has been removed by wet etch to show detail[12]. Reprinted with permission of Solid State Technology, published by Technical Publishing, a company of Dun & Bradstreet.

is performed), the maximum permitted phosphorus content is 6 wt%. This is a precaution taken to again minimize the risk of phosphoric acid formation, particularly when in contact with Al.

PSG has been deposited in PECVD systems by the reaction of SiH_4, N_2, O_2, and PH_3 with Ar as a carrier gas. It is reported that the films are more conformal (Fig. 21), crack resistant, and pinhole free than APCVD films[34]. The phosphorus content is a function of the SiH_4/PH_3 ratio, deposition temperature and $N_2O/(SiH_4 + PH_3)$ ratio[35]. The plasma PSG films can be flowed under conditions similar to those for flowing APCVD PSG films.

Borophosphosilicate Glass

Some VLSI processes would benefit from being performed at lower temperatures than those necessary for PSG reflow (1000-1100°C) because such high temperatures result in excessive diffusion of shallow junctions. Furthermore, radiation-hard MOS gate oxides cannot be exposed to high temperature processing (>900°C). But flowable glass is still very desirable for easing film coverage over abrupt steps in the substrate topography. Glass flow temperatures as low as

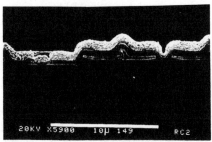
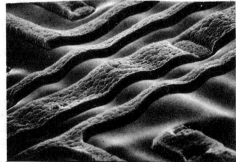

Fig. 22 Micrographs showing BPSG as the interlevel dielectric under the metal. Reprinted with permission of Semiconductor International[2].

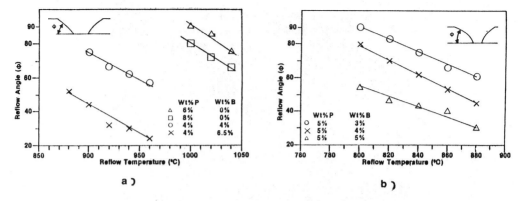

a)   b)

Fig. 23 (a) Reflow angle vs. reflow temperatures in a nitrogen ambient (30 min). (b) Reflow angle vs. reflow temperatures in a steam ambient (30 min). Reprinted by permission of Solid State Technology, published by Technical Publishing, a company of Dun & Bradstreet[37].

700°C can be obtained by adding boron dopant (e.g. B_2H_6) to the PSG gas flow to form the *ternary* (three component) oxide system B_2O_3-P_2O_5-SiO_2, borophosphosilicate glass, or BPSG[36] (Fig. 22).

BPSG flow depends upon film composition, flow temperature, flow time, and flow ambient. It has been reported that an increase in boron concentration of 1 wt% in BPSG decreases the required flow temperature by ~40°C[36]. A plot of required flow temperatures vs. BPSG dopant concentrations in LPCVD films is shown in Fig. 23. In general, increasing the P concentration beyond ~5wt% does not further decrease BPSG flow temperatures. An upper limit on boron concentration is imposed by film stability. That is, BPSG films containing over 5 wt% boron tend to be very hygroscopic and unstable, and if used, should be flowed immediately following deposition. It has also been reported that rapid thermal annealing for 30 s at a temperature 100-175°C higher than that used in a conventional furnace step will result in equivalent BPSG flow[36]. The ambient gas of the flow cycle also affects the flow mechanism. By using a steam ambient instead of N_2, the minimum required flow temperature is reduced by ~70°C[37].

In addition to exhibiting these low temperature flow properties, BPSG (like PSG) is an alkali ion getter and exhibits low stress. Because of its doping, however, BPSG can also be an unwanted diffusion source to underlying silicon. It is found that BPSG is primarily a source of phosphorus, and the phosphorus outdiffusion is increased at higher boron concentrations.

Besides being useful for isolation, passivation, and surface planarization, BPSG is also more attractive for use than PSG in a process known as *contact reflow*. Following anisotropic etching, contact holes have sharp upper corners which make them difficult to fill (Fig. 24a). By successfully rounding these sharp edges with a *second* thermal flow (or *reflow*) cycle (Fig. 24b), contact coverage by a subsequent metal film is significantly improved. (Note that two separate flow cycles are preferred to a single post-etch flow cycle, since the appropriate degree of flow desired in the second cycle is normally less than that required for the first flow step). The ambient is also usually inert, rather than oxidizing, in order to avoid SiO_2 growth in the contacts during the reflow step. Because flow depends on BPSG composition, precise control of dopant uniformity across substrates is necessary to ensure uniform flow and consequent control of contact size.

BPSG deposition reactions for APCVD[4], LPCVD[38], and PECVD[39, 40] processes have all been reported. If hot-wall LPCVD reactors are used, they must be equipped with distributed feed gas systems so that adequately uniform BPSG films can be deposited. During deposition, the source gases for B and P compete with one another for inclusion into the final deposited film. This makes for a complex relation of reactant gas composition to eventual film doping[36].

The chemical composition of BPSG (and PSG), can be determined by several techniques, including the following: a) *wet chemical colorimetry*, which is the most accurate, analyzes the dissolved BPSG film; b) *x-ray photoelectron spectroscopy* (see Chap. 17), which is useful for determining the phosphorus content of the film; c) *Fourier transform infrared spectroscopy* (see Chaps. 1 and 5) which can measure boron levels quite accurately, but phosphorus levels less well accurately (due to the presence of a partially obscured phosphorus-oxygen absorption band); and d) *film etch rates in buffered HF*. Since the etch rate of BPSG in buffered HF depends on the concentration of both B and P in the film, a determination of the etch rate can provide a rapid, qualitative comparison of the composition between BPSG samples.

PROPERTIES AND CHEMICAL VAPOR DEPOSITION OF SILICON NITRIDE

Silicon nitride films are amorphous insulating materials that find three main applications in VLSI fabrication: 1) as final passivation and mechanical protective layers for integrated circuits, especially for parts encapsulated in plastic packages; 2) as a mask for the selective oxidation of silicon; and 3) as a gate dielectric material in MNOS devices. Silicon nitride also has a high dielectric constant (6-9 vs ~4.2 for CVD SiO_2), making it less attractive for interlevel insulation, because of the resultant higher capacitance between conductor layers.

Silicon nitride is highly suitable as a *passivation layer* because of its following properties: a) it behaves as a nearly impervious barrier to diffusion (in particular, moisture and sodium find it very difficult to diffuse through the nitride film); b) it can be prepared by PECVD to have a low compressive stress, which allows it to be subjected to severe environmental stress with less likelihood of delamination or cracking; c) its coverage of underlying metal is conformal; and d) it is deposited with acceptably low pinhole densities.

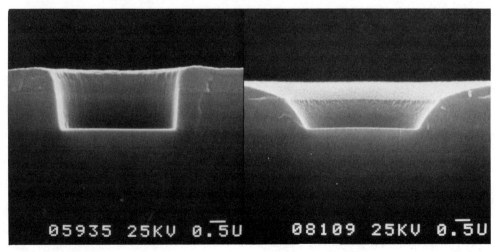

a)   b)

Fig. 24 (a) SEM of dry-etched contact window before reflow. (b) Reflowed BPSG film with 4 wt% P and 4 wt% B. Reflow was 930°C in N_2 for 25 min[37]. Courtesy of Applied Materials, Inc.

Silicon nitride is useful as a *masking layer for selective oxidation*[44], since oxygen also finds it difficult to penetrate silicon nitride. That is, silicon nitride is deposited directly on a silicon substrate (actually onto a thin, thermal stress-relief SiO_2 layer, called a *pad oxide*) and patterned. The wafer is then subjected to an oxidation step. The silicon nitride itself oxidizes very slowly, but is not penetrated by the oxygen. As a result it protects the underlying Si from oxidizing, while allowing a thermal SiO_2 layer to grow on regions of exposed Si. For example, at 1000°C in steam for 10 hours, only a few hundred Å of silicon nitride is converted to SiO_2, while more than 1 μm of SiO_2 can be grown on the exposed silicon. The silicon nitride is removed after the oxidation step.

As shown in Table 3, two techniques are used for depositing silicon nitride. When used as a mask for selective oxidation or as a gate dielectric material in MNOS structures, silicon nitride is generally deposited by high-temperature (700-800°C) LPCVD techniques, for reasons of film uniformity and lower processing cost[8,31]. When used as a passivation layer, the deposition process must be compatible with such low-melting-point metals as aluminum. Thus silicon nitride must be deposited by a low temperature process (300-400°C). For such applications PECVD is the deposition method of choice, as it can deposit nitride films at 200-400°C[32,43,44]. PECVD silicon nitride, however, tends to be nonstoichiometric, and

Table 3. Properties of PECVD Silicon Nitride and High Temperature CVD Nitride

Property		HT-CVD—NP 900°C	PE-CVD—LP 300°C
Composition		Si_3N_4	$Si_xN_yH_z$
Si/N ratio		0.75	0.8–1.0
Density		2.8–3.1 g/cm³	2.5–2.8 g/cm³
Refractive index		2.0–2.1	2.0–2.1
Dielectric constant		6–7	6–9
Dielectric strength		1×10^7 V/cm	6×10^6 V/cm
Bulk resistivity		10^{15}–10^{17} ohms/cm	10^{15} ohms/cm
Surface resistivity		$> 10^{13}$ ohms/square	1×10^{13} ohms/square
Stress at 23°C on Si		1.2–1.8×10^{10} dyn/cm² (tensile)	1–8×10^9 dyn/cm² (compressive)
Thermal expansion		4×10^{-6}/°C	$>4 < 7 \times 10^{-6}$/°C
Color, transmitted		None	Yellow
Step coverage		Fair	Conformal
H_2O permeability		Zero	Low–none
Thermal stability		Excellent	Variable > 400°C
Solution etch rate			
HFB	20–25°C	10–15 Å/min	200–300 Å/min
49% HF	23°C	80 Å/min	1500–3000 Å/min
85% H_3PO_4	155°C	15 Å/min	100–200 Å/min
85% H_3PO_4	180°C	120 Å/min	600–1000 Å/min
Plasma etch rate			
70% CF_4/30% O_2,			
150 W, 100°C		200 Å/min	500 Å/min
Na⁺ penetration		< 100 Å	< 100 Å
Na⁺ retained			
in top 100 Å		> 99%	> 99%
IR absorption			
Si–N max		~870 cm⁻¹	~830 cm⁻¹
Si–H minor		—	2180 cm⁻¹

contains substantial quantities of atomic H (10-30 at%). For this reason it is sometimes chemically represented as $Si_xN_yH_z$. Comparison of the refractive index of CVD silicon nitride with that of thermal silicon nitride gives a quick estimate of the stoichiometry of the CVD film. Higher refractive indices indicate silicon rich films. Table 3 compares the properties of LPCVD and PECVD silicon nitride films.

LPCVD silicon nitride is formed by reacting dichlorosilane ($SiCl_2H_2$) and ammonia (NH_3) at temperatures between 700-800°C according to the overall reaction:

$$3SiCl_2H_2 + 4NH_3 ------> Si_3N_4 + 6HCl + 6H_2 \qquad (9)$$

Silicon nitride depositions by LPCVD are controlled by a large number of deposition parameters including, temperature, total pressure, reactant ratios, and temperature gradients in the reactor. The deposition rate increases with increasing total pressure, or $SiCl_2H_2$ partial pressure, but decreases as the $[NH_3] : [SiCl_2H_2]$ ratio gets larger. A temperature ramp, as discussed in the section on hot-wall tube LPCVD reactors, is required for obtaining uniform depositions. In general, LPCVD silicon nitride films have a high density (2.9-3.1 g /cm^3), a dielectric constant of 6, and are more stoichiometric than PECVD silicon nitride films. Their etch rate in buffered HF is very slow (less than 10Å /min), and their hydrogen content (up to 8 at%) is less than in PECVD films. In addition, they exhibit excellent step coverage and relatively low particulate contamination. Such films, however, have tensile stresses of ~1x10^{10} dynes /cm^2, which is about an order of magnitude higher than that of TEOS-deposited SiO$_2$. This may cause LPCVD silicon nitride films greater than ~2000Å to crack. Sufficient NH$_3$ is also needed to insure that all of the SiH$_4$ is consumed. If insufficient NH$_3$ is available, the films become silicon rich. As a result, depositions are usually carried out with significant excess flow of NH$_3$. It should also be mentioned that, as described in Chap. 7, an anneal in hydrogen is often performed on MOS circuits to reduce the interface trapped charge density, Q_{it}, after the final Al layer is deposited and patterned. Since silicon nitride is such an impervious diffusion barrier, this anneal must be performed on wafers prior to depositing a silicon nitride passivation layer.

Silicon nitride deposition by PECVD was described by Sterling and Swann in 1965. The

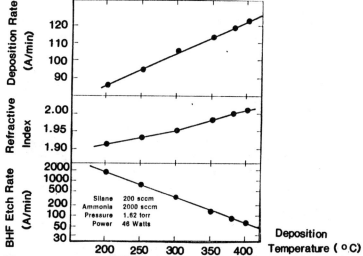

Fig. 25 Effect of deposition temperature on PECVD nitride parameters. Reprinted by permission of Solid State Technology, published by Technical Publishing, a company of Dun & Bradstreet[12].

Table 4. CVD DEPOSITION REACTIONS

PRODUCT	REACTANTS	METHOD	TEMP (°C)	COMMENTS
Polysilicon	SiH_4	LPCVD	580-650	may be *in situ* doped
Silicon Nitride	$SiH_4 + NH_3$	LPCVD	700-900	
	$SiCl_2H_2 + NH_3$	LPCVD	650-750	
	$SiH_4 + NH_3$	PECVD	200-350	
	$SiH_4 + N_2$	PECVD	200-350	
SiO_2	$SiH_4 + O_2$	APCVD	300-500	poor step coverage
	$SiH_4 + O_2$	PECVD	200-350	good step coverage
	$SiH_4 + N_2O$	PECVD	200-350	
	$Si(OC_2H_5)_4$ [TEOS]	LPCVD	650-750	liquid source, conformal
	$SiCl_2H_2 + N_2O$	LPCVD	850-900	conformal
Doped SiO_2	$SiH_4 + O_2 + PH_3$	APCVD	300-500	PSG
	$SiH_4 + O_2 + PH_3$	PECVD	300-500	PSG
	$SiH_4 + O_2 + PH_3 + B_2H_6$	APCVD	300-500	BPSG, low temperature flow
	$SiH_4 + O_2 + PH_3 + B_2H_6$	PECVD	300-500	BPSG, low temperature flow

overall deposition reaction is written as:

$$SiH_4 + NH_3 \text{ (or } N_2) \xrightarrow{\text{(200-400°C, rf)}} Si_xN_yH_z + H_2 \qquad (10)$$

where silane and ammonia or nitrogen are reacted in a plasma at 200-400°C. The use of a high (100-1000:1) N_2: SiH_4 ratio is required to prevent silicon rich film formation (since the N_2 dissociation rate is much lower than that of silane). On the other hand, NH_3 dissociates much more readily, and thus lower NH_3: SiH_4 ratios can be used (e.g. 5-20:1).

The deposition rate depends strongly on rf power, gas flow, chamber pressure and frequency (Fig. 25). If a PECVD deposition process is to be developed, it would be useful to approach the problem with a statistically-based set of experiments as discussed in Chap. 18. A case study of such a process development effort is found in Ref. 44. The stress in PECVD silicon nitride films can also be changed by altering a variety of deposition conditions.

Infrared spectroscopy shows that an appreciable amount of hydrogen in the form of Si-H and N-H is found in PECVD silicon nitride films. The total amount of hydrogen can reach 18-22 at% in films deposited from SiH_4-NH_3 near 300°C. Lower hydrogen content is reportedly found in films from SiH_4-N_2[12]. The presence of large quantities of hydrogen has been reported to be detrimental to IC devices. For example, significant threshold shifts are observed, and etching characteristics (both wet and dry) of the films are impacted. Another advantage of using N_2 rather than NH_3 is that it yields compressive silicon nitride films over a wider range of deposition conditions. Oxygen in the form of Si-O bonds has also been detected in PECVD silicon nitride films. Oxygen concentration increases as the deposition temperature decreases, and is believed to be due to moisture or oxygen released from the reaction chamber walls. Table 4 is a summary of the CVD reactions for poly-Si, SiO_2, PSG, BPSG, as well as silicon nitride.

OTHER DIELECTRIC FILMS DEPOSITED BY CVD

Silicon Oxynitrides

Materials can be prepared with characteristics between those of nitrides and oxides, and these are called silicon oxynitrides [$SiO_xN_y(H_z)$]. They are formed by reacting SiH_4 with N_2O and NH_3. Properties can be tailored for improved thermal stability, low stress, and crack resistance[45]. Such films are also less less permeable to moisture and other contaminants than deposited oxides. The use of helium as a carrier gas is reported to increase the uniformity of the refractive index of the films[46]. Silicon oxynitrides also are reported to exhibit good properties for use as low-temperature deposited insulators between Al conductors. Two-layer films of PECVD silicon oxynitride and silicon nitride have been used in multilevel interconnect planarization schemes utilizing the differential etch rates of the two materials[46].

SIPOS

Silicon rich oxide films (i.e. the so-called semi-insulating polycrystalline silicon, or *SIPOS*) can also be deposited in APCVD, LPCVD, and PECVD reactors. This material has been reported to be excellent for passivating high field transistors[48], and has been studied for use as an injector in electrically-erasable read-only memories (EEROMs)[49] as well as a material for forming emitters in advanced super-gain bipolar transistors[52]. The chemical composition of SIPOS is SiO_x, where x ranges from 0.48 to <2. Such films can be formed by keeping the O_2: SiH_4 flow rate ratio to <3.5. The composition of the films can be analyzed with Auger emission spectroscopy (Chap. 17), with a thermal SiO_2 film as a standard.

REFERENCES

1. A. C. Adams, "Dielectric and Polysilicon Film Deposition," in *VLSI Technology* (S.M. Sze, ed.), Chap. 3, p. 93. McGraw-Hill, New York, 1983.
2. W. Kern, "Deposited Dielectrics for VLSI," *Semiconductor International* 8 (7), 122 (July '85).
3. C. Murray, "Mass-Flow Controllers," *Semiconductor International* 8 (10), 72 (Oct. 1985).
4. W. Kern and V. Ban, "Chemical Vapor Deposition of Inorganic Thin Films," in *Thin Film Processes* (J. L. Vossen and W. Kern, eds.), pp. 257-331. Academic, New York, 1978.
5. M. Hammond, "Intro. to Chemical Vapor Deposition," *Solid State Technol.* Dec. '79, p. 61.
6. W. Kern and G. L. Schnable, "Low Pressure Chemical Vapor Deposition for VLSI Processing - A Review," *IEEE Trans. Electron Devices* ED-26, 647 (1979).
7. P. Singer, "Techniques of Low Pressure CVD," *Semiconductor Intl.*, p. 72 (May 1984).
8. R. S. Rosler, "Low Pressure CVD Production Processes for Poly, Nitride, and Oxide," *Solid State Technol.* 20(4), 63 (April 1977).
9. M. de Fraiteur and J. Goldman, "Pressure Control in LPCVD Systems," *Semiconductor International.*, p. 250, May 84.
10. A. Learn, "Modeling the Reaction of Low Pressure Chemical Vapor Deposition of Silicon Dioxide," *J. Electrochem. Soc.* 132, 390 (Feb. 1985).
11. A. C. Adams, "Plasma Deposition of Inorganic Films," *Solid State Technol.*, p.135, Apr. 83.
12. B. Gorowitz, T. B. Gorczyca, R. J. Saia, "Applications of PECVD in VLSI," *Solid State Technol.*, p. 197 June 1985.
13. A. Weiss, "PECVD: Silicon Nitride and Beyond," *Semiconductor Intl.* 6(7), 88 (July 1983).
14. W. L. Johnson, "Design of Plasma Deposition Reactors," *Solid State Technol.*, 191, Apr. 83.

15. J. Y. Chen, R. Henderson, "Photo-CVD for VLSI," *J. Electrochem. Soc.* **131**, 2147 Sept. 84.

16. R. Solanki, C. Moore, & G. Collins, "Laser Induced CVD," *Solid State Technol.*, 220 June 85.

17. Furumura, *et al.*, "Selective Growth of Poly-Si," *J. Electrochem. Soc.* **133**, 379 (Feb. 86).

18. M. S. Choi and E. W. Hearn, "Stress Effects in Boron-Implanted Polysilicon Films," *J. Electrochem. Soc.* **131**, 2443 (Oct. 1984).

19. M. M. Mandurah, K. C. Saraswat, and T. I. Kamins, "A Model for Conduction in Polycrystalline Silicon, Part I - Theory," *IEEE Trans. Electron Devices* **ED-28**, 1163 (Oct. 1981); "Comparison of Theory and Experiment," *ibid.* **ED-28**, 1171 (Oct. 1981).

20. G. Harbeke, *et al.*, "Growth and Physical Properties of LPCVD Polycrystalline Silicon Films," *J. Electrochem. Soc.* **131**, 675 (March 1984).

21. G. Harbeke, *et al.*, "LPCVD Poly-Si: Growth and Physical Properties of *In Situ* Phosphorus Doped and Undoped Films," *RCA Review* **44**, 287 (June 1983).

22. B. S. Meyerson and W. Olbricht, "Phosphorus-Doped Poly-Si via LPCVD," *J. Electrochem. Soc.* **131**, 2361 (Oct. 1984).

23. T. I. Kamins, "Structure /Properties LPCVD Si Films," *J. Electrochem. Soc.* **127**, 686 Mar 80.

24. M. L. Walker and N. E. Miller, "Control of Polysilicon Film Properties," *Semiconductor Intl.* **7**(5), 90 (May 1984).

25. T. I. Kamins, "Resistivity of LPCVD Poly-Si Films," *J. Electrochem. Soc.* **126**, 833 (1979).

26. N. C. Lu, *et al.*, "The Effect of Film Thickness on the Electrical Properties of LPCVD Polysilicon Films," *J. Electrochem. Soc.* **131**, 898 (April 1984).

27. T. Ohzone, *et al.*, "Ion-Implanted Thin Poly-Si High Value Resistors for Static RAM Applications," *IEEE Trans. Electron Devices* **ED-32**, 1749 (Sept. 1985).

28. S. R. Wilson, *et al.*, "Properties of Ion-Implanted Polycrystalline Si Layers Subjected to Rapid Thermal Annealing," *J. Electrochem. Soc.* **132**, 922 (April 1985).

29. H. Sunami, "Thermal Oxidation of Phosphorus-Doped LPCVD and APCVD Poly-Si Films," *J. Electrochem. Soc.* **125**, 892 (1978).

30. K. Shinada, S. Mori, and Y. Mikata, "Reduction in Poly-Si Oxide Leakage Current by Annealing Prior to Oxidation," *J. Electrochem. Soc.* **132**, 2185 (Sept. 1985).

31. W. Kern and R. S. Rosler, "Advances in Deposition Processes for Passivation Films," *J. Vac. Sci. and Technol.* **14**, 1082 (1977).

32. T. B. Gorczyca and B. Gorowitz, "PECVD of Dielectrics,"in *VLSI Electronics Microstructure Science* (N. Einspruch, ed.), vol. 8, Chap. 4, p. 69. Academic, New York, 1984.

33. A. C. Adams and C. D. Capio, "The Deposition of Silicon Dioxide Films at Reduced Pressure," *J. Electrochem. Soc.* **126**, 1042 (1979).

34. B. Mattson, "CVD Films for Interlayer Dielectrics," *Solid State Technol.*, 60 (Jan. 80).

35. A. Takamatsu, *et al.*, "Plasma Activated Deposition and Properties of PSG Film," *J. Electrochem. Soc.* **131**, 1865 (Aug. 1984).

36. W. Kern and G. L. Schnable, "CVD BPSG for Si Devices", *RCA Review*, p. 423, Sept. 82.

37. J. E. Tong, K. Schertenleib, and R. A. Carpio, "Process and Characterization of PECVD BPSG Films for VLSI Applications," *Solid State Technol.* **27**(1), 161 (Jan. 1984).

38. T. Foster, *et al*, "Low Pressure BPSG Dep. Process," *J. Electrochem. Soc.* **132**, 506 Feb. 85.

39. W. Kern and R. Smeltzer, "BPSG for Integrated Circuits," *Solid State Technol.*, 171 June 85.

40. I. Avigal, "Inter-metal Dielectric and Passivation-Related Properties of Plasma BPSG," *Solid State Technol.* **26**(10), 217 (Oct. 1983).

41. J. A. Appels, *et al.*, "Local Oxidation of Silicon and its Applications in Semiconductor Device Technology," *Phillips Res. Rep.* **25**, 118 (1970).

42. C. Blaauw, "Preparation and Characterization of Plasma-Deposited Silicon NItride," *J. Electrochem. Soc.* **131**, 1114 (May 1984).

43. W. A. P. Claasen, *et al.*, "Influence of Deposition Temp, Gas Pressure, Gas Phase Composition, and RF Freq. on Composition and Mech. Stress of Plasma SiN Layers," *J. Electrochem. Soc.* **132**, 893 (April 1985).

44. P. W. Bohn and R. C. Manz, "A Multiresponse Factorial Study of Reactor Parameters in PECVD Growth of Amorphous Silicon Nitride," *J. Electrochem. Soc.* **132**, 1981 Aug. 85.

45. K. Takasaki, *et al.*, *Electrochem Soc. Ext. Abs.* **80-2**, 260 (1980).

46. V. S. Nguyen, "Highly Reliable High-Voltage Transistors by use of a SIPOS Process," *Electrochem Soc. Ext. Abs.* **83-1**, 216 (1983).

47. D. Barton, C. Maize, "A Two Level Metal CMOS Process for VLSI Circuits," *Semiconductor Intl.* **8**(1), 98 (Jan. 1985).

48. T. Matsushita, *et al.*, "Plasma Deposition of Silicon Nitride and Silicon OxyNitride Using Inert Carrier Gases as Transport Agents," *IEEE Trans. Electron Dev.* **ED-23**, 826 Aug. 76.

49. P. Pan, *et al.*, "The Composition and Properties of PECVD Silicon Oxide Films," *J. Electrochem. Soc.* **132**, 2013 (Aug. 1985).

50. M. Sternheim, *et al.*, "Properties of Thermal Oxides Grown on Phosphorus In Situ Doped Polysilicon," *J. Electrochem. Soc.* **130**, 1735 (Aug. 1983).

51. H.N. Yu, *et al*, "1 µm MOSFET VLSI Tech.: Part 1-An Overview", IEEE Trans. Electron Devices, **ED-26**, 318, (1979).

52. A.R. Reinberg, "Dry Processing for Fabrication of VLSI Devices" in [N.G. Einspruch, Ed.] *VLSI Electronics- Microstructure Science*, Vol. 2, Academic Press, New York, 1981, Chap. 1.

53. J. B. Price, *et al.*, "LPCVD of Tungsten and Tungsten Silicide", Semicon /West, May, 1986.

PROBLEMS

1. Exponential temperature dependence is alternately represented by the form $\exp(-E_a/kT)$, or by $\exp(-\Delta H/RT)$, where E_a is the value of the activation energy in electron volts, and ΔH is its value in Kcal /mole. Show that if $E_a = 1$ eV, the corresponding value of ΔH is 23 Kcal /mole.

2. If a film is deposited conformally into a window that is 0.5 µm high and 1.0 µm wide, sketch the film when it has a thickness of 2000Å, 5000Å, 10,000Å. If the coverage is like that shown in Fig. 19a, what will the deposted film look like when its thickness is 0.5 µm. What does this effect imply for the application of depositing conformal (and nonconformal) films over high aspect ratio windows?

3. Conjecture why diffusion doped polysilicon can exhibit lower resistivities than polysilicon doped with ion implantation.

4. The activation energy of the SiH_4-O_2 reaction is ~0.6 eV for undoped oxides, while that for the undoped TEOS reaction is ~1.9 eV. If the deposition rate of CVD SiO_2 by the SiH_4-O_2 reaction is 250Å /min at 400°C, and is the same for the TEOS reaction at 700°C, find the temperatures for both of these reactions at which the deposition rate will double.

5. As CVD reactors are being designed as single wafer machines (as opposed to batch deposition systems), comment on some of the problems that have to be overcome in single wafer designs.

6. The minimum temperature for flow in phosphorus-doped SiO_2 is ~1000°C. For some VLSI fabrication processes, the maximum allowable temperature is 900°C. Suggest four techniques that are candidates for producing layers of dielectric material between two conducting layers that is planarized, without having to heat the wafers to 1000°C.

7. Explain the difference between the terms *flow* and *reflow*.

8. Summarize the differences in the properties of high-temperature deposited silicon nitride, and plasma-deposited silicon nitride.

7

THERMAL OXIDATION

of

SINGLE CRYSTAL SILICON

The formation of the oxide of silicon (SiO_2) on a silicon surface is termed *oxidation*. The ability to form this oxide (SiO_2), which is stable and tenacious, provides the foundation for planar processing of silicon integrated circuits. Although there are several ways to produce SiO_2 directly on the Si surface, it is most often accomplished by thermal oxidation, in which the silicon is exposed to an oxidizing ambient (O_2, H_2O) at elevated temperatures. Thermal oxidation is capable of producing SiO_2 films with controlled thickness and Si /SiO_2 interface properties. Other techniques used to grow SiO_2 include plasma anodization and wet anodization. These two techniques, however, have not found widespread applicability in VLSI processing.

Thermally grown SiO_2 is found in VLSI processing applications in thicknesses ranging from 60Å to 10,000Å. Some of the functions of these films include: a) masking against ion implantation and diffusion; b) passivation of the silicon surface; c) isolation of individual devices (e.g. local oxidation of silicon, or *LOCOS*); d) use as a gate oxide and capacitor dielectric in MOS devices; and e) use as a tunneling oxide in electrically alterable ROMs (EAROMs). Table 1 lists the range of oxide thicknesses commonly used for these applications.

In this chapter we discuss: a) the properties of silica glass and thermally grown SiO_2; b) the kinetics of oxidation; c) the initial oxidation stage and direct nitridation of silicon and SiO_2; d) the oxidation rate dependence on various growth conditions; e) the properties of the Si /SiO_2 interface; f) dopant impurity redistribution during oxidation; g) oxidation equipment; and h) the measurement of oxide thickness. The oxidation of polysilicon is discussed in Chap. 6. Table 2 lists some typical values for selected properties of thermally grown SiO_2.

Table 1. RANGE OF THERMAL SiO_2 THICKNESSES USED IN VLSI PROCESSING

SiO_2 Thickness	Application
60 - 100 Å	Tunneling Oxides
150 - 500 Å	Gate Oxides, Capacitor Dielectrics
200 - 500 Å	LOCOS Pad Oxide
2000 - 5000 Å	Masking Oxides, Surface Passivation Oxides
3000 - 10,000 Å	Field Oxides

Table 2. SELECTED PHYSICAL CONSTANTS OF THERMAL SILICON DIOXIDE

Dc Resistivity (Ω-cm), 25°C	10^{14}-10^{16}	Melting Point (°C)	~1700
Density (g /cm³)	2.27	Molecular Weight	60.08
Dielectric Constant	3.8 - 3.9	Molecules /cm³	2.3×10^{22}
Dielectric Strength (V /cm)	5-10×10^6	Refractive Index	1.46
Energy Gap (eV)	~8	Specific Heat (J /g°C)	1.0
Etch rate in Buffered HF (Å /min)	1000	Stress in film on Si	2 - 4×10^9
Infrared Absorption Peak	9.3	(dyne /cm²)	compression
Linear Expansion Coefficient (cm /cm°C)	5.0×10^{-7}	Thermal Conductivity (W/cm°C)	0.014

PROPERTIES OF SILICA GLASS

The glass or amorphous state of silicon dioxide (SiO_2) is termed *fused silica*. The misnomer "fused quartz" or "quartz" is often seen in the literature when describing furnace hardware, and such hardware is actually fabricated of *fused silica*. Quartz is the name of one of the crystalline phases of SiO_2. The amorphous state of SiO_2 is thermodynamically unstable below 1710°C. This implies that a tendency for transformation from the amorphous to the crystalline state (*devitrification*) exists at temperatures below 1710°C. The rate of transformation is found to decrease with decreasing temperature, essentially approaching zero at ≤ 1000°C. The white crystalline material that exists on fused silica furnace tubes after long use at high temperature is due to the devitrification process, and can be a source of particulate contamination.

Stevels and Kats[1] have developed a model that allows the structure of fused silica to be more clearly visualized. Although amorphous, and therefore not having long range structure, a short range order does exist. The short range order is centered around the structural formula for the material which is SiO_4^{4-}. The structure can be described as follows: The silicon atom, with a valence of +4 is located at the center of a regular polyhedron (in this case a tetrahedron or triangular polyhedron) with oxygen ions (O^{2-}) at each of its corners. The tetrahedral distance between silicon and oxygen ions is 1.62Å, while the oxygen-oxygen ion distance is 2.27Å.

The polyhedra are then joined to each other by an oxygen ion called a *bridging oxygen*, which is shared between the two touching polyhedra. This is the case for crystalline SiO_2. For silicon dioxide in the amorphous state (fused silica), however, some of the vertices of the

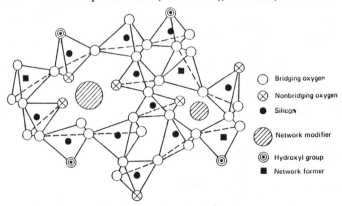

Fig. 1 The structure of fused silica glass[2]. (© 1965 IEEE).

polyhedra have *nonbridging oxygen ions*, which are not shared. The greater the ratio of bridging to non-bridging oxygens, the better the cohesiveness of the glass.

The atomic movement in the glass is more likely to occur by the movement of oxygen atoms rather than silicon atoms, since the rupturing of four Si-O bonds is required to free a silicon atom, while the rupture of only two Si-O bonds are required to free a bridging oxygen atom.   If this scission occurs, an oxygen ion vacancy is formed.   This vacancy has a net positive charge in the network. Both bridging and nonbridging oxygen vacancies may be formed, but the nonbridging is more likely to occur based on binding energy considerations.

SiO_2 films grown by the oxidation of silicon, have an amorphous structure with a random network of polyhedra. The density of thermally grown fused silica (2.15-2.25 g /cm^3) is less than that of crystalline quartz (2.65 g /cm^3). The lower density implies a more open structure. This open structure is conducive to the interstitial diffusion of impurities through the network.

Impurities introduced into fused silica radically change its properties. As is the case in silicon, both *substitutional* and *interstitial* impurities exist. The substitutional impurities replace silicon in the structure. The most important impurities of this type are boron (B^{3+}) and phosphorus (P^{5+}) ions. Another term for these impurities is *network formers*, since they themselves can be the basis of a glassy structure (B_2O_3 or P_2O_5). The missing or extra electrons in the tetrahedra, when these materials are added, are accommodated by the elimination or formation of bridging oxygen ions, respectively. The elimination of bridging oxygens tend to weaken the network (e.g. boron in SiO_2).

The oxides of Na, K, Pb, and Ba enter the structure as *interstitial* impurities. When this occurs the metal ion gives up its oxygen to the network, thereby producing two nonbridging oxygen ions,which replaces the original bridging oxygen. The additional nonbridging oxygen also tends to weaken the structure, allowing the glass to be more porous, and thereby increasing the diffusion rate of other species within the glass. Impurity oxides of this type are termed *network modifiers* since they do not form glasses themselves.

Water vapor is a prevalent impurity in fused silica, and can enter from the atmosphere or be grown-in during wet oxidations (or even in so-called dry oxidations, as described in the section entitled: *Oxidation Growth Rates: Water Dependence*). The water vapor combines with a bridging oxygen to form a pair of stable nonbridging hydroxyl groups (OH$^-$). This reaction can be represented by:

$$H_2O + Si\text{--}O\text{--}Si = Si\text{--}OH + OH\text{--}Si \qquad\qquad (1)$$

The increase in nonbridging oxygens again tends to weaken the silicon network, thereby increasing the diffusivities of many materials in the network. The presence of OH can be detected by IR spectroscopy, since the Si-OH stretching frequency is different than that of Si-O (resulting in absorption peaks at different wavelengths).

Revesz[2] developed a schematic model (Fig. 1) which shows the fused silica structure, including the presence of the *network formers* (P^{5+}, B^{3+}), the *bridging* and *nonbridging* oxygens, and *network modifiers* (Na^+, K^+, Pb^{2+}, and Ba^{2+}).

OXIDATION   KINETICS

Silicon exhibits a propensity to form a stable oxide (SiO_2).   Freshly cleaved Si when exposed to an oxidizing ambient (e.g. O_2, H_2O) will form a very thin (<20Å) oxide layer, even at room temperature.   When Si is exposed to an oxidizing ambient at elevated temperatures, more rapid growth and thicker oxides are produced.

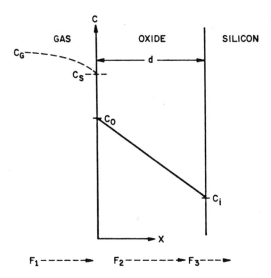

Fig. 2   Silicon dioxide growth by thermal oxidation.

The basic mechanism for the formation of SiO_2 from Si is well understood.   Deal and Grove[3] developed a mathematical model which accurately describes the growth kinetics of oxide films with thicknesses >300Å. They proposed that oxidation proceeds by the diffusion of an *oxidant* (molecular H_2O or O_2) through the existing oxide to the Si /SiO_2 interface, where the molecules react with Si to form SiO_2.   The reactions governing the formation of SiO_2 are given as:

$$Si \text{ (solid)} + O_2 \text{ (vapor)} ----> SiO_2 \text{ (solid)}   \quad Dry\ Oxidation \qquad (2)$$

$$Si \text{ (solid)} + H_2O \text{ (vapor)} ----> SiO_2 \text{ (solid)} + 2H_2 \quad Wet\ Oxidation \qquad (3)$$

The oxidation reaction occurs at the Si /SiO_2 interface. Therefore, as the oxide grows, silicon is consumed and the interface moves into the silicon.   Figure 2 shows the relationship between the thickness of the grown oxide film and the consumed silicon.   Based on the relative densities and molecular weights of Si and SiO_2 it is found that the amount of silicon consumed is 44% of the final oxide thickness. Thus, if 10,000Å of oxide is grown, 4400Å of the Si will be consumed. This relationship is important for calculating step heights that form in the silicon as a result of varying oxidation rates at different locations of the silicon surface.

Linear-Parabolic Model

The linear parabolic model developed by Deal and Grove[3] results in an accurate representation of the growth of SiO_2 on Si over a wide range of thicknesses (~300Å - 20,000Å), temperatures (700–1300°C), and oxidant partial pressures (0.2–25 atm).

The model assumes that the oxidation process occurs as a result of two fluxes that sequentially transport the oxidizing species from the gas to the Si /SiO_2 interface, where a third flux is involved in the consumption of the oxidant by the reaction with the silicon.   These fluxes are shown schematically in Fig. 2. (*Flux* is defined as the number of atoms or molecules crossing a unit area in a unit time, and has the dimensions, number /cm^2sec.) The three fluxes are formally identified as: a) the flux of the oxidizing species moving from the bulk of the gas phase to the gas /oxide interface, F_1; b) the flux of the oxidizing species as it diffuses through the existing oxide to the Si /SiO_2 interface, F_2; and c) the flux of the oxidizing species as it is consumed by reaction at the Si /SiO_2 interface, F_3. In the steady-state condition, the three fluxes will be equal, i.e. $F_1 = F_2 = F_3$. The model continues by approximating each of these fluxes as they relate to physical quantities.

The oxidizing species are introduced into the reactor (e.g. fused silica tube) as a gas, where they are then brought from the *bulk* of the gas stream to the *wafer surface* by a mass-transport process. The flux, F_1, that occurs as a result of this transport, is assumed to arise from the concentration difference between the oxidizing species in the bulk of the gas, C_g, and that at the oxide surface, C_s. It is further postulated that the magnitude of the flux is linearly proportional to the concentration difference, and the proportionality constant, h_g, is defined as the *mass-transfer coefficient*, or:

$$F_1 = h_g (C_g - C_s) \tag{4}$$

The mass transfer coefficient, h_g, is related to the diffusivity of the oxidizing species, D, and the thickness of the stagnant layer. (The concept of a *stagnant layer* is discussed in Chap. 5).

Since the three fluxes are eventually equated, it is neccesary to be able to express them all in terms of the same set of variables. Therefore, it is useful to express Eq. 4 in terms of the equilibrium concentration of gas species in the oxide, instead of in terms of the concentrations in the gas. To do so, we invoke *Henry's law*, which states that the concentration of a species dissolved in a solid at equilibrium is proportional to the partial pressure of the species at the solid surface. That is, the oxidant is absorbed from the gas stream by the existing oxide layer, and the absorption process is described by Henry's law, as:

$$C_o = H P_s \tag{5}$$

and

$$C^* = H P_g \tag{6}$$

where: H is the Henry's law constant; C_o is the equilibrium concentration of the oxidizing species in the oxide at the outer surface; P_s is the partial pressure of the oxidizing species at the oxide surface; C^* is the equilibrium concentration in the bulk of the oxide; and P_g is the partial pressure in the bulk of the gas phase.

Assuming ideal gas behavior (PV = kT) in addition to Henry's law, we can rewrite C_g and

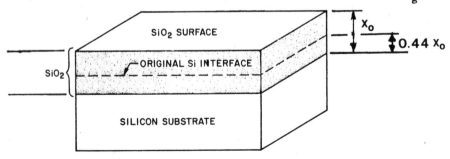

Fig. 3 The thermal oxidation of silicon model[3]. Reprinted with permission of the American Physical Society.

C_s in terms of the partial pressures as follows:

$$C_g = \frac{P_g}{kT}$$
(7)

$$C_s = \frac{P_s}{kT}$$
(8)

Combining Eqs. 4, 5, 6, 7, and 8 we rewrite F_1 as:

$$F_1 = h(C^* - C_o)$$
(9)

where h is a new gas phase mass-transfer coefficient, and is related to h_g by:

$$h = \frac{h_g}{HkT}$$
(10)

Invoking Henry's law for this process implies that the oxidizing species moves through the oxide in molecular form, since the law does not hold under conditions of molecular dissociation.

Once inside the oxide layer the oxidants diffuse towards the Si /SiO$_2$ interface, and this transport effect is described by the second flux, F_2. In steady state, this flux can be expressed as:

$$F_2 = -D\left[\frac{dC}{dx_o}\right] = -D\frac{(C_o - C_i)}{x_o}$$
(11)

in analogy with Fick's law, where D is the diffusion coefficient of the oxidizing species in the oxide, C_i and C_o are the concentrations of the oxidizing species at the growth interface and oxide surface, respectively, and x_o is the thickness of the growing oxide.

Finally, we express the reaction rate flux, F_3, as being proportional to the concentration of the oxidizing species at the interface, C_i:

$$F_3 = k_s C_i$$
(12)

where k_s is the reaction rate constant of the chemical reaction taking place at the interface.

Under steady state conditions (no build-up or depletion of oxidizing species) the three fluxes must be equal ($F_1 = F_2 = F_3 = F$). The result of setting Eqs. 9, 11, and 12 equal is a set of simultaneous equations, with solutions given by:

$$C_i = \frac{C^*}{1 + \left(\frac{k_s}{h}\right) + \left(\frac{k_s x_o}{D}\right)}$$
(13)

$$C_o = \frac{C^*\left(1 + \frac{k_s x_o}{D}\right)}{1 + \left(\frac{k_s}{h}\right) + \left(\frac{k_s x_o}{D}\right)}$$
(14)

We find two important limiting cases for Eqs. 13 and 14. The first arises when the diffusion constant D is very small, and thus the flux of oxidant through the SiO$_2$ is small compared to the flux associated with the reaction at the Si /SiO_2 interface. This is referred to as the *diffusion controlled* case, and under such conditions $C_i \longrightarrow 0$ and $C_o \longrightarrow C^*$.

The second case is encountered when D is very large, and the process is therefore *reaction controlled*. In that case $C_i = C_o = C^* / (1 + k_s /h)$. Under such conditions, a bountiful supply of oxidant is supplied to the Si /SiO_2 interface and the oxidation rate is controlled by k_s and C_i.

We are now in a position to calculate the growth rate (dx_o /dt) of the oxide, by making use of these solutions together with the definition of the number of oxidant molecules incorporated into a unit volume of oxide, N_1. That is, since there are 2.2×10^{22} molecules of SiO_2 per cm^3 in the oxide, and since two oxygen atoms are incorporated per SiO_2 molecule, the value of N_1 for dry (O_2) oxidation is 2.2×10^{22} cm^{-3}, and 4.4×10^{22} cm^{-3} for wet oxidation (H_2O).

If we combine Eqs. 12 and 13 together with the definition of flux, we arrive at:

$$ N_1 \frac{d x_o}{d t} = F_3 = \frac{k_s C^*}{1 + \dfrac{k_s}{h} + \dfrac{k_s x_o}{D}} \tag{15} $$

The differential equation, Eq. 15 can be solved by applying the initial boundary condition, $x_o = x_i$ at t = 0. The value of x_i may include the native oxide that had grown on Si in room ambient, or the oxide present from a previous process. The solution to Eq. 15 is:

$$ x_o^2 + A x_o = B(t + \tau) \tag{16} $$

where;

$$ A = 2D[\frac{1}{k_s} + \frac{1}{h}] \tag{17} $$

$$ B = \frac{2DC^*}{N_1} \tag{18} $$

$$ \tau = \frac{x_i^2 + A x_i}{B} \tag{19} $$

The τ in Eq. 19 represents the time displacement needed to account for the initial oxide layer, x_i.

The relationship predicted by Eq. 16 is obeyed under widely varying conditions of temperature, pressure, and oxidizing species. It fails however, to predict the initial stage of oxidation ($x_o < 300 \text{Å}$) where the growth rate is found to greatly exceed that predicted by Eq. 15. We undertake a discussion of initial oxidation in a subsequent section.

Equation 16 can be simplified by examining its limiting cases. For very long oxidation times, t>>τ, Eq. 16 reduces to:

$$ x_o^2 = Bt \tag{20} $$

This is termed the *parabolic growth law* and B is the *parabolic rate constant* defined by Eq. 18. We see that B is *proportional to the diffusion constant*, D, of the oxidizing species through the existing oxide. This implies that oxide growth in the parabolic regime is *diffusion controlled*. In other words, as the oxide layer gets thicker, the oxidizing species must diffuse through a larger distance to arrive at the Si /SiO_2 interface. The reaction thus becomes limited by the rate at which the oxidizing species diffuses through the oxide, and the oxide growth is proportional to

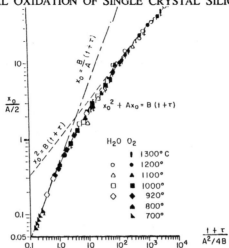

Fig. 4 The relationship governing the thermal oxidation of silicon and its two limiting forms[3]. Reprinted with permission of the American Physical Society.

the square-root of the oxidizing time.

For very short times, on the other hand, $(t + \tau) \ll A^2/4B$. Thus, Eq. 16 reduces to:

$$x_o = \frac{B}{A}(t + \tau) \qquad (21)$$

Equation 20 is termed the *linear growth law* and (B/A) is the *linear rate constant*. From Eqs. 17 and 18 we observe that (B/A) is independent of the diffusion coefficient and is given by:

$$\frac{B}{A} = \frac{k_s h}{k_s + h}[\frac{C^*}{N_1}] \qquad (22)$$

Figure 4 shows a plot of the measured oxide thickness for O_2 and H_2O at various temperatures between 700° and 1300°C as a function of oxidation time (normalized for temperature and oxidizing species). The results cluster very well and follow the curve predicted by Eq. 16. In the limits (short and long times) the results approach the lines given by $x_o = B/A(t + \tau)$ and $x_o^2 = Bt$ (representing the linear and parabolic limits, respectively). We see in Fig. 4 that B/A is most influential at short oxidation times and low temperatures, while B becomes dominant at elevated temperatures and longer times. The transition region between the linear and parabolic dependence is also shown.

The values of B and B/A were obtained as a function of temperature by Deal and Grove[3]. They found that both rate constants followed an exponential dependence on temperature (Arrhenius behavior, see Appendix 3). Figure 5 shows the parabolic rate constant B, as a function of reciprocal temperature for wet and dry oxidations. The behavior of B with temperature, is expressed as:

$$B = B_o e^{\left(\frac{-E_a}{kT}\right)} \qquad (23)$$

where B_o is the pre-exponential constant (which depends on C^* and N_1), E_a is the activation energy for the process (which in this case is the activation energy required for diffusion of O_2 or H_2O through SiO_2), k is Boltzmann's constant, and T is the temperature, in °K.

The values of activation energy obtained from the data in Fig. 5 are 1.24 eV for dry oxidation (O_2) and 0.74 eV for wet oxidation (H_2O). Since the parabolic rate constant depends on the diffusivity, D, of the oxidizing species, we can compare these measured activation energies with those obtained for bulk diffusion of O_2 and H_2O in fused silica (SiO_2). These values have been measured and are reported[3] as 1.17 eV and 0.80 eV for O_2 and H_2O diffusion in fused silica, respectively, agreeing well with the activation energies of the parabolic growth of SiO_2.

Figure 6 shows the temperature dependence of the linear rate constant (B /A) for dry and wet oxidations. Again, an Arrhenius dependence similar in form to Eq. 23 is found. The activation energies calculated from these curves are 1.96 eV and 2.0 eV, for wet and dry oxidations, respectively. These results show that both wet and dry oxidation have similar activation energies. This is indicative that both processes are controlled by the same mechanism. The controlling mechanism believed to operate in the linear region is the breaking of Si-Si bonds, which provides atomic silicon to react with the oxidizing species. The value of the activation energy closely agrees with the energy required to break Si-Si bonds (1.83 eV /molecule). Such bond breaking must occur in order to form SiO_2 from Si and either O_2 or H_2O.

It is also known that the reaction rate constant for oxide growth, k_s, should depend on the Si-Si bond breaking mechanism. Therefore it follows that the linear rate constant (B /A) should depend only on k_s and not h (see Eq. 22). We further confirm this supposition from boundary layer theory, which estimates that $h \approx 10^3 k_s$. Applying this approximation to Eq. 22 we obtain:

$$\frac{B}{A} \cong k_s \left[\frac{C^*}{N_1} \right] \qquad (24)$$

which can be written as:

$$\frac{B}{A} = k_s \left[\frac{C^*}{N_1} \right] = \left[k_o \, e^{\left(\frac{-E_a}{kT} \right)} \right] \left(\frac{C^*}{N_1} \right) \qquad (25)$$

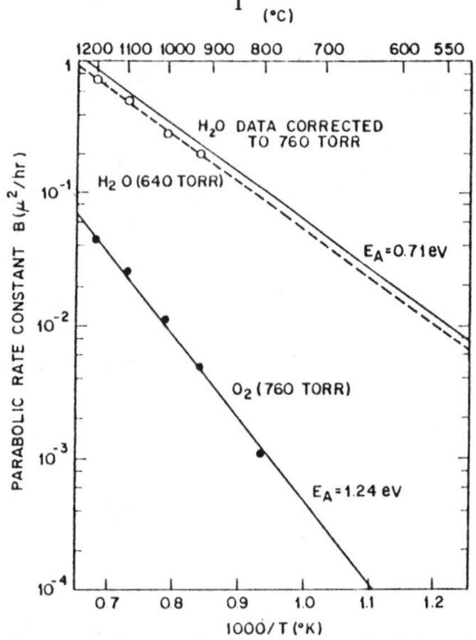

Fig. 5 The temperature dependence of the parabolic rate constant for dry and wet oxidations[3]. Reprinted with permission of the American Physical Society.

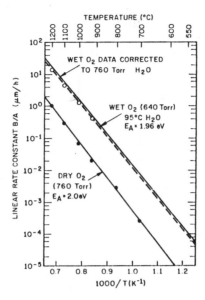

Fig. 6 The temperature dependence of the linear rate constant for dry and wet oxidation[3]. Reprinted with permission of the American Physical Society.

where E_a is the activation energy for Si-Si bond breaking, and k_o is a constant.

The implication of Eq. 24 is that (B /A) is not limited by the supply of oxidizing species, but rather depends only on the rate of the chemical reaction at the Si /SiO$_2$ interface. Evidence shows that significant changes in gas flow do not effect (B /A), thereby confirming its independence from mass-transfer effects.

Tables 3 and 4 list values of A, B, B /A, and τ at several temperatures for wet and dry oxidations, respectively. We note that the values of both B and B /A for wet processes is much greater than for dry processes, leading to higher oxidation rates for the wet case. This result arises even though the diffusivity D for dry oxidation (O$_2$) is higher than D for wet oxidation (H$_2$O). The higher values of the rate constants, B and B /A, result from the fact that the value of C*is much higher for H$_2$O than O$_2$ [C*(H$_2$O) = 3×10^{19} cm^{-3}, and C*(O$_2$) = 5.2×10^{16} cm^{-3}).

THE INITIAL OXIDATION STAGE

One of the most important applications of oxidation to VLSI processing is the formation of

Table 3. RATE CONSTANTS FOR WET OXIDATION OF SILICON[3]

Oxidation temperature (°C)	A (μm)	Parabolic Rate Constant B (μm^2 /hr)	Linear rate constant B /A (μm /hr)	τ (hr)
1200	0.05	0.720	14.40	0
1100	0.11	0.510	4.64	0
1000	0.226	0.287	1.27	0
920	0.50	0.203	0.406	0

Table 4. RATE CONSTANTS FOR DRY OXIDATION OF SILICON[3]

Oxidation temperature (°C)	A (μm)	Parabolic Rate Constant B (μm^2 /hr)	Linear rate constant B /A (μm /hr)	τ (hr)
1200	0.040	0.045	1.12	0.027
1100	0.090	0.027	0.30	0.067
1000	0.165	0.0117	0.071	0.37
920	0.235	0.0049	0.0208	1.40
800	0.370	0.0011	0.0030	9.0
700	...	...	0.00026	81.0

the gate-dielectric of MOS transistors, since in this case, the SiO_2 layer is an active device component. As device dimensions are scaled to one micron or less, oxide layers of 150Å or less will be required. Tunnel oxides for use in *electrically alterable read only memories* (EAROMS) require oxides thinner than 80Å. These thin oxide layers must be producible in a manufacturing environment with high yield and long-term reliability.

The early studies on the kinetics of silicon oxidation have clearly shown that the growth rates for layers thinner than 250Å were higher than predicted by the Deal-Grove linear-parabolic model, when grown in dry oxygen. These thin layers also have properties different than those found for thicker layers.

Massoud, *et al*[4] have recently published the results of a comprehensive study of the oxidation kinetics of SiO_2 formation in the thin regime (<500Å) for dry oxidation, in the temperature range of 800-1000°C. The results of that study are summarized as follows:

1. The total oxidation growth rate in the thin regime is represented by the expression:

$$\frac{dx_0}{dt} = \frac{B}{2x_0 + A} + C_2 e^{[-x_0 / L_2]} \qquad (26)$$

where, B /A is the linear rate constant, B is the parabolic rate constant, C_2 is the pre-exponential factor, and L_2 is the characteristic length. Both C_2 and L_2 are dependent on the conditions of oxidation.

2. For the case of lightly doped silicon oxidized in 1 atm of dry oxygen, it was found that L_2 was independent of temperature and substrate orientation, and had a value of ~70 ± 10Å. The value of C_2 depended exponentially on temperature (Arrhenius behavior) with activation energies of 2.37, 2.32, and 1.80 eV, for <100>,<111>, and <110> orientations, respectively.

3. For lightly doped silicon oxidized at reduced oxygen pressures (O_2-Ar mixtures), the value of L_2 was unchanged from that of 1 atm dry-O_2 oxidations. The activation energies measured for C_2, however, were found to average ~0.8 eV less than the values determined at 1 atm for the three orientations studied. The pressure dependence of C_2 was found to follow a power law dependence ($p^{0.8}$), which is similar to the pressure dependence generally observed for the linear rate constant, B /A.

4. For Si substrates heavily doped with phosphorus (to the solid solubility limit) and oxidized in dry O_2, the value of L_2 was unchanged from that of lightly doped substrates. The pre-exponential, C_2, had activation energies of 1.87 and 1.77 eV for <100> and <111>. The magnitude of C_2 was only moderately affected by heavily doping the substrate.

Several models have been proposed to explain the enhanced oxidation rates for thin oxides. The models can be divided into four groups: a) models based on space charge effects, where the enhancement is electrochemical in nature; b) models based on structural effects in the oxide that result in enhanced oxidizing species transport; c) models based on oxide stress effects, that enhance the diffusivity of the oxidizing species; and d) models which postulate that the enhanced solubility of the oxidant in the oxide results in the O_2 concentration exceeding its solid solubility limits.   All four of these model groups have been critically analyzed by Massoud, *et al*[4]. None of these models, however, was found to apply under all experimental conditions, thus a new model (which was reported to apply universally) was introduced.

The model of Massoud and co-workers, postulates the existence of a *surface layer in the silicon substrate* which contains additional sites for oxidation. These additional sites are responsible for the oxidation rate enhancement observed in the thin film regime. They are believed to have a concentration profile which decays exponentially into the silicon. The profile has a characteristic length of ~30Å. The 30Å length in silicon corresponds to the value of 70Å observed for the characteristic length, L_2 in the thin SiO_2 (0.44 · 70Å ~ 30Å).

Unfortunately, the nature of the excess oxidation sites in the surface layer has not yet been identified. This layer could represent the departure from perfect crystallinity expected in any surface, and oxidation could be enhanced because of changes in the Si-Si bond angles and /or energies. Further investigations are still needed to help delineate features of the Si surface prior to oxidation in order to understand the nature of these additional oxidation sites. Nevertheless, the work is very useful for estimating the growth of SiO_2 in the very thin film regime.

Growth of Thin Oxides

Much work has been done to achieve uniform, thin oxides under controlled processing conditions. In order to *controllably* grow thin oxides, the growth rate must be reduced so that the process entails a reasonable time of growth. Various techniques have been used to achieve this reduced growth rate.   They include: a) dry oxidation; b) dry oxidation with HCl, trichloroethylene, TCE, or trichloroethane, TCA; c) reduced pressure oxidation; d) low temperature high pressure oxidation; and e) rapid thermal processing under oxidizing conditions.

Thin oxides ($\leq$400 Å) have been successfully grown by a number of methods. Irene[5] has

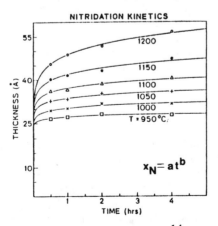

Fig. 7 The kinetics of nitridation in NH_3 for <100> silicon[14]. (© 1985 IEEE).

grown thin oxides (≤ 200Å) in the temperature range of 780 - 980°C in dry O_2. He reported that controlled oxides of 100Å could be grown in 30 min at 893°C. Adams, $et\ al^{6}$ described the growth of thin oxides prepared in LPCVD systems at reduced pressures (0.25-2 torr) in the temperature range of 900-1000°C. It was found that the properties of the oxides were comparable to those grown at 1 atm, but with the additional benefit of excellent thickness control down to 30Å. Oxides grown at high pressures (10 atm) and low temperatures (750°C) have produced 300Å films in 30 min. When these films were used as a gate oxides in NMOS devices (DRAMs), the devices showed significantly improved refresh cycle times[7] when compared with devices with films grown at 1 atm. Oxides in the thickness range of 50-250Å prepared by rapid thermal processing (RTP) have been recently announced (AG Associates). These oxides are also reported to have well controlled thickness and electrical properties.

THERMAL NITRIDATION OF SILICON AND SILICON DIOXIDE

As VLSI devices continue to be scaled-down in size, they will require gate and tunnel dielectric thicknesses in the range of 100Å or less. In spite of the fact that very thin oxides can be grown by a number of techniques as described above, very thin layers of SiO_2 are known to have high defect densities[8], and do not serve as an effective diffusion mask against impurity diffusion[9,10]. Recent studies show that direct *thermal nitridation* of Si and thin SiO_2 appears to be a viable alternative method of growing a good quality dielectric film in this very thin regime.

Thermally grown films of *silicon nitride* (Si_3N_4) have a number of advantages over SiO_2, including: a) they tend to have self-limiting growth kinetics and therefore their thickness is easily controllable; and b) they are effective barriers to impurity diffusion. MOS devices fabricated with these films show large values of gain and reduced hot electron effects[11-13].

Thermal silicon nitride films are generally grown by the high temperature (950-1200°C) nitridation of silicon in pure ammonia (NH_3)[14] or an ammonia plasma[15,16]. They can also be prepared by plasma anodic nitridation using a nitrogen-hydrogen plasma in the temperature range of 700-900°C[17]. An additional, and somewhat novel, technique for producing nitride films on silicon is to use direct nitrogen ion implantation at a dose of 5×10^{16} cm^{-2}, in the energy range of 5-60 keV, followed by an anneal at 1000°C, in a N_2 ambient.

The nitridation kinetics of single crystal silicon have been studied for pure NH_3 ambients over a wide temperature range[18]. There appears to be a slightly higher growth rate for <111> oriented Si than for <100>. This may result from the fact that the nitridation process is diffusion-limited through most (but not all) of its growth time. Growth of the nitride appears to be thermally activated, with activation energies ranging from 0.34-0.51 eV, depending on nitridation time. Growth curves (nitride thickness, x_N vs. time in ambient) are found to follow the generalized relationship:

$$x_N = a\ t^b \tag{27}$$

where **a** and **b** are temperature dependent constants given by: **a** = 920.2 exp (- 0.38 eV /kT)Å, **b** = 0.0183 exp (- 0.238 eV /kT), and t is normalized to 1 min. Figure 7 is a plot of nitride thickness as a function of nitridation time at various temperatures, and also illustrates the self-limiting nature of the growth. Unlike nitridation in NH_3, the growth kinetics of nitridation in pure N_2 at 1200-1300°C are not self-limiting. It is also observed that the initial film grows smooth and amorphous, but becomes rough and crystalline for long times. This data indicates that direct nitridation of silicon in a N_2 atmosphere should be avoided. There is also no observed doping dependence for nitridation similar to that found for oxidation .

The thermal nitridation of SiO_2 films results in the formation of a nitrided-oxide or *nitroxide* film. The nitrogen composition in such films is found to vary with the depth of the film. It has been reported that the nitrogen concentration is highest both at the surface of the oxide film and at the SiO_2/Si interface. It has also been found that the thickness and the index of refraction of the oxide film are increased after nitridation in NH_3[19,20]. Thermal nitridation of SiO_2 in NH_3 results in a small increase in the overall film thickness. The resulting nitroxide film is multi-layer in nature. A nitrogen rich layer is formed both at the surface and at the Si /SiO_2 interface at the very early stages of the process. After a few minutes, the nitridation reaction mainly continues in the bulk, with the surface and interface nitrogen content remaining essentially constant. The Si_3N_4/Si interface is found (by Auger analysis) to move away from the nitrogen rich layer. An oxygen-rich layer is formed underneath the nitrogen-rich layer. The thickness of this oxygen-rich layer increases with time. The oxygen supply used to form this layer is thought to come from the reaction between NH_3 and SiO_2.

Metal-insulator-silicon (MIS) devices have been fabricated[18] with both nitride and nitroxide films serving as the gate dielectric. Electrical measurements, using C-V and I-V measurements, time dependent breakdown, dielectric breakdown, and trapping techniques, showed good device characteristics for nitride films thicker than 40Å. Extremely low levels of trapping are found in these films, and this results in better long term reliability than devices built with SiO_2 films. The Si_3N_4/Si interface is also found to be as abrupt as that observed for the SiO_2/Si interface. The results of this work show that highly scaled metal-insulator-silicon field-effect-transistor (MISFET) devices which require ultra-thin dielectrics will find thermal silicon nitride and nitroxide films very useful as gate dielectrics. Significant work nevertheless remains to be done to further characterize and understand these films.

FACTORS WHICH AFFECT OXIDATION RATE

The thermal oxidation rates of silicon depend on many parameters, the following being the most important: a) the crystallographic orientation of the Si; b) the Si doping level; c) the presence of halogen impurities (Cl, HCl, TCA, TCE) in the gas phase; d) the growth pressure; e) the presence of a plasma during growth; and f) the presence of a photon flux during growth. In this section we discuss these effects and relate them to VLSI processing technology. Since the presence of chlorine and its compounds during oxidation is also used to improve MOS device characteristics, this phenomenon will also be discussed (in the section *Oxidation Growth Rates: Chlorine Dependence*).

Oxidation Growth Rates: Crystal Orientation Effects

The parabolic rate constant, B, has been observed to be relatively independent of the crystal orientation of the Si substrate. This is not surprising since, as we have seen, the parabolic rate is tied to the diffusivity of the oxidizing species through the existing SiO_2, and not to the reaction at the Si /SiO_2 interface. The linear rate constant, B /A, on the other hand, has been found to depend strongly on the crystal orientation. Table 5 shows the effect of orientation and temperature on the rate constants.

Since B /A depends on the surface reaction rate constant, k_s, this fact has been used by Ligenza[21] to develop a model that accounts for the relationship between B /A, and the orientation dependence. This model is based on the oxidation reaction between an H_2O molecule

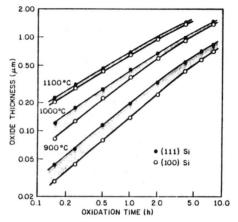

Fig. 8   Silicon dioxide thickness as a function of oxidation time for silicon in H_2O at 640 torr[66]. Reprinted with permission of the publisher, the Electrochemical Society.

in the SiO_2 and a Si-Si bond at the Si /SiO_2 interface. We summarize Ligenza's model here, but the reader is referred to the original paper for a more detailed description.

It is observed that the orientation dependence of oxidation depends on two factors: (a) the number of Si bonds available per cm^2 of Si surface for a particular crystal plane; and (b) the orientation dependence of the activation energy that is required to initiate the reaction. Although each Si atom at the surface has two bonds available for reaction, the number of Si-Si bonds capable of reacting is actually *less* than twice the number of surface atoms, since reacting H_2O molecules are sufficiently large that they can shield other H_2O molecules from reacting with adjacent Si-Si bonds. This geometric effect is called *steric hindrance*. The steric hindrance leads to higher activation energies since Si-Si bonds that make an angle with the surface have a less favorable position for the formation of SiO_2, and therefore have a higher activation energy than those that are parallel to the surface.

The average value of B /A for <111> silicon is measured to be 1.68 times that of <100>

Table 5. RATE CONSTANTS FOR THE THERMAL OXIDATION OF SILICON IN H_2O AS FUNCTION OF TEMPERATURE AND SUBSTRATE ORIENTATION[66].

Oxidation temperature (°C)	Orientation	A (μm)	Parabolic Rate Constant B (μm^2 /hr)	Linear rate constant B /A (μm /hr)	B/A ratio (111)/(100)
900	(100)	0.95	0.143	0.150	
	(111)	0.60	0.151	0.252	1.68
950	(100)	0.74	0.231	0.311	
	(111)	0.44	0.231	0.524	1.68
1000	(100)	0.48	0.314	0.664	
	(111)	0.27	0.314	1.163	1.75
1050	(100)	0.295	0.413	1.400	
	(111)	0.18	0.415	2.307	1.65
1100	(100)	0.175	0.521	2.977	
	(111)	0.105	0.517	4.926	1.65

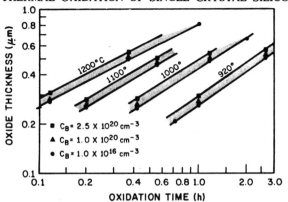

Fig. 9 Silicon dioxide thickness as a function of oxidation time in wet oxygen (95°C H_2O) for boron-doped silicon[67].   Reprinted by permission of publisher, the Electrochemical Society.

silicon. Figure 8 shows oxide growth curves for <100> and <111> Si in H_2O (at a partial pressure of H_2O of 640 torr) at several growth temperatures. Two important phenomena are illustrated in this figure: a) at high temperatures the growth rate difference between the two orientations is smaller than at low temperatures. This results from the fact that at high temperatures, the percentage of time where linear kinetics dominate is smaller than at low temperatures (i.e. the parabolic rate constant is predominant, and this constant is not orientation dependent); and b) at long times, the difference in SiO_2 thickness obtained for the two orientations is reduced, since the linear portion of the growth is again a smaller fraction of the total growth.

The linear portion of the dry oxidation (O_2) curve also depends on the crystal orientation, because the O_2 molecule reacts directly with Si-Si bonds at the Si /SiO_2 interface to form the oxide. This reaction is similarly impacted by steric hindrance. Irene found that the linear oxidation rate in dry O_2 (≤1 ppm H_2O) followed the oxidation growth rate dependence sequence[22], $(dx_o /dt)_{<110>} > (dx_o /dt)_{<111>} > (dx_o /dt)_{<100>}$, which agrees well with the sequence found for H_2O oxidations by both Ligenza[21] and Pliskin[23] .

Oxidation Growth Rates: Dopant Effects

The commonly used Group III and Group V dopants (boron, phosphrous, arsenic, and antimony) are known to enhance the oxidation rate when present in high concentration in silicon. For diffusion limited oxidations (where the parabolic rate constant dominates), the oxidation rate of the heavily doped silicon is primarily dependent on the impurity concentration in the SiO_2. For reaction-rate limited oxidation, B /A dominates, and the growth rate depends on the dopant concentration at the Si surface.

For example, consider the the case of boron, which segregates into, and then remains in, the SiO_2 during oxide growth. A weakening of the bond structure of the SiO_2 (glassy network) occurs as a result of the presence of the large boron concentration. The weakened bond structure allows both O_2 and H_2O molecules to enter the SiO_2 more easily, and also to diffuse through it more rapidly, thereby leading to enhanced oxide growth. Figure 9 is a plot of the wet oxide thickness as a function of oxidation time at several temperatures and boron concentrations. For high concentrations of boron (~1×10^{20} cm^{-3}) an increase in oxidation rate at all temperatures and for all thicknesses is measured.

The cause of the oxidation rate dependence on phosphorus concentration is different than that for boron, since phosphorus tends to pile up at the Si surface and very little is incorporated into

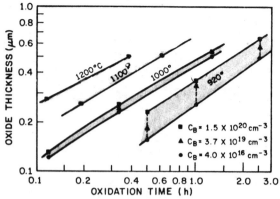

Fig. 10 Silicon dixoide thickness as a function of oxidation time in wet oxygen (95°C H$_2$O) for phosphorus-doped silicon[68]. Reprinted with permission of the Electrochemical Society.

the SiO$_2$. Thus, there is no significant increase in growth rate for high phosphorus concentrations when the oxidation is primarily diffusion controlled. On the other hand, there is a strong effect when the oxidation is reaction limited. Figure 10 shows the oxide growth curves for P-doped Si during wet oxidation. The effect of phosphorus concentration is much stronger at lower temperatures (~920°C) and short times, where linear kinetics control the oxidation. Above 1000°C, where the diffusivity of H$_2$O in SiO$_2$ increases so that parabolic rate constant, B, predominates, phosphorus presence has little or no effect on the oxidation rate. Figure 11 shows the variation of the rate constants with P-concentration for the dry oxidation case.   We can see that there is a strong dependence of the linear rate constant, B /A, on phosphorus concentration.

An atomistic model has been developed to explain this effect [24]. The model states that the high surface concentration of phosphorus shifts the position of the Fermi level, which in turn increases the surface vacancy concentration. These vacancies are thought to provide additional sites for the reaction of the oxidizing species with the silicon, thereby increasing B /A and the attendant growth rate in reaction-rate controlled regimes.

Other dopants such as gallium and aluminum (which accumulate in the oxide, but then, unlike boron, rapidly diffuse through it) do not show any enhanced oxidation effects.

The significance of enhanced oxidation rates due to doping variations in VLSI processing is that there are regions on a circuit that have different dopants, which are also present in

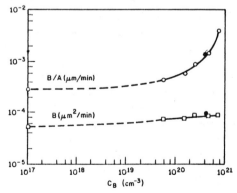

Fig. 11 Oxidation rate constants for dry oxidation of silicon as a function of phosphorus doping level at 900°C[68].   Reprinted with permission of the publisher, the Electrochemical Society.

varying concentrations. During oxidation, the oxide thickness over the heavily doped regions may be greater than over the lightly doped regions. Such variations in oxide thickness must be taken into account when specifying etch processes to insure that the thickest oxide regions are completely removed. Another concern is that unwanted steps will appear in the silicon, as a result of the thicker oxide consuming more silicon. The following example calculates the silicon step height formed between a lightly and heavily doped region on the silicon surface.

Example: Calculate the step height formed in the silicon at the edge of a phosphorus doped emitter $(C_s = 1.5 \times 10^{20} \text{ cm}^{-3})$ and a boron doped base $(C_s = 4 \times 10^{18} \text{ cm}^{-3})$ after a 920°C wet oxidation of 60 min.

Solution: The oxide thickness grown over the base region is determined from Fig. 9 as 2800Å. During the same time the oxide thickness grown over the emitter is determined from Fig. 10 to be 3700Å.
The difference in oxide thickness across the two regions is given by:
3700Å - 2800Å = 900Å. Since the amount of silicon consumed during oxidation is 44% of the oxide thickness, the step height in the Si is given by: 0.44 (3700Å - 2800Å) = 0.44 (900Å) = 396Å.

Oxidation Growth Rates: Water (H_2O) Dependence

As the presence of H_2O in the oxidation growth enivronment influences the growth rate, it must be be treated as an impurity during dry oxidation. In practice, it is extremely difficult to remove all traces of H_2O from a dry oxidizing ambient. Irene[25] has shown that if as little as 25 ppm of H_2O is deliberately added to the oxygen stream at temperatures between 800 and 1000°C, both the linear and parabolic rate constants are measurably increased when compared to 1 ppm of H_2O contamination. If it is desired to obtain oxidation curves for pure oxygen (less than 1 ppm H_2O), all water vapor must be removed from the oxidizing ambient. Such residual water vapor has been found to come from three sources: 1) H_2O in the O_2 gas; 2) H_2 and hydrocarbons which oxidize at elevated temperatures to form H_2O and CO_2; and 3) the diffusion of H_2O molecules from the room ambient through the fused silica furnace tube. The H_2O, H_2, and hydrocarbons can be removed by passing the O_2 through a pre-burner in order to oxidize the H_2 and hydrocarbons, and subsequently through a drier or cold trap, to remove the H_2O. The third H_2O component can be minimized by using a double-wall tube with an inert gas flowing between the walls to sweep out the water vapor. It should be noted that in production environments it is not necessary to employ such water reduction techniques. As long as reasonable care is exercised to maintain repeatable levels of H_2O concentrations, and as long as the maximum level remains in the 25-50 ppm range, repeatable oxide thicknesses can be obtained.

Oxidation Growth Rates: Chlorine Dependence

The presence of chlorine (Cl) mixed with the O_2 gas stream during all or part of the oxidation cycle, has two important effects: 1) during dry oxidation the Cl has a signficant effect on the oxidation rate of the silicon; and 2) the introduction of the Cl enhances several material properties of the SiO_2, and this results in improved device characteristics. The reported improvements include: a) a reduction of the mobile charge in the SiO_2; b) an increase in the lifetime of minority carriers in the underlying silicon; c) a reduction in the number of oxide defects, which results in increased SiO_2 breakdown strength; d) a reduction in the interface and fixed charge density; and e) a reduction in the number of oxidation induced stacking faults in the underlying silicon. It is therefore common practice in VLSI processing to intentionally add Cl during dry oxidation. The chlorine can be introduced by a variety of chlorine-containing gases, including: anhydrous chlorine (Cl_2), anhydrous hydrogen chloride (HCl), trichloroethylene (TCE), or trichloroethane (TCA). A potential deleterious effect of the presence of Cl is the anisotropic etching of the silicon that can occur under conditions of low O_2 flow.

Hess and Deal[26] have shown that the parabolic rate constant, B, gradually increases as the HCl concentration is increased in the O_2 ambient gas. They also determined that B /A increases by approximately a factor of 2 for an increase in the HCl concentration from 0 to 3%. Although H_2O is formed when HCl is added to O_2 at the oxidation temperature (i.e. $2 \ HCl + 1/2 \ O_2 \rightarrow H_2O + Cl_2$) its presence is not adequate to explain the observed increase in oxidation rate. It is also found that chlorine additions increase the oxidation rates without the presence of H_2O. These results suggests that the chlorine itself plays a major role in increasing the oxidation rates, but a detailed understanding of the mechanism is not yet available.

Although significant effort was expended to develop processes which utilized HCl in O_2 as the Cl species, the corrosive nature of HCl led to consideration of other Cl containing materials as alternatives. These gases, however, introduce their own problems. For example, TCE is a known cancer causing agent (carcinogen), while TCA reacts to form phosgene, $COCl_2$, a deadly gas. The phosgene formation can be suppressed by the use of excess oxygen, and this precaution must be taken during processing with this gas. Once it was demonstrated that such safeguards could be implemented, TCA rapidly replaced HCl for many oxide growth applications. Oxidation furnaces are now sold with plumbing to accommodate TCA delivery systems[27].

Oxide films grown in the presence of TCA or TCE are found to exhibit properties comparable to those grown in HCl[28,29]. Excellent gate oxide films have been prepared using a two-step TCE[30,31] or TCA process. The initial step involves growing the oxide at low temperatures ($\leq 900°C$) to provide a dense, high breakdown, low defect density oxide. This is followed by a high temperature oxidation to provide good interface properties and passivation against mobile ions. An example of a process that yields such films is as follows: a) a low temperature (850°C) dry oxidation; b) next a low temperature (850°C) TCE or TCA oxidation; c) a high temperature (1050°C) TCE or TCA oxidation, with low oxygen partial pressure; and d) finally, a 1050°C, N_2 anneal is used to reduce the fixed oxide charge, Q_f (see Section entitled *Fixed Oxide Charge*).

Chlorine atoms have been detected in oxides grown in a Cl environment[32,33]. Their concentration was found to increase as the oxidation temperature, oxide thickness, and chlorine partial pressure increased. Chlorine tends to accumulate in the oxide within 30 Å of the Si /SiO_2 interface and also moves rapidly in the oxide to follow the advancing interface as the oxide grows.

Oxidation Growth Rates: Pressure Effects

VLSI technology requires shallow diffusions and thin oxides. Shallow diffusions are achieved by use of ion implanted predepositions followed by processes that are restricted to low temperatures (<950°C). In such processes the movement of dopants introduced near the surface is minimized. One technique for achieving low temperature processing is to grow the SiO_2 in a high pressure environment. Pressures of up to 25 atmospheres of steam (H_2O) are routinely achieved, and several commercial high pressure oxidation system are available (HiPOx® manufactured by GaSonics and the FOX® manufactured by Thermco Systems).

The use of *high pressure* significantly increases the oxide growth rate, by increasing the linear and parabolic rate constants. This allows growth rates comparable to those obtained at

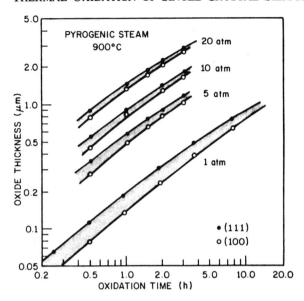

Fig. 12 Silicon diode thickness as a function of oxidation time and pyrogenically produced steam pressures at 900°C[69]. Reprinted by permission of the publisher the Electrochemical Society.

1 atm and high temperatures to be achieved at lower temperatures or to grow oxides more rapidly at the same temperature. The increase in the parabolic and linear rate constants arises from the increase in C^*, the equilibrium bulk concentration of the oxidizing species in the oxide, which in turn is proportional to the partial pressure of the oxidizing species. Figure 12 shows the oxide thickness as a function of time at various steam pressures at 900°C for <100> and <111> silicon. For any growth time the oxide thickness increases rapidly with the steam pressure. The increased growth rate can be traded off for temperature and a good rule of thumb is: *a constant growth rate can be maintained, if for each increase in pressure of 1 atm, the temperature is decreased 30°C*. Analysis of the data in Fig. 12 reveals that both B/A and B increase nearly linearly with pressure (actually B/A increases less than linearly).

High pressure can be achieved by pressurizing water-pumped equipment up to 25 atm or by using a pyrogenic ($H_2 + 1/2\ O_2 \rightarrow H_2O$) system to produce water at 25 atm. The pyrogenic system requires pressurizing both the H_2 and O_2 to more than 25 atm, and great care must be exercised in handling these materials at such pressures. The water pumped system depends on the quality of the water and the pump system for the quality and cleanliness of the high pressure oxide. Figure 13 shows a schematic drawing of a high pressure oxidation system. The high pressure exists on both sides of the quartz tube. This eliminates any pressure differential that the tube must support at elevated temperatures. Significant effort is expended to design automated safety features into commercially available systems, because of the hazards associated with such high pressures. The safety features are incorporated into microprocessor based control systems.

Applications of high pressure oxidation in VLSI processing have not been widely implemented. Although in principle, high pressure affords several advantages over atmospheric oxidations, problems with the process and equipment have slowed its acceptance. These problems include: a) concern for safety of gases at very high pressure; b) the equipment requires significant floor space relative to conventional 4-stack furnaces, yet cannot accommodate as many wafers as a single conventional tube; c) poor thickness uniformity results from temperature inhomogenities

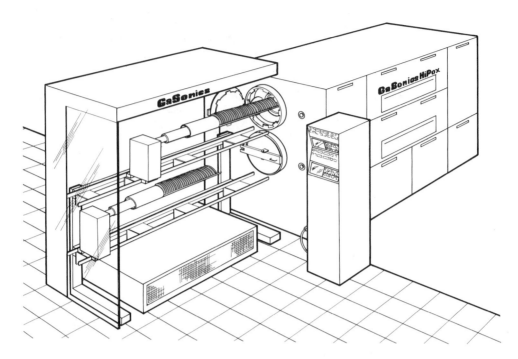

Fig. 13 Schematic representation of a commercial high pressure oxidation system. Courtesy of Gasonics, Inc.).

which arise from convection currents at high pressures; and d) particulates often become embedded in the oxide, and such particulate generation appears to be inherent to the process.

Improvements in the equipment, such as a 2-stack configuration, are making the systems more economical, and applications are slowly beginning to find their way into VLSI processing. High pressure gate oxides have been employed to produce high density dynamic RAMs with improved refresh cycle time[34]. Bipolar devices have shown reduced emitter pipe defects when the LOCOS (Local Oxidation of Silicon) field oxidation was performed at high pressure[35]. Finally higher performance bipolar devices fabricated at lower temperatures by the use of high pressure oxidation, have been reported[7]. The lower temperature allows the use of a thinner epitaxial layer since the buried (n^+) collector undergoes less up-diffusion. It is interesting to note, that high pressure oxidation furnaces have found a VLSI application other than enhanced oxidation. This application is the use of high pressure furnaces to achieve significantly reduced reflow temperatures for PSG and BPSG glasses (see Chap. 6).

Oxidation Growth Rates: Plasma and Photon Effects

It has been known for some time that a *plasma environment* can enhance the oxidation rate[36,37]. This technique allows oxides to be grown at temperatures even below those achieved by high pressures (e.g. <600°C). The technique is usually carried out in an oxygen discharge generated by rf or a dc electron source. By biasing the silicon substrate to a potential below that of the plasma potential, charged active oxygen is attracted to, and collects on, the substrate. This results in rapid oxidation of the silicon. The SiO_2 growth rate is found to increase with increasing temperature, plasma density, and substrate doping concentration.

The plasma-enhanced oxide growth process, however, suffers from several disadvantages including: a) as a result of the high energy density in the plasma region and the large number of reactive species in the plasma, it is possible for unwanted contaminants to also be deposited on the substrate. These can arise from material that is sputtered off the reactor walls, or from gas-phase reactions of the active species generated by the plasma; and b) as with other low temperature techniques for preparing SiO_2 (CVD, PECVD) these plasma oxidized films have inferior electrical properties compared to thermal SiO_2 grown at 1000°C. Note that these properties can be substantially improved by post-oxide thermal treatment in an oxidizing ambient[38].

It has also been observed that by directing light (photons) on silicon in an oxidizing environment the oxidation rate of the silicon can be increased. This effect is called *photon-assisted oxidation*. Early studies[39] of this mechanism used Hg and I_2 vapor lamps to produce abundant ultraviolet (UV) radiation. Under these conditions an increase in oxidation rate was observed only during the initial growth (<300Å) and at the lowest temperatures. More recently, workers[40-42] have utilized CW lasers which produce IR and UV radiation in attempts to increase the rate of oxide formation. Again, enhancement of oxidation rates under the photon flux was observed. Although substrate heating due to the photon flux was involved, the oxide growth rates that were recorded were greater than those that could be attributed to thermal effects alone.

The actual mechanism for the rate enhancements in photon-enhanced growth are as yet not known. Nevertheless, the technique has promise for lower temperature VLSI processing. A particularly useful feature is that local oxidation can be performed by illuminating desired wafer regions, thereby causing oxidation to occur. Other regions of the circuit can be kept at temperatures lower than that required to cause thermal oxidation, and thus they remain oxide free.

MASKING PROPERTIES OF THERMALLY GROWN SiO$_2$

The use of SiO_2 to mask against the diffusion of the common dopants in silicon is a cornerstone of planar technology. Devices are formed by etching windows in selected areas of the SiO_2 grown on the silicon. Junctions are then formed by diffusing or ion implanting impurities into these selected regions. In order for the SiO_2 to be effective in preventing penetration of dopants into the silicon under the SiO_2, the diffusivities of the common dopants (B, P, As, Sb, Ga) in SiO_2 should be significantly smaller than in silicon. Table 6 lists the diffusion coefficients of the dopants in SiO_2[43] and silicon. As shown in Table 6, SiO_2 serves as an excellent masking material for all of these impurities except Ga.

Table 6.   Diffusion Coefficients in SiO$_2$ and Si of the Common Dopants

Dopant	SiO$_2$ D (900°C) cm^2/sec	Si D$_i$ (900°C) cm^2/sec[*]
Boron	2.2×10^{-19} - 4.4×10^{-16}	~ 1.5×10^{-15}
Gallium	1.3×10^{-13}	~ 6×10^{-14}
Phosphorus	9.3×10^{-19} - 7.7×10^{-15}	~ 4×10^{-17}
Arsenic	4.8×10^{-18} - 4.5×10^{-19}	~ 2×10^{-16}
Antimony	3.6×10^{-22}	~ 8×10^{-17}

[*] These are estimated values for intrinsic diffusion, the values for extrinsic diffusion are higher.

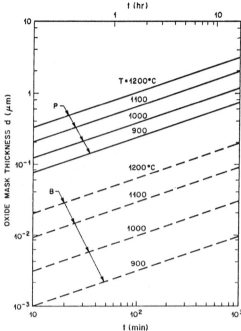

Fig. 14 Minimum silicon dioxide thickness required to mask against phosphorus and boron diffusions as function of diffusion time & temperature[70]. Reprinted by permission Pergamon Press.

Nevertheless, to insure that dopants do not penetrate the SiO_2 and reach the silicon during a diffusion process, a minimum thickness of SiO_2 is required. Figure 14 shows the minimum thickness for dry O_2 grown SiO_2 required to mask against phosphorous and boron predepositions, as a function of time and temperature. These curves are conservative since they are determined from the highest diffusivities values, and concentrations of boron and phosphorus. For VLSI applications in which temperatures are generally kept below 1000°C, masking oxide thicknesses can be kept below 0.4-0.5 µm.

To mask against ion implantation, the oxide should be sufficiently thick so that no ions will have penetrated through the oxide after both the ion implantation and all high temperature processing steps have been completed. The stopping ability of the SiO_2 to the ions depends on the atomic weight and acceleration energy of the ions. Since light ions (e.g. boron) have deeper penetration depths than do heavier ions (e.g. As), they require thicker masking layers. The penetration depths of common dopant ions in SiO_2 are given in Chap. 9. Note that it is usually good practice to etch off the heavily implanted SiO_2 surface regions prior to any subsequent thermal processing. This minimizes the chance that implanted ions will reach the Si surface.

PROPERTIES OF THE Si /SiO$_2$ INTERFACE AND OXIDE TRAPS

The SiO_2 and Si /SiO$_2$ interface both contain various charges and traps. These charges have profound effects on the properties of the devices fabricated in the underlying silicon. In this section we describe the nature of the Si /SiO$_2$ interface, the types of oxide traps and charges, and the techniques for measuring their levels in the oxide.

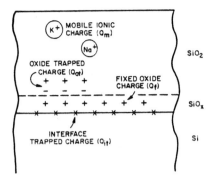

Electrical measurements on various device structures is the most important method used to characterize the properties of the SiO_2 and the Si /SiO_2 interface. The device structures most widely employed to evaluate such properties are the metal-oxide-semiconductor (MOS) capacitor and the MOS transistor. The electrical measurements provide direct information about such device structures. The device characteristics are then analyzed to yield information about the charges and traps, in the SiO_2 and at the Si /SiO_2 interface. It is necessary to obtain information about the SiO_2 in this manner because, for most of these charges, the defect densities in a device quality oxide are too low to be detected by chemical means.

There are four types of charges that exist in the oxide or near the Si /SiO_2 interface: a) interface trap charge, Q_{it}; b) fixed oxide charge, Q_f; c) mobile ionic charge, Q_m; and d) bulk oxide trapped charge, Q_{ot}. Figure 15 schematically shows these charges in oxidized silicon[44].

Interface Trap Charge

The *interface trap charge*, Q_{it}, refers to charge which is localized at the Si /SiO_2 interface on sites that can change their charge state by exchange of mobile carriers (electrons or holes) with the silicon. The charge state of the interface trap site changes with gate bias if the interface trap is moved past the Fermi level, causing its occupancy to change. These traps have energy levels distributed throughout the silicon bandgap with a U-shaped distribution (see Fig. 16) across the bandgap[45]. The minimum level at mid-gap is typically the concentration level used to characterize their presence. The density of these charges is expressed as the number /cm^2 eV.

In as-grown oxides, Q_{it} depends on the oxidation temperature, furnace ambient (wet vs. dry), oxygen partial pressure, and Si substrate orientation. Figure 17 shows the value of Q_{it} measured at mid-gap as a function of dry oxidation temperature, and crystal orientation (<100> vs. <111>)[46]. The value of Q_{it} is found to decrease with increasing temperature and <100> surfaces are found to have a lower density of traps than <111> surfaces. For example, VLSI gate oxides grown at 950°C on <100> material will exhibit Q_{it} densities of about 10^{12} cm^{-2} eV^{-1}. This value is too high to fabricate good devices and must be reduced by additional processing. Two annealing techniques found to be effective in reducing Q_{it} to acceptably lower values are: a) the low temperature post-metallization anneal; and b) the high temperature post-oxidation anneal.

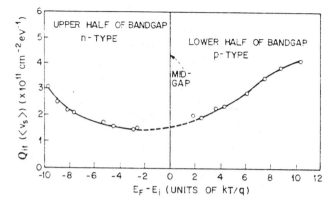

Fig. 16 Typical interface trap density as a function of position in the band gap for <111> silicon[45]. Copyright, 1967, Bell Telephone Laboratories, Incorporated, reprinted by permission.

In the *low temperature post-metallization* case, the sample is annealed in H_2 or other non-oxidizing ambients, such as forming gas (H_2-N_2) or Ar, in the temperature range of 350-500°C for up to 30 min. Aluminum must be present for the anneal to be effective. Before such an anneal, a steam grown oxide may exhibit Q_{it} densities in the low 10^{11} cm^{-2} eV^{-1} range, while after anneal, the densities are decreased to the low 10^{10} cm^{-2} eV^{-1} regime. This is an acceptable level for most device applications. It is believed that water in SiO_2, present even in dry oxides, reacts with the aluminum to form Al_2O_3 and atomic hydrogen (H). The H diffuses to the Si /SiO_2 interface, where it reacts chemically with traps, rendering them electrically inactive. If silicon nitride is present, it serves to block the diffusion of atomic H, and no reduction in Q_{it} is observed. Thus post-metallization anneals should be performed prior to

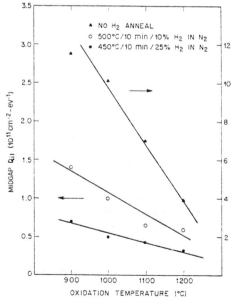

Fig. 17 The interface trap density near midgap as a function of the oxidation temperature in dry oxygen[46]. Reprinted with permission of the publisher, the Electrochemical Society.

passivation, if Si_3N_4 is used as the passivating overcoat.

In the case of the *post-oxidation high temperature anneal*, the oxide is grown and then annealed at the growth temperature (*in-situ*) in an H_2 or an inert ambient (N_2 or Ar). In this case, the reduction in Q_{it} probably results from H_2 or trace quantities of H_2O in the ambient, again promoting the reaction leading to atomic hydrogen.

The interface trap density, Q_{it}, can be measured by several techniques including: a) the conductance (I-V) method on a MOS transistor; b) high frequency (>100 kHz) capacitance-voltage (C-V) measurement; c) low frequency (<1 kHz) C-V measurement; d) and emission time spectroscopy, which is also termed *deep level transient spectroscopy*, or DLTS. These measurement techniques, as well as their underlying principles are throughly treated in the book *MOS Physics and Technology* by E.H. Nicollian and J.R. Brews[47], to which the reader is referred for details.

Fixed Oxide Charge

The *fixed oxide charge*, Q_f, is located within 35Å of the Si /SiO_2 interface in the so-called *transition region* between silicon and SiO_2. These charges do not change their charge state by exchange of mobile carriers with the silicon, as occurs with Q_{it}, hence the name fixed charge. The Q_f charge centers are predominately positive, although some negative, compensating centers may also be present. The fixed charge, Q_f, is considered to be a sheet of charge at the Si /SiO_2 interface and is expressed as the number of charges per unit area (number /cm^2).

The value of Q_f depends on the oxidizing ambient (H_2O or O_2), oxidizing temperature, silicon orientation, cooling rate from elevated temperature, cooling ambient, and subsequent anneal cycles. Although, it is desirable to minimize the value of Q_f, current VLSI MOS technology uses ion implantation to control the device threshold voltage, V_T, which is the device parameter most impacted by Q_f. Control of V_T by ion implantation allows for precise and uniform tailoring of the threshold voltage, rather than having to achieve uniform and low values of Q_f. In spite of the ability to override small variations in Q_f, a goal of maintaining a low and reproducible value of Q_f is nevertheless still desirable during fabrication.

Figure 18 shows the effect of oxidation temperature on Q_f for both n and p-doped <111> Si^{48}. The lowest values of Q_f are obtained at the highest oxidation temperatures. For wet oxidation, Q_f decreases much more slowly with increasing temperature than for dry oxidation.

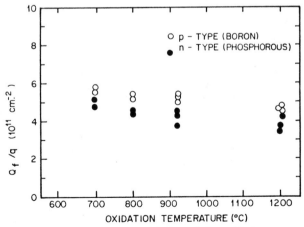

Fig. 18 The oxide fixed charge density as a function of the oxidation temperatures for p-type and n-type <111> silicon[48]. Reprinted with permission of the publisher, the Electrochemical Society.

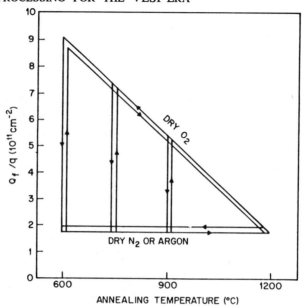

Fig. 19 Oxide fixed charge density as a function of annealing temperature for times one hour or less (The Deal Triangle)[48]. Reprinted by permission of the publisher, the Electrochemical Society.

The rate of cooling from the oxidation temperature also affects the value of Q_f. The fastest cooling rates result in the lowest Q_f values. Rapid cooling prevents oxidation at low temperature, and simultaneously high values of Q_f. The use of rapid cooling (quenching), however, is not practical for large diameter wafers (>100 mm), where warpage and slip are known to occur. Converting from oxygen to an inert gas (Ar, N_2) at the oxidation temperature, however, avoids this problem by stopping oxidation while wafers are still at their highest temperature. Deal[48] summarized the relationship of Q_f for <111> material to oxidation conditions and to cooling or annealing in inert ambients, by the so-called *Deal Triangle* (Fig. 19). The hypotenuse (AC) of the triangle represents the effect of temperature on Q_f, while high temperature anneals in inert gas are represented by the vertical bars. The base of the triangle represents an inert gas cool down.

Figure 20 shows the dependence of Q_f and Q_{it} on Si crystallographic orientation[49], with <100> Si resulting in the lowest values for each type of charge. Consequently, this orientation is almost exclusively used for MOS VLSI circuits. Values of Q_f in the low 10^{10} cm^{-2} range are achievable in <100> material after careful oxidation, anneal, and cooling in an inert environment.

The value of Q_f is determined by measuring the voltage shift of a high frequency capacitance-voltage (C-V) curve of a MOS capacitor test device. The details of the measurement are discussed in Ref. 47. For the case of oxide fixed charge at the interface, the flat-band voltage, V_{FB}, of the C-V curve is related to the oxide charge, Q_f, oxide thickness, x_o, and work function difference between the gate electrode and the silicon, ϕ_{ms}. The relationship is given by:

$$\frac{Q_f}{q} = \frac{(-V_{FB} + \phi_{ms})\, \varepsilon_{ox}\, \varepsilon_o}{q\, x_o} \qquad (28)$$

where ε_{ox} is the dielectric constant of the SiO_2 ($\varepsilon_{ox} = 3.9$), and ε_o is the permittivity of free space. The capacitance per unit area of oxide is given by $C_{ox} = \varepsilon_o \varepsilon_{ox}/x_o$. Figure 21 shows C-V curves for n-type Si and an oxide with positive fixed charge. The presence of positive oxide charge shifts the C-V curve from its ideal (no charge) value to more negative voltage values.

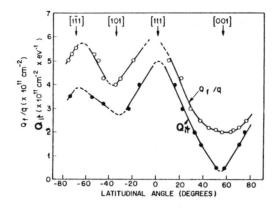

Fig. 20 Oxide fixed charge density and interface trap density as a function of the silicon orientation[49]. Reprinted with permission of the American Physical Society.

Mobile Ionic Charge

The *mobile ionic charge*, Q_m, is commonly caused by the presence of ionized alkali metal atoms (Na^+, K^+). This type of charge is located either at the gate (metal or polysilicon) /SiO_2 interface (where it often originally enters the oxide layer), or at the Si /SiO_2 interface, to where it drifts under the presence of an applied positive field to the gate. The field-assisted drift across the oxide occurs even at room temperature, since these ions are extremely mobile in SiO_2. The amount of mobile charge incorporated into the SiO_2 depends on the *cleanliness* of the oxidation process, and involves such components of the process as: a) the furnace; b) processing chemicals; c) oxidizing ambient; d) gate electrode material; and e) wafer handling, all factors which ultimately amount to keeping Na out of the SiO_2. The density of mobile charge varies between 10^{10} and 10^{12} cm^{-2}. The presence of mobile charge can result in long term changes in the device threshold voltage, V_T, as the ions drift from the gate to the Si /SiO_2 interface. For this reason, low values of Q_m are necessary. A value of Q_m in the low 10^{10} cm^{-2} range will result in a V_T shift of only a few tenths of a volt for an MOS device with a gate oxide of 1000Å. For VLSI gate oxides (<250 Å) the shift will be less than 0.1 V. On the other hand, values of Q_m in the 10^{12} cm^{-2} range will result in V_T shifts of several volts.

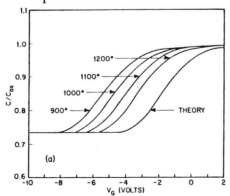

Fig. 21 High frequency C-V curves showing the effect of the presence of a positive oxide charge.[48] Reprinted with permission of the publisher, the Electrochemical Society.

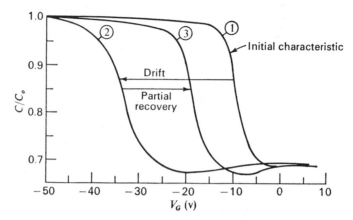

Fig. 22 High frequency C-V curves showing the shift after B-T stressing as a result of ionic contamination[50]. Reprinted with permission of the American Physical Society.

The mobile ionic charge density is also measured with the use of the MOS C-V technique. In this method a C-V curve of a MOS capacitor is initially measured. The MOS capacitor is then heated to 300°C, while the gate is subjected to a positive bias (i.e. an electric field of ~2 x10^6 V /cm). This bias-temperature stress condition is maintained for ~30 min to insure that all the mobile charge has drifted to the Si /SiO$_2$ interface. The capacitor is then cooled to room temperature with the bias still applied, and another C-V curve taken. The flat band voltage shift, ΔV_{FB}, that is measured between the C-V curves before and after bias-temperature treatment is used to calculate the value of Q_m drifted at 300°C. Since drifted mobile ions pile-up as a sheet of charge at the interface, the value of Q_m is given by ΔV_{FB} x C_{ox}. Figure 22 shows the effect of mobile ionic charge on a C-V curve after B-T testing[50]. Another technique for measuring Q_m is the *triangular voltage sweep* (TVS)[51-53] method.

To eliminate mobile ionic charge, it is necessary to minimize the introduction of Na into the oxide. The major sources of Na during processing are: a) gate or contact metallization; b) oxidation and annealing furnaces and gases; c) diffusion furnaces and gases; d) photoresist bake and incomplete resist stripping; and e) contaminated chemicals used for wafer cleaning and /or oxidation. Oxidation tubes may also be a source of Na. Cleaning tubes with an O$_2$-HCl mixture at elevated temperatures reduces Na to very low levels (e.g. 6 mole % HCl in O$_2$, at 1150°C for 2 hr). Growing SiO$_2$ in HCl-O$_2$ or other Cl containing gases, such as TCE or TCA, also results in stable oxides. This occurs since Cl is incorporated in the oxide at the Si /SiO$_2$ interface, neutralizing Na ions as they arrive (if their concentration is not excessive).

Even if the oxide is initially free of Na, contamination can occur during the device life. In order to protect the devices from external Na contamination, several types of protective layers have been developed. A phosphosilicate glass (PSG) can act as a getter of Na (see Chap. 2), and is often used as a interlevel dielectric between the gate and the metal level and /or as a final passivation in silicon gate MOS devices. Plasma deposited silicon nitride (Si$_3$N$_4$) serves as an impervious layer to Na, thereby preventing Na penetration to the gate oxide, and is often used as a final passivation. The deposition and properties of these materials are discussed in Chap. 6.

Oxide Trapped Charge

The last of the oxide charges to be discussed is *oxide trapped charge*, Q_{ot}. Bulk oxide traps can be located at the gate $/SiO_2$ interface, the Si $/SiO_2$ interface, as well as deep in the oxide. The traps are associated with defects in the SiO_2, such as impurities and broken bonds. They can be reduced to low values by annealing conditions similar to those established to reduce Q_{it}. Such traps are usually uncharged, but can become charged when electrons and holes are introduced into oxide and trapped at the trap site. One way that traps become charged is by avalanche injection of highly energetic electrons or holes (*hot carriers*) into the oxide. These carriers develop when fields in the silicon exceed $\sim 3x10^5$ V /cm (depending on doping level). Exposing the oxide to radiation environments can also produce significant levels of trapped charges.

There are two distinct ways in which devices are usually subjected to a radiation environment: 1) the devices operate in a space environment, or in proximity to a nuclear reactor where radiation is readily available; and 2) the VLSI processing itself leads to a radiation environment. VLSI processing that results in radiation damage to silicon devices includes: a) e-beam evaporation; b) sputtering; c) plasma etching; d) direct write e-beam and x-ray lithography; and e) ion implantation.

Highly energetic particles, such as γ–rays from space, can enter the oxide and create electron-hole pairs which can be trapped by oxide defects, thereby charging the defect. Heavier particles (e.g. As^+ ions from ion implantation) can possess sufficient energy to dislodge atoms from their sites, thereby creating defect centers and potential oxide trapping sites. The value of Q_{ot} can range from 10^9 to 10^{13} charges cm^{-2} depending on the environment the oxide has seen. The density of Q_{ot} is also measured using previously described C-V techniques[47].

The Nature of the Si $/SiO_2$ Interface

Numerous studies have been performed in order to elucidate the nature of the Si $/SiO_2$ interface. These studies have investigated the following interface characteristics: a) the spatial location of the interface traps and the oxide fixed charge; b) the morphology of the boundary; c) the chemical composition of the SiO_2 near the boundary; and d) the width of the interface. Electrical measurements have shown that fixed charge and interface traps exist in the oxide within 10-35Å of the Si $/SiO_2$ interface[54]. The interface traps occupy energy levels within the bandgap of the silicon, while the fixed charge occupies energy levels outside the bandgap. These charges and traps exist within a very narrow transition region between the Si and SiO_2.

Grinthaner and Maserijian[55] depict the interfacial region as consisting of single crystal silicon followed by: a) a monolayer in which there is some SiO_2, Si_2O_2, and Si_2O (the last two are incompletely oxidized silicon or sub-stoichiometric structures); b) a strained region of SiO_2 approximately 10-40Å deep; and c) the remaining stoichiometric, strain-free, amorphous silicon dioxide. (Stoichiometric silicon dioxide is SiO_x where x = 2.) During subsequent oxidations and anneals, defects form and are accumulated in the strained region. These defects act as sinks for impurities such as heavy metals (Au, Ni, Fe), alkali metals (Na, K), and chlorine.

Interface traps can be formed in several ways and three different models have been proposed to explain their origin. In the *Coulombic model*[56], charges in the oxide are believed to induce potential wells in the silicon, and quantum levels within these wells are associated with interface trap levels. The *bond model*[57] proposes that the interface trap distribution is a result of the distribution of bond angles, or stretched bonds at the silicon surface. Local strain or local nonstoichiometry at the Si $/SiO_2$ interface could cause these bond distortions. The third model assumes that *defects within or near the interfacial region cause interface trap levels*. Defects responsible for causing interface traps include stacking faults and micropores, molecular fragments remaining from imperfect oxidation, excess silicon, excess oxygen, or impurities.

The positive fixed charge could result[45] from the loss of an electron from non-bridging oxygen centers near the Si /SiO$_2$ interface to the silicon, making the center positively charged. All of these proposal are tentative and work continues toward a more complete understanding of the interface and the origin of the charges and traps.

STRESS IN SiO$_2$

It is important to establish the level of stress in oxide films, since it can contribute to wafer warpage and possibly to defect (slip) generation in the substrate. When the stress in a thermally grown SiO$_2$ film is measured at room temperature it is found[58] to be *compressive* and to have a relatively small magnitude ($\sim$3x10^9 dyne /cm^2). The residual stress in the oxide results from the difference between the thermal coefficients of expansion of Si and SiO$_2$ (see Chap. 4).

When the stress is measured at the growth temperature (e.g. 950°C), it is found[59] to be somewhat more compressive ($\sim$7x10^9 dyn /cm^2). The lower value observed at room temperature implies that some stress relief occurs as the film cools down. At growth temperatures above 975°C, measurements indicate that *viscous flow* apparently occurs, producing stress free films. Upon cooling, however, the same thermal stress (3x10^9 dyn/cm^2) as in lower temperature grown films is found to persist in the oxide.

The stress at the corner of a window cut into the oxide can be sufficiently large to induce plastic deformation in the silicon at processing temperatures above 800°C. In addition to plastic deformation, at these regions of concentrated stress enhanced diffusion, and selective tungsten deposition *wormholes* can occur (see Chap. 11 for data on selective tungsten deposition).

DOPANT IMPURITY REDISTRIBUTION DURING OXIDATION

The dopant impurities (boron, phosphorus, arsenic, and antimony) near the silicon surface become redistributed during thermal oxidation. Such *dopant redistribution* is important to consider since the surface concentration controls a variety of device parameters. The details of redistribution depend on three factors. The first is called the *segregation coefficient*, m, which is defined as:

$$m = \frac{\text{equilibrium concentration of impurity in Si}}{\text{equilibrium concentration of impurity in SiO}_2} \qquad (29)$$

Redistribution from this factor occurs because the impurities have different concentrations at

Table 7. Values of Segregation Coefficents of Impurities at Si /SiO$_2$ Interface[22]

Impurity	Thermodynamic Estimate	Redistribution Experiments	Oxide Masking Experiments
		m	
Ga	> 10^3		~20
B	10^{-3}-10^3	~0.3, 0.1	~10^{-2}
In	> 10^3		
P, Sb, As	> 10^3	~10	

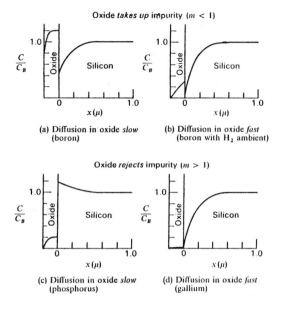

Fig. 23 The effect of thermal oxidation of silicon on the impurity segregation at the Si /SiO$_2$ interface: (a) slow diffusion in oxide, m<1 (boron); (b) fast diffusion in oxide, m<1 (boron in H$_2$ ambient); (c) slow diffusion in oxide, m>1 (phosphorus); and (d) fast diffusion in oxide, m>1 (gallium)[60]. Reprinted with permission of the American Physical Society.

equilibrium in each of the two solid phases (Si and SiO$_2$).

The second factor is *diffusivity of the impurity in the oxide.* That is, even if impurities segregate into the oxide, they will escape through the SiO$_2$ surface if they are able to rapidly diffuse. Boron in an H$_2$ ambient, and gallium exhibit a high diffusivity in SiO$_2$.

The third factor is the *rate of the Si /SiO$_2$ interface movement* into the underlying silicon. That is, the volume of the SiO$_2$ is more than twice that of the silicon consumed in growing it. Thus if the oxide has the same impurity concentration as the silicon, some impurities must be depleted from the Si to compensate for the volume gained by the oxide growth.

Grove, *et al*[60] developed a pictorial representation of four impurity redistribution cases, which are shown in Fig. 23. Let us discuss them by considering two situations: a) the oxide takes up the impurity (m<1); and b) the oxide rejects the impurity (m>1). The redistribution profile then depends, to a large extent, on how rapidly the impurity can diffuse through the oxide. In the first case (Fig. 23a), m<1 and the diffusion through the oxide is slow (e.g. boron). As a result the silicon surface is depleted of boron. If in subsequent processing however, the SiO$_2$ is subjected to an H$_2$ atmosphere, where the diffusivity of boron in SiO$_2$ increases significantly, we find the boron at the silicon surface is further depleted (i.e. case 2 as shown in Fig. 23 b). In the third case, in which the oxide rejects the impurities (m>1), and the impurity diffuses slowly (e.g. phosphorus), an accumulation of the impurity in the Si rapidly occurs. In the fourth case, where m>1 but the impurity rapidly diffuses through the oxide (e.g. gallium), it is found that the impurity leaves the solid, and the surface may actually be depleted (Fig. 23d).

The values of the segregation coefficients for the common impurities are listed in Table 7. The value of m can be calculated by fitting the measured diffusion profiles to a model for redistribution. The model involves the solution of the diffusion equations with a moving boundary (i.e. taking into account that the silicon is being consumed). The details of the model are beyond the scope of this section and are discussed in Ref. 60.

The value of m, however, may be experimentally measured. Secondary ion mass spectroscopy (SIMS), as discussed in Chap. 16, can be used to directly determine the segregation coefficient by measuring the concentration of the impurity both in the oxide, and in the silicon near the Si $/SiO_2$ interface. Using this procedure it has been found[61], for example, that the value of m for boron increases exponentially with temperature, with an activation energy of 0.33 eV for dry oxidations and 0.66 eV for near dry or wet oxidations. It is also found that m is greater for <100> than <111> oriented silicon.

OXIDATION SYSTEMS

The hotwall (see Chap. 6) horizontal *diffusion furnace* has been a workhorse of the semiconductor industry for over 30 years. Such furnaces were originally designed to perform diffusions (i.e. chemical predepositions and drive-ins, described in Chap. 8), as well as oxidations. More recently these furnaces have also been used for low pressure chemical vapor deposition *(LPCVD)* processes (Chap. 6). As more predepositions are being accomplished by ion implantation rather than by chemical means, furnace systems are primarily being used for oxidation, ion implant activation, and annealing. In addition, such VLSI fabrication demands as accurate control of oxide thickness, reduction in defect generation, and the necessity of processing large diameter wafers, has resulted in major modifications to the basic furnace. In this section we describe the basic hotwall diffusion furnace system, and some recent advances that

Fig. 24 A commercial 4-stack diffusion furnace system. Courtesy Thermco Systems.

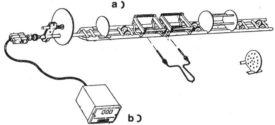

Fig. 25 (a) Wafers in an angle-slotted quartz boat. (b) Wheeled carrier. Courtesy of QBI, Inc.

have allowed furnaces to remain compatible with the needs of VLSI processing.

The basic diffusion (oxidation) system consists of the following modules: a) a cabinet; b) heating elements; c) measuring thermocouples; d) process tubes; e) quartz paddles and boats; f) a source cabinet; g) a temperature control system; and h) a load station.   A typical furnace system is shown in Fig. 24.

The standard furnace *cabinet* is constructed to house four horizontally oriented tubes. This allows high wafer throughput in a reduced floorspace, and such configurations are known as *4-stacks*.   Furnace heating elements are constructed of high purity, high temperature ceramic materials, wound with wire resistance heaters.   The windings are spaced along the tube length so that three heating zones can be provided.   Each of these zones is monitored by a Pt /Pt-Rh thermocouple, which allows for temperature measurement and accurate profiling of the furnace (i.e. determining the temperature as a function of distance along the length of the furnace tube). The process tubes which fit inside each of the four heating elements are most often fabricated of clear fused silica with special terminations.   These terminations are designed to allow for the introduction of the process gases on one end, and to provide for the exhaust of process gases at the other.   A fused silica *paddle* is used to support the quartz *boats*, which in turn, hold the wafers during the process.   Figure 25a shows a photograph of wafers loaded in such a boat.

The *source cabinet* contains the gas distribution system (valves, mass flow controllers, etc.) required for precise delivery of the process gases.   In an oxidation system the process gases include: O_2, H_2, N_2, TCA or HCl, and Ar.   The gas distribution system is constructed from stainless steel components, to minimize contamination.

The *temperature control* system receives data from the thermocouples, and adjusts power to the heater windings to maintain the temperature at a predetermined set point . In the past this function was performed by analog control circuitry. With tighter temperature control requirements, larger wafer loads, and controlled ramping (i.e. heating and cooling of the furnace at pre-determined rates), digital techniques are required to maintain adequate control.

The *load station* provides a clean environment (Class 10 or better) for loading and unloading wafers into the furnace tube. An automatic loading system is provided to push and pull the wafer boats into the furnace tube at a controlled rate.

Modern diffusion systems are capable of controlling temperature over the range of 300-1200°C to an accuracy of ±0.5°C over a length up to 40 inches, the so-called *flat-zone*. The use of 150 mm (6 inch) wafers requires process tube diameters greater than 10 inches. In such large tubes, heat losses from the ends are so high that adjustments of end zone temperatures cannot effectively maintain flat zones of adequate length. To help alleviate this problem some new systems employ five-zone furnaces. In addition to achieving improved flat-zones, the five-zone furnaces provide for more linear temperature ramping, with no overshoot[62].

Another key issue in temperature control is *recovery time*. This is the time required to return the furnace temperature to the set point after a cold load of wafers has been pushed into the heated zone. The thermal load generated by 200-150 mm wafers in a fused silica boat on a cantilever system (to be described below) is significant. Temperature reductions of as much as 50°C have been observed when such loads are inserted. With five-zone furnaces, each zone can be programmed to accelerate recovery of the flat zone temperature by anticipating the thermal load.

Most modern furnace systems utilize either microprocessors or minicomputers to control the operation of the tube, and to provide for status reports and data logging capability for each tube. Such temperature control modules can monitor the five furnace zones and provide real time corrections to temperature deviations. As mentioned previously, the zones can be preprogrammed to anticipate such large heat loads as a boat full of wafers entering the hot zone, or the ignition of an injector (which burns hydrogen and oxygen to form H_2O). The process sequence, which includes the type, quantity, and time that a gas is on, is controlled and monitored

Fig. 26 CRT output of computer controlled furnace showing tube status.   Courtesy Thermco Systems.

Fig. 27 Open cantilever wafer loading system. Courtesy Thermco Products.

through the *program module* of the micrcomputer.

The *boat loader module* controls the speed and position of the boat loader carriage assembly, and prevents the boat from traveling beyond preset limits. Many safety features are normally also incorporated into the system. For example, the controller may trigger alarms for tube overtemperature, stack overtemperature, scavenger exhaust failure, and loss of purge N_2 gas pressure. The programming and monitoring functions are handled by a keyboard /CRT system. Several process programs can be stored in the computer to run the furnace, allowing for single button operation thereby minimizing the possibility for operator error. The individual furnace computers may be tied to a *fabrication management computer,* which controls several 4-stack systems or perhaps the complete wafer fabrication facility. Figure 26 shows a photograph of the output of a CRT used in a furnace control system.

Considerable particle generation occurs when boat-laden fused silica paddles are dragged along the furnace tube during loading and unloading, as evidenced by measurements taken of the particle environment in the furnace tube during such loading. These generated particles can land on wafers and result in defects if they become embedded in the growing oxide. Such particle contamination can be somewhat reduced by the use of wheeled carriers (Fig. 25b), but friction at the wheel bearings and movement of the wheels over the tube still generates particles. Thus, greater reductions in particle generation can be achieved by use of *suspended loading* systems.

Two types of suspended loading systems are currently available. They are: a) fully-suspended loading systems (*cantilever*); and b) soft-landing systems. In *fully-suspended systems* the boats and wafers are suspended at the end of a motor-driven rod and pushed into the furnace without touching the process tube walls. During processing the wafers remain suspended, and upon completion, are removed from the tube, again without touching the walls. The materials of construction for such systems must be strong enough to support the load (up to 20 lbs) at processing temperatures of up to 1200°C. Such materials must also be inert to the gas environment (O_2, H_2, H_2O, HCl, etc.). Favored materials for these systems are silicon carbide (SiC), sapphire (Al_2O_3), and fused silica (SiO_2). A shortcoming of fully suspended systems is that variable deflections of the rod may cause problems in centering the load in the process tube. Figure 27 shows a photograph of a cantilever system.

The *soft landing system*, on the other hand, overcomes many of the problems associated with the suspended system. The early soft landing system set the paddles down within the process tube, resulting in a sag-free support for the wafer load. More recently *boats only* soft landing systems have become available. These systems carry the boats into the tube, lower them until they are supported by the process tube, and then withdraw, leaving behind the boats and wafers. An automatic tube door closes to permit processing. Upon process completion, the door opens and the paddle re-enters, picks up, and removes the boats from the process tube.

Vertical Furnaces

A recent innovation in furnace technology is the *vertical furnace* (Fig. 28). The major advantages that vertical furnace systems offer over conventional horizontal systems are: a) no cantilever or soft-landing is required since the wafers are held in a quartz boat which does not touch the process tube walls; b) the wafers can be loaded and unloaded automatically, and in a cassette-to-cassette manner (allowing the furnace operation to be performed unattended); and c) the clean room footprint of the system is somewhat smaller than for the horizontal configuration[71].

MEASUREMENT OF OXIDE THICKNESS

The accurate measurement of oxide film thickness is an important process control tool used

Fig. 28 Vertical diffusion furnace. Courtesy of the Silicon Valley Group.

during VLSI wafer fabrication. For example, the thickness of gate oxides in VLSI MOS devices (which is typically ≤ 250Å) must be tightly controlled, as it is one of the parameters that directly determines the device threshold voltage, V_T. Various techniques are available for measuring oxide thickness, the most important being: a) optical interference; b) ellipsometry; c) capacitance; and d) use of a color chart. Microprocessor controlled optical interference equipment is the most widely used for thickness measurements in a production environment. Automated ellipsometry equipment has also become available, and its use is gaining in popularity.

The *optical interference* method is a simple and nondestructive technique which can be used to routinely measure thermal oxide thickness from less than 100Å to more than 1 µm. The method is based on the interference that occurs between light reflected from the air $/SiO_2$ interface and the Si $/SiO_2$ interface. The equation governing this interference is:

$$x_o = \frac{\lambda (g - P)}{2 n^*}$$

(30)

where $n^* = (n_i^2 - \sin^2\theta)^{1/2}$; n_i is the index of refraction of the oxide film; θ is the angle of incidence of the light relative to the substrate; g is the order of the interference; λ is the wavelength of the incident radiation; and P is the net phase shift (i.e. $P = [\phi_s - \phi_f]$, where ϕ_f is the phase shift at the air $/SiO_2$ interface, and ϕ_s is the phase shift at the Si $/SiO_2$ interface). Equation 30 assumes that n_i is independent of wavelength, and that multiple reflections can be neglected. It also predicts that energy minima in the reflected light occur when $g = m + 1/2$, where m is an integer (m = 1, 2, 3, ...), and energy maxima result when $g = m$.

The most widely used implementations of this technique vary the wavelength of the light, while holding the incident angle constant. A spectrophotometer that supplies the incident light in the ultra-violet-to-visible range is used in performing the measurement. The reflected light intensity is measured as a function of wavelength. The minima in intensity are then fitted to Eq. 30 to calculate the film thickness.

Riezman and Van Gelder[63] developed a *fringe chart*, in which film thickness is plotted as a function of wavelength minima for various orders of reflection, g. The film thickness was measured independently by multiple-beam interferometry, an accurate but complex measurement technique. Therefore, such charts account for the phase shift terms (ϕ_f and ϕ_s), multiple reflections, and dispersion effects. The precision of measurements using optical interference and the fringe chart is ±25Å in the 500-10,000Å range. For thin gate oxide films (<500Å) the chart does not provide adequate results, and thickness data must be calculated directly from Eq. 30.

The development of microprocessor controlled commercial instrumentation has replaced the use of the fringe chart. Such automated systems include a microspectrophotometer, which measures reflected light intensity in the wavelength range of 480-790 nm, together with the use of a grating monochromator and photomultiplier tube detector. This instrument is mounted on a customized microscope with vertical reflected light illumination, and turret mounted 10X and 40X parfocal objectives (*parfocal* means that the field of view essentially remains in focus, and only fine focus adjustments are needed when switching between objective lenses). The viewer has a 10X eyepiece for precisely locating the area on the circuit to be measured. Measurements in the active region of the circuit (e.g. contact holes) can be made with this equipment. The thickness is automatically calculated from Eq. 30 using preassigned values of index of refraction.

Such automated optical interference instruments (Fig. 29) are widely used to measure the thickness of SiO_2 on Si, Si_3N_4 on Si, photoresist on Si, and poly-Si on SiO_2 (300-1200Å thick SiO_2). In the case of SiO_2 on Si, such equipment can readily measure film thicknesses in the range of 400Å to 30,000Å, with an accuracy of ±2%. For thinner oxides, special algorithms are employed by the instruments to determine thicknesses as small as 100Å. Below this

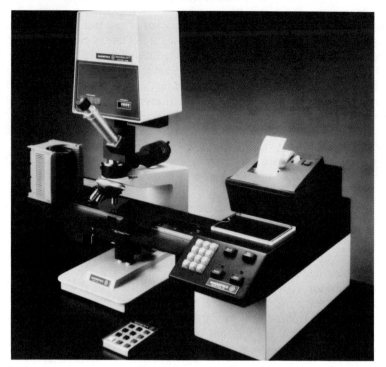

Fig. 29   Automated thin film thickness measuring system based on interference.   Courtesy of Nanometrics, Inc.

thickness a technique such as ellipsometry should be employed.

Ellipsometry   provides a non-destructive optical technique for measuring the oxide thickness, as well as the optical index of refraction at the measuring wavelength.   Archer was the first to use this method, and a detailed analysis of the method is found in Ref. 64.   In general, the index of refraction, n_i, of thermally grown SiO_2 is constant at a value of 1.46 at a wavelength of 5460Å.   On the other hand, the value of n_i for deposited films depends strongly on deposition conditions.   As a result, thermal oxides usually require only a thickness measurement.

The ellipsometry technique makes use of the change of state of the polarization of light when it is reflected from a surface (in this case the oxidized Si/SiO_2 interface).   The state of polarization is determined by the relative amplitude of the parallel and perpendicular components of the radiation, and by the phase difference between these two components.   The polarization change depends on the optical constants of the silicon, the angle of incidence of the light, the optical constants of the film (i.e. n_i, and the extinction coefficient, k), and the film thickness.   If the optical constants (n_i, k) of the substrate are known and the film is non-absorptive at the wavelength being used (k = 0), the state of polarization of the reflected beam depends on n_i and the thickness of the transparent film.

In the past, ellipsometry measurements required a delicate and experienced skill to determine the extinction points (which allow the state of polarization to be directly calculated).   Automated equipment, which directly outputs values of n_i and thickness has since been developed (Fig. 30). Such instruments are capable of measuring the oxide thickness and n_i in 4 to 20 sec, in films which are 20Å to 60,000Å thick.   Thickness measurement of ±3Å in accuracy and ±1Å in repeatabilty are possible, and the index of refraction can be determined to an accuracy of ±0.005

Table 8. COLOR CHART FOR THERMALLY GROWN SiO_2 FILMS OBSERVED PERPENDICULARLY UNDER DAYLIGHT FLUORESCENT LIGHTING[65]

Film Thickness (μm)	Color and Comments	Film Thickness (μm)	Color and Comments
0.05	Tan	0.63	Violet red
0.07	Brown	0.68	"Bluish" (Not blue but
0.10	Dark violet to red violet		borderline between violet
0.12	Royal blue		and blue green. It appears
0.15	Light blue to metallic blue		more like a mixture
0.17	Metallic to very light		between violet red and blue
	yellow green		green and looks grayish)
0.20	Light gold or yellow—	0.72	Blue green to green (quite
	slightly metallic		broad)
0.22	Gold with slight yellow	0.77	"Yellowish"
	orange	0.80	Orange (rather broad for
0.25	Orange to melon		orange)
0.27	Red violet	0.82	Salmon
0.30	Blue to violet blue	0.85	Dull, light red violet
0.31	Blue	0.86	Violet
0.32	Blue to blue green	0.87	Blue violet
0.34	Light green	0.89	Blue
0.35	Green to yellow green	0.92	Blue green
0.36	Yellow green	0.95	Dull yellow green
0.37	Green yellow	0.97	Yellow to "yellowish"
0.39	Yellow	0.99	Orange
0.41	Light orange	1.00	Carnation pink
0.42	Carnation pink	1.02	Violet red
0.44	Violet red	1.05	Red violet
0.46	Red violet	1.06	Violet
0.47	Violet	1.07	Blue violet
0.48	Blue violet	1.10	Green
0.49	Blue	1.11	Yellow green
0.50	Blue green	1.12	Green
0.52	Green (broad)	1.18	Violet
0.54	Yellow green	1.19	Red violet
0.56	Green yellow	1.21	Violet red
0.57	Yellow to "yellowish" (not	1.24	Carnation pink to salmon
	yellow but is in the position	1.25	Orange
	where yellow is to be	1.28	"Yellowish"
	expected. At times it	1.32	Sky blue to green blue
	appears to be light creamy	1.40	Orange
	gray or metallic)	1.45	Violet
0.58	Light orange or yellow to	1.46	Blue violet
	pink borderline	1.50	Blue
0.60	Carnation pink	1.54	Dull yellow green

units. Such film thickness measuring systems are quite versatile, and they are routinely used to measure the thickness and index of refraction of other dielectric films, including: a) silicon nitride (Si_3N_4); b) silicon oxynitride (SiO_xN_y); and c) aluminum oxide (Al_2O_3) on silicon substrates. They can also be used to determine the thickness of multilayer film structures, such as polysilicon on SiO_2 on Si, or photoresist on SiO_2 on Si.

The *capacitance method* can be used for measuring oxide thickness, but requires the fabrication of a MOS capacitor. The oxide thickness is given by $x_0 = C_{ox} A_g / \varepsilon_{ox} \varepsilon_0$, where C_{ox} is the measured oxide capacitance, A_g is the area of the gate, ε_{ox} is the dielectric constant of SiO_2, and ε_0 is the permittivity of free space. The metal electrode of the capacitor must be large compared to the oxide thickness in order to minimize capacitive fringing effects, which introduce errors in the measurement. The area of the metal electrode needs to be accurately known since it also enters into the calculation. The capacitance is measured with the capacitor biased so that the Si

Fig. 30 Automated ellipsometer measuring system. Courtesy of Gaertner Scientific Corporation.

is in accumulation (negative voltage on gate for p-type and vice versa). This insures that the silicon capacitance does not contribute to the measurement.

The *oxide film color chart* provides a rapid but less accurate procedure of determining SiO_2 film thickness. To prepare a color chart, oxides of various thicknesses are first grown and then measured by ellipsometry or optical interference, and these films become a *measurement standard*. Each oxide thickness has a specific color when it viewed in white light perpendicular to its surface. The colors are cyclically repeated for different orders of reflection. Sample wafers are then compared to the known oxide films. The sample and standard are tilted to vary the angle of incidence. If, as the sample and the standard films are tilted through various angles of incidence, both pass through the same sequence of colors, then the reflection order is correct and the thickness identified. Such a comparison technique is used only for rapidly determining oxide thicknesses where great accuracy is not required. Table 8 is a color chart that lists the colors observed for SiO_2 on Si, when viewed in normal incidence fluorescent lighting, as a function of the oxide thickness[65].

REFERENCES

1. J.M. Stevels and A. Kats, "The Systematics of Imperfections in Silicon-Oxygen Networks", *Philips Research Repts.*, **11**, 103 (1956).
2. A.G. Revesz, "The Defects Structure of Grown Silicon Dioxide Films"; *IEEE Trans. Electron Dev.*, **ED-12**, 97, (1965).
3. B.E. Deal and A.S. Grove, "General Relationship for the Thermal Oxidation of Silicon", *J. Appl. Phys.*, **36**, 3770, (1965).
4. H.Z. Massoud, J.D. Plummer, and E.A. Irene, "Thermal Oxidation of Silicon in Dry Oxidation: Growth Rate Enhancement in the Thin Regime, I. Experimental Results, II. Physical Mechanisms", *J. Electrochem. Soc.*, **132**, 2685, (1985).
5. E.A. Irene, "Silicon Oxidation Studies: Some Aspects of the Initial Oxidation Regime", *J.*

Electrochem. Soc., **125**, 1708 (1978).

6. A.C. Adams, T.E. Smith, and C.C. Chang, "The Growth and Characterization of Very Thin Silicon Dioxide Films", *J. Electrochem. Soc.*, **127**, 1787 (1980).

7. M. Hirayama, H. Miyashi, N. Tsubouchi, and H. Abe, "High Pressure Oxidation for Thin Gate Insulator Process", *IEEE Trans. Electron Devices*, **ED-29**, 503 (1982).

8. A. Kovchavtsev and A. Frantsuzov, *Mikroelectronika*, **8**, 439 (1979), p. 12.

9. H. Grinolds, *et al*, "Nitrided-Oxides for Thin Gate Dielectrics in MOS Devices", *IEDM Tech. Dig.*, p. 42 (1982).

10. I. Kato, *et al*, "Ammonia-Annealed SiO_2 Films for Thin Gate Insulators", *Japan J. Appl. Phys.*, **21**, Suppl. 21-1, 153 (1982).

11. T. Ito, *et al*, "Thermally Grown Silicon Nitride Films for High Performance MNS Devices", *Appl. Phys. Lett.*, **32**, 330 (1978).

12. T. Ito, H. Ishikawa, and M. Shinoda, "Thermal Nitridation of Silicon in Advanced LSI Processing", *Japan J. Appl. Phys.*, **20**, Suppl. 20-1, 33 (1981).

13. T. Ito, *et al*, "Thermal Nitride Gate FET Technology for VLSI Devices", *ISSCC Digest Tech. Papers*, p. 79 (1980).

14. M.M. Moslehi and K.C. Saraswat, "Thermal Nitridation of Silicon and Silicon Dioxide in a Cold-Wall RF-Heated Reactor" in *Proc. Symp. Silicon Nitride Thin Insulating Films, Electrochem. Soc.*, V. 83-8, pp. 324-345 (1983).

15. S. Wong and W. Oldham, "A Multiwafer Plasma System for Anodic Nitridation and Oxidation", *IEEE Electron Device Lett.*, **EDL-5**, 175 (1984).

16. T. Ito, *et al*, "Thermal Nitride Thin Films for VLSI Circuits" in *Proc. Electrochem. Soc. 163rd Meeting*, p. 295, (1983).

17. M. Hirayama, *et al*, "Plasma Anodic Nitridation of Silicon in N_2-H_2 Systems", *J. Electrochem. Soc.*, **131**, 663 (1984).

18. M.M. Moslehi and K. Saraswat, "Thermal Nitridation of Si and SiO_2 for VLSI", *IEEE Trans. on Electronic Devices*, **ED-32**, 106 (1985).

19. Y. Hayafuji and K. Kajiwara, "Nitridation of Silicon and Oxidized Silicon", *J. Electrochem. Soc.*, **129**, (1982).

20. Y. Hayafuji, K. Kajiwara, and S. Usui, "Shrinkage and Growth of Oxidation Stacking Faults During Thermal Nitridation of Si and Oxidized Silicon", *J. Appl. Phys.*, **63**, 8639 (1982).

21. J.R. Ligenza, "Effect of Crystal Orientation on Oxidation Rates in High Pressure Steam", *J. Phys. Chem.*, **65**, 2011 (1961).

22. E.A. Irene, "The Effect of Trace Amounts of Water on the Thermal Oxidation of Silicon in Oxygen", *J. Electrochem. Soc.*, **121**, 1613 (1974).

23. W.A. Pliskin, "Separation of the Linear and Parabolic Terms in Steam Oxidation of Silicon", *IBM J. Res. Dev.*, **10**, 198 (1966).

24. C. Ho, J.D. Plummer, "Si-SiO_2 Interface Oxidation Kinetics: Physical Model for the Influence of High Substrate Doping Levels. I Theory", *J. Electrochem. Soc.*, **126**, 1516 (1979), "II Comparison with Experiments and Discussion", *J. Electrochem. Soc.*, **126**, 1523 (1979).

25. E.A. Irene, *J. Electrochem. Soc.*, **120**, 1613 (1974).

26. D.W. Hess and B. Deal, "Kinetics of the Thermal Oxidation of Silicon in O_2/HCl Mixtures", *J. Electrochem. Soc.*, **124**, 735 (1977).

27. e.g. Thermco Systems, Orange, CA.

28. R. Cosway and S. Wu, "Comparison of Thin Thermal SiO_2 Grown Using HCl and 1,1,1 Trichloroethane (TCA)", *J. Electrochem. Soc.*, **132**, 151 (1985).

29. M.B. Das, R.E. Tressler, and W.H. Grubbs, "A Comparison of HCl and Trichloroethylene-Grown Oxides on Silicon", *J. Electrochem. Soc.*, **131**, 389 (1984).

30. B.Y. Liu and Y.C. Chang, "Growth and Characterization of Thin Gate Oxide by Dual TCE Process", *J. Electrochem. Soc.*, **131**, 683, (1984).

31. Y.C. Cheng and B.Y. Liu, "Oxidation Characteristics and Electrical Properties of Low Pressure Dual TCE Oxides", *J. Electrochem. Soc.*, **131**, 354 (1984).

32. H.L. Tsai, *et al.*, "Cl Incorporation at the Si /SiO$_2$ Interface during Oxidation of Si in HCl /O$_2$ Ambients", *J. Electrochem. Soc.*, **131**, 411 (1984).

33. J.W. Rause, *et al.*, "Auger Sputter Profiling and Secondary Ion Mass Spectroscopy Studies of SiO$_2$ Grown in O$_2$ /HCl Mixture", *J. Electrochem. Soc.*, **131**, 887 (1984).

34. N. Tsubouchi, *et al.*, "Application of the High-Pressure Oxidation Process to the Fabrication of MOS LSI", *IEEE Trans. Elec. Dev.*, **ED-26**, 618 (1979).

35. D. Craven and J.B. Stimmel, "The Silicon Oxidation Process-Including High Pressure Oxidation", *Semiconductor International*, 59, June, 1981.

36. J.R. Ligenza, *J. Appl. Phys.*, **36**, 2703 (1965).

37. D.L. Pielfrey and J. Reche, *Solid State Elect.*, **17**, 627 (1974).

38. A. Ray and A. Reisman, *J. Electrochem. Soc.*, **128**, 2424 (1981).

39. R. Oren and S.K. Ghandi, *J. Appl. Phys.*, **42**, 752 (1971).

40. S.A. Schafer and S.A. Lyon, *J. Vac. Sci. Technol.*, **19**, 494 (1981).

41. E. Young, W. Tiller, "Photon Enhanced Oxidation of Si", *Appl. Phys. Lett.*, **42**, 63, (1983).

42. I.W. Boyd, "Laser-Enhanced Oxidation of Si", *Appl. Phys. Lett.*, **42**, 728 (1983).

43. W.E. Beadle, J.C.C. Tsai, and R.D. Plummer, eds., *Quick Reference Manual for Silicon Integrated Circuit Technology*, J. Wiley and Sons (1985), p. 6-33.

44. B.E. Deal, "Standardized Terminology for Oxide Charges Associated with Thermally Oxidized Silicon", *IEEE Trans. Electron Dev.*, **ED-27**, 606 (1980).

45. E.H. Nicollian and A. Goetzberger, *Bell System Tech. J.*, **46**, 1055 (1967).

46. R.R. Razouk and B.E. Deal, *Extended Abstracts of the May 1979 Electrochemical Soc. Meeting*, Abs. 135, pp. 363-365; *J. Electrochem. Soc.*, **126**, 1573 (1979).

47. E.H. Nicollian and J.R. Brews, *MOS (Metal Oxide Semiconductor) Physics and Technology*, Wiley, New York, 1982, Chap. 12.

48. B.E. Deal, M. Sklar, A.S. Grove, and E.H. Snow, "Characteristics of the Surface State Charge of Thermally Oxidized Silicon", *J. Electrochem. Soc.*, **114**, 266 (1967).

49. E. Arnold, J. Ladell, and G. Abowitz, *Appl. Phys. Letts.*, **13**, 413 (1968).

50. E.H. Snow, A.S. Grove, B.E. Deal, and C.T. Sah, "Ion Transport Phenomena in Insulating Films", *J. Appl. Phys.*, **36**, 1664 (1965).

51. M. Yamis, *IEEE Trans. Electron Dev.*, **ED-12**, 88 (1965).

52. M. Kuhn and D.J. Silversmith, *J. Electrochem. Soc.*, **118**, 966 (1971).

53. N.J. Chou, *J. Electrochem. Soc.*, **118**, 601 (1971).

54. E.H. Nicollian and J.R. Brews, *MOS (Metal Oxide Semiconductor) Physics and Technology*, Wiley, New York (1982), Chap. 16.

55. F.J. Grunthaner and J. Maserjian, *Physics of SiO$_2$ and its Interfaces*, S.T. Pantelides, Ed., Pergamon, New York, 1978, Chap. 7.

56. A. Goetzberger, V. Heine, and E.H. Nicollian, *Appl. Phys. Lett.*, **12**, 95 (1968).

57. R.B. Lauglin, MIS Gordon Conference, Tilton, N.H., 1980.

58. R.J. Jaccodine and W.A. Schlegal, "Measurement of Strains at Si-SiO$_2$ Intefaces", *J. Appl. Phys.*, **37**, 2429 (1966).

59. E.P. EerNisse, "Stress in Thermal SiO$_2$ During Growth", *Appl. Phys. Lett.*, **35**, 8 (1979).

60. A.S. Grove, O. Leistiko, and C.T. Sah, "Redistribution of Acceptor and Donor Impurities During Thermal Oxidation of Silicon", *J. Appl. Phys.*, **35**, 2629 (1964).

61. R.B. Fair, and J.C.C. Tsai, "Theory and Direct Measurement of Boron Segregation in SiO$_2$

during Dry, Near Dry, and Wet O_2 Oxidations", *J. Electrochem. Soc.*, **125**, 2050 (1978).

62. J.C. Malaikal, D. Fisher, Jr., and A. Waugh, "Trends in Automated Diffusion Furnace Systems for Large Wafers", *Solid State Tech.*, p. 105, Dec. 1984.

63. F. Reizonan and W. VanGelder, *Solid State Electronics*, **10**, 625 (1967).

64. R.J. Archer, "Determination of the Properties of Films on Silicon by the Method of Ellipsometry", *J. of the Optical Soc. of America*, **52**, 970 (1962).

65. W.A. Pliskin and E.E. Conrad, "Nondestructive Determination of Thickness and Refractive Index of Transparent Films", *IBM J. Res. and Dev.*, **8**, 43 (1964).

66. B.E. Deal, *J. Electrochemical Soc.*, **125**, 576, 1978.

67. B.E. Deal and M. Sklar, "Thermal Oxidation of Heavily Doped Silicon", *J. Electrochem. Soc.*, **112**, 430, (1965).

68. C.P. Ho, *et al*, "Thermal Oxidation of Heavily Phosphorus Doped Silicon"*J. Electrochem. Soc.*, **125**, 665, (1978).

69. R.R. Razouk, L.N. Lie, and B.E. Deal, "Kinetics of High Pressure Oxidation of Silicon in Pyrogenic Steam", *J. Electrochem. Soc.*, **128**, 2214, (1981).

70. H.F. Wolfe, *Semiconductor Silicon Data*, Pergamon Press, 1969, New York, p. 601.

71. P. Burggraaf, "Vertical Furnaces", *Semiconductor International*, April, 1986.

PROBLEMS

1. Explain the difference between mass-transport limited oxidation, and reaction-rate limited oxidation.

2. Explain why steam oxidation proceeds at a faster rate than dry oxidation.

3. A silicon wafer of 150 mm diameter has a 0.2 μm thick layer of SiO_2. Calculate the weight of the the SiO_2 in this layer.

4. Show that in Eq. 16, $x_o^2 + Ax_o = B (t + \tau)$, reduces to $x_o^2 = Bt$ for log times, and $x_o = B/A (t + \tau)$ for short times.

5. A silicon wafer has a 0.3 μm layer of thermally grown SiO_2 on its surface. Find the time required to increase the SiO_2 to a thickness of 0.5 μm by oxidation in wet oxygen at 1100°C. Repeat the calculation for dry oxygen at 1100°C.

6. If the wet oxidation of problem 5 is conducted with 3%HCl added to the oxidizing ambient, determine the time that would be necessary for the film thicness to reach 0.5 μm.

7. Determine the SiO_2 thickness which would result from the following oxidation sequence: 30 min dry O_2, followed by 30 min in wet O_2, both at 1050°C. Assume (100) Si is being used.

8. A cut is made in an SiO_2 film that is 0.4 μm in thickness. An oxide of 1500 Å is grown over this cut. Sketch a cross section of the resulting Si /SiO_2 structure, showing the thickness of the oxide in all locations, and Si /SiO_2 interface at all locations.

9. Estimate the thickness of an SiO_2 layer needed to mask against a boron diffusion at 1000°C for an hour, if the solid solubility of boron is three times that in silicon. Compare your answer with the data given in Fig. 14.

10. List several techniques that have been found to allow an oxide to be initially grown on a silicon surface without forming oxidation-induced stacking faults.

11. Give some reasons why vertical furnaces have advantages over horizontal furnaces for growing thermal oxides, and also list some possible disadvantages.

8

DIFFUSION in SILICON

Diffusion is the process by which a species moves as a result of the presence of a chemical gradient. The diffusion of controlled impurities or dopants into Si is the basis of p-n junction formation and device fabrication in VLSI processing. In the early days of transistor and IC processing, dopants were supplied to the silicon by *chemical sources*. These dopants were then diffused to the desired depths by subjecting the wafers to elevated temperatures (900-1200°C). Such dopant sources are available in gas, liquid, or solid form. More recently, ion implantation has become the major means for the introduction of impurities into silicon (see Chap. 9). Ion implantation is capable of placing the impurities near the silicon surface, but a diffusion step is still required to drive the impurities to specified depths. In addition, ion implanted devices must be subjected to elevated temperatures in order that the implanted impurities become electrically active, and so that defects from the implantation are removed by annealing (see Chap. 9). When these high temperature steps are performed, diffusion of the impurities occurs at the same time.

In the past, junction depths in devices were often in the range of 1-3 μm. In VLSI devices, however, these depths have decreased to 0.15-1.0 μm. For example, in advanced CMOS devices it is common for source and drain diffusions to be as shallow as 0.2 μm. Control of such shallow diffusions is more difficult to achieve, and their measurement requires special skill and care.

In this chapter we present: a) the mathematics which govern the mass transport phenomena of diffusion (Fick's equations and their solutions); b) the diffusion constants (diffusivities) of the common dopants in silicon; c) some current models which attempt to explain the atomistic mechanisms that give rise to diffusion in silicon; d) anomalous diffusion effects; e) diffusion systems and chemical sources; and f) measurement of diffusion depth (junction depth, diffused sheet resistance, and diffused profile).

MATHEMATICS OF DIFFUSION

In this section we describe the basic differential equations of diffusion and their solutions for special cases. The *diffusion coefficient* (the terms *diffusion coefficient*, *diffusivity*, and *diffusion constant*, D, are used interchangably throughout this text) is also defined and its experimental determination presented. It should be pointed out that the atomic nature of the silicon matrix, as related to the diffusion of atoms, is discussed in a subsequent section. This area of study is very important, and a significant research effort continues in developing atomistic models of the diffusion of substitutional impurities in silicon.

Fick's First Law

If there is an impurity concentration gradient present, $\partial C / \partial x$, in a finite volume of a matrix

242

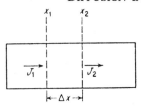

Fig. 1 Schematic showing an element of volume with the flux J_1 entering and J_2 leaving.

substance, there will be a tendency for the impurity material to move so that a decrease in the gradient is produced. If the flow persists for a sufficiently long time, the material will become homogeneous and the net flow of matter will cease. In 1855 Fick formalized the mathematics of this mechanism by postulating that the flux of the matter across a given plane is proportional to the concentration gradient across the plane. This material flow is expressed, in the one dimensional case, as a flux J, which is given by:

$$J = -D \frac{\partial C(x,t)}{\partial x} \tag{1}$$

where D, the constant of proportionality, is the *diffusion constant* for the material that is diffusing (diffusant) in the specific solvent. Equation 1 is called *Fick's First Law*. The dimensions of the terms in Eq. 1 are given in parentheses below:

$$J \text{ (mass) / (length}^2 \text{ time)} = -D \text{ (length}^2 \text{ /time)} \, \partial C / \partial x \text{ (mass /length}^3 \text{ /length)} \tag{1a}$$

The units of the diffusion constant D are cm^2/sec, but they are often given in the more convenient form of μm^2/hr.

Fick's Second Law

We now consider the case of a finite volume of material in which the impurity gradient decreases with increasing time. In this case, the impurity concentration is also changing with time. Since this is a condition frequently encountered during a diffusion process, Eq. 1 must be modified to include the effect. For example, if we consider a bar of unit cross-sectional area (Fig. 1) with the x-axis along its center, and apply a material balance to it, we obtain the following: An element, Δx thick along the x-axis has a flux J_1 entering on one side and J_2 leaving on the other. If Δx is very small, J_1 can be accurately related to J_2 by the expression:

$$J_1 = J_2 - \Delta x \left(\frac{\partial J}{\partial x} \right) \tag{2}$$

Since the amount of material that entered into the element in unit time, J_1, is different than that which left, J_2, the concentration within the element must change with time, $\partial C / \partial t$ (assuming also that no impurity materials are formed or consumed in the matrix). The volume of the element can be expressed as $1 \cdot \Delta x$ (unit area $\cdot$ thickness), and thus the net increase in matter in the element per unit time is given by:

$$J_2 - J_1 = \Delta x \left(\frac{\partial C}{\partial t} \right) = -\Delta x \left(\frac{\partial J}{\partial x} \right) \tag{3}$$

or,

$$\frac{\partial C(x,t)}{\partial t} = - \frac{\partial J}{\partial x} \qquad (4)$$

Substituting Eq. 1 in Eq. 4 we obtain *Fick's Second Law* which is written as:

$$\frac{\partial C(x,t)}{\partial t} = \frac{\partial}{\partial x} (D \frac{\partial C}{\partial x}) \qquad (5)$$

If the diffusion coefficient is independent of position, as is the case when the solute concentration is low, then Eq. 5 reduces to:

$$\frac{\partial C(x,t)}{\partial t} = D [\frac{\partial^2 C(x,t)}{\partial x^2}] \qquad (6)$$

Solutions To Fick's Second Law

There are two technologically important ways to introduce impurities into the silicon using thermal diffusion. In the first, a flux of impurities constantly arrives at the surface from a source via vapor transport, and diffuses into the silicon. The source effectively maintains a constant value of surface concentration during this diffusion process. In diffusion technology this type of dopant incorporation is known as *predeposition*, and is used to introduce a uniform and repro-ducibile quantity of impurities into the silicon, generally at lower temperatures (800–900°C). For this case, the initial condition, which is represented by the concentration at t = 0 is:

$$C(x, 0) = 0 \qquad (7)$$

and the boundary conditions are;

$$C(0, t) = C_s \qquad (8)$$

and;

$$C(\infty, t) = 0 \qquad (9)$$

where C_s is the surface concentration in atoms $/cm^3$. The solution to Fick's second law (Eq. 6) under these initial and boundary conditions is:

$$C(x, t) = C_s \text{ erfc } [x / 2\sqrt{Dt}] \qquad (10)$$

where D is the diffusion coefficient, x is the distance coordinate, t is the time of the diffusion, erfc is the *complementary error function*; and $\sqrt{Dt}$ is the *diffusion length*. The complementary error function is tabulated[1] for values of z = x $/2\sqrt{(Dt)}$. Note that with increasing time, the impurity penetrates deeper into the silicon.

The total amount of impurity in the diffused layer Q(t) is obtained by integrating Eq. 10 over all of space (-∞ to +∞). The result of this integration is given by:

$$Q(t) = \frac{2}{\sqrt{\pi}} C_s \sqrt{Dt} \qquad (11)$$

We define the junction depth, x_j, as the distance in the silicon at which the diffused profile is equal to the substrate background concentration, C_{sub}. Therefore, x_j may be calculated from Eq. 10 when $C(x_j, t) = C_{sub}$ or:

$$x_j = 2 \sqrt{Dt} \ \text{erfc}^{-1} \frac{C_{sub}}{C_s} \qquad (12)$$

In the second method for impurity introduction by thermal diffusion, we assume that a thin layer of dopant is present at the surface, either through chemical predeposition or ion implantation. The total quantity of impurity present is fixed at Q_0 (atoms /cm^2). The initial condition for this case is represented by:

$$C(x, 0) = 0 \qquad (13)$$

and the boundary conditions are;

$$\int_0^\infty C(x, t) \, dx = Q_0 \qquad (14)$$

and;

$$C(x, \infty) = 0 \qquad (15)$$

The solution to Fick's second law (Eq. 6) under these conditions is:

$$C(x, t) = \frac{Q_0}{\sqrt{\pi D t}} e^{[-x^2 / 4Dt]} \qquad (16)$$

This solution has the form of a *Gaussian distribution*. The surface concentration, C_s, is obtained by solving Eq. 16 when x = 0:

$$C_s = C(0, t) = \frac{Q_0}{\sqrt{\pi D t}} \qquad (17)$$

The junction depth for this case is determined in a manner similar to that used to obtain Eq. 12 and is given by:

$$x_j = \{4 D t \ \ln \frac{Q_0}{C_{sub} \sqrt{\pi D t}} \}^{1/2} \qquad (18)$$

This process is termed the *redistribution* or *drive-in* diffusion, and is used to move the impurities to the desired junction depth subsequent to the predeposition. Equation 16 applies for the Q_0 existing as a δ-function, a very shallow rectangular distribution of dopant near the surface.

Figure 2 is a plot of normalized concentration (C /C$_0$) as a function of normalized distance (x /$\sqrt{4Dt}$) in the silicon for the complementary error function (erfc), and for the Gaussian distributions. The following examples explain how to obtain the diffusion profile and the junction depth for the two cases discussed above.

Example 1: Calculate: a) the diffusion profile; b) the junction depth; and c) the total amount of dopant introduced after boron predeposition performed at 950°C for 30 min, in a neutral ambient. Assume the substrate is n-type with a background doping level of 1.5×10^{16} atoms /cm^3 and the boron surface concentration reaches solid solubility ($C_s = 1.8 \times 10^{20}$ atoms /cm^3).

Solution: Under the conditions of the predeposition, the diffusion profile is controlled by the erfc of Eq. 10. The profile can be obtained by using the erfc curve of Fig. 2. In order to use

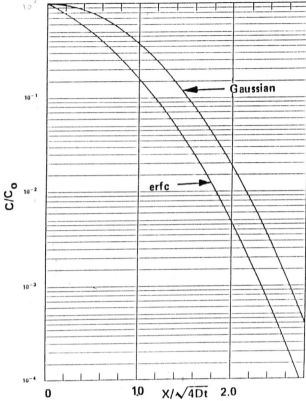

Fig. 2 Normalized concentration (C/C_0) as a function of normalized distance (dC/dt) for the Gaussian and erfc curves.

this curve, values of $x/(2\sqrt{Dt})$ must be calculated for values of x. Corresponding values of $C(x)/C(0)$ [where $C(0)$ is the B surface concentration C_s], are found from Fig. 2, and finally values of $C(x)$ are calculated. First, the value of the B diffusion coefficient at 950°C is found from:

$$D_B(950°C) = 0.76 \exp(-3.46 \text{ eV}/k\ 1223°K) [\text{cm}^2/\text{sec}] = 3.0\text{x}10^{-15} \text{ cm}^2/\text{sec}.$$

We next calculate the diffusion length, $z = 2\sqrt{Dt}$ for the process:

$$z = 2 (3\text{x}10^{15} \text{ cm}^2/\text{sec} \times 1.8\text{x}10^3 \text{ sec})^{1/2} = 4.65\text{x}10^{-6} \text{ cm}.$$

We now set up the following table to assist in the calculation of the diffusion profile:

x, cmx10^{-4}	x $/(2\sqrt{Dt})$	C(x) /C (0)	C(x) atoms /cm^3
0	0	1	1.8 x 10^{20}
0.05	1.07	0.13	2.3 x 10^{19}
0.075	1.67	0.018	3.2 x 10^{18}
0.10	2.15	0.0022	3.9 x 10^{17}
0.15	3.23	0.000005	9.0 x 10^{14}

b) The junction depth is defined as the value of x where the concentration of the diffusion profile equals the substrate doping (1.5×10^{16} atoms /cm^3). When C (x) = 1.5×10^{16} atoms /cm^3, then C (x) /C (0) = 8.3×10^{-5}. We next determine x_j /(2$\sqrt{Dt}$) from Fig. 2, which results in:

$$x_j / (2\sqrt{Dt}) = 2.8; \quad \text{or} \quad x_j = 2.8 \times 4.65 \times 10^{-6} \text{ cm} = 1.3 \times 10^{-5} \text{ cm} = 0.13 \text{ } \mu\text{m}$$

c) The total amount of dopant is calculated from Eq. 11:

$$Q_o = (2\sqrt{Dt}) / \sqrt{\pi} = 4.62 \times 10^{-6} / 1.77 = 4.7 \times 10^{14} \text{ atoms /cm}^2.$$

Example 2: Calculate the diffusion profile and junction depth after the predeposition of Example 1 is subjected to a neutral ambient drive-in at 1050°C for 60 min.

Solution: For the drive-in conditions of this example the Gaussian distribution of Eq. 16 (curve labeled Gaussian in Fig. 2) controls the profile. First calculate the diffusion length (2$\sqrt{Dt}$) for the new diffusion conditions:

$$D_B \text{ (1050°C)} = 0.76 \exp (-3.46 \text{ eV } /k \text{ 1323°K}) = 3.3 \times 10^{-14} \text{ cm}^2 \text{ /sec.}$$

We next calculate the value of the surface concentration C (0) from Eq. 17:

$$C (0) = Q_o / \sqrt{\pi Dt} = 4.7 \times 10^{14} \text{ atoms /cm}^2 / (\pi \text{ 3.3 x } 10^{-14} \text{ cm}^2 \text{ /sec x 3600 sec})^{1/2}$$

$$= 2.44 \times 10^{19} \text{ atoms /cm}^3$$

We set up the following table to assist in the calulation.

x, cm 10^{-4}	x /(2$\sqrt{Dt}$)	C (x) /C (0)	C (x) atoms /cm^3
0	0	1	2.44 x 10^{19}
0.1	0.45	0.75	1.83 x 10^{19}
0.2	0.91	0.60	1.46 x 10^{19}
0.3	1.36	0.17	4.10 x 10^{18}
0.4	1.82	0.038	9.20 x 10^{17}
0.5	2.27	0.006	1.46 x 10^{17}
0.6	2.73	0.0004	9.80 x 10^{15}
0.7	3.18	0.00004	9.80 x 10^{14}

We calculate x_j in a manner similar to the one used in Example 1.

$$C (x_j) / C (0) = 1.5 \times 10^{16} / 2.44 \times 10^{19} = 6 \times 10^{-4}.$$

We next find the corresponding value of x /(2$\sqrt{Dt}$) from Fig. 2, which is = 2.7. Hence, $x_j / \sqrt{Dt}$ = 2.7, or:

$$x_j = 2.7 \times 2.2 \times 10^{-5} \text{ cm} = 5.9 \times 10^{-5} \text{ cm} = 0.59 \text{ } \mu\text{m.}$$

If the predeposition is from an ion implanted source, which produces a doping concentration with a near-Gaussian distribution close to the surface, the Gaussian profile will move when subjected to a high temperature anneal. The solution to Eq. 6 with an initially implanted Gaussian profile has been reported for the cases of redistributions performed in neutral ambients[2], and oxidizing ambients[3]. For redistributions in oxidizing ambients, the solution to Eq. 6 is difficult to obtain in closed form, since it involves a moving boundary problem. As a result, numerical methods must be employed to obtain these solutions. For redistributions in a neutral atmosphere from an implanted Gaussian, where the concentration of dopant C (x) is:

$$C(x) = \frac{\phi}{\sqrt{\pi} \Delta R_p} \exp\left[\frac{-(x - R_p)^2}{(2\Delta R_p^2)}\right] \qquad (19)$$

(where ϕ is the ion dose [ions /cm^2]; R_p is the projected range, in cm; and ΔR_p is the projected range straggle, in cm), the following expression is obtained for the diffusion profile:

$$C(x,t) = \frac{\phi}{\sqrt{\pi} (2\Delta R_p^2 + 4 Dt)^{1/2}} \exp\left(\frac{-(x - R_p)^2}{(2\Delta R_p^2 + 4Dt)}\right) \qquad (20)$$

where x is the distance along the diffusion profile, t is the diffusion time, and D is the diffusion coefficient of the impurity (see Chap. 9 for a definition of R_p and ΔR_p).

Concentration Dependent Diffusion Coefficients

The diffusion profiles discussed earlier are valid if the diffusion coefficients are constant in value. This is implied by the fact that Eq. 6, rather than Eq. 5, was used to obtain solutions to Fick's second law. This situation of constant diffusion coefficients prevails when the doping concentration is lower than the intrinsic carrier concentration n_i, at the diffusion temperature (e.g. $n_i = 5 \times 10^{18}$ /cm^3 at T = 1000°C) .

Under conditions of high dopant concentration it is observed that the diffusion profiles predicted by Eqs. 10 and 16 do not match measured profiles. This arises because under high concentration conditions, D becomes a function of concentration [D = D(C)], and Eq. 5 must be used to obtain correct solutions to Fick's second law. Since the value of D changes with concentration, and since the concentration changes with distance (i.e. a concentration gradient exists), D also changes with distance into the sample. Under these conditions Eq. 5 may be expanded to:

$$\frac{\partial C}{\partial t} = \frac{\partial}{\partial x}\left(D \frac{\partial C}{\partial x}\right) = \frac{\partial D}{\partial x}\frac{\partial C}{\partial x} + D \frac{\partial^2 C}{\partial x^2} \qquad (21)$$

The $\partial D /\partial x$ term, however, makes Eq. 21 an inhomogeneous differential equation, and obtaining solutions in closed form is difficult at best (for special cases), and often even impossible.

Fortunately, a methodology termed the *Boltzmann-Matano analysis*[4] has been developed, which provides for a solution of D = D (C). It does not, however, result in a solution for C (x,t) as previously obtained, but rather allows D (C) to be calculated from an experimentally determined concentration profile (C vs. x plot). The mathematical argument proceeds as follows:

When D is a function of concentration only (and not time), and the surface concentration is held constant, Eq. 21 can be transformed into an ordinary homogeneous differential equation by using a new variable η, which is defined as:

$$\eta = x / t^{1/2} \qquad (22)$$

Applying this transformation of variables to Eq. 21 we obtain:

$$-\frac{\eta}{2} \frac{dC}{d\eta} = \frac{d}{d\eta} [D \frac{dC}{d\eta}] \qquad (23)$$

Next, the situation of diffusing impurities into an undoped background is considered. The initial conditions are given by:

$$C = C_0 \qquad \text{for} \qquad x < 0, \ t = 0 \qquad (24)$$

and,

$$C = 0 \qquad \text{for} \qquad x > 0, \ t = 0 \qquad (25)$$

Integrating Eq. 23 between the limits $C = 0$ to $C = C_1$, where C_1 is the concentration at which we desire the value of the diffusion coefficient, we obtain:

$$\frac{-1}{2} \int_0^{C_1} \eta \, dC = \left[D \frac{dC}{d\eta} \right]_{C=0}^{C=C_1} \qquad (26)$$

Applying the boundary condition, D [dC /dη] = 0, which is valid when C = 0, and substituting back in Eq. 23 for that boundary condition and for η at a fixed time, an expression for the concentration dependent diffusion coefficient is obtained:

$$D(C_1) = \frac{-1}{2t} (\frac{dx}{dC})_{C_1} \int_0^{C_1} x \, dC \qquad (27)$$

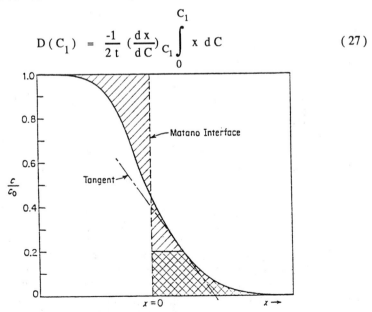

Fig. 3 The Matano interface is positioned to make the hatched areas on either side equal. The cross-hatched area and tangent show the quantities involved in calculating D at C= O, C = 0.2 C_0^4.

Thus, D (C) can be obtained from graphical integration and differentiation of C (x) using Eq. 27.

The quantities that are required to calculate a value of D are shown in Fig. 3, which is a plot of normalized concentration C /C_o, as a function of distance. The Matano interface is the plane at which x = 0. Graphically it is the line that makes the two cross-hatched areas of Fig. 3 equal. For example, the value of D (C_1) at C_1 = 0.2 C_o, would be calculated by measuring the area under the curve between C /C_o = 0 and C_1 /C_o = 0.2 (cross-hatched) and the reciprocal of the slope at that point. The cross-hatched area is equal to the integral term in Eq. 27, while the reciprocal of the slope at C_1 is equal to the differential term in the same equation. This process is repeated for several values of C_1, until the diffusion constant is obtained as a function of concentration, D (C). The errors in the calculated values of D (C) are the largest when (C /C_o) ~ 1 and (C /C_o) ~ 0, since in these regions the integral is small, and the slope large. To minimize these errors, the original data are often plotted on probability paper. The methods for obtaining C (x) are discussed in a later section entitled *Doping Profile Measurements*.

Equation 27 assumes that for large negative values of x, the concentration remains constant with time. In most diffusions, however, the concentration gradient is measured after the predeposition and drive-in. Therefore Eq. 27 is not applicable to redistribution (drive-in) from a high concentration predeposition. The constant surface concentration requirement may be removed as long as the total dopant concentration does not change with time. This is expressed as:

$$Q = \int_0^\infty C (x, t) \, dx = \text{constant} \quad (28)$$

Under these conditions the diffusion coefficient is given by:

$$D \left[\frac{C (x_o, t)}{C_s} \right] = \frac{-C (x_o, t) x_o}{2 t [\frac{dC}{dx}]_{x = x_o}} \quad (29)$$

where C_s is the surface concentration, x_o the location where D is determined and (dC /dx)$_{x = xo}$ is the concentration gradient at x = x_o.

THE TEMPERATURE DEPENDENCE OF THE DIFFUSION COEFFICIENT

Empirical measurements of diffusion coefficients, D, as a function of temperature show that the data can be represented by:

$$D = D_0 \, e^{[- E_a /kT]} \quad (30)$$

where D_0 and E_a may vary with composition, but are independent of temperature. Experimentally, D_0 and E_a are obtained from Arrhenius plots of ln D vs. 1 /T, which are linear when Eq. 30 is valid. The slope of this plot gives the value of -E_a /k, while ln D_0 is given by the intercept at 1 /T = 0. D_0, E_a, k, and T in Eq. 30 are the frequency factor, the activation energy, Boltzmann's constant, and the absolute temperature, respectively.

The value of D_0, the *frequency factor*, is related to the frequency of lattice vibrations (i.e. the frequency at which atoms strike the potential barrier that they must overcome to move in the lattice). This number is typically 10^{13}-10^{14} Hz. The value of E_a is related to the height of the energy barrier that the impurity must overcome in order to move within the lattice.

Table 1. INTRINSIC DIFFUSION COEFFICIENTS OF B, As, and Sb

	Unit	Boron[6]	Phosphorus[6]	As[7]	As[8]	As[9]	Antimony[6] (See Refs. 6, 7, 8, 9)
		$(D_i^+)_B$	$(D_i^x)_P$		$(D_i^{--})_{As}$		$(D_i^x)_{Sb}$
D_o	cm^2/s	0.76	3.85	24	22.9	60	0.214
E_a	eV	3.46	3.66	4.08	4.1	4.2	3.65

The measured values of activation energy for diffusing atoms in silicon vary from 0.2-2 eV for the fast diffusing elements (such as He, H_2, O_2, Au, Pt, Na, Fe, and Ni), to 3-4 eV for the substitutional impurities (such as B, Ga, P, As, and Sb). It was believed that the low values of activation energy implied interstitial diffusion, while the high values were due to vacancy diffusion. Recent studies, however, show that this analysis is not necessarily correct, and that both vacancies and interstitials play a role in substitutional diffusion in silicon (see Section entitled *Interstitial-Vacancy Model*).

DIFFUSION CONSTANTS OF THE SUBSTITUTIONAL IMPURITIES   Boron, Arsenic, Phosphorus, and Antimony

As previously discussed, the diffusion coefficient is concentration dependent at high concentrations. The upper limit of concentration at which the diffusion coefficient is constant, can be estimated from the intrinsic carrier concentration, n_i, of the silicon at the diffusion temperature. The intrinsic carrier concentration as a function of temperature for silicon is given by[5]:

$$n_i \,(cm^{-3}) \;=\; 3.87 \times 10^{16} \, T^{3/2} \, \exp\left[-\,(0.605 + \Delta E)\,/kT\right] \tag{31}$$

where ΔE (eV) $= 7.1 \times 10^{-10} \,(n_i \,/T)^{1/2}$. When the diffusing impurity concentration C (x) is less than that of n_i, it is assumed that D is independent of concentration. In that case, the value of D can be determined from the measured concentration profiles applied to Eqs. 10 or 16, depending on the applicable boundary conditions. The diffusion coefficient at low concentrations is termed the *intrinsic diffusion coefficient*, D_i. When the total impurity concentration (diffusion plus substrate background) exceeds n_i, we refer to the *extrinsic diffusion coefficient*, D_e, which then depends on the ratio of the extrinsic to intrinsic carrier concentrations.

The values of the intrinsic diffusion coefficients have been tabulated by Fair[6] as a function of temperature. These data are presented by listing values of D_o and E_a for the common dopants, and are repeated here in Table 1. To determine the value of D at the diffusion temperature, the tabulated values of D_o and E_a are inserted into Eq. 30. In the same report, Fair attempts to explain the values of the diffusion coefficients by postulating the existence of a multi-charge state impurity-vacancy interaction mechanism. He proposes, that the intrinsic diffusivities are dominated by interactions with neutral vacancies, V^x; singly charged acceptor vacancies, V^-; doubly charged acceptor vacancies, V^{-2}; and singly charged donor vacancies, V^+. He establishes a corresponding set of diffusivities, which are labeled as D^x_i, D^-_i, D^{-2}_i, and D^+_i, respectively. The values of the D_i for these various defects are fitted well to the collected diffusion data.

Figures 4 a-d summarize the diffusion coefficients of boron, phosphorus, arsenic, and

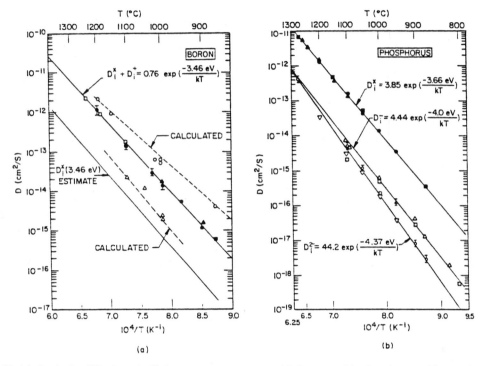

Fig. 4 Intrinsic diffusion coefficients vs. temperature: (a) boron; (b) phosphorus; (c) arsenic; and (d) antimony[6]. Reprinted with permission of North-Holland Publishing Co.

antimony as a function of temperature. For a detailed description of the data and its application, the reader is referred to Ref. 6. Although the model accurately represents the diffusion coefficients as a function of temperature, a question remains as to the physical correctness of Fair's vacancy model. The current status of the understanding of the diffusion models in silicon is discussed in the section entitled *Atomistic Models of Diffusion in Silicon*.

Table 2 lists the values of D_o for the fast diffusing elements in silicon. Most of these elements degrade the properties of devices, and as a consequence their presence in VLSI applications is usually undesirable.

Fair[6] has also developed expressions for high impurity concentration diffusion coefficients in Si. These models are valuable, since they are the result of careful experimentation, and they

Table 2. DIFFUSION COEFFICIENTS OF THE FAST DIFFUSANTS IN SILICON

Element	D_o (cm^2/sec)	Element	D_o (cm^2/sec)
Na	1.6×10^{-3}	Ag	2×10^{-3}
K	1.1×10^{-3}	Pt	1.6×10^{2}
Cu	4×10^{-2}	Fe	6.2×10^{-3}
Au	1.1×10^{-3}	Ni	0.1
Au_i	2.4×10^{-4}	O_2	7×10^{-2}
Au_s	2.8×10^{-3}	H_2	9.4×10^{-3}

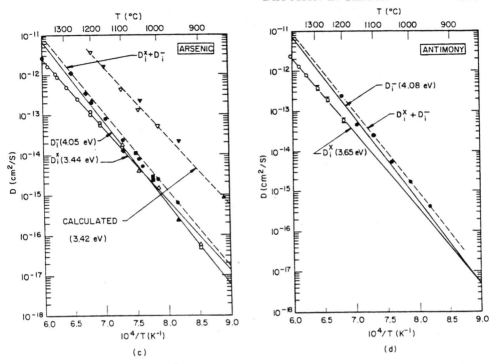

Fig. 4 (c) Arsenic. (d) Antimony [6]. Reprinted with permission of North-Holland Publishing Co.

can be used to provide analytical expressions for diffusion profiles. The validity of the basic vacancy model on which they are based however, is currently being debated. The following sections provide a summary of the concentration-dependent diffusion coefficients for As, B, and P.

Arsenic Diffusion

The diffusion coefficient for *arsenic* (As), according to the multi-charge state vacancy mechanism is given as:

$$D\,(C)_{As} \;=\; (\,2n\,/n_i\,)\,(\,D_i\,C_{As}\,) \tag{32}$$

for $C_s \gg n_i$. The factor of 2 results from the electric field effect (see the section entitled *Anomalous Diffusion Effects in Silicon*). The approximate solution (empirical) for the measured As impurity profile can be represented by the following polynomial[10]:

$$C \;=\; C_s\,(1.00 \,-\, 0.87\,Y \,-\, 0.45\,Y^2) \tag{33a}$$

where;

$$Y \;=\; x\,\frac{(\,8\,C_s\,D_i\,t\,)^{-1/2}}{n^{-1/2}} \tag{33b}$$

and the intrinsic diffusion coefficient is given by: $D_i = 22.9 \exp(-4.1\,eV\,/kT)\,cm^2/sec$.

Equation 32 predicts a very abrupt profile (see Fig. 5), and a junction depth which is relatively insensitive to the substrate (background) doping. For constant surface concentration the junction depth is given by:

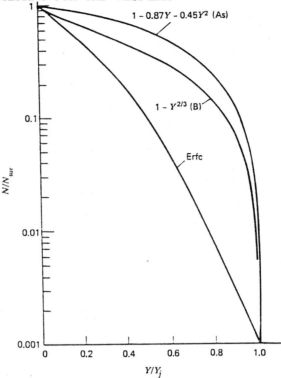

Fig. 5 Normalized diffusion profiles for arsenic (Eq. 32) and boron (Eq. 40). The erfc is shown for comparison. From S.K. Ghandi, *VLSI Fabrication Principles*, Copyright © 1983 John Wiley & Sons. Reprinted with permission of John Wiley & Sons.

$$x_j = 2.3 \, (\, C_s \, D_i \, t \, / \, n_i \,)^{1/2} \tag{34}$$

For ion implanted As layers that are subsequently diffused[11] the following expressions give: a) junction depth, x_j; b) surface concentration, C_s; c) total dopant (also the ion dose) Q; and d) sheet resistance, R_s:

$$x_j = 2 \, (\, Q \, D_i \, t \, / \, n_i \,)^{1/3} \tag{35}$$

$$C_s = 0.91 \, (\, Q^2 \, n_i \, / \, D_i \, t \,)^{1/3} \tag{36}$$

$$Q = 0.55 \, C_s \, x_j \tag{37}$$

$$R_s = (1.76 \times 10^{10} \, / \, Q^{7/9}) \, (n_i \, / \, D_i \, t \,)^{1/9} \tag{38}$$

Boron Diffusion

The diffusion coefficient for *boron* is given by:

$$D \, (C) = D_i \, C_B \, / \, n_i \tag{39}$$

An approximate solution for the impurity profile for $p \gg n_i$ is[12]:

$$C = C_s (1 - Y^{2/3}) \qquad (40)$$

where;

$$Y = (x^2 / 6 D_i t)^{3/2} \qquad (41)$$

is a dimensionless parameter (see Fig. 5).

The junction depth of boron diffusions in n-type material with background doping of $<10^{18}$ atoms $/cm^3$ is given by:

$$x_j = 2.45 (C_s D_i t / n_i)^{1/2} \quad \text{at } C_B \sim 10^{18} \qquad (42)$$

where;

$$C_s = (2.78 \times 10^{17}) / (R_s x_j) \qquad (43)$$

For an implanted, diffused B layer:

$$Q = 0.4 C_s x_j \qquad (44)$$

$$C_s = 0.53 (Q^2 n_i / D_i t)^{1/3} \qquad (45)$$

where:

$$D_i = 3.17 \exp (- 3.59 \text{ eV } /kT) \quad cm^2 /sec \qquad (46)$$

Phosphorus Diffusion

The diffusion of *phosphorus* under high concentration conditions in silicon is a complex phenomenon, and has a shape which is schematically represented in Fig. 6. It can be seen that there are three distinct regions: a) the *high concentration* region; b) the *transition* or *kink* region; and c) the low concentration, or *tail* region. No simple equation has been found to fit this curve. A relationship exists, however, between the total P concentration, C_T, and the electrically active concentration, which is represented by:

$$C_T = n + 2.4 \times 10^{-41} n^3 \qquad (47)$$

for temperatures between 900° and 1050°C.

Fair and Tsai[13] developed a phosphorus diffusion model based on differing diffusion constants for the high concentration and tail regions of the profile. They describe the existence

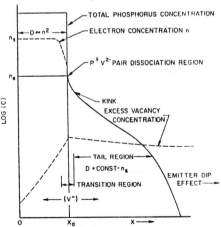

Fig. 6 A model describing high concentration phosphorus diffusion in silicon[6]. Reprinted with permission of North-Holland Publishing Co.

of a *phosphorus ion-vacancy pair*, $(PV)^{-1}$, as the controlling defect for phosphorus diffusion. This defect is believed to have a negative charge, and is present in the high concentration region, but dissociates in the tail region, thereby increasing the vacancy concentration. Hu, *et al*[14] described the phosphorus profile by considering a two-stream diffusion. They claimed that the streams are substitutional (P_s) and interstitial (P_i) phosphorus streams. The presence of a phosphorus interstitial donor P_i has not been established, and other assumptions of the model are difficult to verify. Tan and Gosele[15] have proposed a model based on ternary equilibrium among Si-P-O, the constituents at the silicon surface, which provides for a higher solubility of P in Si then observed otherwise. In the interior of the crystal, the binary equilibrium conditions are valid, and SiP precipitates form. These three theories are indicative that the origin of the P diffusion curve is not well understood at this time. Work continues to clarify our understanding.

ATOMISTIC MODELS OF DIFFUSION IN SILICON

The mechanisms that occur on the atomic scale control the phenomena of diffusion in silicon, and a great deal of effort has been expended to identify these mechanisms. Atomic diffusion in silicon has been found to proceed by *direct* and *indirect* mechanisms. Impurity atoms that do not have a strong bonding interaction with Si atoms, and are located exclusively in the interstices of the lattice, jump directly between the interstices. As discussed previously, species such as H_2, He, Ar, and Au^+ (the Au interstitial) are believed to diffuse in this manner. Oxygen atoms, which possess strong bonding interactions with Si atoms, are also believed to diffuse by jumping between the bond-centered interstitial positions. This type of diffusion is schematically depicted in Fig. 7a

The diffusion of the substitutional impurities and the self-diffusion of silicon requires the presence of intrinsic point defects (see Chap. 2) as the diffusion vehicle. The *vacancy* [V] mechanism is known to be the controlling mechanism in the self-diffusion of fcc metals[4] and in Ge[16], and has also been proposed by Fair[6] to be the controlling mechanism in Si. Figure 7b depicts the movement of atoms via the vacancy mechanism. Other studies have shown, however, that *interstitials* [I] also plays a role in Si self-diffusion, and in the substitutional diffusion of P, B, Al, and Ga, particularly above 1000°C.

The Vacancy Model

The *vacancy model* of Fair[6] is based on the premise that the diffusion coefficient of an impurity is dominated by the interaction of the impurity with vacancies. If the impurity

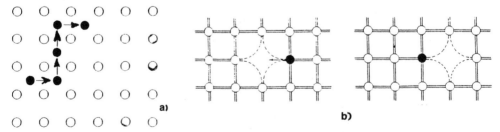

Fig. 7 (a) Interstitial diffusion by jumping. (b) Substitutional diffusion to via the vacancy mechanism. From S.K. Ghandi, *VLSI Fabrication Principles*, Copyright © 1983 John Wiley & Sons. Reprinted with permission of John Wiley & Sons.

diffusion is dominated by an acceptor type monovacancy mechanism, the diffusion coefficient is approximately proportional to the acceptor monovancancy concentration.   This is described by:

$$\frac{D}{D_i} = \frac{n}{n_i}$$ (48)

Since the defects, however, can have various charge states, Eq. 48 can be generalized to include all combinations of point defect interactions[17] and is given by:

$$D = D_{Si}^x + \sum_{r=1}^{m} (D^{-r})[\frac{n}{n_i}]^r + \sum_{r=1}^{m} (D^{+r})[\frac{n}{n_i}]^r$$ (49)

where D^x, refers to the diffusion coefficient associated with neutral defects, and D^{-r} and D^{+r} refer to the intrinsic impurity diffusion coefficients associated with the particular charge state, r, of the defect (r = 1, 2, 3 ... m) .   Equation 48 can be used to fit the experimental diffusion profiles without specifying the nature of the defects of different charge states, and does not imply, by itself, a dominant mechanism (vacancy or interstitial). Fair, however, attempts to explain self-diffusion of silicon and diffusion of the substitutional impurities in Si (Group III and V), by a predominant vacancy mechanism. For example, he includes the existence of neutral vacancies $[V^x]$, singly charged acceptor vacancies $[V^-]$, doubly charged acceptor vacancies $[V^{2-}]$, and singly charged donor vacancies $[V^+]$, as all contributing to Si self-diffusion. The diffusion associated with each defect has a specific value of D_o and E_a. Fair described the self-diffusion of Si by:

$$D_{si} = D_{Si}^x + D_{Si}^-(n/n_i) + D_{Si}^{2-}(n/n_i)^2 + D_{Si}^+(n/n_i)$$ (50)

where;

$$D_{Si}^x = 0.015 \exp(-3.89 \text{ eV}/kT) \text{ cm}^2/\sec$$ (51)

$$D_{Si}^- = 16 \exp(-4.54 \text{ eV}/kT) \text{ cm}^2/\sec$$ (52)

$$D_{Si}^+ = 1180 \exp(-5.09 \text{ eV}/kT) \text{ cm}^2/\sec$$ (53)

$$D_{Si}^{2-} = 10 \exp(-5.1 \text{ eV}/kT) \text{ cm}^2/\sec$$ (54)

The Interstitial-Vacancy Model

In the last few years, studies in two areas have helped clarify our understanding of the nature of point defects and diffusion processes in silicon. These studies have indicated that silicon vacancies, [V], alone do not control diffusion, but that silicon interstitials, [I], also play a role, particularly for temperatures above 1000°C. These results have been obtained: a) from the studies on the effect of oxidation on stacking fault growth kinetics and diffusion; and b) from the analysis of Au diffusion in dislocation-free Si. The stacking fault studies showed that [I] and [V] coexist in Si at high temperatures, under thermal equilibrium, as well as under oxidizing conditions. The Au diffusion studies showed that [I] play a role, and that it is possible for both [I] and [V] to coexist. The value of the [I] component for self-diffusion in Si has been obtained, and an estimate of the [V] component has been made. The arguments for these conclusions, which we present here, are in outline form only. Readers are referred to the Tan and Gosele paper[15], and to the numerous references they give, for more detailed descriptions of these studies.

First we consider the effects of oxidation on diffusion. The oxidation of silicon is known to lead to *oxidation-enhanced* or *retarded* diffusion (OED or ORD) of dopants, as well as generation of oxidation induced stacking faults (OISF, Chap. 2). As pointed out by Tan and Gosele, the enhancement or retardation effects, as well as stacking fault generation, probably result from the same cause, oxidation injection of [I] into Si. Tan, *et al*[18] have observed that OED /ORD cannot be explained if either [I] or [V] are assumed to exist alone under thermal equilibrium. It is assumed that both [I] and [V] are present in Si at high temperatures, and that they achieve local equilibrium, which is expressed as:

$$C_I \, C_V \; = \; C_I^{eq} \, C_V^{eq} \qquad (55)$$

where C_I and C_V are the [I] and [V] concentrations, respectively, and the superscript "eq" denotes their thermal equilibrium values. Under oxidation conditions, we define supersaturation ratios for [I] and [V] as:

$$S_I \; = \; (C_I / C_I^{eq}) \; - \; 1 \qquad (56)$$

and

$$S_V \; = \; (C_V / C_V^{eq}) \; - \; 1 \qquad (57)$$

Thus, when $C_I = C_I^{eq}$, $S_I = 0$ which implies no supersaturation, and similarly for S_V.
Using Eqs. 55, 56, and 57, a relationship between S_I and S_V can be established:

$$S_V \; = \; - \, S_I \, / \, (1 + S_I) \qquad (58)$$

Next consider the case when both [I] and [V] contribute to diffusion. Under such conditions we can write the diffusion coefficient of the substitutional impurities D^S, as consisting of an (I) and a (V) component:

$$D^S \; = \; D_I^{\,S} \; + \; D_V^{\,S} \qquad (59)$$

where $D_I^{\,S}$ and $D_V^{\,S}$ are the [I] and [V] components of the diffusion coefficients, respectively. Under oxidation conditions, however, the thermal equilibrium between [I] and [V] is perturbed, and the value of D^S is now given by D^S_{ox}, which is written as:

$$D^S_{ox} \; = \; D_I^{\,S} \, (C_I / C_I^{eq}) \; + \; D_V^{\,S} \, (C_V / C_V^{eq}) \qquad (60)$$

It should be noted that each of the diffusion coefficient components is multiplied by the ratio of the concentration of the point defect to the equilibrium concentration of that point defect.
We now define the diffusion coefficient *enhancement* term as:

$$\Delta_{ox}^{\,S} \; = \; (D_{ox}^{\,S} / D^S) \; - \; 1 \qquad (61)$$

so when $D_{ox}^{\,S} = D^S$, $\Delta_{ox}^{\,S} = 0$, implying that no diffusion enhancement is occurring.
A final definition required to complete this discussion is that of $G_I^{\,S}$, which is the fractional [I]-component of the dopant diffusion constant under thermal equilibrium conditions:

$$G_I^{\,S} \; = \; D_I^{\,S} \, / \, D^S \qquad (62)$$

Substituting Eqs. 55, 60, and 62 into Eq. 61 we arrive at an expression for the diffusion enhancement which is related to measurable factors, and is given by:

$$\Delta_{ox}^{\,S} \; = \; (2G_I^{\,S} + G_I^{\,S} \, S_I \, - 1) \, S_I \, / \, (1 + S_I) \qquad (63)$$

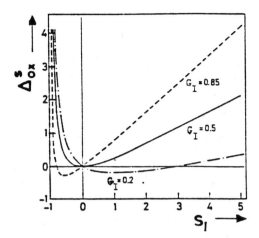

Fig. 8 Predicted values of diffusion coefficient enhancement during oxidation as a function of the supersaturation of interstitials[15]. This figure was originally presented at the Spring, 1984 Meeting of the Electrochemical Society. Reprinted by permission.

Figure 8 shows a plot of Δ_{ox}^S as a function of S_I for three different values of G_I^S (Eq. 63).

The values of S_I used in Fig. 8 were obtained from measurements of the growth of oxidation stacking fault sizes. The interstitial supersaturation ratio S_I has been determined from the stacking fault growth measurements, to depend exponentially on temperature, and is given by[15,19] (where t is the time of oxidation):

$$S_I = 6.6 \times 10^{-9} \exp (2.52 \text{ eV } /kT) \, t^{-1/4} s^{1/4} \qquad (64)$$

The values of Δ_{ox}^S can be either positive or negative (i.e. OED or ORD), depending on the values of G_I^S and S_I. Figure 9 shows the fitting of the available data on Sb ORD to Eq. 63. The fit is quite good for values of $G_I^S = 0.02$, implying that there is an interstitial component to Sb diffusion under oxidizing conditions. It is found experimentally,[19] however, that for short Sb diffusion time (<10 min), OED is observed. This is possible if dynamical equilibrium is not reached between [I] and [V], but rather C_I increases while C_V remains constant, thereby increasing G_I until equilibrium is achieved. Such a close fit between theoretical and experimental data indicates that the model, which postulates the coexistence of [I] and [V], and which holds

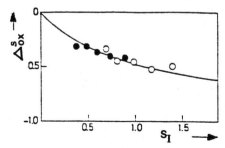

Fig. 9 Quantitative fitting of available Sb ORD data for <100> Si to Eq. 63 for $G_I = 0.02$[15]. This figure was originally presented at the Spring, 1984 Meeting of the Electrochemical Society. Reprinted by permission.

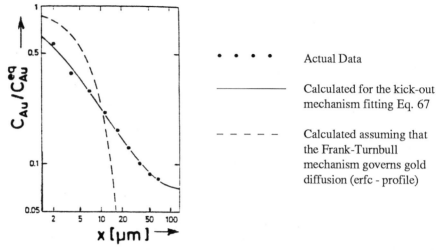

Fig. 10 Gold concentration profile for a 1 hr. diffusion in dislocation-free silicon at 900°C. Dash: erfc fitting, solid: data fitting Eq. 67[15]. This figure was originally presented at the Spring, 1984 meeting of the Electrochemical Society. Reprinted by permission.

that they attain local equilibrium, is essentially correct.

We next consider the diffusion of Au in dislocation-free Si. Gold atoms can be incorporated in Si as either substitutional impurities, Au_s, or interstitial impurities, Au_i, and these may diffuse either by the kick-out mechanism, or the Frank-Turnball mechanism. The long range transport of Au atoms in both mechanisms occurs via migration of Au_i, which may jump from one interstitial site to another, or from an interstitial site to a lattice position, to become Au_s. In the *kick-out mechanism* the interchange involves [I], and the equilibrium between the two types of Au sites is given by:

$$Au_i \Leftrightarrow Au_s + I \qquad (65)$$

This creates [I] supersaturation in the crystal, which is balanced by [I]-outdiffusion. In the *Frank-Turnball mechanism*, the interchange involves [V] and the equilibrium is given by:

$$Au_i + V \Leftrightarrow Au_s \qquad (66)$$

This creates vacancy [V] undersaturation in the crystal, which is balanced by [V] in-diffusion.

If Eq. 65 is controlling, it means that [I] are involved in Si self-diffusion, but does not imply that [V] are not also contributing. If Eq. 66 is controlling it means that [V] are involved in the Si self-diffusion, but again there is no implication regarding the contribution of [I]. If only [V] and no [I] are contributing it can be shown that Au should have a diffusion coefficient that is independent of the Au concentration. Under these conditions the diffusion profile should follow a erfc dependence.

Figure 10 shows an experimental diffusion profile for Au diffusion in dislocation-free silicon at 900°C[20]. The erfc type profile does not fit the data, implying that vacancy controlled diffusion is not operative.

The data, on the other hand, results in an excellent fit to the expression:

$$D_I^{eff} = C^{-2} D_I C_I^{eq} / C_S^{eq} \qquad (67)$$

where C is the concentration of Au, C_I^{eq} is the equilibrium concentration of [I], and C_S^{eq} is the equilibrium concentration of substitutional Au.

The implication of this particular concentration dependence is that [I] are indeed contributing to Si self-diffusion, but does not preclude the contribution of [V]. At temperatures below 800°C, when the equilibria expressed in Eqs. 55 and 65 cannot be attained, the erfc-profile should be controlling (vacancy controlled diffusion). Wilcox, et al[21] found that at 700°C, the diffusion profile of Au does have the erfc dependence.

These experiments clearly show that [V] alone does not control self-diffusion or substitutional impurity diffusion in Si. They do, however, show that [I] contributes to diffusion, and may be the predominant point defect at temperatures in excess of 1000°C. Over the next few years we anticipate continued progress in our understanding of the nature of point defects and their roles in impurity diffusion in silicon.

DIFFUSION IN POLYCRYSTALLINE SILICON

Impurity diffusion in polycrystalline thin films (such as polysilicon) is quantitatively different from that observed in single crystal material. The difference arises because grain boundary diffusion predominates in polycrystalline films, while bulk diffusion is the diffusion mechanism in single-crystal material. As discussed in Chaps. 4 and 6, the diffusivity of elements along grain boundaries is ~100 times faster than in bulk material.

Polycrystalline Si films are made up of grains with sizes ranging from 0.1 to 0.3 μm, that are separated by grain boundaries. The diffusion within the grains is comparable to that of single crystal silicon. The impurity atoms diffuse along the grain boundaries and then diffuse into the grains. Since the grains are small, only a short time is required for the dopant, which is entering from all sides of the grain, to diffuse therein. As a result, the diffusion is controlled by the grain boundaries, the grain structure, and the preferred orientation of the film (see Chap. 6). These quantities, in turn, depend upon the deposition conditions of the film, such as temperature, deposition rate, thickness, and post-deposition annealing treatments.

Each measurement of the diffusion constant, D, in poly-Si depends very much on the film's history and published results cannot be taken universally. Data have been obtained for phosphorus[22], boron[23,24], and arsenic[25,26] diffusivities in poly-Si, and are listed in Table 3. It has been found that P and As precipitate at grain boundaries, resulting in reversible changes of resist-

Table 3. SOME REPORTED DIFFUSION COEFFICIENTS IN POLYSILICON FILMS

Elements	D_o (cm^2/s)	E (eV)	D (cm^2/s)	T (°C)
As	8.6×10^4	3.9	2.4×10^{-14}	800
As	0.63	3.2	3.2×10^{-14}	950
B	$(1.5-6) \times 10^{-3}$	2.4-2.5	9×10^{-14}	900
B			4×10^{-14}	925
P			6.9×10^{-13}	1000
P			7×10^{-13}	1000

Table 4. DIFFUSION COEFICIENTS OF FAST DIFFUSERS IN AMORPHOUS SiO$_2$

Material	D_o (cm^2/s)	D at 1000°C (cm^2/sec)	D at 300°C (cm^2/sec)	Material	D_o (cm^2/s)	D at 1000°C (cm^2/sec)	D at 300°C (cm^2/sec)
H$^+$	1	1.01×10^{-3}	2.05×10^{-7}	P	5.3×10^{-8}	8.1×10^{-14}	1×10^{-25}
He	3.0×10^{-4}	5×10^{-7}	3×10^{-8}	B	1.7×10^{-5}	1.6×10^{-18}	4.0×10^{-30}
H$_2$	5.65×10^{-4}	9×10^{-6}	6.3×10^{-8}	Ga	3.8×10^5	8.0×10^{-12}	2.5×10^{-31}
H$_2$O	1.0×10^{-6}	7×10^{-10}	1.1×10^{-13}	Au	8.2×10^{-10}	5.54×10^{-15}	2.5×10^{-19}
OH	9.5×10^{-4}	2×10^{-7}	9.5×10^{-10}	Na	6.9	5.2×10^{-5}	2.8×10^{-11}
O$_2$	2.7×10^{-4}	1×10^{-4}	1.6×10^{-14}	Pt	1.2×10^{-13}	4.5×10^{-17}	2.0×10^{-20}

ivity upon annealing. Boron has not shown any evidence of such grain boundary segregation.

DIFFUSION IN SILICON DIOXIDE

Silicon dioxide serves as a diffusion mask for both chemical and ion implanted predepositions. Its sole purpose, in that case, is to keep diffusants away from regions where they are not desired. Silicon nitride (Si$_3$N$_4$) can be used for similar purposes. The diffusivities of the common dopants in SiO$_2$ can be calculated from measured profiles assuming solutions to Fick's second law, with the appropriate boundary conditions.

The Group III and V elements are known to form glassy networks with SiO$_2$ and as a result their diffusivity strongly depends on their concentration. The diffusivities of these elements, are very low for concentrations less than 1%, and do not need to be considered in any detail. The diffusivities of the common dopants in SiO$_2$ are listed in Chap. 7, Table 5.

There is a class of materials that are fast diffusants in SiO$_2$, including H$_2$, He, OH^{-1}, Na, O$_2$, and Ga. Values of D greater than 10^{-13} cm^2/sec have been determined for these elements. Table 4 lists fast diffusing materials and their diffusivities in SiO$_2$ for various conditions.

ANOMALOUS DIFFUSION EFFECTS IN SILICON

Electric Field Enhancement

During diffusion at elevated temperatures, the diffusing species generally become ionized. As a result, an electric field is set up within the crystal, from the gradient of charge that exists along the concentration gradient. This electric field operates in a direction to enhance the

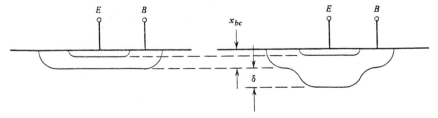

Fig. 11 Enhanced base diffusion under the emitter resulting from a heavy phosphorus diffusion. (Emitter-dip effect).

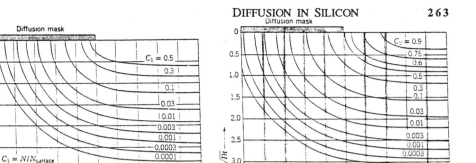

Fig. 12 Diffusion contours at the edge of an oxide window: a) Constant source diffusion; b) Limited source diffusion. From S.K. Ghandi, *VLSI Fabrication Principles*, Copyright © 1983 John Wiley & Sons. Reprinted with permission of John Wiley & Sons.

diffusion of the impurities. It can be shown[27] that the diffusion flux of dopants in an electric field is given by:

$$ J = - q D h \left(\frac{\partial N_D}{\partial x} \right) \tag{68} $$

where N_D is the donor concentration, and h is given by;

$$ h = 1 + \frac{\dfrac{N_D}{2 n_i}}{\sqrt{\left(\dfrac{N_D}{2n_i}\right)^2 + 1}} \tag{69} $$

For the case of high donor concentration (i.e. $N_D \gg n_i$), Eq. 69 results in a value of h = 2. Hence the maximum obtainable electric field enhancement of the diffusion constant is 2.

Emitter Push Effect

The *emitter push* (or dip effect) was first noted in npn transistors with heavily phosphorus doped emitters. It was observed that the diffusion of the base (boron diffusion) under the emitter was enhanced relative to the base diffusion outside the emitter. The effect is depicted schematically in Fig. 11.

Since the observation of this effect, numerous physical mechanisms have been proposed to explain the behavior. Fair[28] proposed that the enhanced boron diffusion results from a combination of: a) bandgap narrowing, resulting from strain induced by the phosphorus diffusion, which reduces the concentration of the (P^+V^{2-}) pairs; and b) the (P^+V^{2-}) dissociation then increases the vacancy concentration in the boron region, thereby increasing the diffusion rate in that region. Other experiments[29-32] have shown that there is a net interstitial, [I], supersaturation below the phosphorus diffused layer, indicating that perhaps the interstitials play a role in the emitter push effect.

The technological importance of this effect is diminishing, as arsenic has become the dopant of choice for both VLSI bipolar emitter and NMOS source and drain diffusions. No extended push effect has been observed for arsenic diffusion.

Lateral Diffusion Under Oxide Windows

Since most diffusions are carried out through an SiO_2 mask, the lateral diffusion under the mask is an important parameter. In concentration-independent diffusion cases, under both constant-source and limited-source conditions, it has been determined that lateral penetration is ~75-85% of the vertical diffusion depth (Figs. 12a and 12b)[33]. In high concentration and shallow diffusion cases, in which values of the diffusion coefficients are concentration dependent, lateral penetration is found to extend ~65-70% of the vertical diffusion distance[34]. A second lateral diffusion effect exists, and is manifested as the enhancement or retardation of junction depths near the edge of SiO_2 or Si_3N_4 windows resulting from elastic strain near the edge[35].

Diffusion in an Oxidizing Ambient

It is well known that diffusion in an oxidizing ambient can result in oxidation enhanced diffusion (OED) or oxidation retarded diffusion (ORD). The technique for measuring OED /ORD is discussed in Ref. 36. It has been observed that OED /ORD follows these general rules: a) in dry oxidation, OED predominates for As and P in <100> wafers, and for P in <111> wafers (with only small effects for As in <111> silicon); b) the enhancement for <100> silicon is greater than for <111> silicon; c) for boron diffusion in <100> silicon, OED is observed under oxidizing ambients; and d) Δ_{ox}^S can be expressed in terms of an effective oxidation rate $(x_{ox}/t)^n$.

DIFFUSION SYSTEMS AND DIFFUSION SOURCES

Diffusion processes are conducted in systems known as *diffusion furnaces*, which provide controlled high temperature and gas flow conditions. A typical *diffusion system* consists of a heating element, diffusion tube, boat, and dopant delivery system. The components of diffusion furnaces are essentially identical to those used to grow thermal SiO_2 films, and the reader is directed to Chap. 7 for a discussion on their characteristics. In the past, most predepositions of impurities in device fabrication were done with chemical sources. More recently, ion implantation has taken over as the preferred means of introducing dopants into Si (Chap. 9). Regardless of the predeposition process, drive-in steps which redistribute impurities or achieve electrical activation, are still performed in diffusion furnaces. Ambient gases in furnace tubes during drive-in can be oxidizing (O_2, H_2O), reducing (H_2), or neutral (Ar, He, N_2).

In some of the most advanced MOS and bipolar technologies, ion implantation is used to predeposit all layers of impurities. There are, however, many other VLSI fabrication processes that still use chemical procedures for introducing dopants. For example, backside phosphorus gettering (Chap. 2) uses $POCl_3$ sources and diffusion furnace processes, while bipolar base diffusions may be predeposited in diffusion furnaces sourced with boron nitride (BN) discs. In the following sections, we describe the major classes of chemical sources available to the process engineer. It should be noted that these sources are used to provide dopant materials to silicon wafers both in diffusion furnaces and ion-implanted processes. These sources are: a) gaseous; b) liquid (bubbler and spin-on); and c) solid (tablet, powder, or disc). Table 5 shows the various types of sources available for each of the major dopant atoms (As, P, B, and Sb).

Gaseous Sources

Gaseous dopant sources are the most widely used types throughout the semiconductor industry. This is particularly true for ion implantation applications, where 90% of the sources

Table 5. SELECTED SOURCES FOR CHEMICAL DIFFUSION IN SILICON

Dopant	Gaseous Source	Liquid Source	Solid Source
As	AsH_3, AsF_3	arsenosilica^S	$AlAsO_4^d$
P	PH_3, PF_3	$POCl_3$, phosphosilica^S	$NH_4H_2PO_4^d$, $(NH_4)_2H_2PO_4^d$
B	B_2H_6, BF_3, BCl_3	BBr_3, $(CH_3O)_3B$ borosilica^S	BN^d
Sb	SbH_3^I	Sb_3Cl_5, antimonysilica^S	Sb_2O_3, Sb_2O_4

d = disc source   s = spin on source   I = ion implantation source only

used are gaseous[37]. The advantages of the gaseous sources are their convenience, ease of delivery from a pressurized cylinder, and the ability to accurately measure their flow rate with a *mass flow controller* (See Chap. 6). Gases also have the advantage of retaining their purity for longer time periods than corrosive liquid sources.

The gas species most widely used for sources of boron, particularly for ion implantation, but also for chemical predeposition, are boron trifluoride (BF_3), boron trichloride (BCl_3), and diborane (B_2H_6). Gas types used as phosphorus sources include phosphine (PH_3), and phosphorous pentafluoride (PF_5). Arsenic sources include arsine (AsH_3), and arsenic trifluoride (AsF_3). It should be noted that many of these gases are extremely toxic and /or corrosive.

When these materials are used for chemical predeposition the active source gas is often mixed with a carrier or diluent gas, such as N_2, and a reactive gas such as O_2. Typically, the source gas reacts with the oxygen (e.g. $2PH_3 + 4O_2 \text{---}> P_2O_5 + 3H_2O$) at the wafer surface to form a dopant oxide (e.g. P_2O_5, B_2O_3). The dopant then diffuses from the oxide into the Si, resulting in a uniform dopant concentration across the surface. The depth profile of such pre-deposited layers of dopant in Si can be represented by complementary error functions.

Liquid Sources

Liquid sources are available in two forms: a) those used with *bubblers*; and b) those used as *spin-on dopants*. Bubblers convert the liquid source to vapor, by flowing a carrier gas through the liquid dopant held at a constant temperature, and dopant vapor is swept from the bubbler into the diffusion furnace by the carrier gas. Bubbler temperatures determine the vapor pressure of the liquid source, and thus the concentration of dopant reaching the wafer.

In a manner similar to that for gas sources, the vapors from the liquid sources react with oxygen to form dopant oxides on the wafers (e.g. $4POCl_3 + 3O_2\text{---}> 2P_2O_5 + 6Cl_2$). The process can be easily controlled by starting and stopping the gas flow to the bubbler.

Commercial vendors market liquid dopants in bubblers. The liquids provide good uniformity and purity. Many of the liquid sources are halogen compounds, which play an additional role: that of a gettering agent for removal of heavy metals contamination.

Phosphorus liquid sources include phosphorus oxychloride, or *pocl* ($POCl_3$). The boron sources are boron tribromide (BBr_3), and trimethylborate or TMB ($(CH_3O)_3B$). The antimony source is antimony pentachloride (Sb_3Cl_5).

The *spin-on* dopants are liquid solutions which, upon drying, form doped SiO_2 layers. Most of the formulations are held proprietary by their manufacturers. The spin-on dopants are applied to the wafers in a similar fashion as photoresist (see Chap. 12). The thickness of the deposit depends on the solution viscosity and the spin speed. The dopant concentration in the

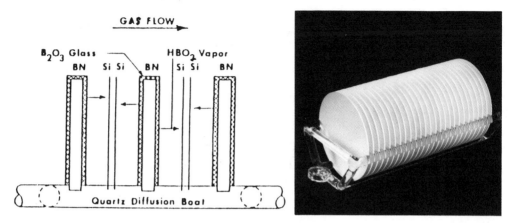

Fig. 13 Stacking pattern for solid disc type dopants in a diffusion furnace. Courtesy of Standard Oil Engineered Materials.

film can be varied by dilution with organic solvents (e.g. ethyl alcohol). It is often necessary to bake the wafers at 200°C for 15 minutes to densify the film, and to prevent absorption of water vapor prior to driving the dopant into the silicon. The diffusion is performed over a range of temperatures and times depending on the desired sheet resistance, and junction depth. Spin on dopants for the following diffusants are commercially available: As (arsenosilica); Sb (antimonysilica); B (borosilica); P (phosphorosilica); and Ga (galliumsilica).

Solid Sources

Solid sources also come in two forms: a) *tablet* or *granular* form; and b) *disc* or *wafer* form. The granular or powdered forms are generally heated (500-700°C) in either a separate furnace or in a cooler portion of the diffusion furnace, to convert the material to its vapor form. Vapors from the heated tablet or powder source are transported to the wafers, where they react with Si and release the dopant, which then diffuses into the Si (e.g. $2Sb_2O_3 + 3 \ Si \ ---> 3SiO_2 + 4Sb$).

When diffusing Sb or As care should be exercised to exclude oxygen from the system since a very thin layer of SiO_2 will stop the diffusion of Sb and As. The solid sources used for this process are antimony trioxide (Sb_2O_3), and arsenic trioxide (As_2O_3).

Disc type sources now are available for boron, phosphorus, and arsenic diffusion. Boron nitride (BN) discs are the most common variety in use, and may contain some B_2O_3 as a binder. Such discs are typically slightly larger than the Si wafers. Figure 13 shows the proper stacking arrangement of this type of dopant in quartz diffusion boats. The discs are oxidized at 750-1100°C to form a thin skin of B_2O_3, which serves as the diffusion source. A small flow of N_2 is used to prevent backstreaming of airborne contaminants into the diffusion tube.

High concentrations of boron can be transported from the BN disc using a process termed *hydrogen injection*. Hydrogen is streamed to the discs to form a HBO_2 vapor. The HBO_2 has a high vapor pressure, therefore significant surface concentrations of boron on the Si can be achieved with the technique.

Discs in the form of hot pressed ammonium monophosphate ($NH_4H_2PO_4$) and ammonium diphosphate [$(NH_4)_2H_2PO_4$] in inert ceramic binders are also available as phosphorus disc sources. Recently, planar discs of aluminum arsenate ($AlAsO_4$), in a proprietary substrate, have become available as arsenic predepositon sources[38]. The $AlAsO_4$ decomposes to form As_2O_3, which is transported to the wafers. Once again, care must be exercised to avoid oxygen in the diffusion ambient, as even a thin layer of SiO_2 will act as an effective mask against As diffusion.

MEASUREMENT TECHNIQUES FOR DIFFUSED (AND ION IMPLANTED) LAYERS

The diffusion coefficient, D, is experimentally determined from the concentration profile in the silicon subsequent to predeposition and drive-in.   In order to determine accurate values of D, reliable measurements of junction depths, sheet resistances, and diffusion profiles must be obtained. In addition, it is necessary to monitor the diffusion process on the production line to assure process control. This can generally be achieved by measuring the diffused junction depth and sheet resistance after each of the diffusion steps. In this section we describe the measurement techniques for determining these quantities, as well as doping profile measurements.

Sheet Resistance Measurements

The *sheet resistance*, R_s, of a diffused layer is the resistance exhibited in a square (i.e. a region of equal length and width) of diffused material which has a thickness x_j (junction depth). The value of R_s is expressed in units of (Ω/sq), and is related to the resistivity of a diffused layer by: R_s (Ω/sq) = ρ (ohm-cm) /x_j (cm), where ρ is the volume resistivity of the diffused layer.

The value of R_s is obtained by measuring the resistance of the diffused layer using a *4-point probe*. Figure 14 shows the colinear probe arrangement that is the basis for the ASTM standard[40], which has been developed for conducting measurements of R_s.   To make the measurement a current, I, is forced between the outer two probes, and the voltage drop, V, between the two inner probes is measured.   To prevent erroneous readings, due to thermoelectric heating and cooling, the measurement is performed first with current in the forward direction, then with current in the reverse direction. The two voltage readings are averaged. To achieve accurate results the measurements should be made at several current levels, until the proper level is found.   That is, if the current is too low, the values of the forward and reverse readings will differ, while if it is too high, I^2R heating will cause the measured reading to drift with time. The ASTM standard also recommends best current levels, which depend on the resistivity range of the diffused layer.

The sheet resistance R_s is calculated from the measured values of current and voltage by the

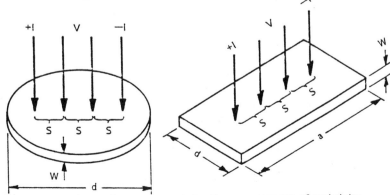

Fig. 14 Co-linear four-point probe arrangement for the measurement of resistivity.

Table 6. CORRECTION FACTOR F$_1$ AS A FUNCTION OF AVERAGE PROBE SEPARATION, s, TO AVERAGE SPECIMEN DIAMETER, D.

s / D	F$_1$	s / D	F$_1$
0	4.532	0.05	4.436
0.005	4.531	0.06	4.395
0.01	4.528	0.07	4.348
0.02	4.517	0.08	4.294
0.03	4.497	0.09	4.235
0.04	4.470	0.10	4.171

expression:

$$R_s = (V/I)\ F_1 \tag{70}$$

where F$_1$ is a correction factor depending on the geometry of the probes (spacing between probes) relative to the specimen thickness, and distance of the probes from the edge of the specimen. Table 6 lists some of the correction factors[40] for circular wafers, and differing s /d values, where s is the interprobe spacing (usually 30-70 mils), and d is the wafer diameter. For small values of s /d (<.01) on circular wafers, F$_1$ approaches a value of 4.53. This is the case for readings taken away from the edge of ≥ 100 mm wafers.

For a diffused layer the average value of R$_s$ depends on the junction depth x$_j$, the impurity profile C (x), and carrier mobility μ (C). The relationship among these parameters is:

$$R_s = \left[q \int_0^{x_j} \mu\ C(x)\,dx \right]^{-1} \tag{71}$$

Mobility is a function of concentration, and hence depth in the wafer. Equation 71 can be simplified if the effective layer mobility, μ$_{eff}$, is used. The effective mobility is expressed as:

$$\mu_{eff} = \frac{\displaystyle\int_0^{x_j} \mu[C(x)]\ C(x)\,dx}{\displaystyle\int_0^{x_j} C(x)\,dx} \tag{72}$$

Utilizing μ$_{eff}$, Eq. 72 is then rewritten as:

$$R_s = \frac{1}{q\,\mu_{eff}\displaystyle\int_0^{x_j} C(x)\,dx} \tag{73}$$

For a given diffusion profile C(x) (e.g. erfc, Gaussian, or even the complex profiles discussed previously), the surface concentration C$_s$ is related to the average resistivity

ρ (av) = R_s x_j, and the substrate doping, C_{sub}. Irvin[41] developed a set of curves relating C_s to ρ (av) for the simple diffusion profiles (see Chap. 1, Fig. 21). These curves, however, are not accurate for high concentrations or shallow junctions, conditions which are often found in VLSI applications, and hence are of limited use. Empirical curves of sheet resistance as a function of junction depth, and sheet resistance as a function of time, temperature, and As dose have been obtained for ion implanted and diffused As^2. Similar curves have also been obtained for P implantation and diffusion[42].

Automated 4-point probe equipment is currently available from several vendors. This equipment is capable of measuring the sheet resistance at many places on a wafer, thereby providing a contour map of the sheet resistance. Figure 15 shows an example of such a sheet resistance wafer map. Sheet resistance contour maps are described in more detail in Chap. 9.

Junction Depth Measurements

VLSI junction depths, x_j, are often shallow (<0.5 μm), and thus, in order to be useful, measurement techniques used to determine x_j should be accurate to ± 200Å.

Angle-Lap and Stain

The *angle-lap and stain method* for measuring junction depth is an effective technique that has been utilized for such a long time that it has been standardized in ASTM Standard Test Method F-110-84[43]. The ASTM method works well on relatively deep, highly doped junctions, but is accurate to only ± 0.5 μm. Wu, *et al*[44] have reported that the accuracy of this technique can be significantly improved, to ± 200Å of the true metallurgical junction, if the proper lapping and staining precautions are taken.

The angle-lapping technique requires the grinding of a small angular bevel on the silicon wafer. This is accomplished by mounting the wafer on an angle block, and moving the assembly over a flat glass onto which the lapping compound slurry has been poured. To obtain small angles ($\leq 2°$) of good quality, the following practices should be followed:

1. The angle lap assembly should be massive, to eliminate rocking during the lapping process. Small pieces of silicon affixed to the bottom of the holder, where

Fig. 15 Four-point probe contour map of a wafer after a furnace phosphorus predeposition.

they act as runners, are found to be helpful in minimizing the rocking, as well.

2. A slurry consisting of the lapping compound (alumina of 0.3 μm particle size mixed with water) is used. The slurry should be changed often to keep it free from dirt and silicon chips.

3. Any metallization on the wafer surface should be removed prior to lapping, since it smears when lapped, and will interfere with the subsequent staining process.

4. If the region of interest does not have a reasonably thick (>3000Å) protective oxide or nitride layer, edge rounding can occur. This can make it difficult to locate the position of the surface with certainty. Thus, it is good practice to chemically deposit ~5000Å of low temperature silicon dioxide onto such surfaces prior to lapping.

After lapping has been performed to bring the junction to the surface, a staining solution is used to delineate either the n-type area or the p-type area.

A copper sulphate ($CuSO_4$) solution is recommended by Wu for effectively staining the n-region. The staining occurs as a result of the following events: a) a drop of $CuSO_4$ is dispensed onto the junction; b) at the same time the junction is illuminated by an intense light source which causes the junction to be forward-biased; and c) as a result of the voltage drop across the junction, Cu atoms plate onto the n-side of the junction ($Cu^{++} \rightarrow Cu^{\circ}$ + 2 holes), thereby delineating the region. The staining effect can be optimized by heeding the following recommendations:

1. If the $CuSO_4$ concentration is too high, excessive Cu plating will occur, and this will result in poor junction delineation. If it is too low, very little plating will occur, especially on low doped n-regions. A recommended recipe is 8g of ($CuSO_4$:$5H_2O$) and 10 cc of 49% HF, in one liter of DI water. The HF dissolves any native oxide from the lapped surface, and thus exposes fresh Si to the $CuSO_4$ solution.

2. It is recommended that a Sun Gun® be used to obtain sufficiently strong illumination (which is vital for reproducibility). The light should be incident for 5-10s immediately after the staining solution is applied to the junction area. The stained junction should subsequently be rinsed with water and dried.

3. If, upon examination, the stain is not clear, it can be removed in a mixture of NH_4OH and H_2O_2, and the junction can be restained. The use of UV illumination (which is absorbed near the surface of the silicon) is helpful in delineating diffusions with low doping concentration (10^{14} to 10^{15} atoms /cm^3).

Note that a second staining solution consisting of 100 cc of HF and a few drops of nitric acid (HNO_3) has also been widely used. Such solutions stain the p-region of the junction for reasons that are not well understood. Wu also describes their applicability and limitations.

For accurate measurements of shallow junctions, an angle of ~1° is required to achieve adequate magnification. A large source of error in junction depth measurements, however, arises from the uncertainty of the lapping angle on the chip or wafer. Since the chip is mounted to the block by wax, the angle on the wafer may not coincide exactly with that on the block. Therefore, a technique to determine the angle on the wafer after lapping (and staining) is required. Such a technique is provided by the laser reflection system shown schmatically in Fig. 16. In this technique the laser beam is collimated, and then made to pass through a hole in a screen. The screen is typically ~140 inches from the sample. The sample is mounted such that reflections from the beveled edge and the surface are occurring at the same time. The sample

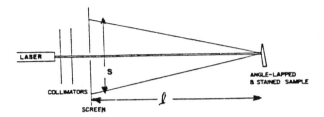

Fig. 16 Schematic diagram of the set up for small angle measurements with a laser[17]. Reprinted with permission of the publisher, the Electrochemical Society.

orientation is then adjusted to achieve a condition such that reflections from these two surfaces are roughly equidistant from the primary beam. Under these conditions the angular magnification, **m**, from the lapped surface is given by:

$$ m = [(l^2 + (s^2/4)] / (s/2) \qquad (74) $$

where l and s are defined in Fig. 16.

If the beveled angle is small, so that l >> s then:

$$ m \cong 2l/s \qquad (75) $$

Typically, l = 140 inches and s = 5 ± 0.1 inches, so m can be measured to 2% accuracy for angles about 1°, and better for smaller angles. After obtaining the magnification factor, **m**, the junction depth can be found from:

$$ x_j = x/m \qquad (76) $$

where x_j is the depth of the stained junction measured along the beveled surface. The value of x_j

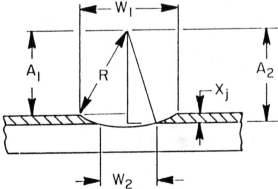

Fig. 17 Arcuic trigonometric method for determining junction depth. Reprinted with permission of Semiconductor International.

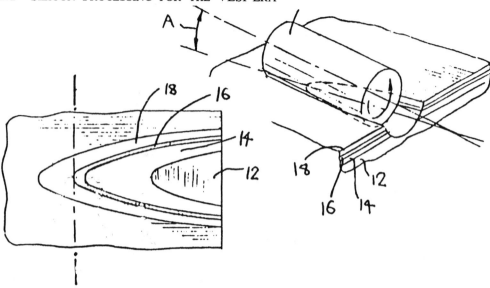

Fig. 18 Tapered groove, stained to delineate the function present in a p/n/p/n+ structure[46]. Reprinted with permission of the publisher, the Electrochemical Society.

is typically determined from photographs taken at ~1000x magnification. The accuracy to which x_j can be measured on the photograph varies between ±0.5-1 mm. For a 1° angle lap, this corresponds to an experimentally uncertainty in x_j between 100-200Å. The angle-lap and stain accuracy is also limited by the resolution of the staining procedure, which in some cases may not accurately delineate the metallurgical junction. Strong illumination helps to assure that staining provides an accurate indication of the junction edge.

A factor that influences stained junctions is the *wedge effect*.   For small angle laps (≤ 1°) there is some question about whether enough dopant ions exist in the 1° wedge to truly stain the p-n junction, particularly if the upper layer is lightly doped.   Again, strong illumination will help uniformly plate the Cu on the n-side of the depletion width to within ~200Å of the metallurgical junction.

Groove and Stain

A technique related to the angle-lap and stain method for measuring x_j is that of *groove and stain*. This technique has the advantage over angle lapping in that it rapidly exposes the junction. That is, by rotating a diamond grit impregnated wheel or ball, ranging in diameter from 0.75-1.5 in, against the wafer at a predetermined location, a groove will be cut in the silicon. If the groove is sufficiently deep, the junction will be exposed. After the groove is formed, the wafer is cleaned and the junction stained using the same methods discussed in the angle-lap section.

The *arcuic trigonometric* method uses a groove of precise radius, R. The groove width at the surface of the wafer, W_1, is measured, as is the width at the junction, W_2 (see Fig. 17). The junction depth, x_j, is then calculated from the following trigonometric relationship:

$$x_j = A_2 - A_1 \tag{77}$$

or;

$$x_j = [\, R^2 - (W_1/2)^2\,]^{1/2} + [\, R^2 - (W_2/2)\,]^{1/2} \tag{78}$$

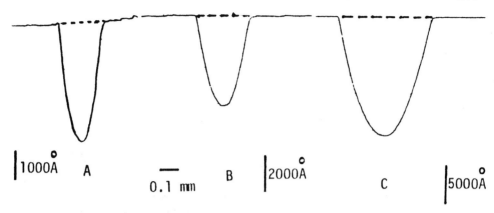

$1000\overset{o}{A}$ A $\overline{}$ B $2000\overset{o}{A}$ C $5000\overset{o}{A}$
0.1 mm

Fig. 19   Profilometer tracings at points A, B, and C providing direct measurements of junction depth[46]. Reprinted with permission of the publisher, the Electrochemical Society.

The multiplying ratio of the groove's width to depth expands the layer thickness into the easy measuring range of an optical microscope.   Commercial grooving instruments manufactured by Philtec Instruments Company[45] are available.   Their instrument is advertised as being precise to ±1% for junction depths as shallow as 500Å.   The major sources of error in this method arise from the assumptions that: a) the groove is exactly cylindrical; and b) the edge of the groove can be accurately determined when viewed in the microscope.

Tapered Groove and Profilometry
A grooving and staining technique has been developed by Prussin[46], which directly measures the junction depth.   This technique involves the use of a *tapered groove*.   The taper is achieved by tilting the wafer relative to the groover to insure that the groove reaches the surface in a gradual manner.   Figure 18 shows an example of such a groove, which has been stained to delineate the presence of the junctions.   In this example a 4-layer structure (p /n /p /n$^+$) is shown, and points A, B, and C show the location of the three junctions.   The depth of the junction is determined by measuring the height of the groove at the junction position.   This is accomplished by running the stylus of a surface profilometer (see Chap. 10) across the junction at right angles to the groove.   Figure 19 shows the profilometer tracings taken at points A, B, and C.   The vertical distance from the surface of the silicon to the junction is a direct measurement of the junction depth.   A precision of ±200Å has been attributed to this measurement.
Finally, Sheng and Marcus[47] have developed on extremely accurate measurement for junction depth using *transmission electron microscopy* (TEM).   They delineate the n$^+$ /p junction using a solution of 0.5% HF in HNO$_3$ to selectively remove the n$^+$ material.   The sample is then prepared for cross-section TEM observation.   A direct measurement of x_j to within ±50Å is obtained from the TEM image, but the technique suffers from a tedious and time consuming sample preparation procedure.

Doping Profile Measurements

Accurate measurements of the doping profile are required to determine both diffusion coefficients at high concentrations and anomalous diffusion effects.   The techniques available for determining doping profile are: (a) capacitance-voltage (C-V); (b) spreading resistance (SR); and (c) secondary ion mass spectroscopy (SIMS).

Capacitance-Voltage Measurement

The use of *capacitance-voltage* (C-V) measurements to determine the diffusion profile is predicated on the measurement of the reverse bias capacitance of a n^+/p or p^+/n junction as a function of voltage. The capacitance of a junction is given by:

$$C(V) = [q \, \varepsilon_s \, C_B /2)^{1/2} [V_{bi} \pm V_R - (2 \, k \, T /q)]^{1/2} \qquad (79)$$

where ε_s is the permittivity of silicon, C_B is the substrate doping concentration, V_{bi} is the built-in potential of the junction, and V_R is the reverse bias voltage.

The technique requires the formation of a shallow n^+ or p^+ diffusion over the surface of the diffused layer of interest. The capacitance is measured while applying a ramped reverse voltage to the junction. The instantaneous value of C_B is then calculated from Eq. 79. The depth at which C_B is measured is obtained from the value of the applied reverse voltage. The major limitation of the technique is that the initial values of C_B that are measured are at least as deep as the value of x_j for the n^+ or p^+ diffusion, plus the zero bias depletion width (which extends into the substrate). The technique also falls short when abrupt profiles are measured since fictitious tails are often observed near the steep edge[48].

Spreading Resistance Measurements

The use of *spreading resistance*, SR (i.e. the resistance associated with the divergence of the current lines which emanate when a small-tipped electrical probe is placed onto Si), was proposed by Mazur and Dickey[49] for measuring the thickness of diffused or epitaxial layers, and for establishing the impurity profile for these structures. In the technique, the spreading resistance of a reproducibly formed point-contact is measured. This can be accomplished by using one, two, or three probe configurations. The two probe method has met with the most success on commercially available equipment.

To make a spreading resistance measurement, a known current is applied between two probes, and the voltage drop is measured across these probes to obtain a spreading resistance ($R_{SR} = V /I$). The value of R_{SR} is measured in a very small volume of Si immediately under the probe. In the two probe method, the resistivity is related to the R_{SR} by the relationship:

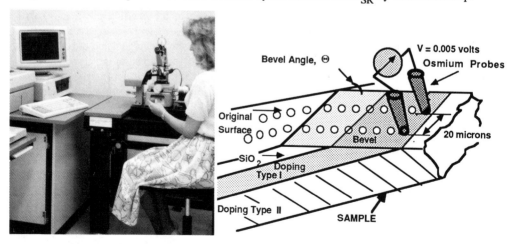

Fig. 20 Commercial spreading resistance measuring system. Courtesy of Solid State Measurements, Inc.

$$\rho = 2 R_{SR} \mathbf{a} \qquad\qquad (80)$$

where $\mathbf{a}$ is an empirical quantity, which is related to the effective electrical contact radius (it is not however the probe tip radius). The value of $\mathbf{a}$ is determined by measuring R_{SR} on a well-characterized material of known resistivity. An ASTM Standard[50] has been developed for conducting SR measurements, and the reader is referred to that reference for details. Figure 20 shows a photograph of a commercially available SR measuring system.

In order to use such equipment the probe tips must be well conditioned. This is achieved by stepping the probes at least 500 times on a silicon substrate that has been ground with 5 µm grit. The probe tips must be calibrated (i.e. to obtain the value of $\mathbf{a}$) on a wafer that has a well documented resistivity (determined with a 4-point probe). It is important that the surface of the calibration wafer and the surface of the unknown sample are identically prepared.

The main use of SR is to determine doping profiles of diffused or epitaxial layers. This is accomplished by angle lapping a wafer and then making SR measurements along its length. Knowing the angle of the taper gives the depth as a function of distance from the surface. The same precautions used to lap a wafer for junction depth measurements must be obeyed when beveling a specimen for SR measurements, including the use of the laser measurement to determine the angle. Elaborate correction techniques and correction factors (CF) have been developed to convert the values of R_{SR} into carrier concentration levels, particularly in multilayer samples. The values of the CF depend on the nature of the layer and the proximity of the measurement to the junctions.

Secondary Ion Mass Spectroscopy

Secondary Ion Mass spectroscopy (SIMS) provides a direct method for measuring the impurity profile of a diffused (or epitaxial) layer. Details of the method are presented in Chap. 17.

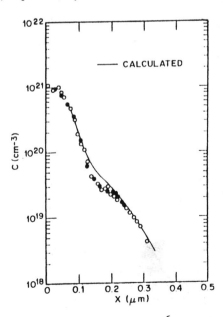

Fig. 21 Phosphorus diffusion profile in silicon, SIMS data[6]. Reprinted with permission of North-Holland Publishing Co.

The technique directly measures the amount of impurity in a Si matrix as the sample surface is slowly sputtered. Unlike the techniques discussed previously, SIMS measures both electrically active and inactive impurity levels. To obtain quantitative data about impurities present, either external standards or a self-standard are required. The ion yields of the external standard are measured at the same time as the ion yields from the test sample, and a linear dependence between the ion yield and concentration is assumed to determine the unknown concentration. The self-standard uses the ion yield of the Si matrix to provide a standard against which the impurity is measured. This works quite well for high impurity concentrations ($\sim 10^{20}$ atoms /cm^3). The SIMS technique can also be used to profile impurities remaining in SiO$_2$ diffusion or implant masks. Data about impurity concentration from SIMS in the SiO$_2$ and in the Si near the Si /SiO$_2$ interface allows the segregation coefficient to be readily measured. Figure 21 shows the impurity profile for P diffusion in Si determined by both SIMS analysis and differential conductivity. The three regions of the high concentration P profile are apparent in Fig. 21.

REFERENCES

1. H. Carslaw and J. Jaeger, *Conduction of Heat in Solids*, Appendix II, Oxford Univ. Press. Fairlawn, N.J., 1959.
2. W.E. Beadle, J.C.C. Tsai, and R.D. Plummer, Eds. *Quick Reference Manual for Silicon Integrated Circuit Technology*, John Wiley & Sons, New York, 1985, p. 723.
3. C.P. Wa, E. Douglas, and C. Mueller, "Redistribution of Ion-Implanted Impurities in Si During Diffusion in Oxidizing Ambients", *IEEE Trans. on Electron Dev.*, **ED-23**, 1095 (1976).
4. P.G. Shewmon, *Diffusion in Solids*, McGraw-Hill, New York, 1963.
5. F.J. Morin and J.P. Maita, "Electrical Properties of Silicon Containing Arsenic and Boron", *Phys. Rev.*, **96**, 28 (1954).
6. R.B. Fair, "Concentration Profiles of Diffused Dopants in Silicon" in F.F.Y. Wang, Ed., *Impurity Doping Processes in Silicon*, North-Holland, New York, 1981, Chapter 7.
7. T.L. Chui and H.N. Ghosh, "Diffusion Model for Arsenic in Silicon", *IBM J. Research Develop.*, **11**, 472 (1971).
8. R.B. Fair and J.C.C. Tsai, "The Diffusion of Ion-Implanted Arsenic in Silicon", *J. Electrochem. Soc.*, **122**, 1689 (1975).
9. B.J. Masters and J.M. Fairfield, "Arsenic Isoconcentration Diffusion Studies in Silicon", *J. Appl. Phys.*, **40**, 2390 (1969).
10. R.B. Fair, "Profile Estimation of High Concentration Arsenic Diffusion in Silicon", *J. Appl. Phys.*, **43**, 1278 (1972).
11. R.B. Fair and J.C.C. Tsai, "Profile Parameters of Implanted-Diffused Arsenic Layers in Silicon", *J. Electrochem. Soc.*, **123**, 583 (1976).
12. R.B. Fair, "Boron Diffusion in Silicon-Concentration and Orientation Dependence, Background Effects, and Profile Estimation", *J. Electrochem. Soc.*, **122**, 583 (1976).
13. R.B. Fair and J.C.C. Tsai, "A Quantitative Model for the Diffusion of Phosphorus in Silicon and the Emitters Dep Effect", *J. Electrochem. Soc.*, **124**, 1107 (1977).
14. S.M. Hu, P. Fahey, and R.W. Dulton, *J. Appl. Phys.* in Press.
15. T.Y. Tan and V. Gosele, "Point Defects and Diffusion Processes in Silicon", *Proceedings of VLSI Science and Technol,1984*; Electrochem. Society, N.J. (1984). Vol. **84-7**, 151 (1984).
16. A. Seeger and K.P. Chik, *Phys. Stat. Solidi*, **29**, 455 (1968).
17. T.Y. Tan, U. Gosele, and F. Morehead, "Point Defects, Diffusion Processes, and Swirl Defect Formation in Silicon"*J.Appl. Phys.* **A31**, 97 (1983).
18. T.Y. Tan and U. Gosele, "Oxidation-Enhanced or Retarded Diffusion and the Growth or

Shrinkage of Oxidation-Induced Stacking Faults"*Appl. Phys. Letts*, **40**, 616 (1982).

19. D.A. Antoniadis and I. Moskowitz, *J. Appl. Phys.*, **53**, 9214 (1982).

20. N.A. Stowlwijk, B. Schuster, J. Holzl, H. Mehrer, and W. Frank, "Diffusion and Solubility of Gold in Silicon" *Physica*, **116B**, 335 (1983).

21. W.R. Wilcox, T.J. LeChappele and D.H. Forbes, "Gold in Silicon: Effect on Resistivity and Diffusion in Heavily-Doped Layers", *J. Electrochem. Soc.*, **111**, 1377 (1964).

22. T.I. Kamins, J. Manolin, and R.N. Tucker, "Diffusion of Impurities in Polycrystalline Silicon", *J. Appl. Phys.*, **43**, 83 (1972).

23. S. Horuichi and R. Blanchard, "Boron Diffusion in Polycrystalline Silicon Layers", *Solid State Electron.*, **18**, 529 (1975).

24. C.J. Coe, "The Lateral Diffusion of Boron in Polycrystalline Silicon and its Influence on Fabrication of Sub-Micron Most's", *Solid State Electron.*, **20**, 985 (1977).

25. B. Swaminathan, K.C. Saraswat, R.W. Dulton, and T.I. Kamins, "Diffusion of Arsenic in Polycrystalline Silicon", *Appl. Phys. Lett.*, **40**, 795 (1980).

26. K. Tsukamoto, Y. Aklasaka, and K. Houe, "Arsenic Implantation into Polycrystalline Silicon and Diffusion to Silicon Substrates", *J. Appl. Phys.*, **48**, 1815 (1977).

27. S.M. Hu and S. Schmidt, "Interaction in Sequential Diffusion Processes in a Semiconductor", *Phys. Rev.*, **107(2)**, 392 (1957).

28. R.B. Fair, "The Effect of Strian-Induced Bandgap Narrowing on High Concentration Phosphorus Diffusion in Silicon", *J. Appl. Phys.*, **50**, 860 (1979).

29. R.J. Jaccodine, in *Defects in Semiconductors II*, S. Hahayan and J.W. Corbett, eds, North-Holland, N.Y. (1983), p. 101.

30. C.L. Claeys, G.J. Declerck, and R.J. VanOverstaeten, "Influence of Phosphorus Diffusion on the Growth Kinetics of Oxidation-Induced Stacking Faults in Silicon" in *Semiconductor Characterization Techniques*, P. Barnes, and G.A. Rozgonyi, eds, Electrochem Soc., Princeton, (1978), p. 336.

31. A. Amigliato, M. Servidori, S. Solmi, and I. Vecchi, "On the Growth of Stacking Faults and Dislocations Induced in Si by Phosphorus Predeposition", *J. Appl. Phys.*, **48**, 1806 (1977).

32. H. Strunk, U. Gosele, and B.O. Kolbesen, "Interstitial Supersaturation near Phosphorus-Diffused Emitter Zones", *Appl. Phys. Lett.*, **34**, 530 (1979).

33. D.P. Kennedy and R.R. O'Brien, "Analysis of the Impurity Atom Distribution Near the Diffusion Mask for a Planar p-n Junction", *IBM J. Res. Dev.*, **9**, 179 (1965).

34. D.D. Warner and C.L. Wilson, "Two Dimensional Concentration Dependent Diffusion", *Bell Sys. Tech. J.*, **59**, 1, (1980).

35. C.F. Gibbons, E.I. Povelonis, and D.R. Ketchow, "The Effect of Mask Edge on Dopant Diffusion into Semiconductors", *J. Electrochem. Soc.*, **119**, 767 (1972).

36. K. Taniguchi, K. Kurosawa, and M. Kashiwagi, "Oxidation Enhanced Diffusion of Boron and Phosphorus in <100> Silicon", *J. Electrochem. Soc.*, **127**, 2243 (1980).

37. A.A. Weiss, "Semiconductor Dopant Status Report", *Semiconductor Int'l*, p. 66 (Aug. 83).

38. R.E. Tressler, *et al*, "Present Status of Arsenic Planar Diffusion Sources", *Solid State Tech.*, p. 165 (Oct. 84).

40. "Standard Method for Measuring Resistivity of Silicon Slices with a Colinear Four-Probe Array"; *1984 Annual Book of ASTM Standards*, F-84-84a; Vol. 10.05, p. 191 (1984).

41. J.C. Irvin, "Resistivity of Bulk Silicon and of Diffused Layers in Silicon", *Bell Syst. Tech. J.*, **41**, 387 (1962).

42. R.B. Fair, "Analysis of Phosphorus-Diffused Layers in Silicon," *J. Electrochem. Soc.*, **125**, 323 (1978).

43. "Standard Test Method for Thickness of Epitaxial or Diffused Layers in Silicon by the Angle

Lapping and Staining Technique"; *1984 Annual Book of ASTM Standards*, F110-84; Vol. 10.05, p. 230 (1984).

44. C.P. Wu, E.C. Douglas, C.W. Mueller, and R. Williams, "Techniques for Lapping and Staining Ion-Implanted Layers", *J. Electrochem. Soc.*, **126**, 1982 (1979).

45. Philtec Instruments Co., Philadelphia, PA.

46. S. Prussin, "Junction Depth Measurements for VLSI Structures", *J. Electrochem. Soc.*, **130**, 184 (1983).

47. T. Sheng and R.B. Marcus, "Delineation of Shallow Junctions in Silicon by Transmission Electron Microscopy", *J. Electrochem. Soc.*, **128**, 883 (1981).

48. J. Hilibrand and R.D. Gold, *RCA Review*, **21**, 245 (1960).

49. R.G. Mazur and P.H. Dickey, "A Spreading Resistance Technique for Resistivity Measurements on Silicon", *J. Electrochem. Soc.*, **113**, 255 (1966).

50. "Standard Method for Measuring Resistivity of Silicon Wafers Using a Spreading Resistance Probe", *1984 Annual Book of ASTM Standards*, F525-84, Vol. 10.05, p. 455 (1984).

PROBLEMS

1. Determine the solid solubility and the diffusion coeeficient of :
    a) boron at 950°C;
    b) phosphorus at 1050°C; and
    c) arsenic at 1100°C.

2. During predeposition, what parameter controls the concentration of dopant at the surface of the wafer?

3. Predeposition is carried out on a silicon wafer at 975°C for 30 minutes in the presence of an excess of phosphorus. The substrate is boron doped to a concentration of 10^{17} atoms /cm^3:
    a) Find C_s for this procedure;
    b) Calculate $\sqrt{Dt}$ for this procedure;
    c) Calculate the dopant concentration in the silicon at the following depths: x = 0, x = 0.1 µm, x = 0.2 µm, x = 0.3 µm, x = 0.4 µm, and x = 0.5 µm. Plot this concentration on log-linear graph paper;
    d) Determine the depth at which the metallurgical p-n junction occurs;
    e) Determine the junction depth if a 2 Ω-cm p-type substrate is utilized;
    f) Find the total amount of dopant incorporated into the wafer per unit area.

4. Assume the wafer of Problem 3 is used in this problem. The objective will be to determine the doping profile as a function of depth after a drive-in step of 50 minutes at 1100°C has been carried out:
    a) Calculate $\sqrt{Dt}$;
    b) Find the final surface concentration, C_s;
    c) Find the values of the dopant concentration at the following depths: x = 0, x = 0.5 µm, x = 1.0 µm, x = 1.5 µm, and x = 2.0 µm. Plot this data on log-linear graph paper;
    d) Determine the depth at which the metallurgical p-n junction occurs.

5. A diffusion furnace is operated at 950°C ± 1°C. Assuming a diffusion profile described by a Gaussian distribution, what is the corresponding spread in the diffusion lengths due to the uncertainty in the temperature?

6. Show that the logarithm of the sheet resistance after a predeposition is approximately a linear function of the reciprocal. This indicates an Arrhenius behavior (see Appendix 3).

7. Verify that equations 10 and 16 satisfy Ficks second law (Eq. 6) and the appropriate initial and

boundary conditions.

8. The base region of a pnp transistor is fabricated according to the following process sequence:

Starting material: p-type, 10^{15} boron atoms /cm^3;

Oxidation: 80 minutes at 1200°C , wet oxygen

Open base region window

Predeposition of phosphorus, 30 minutes at 800°C;

Drive-in diffusion: 50 minutes at 1200°C, dry oxygen.

The measured V /I after the drive-in step is 1.4 Ω, and the junction depth is measured to be 4 μm deep. Calculate:

   1) The oxide thickness over the base window and in the field regions;

   2) The phosphorus dopant profile after the drive-in diffusion;

   3) The total number of impurities per unit area of base.

9.   When determining if the intrinsic diffusivity of an impurity is applicable at a given temperature, it is necessary to know the intrinsic carrier concentration, n_i.   Using Eq. 31 and the expression for ΔE given immediately after Eq. 31, plot the value of n_i versus T.

10.   To prevent wafer warpage that would occur due to a rapid decrease in temperature, the temperature in a diffusion furnace is linearly reducd from 1100°C to 600°C over a period of 30 minutes.   Calculate the effective diffusion time at the initial diffusion temperature for an arsenic diffusion in silicon.

9

ION IMPLANTATION

FOR VLSI

Ion implantation is a process in which energetic, charged atoms or molecules are directly introduced into a substrate. Acceleration energies range between 10-200 keV for most ion implanters (although energies as high as several MeV are being utilized in high-energy implant systems). In VLSI fabrication, ion implantation is primarily used to add dopant ions (most often selectively) into the surface of silicon wafers. The superiority of ion implantation over chemical (diffusion) doping methods for this purpose, has caused it to steadily replace diffusion doping in an increasing number of applications. In order to discuss this important but complex technology, it is useful at the outset to identify the goals that a successful ion implantation process should achieve. This will allow readers to see the direction that our discussion will take.

The overall objective of an ion implantation process is to *introduce a desired atomic species into a target material* (e.g. in VLSI fabrication into single-crystal silicon substrates), and in order to do so, the following goals need to be be attained:

1) The species should be implanted in the exact quantity specified;
2) The implanted species should end up located at the correct depths below the surface;
3) The implantation should be limited to only the designated areas of the substrate;
4) When required, it should be possible to electrically activate all the implanted impurities;
5) As much as possible, the silicon lattice structure should be unchanged by the dopant incorporation process.

In order to achieve the above goals, it must be possible to successfully satisfy the following aspects of the implantation process:

a) To meet *Goal 1*, equipment is needed that can accurately implant and monitor the quantity of the species being implanted;
b) To meet *Goal 2*, models must be available to predict the distribution of the ions in the implanted layer following the implantation and annealing steps. Such models must correctly account for implant species (e.g. B, P, As, Sb), energy, dose (i.e. number of ions implanted $/cm^2$), substrate structure (e.g. amorphous or single-crystal [including crystal orientation]), and effects of any thin film structures on top of the Si substrate;
c) To satisfy *Goal 3*, adequate masking structures against implantation must be available;
d) To reach *Goal 4*, we need to understand how implanted atoms are electrically activated in

the silicon lattice;

e) To achieve *Goal 5*, a model that describes how the implantation disturbs the lattice (i.e. including the type of damage, and its location in the substrate) must be available. This model must also be able to correctly account for implantation species, energy, dose, crystal structure and orientation, and temperature of the substrate. Furthermore, a related model (for predicting how the lattice can be restored most closely to its pre-implantation condition, and describes the residual defects that exist after various annealing treatments are applied), is also needed.

Many of the tasks spelled out in the second list above have been accomplished. Work is continuing in attempts to increase the understanding and equipment capabilities where gaps still exist. In this chapter we also describe the current state of progress for meeting the above goals.

In our presentation the models used to portray aspects of implantation will first be discussed. The approach will be to describe the physical mechanisms that are the basis of the theoretical models. Then a comparison among the models derived from theory and experimental observations will be given (i.e. to illustrate how well actual implantations are described by the models). Next, a description of ion implantation systems is undertaken (including safety aspects and operational limitations). This is followed by a presentation of the methods used to monitor and evaluate various aspects of ion implantation, including: a) implantation doses; b) implantation dose uniformity across a wafer; c) implantation profiles; d) implantation damage; and e) degree to which damage is removed (and /or remains) after annealing. Finally a group of topics relating to practical aspects of ion implantation processes is included. Table 1 lists some of the important applications of ion implantation in VLSI.

It is useful to begin the discussion with a definition of *implantation dose*. The ion beam

Table 1. ION IMPLANTATION APPLICATIONS IN VLSI

Junction Formation
   (used to fabricate both MOS and bipolar devices)

CMOS Fabrication (Vol. 2, this text):
- Threshold Voltage Control /Adjustment
- Channel Stop Implantation
- Source /Drain Formation
- Well Formation
- Punchthrough Stopper Implantation
- Graded Source and Drain Formation
- High Energy Implantation to Form Retrograde Well

Bipolar Fabrication (Vol. 2 of this text):
- Predeposition
- Base Formation Implantation
- Arsenic-Implanted Poly-Si Emitter
- High Value Resistor Formation Implantation

Miscellaneous Process Applications:
- Backside Damage Layer Formation for Gettering (Chap. 2)
- Implant-Induced Damage of SiO_2 and Si_3N_4 to Enhance Etching (Chap.15)
- Photoresist Hardening to Improve Dry-Etch Resistance (Chap. 12)
- Buried Insulator Layer Formation (Chap. 5 and Vol. 2 of this text)
- Silicon-on-Insulator Using Oxygen or Nitrogen Implantation (Chap. 5)
- Ion Beam Mixing to Promote Silicidation Reactions

- Polysilicon Resistor Formation
- High Energy Implantation to Form Buried Collector
- Emitter Formation Implantation

---

current in implanters ranges between about 10 μA and 30 mA, depending on the implant species, energy, and model of implanter. The number of implanted ions per unit area is termed the *dose*, φ. Typical doses range from 10^{11}-10^{16} atoms /cm^2. The dose (in atoms /cm^2) is related to beam current I (in amperes), beam area A (in cm^2), and implantation duration, t (in sec) by:

$$\phi = \frac{I\,t}{q_i\,A} \tag{1}$$

where q_i is the charge per ion (normally equal to one electronic charge, 1.6×10^{-19} C).

ADVANTAGES (AND PROBLEMS) OF ION IMPLANTATION

Advantages

- *Most important advantage:* Ability to more precisely control the number of implanted dopant atoms into substrates (e.g. to within ±3%). For dopant control in the 10^{14}-10^{18} atoms/cm^3 range, ion implantation is clearly superior to chemical deposition techniques.
- Implanted impurities are introduced into the substrate with much less lateral distribution than from diffusion doping processes. The reduced lateral impurity penetration allows devices to be fabricated with features of smaller dimensions.
- Mass separation by the ion implanter results in extremely pure dopant beams, even if the source material is relatively impure.
- A single ion implanter can be used for a variety of implant species, with little cross contamination by other species. Note that in theory this is an advantage, but in many implanters, some cross contamination still occurs. New generation implanters are being built to minimize this problem through innovative equipment designs.
- Ion implantation can inject dopant atoms into a semiconductor by implantation through a thin surface layer (e.g. SiO$_2$). The threshold voltage of MOS transistors can be adjusted with this technique, a procedure not easy with any other process method. The surface layer can also serve as a protective screen against contamination by metals and other dopants during the introduction of the desired impurities.
- The predicted distribution of implanted ions in silicon fits experimental data very well (e.g. predicted ranges for low-energy heavy ions are within ±10% of the measured values, while for high-energy light ions the agreement is within ± 2%).
- A variety of doping profiles can be produced by superimposing multiple implants.
- Under some implantation conditions, highly abrupt junctions can be formed.
- Uniform doping across large wafers can be achieved.
- A variety of materials is suitable for creating masks to keep ions from being implanted into unwanted regions. The masking layer prevents ions from being introduced into regions under the mask (except near the mask edge). Lateral scattering effects do occur as a result of the ion implantation process, but are smaller than lateral diffusion distances.
- Ion implantation is a low-temperature process, allowing materials such as photoresist to routinely be used as implant mask layers.
- The inherent purity of processing under high vacuum reduces problems from a variety of contamination and defect causing mechanisms.
- The parameters that control an ion implantation process are amenable to automatic control, and major advances toward cassette-to-cassette operation have been achieved.

Problems /Limitations of Ion Implantation

- Ion implantation causes damage to the material structure of the target. In crystalline targets (e.g. single-crystal Si wafers), crystal defects and even amorphous layers are formed. To restore the target material to its pre-implantation condition, thermal processing after implantation must be performed. In some cases, significant implantation damage cannot be removed.
- The maximum implantation depth achievable with conventional (i.e. non-high energy) implanters is relatively shallow, especially for heavy atoms (e.g. arsenic).
- The lateral distribution of implanted species (although smaller than lateral diffusion effects), is not zero. This is a fundamental limiting factor in fabricating some minimum sized device structures, such as the electrical channel length between source and drain in self-aligned MOS transistors.
- Throughput is typically lower than diffusion doping processes. High implantation doses may require undesirably long implantation times, depending on the implanted species.
- Ion implanters are complex machines, among the most sophisticated systems in wafer manufacturing. In order to be effectively utilized they must be conscientiously operated, monitored, and maintained by well-trained personnel.
- Ion implantation equipment contains many safety hazards (e.g. high voltage and toxic gases) to the personnel which operate and (especially) that service the machines. To minimize the likelihood of accidents from operating and maintaining such equipment, careful safety procedures must be established and strictly followed.
- Some older-style implanters use diffusion pumps in the beam column, and this may cause contamination due to oil backstreaming. The effects of such contamination are described in Chap. 3.

IMPURITY PROFILES OF IMPLANTED IONS

In order to benefit from the ability to control the number of impurities implanted into a substrate, it is necessary to know where the implanted atoms are located after implantation (i.e. it must be possible to predict the *depth distribution*, or *profile* of the as-implanted atoms). For example, this information is necessary for selecting appropriate doses and energies when designing a fabrication process sequence for new or modified integrated circuit devices. What is needed to make accurate predictions of implantation profiles is a theoretical model (or models) based on the energy interaction mechanisms between the impinging ions and the substrate. In this section we will address the topic of how such theoretical models have been developed, and under what conditions they provide accurate predictions of implantation profiles. Figure 1 outlines the evolution of the models developed for predicting implantation profiles and indicates the conditions under which they can be used to provide useful predictions.

Despite the fact that the derivation of the models is quite mathematical and complex, the scope of this text limits us to a more qualitative discussion. Even on a largely qualitative level such a presentation is valuable. That is, it provides the reader with an appreciation of the intellectual underpinning of ion implantation profile prediction, and also serves as an introduction to other physical mechanisms associated with ion implantation. These include channeling effects during implantation, substrate damage from implantation, and recoil effects that occur when implantations are done through thin layers present on the substrate surface. For readers interested in gaining a deeper and /or more quantitative understanding of implantation profile

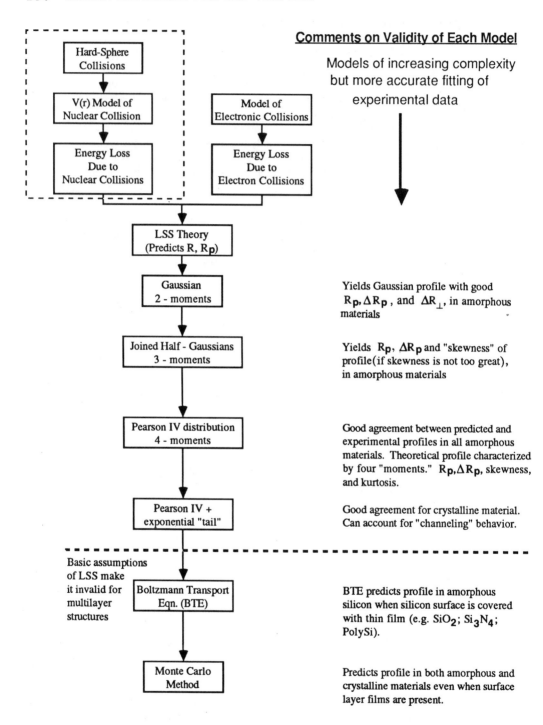

Fig. 1 Mathematical models for predicting ion implantation concentration profiles.

models, several references are given at the end of the chapter.   Some are comprehensive surveys[1,2], while others are papers in which the models were originally published (and which thus discuss more fully the assumptions underlying the model, and details of the derivations and associated calculations[3,4,5]).

Definitions Associated with Ion Implantation Profiles

As energetic ions penetrate a solid target material, they lose energy due to collisions with atomic nuclei and electrons in the target, and eventually come to rest. The total distance that an ion travels in the target before coming to rest is termed the *range*, R.   As a result of the collisions between the ions and the target material nuclei, this trajectory is not a straight line (and in fact, the value of the total distance traveled is not even the quantity that is of highest interest).   What is of greater interest than R is the projection of this range on the direction parallel with the incident beam, since this represents the penetration depth of the implanted ions along the implantation direction.   This quantity is called the *projected range*, R_p (Fig. 2a). As the number of collisions (and the energy lost per collision experienced by the penetrating ion) are random variables, ions having the same initial energy and mass will end up spatially distributed in the target.   An average ion will stop at a depth below the surface given by R_p. Some ions, however, undergo fewer scattering events than the average, and come to rest more deeply into the target.   Others suffer more collisions, and come to rest closer to the surface than R_p.   As a result, when large numbers of ions are implanted, R_p corresponds to the depth at which most ions stop, and this is the distance at which the profile has maximum value.   The statistical fluctuation along the direction of the projected range is given by the quantity known as the *projected straggle*, ΔR_p (Fig. 2b).

The ions are also scattered to some degree along the direction perpendicular to the incident direction, and the statistical fluctuation along this direction is called the *projected lateral straggle*, $\Delta R_\perp$ (Fig. 2b).   Information about the lateral straggle is important because it describes the extent

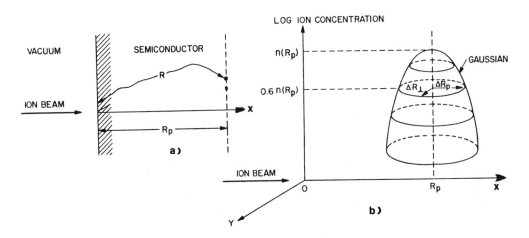

Fig. 2 a) Schematic of the ion range, R, and projected range, R_p. b) Two-dimensional distribution of the implanted atoms. From S.M. Sze, *Semiconductor Devices, Physics and Technology*, Copyright © 1985 John Wiley & Sons. Reprinted with permission of John Wiley & Sons.

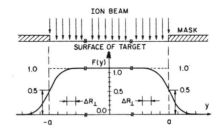

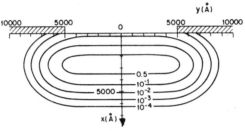

Fig. 3 Illustration of lateral profiles. (a) Ion concentration along the lateral direction (y) for a gate mask with $a \gg \Delta R_\perp$ and infinite extension in the x-direction. (b) Contours of equal-ion concentrations for 70 keV B^+ (R_p = 2710Å, ΔR_p = 824Å, and $\Delta R_\perp$ = 1006Å) incident into silicon through a 1 µm slit[6]. Reprinted with permission of Japanese Journal of Applied Physics.

of lateral penetration of ions under the edges of implant masks. Such lateral penetration may represent a limiting dimension in some integrated circuit device structures. In general, values of ΔR_p and $\Delta R_\perp$ are within ±20% of one another. Figure 3a shows an opening in a "thick" mask material, and resultant ion concentrations in the lateral direction in the target. It is seen that lateral straggle effects cause implanted ions to be distributed under edges of the mask. In Fig. 3b contours of equal-ion concentration for 70-keV B atoms implanted into a 1-µm slit are shown[6].

Theory of Ion Stopping

As an ion moves through a solid target, it transfers energy by collisions with the target nuclei (*nuclear collisions*) and by Coulombic interaction with the electrons in the target material. In the latter mechanism, the energy transferred to the electrons can lead to exciting the electrons to higher energy levels (*excitation*), or to the ejection of electrons from their atomic orbits (*ionization*). The energy loss due to such target interactions gradually slows the ion, eventually bringing it to a stop. If the energy of the ion at any point along its trajectory in the target is given by E, the process of *energy loss through nuclear collisions* can be characterized by an *energy loss per unit length due to nuclear stopping*, $S_n(E)$, and *energy loss from interactions with target electrons* by an *energy loss per unit length due to electronic stopping*, $S_e(E)$. The total rate of energy loss $(dE/dx)_{total}$ is given by the sum of these stopping mechanisms:

$$\left[\frac{dE}{dx}\right]_{total} = S_n(E) + S_e(E) \qquad (2)$$

If the total distance that the ion travels before coming to a complete stop is given by R, then:

$$R = \int_{0}^{E_{o}} \frac{dE}{(-dE/dR)} \qquad (3)$$

where E_o is the initial incident ion energy.

The *nuclear stopping process* can be visualized with the aid of the extreme simplification that treats the event as a collision between two hard spheres (Fig. 4), or by a more correct approximation that assumes that the scattering is described by a Coulombic force-at-a-distance interaction. In the latter description, an appropriate atomic scattering potential, $V(r)$, must be determined. The most successful model for predicting implantation profiles based on the ion stopping approach is the so called *LSS model*, which is discussed in further detail in the following section. The LSS model utilizes a modified Thomas-Fermi screened potential for $V(r)$. Calculations based on this model show that nuclear stopping increases linearly in effectiveness at low energies, reaches a maximum at some intermediate energy, and decreases at higher energies, because at high velocity, ions move past target nuclei too quickly to efficiently transfer energy to them. Values of $S_n(E)$ for boron, phosphorus, and arsenic are shown in Fig. 5. It is important to note that $S_n(E)$ increases with the mass of the implanted ion, and thus heavy ions such as As will transfer much more of their energy through nuclear collisions than will B atoms.

The *electronic stopping process* can be considered as similar to the stopping of a projectile in a viscous medium, and the stopping magnitude can be approximated to be proportional to the square root of the ion velocity:

$$S_e(E) = k_e (E)^{1/2} \qquad (4)$$

where k_e is a constant that depends weakly on the ion and target atomic masses and numbers. Figure 5 also shows $S_e(E)$ for B, P, and As. As can be seen in Fig. 5 the crossover energy at which electronic stopping becomes more effective than nuclear stopping is higher for heavier ions. For example, for boron $S_e(E)$ is the predominant energy loss mechanism down to ~10 keV, while for P and As, the energy loss due to nuclear stopping predominates for energies up to 130 keV and 700 keV, respectively.

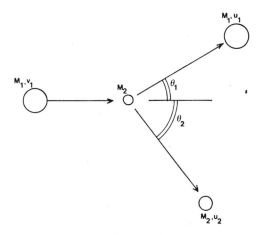

Fig. 4 Collision between two particles treated as if they are hard spheres.

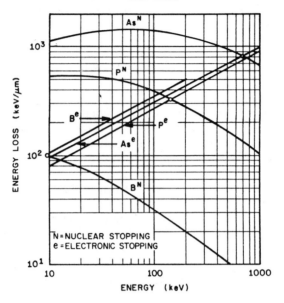

Fig. 5 Calculated values of dE/dx for As, P, and B at various energies. The nuclear (N) and electronic (e) components are shown. Note the points (o) at which nuclear and electronic stopping are equal. (After Smith, Ref. 7, redrawn by Seidel in Ref. 63, Copyright, 1983, Bell Telephone Laboratories, Incorporated, reprinted by permission.)

Models for Implantation Profiles in Amorphous Solids

The range-energy relation given by Eq. 2 was reformulated by Lindhard, Scharff, and Schiott (LSS) for implantation into amorphous material in terms of the reduced parameters, ε and ρ, as:

$$\frac{d\varepsilon}{d\rho} = (\frac{d\varepsilon}{d\rho})_n + k_\varepsilon (\varepsilon)^{1/2} \qquad (5)$$

where ρ and ε are dimensionless variables related to the range, R, and incident energy, E_0, by:

$$\rho = \frac{(R\,N\,M_1\,M_2\,4\,\pi\,a^2)}{(M_1 + M_2)} \qquad (5a)$$

and;

$$\varepsilon = \frac{E_0\,a\,M_2}{[Z_1\,Z_2\,q^2\,(M_1 + M_2)]} \qquad (5b)$$

where: M_1 and M_2 are the mass of the incident ions and target atoms, respectively; N is the number of atoms per unit volume; a is the screening length, equal to $0.88a_0 /(Z_1^{1/3} + Z_2^{2/3})^{1/2}$ and a_0 is the Bohr radius; and Z_1 and Z_2 are the atomic numbers of the ion and target.
LSS used a modified Thomas-Fermi screened potential to calculate the energy loss due to nuclear stopping, together with the assumption that the energy loss due to electronic stopping is given by Eq. 4 (Fig. 6). Using this approach, they calculated values of ρ for different values of ε. (Note that these calculations are quite complex, but can be found in the classic paper by

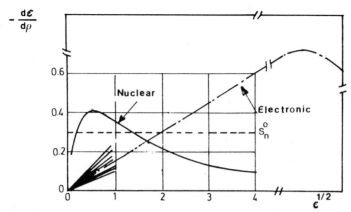

Fig. 6 (a) Nuclear stopping power for Thomas-Fermi potential (solid line) and electronic stopping power (dash-dot line) for k = 0.15 in terms of the reduced variables ε and ρ, based on LSS theory[3]. (b) Representation of as-implanted impurity profiles by joined half-Gaussian distributions.

LSS[3].) The value of ρ was then converted to R (using Eq. 5a), and finally a value for R_p was obtained from the approximate expression[8]:

$$R_p \cong \frac{R}{1 + \left[\dfrac{M_2}{3 M_1} \right]} \qquad (6)$$

LSS assumed that the distribution of the implanted ions in amorphous materials could be described by a symmetrical Gaussian curve (see Chap. 18, Fig. 2). If this assumption is valid, then the implanted ion concentration, n, as a function of depth, x, can be described by:

$$n(x) = \frac{\phi}{\sqrt{2\pi}\, \Delta R_p} \exp \left[\frac{-(x - R_p)^2}{2\Delta R_p^2} \right] \qquad (7)$$

where: ϕ is the dose (in number of implanted ions /cm^2) , and ΔR_p is the standard deviation of the Gaussian distribution (or *projected straggle* of the distribution in the direction of incidence of the beam). The value of ΔR_p is calculated in terms of R_p and the mass of the implanted ions M_1 and the target atoms M_2, by the approximate expression[8]:

$$\Delta R_p \cong \frac{2 R_p}{3} \left[\frac{\sqrt{M_1 M_2}}{M_1 + M_2} \right] \qquad (8)$$

The concentration is maximum at R_p, and Eq. 7 at x = R_p reduces to:

$$n(x = R_p) = \frac{\phi}{\sqrt{2\pi}\, \Delta R_p} \cong \frac{0.4\,\phi}{\Delta R_p} \qquad (9)$$

The assumption by LSS that the distribution of implanted atoms in amorphous materials is well approximated by a Gaussian curve is not completely correct, but is nevertheless very useful as a first order description. Indeed, the fit to experiment is almost always good for all implantations near the peak. For example, the peak value predicted by Eq. 9 is generally within

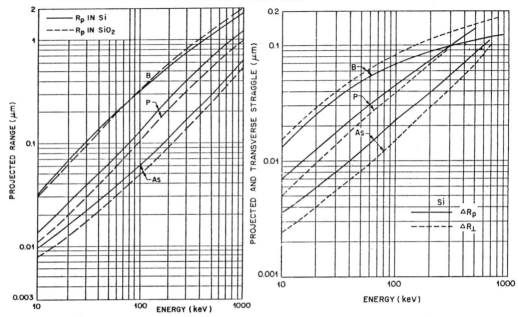

Fig. 7 (Left) Projected range for B, P, and As in Si and SiO_2 at various energies. The results pertain to amorphous silicon targets and thermal SiO_2 (2.27 g /cm^3). After Smith, Ref. 7, redrawn by Seidel in Ref. 63, Copyright, 1983, Bell Telephone Laboratories, Incorporated, reprinted by permission. **Fig. 8 (Right)** Calculated ion projected straggle ΔR_p, and ion lateral straggle $\Delta R_\perp$ for As, P, and B ions in silicon. After Smith, Ref. 7, redrawn by Seidel in Ref. 63, Copyright, 1983, Bell Telephone Laboratories, Incorporated, reprinted by permission.

1% of the measured value, except for very shallow [low energy] implants]. On the other hand, significant asymmetries begin to appear in experimental implant profiles in amorphous targets once the concentration levels drop by a factor of 10 below the peak value (and which are not taken into account by a symmetrical Gaussian approximation). Therefore, various distribution curves with a higher number of moments than a Gaussian (which can be described with only two moments, R_p and ΔR_p), have been examined for their ability to fit the experimental implant profiles (Fig. 1). In addition, LSS theory and the Gaussian curve fail to account for several effects that occur when implants are made into *single-crystal* material. Thus, modifications or even other analytical models must be used to obtain a good fit to data obtained in such situations.

Nevertheless, in practice the Gaussian distribution is still commonly used to provide quick estimates of doping distributions into amorphous and single-crystal targets. The higher moment distributions (or alternate models) are subsequently utilized to fine tune the dose or energy to obtain better results. Several workers have calculated R_p and ΔR_p values, using LSS theory to perform the calculations, for many of the elements that are commonly implanted into Si and SiO_2. Figure 7 is an example of such data for B, P, and As into silicon. Values for projected lateral straggle, $\Delta R_\perp$, have also been compiled, and are given in Fig. 8 (together with ΔR_p).

> **Example:** A 150 mm wafer is to be implanted with 100 keV boron atoms to a dose of 5×10^{14} ions /cm^2. 1) Determine the projected range, projected straggle, and peak concentration using Figs. 7 and 8. 2) If the implantation time is 1 min, calculate the required ion beam current.

Solution: 1) From Figs. 7 and 8, it is found that the *projected range* and *projected straggle* are 0.32 µm and 0.07 µm, respectively. The peak concentration, $N(x = R_p)$, can be calculated from Eq. 9, or:

$$N(x = Rp) = \frac{0.4\,\phi}{\Delta R_p} = \frac{0.4\,(5 \times 10^{14})}{(0.07 \times 10^{-4}\,cm)} = 2.8 \times 10^{19}\ ions\ /cm^3$$

2) To find the beam current, first calculate the total number of implanted ions (Q = dose * wafer area:

$$Q = (5 \times 10^{14}\ ions\ /cm^2) * [\ \pi\ (15\ /2)^2\ ] = 8.8 \times 10^{16}\ ions$$

Then, the required scanned beam current is determined by dividing the total charge, qQ, by the time of implantation:

$$I = (qQ\ /t) = [(1.6 \times 10^{-19}\ C)\ (8.8 \times 10^{16})]\ /60\ sec = 0.23\ mA.$$

Higher Moment Distributions for Implant Profiles in Amorphous Material

As noted earlier, even when implanting into amorphous material, the experimental profiles exhibit some asymmetry, or *skewness*. This is not surprising if one considers the forward momentum of the ions. That is, when relatively light atoms make collisions with target atoms (e.g. B in Si), they experience a significant degree of backscattering. Hence more will come to rest at a distance closer to the surface than R_p (causing the concentration near the surface to be higher). On the other hand, heavier atoms will undergo little backscattering, and the concentration on the deep side will be higher. Thus, even if a Gaussian distribution is used to approximate the implantatation profile, such non-Gaussian effects on the behavior of devices can be anticipated. That is, when boron is utilized to implant deep p-wells (in CMOS technology), higher doping close to the surface is expected (than predicted by a Gaussian distribution). On the other hand, the skewness in arsenic implants will produce deeper junctions than predicted when implanting n⁺ source /drain, or emitter regions with arsenic.

To theoretically account for the skewness found in measured profiles, probability

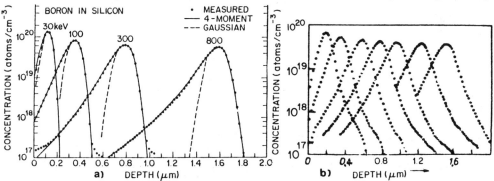

Fig. 9 (a) Boron implanted atom distributions, showing 1) measured data points, 2) four-moment (Pearson-IV) curves, and 3) symmetric Gaussian curves. The boron was implanted into *amorphous silicon* without annealing. (b) Depth distributions of boron implanted in *crystalline silicon* in a <763> direction - a dense crystallographic direction at various energies. *Tails* occur because of channeling effect[5]. Reprinted with permission of Philips Research Reports.

distributions with higher order moments must be used to approximate impurity distributions. Gibbons and Mylroie[9] have demonstrated that use of a distribution with a third central moment can accurately approximate depth profiles if the asymmetry of the profile is not excessive (i.e. the value of the third central moment is less than ΔR_p). When this approach is undertaken, the distribution is actually represented by two half-Gaussian profiles, each with their own projected straggle ΔR_{p1} and ΔR_{p2} (Fig. 6b), joined together at the depth of their modal range, $R_M = R_p - 0.8 (\Delta R_{p1} - \Delta R_{p2})$. The joined half-Gaussian approximation produces a good fit to the profiles of phosphorus and arsenic atoms implanted into Si. The concentration values of such distributions as a function of depth can be calculated from:

$$ n(x) = \frac{2\phi}{\sqrt{2\pi}(\Delta R_{p1} + \Delta R_{p2})} \exp\left[\frac{-(x - R_M)^2}{2\Delta R_{p1}^2}\right] \quad x \geq R_M \qquad (10a)$$

$$ n(x) = \frac{2\phi}{\sqrt{2\pi}(\Delta R_{p1} + \Delta R_{p2})} \exp\left[\frac{-(x - R_M)^2}{2\Delta R_{p2}^2}\right] \quad x \leq R_M \qquad (10b)$$

and the values for ΔR_{p1} and ΔR_{p2} are found in the tabulated data of Gibbons, et al[2].

An approach that represents the implanted profile with a distribution described by *four moments* is more exact than the three moment approach (and is applicable even when the third central moment is large). In fact, Hofker[5] has demonstrated that excellent agreement with measured B profiles in amorphous Si is obtained (Fig. 9), by assuming that the implantation distribution can be described by a *Pearson IV distribution function,* a type of distribution function which can be specified by four moments. The four moments describe various character- istics of the implant profile curve: 1) μ_1 (mean range); 2) μ_2 (straggle); 3) γ_1 (skewness); and 4) β (*kurtosis* - which characterizes the *tail* aspect of the distribution). It was later shown by Ryssel, et al[7] that equally good agreement is found using a Pearson IV curve for other implanted species into amorphous targets. The mathematical expressions for the Pearson IV curve are quite lengthy and fall beyond the scope of this text. They can be found in Refs. 1 and 10.

Implanting Into Single Crystal Materials: Channeling

LSS theory matched with an appropriate distribution curve can accurately predict concentration profiles for implantation into amorphous materials. In practice when fabricating VLSI, however, most implants are made into single-crystal Si. In single-crystal lattices there are some crystal directions (known as *channels*) along which the ions will not encounter any target nuclei, and will be *channeled*, or steered along such open channels of the lattice. Figure 10 illustrates that the most likely channeling directions, in order, are <110>, <111>, and <100>, as these are directions that exhibit channels of decreasing "openness". As the implanted atoms travel down channels, the slowing down is accomplished mainly by electronic stopping, and the ions can penetrate the lattice several times more deeply as in amorphous targets (Fig. 11a). At first, this might appear to offer the advantages of being able to produce deeper implanted junc- tions and less lattice damage. It turns out, however, that the large sensitivity to incident beam direction, and the unpredictable effects of *dechanneling* (leading to anomalous profile tails), make channeling effects difficult to control, and has kept these apparent benefits from being exploited.

As a result, instead of seeking to harness the effect of channeling, techniques have been sought to avoid its occurrence. The most widely adopted procedure to minimize channeling has

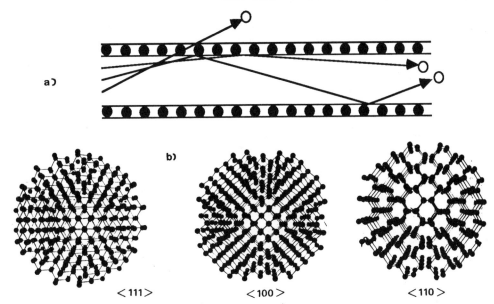

Fig. 10 (a) Schematic representation of ion trajectories in an axial channel for various entrance angles. (b) Ball model showing relative degree of "openness" of the diamond (Si) lattice when traversing in <111>, <100>, and <110> directions.

been to tilt the wafer surface relative to the incident beam direction (most commonly by ~7°), so that the lattice presents a dense orientation to the incident beam (i.e. the approximate <763> direction). There are two reasons why this technique *reduces*, but does not eliminate channeling:

1) Even when atoms enter the lattice in an apparently random, "dense-appearing" direction, they can be diverted into one of the open channels after entering the lattice (Fig. 11b). This effect leads to the *tails* that are observed in implanted profiles into single-crystal lattices. Figure 9b shows the depth profiles of boron implanted into crystalline Si in a <763> direction. Compared to Fig. 9a, it can be seen that tails in the profile exist. Based on experimental results, an empirically modified Pearson IV distribution has been constructed to account for this effect. This is done by adding an exponential tail with a fixed characteristic length (of 0.045 μm), which is independent of dose energy and crystalline surface orientation, to the Pearson IV distribution. The tail is appended to the shoulder of the distribution where the concentration has decreased to 50% of the peak value. This modification has been applied in SUPREM II (a process simulation computer program) for simulating boron implant profiles, and the resulting calculated profiles are compared to experimentally implanted profiles into Si in the <763> direction in Fig. 12.

2) If the 7° tilt is performed such that the Si atoms are still inadvertently aligned in a highly symmetric array of planes, the phenomenon of *planar channeling* can still produce channeling effects (Fig. 13). Turner, *et al*[13] reported that in order to insure that both planar and axial channeling effects are avoided, wafers must also be oriented with an appropriate azimuthal (or *twist*) direction, in addition to a proper tilt angle. Recommended optimum twist angles for various implantation species, energy, and orientation are shown in Fig. 14. It was also noted that control of the twist angle is not possible with many of the gravity-fed wafer handling systems used on medium-current implanters. More recently, however, new end-stations have been introduced by most implanter suppliers which make provision for controlling twist angle.

In addition to tilting and twisting wafers, other methods investigated to further reduce

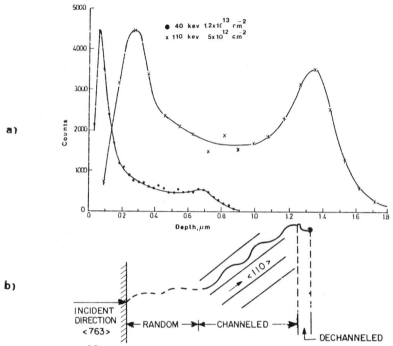

Fig. 11 Variation of ^{32}P-concentration profiles with ion energy for implantation along <110> axis, plotted on a linear scale[11]. Reprinted with permission of Canadian Journal of Physics. (b) Schematic diagram of an ion path in a single crystal for an ion incident in a "dense" <763> direction. The path shown has non-channeled and channeled behavior[63]. Copyright, 1983, Bell Telephone Laboratories, Incorporated, reprinted by permission.

channeling effects include: a) pre-amorphizing the lattice by a prior implantation (e.g. with Si or Ge)[14,15]; b) implanting through a surface oxide, to randomize the directions of the ions as they enter the lattice[14,15,16]; and c) using heavy ions for the implantation (e.g. BF_2 rather than B).

Boltzmann Transport Equation and Monte Carlo Approaches for Calculating Implantation Profiles

When fabricating integrated circuits, the ion implantation is frequently performed through a thin surface layer (e.g. SiO_2, Si_3N_4, or even a composite layer). As is discussed in a later section, if the stopping powers of the various layers are not too different, simple approximations can be used to fit actual profiles fairly well. More accurate predictions of profiles into multilayered materials, however, requires either a numerical solution of the Boltzmann transport equation (BTE)[17,18] or Monte Carlo simulation[19,20,21]. The assumptions used by the LSS theory that the target material is homogeneous and isotropic, make it invalid for predicting implantation ranges in multilayer structures. In both of these methods the predicted distribution is calculated directly, rather than inferred from a range value and an assumed distribution curve.

In the Monte Carlo approach, ion implantation is simulated by following the history of an energetic ion through succesive collisions with target atoms using the binary collision assumption. The calculation of each trajectory begins with a given energy, position, and direction. A large number of ion trajectories are calculated (>10^3), and the depth at which each ion stops is determined. The predicted profile is generated by plotting histograms of the number of ions stopped within each depth interval. Monte Carlo simulations can be performed for either amorphous or crystalline targets. In the amorphous material simulation, the position of the target atoms follows a Poisson distribution, while in crystalline target simulation the atom positions are specified to correspond to the positions that they would assume on a lattice. Both the BTE and the Monte Carlo simulation calculations are complicated and time consuming numerical procedures, and they are probably not justified for many typical integrated circuit process simulations. Nevertheless, they are valuable in some applications, as they can provide distributions of recoil atoms (e.g. the O atoms recoiled into the substrate from a surface SiO_2 layer).

The Monte Carlo technique is the more general of the approaches and has two advantages over the BTE technique: a) it is a three-dimensional technique, allowing lateral spreading effects to be calculated; and b) it can be used to simulate profiles into crystalline material. The Monte Carlo approach, however, is computationally very intensive. For example, a typical VLSI calculation of a single profile takes 50-500 sec on a Cray-1 supercomputer[22]. If crystalline effects are to be modeled, the required computer time becomes even greater (e.g. 4 hours on an IBM mainframe). As a result, work is being pursued to enhance the capability of the less-computationally intense method of BTE. Recent work by Giles and Gibbons indicates that channeling effects can be successfully modeled using the BTE technique[22].

ION IMPLANTATION DAMAGE and DAMAGE ANNEALING IN SILICON

As outlined earlier, ion implantation has many advantages for VLSI processing, the most important being the ability to control the number of impurity atoms introduced into a substrate

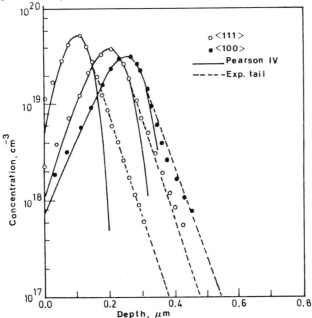

Fig. 12 Experimental profiles of boron implanted into <111> (closed circles) and <100> (open circles) silicon. The solid lines represent the Pearson IV distribution and the dashed lines the exponential tail[12]. (© 1979 IEEE).

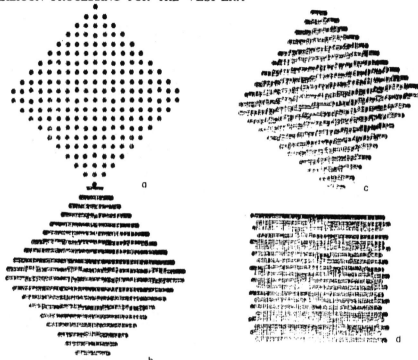

Fig. 13 A ball model of representation of Si (100) lattice: (a) viewed along the direction of direct axial channeling; (b) planar channeling along (110) for a 7° wafer tilt; (c) near-channeling at a twist of 10° away from planar channeling; (d) "random" conditions for 7° tilt, 30° twist[13]. Reprinted with permission of Solid State Technology, published by Technical Publishing, a company of Dun & Bradstreet.

more precisely. The price for such benefits is high, however, as implantation cannot be achieved without damage to the substrate material. That is, high-energy ions collide with substrate atoms and displace them from their lattice sites in large numbers. Furthermore, only a small percentage of the as-implanted atoms end up on electrically active lattice sites. In order to successfully fabricate devices, the damaged substrate regions must be restored to their pre-implanted structure, and the implanted species must be electrically activated. In this section we will describe the aspects of implantation damage, mechanisms of damage annealing, and electrical activation of implanted dopants. Figure 15 illustrates the subjects that will be covered. At the conclusion of the section, it will be noted that as devices are designed with very shallow junctions (e.g. $\leq 2500\text{Å}$), the remaining residual implantation damage (even after the annealing process), represents a mechanism that degrades device operation. An exmple is the excess reverse-bias leakage current observed when such shallow junctions are fabricated.

Implantation Damage in Silicon

When energetic ions strike a silicon substrate, they lose their energy in a series of nuclear and electronic collisions, and rapidly come to rest some hundreds of atom layers below the surface. Only the nuclear collisions result in *displaced silicon atoms* (also referred to as *damage*, or *disorder*). An individual nuclear collision can result in different types of displacement events,

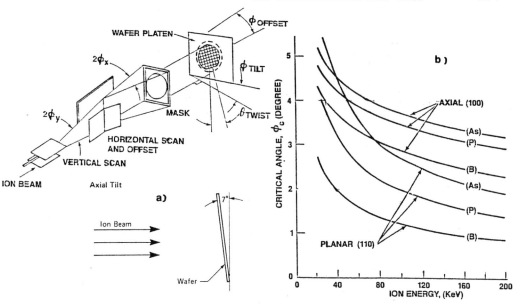

Fig. 14 (a) Channeling avoidance by using axial tilt and azimuthal twist. (b) Calculated critical angles vs. ion energy for channeling along the axial Si <100> direction and planar <110> direction for B^+, P^+, and As^+ ions[13]. Reprinted with permission of Solid State Technology, published by Technical Publishing, a company of Dun & Bradstreet.

depending on the magnitude of the energy transferred. If the energy transferred to a silicon atom (ΔE_n) is less than the energy required to displace it from its lattice site, E_{di}, no displacement event results. If $2E_{di} \geq \Delta E_n \geq E_{di}$, a single displacement and simple isolated point defects are created. If $\Delta E_n \geq 2E_{di}$, point defects and *secondary displacements* (i.e. recoiled lattice atoms with enough energy to generate additional lattice disorder) are produced. Finally, if $\Delta E_n >> E_{di}$, multiple secondary displacements and defect clusters are created. (Note, E_{di} [Si] ~ 15 eV.)

Because of their extremely small size, the exact nature of isolated defects and defect com- plexes from ion implantation are hard to characterize. However, each displacement of a lattice atom, whether by a primary beam ion, or by an energetic recoil lattice atom, produces a *Frenkel defect* (see section on *Point Defects* in Chap. 2). In addition, it is widely agreed that the defects include: a) vacancies [V]; b) divacancies, [V^2] (i.e. two vacancies bound together); c) higher order vacancies and (vacancy-impurity complexes); and d) interstitials [I]. The ions create zones of gross disorder, populated by such defects, in regions where they deposit their kinetic energy (Fig. 16). These zones are vacancy-rich at the center, and are surrounded by [I], since each displaced Si atom moves into the lattice with velocity components perpendicular to the ion track.

The lattice in these disordered regions exhibits several different damage configurations: a) *isolated point defects or point defect clusters in essentially crystalline silicon* (i.e. the type of damage that results from implanting light ions, or when $\Delta E_n \cong 2E_{di}$); b) *local zones* of completely *amorphous material* in an otherwise crystalline layer (i.e. an *amorphous region* is defined as a region in which the displaced atoms per unit volume approach the atomic density of the semiconductor). This form of damage configuration is associated with low-dose implants of heavy ions (i.e. $\Delta E_n >> E_{di}$); and c) *continuous amorphous layers* which form as the damage from the ions accumulates. That is, as the dose of ions (typically heavy ions) increases, the locally amorphous regions eventually overlap, and a continuous amorphous layer is formed.

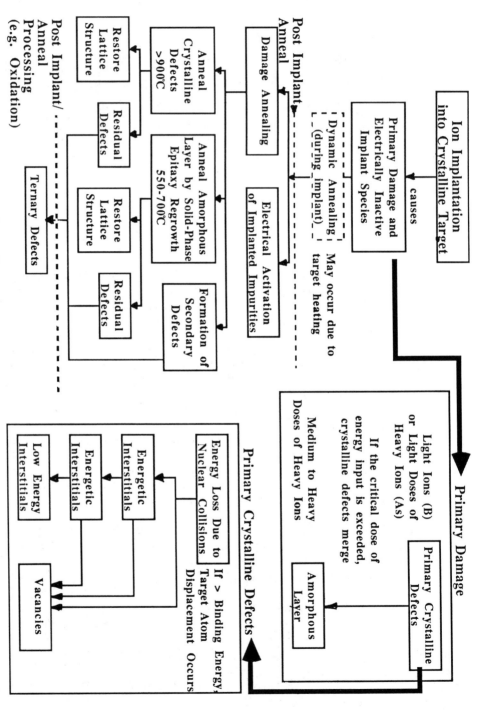

Fig. 15   Ion implantation damage and annealing.

In our discussion we will group damage types a) and b) into the category of *primary crystalline-defect damage*. Type c) will be referred to as *amorphous layer damage*. The basis for this grouping, as we shall see in the sections that discuss damage annealing, is that the annealing strategy for damage types a) and b) are the same, but a different annealing procedure is employed for type c) damage.

Regardless of the form of the damage configuration, the number of displaced atoms after implant is almost always larger than the number of implanted atoms. These displaced atoms reduce the mobility in the damaged regions and produce defect levels in the band gap of the material (i.e. deep-level traps, for both electrons and holes) which have a strong tendency to capture free carriers from the conduction and valence bands. Due to the relatively large number of displaced atoms, and the fact that few as-implanted impurities occupy substitutional sites, non-annealed implanted regions typically exhibit high-resistivity.

Primary Crystalline Defect Damage

Primary crystalline defect damage is observed after implanting a crystal with relatively light ions (e.g. B^{11} in Si^{28}), or with light doses of heavier ions (e.g. P^{31}, As^{75}, Ar^{40}). As noted in the previous section, the damage configurations from light ions are in fact quite different than those from heavy ion implantation. Let us discuss the reason for the difference.

As shown in Fig. 5, at the initial impact energies, most of the energy loss for *light ions* (in this example, boron), is due to electronic collisions (which do not produce displacement damage). However, as B ions penetrate deeper into the lattice, they lose energy. Eventually, the cross-over energy is reached, below which nuclear stopping predominates (~10 keV for B). As shown in Fig. 16, *most lattice damage occurs in the part of the light ion trajectory beyond that point.*

An estimate of the damage can be calculated by considering the case of an 80 keV boron ion. Its projected range is ~2500Å (Fig. 7), and its initial *nuclear energy loss* is ~3.5 eV /Å (Fig. 5). The boron ion will thus lose ~8.75 eV from nuclear collisions for each lattice plane that it passes (as the lattice spacing in Si is ~2.5Å). Since 8.75 eV is less than E_{di} [Si], the B atom

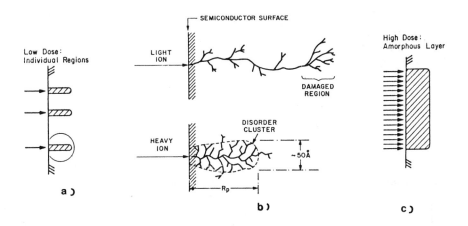

Fig. 16 Schematic representation of the disorder produced by ion implantation. (a) Low dose; b) light ions – individual regions with degree of disorder increasing as ions penetrate deeper into substrate,   heavy ions - individual regions of more uniform disorder along entire ion trajectory. (c) Heavy doses - formation of amorphous layer.

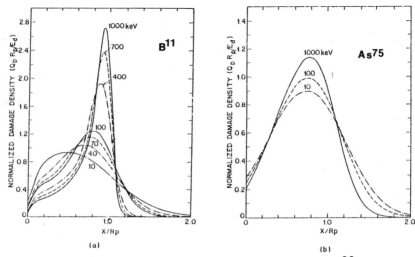

Fig. 17 Calculated damage density profiles of (a) boron, and (b) arsenic[23]. Reprinted with permission of Plenum Publishing Company.

upon initially entering the Si, does not transfer enough energy per nuclear collision to displace silicon atoms from their lattice sites. When the ion energy has decreased, however, to 40 keV (at a depth of ~1300Å in the lattice), the energy loss due to nuclear collisions rises to ~15 eV per lattice plane (i.e. 6 eV /Å), which is sufficient to displace Si lattice atoms. (Note that although electronic stopping is still predominant, nuclear stopping between 10-40 keV can still displace Si atoms.) Assuming that one Si atom is displaced per lattice plane for the remainder of the ion trajectory, 480 lattice atoms are displaced (i.e. 1200Å /2.5Å) by the time the boron atom comes to complete rest. If each displaced atom is moved roughly 25 Å by such collisions, the damage volume is found from $V_{dam} \cong \pi (25\text{Å})^2 (1200\text{Å}) = 2.4 \times 10^{-18}$ cm^3. The damage density is $480 /V_{dam} \cong 2 \times 10^{20}$ cm^{-3}, which amounts to only ~0.2% of the atoms. This calculation implies that very large doses of light ions are thus required to produce an amorphous layer, and for the most part, each ion produces a trail of well separated primary recoiled Si atoms in the wake of the implanted ion. Furthermore, displaced atoms will be separated by short distances from the vacancies they leave, because the energies of their recoils are low. This suggests that only a relatively small input of energy to the lattice could cause such separated pairs to rejoin. In fact, as we shall explain in the following section on annealing, a large fraction of the disorder produced during boron implantations is *dynamically annealed during the implantation*, and therefore at room temperatures even high-dose boron implantations may not produce an amorphous layer. The damage from boron implantations is thus characterized by *primary crystalline defects*. Damage density is distributed versus depth as shown in Fig. 17a, which shows a sharp buried peak concentration, and qualitatively fits our description of the damage-creation process.

When *heavy ions* are implanted, the energy loss is predominantly due to nuclear collisions over the entire range of energies experienced by the decelerating heavy ions (Figs. 5 and 16). Thus, substantial damage is expected. Let us examine the case of 80 keV arsenic atoms, which will have a projected range of ~500Å. The average energy loss due to nuclear collisions will be ~120 eV /Å over the entire range. As a result, the As atoms lose ~300 eV for each Si atomic plane that they pass. Most of this energy is transferred to a single lattice atom. The recipient Si atom, however, will subsequently produce ~20 displaced lattice atoms. The total number of displaced atoms is thus 4000. Again, assuming an average distance moved for each displaced atom of ~25Å, the damage volume is $V_{dam} \cong \pi \ (25Å)^2(500Å) = 0.8 \times 10^{-18}$ cm^3. The damage density is then $4000 /V_{dam} \cong 5 \times 10^{21}$ cm^{-3}, or ~10% of the number of atoms in the lattice within the damage volume. Since a single ion is capable of producing such heavy damage, it is reasonable to expect that some local regions of a silicon substrate (which when subjected to even light doses of heavy-ion bombardment) will suffer enough damage to become amorphous. Some damage density distributions due to heavy-ion (e.g. As) implants are shown in Fig. 17b. They exhibit a *broad* buried peak which is a replica of recoiled range distribution.

Amorphous Layer Damage

Simple qualitative concepts illustrate how continued bombardment by heavy ions will lead to the formation of continuous amorphous layers. That is, heavy-ion damage accumulates with ion dose, through an increase in the density of localized amorphous regions. Eventually these regions overlap, and a continuous amorphous layer is the result. The evolution of a continuous amorphous layer from the accumulation and overlap of damage formed by individual atoms has been observed (Fig. 18) using Rutherford backscattering spectroscopy in a channeling mode (see Chap. 17 for more information on this technique). In Fig. 18 it can be seen that damage produced by 1.7 MeV Ar$^+$ ions in Si builds up to an initial damage distribution with a peak at a depth of ~1.3 μm. At that depth, individual amorphous zones are likely to be created by each ion (Fig. 18b). Closer to the surface, the damage consists predominantly of isolated defect clusters (akin to the damage caused by light ions). As the dose is increased, both types of damage increase, and finally a 1.5 μm thick (almost continuous, as in the example of Fig. 18a) amorphous layer extends to the surface. This also illustrates another aspect of the formation of amorphous layers. That is, amorphization begins at the depth of the maximum nuclear collision

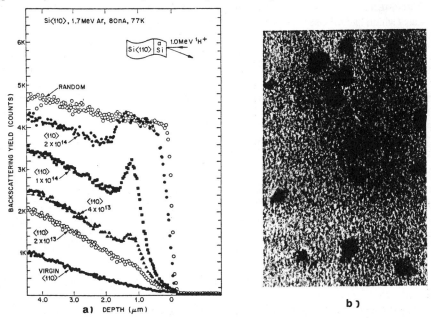

Fig. 18a) RBS channeling spectra of damage accumulation in Si at liquid nitrogen temperature following bombardment with 1.7 MeV Ar$^+$ ions[24]. By permission of the Harvard University Archives. (b) Bright-field electron micrographs of amorphous regions in Si produced by bombarding to doses of 3×10^{11} cm^{-2} with 10 keV bismuth ions[25]. Courtesy Inst. Phys. Conf. Ser.

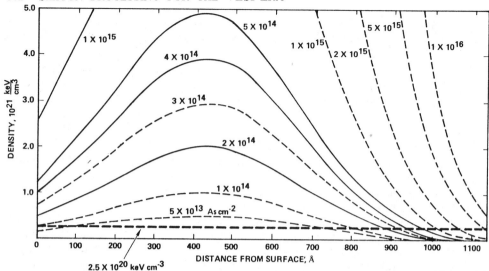

Fig. 19 Damage density distribution for 100-keV As implantations for doses of 5×10^{13}-1×10^{16} As cm^{-3} calculated from Brices[23] curves[27]. Reprinted by permission of American Physical Society.

energy deposition, at slightly less than the projected range, and spreads towards the surface, as well as towards deeper positions in the target. In addition, the interface with the single crystal region below is not a well defined plane, due to the statistical nature of the penetration. Beyond the interface, primary crystalline defect damage, as well as a considerable concentration of Si interstitials (which had diffused out of the damage clusters during implantation) is expected.

The minimum (or *threshold*) dose required to convert a crystalline material to an amorphous layer, can be arrived at from a number of different viewpoints: a) First, it is certain that when the number of stably displaced Si atoms reaches the number of Si atoms /unit volume (e.g. 5×10^{22} cm^{-3}), the material has become "amorphous". When the damage density reaches this value, an amorphous condition is reached; b) Another view holds that there is a critical energy density that must be placed into the crystal to make it amorphous[26]. This critical energy, E_c, is given by:

$$E_c \cong f \ (10^{24} \ eV) \ /cm^3 \qquad (11)$$

and the pre-factor, f is 0.1-0.5 for Si. Based on this model, Brice published tables which predict the extent of the amorphous layer formation for different doses (and constant implant energy). Prussin, *et al*[27], plotted the data from these tables for As implanted into Si (Fig. 19); and c) a simple estimate can be obtained at by assuming that if enough energy is applied to the crystal to cause melting (for Si, 10^{21} keV /cm^3), that this would also produce an amorphous layer. For 100-keV As ions, the dose, D_{crit}, calculated from this approach (where E_o is the beam energy in keV, and R_p in cm), is:

$$D_{crit} = [(10^{21} \ keV \ /cm^3) \ R_p] \ /E_o = 6 \times 10^{13} \ ions \ /cm^2 \qquad (12)$$

From the discussion on minimum implant dose for creating an amorphous layer, three more important aspects of amorphous layer structure can be inferred: 1) From Brice's model (Fig. 19), it can be seen that some implant doses will cause an amorphous layer to form below the surface of the Si, but that near the surface not enough energy has been transferred to the lattice to cause the material to become amorphous (e.g. in the case of As at 100 keV, for doses less than

1×10^{14} cm^{-2}, the energy density at the surface drops below the threshold required to create an amorphous layer). Thus the amorphous region does not extend all the way to the surface, but is a *buried amorphous layer*. Such buried amorphous layers have been found to exhibit profoundly different annealing characteristics than those which extend all the way to the surface; 2) In the crystalline layer just below the amorphous layer, the silicon has still been the recipient of impurity atoms, albeit in lower quantities than necessary to cause an amorphous region. Yet, heavy primary crystalline defect damage from the implanted ions exists in this zone, as well as impurities that must be electrically activated in order to achieve full electrical activation of the dose. These cannot be ignored when considering an annealing strategy for amorphous layers; and 3) Prussin, *et al*[27] have shown that the critical energy, E_c, required to create an amorphous layer depends on the implanted species, and is 2.5×10^{20} keV /cm^2, 1.0×10^{21} keV /cm^2, and 5.0×10^{21} keV / cm^2, for As, P, and B, respectively.

The temperature of the substrate during implantation also impacts amorphous layer formation. Figure 20 shows a plot of the critical dose versus reciprocal temperature for various ions[26]. It can be seen that for light ions such as B^{11}, a 50°C rise above room temperature prevents the formation of an amorphous layer at any dose. This is due to the fact that B implanted at room temperature generates only a few stably displaced atoms during implantation, and a small increase in temperature allows such closely spaced vacancy-interstitial pairs to recombine.

Electrical Activation and Implantation Damage Annealing

In this section we consider the methods that can be used in attempts to restore the silicon to its pre-implanted structure, and to electrically activate the implanted impurity atoms. The subjects will be covered in the following order: a) electrical activation of implanted impurities; b) annealing primary crystalline defect damage; c) annealing of amorphous layers; d) dynamic annealing effects; and e) diffusion of implanted impurities.

Electrical Activation of Implanted Impurities - Since most as-implanted impurities do not occupy substitutional sites, a subsequent thermal step is employed to bring about electrical activation. The degree to which the thermal procedure is effective in electrically activating impurities is commonly determined by Hall effect measurements, but can also be checked more simply by measuring the sheet resistance, R_s[28]. To a first approximation, R_s

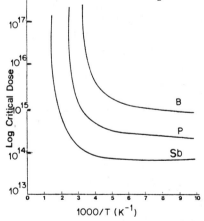

Fig. 20 Theoretical relationship between amorphous threshold and temperature for B$^+$, P$^+$, and Sb$^+$ into Si[26]. Reprinted with permission of Gordon and Breach Publishing Company.

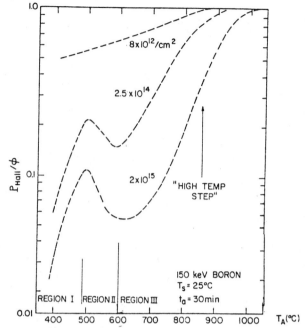

Fig. 21 Isochronal annealing behavior of boron. The ratio of free-carrier content, p_{Hall}, to dose, ϕ, is plotted against anneal temperature, T_A, for three doses of boron[30]. Reprinted with permission of Gordon and Breach Publishing Company.

is inversely proportional to the mobility, μ, and the implanted dose, ϕ (cm^{-2}):

$$R_s = 1 / (q \mu \phi) \qquad (13)$$

where q is the electronic charge. Although μ depends strongly on the concentration of doping atoms and implantation damage, values of R_s have been tabulated utilizing known mobility data[29]. For a known dose, full electrical activity is reached when the predicted R_s is reached.

Electrical activation of implanted impurities in amorphous layers proceeds differently than in layers with primary crystalline damage. As we will discuss, electrical activation in amorphous layers occurs as the impurities in the layer are incorporated onto lattice sites during recrystallization. Electrical activation in crystalline damaged regions exhibits more complex behavior.

For example, Fig. 21 shows the *isochronal* electrical activation behavior of implanted boron (i.e. anneals performed at varying temperatures, but for identical times). In this curve, the measured surface carrier concentration (normalized to the dose, in cm^{-2}) is used to indicate the degree of activation. That is, when $p_{Hall} / \phi = 1$, full activation is reached. Note that other impurities exhibit similar behavior to that shown in Fig. 21, provided the implantation does not cause a continuous amorphous layer to be formed.

The temperature range up to 500°C (Region 1 of Fig. 21) shows a monotonic increase in electrical activity. This is due to the removal of trapping defects and the concommitant large increase in free carrier concentration, as the traps release the carriers to the valence or conduction bands. In Region 2 (500-600°C), substitutional B concentration actually decreases. This is postulated to occur as a result of the formation of dislocations at these temperatures. Some substitutional boron atoms are believed to precipitate on or near these dislocations. In Region 3

(>600°C), the electrical activity increases until full activation is achieved at temperatures ~800-1000°C. The higher the dose, the more disorder, and the higher the final temperature required for full activation. At such temperatures, Si self-vacancies are generated. They migrate to the B precipitates, allowing boron to dissociate and fill the vacancy (i.e. a substitutional site).

Activation of implanted impurities by rapid thermal processing (RTP) has also been studied. The time-temperature cycle to reach minimum sheet resistance for As, P, and B is ~5-10 sec at 1000-1200°C, the exact condition being dependent on implanted species, energy, and dose[31,32].

b) **Annealing of Primary Crystalline Damage** - Isolated point defects and point defect clusters (as predominantly occur during light ion implantation), and locally amorphous zones, that are typically observed from light doses of heavy ions, are both regions of primary crystalline damage that exhibit comparable annealing behavior. At low temperatures (up to ~500°C), vacancies and self-interstitials that are in close proximity undergo recombination, thereby removing trapping defects. At higher temperatures (500-600°C), as described above, dislocations start to form, and these can capture impurity atoms. Temperatures of 900-1000°C are required to dissolve these dislocations. Note that the activation energy of impurity diffusion in Si is always smaller than that of Si self-diffusion. Therefore, the ratio of defect annihilation to the rate of impurity diffusion becomes greater as the temperature is raised. This implies that *the higher the anneal temperature the better*, with the upper limit being constrained by the maximum allowable junction depth dictated by the device design[28].

It is also important that the steps used to anneal implantation damage be conducted in a neutral ambient, such as Ar or N_2. That is, dislocations which form during annealing[33] can serve as nucleation sites for oxidation induced stacking faults (OISF, see Chap. 2) if oxidation is carried out simultaneously with the anneal (i.e. the annealing is performed in an oxygen ambient).

c) **Annealing of Amorphous Layers** - The annealing of continuous amorphous layers that extend to the Si surface has been found to occur by solid-phase epitaxy (SPE)[34], at temperatures between 500-600°C (Fig. 22). That is, a recrystallization process occurs on the underlying crystalline substrate, and regrowth proceeds toward the surface. The amorphous layer regrows at varying rates, depending on the crystal orientation and the implanted species, although at 600°C regrowth is generally completed in a matter of minutes. Regrowth is faster on (100) than (111) Si, and impurities such as B, P, and As enhance regrowth, while O, C, N, or Ar retard regrowth. In practice, 550°C is favored, since (100) regrowth at 600°C is too rapid.

The impurity dopant atoms are swept into substitutional lattice sites during the SPE regrowth, and thus full electrical activation within the amorphous layer can be obtained at relatively low temperatures. The impurities that are implanted into the region beyond the amorphous layer, however, must be subjected to the higher temperatures needed to electrically activate impurities in regions of primary crystalline damage (Fig. 23). Therefore, to fully

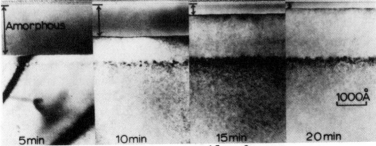

Fig. 22 Solid phase regrowth of a 200 keV, 6×10^{15} /cm^2 antimony implantation at 525°C. TEM cross section micrograph. Courtesy of Institute of Physics, Conference Series.

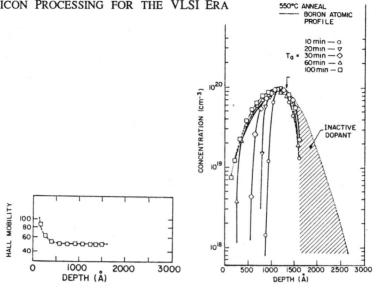

Fig. 23 Free-carrier concentration and mobility for implanted layers which illustrate dopant incorporation by solid phase epitaxy (SPE). Profiles for BF_2^+ implanted into <100> Si at 150 keV and 10^{15} /cm^2 after different isothermal anneals. The dotted curve is the as-implanted atomic profile from SIMS analysis. The original amorphous-crystalline interface is denoted by the arrow, SPE is complete after ~100 minutes. The hatched region is electrically inactive.[64] Reprinted with permission of the American Physical Society.

activate an amorphous layer, and the region of heavy primary crystalline damage behind it, higher temperature anneals than 600°C must be used (normally 800-1000°C). Note that specific *minimum* times and temperatures depend on the particular implanted species and dose.

If the amorphous layer does not extend all the way to the surface (i.e. a buried amorphous layer), the annealing proceeds differently[36]. That is, SPE occurs at both amorphous-single crystal interfaces, and the regrowing interfaces meet below the surface. This meeting point has been found to be a heavily damaged layer, with properties likely to cause device degradation. Thus, it is prudent to select implantation conditions which avoid the formation of buried amorphous layers. Note that if a wafer exhibits color bands after implantation and anneal, they are probably symptomatic of a subsurface damage layer left behind by a buried amorphous layer. The gives rise to optical interference effects from the light reflected from the subsurface damage layers and the surface.

Some of the crystalline defects in the region beyond the amorphous layer are annealed out during subsequent thermal cycles, but others give rise to extended defects, such as dislocation loops and stacking faults, which then grow and interact. Under some implantation and annealing conditions, these defects move to the surface and eventually disappear, while in others they grow into larger structures which intersect the surface or remain in the bulk. Details of these mechanisms are not well understood at this time, and research is continuing in efforts to increase this understanding. Reference 28 describes one model of the events associated with such phenomena.

d) **Dynamic Annealing Effects** - The heating of the wafer during implantation can impact the implantation damage and the effects of subsequent annealing. A rise in temperature increases the mobility of the point defects caused by the damage, and this gives rise to healing of damage even as the implant process is occurring, hence the name *dynamic annealing*[36]. In the case of light ions, sufficient damage healing may occur to prevent the formation of amorphous layers, even at very high implantation doses. In the case of heavy ion implantations, dynamic annealing can cause amorphous layer regrowth during the implantation step.

A study by Prussin, *et al* [37] showed that the wafer cooling capability of an ion implanter can impact the structure of the damage following implantation because of dynamic annealing effects. That is, if a wafer is prevented from being significantly heated above room temperature by adequate heat sinking during implantation, dynamic annealing is minimized.   On the other hand, if no heat sinking is provided and wafers are allowed to rise to temperatures ~150-300°C, dynamic annealing effects can produce changes in implantation damage structures. This typically occurs in unreproducible and unwanted ways, such as the formation of buried amorphous layers, or crystalline layers containing high densities of dislocation loops.

e) **Diffusion of Implanted Impurities** - As described in Chap. 8, the diffusion of impurities in single-crystal Si is a complex phenomenon. The diffusion of impurities in implanted Si is even more complicated as a result of the presence of implantation damage. As an example, empirical studies of the *diffusion of B* in implanted single crystal Si indicate that at high temperatures (≥ 1000°C), the data seems to obey ordinary diffusion theory (Fig. 24). At lower temperatures, however, ordinary diffusion theory does not accurately predict the diffusion behavior.   That is, at 900°C the boron profile can be closely fit only if a diffusion constant is used that is about three times the value observed under a chemical 900°C diffusion. In addition, at 700-800°C, the depth of profile peak remains fixed, but the concentration in the tail is anomalously higher than is predicted from published boron diffusion constants. Furthermore, the presence of an amorphous layer has been observed to enhance the diffusivity of B by a factor of four.

In the past diffusion of impurities during annealing was often used to drive them to depths beyond the range of implantation damage, thereby producing junctions not degraded by such damage. Present demands for shallow junctions, and narrow-base and emitter regions in bipolar devices, no longer allow extensive dopant redistribution during anneal. Therefore rapid thermal processing (RTP) is being pursued to anneal implantations with minimal impurity redistribution.

As discussed earlier, several papers on the application of RTP to annealing implantations have been published[31]. They indicate that RTP cycles of ~1000°C for 10 sec can activate implanted layers as effectively as 30 minute furnace anneals at 1000°C (Fig. 25), but with impurity redistribution distances of a only few hundred angstroms (compared to several thousand angstroms for furnace anneals). Shallow junctions have been fabricated by pre-amorphizing with a prior implantation of Si or Ge, to reduce the depth of the tail from channeling, and annealing

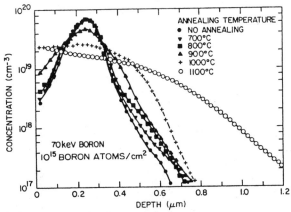

Fig. 24 Boron concentration as a function of annealing at various temperatures. The anneal time is 35 minutes[5]. Reprinted with permission of Philips Research Reports.

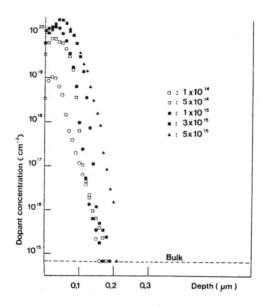

Fig. 25 Dopant concentration profiles for 50 keV $BF_2{}^+$ RTP (1000°C, 10 sec).[32]   Reprinted with permission of the publisher, the Electrochemical Society, Inc.

with RTP after implanting the desired impurity (see also section on *Shallow Junctions*).

The redistribution of impurities in polysilicon should also be considered. It is observed that dopants redistribute themselves much more rapidly in polysilicon than in single-crystal Si, as a result of grain boundary diffusion. Thus, even short RTP cycles that anneal implantations in single-crystal Si without appreciable redistribution, are likely to uniformly distribute impurities throughout a thin film of polysilicon. For some impurities (e.g. As), a capping oxide must be present to prevent significant As outdiffusion from the polysilicon by such cycles.

ION IMPLANTATION EQUIPMENT

Components of an Ion Implantation System

Ion implanters are among the most complex systems used in the fabrication of VLSI. They contain many subsystems (Fig. 26), including the following:

a) A *feed source* of material containing the species to be implanted. Since the most commonly used ions for implanting silicon, B, P, and As, are not gases at room temperature, they must be supplied to the ion source in some other form.   In most semiconductor applications, gases are the favored feed materials (e.g. BF_3, BCl_3, PH_3, AsH_3, and $SiCl_4$) because of ease of operation. An adjustable valve allows flow control of the feed gas into the ion source. In some high-current implanters, however, which require large quantities of feed gases, such highly toxic gases as AsH_3 and PH_3, may be replaced by less toxic solid sources;

b) An *ion source*, with its own power supply and vacuum pump, to ionize the feed gas, and produce a plasma at pressures of ~10^{-3} torr. That is, ions are typically formed through collision with electrons from an arc discharge, or cold cathode source. Figure 27 shows

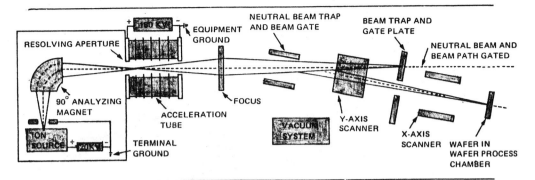

Fig. 26 Configuration of a typical medium-current implanter. Courtesy Varian Associates-Extrion.

a Freeman type hot-cathode ion source, commonly used in medium-current implanters;

c) An *ion extraction and analyzing device*, that selects only the ion species of interest according to their mass, and rejects all others. In most modern implanter designs the ions are extracted from the ion source with voltages of 15-40 keV, and then analyzed. The extracted ion beam must be analyzed (i.e. spatially separated into several beams according to their ionic mass), because the ion beam is a mixture of different fractions of molecules and atoms of the source feed material. For example, BF_3 gas will dissociate into B^{++}, B^+, and BF_2^+, and only one of these ionic groups is be implanted at any one time. The separation is performed with an analyzing magnet, as the radius of the arc of of a charged particle (e.g an ion), r_0 (in cm), accelerated at a voltage, V, within the field of the analyzing magnet, will depend on the mass and charge state of the ion according to:

$$r_0 = (143.95 / H) (M V / n)^{1/2} \qquad (14)$$

where H is the magnetic field (in gauss), M is the ionic mass (in a.m.u.), V is the accelerating voltage; and n is the charge state of the ion. By adjusting the magnetic field strength, only the ionic specie of interest will be given the radius of curvature that will

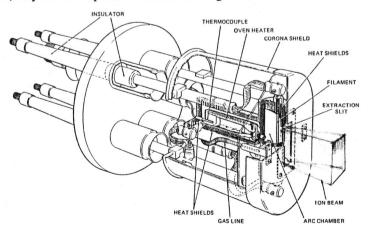

Fig. 27 Sectioned drawing of a Freeman Ion Source (after Freeman)[65].

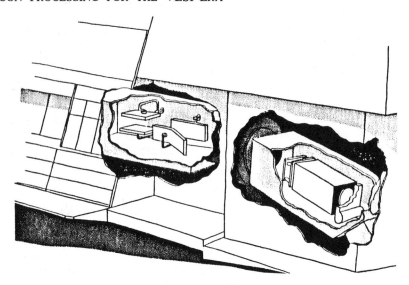

Fig. 28   Illustration of the location of the scanner and end station.

allow it to pass through the resolving slit (aperture) and into the acceleration tube. The analyzing magnet can effectively separate 1 a.m.u., which is useful for separating isotopic species (e.g. $^{10}B^+$ from $^{11}B^+$). Aperture material may be sputtered onto the target. Therefore apertures are fabricated from such materials as graphite (which have low sputtering yields, and whose presence is electrically less damaging in Si than metals);

d) An *acceleration tube* (and its power supplies), which creates the acceleration field to increase the ion energy to the desired energy level, or to decelerate the ion if an energy less than the extracted voltage is required. The ion beam is also focused to a particular size and shape. Round or ribbon-shaped beams are used, depending on equipment design;

e) A *scanning system* to distribute the ions uniformly over the target. The various types of scanning systems used in implanters are discussed in more detail in the section that describes medium- and high-current implanters;

f) A *system end station*, which includes an area defining aperture, a Faraday cup and current integrator (which directly measures the implant dose by collecting the beam current and integrating over the implant time), and a subsystem that loads, holds, and positions the target (e.g. wafers). Secondary electron emission arises from the ion beam striking the target. If these electrons are lost to the walls a dose error results. The Faraday cage nearly surrounds the target and has an opening around the aperture. A negative bias of several hundred volts is applied to the cage. This arrangement returns most of the secondary electrons to the target, reducing the dose error;

g) A *high vacuum system* for evacuating the source, acceleration column, beam chamber, and end station, to typically $<10^{-6}$ torr. A high vacuum is needed to minimize the formation of neutrals by collision of beam ions with residual gas atoms. Neutral beam effects are further reduced by offsetting the slit and target by ~5-10°, and trapping the neutral beam at a beam stop. Cryopumps are used for the beam chamber and end-station to avoid oil backstreaming from diffusion pumps. Turbomolecular pumps (in modern implanters) or diffusion pumps are used to pump the ion source, as the high gas load of the ion source would rapidly saturate a cryopump.

h) A *computer and control system*, which allows for varying degrees of automation in the operation of the implanter[39].

Ion Implanter Types

Due to the variety of implantation applications that are specified in VLSI fabrication, different implantation equipment types have evolved. There are four types of ion implanters encountered in semiconductor research and production environments[40,41]:

a) *medium-current implanters*, MCI, with a total beam current of up to ~2 mA, and maximum energies of 200 keV;

b) *high-current implanters*, with maximum beam current up to 30 mA, and beam energy ranges from 10-160 keV;

c) *pre-deposition implanters* (low beam-energy, high current machines, ~10-30 keV);

d) *high-energy implanters* (up to several MeV), which are the most recently introduced type.

a) **Medium-Current Implanters** - are distinguished not only by their maximum beam-current output, but also by the fact that they are *serial processing machines* (or, machines that implant one wafer at a time). The maximum beam-current is ~2 mA and two or three sets of deflection plates are utilized to electrostatically raster scan the focused beam in a square pattern in the x and y directions (Fig. 14). The maximum deflection angle for 125 mm wafers is $\leq 2°$, and the wafer tilt is adjustable from 0-10°. The scan frequencies range from 10 Hz-1 kHz. MCI are typically used in MOS applications to perform threshold voltage adjust, well, and isolation implantations, and in bipolar technology to perform resistor, base, and isolation implantations. The maximum practical dose in these machines is in the range of 10^{14}-10^{15} ions /cm^2.

In MCI, the temperature of the wafers may rise to undesirably high values if radiant heat loss is the only form of heat dissipation. For example, in some implantation processes, wafer temperatures may reach several hundred °C. That is, for a 0.5 mA beam at 150 keV, the wafer is absorbing 75W, and for a 10 sec implantation the corresponding energy absorbed is 750 J. The two problems associated with such excessive heating involve undesirable partial annealing effects of damage during implant, and degradation of the resist masking layer. As a result, heat must be removed to a cooled wafer holder or platen. The Varian Extrion Model CF-3000 shown in Fig. 26 can be equipped with a cooled platen called Waycool®, in which the wafer is clamped against a Freon-cooled heat sink. It is reported that without Freon cooling, 8W of input power will raise the temperature of the wafer to 105°C, but with Freon cooling, it takes 100 W to raise the temperature to 105°C. Single wafers are gravity-fed through vacuum-locks to a movable platen, and leave the chamber via exit locks.

The *throughput*, in wafers per hour, of an MCI for a specific process and wafer size, can be calculated by determining the *flux density of the machine*, Φ, in [ions /(cm^2 sec)]:

$$\Phi = \frac{6.24\times10^{15} \ (\text{ions /sec mA})}{I_s A_s} \tag{15}$$

where: I_s is the scanned beam current in mA; and A_s is the scanned area in cm^2. Then the *throughput per hour*, θ, is found from:

$$\theta = \frac{3600}{T_i + T_t} \tag{16}$$

where: T_i is the implantation time (in sec), given by dose /Φ; and T_t is the wafer transfer time (or the time from the end of one implantation to the beginning of the next) in sec, as specified by the manufacturer. A typical implantation time in an MCI is ~10 sec. Nearly all MCIs use vacuum loadlocks with cassette-to-cassette loading, and wafer handling time is kept to ~2-4 sec. Note that as the wafer size increases, the flux density decreases, and hence for a given dose and beam current the implantation time is longer.

b) **High-Current Implanters** - have been developed for applications that require doses greater than 10^{14}-10^{15} ions /cm^2 (e.g. to doses >10^{16} /cm^2). Some of these include: source-drain implantations and polysilicon doping implantations in MOS devices; emitter formation or buried layer doping in bipolar devices; and damaged layer formation on wafer backsides for impurity gettering (see Chap. 2). The high-current output (up to 30 mA) of these machines enables them to perform these doping tasks with adequate throughput by processing the wafers in a batch mode[42]. That is, the wafer heating of the beam is distributed throughout the entire load.

There have been a variety of scanning methods and end stations developed for high-current implanters, all of which are also quite different than those utilized in medium-current machines (Fig. 29). In the latest generation systems, the wafers are mounted on the periphery of a disk (Fig. 30) rotating at ~1000 rpm, to produce a rotating mechanical scan. Wafers pass in front of the large defocused ion beam once during each revolution. The ion beam is also scanned linearly across the wafer, either by electrostatic scanning of the beam or by applying an additional linear motion to the rotating disk, to provide a uniform implant dose across the wafer (Fig. 31a). Note

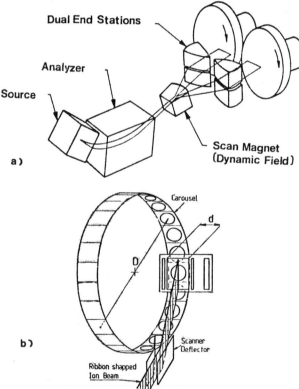

Fig. 29 Some of the various scanning methods used in high-current implanters. (a) Mechanical and magnetic scanning on the Varian model 80-10. (b) Schematic of hybrid scanning (mechanical and electrostatic). Reprinted with permission of Semiconductor International.

that it is argued that electrostatic scanning is less effective than linear mechanical scanning. That is, if a beam is electrostatically scanned, its angular deflection relative to the wafer surface changes. With a change in angular deflection, implantation depth can vary, and can also cause shadowing at steep wafer features (Fig. 31b). The rectilinear motion of the disk in front of a stationary beam eliminates such problems. The dose received in high current implanters is defined by the integrated charge, corrected by the fraction of time that the ion beam hits the wafer.

The handling time of high current implanters is ~10 sec per wafer (300-400 wafers /hr). This time is longer than in single-wafer MCI machines for several reasons, including: a) when the batch of wafers and its fixturing is loaded into the load-lock, much more moisture is introduced into the lock chamber than during loading of a single wafer, as the chamber may be exposed to atmosphere for several minutes. This water must be desorbed before adequate vacuum pressures are reached (see Chap. 3); and b) in order to start and stop the large rotating mass of disk and wafers (without damaging the wafers), acceleration and deceleration must be gra- dual. The throughput, θ, in wafers per hour, of high-current implanters, can be calculated from:

$$\theta = \frac{3600\,H}{T_i + T_l} \qquad (17)$$

where: H is the wafer capacity /run /end station (i.e. the machine may have multiple end stations); T_i is the batch implant time for the dose being implanted (in sec); and T_l, the loading and vacuum pump time (in sec).

Wafer heating in high-current implanters is comparable to that in MCI, and thus wafer-cooling heat sink methods must also be utilized to maintain wafer temperatures within acceptable limits. In the latest models, maximum wafer temperature is limited to 80°C under fully-rated beam current conditions, by use of water-cooled heat-sinking.

c) **Pre-deposition Implanters** - are relatively inexpensive machines dedicated to processes in which an ion deposition is followed by a thermal drive-in to obtain a desired dopant

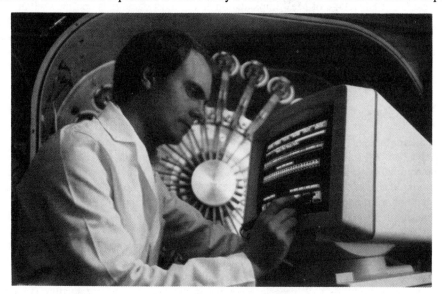

Fig. 30 Rotating disc scanning system and light pen /CRT control unit of the Applied Materials 9000 high-current implanter. Courtesy of Applied Materials, Inc.

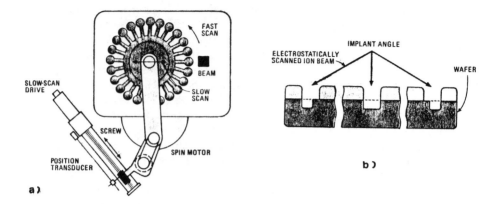

Fig. 31 (a) Dual mechanical scan used in the Applied Materials 9000 high-current implanter. (b) Scan angle errors and shadowing effects that can occur in electrostatically scanned systems. A dual mechanical scan avoids this problem. Courtesy of Applied Materials, Inc.

distribution. Their system design differs from that of the MCI and high-current implanters in that only acceleration of ~10-30 keV is performed during ion extraction. The beam is then analyzed at that point, and directed at the target without further acceleration.

d) **High-Energy Implanters** - are machines whose maximum beam energy is as high as 400 keV-several MeV, but with relatively low beam currents (0.005-0.05 mA). A number of applications of high-energy implantations have been investigated, including formation of buried grids as protection against "soft errors" in memories[56], direct formation of buried collectors in bipolar transistors[57], and retrograde well and buried layers for control of latchup in CMOS devices[58]. An advantage of MeV implantation is that most damage is confined within the ion range distribution, and near surface regions are relatively undisturbed. Much of the work is still being performed in research laboratories, but with commercial introduction of several high-energy implanters, development of high-energy implantation processes is expected to increase.

Ion Implantation Equipment System Limitations

Ion implantation systems can introduce a number of non-uniformities and defects that can limit the yield of a process. In the previous sections we discussed the impact of maximum beam current and vacuum pumping issues on wafer throughput, and planar channeling and wafer heating as process limitations associated with implantation equipment. Some other limitations of the equipment will be discussed here, including: a) contamination and particulates; b) dose monitoring inaccuracies due to beam neutralization effects and secondary-electron emission; c) implantation mask problems; d) wafer charging during implantation; e) poor dose uniformity from machine-to-machine in production implanters; and f) tilt-angle and "scan lock-up" non-uniformities in electrostatically scanned machines.

a) **Contamination and Particulates** - Although ion implantation is considered to be a clean process compared to other doping techniques, several sources of contamination have been nevertheless been observed. These include the following: 1) heavy metals from sputtering of the apertures or wafer holders; 2) cross-contamination, from previous implant runs; 3) pump oil contamination from diffusion pumps; 4) Na contamination; and 5) particulates.

The steps which have been suggested to minimize problems from the above contamination sources are[41,43]: 1) use of graphite or copper free aluminum apertures and target holders has effectively reduced heavy metal contamination due to sputtering; 2) cross-contamination is difficult if not impossible to completely eliminate in high-current implanters. Thus, atoms which can significantly reduce lifetime, such as Au or Cu, should never be used in an implanter. Cross-contamination in medium-current implanters is not considered to be a problem, since a neutral trap is an integral component of the electrostatic scanning system; 3) Contamination from backstreaming pump oil was a problem associated with implanters equipped with diffusion high-vacuum pumps for the beam chamber. Newer implanter designs use cryopumps, which avoid such contamination; 4) Na contamination is minimized by keeping the beamline components carefully cleaned, so that no Na can be sputtered or evaporated from their surfaces. Use of screening oxides has also found to be helpful, although the short wafer-exposure time and neutral trap keep this from being a serious problem in medium-current implanters; and 5) particles present on a wafer surface during implantation mask the regions covered by the particle, while those that fall on a wafer in the implanter after implantation can contaminate the equipment of following processes. Thus, minimal particulate generation in implanters is critical. The concept of requiring equipment suppliers to add a *maximum particles per cm²* clause to their system specifications has even been proposed by some users. A variety of automated wafer handling systems in modern implanters are designed to minimize particle contamination.

b) **Dose Monitoring Inaccuracies due to Beam Neutralization and Secondary-Electron Emission** - Ion implanters measure dose by integrating the total current collected by the electrically isolated sample located within a Faraday cup. It is assumed that the measured charge quantity accurately reflects the dopant flux delivered to the wafer. In practice this assumption may not be valid, due to such mechanisms as secondary electron emission and beam neutralization. Dose monitoring inaccuracies due to *secondary electron emission* are reduced by biasing the Faraday cup as described in the section on *Components of an Ion Implantation System.*

Beam neutralization occurs as a result of ion collisions with residual gas atoms in the beam chamber, and by neutralization from thermal electrons caught in the beam[45]. The sources of the residual atoms are poor beamline vacuums, and vaporization of photoresist masking material by the ion beam. The degree of such resist outgassing effects generally depends on the vacuum capacity of the implanter. Electrostatically-scanned systems are particularly vulnerable to neutralization non-uniformity effects, as the center of the wafer receives a higher implantation dose than the periphery. That is, the neutral beam continues to strike the wafer center during the entire implant, even when the the beam is scanned elsewhere. In some Faraday cup designs, the beam is de-neutralized by a suppression electrode (biased to -500 V) prior to entering the cup, and this electrode also keeps thermal electrons from entering the cup and adding to the dose error.

c) **Implantation Mask Problems** - The impact of the ion beam on photoresist material, which is a commonly utilized mask against implantation into unwanted areas on a wafer, is significant. That is, the resist is vaporized, leading to increased neutralization of the ion beam as mentioned in the previous section, and the resist layer thickness is thus decreased. In addition, the remaining resist becomes increasingly difficult to strip as the dose and temperature increase. In extreme cases the resist flows, cracks, or blisters. Wafer cooling during implantation reduces these effects. Use of polyimide as an alternative mask material has been suggested to solve resist flow and removal problems following heavy implantations. Polyimides are able to tolerate much higher temperatures without flow than resist materials.

d) **Wafer Charging During Implantation** - As a wafer is being implanted, many types of charged species are ejected, primarily secondary electrons[46]. During high current implantation, this can lead to positive charge build-up on the wafer, particularly when implanting into insulating layers, or poly-Si on SiO_2. Such build-up can alter the charge balance in the ion beam and lead to significant dose variations across the wafer. In other instances this may even lead to destructive effects on the surface of the integrated circuit (e.g. microscopic craters, which are a manifestation of oxide breakdown when the surface potential exceeds the breakdown voltage of the oxide surface layer). Many implanter systems direct a stream of low-energy electrons at the wafer during implantation from an *electron flood gun*, to neutralize positive charge build-up. Use of such flood guns helps control the charge effects on doping uniformity[46].

e) **Machine-to-Machine Non-Uniformities** - A study by Current, *et al*[47] compared the dose accuracy and uniformity of eight different models of production implanters from four implanter suppliers. Differences in dose accuracy were surprisingly high (within 14% for a dose of $5x10^{13}$ ions /cm^2, and within 6.5% for $5x10^{15}$-$1x10^{16}$ ions /cm^2), but the dose uniformity across a wafer was significantly better (the 1σ uniformities were within ~1.0-1.75%). The range of dose accuracy variations is wider than can be considered acceptable for production purposes.

The variations from implanter-to-implanter were thought to be caused by a variety of equipment related discrepancies, including: a) acceleration energy drift; b) current integration accuracy; c) implanter leakage currents; d) scan area shift; e) scanning uniformity problems; f) incomplete collection in the Faraday cup; g) electron bias suppression failure; h) sputter effects; i) whole wafer charging; j) neutral ion formation; and k) channeling. This study also demonstrates that only by being aware of the large number of potential parameters that can fall out of specification, can the important task of insuring that wafers repeatedly receive the required dose, and uniformity of dose, be undertaken with high probability of longterm success.

f) **"Scan Lock-Up" and Tilt Angle Non-Uniformities in Electrostatically Scanned Machines** - In order to obtain the specified dose uniformity across a wafer, a minimum implantation time must be allowed to produce a dense scan pattern for adequate uniformity. That is, the maximum scan frequency of a machine sets a lower limit on minimum scan time. Since the ion beam spot has a Gaussian shape, the minimum overlap of the individual Gaussian beam traces must be better than 1.5σ of the beam diameter for 1% uniformity. An optimum scan pattern for two upper scan frequency conditions in an electrostatically-scanned system: a) vertical scan frequency of 64.935 Hz, vertical frequency of 500 Hz; and b) horizontal

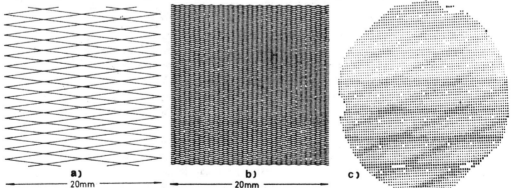

a) ⎯⎯⎯ 20mm ⎯⎯⎯ b) ⎯⎯⎯ 20mm ⎯⎯⎯ c)

Fig. 32   Schematic showing the scan pattern utilized in x-y electrostatically scanned implanters.   Optimum scan pattern for a standard MCI.   **(a)** the horizontal frequency is 64.935 Hz, and the vertical frequency 500Hz, and   **(b)** the horizontal frequency is 500 Hz and the vertical frequency is 1243.57 Hz[45]. (c) Doping variations due to a scan lock-up condition with an electrostatic-scan system (from C-V measurements at 4300 test sites)[48]. Reprinted with permission of Academic Press.

frequency 500 Hz, vertical frequency 1234.57 Hz is shown in Fig 32a. The fact that both horizontal and vertical scans are used, implies that values of these scan frequencies must be selected.

A common problem, however, from such selection, is that partial scan synchronization may inadvertently occur between the x and y deflection waveforms. This phenomenon, termed *scan lock-up*, can cause large implantation non-uniformities. The effects of scan lock-up are most pronounced at low doses, where decreased beam currents and scan times lead to significant overlap in some areas, and doping level gaps in others. Use of wafer mapping measurement tools can detect this effect (Fig. 32b) after the fact. Instruments that constantly monitor and adjust scan frequencies so that lock-up does not occur, are also commercially available (e.g. Brookhaven Instruments).

The effect of the 7° tilt angles at which wafers are held to minimize channeling, is to change the scan rate at the wafer surface so that a linear gradient in the implant dose across the wafer is created (in machines designed with electrstatic x-y scanning). The half of the wafer closest to the beam scan plates receives a larger dose. Such tilt-angle effects become more pronounced as wafer size, and hence scan angle, increases.

Ion Implantation Safety Considerations

The operation of ion implanters involves a variety of potentially severe safety hazards. A brief discussion of these hazards is presented here, and a more detailed description of safety issues involving ion-implanters can be found in Ref. 49.

First, implanters utilize *lethal substances* (e.g. PH_3 and AsH_3) as their source gases. Safe use of such gases requires carefully interlocked exhaust systems, which remove any toxic gases that may be present from small leaks in the gas system. Flow or pressure sensors are installed in the vent to sound an alarm if the vent becomes inoperative. Maintenance personnel must use special exhaust and respiration precautions (e.g. breathing apparatus) when changing toxic gas bottles, cleaning sources and source housing, or when changing the roughing pump oil.

Second, personnel must be protected against an accidental encounter with the high-voltages generated by the implanter. Note that there are several power supplies used in an implanter besides the one used to produce the acceleration voltage. Auxiliary supplies used for beam extraction, scanning, etc., also produce maximum voltages of several keV, and can therefore all produce lethal electric shocks. If contact with even low voltage conductors occurs, nerve conductance may be blocked, and an individual may be thereby disabled and unable to disconnect the electrical power, or even push the "emergency off" button. At least two-levels of safety interlocks, utilizing keys, door switches, and grounding bars should therefore be designed into the systems to prevent such an event. Since such interlocks are breeched during maintenance procedures, personnel are invariably exposed to some risk. In such situations there is no substitute for precaution and common sense, including the rule that no work should be performed alone on hazardous equipment!

Third, x-rays are produced as high energy electrons strike surfaces in the ion column (e.g. the extraction electrodes or beam-defining aperture) of the implanter. Such electrons are produced as energetic ions strike other atoms in the beam extractor. X-ray radiation production in implanters can be reduced by design techniques. Proper shielding, however, must also be employed to absorb whatever x-rays are produced, so that their radiation level external to implanters is below the recommended limit of 0.25 mrem /hr. *Lead shielding* of >6.5 mm thickness is needed to reduce radiation to such levels in medium and high-current implanters. Operators should also wear x-ray monitor badges.

CHARACTERIZATION OF ION IMPLANTATIONS

Extensive utilization of a variety of characterization tools is required to successfully benefit from the advantages of ion implantation.   Such characterization techniques must be employed both during process development as well as in production.   The quantities determined by such measurements include: a) implanted dose; b) uniformity of implanted dose across the wafer; c) implantation depth profiles; d) implantation damage; and e) the efficacy of implantation damage annealing methods.   In this section we discuss methods available to perform such measurements.

Measurement of Implantation Dose and Dose Uniformity

Although monitoring *during* a process is used to determine the total dose received by each wafer, it is also important to measure the wafer after implantation to insure that equipment problems discussed in previous sections have not resulted in dose errors. Post-implantation dose measurements are performed with the following techniques: a) four-point probe sheet resistance; b) optical dosimetry measurements; and c) capacitance-voltage measurements.   General information on techniques a) and c) are found in Chap. 8.   In this section we introduce technique b), and give information on technique a) related to ion implantation dose measurement.

In addition to total dose checking, it must be verified that the process has produced a uniform deposit across the wafer. Dose uniformity is monitored by wafer mapping techniques, which can be based on any of the above measurements[48].   The advent of low-cost computers has made it feasible to perform a large number of measurements per wafer, and to display the results in the form of *two-dimensional wafer maps*.   The variations of the parameter being measured can be represented by a number of formats, including the actual values measured at each location, gray-scale shading, or contour maps.   The *contour map* format has gained wide acceptance, as it provides an easily understood representation of process variation trends, as well as a useful record of the local values. That is, the value at any point is quickly determined by counting the contour lines from the average value contour.   The contour lines represent locations of constant value, and closely spaced contours indicate regions where the mapped parameter is sharply varying. Examples of using contour maps to diagnose implanter problems are given in later paragraphs.

It should be noted that post-implantation total dose and dose uniformity measurements are normally conducted on special monitor or test wafers. *Monitor wafers* are usually devoid of device structures, while *test wafers* contain arrays of test structures designed for measuring the parameters of interest.   It is important to be able to rapidly and accurately characterize such wafers before allowing product wafers to move on to subsequent processing steps.

Implantation Dose Measurements
Modern VLSI processes employ doses ranging from $\sim 10^{11}$-10^{16} ions /cm^2.   Each of the dose measurement techniques listed above find best use over specific ranges. The *four-point probe sheet resistance method* is the most commonly utilized technique because of its superior versatility[50].   Doses from 10^{12} ions /cm^2 upwards can be directly measured with routine accuracy of $\sim 0.25\%$, and use of a special double-implant technique extends its capability down to the mid-10^{10} ions /cm^2 range[50]. The key to high precision measurements with this technique is the use of two measurement configurations at each probe location.   It is also very amenable to computer data acquisition and contour map plotting.   To provide rapid turnaround, RTP techniques are widely used to electrically activate implantations prior to sheet resistance measurement.

The *optical dosimetry technique* is based on measurements of the darkening of photoresist that occurs as a result of exposure to ion beams[51].   Doses down to the 10^{10} ions /cm^2 range

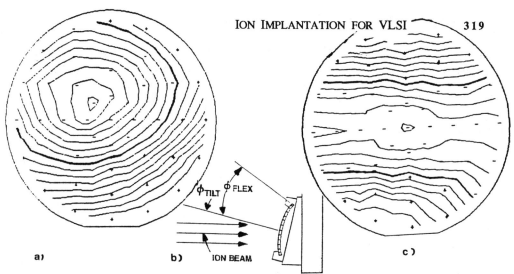

Fig. 33 (a) Example of a contour map of the sheet resistance of a wafer that was implanted in an electrostatically-scanned implanter while mounted on a fixture that slightly flexes the wafer to insure good thermal contact. No wafer tilting was utilized. (b) Schematic of the wafer mounting arrangement. (c) Contour map of the sheet resistance of a wafer implanted under the same conditions as (a) but with a 7° wafer tilt. See text for explanation of the contour configurations.

can be monitored with this method. Test wafers are fabricated by spinning resist onto transparent substrates. Prior to implanting, the resist is scanned by a dosimeter, and its background optical absorption stored in computer memory. After implantation, the substrate is scanned again, and the background subtracted to yield contour maps. These reveal the spatial distribution of ions and the regions over which the dose is within design tolerance. The scanning of the substrates can be done with a 3 mm beam, permitting greater resolution than with the four-point probe technique. This allows sensitive diagnosis of such implanter problems as scan lock-up, non-linear scanning, and loss of beam diameter control during low dose or low energy implantations. In addition, although the precision of optical dosimetry methods are somewhat lower those based on the four-point probe (~1% vs <0.25%), this is adequate for some applications (e.g. qualitiative checks), and no additional post-implantation processing steps (e.g. annealing) are required.

Capacitance-voltage (C-V) techniques are most useful for monitoring the low-dose implantations used for such applications as threshold voltage adjustment in MOS devices. New advances in both the sheet resistance and optical dosimetry methods, however, have eroded some of these advantages of the C-V techniques.

Implantation Dose Uniformity and Diagnosis of Implanter Performance

The contour mapping technique is very useful for routinely monitoring implantation dose uniformity. The causes of non-uniformity are related to limitations of implantation equipment design and to machine malfunction, and contour map data can frequently be used to track down their origin. That is, such malfunctions as scan non-uniformities, scan lock-up, beam neutral-ization problems, charging, and channeling effects, can be identified. Two such examples are:

1) In Fig. 33a, a contour map of a <100> wafer that was mounted in an electrostatically-scanned implanter, on a fixture that slightly flexes the wafer (to insure good thermal contact to the heat sink pad (Fig. 33b). The wafer in this example, however was not tilted 7° off axis. Thus, the flexing and beam scan angles together cause the edges of the wafer to appear to be tilted at a small angle off perpendicular to the beam, while the center remains perpendicular to the beam

axis. The sheet resistivity contour map shows a "bulls-eye" pattern, with 15% higher resistivity near the edges. This results because in the center of the wafer there is a deeper implant due to channeling. Thus, a lower average carrier concentration over the depth profile exists at the center than at the edges, where the non-perpendicular wafer surface reduces channeling. Since the mobility of charge carriers increases with decreasing doping concentration, the implant layer at the wafer center exhibits lower sheet resistance, as revealed by the contour patterns. Figure 33c shows a contour map of a wafer that is tilted 7°, to minimize axial channeling, but with no attempt to avoid planar channeling by deliberate twisting. The contour map reveals horizontal regions of uniform sheet resistivity, which can be attributed to planar channeling.

2) Figure 34 shows a contour map of a wafer in which neutrals in the incident beam are significant. Since neutrals are not deflected by the electrostatic scan plates, they end up striking the center of the wafer, causing a higher dose (and consequent lower resistivity) at that location.

Measurement of Implantation Depth Profiles

The concentration of the implanted species as a function of depth (atomic depth profiles), and the depth profiles of the impurities that have been electrically activated must both be determined. These two profiles are each found with a different set of techniques. The *atomic profiles* are measured with secondary ion mass spectroscopy (SIMS), Rutherford backscattering spectroscopy (RBS), and neutron activation analysis (Chap. 17). The *profiles of the electrically activated dopants* are characterized with spreading resistance and capacitance voltage measurements (Chap. 8). The *junction depth* of implanted layers can also be measured by lapping and staining techniques (Chap. 8).

Measurement of Implantation Damage and Annealing Efficacy

Both the as-implanted damage and the damage that remains after various annealing cycles is studied with the same set of characterization methods. Damage resulting in defects that intersect the surface (e.g. oxidation-induced stacking faults and dislocations that intersect the surface) can be detected using *chemical etches* and inspection with optical microscopy or SEM (Chap. 2 and 15). SEM analysis may also be conducted in the *electron-beam-induced-conductivity*, or EBIC

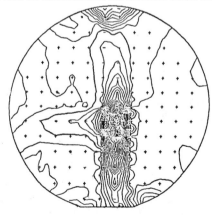

Fig. 34 Sheet resistance contour map of an ion implanted wafer in which neutrals of the incident beam are not deflected by the implanter's electrostatic scan plates, and hence implant the center of the wafer with a higher dose than the periphery[36]. Reprinted by permission of Academic Press.

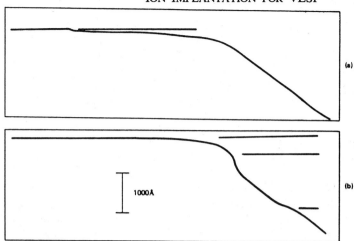

Fig. 35 Profilometer traces of angle-lapped bevel for 0.5° fixture, as-implanted - 2.5x10^{15} atoms /cm^2 at 190 keV, Wayflow® fixture. (a) as-lapped. (b) after 3 h etch in 48% HF, amorphous layer is delineated since it etches more rapidly than the single crystal regions. Surface crystal film is 450Å thick[27]. Reprinted with permission of the American Physical Society.

mode, to identify the location of dislocations (since enhanced electron-hole recombination occurs near their cores and surrounding strain fields, Chap. 17).

Damage that produces defects and amorphous regions in the shallow layers below the surface can be detected by several techniques, including: a) *transmission electron microscopy*, TEM, since the resolution capabilities of TEM allow defect features and amorphous regions [whose dimensions are on the order of nm to μm] to be directly imaged. Cross-sectional TEM provides data on the type and depth of the damage below the surface, while planar TEM gives information on the areal location of the damage (Chap. 17); b) *x-ray topography*, which also images distortion of lattice planes. For example, dislocations and slip lines present in the entire volume of the wafer are revealed; c) Rutherford backscattering spectroscopy, which can yield information on the location of some impurity atoms in the Si lattice, as well as the degree of crystalline perfection present after an annealing step (Fig. 18); and d) *tapered groove profilometry* (Chap. 8) and *angle-lapping edge profilometry* (ALEP) which are rapid techniques for locating the depth of amorphous layers and buried amorphous layers, respectively. In the ALEP technique, wafers are beveled, and then etched with a chemical etchant that attacks the amorphous Si more quickly than the damaged single-crystal regions, thereby delineating the buried amorphous layer (Fig. 35).

SPECIAL ION IMPLANTATION PROCESS CONSIDERATIONS

Selecting Masking Layer Materials and Thicknesses

To restrict the ionic species from being implanted into unwanted substrate regions, an appropriate mask layer needs to be present on the wafer surface. Many materials are used for such masking purposes in IC fabrication including photoresist, SiO_2, Si_3N_4, polysilicon, metal films, and polyimide. The desirable features of an implantation mask material include: a) the material should have good ion stopping power in the smallest thickness; b) the material should be compatible with microlithographic techniques; c) the mask should not contaminate the wafer surface; and d) the material should be easily removable, even after a heavy implantation dose.

The minimum thickness of masking material required to stop a given percentage of incident ions can be estimated from the range parameters for ions. Using a Gaussian distribution to approximate the implantation profile, a mask layer of thickness $R_p + 3\Delta R_p$ is required to achieve a masking effectiveness of 99.99%. The minimum thickness for stopping this percentage of incident ions is illustrated as a function of ionic type and energy in Fig. 36a for SiO_2 and Si_3N_4, and in Fig. 36b for photoresist.

Several useful comments can also be mentioned about these materials as masking layers: a) When SiO_2 and Si_3N_4 are used as masking layers, they may not always be removed after the implantation step. The ions implanted into such layers can, however, diffuse during subsequent thermal steps. Thus, an adequate masking layer thickness, to protect the underlying substrate against introduction of unwanted impurities from the combined effects of implantation and subsequent diffusion, must be selected; b) SiO_2 and Si_3N_4 mask layers can be significantly damaged by implanting, with the result that they etch faster than unimplanted layers. If not annealed out, this damage may impact the thickness of these layers after subsequent etching; c) After heavy doses the resist masking layers may become difficult to remove. Heating of the resist during heavy implantations is the main cause of this problem, and the heating may also cause unwanted resist flow. Use of adequate cooling of the wafers during implantation can reduce the severity of the effect. As mentioned in an earlier section, use of polyimide as an alternative to resist for extreme cases of resist removal difficulty, has been suggested.

The profile of the mask edges may also be important in many applications. In some cases, it is necessary that the mask layer should be sharply defined at the edge (as shown in Fig. 3). An example of an application where this is important is in the formation of self-aligned source /drain regions in MOS devices. That is, by using a polysilicon (or polycide) gate structure with vertical walls as the masking material, the implanted regions of the source and drain provide a well-defined gate length. In other cases, it may be more useful to have the masking layer edges sloped. The ions then penetrate the thin regions of the sloping edges, and cause implantation under the mask as shown in Fig. 37. If the mask is SiO_2, the pn junction will be thus be passivated by the oxide, and the radius of curvature of the junction will be larger than for a steep-edged mask. The larger radius of curvature produces junctions with higher breakdown voltages, which is a desirable device property in some circuit applications.

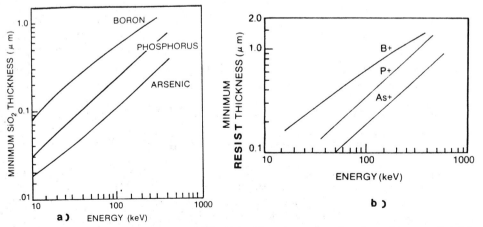

Fig. 36 Minimum thickness to stop 0.9999 of incident ions as a function of energy for (a) silicon dioxide,   and   (b) photoresist[61]. Reprinted by permission of Academic Press.

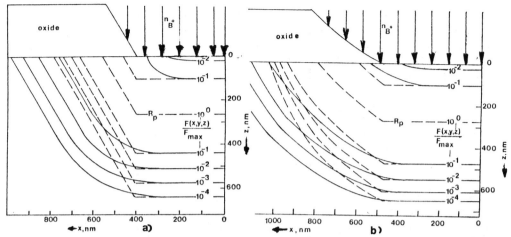

Fig. 37 (a) Equidensities calculated from implanting 70-keV B-ions into the structure shown. The mask is infinitely thick and has a taper of 60°. The dashed lines give the equidensities without lateral spread, and the solid lines include lateral spread. (b) Equidensities from implanting 70 keV B-ions into the structure shown. The mask edge has a curvature of 1.1 μm as might be obtained by chemical etching. The dashed lines give the equidensities without lateral spread and the solid lines include the lateral spread[54].

In an earlier section, it was shown how the projected lateral straggle, $\Delta R_\perp$, is estimated using LSS theory to produce information about the lateral distribution of ions under a mask edge. Figure 3 illustrated such calculated distributions for a "thick" (i.e. completely opaque) mask with vertical-walled edges. If the mask is not uniformly opaque (i.e. its thickness t_0 at an edge varies gradually between 0 and t_0), the ion distribution under the mask is typically calculated numerically by using a piece-wise quasi-planar approximation (e.g. for cases shown in Fig. 37).

Implanting Through Surface Layers

In some applications, the implanted ions first pass through a thin layer on the wafer surface before reaching the Si substrate. The layer may: a) be deliberately added prior to the deposition in order to achieve some improvement in the implant process; or b) the layer may be present as an integral part of the final device structure (e.g. the gate oxide present during a threshold voltage adjust implant, as is described in the next section). Layers are usually deliberately added for two reasons: 1) implantation through a thin layer of SiO_2 (e.g. 200-300Å thick) provides a protective screen against contamination by metals or other impurities during the implant. The risk of contamination of the Si material is always present when bare Si is exposed during processing, and the problem is especially acute during high-current implant processes; and 2) thin dielectric surface layers have been shown to reduce the fraction of channeled ions for both axial and planar channeling conditions, since the incident ions are scattered in random directions (and the number of channeled particles is thereby is reduced). Regardless of the reason that the implant is done through a thin surface layer, it has the following implications on the implant process:

a) Higher energy implanters are needed, especially for As implants, as the beam must first penetrate the screening layer (during which time it loses energy), before arriving at the substrate;

b) A significant number of surface layer atoms (e.g. oxygen, if the layer is an oxide), are also *recoil implanted* (or *knocked-on*) into the Si (Fig. 38). The effect increases with implant

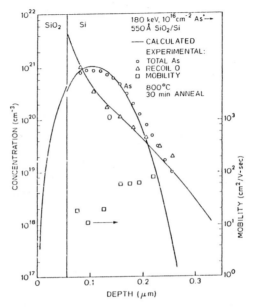

Fig. 38 Comparison of a Boltzmann transport equation calculation and experimental SIMS results for the recoil oxygen distribution resulting from an implantation of As into 550Å of SiO_2 from a dose of 10^{16} cm^{-2}. Reprinted with permission of the American Physical Society[18].

ion mass, and is thus much greater for As than for B. In As implants the number of knock-on oxygen atoms is comparable to the As dose. The effect is also greatest when the oxide thickness approaches the R_p of the implanted ions. The knock-on oxygen atoms have a significant effect on the residual defects that remain after annealing. Several research efforts are underway to unravel the interactions of implanted dopants, knock-on oxygen (and the resulting defect structures), and their effects on device characteristics. At this time it is postulated that the oxygen atoms pin dislocation loops and prevent them from annealing by climb to the surface, and that the oxygen atoms also move into the damaged crystalline regions below the amorphous layer (normally still the device region) and form unwanted oxygen precipitates;

c) The range distribution in such multilayered media must be determined. Earlier it was discussed how Monte Carlo methods can be used to accurately determine such range distributions. A simpler interpolation rule has been proposed by Schiott which yields agreement with measured data to within ~10%, if the range and standard deviation in the mask layer is known[55]. Such data has been tabulated for such mask materials as SiO_2, Si_3N_4, photoresist, poly-Si, and refractory metal silicides. Schiotts rule can be expressed as:

$$\frac{1}{R_p} = \sum_i \frac{C_i}{R_{pi}} \qquad (18)$$

and

$$\frac{R^2}{\sigma_p^2} = \left[\sum_i \gamma_i \frac{C_i}{R_i} \frac{R_i^2}{\sigma_{pi}^2} \right] \left[\sum_i \gamma_i \frac{C_i}{R_i} \right]^{-1} \qquad (19)$$

where

$$\gamma_i = 4 M_1 M_i / (M_1 + M_i)^2 \qquad (20)$$

and C_i represents the fraction of the total mass of the ith component, and R_i and σ_i, the range and projected range of the ith component in mass units per cm^2.

Predeposition and Threshold Control

Ion implantation is widely used for predeposition, since the amount of the total dopant introduced into the Si can be precisely controlled. As mentioned earlier, special implanters are sometimes dedicated to predeposition processes. A high dose implant at low energies is first deposited. Next, the dopant impurities are driven by a high temperature diffusion step to the desired final position. The diffusion step anneals some of the damage, but also has the effect of moving the pn junction location away from the region of unannealed residual or secondary implantation damage.

One of the most important applications of ion implantation is the *control of the threshold voltage* of MOS devices. For example, a precise quantity of B atoms is implanted through the thin gate oxide of NMOS devices to adjust their threshold voltage. If the deposited dopants are implanted into the channel region, the threshold gate voltage, V_T, changes by $\Delta V_T = - \Delta Q_B / C_{ox}$, where ΔQ_B is the change of the sheet ionized dopant charge in the channel, and C_{ox} is the gate oxide capacitance per unit area. By selecting a suitable energy, the B atoms just penetrate the thin gate oxide (~200-500Å) of the device regions, but not the thicker oxide (~7500Å) of the field regions, the projected ranges of B in Si and SiO_2 are about the same, making the correct selection of the implantation energy a relatively simple task.

Shallow Junction Formation By Ion Implantation

As VLSI device structures shrink laterally in dimension, it also becomes necessary to scale the vertical dimensions of the devices as well. One important vertical dimension is the depth of pn junctions. For example, CMOS VLSI devices require the formation of shallow, low resistance source /drain regions (<0.5 μm deep). The preferred technology for forming such shallow junctions is ion implantation. Arsenic can be used to form shallow n^+ source /drain regions (0.25-0.5 μm), while formation of shallow p^+ regions entails the implantation of BF_2. Implantations in the 10^{15}-10^{16} atoms /cm² range are utilized to produce low-resistance layers.

The molecular species BF_2 is preferred over B for the formation of shallow p^+ regions. Boron is a light atom (B^{11}), and as discussed earlier, when implanted at room temperature, does not cause amorphous layer formation. Thus, the crystalline-defect damage caused by boron must be annealed at temperatures >900°C, and this leads to significant diffusion of the implanted ions. In addition, in most machines the lowest practical energy for obtaining adequate beam currents is ~30 keV, and this energy yields an R_p for B of 1000Å, which may be undesirably deep.

The use of BF_2 alleviates these problems. Upon undergoing its first collision after striking the target, the BF_2^+ ion dissociates, releasing a low energy B atom. Since the mass of BF_2, however, is ~5 times that of B, it yields an R_p for boron in the substrate that is much lower than the R_p from a beam of B^+ ions at equal energy (e.g. R_p of boron atoms from a 50 keV BF_2^+ implantation is ~300Å). In addition, at such high doses, the Si atoms displaced by the initial BF_2^+ collision create an amorphous layer, which can be annealed at temperatures of 550-700°C. Finally, since the implantation forms an amorphous layer, this reduces the channeling phenomenon that occurs when implantations are made into crystalline material.

As noted in the sections that discussed the annealing of implantation damage, earlier applications of ion implantation to form source /drain junctions involved the deposition of the impurities, followed by a thermal drive-in step that located the junction at a substrate depth beyond the residual implantation damage. The requirements for shallow junctions no longer allow the option of such a drive-in step, as the junction becomes too deep. As a result, the damage from the implantation remains at the junction location, and this leads to high junction leakage currents. Extensive research efforts are being conducted to better understand the formation of damage, and the kinetics of damage transformation during annealing, in an attempt to satisfactorily fabricate shallow junctions. Rapid thermal processing (RTP) and sophisticated materials characterization techniques (e.g. TEM, RBS, SIMS) are some of the tools being most widely utilized in the efforts to attack this problem. For example, Tung combined the use of BF_2 implants with RTP (1000°C, 30 sec), and achieved junction depths of 0.07, 0.19, and 0.21 μm for doses of 1, 3, and 5×10^{15} atoms /cm^2, respectively[59]. These junction depths are 3 to 4 times less than those obtained from conventional annealing techniques. Successful shallow junction implantations have also been reported using BCl_3[60].

Arsenic has a very shallow projected range, R_p, in Si (e.g. ~300Å at 50 keV). This energy is convenient in that it allows the use of relatively high beam currents to carry out the implantation. Since As^+ is a heavy ion (As^{75}), it produces an amorphous layer which can be annealed by solid-phase epitaxy regrowth at relatively low temperatures, and which therefore results in very little As diffusion. It has been found, however, that the free carrier concentration in heavily doped As layers saturates at $\sim 3 \times 10^{20}$ cm^{-3}, even though the maximum equilibrium solubility is $\sim 1.5 \times 10^{21}$ cm^{-3}. This effect, thought to be due to the formation of neutral As clusters at high As concentrations, is a limitation of As for this application. The lowest sheet resistance reportedly achievable in a 0.2 μm deep As implanted junction is ~26 Ω /sq.

It should be noted that as device dimensions shrink, the series resistance of the source /drain regions becomes increasingly important. To reduce this parameter to values lower than are possible though implantation, alternative processes are being developed. In Chap. 11 the formation of self-aligned source /drain silicide (or *salicide*) films for this purpose is described.

The formation of shallow junction emitters in bipolar technologies can also be effected by implanting arsenic into a polysilicon film and then diffusing it afterward into the single-crystal

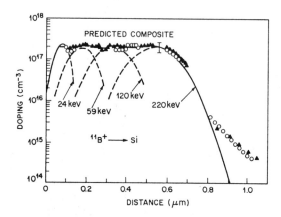

Fig. 39 Uniform B-profile (10-150 keV) calculated using minimization procedure showing nine discrete components[62].

silicon beneath. In this technique several benefits are derived, including: a) no implantation-induced damage occurs in the substrate Si; b) no dopant supersaturation occurs in the emitter (i.e. shallow [<0.3 μm] emitters are formed, with superior performance and extremely low densities of collector-emitter shorts); c) if a PtSi is formed as part of the contact metallurgy, none of the emitter-Si is consumed by the PtSi formation, as the contact is made to poly-Si, allowing improved control of the emitter thickness. Formation of As-doped poly-Si emitters by RTP has also been reported, with the junction depth being ~0.1 μm[60].

Multiple Implantations

In some applications a dopant distribution profile other than a Gaussian is desired. For example, it may be desired to pre-implant a region with inert element (or Si or Ge) ions to make it amorphous prior to implanting the dopant species. To insure that the entire layer into which dopant species are implanted will be amorphous, a sequence of implantations at different energies can be performed. Multiple implantations can also be employed to produce flat doping profiles, linear doping profiles, etc. Freeman and Booker developed a technique for predicting the profile from multiple implantations by combining multiple discrete dose implantations[62]. As shown in Fig. 39, the result of such a procedure for a uniform boron profile, produced by 9 discrete components with energies ranging from 10-150 keV, gives good agreement with measured data.

Multiple implantations may, however, interact. For example, in npn bipolar devices, double-implanted structures have been fabricated for some time. Enhanced performance has been reported as a result of the narrow base fabricated with a double-implanted process. In order to obtain adequate current gain, the emitter (formed from an As implantation) had to be diffused much deeper than its original implanted profile. This leads to an interactive diffusion effect between As and B. That is, after a double implantation (i.e. B followed by As), and a 950°C heat treatment, a "dip" in the B profile occurs[64]. This is due to the transport of B ions in the electric field caused by the steep gradient of As ions. This is but one example that indicates that such interaction effects in multiple implantations can be significant.

Implanting Doubly Ionized Species

A technique for increasing the energy capability of an ion implanter by a factor of two is to implant *doubly ionized species*. That is, the ion energy of a doubly-ionized atom, E_{dd}, for an acceleration voltage V, is $E_{dd} = 2\,qV$. When the atom dose due to a beam of doubly-ionized species is calculated, it must be remembered that two electrons are counted for each implanted atom. Normally doubly ionized species are produced less abundantly by the implanter ion source, and consequently beam currents are typically smaller, and implantation times longer.

REFERENCES

1. H. Maes, W. Vandervorst, and R. van Overstraten, "Impurity Profile of Implanted Ions in Silicon", in F. Y. Wang. Ed. *Impurity Doping Processes in Silicon*, North-Holland, Amsterdam, 1981, Chap. 7, p. 443.

2. J.F. Gibbons, "Ion Implantation in Semiconductor-Part I: Range Distribution Theory and Experiments", *Proc. IEEE*, **56**, 295 (1968).

3. J. Lindhard, M. Scharff, and H. Schiott, "Range Concepts and Heavy Ion Ranges", *Mat-Fys. Med. Dan. Vid. Selsk 33*, No. 14, 1 (1963).

4. J.F. Gibbons, W.S. Johnson, and S.W. Mylroie, in Dowden, Hutchinson, and Ross, Eds.

Projected Range in Semiconductors, Academic, New York, 1975, Vol. 2.

5.   W.K. Hofker, "Implantation of Boron in Silicon", *Philips Res. Repts. Suppl.,* No. 8 (1975).

6.   S. Furukawa, *et al,* "Theoretical Considerations on Lateral Spread of Implanted Ions", *Jap. Jnl. Appl. Phys.,* **11**, 2, p. 134 (Feb. '72).

7.   B. Smith, "Ion Implantation Range Data for Silicon and Germanium Device Technologies", *Research Studies,* Forest Grove, Oregon, 1977.

8.   J. Lindard and M. Scharff, Phys. Rev. **124**, (1961) 128.

9.   J.F. Gibbons and S. Mylroie, *Appl. Phys. Lett.* **22**, (1973) 568.

10. D.A. Antoniadis, S. Hansen and R. Dutton, *Tech. Report No. 5019-2,* Stanford Univ. (1978).

11. G. Dearnaley, *et al,* Canadian J. of Phys., **46**, (1968) 587.

12. D.A. Antoniadis, S.E. Hansen, and R.W. Dutton, "SUPREM II", *IEEE Trans. Electron Dev.,* **ED-26**, 490 (1979).

13. N.L. Turner, *et al.,* "Effects of Planar Channeling Using Modern Ion Implantation Equipment", *Solid State Technol.,* Feb. '85, p. 163.

14. T.E. Seidel, *IEEE Electron Device Lett.,* Vol. **EDL-4**, p. 353 (1983).

15. C. Carter, *et al, Appl. Phys. Lett.* 44, 459 (1984).

16. T.M. Liu, W.G. Oldham, *IEEE Electron Dev. Lett.* **EDL-4**, 59 (1983).

17. D.H. Smith and J.F. Gibbons, "Application of the Boltzmann Transport Equation to the Calculation of Range Profiles and Recoil Implantation in Multilayer Media", in F. Chernow, *et al.,* Eds., *Ion Implantation in Semiconductor 1976,* Plenum, New York, 1977.

18. L.A. Cristel and J.F. Gibbons, "An Application of the Boltzmann Transport Equation to Ion Range and Damage Distribution in Multilayer Targets", *J. Appl. Phys.,* **51**, 6176 (1980).

19. W. Fichtner, "Process Simulation-Ion Implantation" in S.M. Sze, Ed., *VLSI Technology,* McGraw-Hill, New York, 1983, Chap. 10, Section 3, p. 390.

20. J.P. Biersack and L.G. Haggmark, "A Monte Carlo Computer Program for the Transport of Energetic Ions in Amorphous Targets", *Nucl. Instrum. and Methods,* **174**, 257 (1980).

21. A. DeSalvo and R. Rosa, "A Comprehensive Computer Program for Ion Penetration in Solids", *Radiat. Eff.,* **31**, 41 (1976).

22. M.D. Giles and J.F. Gibbons, "Calculation of Channeling Effects During Ion Implantation Using Boltzmann Transport Equation", *IEEE Trans. Electr. Dev.,* **ED-32**,10,p.1918 (Oct. '85).

23. D.K. Brice, *Ion Implantation Range and Energy Deposition Distributions,* IFI /Plenum, New York, 1975.

24. E.P. Donovan (1982), *Ph.D. Thesis,* Harvard University.

25. L.M. Howe and M.H. Rainville, *Nucl. Instrum. Methods* **182** /**183**, 143 (1981).

26. F.F. Morehead and B.L. Crowder, "A Model for the Formation of Amorphous Si by Ion Implantation", F. Eisen and L. Chadderton, Eds. *1st Int'l. Conference on Ion Implantation, Thousand Oaks,* Gorden & Breach, New York, 1971.

27. S. Prussin, D. Margolese, and R. N. Tauber, "Formation of Amorphous Layers by Ion Implantation", *J. Appl. Phys.,* **57** (2), 15 Jan. '85, p. 180.

28. S. Mader, "Ion Implantation Damage in Silicon", in J.F. Ziegler, Ed. *Ion Implantation Science and Technology,* Academic Press, Orlando, Fla., 1984, p. 109.

29. G. Dearnaley, *et al, Ion-Implantation,* North-Holland, Amsterdam, 1973.

30. T.E. Seidel and MacRae, in F. Eisen and L. Chadderton, Eds., *First Intl. Conf. on Ion Implantation,* Thousand Oaks, Gordon and Breach, New York, 1971.

31. T.E. Seidel, *et al,* "Rapid Thermal Annealing (RTA) of Dopants Implanted into Pre-Amorphized Silicon", in *VLSI Science and Technology,* 1984, Electrochem. Soc., Pennington, N.J.

32. N.C. Tung, "Application of Rapid Isothermal Annealing to Shallow p-n Junctions via BF_2 Implants", *J. Electrochem. Soc.,* **132**, No. 4, p. 914 (Apr. '85).

33. S. Prussin, "Role of Sequential Annealing, Oxidation and Diffusion Upon Defect Generation in Ion-Implanted Silicon Surfaces", *J. Appl. Phys.*, **45**, 1635 (1974).

34. L. Csepregi, J.W. Mayer, and T.W. Sigmon, *Appl. Phys. Lett.* **29**, 92 (1976).

35. J. Fletcher, J. Narayan, and O.W. Holland in *Inst. Physics Conf. Series*, **60**, 295 (1981).

36. M.I. Curent and D.K. Sadana, "Materials Characterization for Ion Implantation", in N.G. Einspruch, Ed., *VLSI Electronics-Microstructure Science*, Vol. 6, Academic Press, New York, 1983, Chap. 6, p. 466.

37. *ibid.*, Ref. 27.

38. R.A. Powell and R. Chow, "Dopant Activation and Redistribution in As-Implanted Poly-Si by RTP", *J. Electrochem. Soc.*, **132**, No. 1, p. 195, Jan. '85.

39. O.C. Woodard & R. Pipe, "Automation of Implanter Equipment", *Solid StateTech*, Feb.'85, p.177.

40. P. Burggraaf, "Production Issues in Ion Implantation", *Semi. International*, May '84, p. 128.

41. P. Burggraaf, "Ion Implantation in Wafer Fabrication", *Semi. International*, Nov. 1981, p. 39.

42. "Here Comes a New Breed of Implanter", *Electronics*, Sept. 9, 1985, p. 86.

43. D. Pramanik and M.I. Current, "MeV Implantation for Silicon Device Fabrication", *Solid State Tech.*, May 1984, p. 211.

44. M.I. Current, *et al*, "MeV Implantation for CMOS Applications", *Semi. International*, June 1985, p. 106.

45. H. Glawishnig and N. Noack, "Ion Implantation System Concepts", in J.F. Ziegler, Ed. *Ion Implantation, Science and Technology*, Academic, Orlando, Fla., 1984, p. 313.

46. N. White, *et al*, "Wafer Charging and Beam Interactions in Ion Implantation", *Solid State Technol.*, Feb. 1985, p. 151.

47. M.I. Current and W.A. Keenan, "A Performance Survey of Production Ion Implanters", *Solid State Technol.*, Feb. 1985, p. 139.

48. M.I. Current and M.J. Markert, "Mapping of Ion Implanted Wafers", in J.F. Ziegler, Ed. *Ion Implantation, Science and Technology*, Academic, Orlando, Fla., 1984, p. 487.

49. H. Ryssel and K. Haberger, "Ion Implantation: Safety and Radiation Considerations", in J.F. Ziegler, Ed. *Ion Implantation, Science and Technol.*, Academic, Orlando, Fla., 1984, p. 313.

50. W.A. Keenan, *et al.*, "Advances in Sheet Resistance Measurements for Ion Implant Monitoring", *Solid State Technol.*, June 1985, p. 143.

51. J.R. Golin, *et al*, "Advanced Methods of Ion Implant Monitoring Using Optical Dosimetry", *Solid State Technol.*, June 1985, p. 155.

52. S. Prussin, "The Nature of Defect Layer Formation for Arsenic Ion Implantation", *J. Appl. Phys.*, **54** (5), May 1983, p. 2316.

53. K.A. Pickar, "Ion Implantation in Silicon-Physics, Processing, and Microelectronic Devices", *Applied Solid State Science 5*, Academic, New York, 1975.

54. H. Runge, *Phys. Stat. Sol.*, (a) **39** (1977) 595.

55. H.E. Schiott, *Proc. 1st Int. Conf. on Ion Implantation*, Eds. L. Chadderton and F.H. Eisen (Gordon and Breach, New York, 1971), p. 197.

56. M.R. Wordman, R. Dennard, G. A. Sai-Halasz, *IEDM-81*, p. 40, 1981

57. M. Doken, *et al.*, *IEDM Tech. Digest-81*, p.586, 1981.

58. J.Y.T. Chen, *IEEE Trans. Electron Devices*, **ED-31**, p. 910, 1984.

59. M. Delfino and M.E. Lunnon, "A Structural and Electrical Comparison of BCl and BF_2 Ion-Implanted Silicon", *J. Electrochem. Soc.*, **132**, No. 2, Feb. '85, p. 437.

60. H.S. Rupprecht and A.E. Michel, "Trends of Ion Implantation in Silicon Technology", in J.S. Williams and J. Poate, Eds., *Ion Implantation & Beam Tech.*, Academic, 1984, Ch. 9, p. 311.

61. K. Pickar, *Ion Implantation in Silicon, Applied Solid State Science 5*, Academic Press, 1975.

62. J.H. Freeman and D.V. Booker, *Nucl. Instr. and Meth.*, **144** (1977), 175.

63. T.E. Seidel, "Ion Implantation", in S.M.Sze, Ed., *VLSI Technology*, McGraw-Hill Book Co., 1983, Chap. 6, pp. 219-265.
64. M.Y. Tsai and B.G. Streetman, "Recrystallization of Implanted Amorphous Si Layers, I. Electrical Properties of Si Implanted with BF_2^+ or $Si^+ + B^+$", *J. Appl. Phys.*, **50**, 183, (1979).
65. J. H. Freeman, *Nuclear Instrumentation and Methods*, **22**, 306, (1963).

PROBLEMS

1. Plot the vertical implanted ion profiles about the value of R_p for a dose of 3×10^{14} P atoms /cm^2 and an implant energy of 80 keV. Assume a Gaussian distribution for the implanted ions and a thick mask layer with a vertical mask edge profile.

2. Describe the difference between *range* and *projected range*.

3. Explain why light ions penetrate more deeply into a target than do heavier ions.

4. A p-n junction is formed by implanting As through an opening in a layer of thermal SiO_2. If a dose of 5×10^{15} ions /cm^2 is used and a background concentration of 1×10^{15} cm^{-3} exists in the substrate, calculate the depth of the junction. Assume a Gaussian distribution for the impanted dose. Sketch the junction, and also sketch the junction that would be created if a diffusion were used to form the junction. In what way are these junctions different?

5. An high current ion implanter has a beam current of 30 mA. The wafer holder can accommodate 30 100 mm diameter wafers. Assume a 130 keV energy and a total implant time of 5 minutes. What is the dose received by the wafers?

6. A threshold voltage adjust implant is made by implanting through a 300Å SiO_2 gate oxide. If the field oxide is 5000Å thick, what percentage of the implanted atoms penetrates the field oxide. Assume that $\Delta R_p \sim 0.3\ R_p$.

7. Select an implant dose of 30 keV boron atoms, and a drive-in time and temperature reqired to produce 200 Ω /square diffused layer and a 5000Å junction depth.

8. If it is desired to form a shallow defect-free junction, describe how the following concepts will impact the implantation process selected to produce such junctions: channeling (axial and planar), pre-amophization, solid phase epitxy, thermal cycles, and annealing ambients.

9. Describe (a) planar channeling, (b) dynamic annealing, (c) buried amorphous layers, and (d) secondary defects.

10. Explain why the center of a wafer is likely to receive a higher dose than the periphery in medium current implanters that use a electrostatic scanning system.

10

ALUMINUM THIN FILMS and

PHYSICAL VAPOR DEPOSITION in

VLSI

This chapter is divided into four major topics involved in the formation of interconnect structures in VLSI. They are:

1. *The Properties of Aluminum (and Aluminum Alloy) Thin Films in VLSI.* This section considers the advantages and limitations of such films for VLSI applications. It should be noted, however, that many of the fabrication processes that involve aluminum are also discussed elsewhere in the chapter (e.g. deposition of aluminum films by evaporation or sputtering), or in other chapters of this volume (e.g. wet etching of aluminum in Chap. 15). A variety of issues associated with the use of thin film aluminum are addressed in other parts of this text as well, including: a) step coverage (this chapter); b) effects of gas incorporation in the deposited film (this chapter); c) electromigration (Vol. 2 of this text); d) hillocks in Al thin films and their suppresion (Vol. 2); and e) contact formation to silicon, including spiking, contact electromigration failure, and contact structures /barrier layers (this chapter and Vol. 2).

2. *Physical Vapor Deposition (PVD) by Sputtering.* The discussion on sputtering is quite extensive, as this is the primary PVD method utilized in VLSI applications. The sputtering presentation covers the following subjects: a) glow-discharge physics; b) the physics of sputtering; c) the deposition kinetics of sputtered films; d) sputter system configurations (dc, rf, and magnetron); e) commercially available sputtering equipment for microelectronic applications; and f) sputter deposition processing issues.

3. *Physical Vapor Deposition of Thin Films by Evaporation.* This section is less detailed, as evaporation is not as extensively used in VLSI applications as sputtering.

4. *Thickness Evaluation of Opaque (Especially Metallic) Thin Films.*

Since the key topic of this chapter is *physical vapor deposition* (PVD) it is useful to define the concept and introduce the common characteristics shared by PVD processes. All PVD processes proceed according to the following sequence of steps:

    a) The material to be deposited (solid or liquid source) is physically converted to a vapor phase;

    b) The vapor is transported across a region of reduced pressure (from the source to the substrate); and

    c) The vapor condenses on the substrate to form a thin film.

The conversion to a gaseous phase (Step **a**) is done by the addition of heat (in evaporation deposition), or through the physical dislodgement of surface atoms by momentum transfer (in

Table 1. PROPERTIES OF ALUMINUM AND ALUMINUM ALLOY THIN FILMS

Name	Symbol	Melting Point (°C)	Al /Si Eutectic (°C)	Density (g/cm^3)	Resistivity (μΩ–cm)
Aluminum	Al	660	577	2.70	2.7
Aluminum/ 4% Copper	Al 4%Cu	650	~577	2.95	3.0
Aluminum/ 2% Silicon	Al 2%Si	640	~577	2.69	2.9
Aluminum/ 4% Copper 2% Silicon	Al 4%Cu 2%Si		~577	2.93	3.2

sputter deposition). It is this step which will receive the most attention in the chapter, as some of the aspects of film condensation and formation are discussed in the earlier chapters dealing with Thin Films, Epitaxy, and CVD. It is also worth mentioning that many concepts used to describe PVD processes, such as vacuum pressure, mean free path, impingement rate, monolayer time, vapor pressure, molecular and viscous flow, are presented in Chap. 3.

ALUMINUM THIN FILMS IN VLSI

Aluminum is the third material in the trinity of matrix substances employed to fabricate silicon solid-state components (the other two being silicon and silicon dioxide). Aluminum is primarily utilized in these applications in thin-film form, and it functions as a material which interconnects the device structures formed in the silicon substrate. It has emerged as the most important material for such applications because of its low resistivity (ρ_{Al} = 2.7 μΩ-cm), and its compatibility with the other two matrix substances[1,49]. For example, aluminum thin films adhere well to SiO_2. (During the thermal step that sinters the metal silicon contacts, the aluminum atop SiO_2 forms a thin layer of Al_2O_3 at the Al/SiO_2 interface, and this promotes good adhesion.) The material properties of aluminum films which are of most interest for silicon device fabrication are listed in Table 1. Aluminum alloys are also often utilized in microelectronic applications, including: a) Al-1wt%Si); b) Al-2wt%Cu); c) Al-4wt%Cu); d) Al-1wt%Si-2wt%Cu); and e) Al-1.2wt%Si-0.15wt%Ti. Note that the relatively low values of the melting point of Al (660°C) and the Al-Si eutectic temperature (577°C), restrict the maximum value of subsequent processing temperatures once the Al film has been deposited.

Aluminum thin-films are deposited as polycrystalline materials, usually in the 0.5 μm-1.5 μm thickness range. Early microelectronic devices utilized evaporation to deposit pure Al, Al-Si, or Al-Cu alloys, but the stringent alloy-composition requirements of VLSI (and other limitations of evaporation) have made sputtering the dominant PVD technology for aluminum films. The development of magnetron sputtering (which allows aluminum to be deposited at deposition rates of up to ~1μm /min), further decreased the use of evaporated Al film deposition. Aluminum alloys are used more frequently than pure aluminum because they possess enhanced properties for certain interconnect requirements, including superior contact formation characteristics and better resistance to electromigration. A brief introduction to these topics follows, but the reader is directed to Vol. 2 for a more detailed discussion of them.

The use of *Al-Si* in place of *pure Al* stems from problems of *junction spiking* that occur at the interface of pure Al and Si, when the interface is heated. That is, the solubility of silicon in aluminum (Fig. 1) rises as the temperature increases (e.g. to about 0.5% at 400°C). Thus, when the silicon substrate is put into intimate contact with a pure aluminum film, the silicon substrate acts as a source of the silicon that the aluminum can take up in solution. In addition, the diffusivity of silicon along the grain-boundaries of the aluminum film at 400°C is quite high. Thus, a significant quantity of Si can move from the region below the metal-substrate interface and into the aluminum film. Simultaneously, aluminum from the film will move to fill the voids created by the departing Si (Fig. 2). If the penetration by the aluminum is deeper than the p-n junction depth below the contact, the junction will be electrically shorted (junction spiking). To suppress this mechanism, an alloy of Al which contains a concentration of Si in excess of the Si solubility at the maximum process temperature, can be used in lieu of pure aluminum. Thus, when the (Al-Si)-silicon interface is heated, the aluminum alloy film does not need to receive silicon into solution from the substrate, and junction spiking is avoided.

Although this remedy has been successfully applied in the past, it appears to be inadequate for VLSI applications. The shallower junctions, narrower metal lines, and more severe topography of VLSI are less tolerant of Si precipitation in the Al-Si film that occurs when such films are cooled down from 400°C temperature steps. In addition, electromigration failure effects at Al-Si-silicon contacts become more severe as the contacts get smaller. More elaborate contact structures employing barrier materials, and other layers, appear to be necessary replacements (e.g. Si/W/Al, or Si/PtSi/Ti:W/Al contact structures). Volume 2 discusses the topic of metal-silicon contacts and the inadequacies of Al-Si for VLSI in substantially more detail.

It has also been found that the addition of small quantities (0.1-4%) of other materials [e.g.

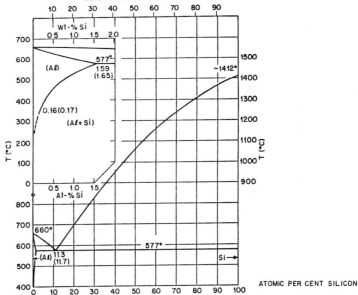

Fig. 1 Phase diagram of the aluminum-silicon system. From M. Hansen and A. Anderko, *Constitution of Binary Alloys*, 1958.   Reprinted with permission of McGraw-Hill Book Co.

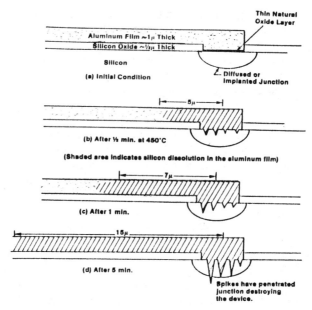

Fig. 2 Junction spiking and silicon migration during contact sintering.

Cu (0.5-4wt%), and more recently Ti (0.1-0.5wt%)] to the aluminum, improves the electromigration resistance and reduces the propensity of aluminum thin-films to form hillocks (i.e. protrusions on the Al film surface, see Fig. 36). One penalty for such improvements is that the resistivity of such films is increased (by ~10-30%) over that of pure aluminum films. Other drawbacks that must be accepted in exchange for the enhanced behavior of such alloys include added process complexity, and more stringent process controls. That is, an appropriate combination of alloy composition, film deposition, annealing, and etch processes must be selected, developed, evaluated, and maintained. Furthermore, some processes are more difficult to perform on alloy films than on pure Al films, such as dry etching of Al-Cu films (Chap. 16). Details of the issues that must be considered in developing and evaluating processes that utilize Al alloy films are discussed elsewhere, both in this volume and Vol. 2 of this text.

Some other characteristics of aluminum thin films also need to be described at this point. Aluminum readily forms a thin native oxide (Al_2O_3) on its surface upon exposure to oxygen, even at room temperature. The presence of this oxide can effect the contact resistance when another metal layer is deposited on the Al, and can inhibit both the sputtering of an aluminum target and the etching of aluminum films. Aluminum thin films can also suffer corrosion problems as a result of some fabrication processes. For example, if phosphorus-doped SiO_2 is deposited onto Al films, phosphoric acid (HPO_3) can be formed if moisture is absorbed by the glass. This acid will attack Al and cause corrosion. In addition, dry etching of Al may leave chlorine residues on the Al surfaces. Exposure to ambient moisture can lead to the formation of HCl. If Cu is present as an alloy in the Al film, severe corrosion can occur (as the $CuAl_2$ compound and the Al form microelectrodes of a battery, and HCl acts as the electrolyte). Both of these problems are discussed elsewhere in the text in more detail (see Index, under *Al, corrosion*).

SPUTTER DEPOSITION FOR VLSI

Sputtering is a term used to describe the mechanism in which atoms are dislodged from the surface of a material by collision with high energy particles. It has become the most widely utilized deposition technique for a variety of metallic films in VLSI fabrication, including aluminum, aluminum alloys, platinum, gold, titanium:tungsten and tungsten. It is also used in some applications to deposit molybdenum, Si, SiO_2 (silica glass), and refractory metal silicides, although CVD or evaporation may be more frequently used to deposit this group of materials.

Sputtering has displaced evaporation as the workhorse PVD method for VLSI because of the following advantages:

a) Sputtering can be accomplished from large-area targets, which simplifies the problem of depositing films with uniform thickness over large wafers;

b) Film thickness control is relatively easily achieved by selecting a constant set of operating conditions and then adjusting the deposition time to reach it;

c) The alloy composition of sputter-deposited films can be more tightly (and easily) controlled than that of evaporated films;

d) Many important film properties, such as step coverage and grain structure, can be controlled by varying the negative bias and heat applied to the substrates. Other film properties, including stress and adhesion, can be controlled by altering other process conditions, such as power and pressure;

e) The surface of the substrates can be sputter-cleaned in vacuum prior to initiating film deposition (and the surface is not exposed again to ambient after such cleaning);

f) There is sufficient material in most sputter targets to allow many deposition runs before target replacement is necessary, and;

g) Device damage from x-rays generated during electron-beam evaporation is eliminated (although some other radiation damage may still occur).

As is true with other processes, however, sputtering also has its drawbacks. They are:

a) Sputtering processes involve high capital equipment costs;

b) The deposition rates of some materials are quite slow (e.g. SiO_2);

c) Some materials (e.g. organic solids) are frequently unable to tolerate ionic bombardment, and degrade in a sputter environment; and,

d) Since the process is carried out in low-medium vacuum ranges (compared to the high vacuum conditions under which evaporation is conducted), there is greater possibility of incorporating impurities into the deposited film.

In general, the sputtering process consists of four steps: 1) ions are generated and directed at a target; 2) the ions sputter target atoms; 3) the ejected (sputtered) atoms are transported to the substrate, where; 4) they condense and form a thin film. Although it is of interest to note that sputtering can be conducted by generating the energetic incident ions by other means (e.g. ion beams), in virtually all VLSI sputtering processes their source is a glow-discharge. The discussion of sputtering in this section will be limited to *glow-discharge sputtering*[4,5,12].

Properties of Glow Discharges

The energetic particles used to strike target materials to be sputtered in VLSI sputter deposition systems are generated by glow-discharges[4,5]. A *glow-discharge* is a self-sustaining

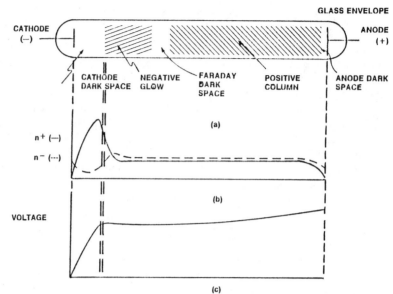

Fig. 3 (a) Structure of a glow discharge in a dc diode system. (b) Charged particle concentration in a glow discharge. (c) Voltage variation in a dc diode glow discharge. Courtesy of MRC.

type of plasma (i.e. a *plasma* is defined as a partially ionized gas containing an equal number of positive and negative charges, as well as some other number of non-ionized gas particles). In Fig. 3 we show a simple *dc diode type* system that can be employed to study the properties of glow-discharges used in sputtering. It consists of a glass tube which is evacuated and then filled with a gas at low pressures. Within the tube there are two electrodes (a positively charged *anode* and a negatively charged *cathode*) between which a dc potential difference is applied.

Creation of Glow Discharges

Let us consider this system to examine the situation when the tube is filled with argon at an initial pressure of 133 Pa (1 torr), the distance between the electrodes is 15 cm, and a 1.5 kV potential difference is applied between the cathode and anode. At the outset no current flows in the circuit, as all the argon gas molecules are neutral and there are no charged particles in the gas. The full 1.5 kV is thus dropped between the electrodes. If a free electron is introduced into the tube (most likely created from the ionization of an argon atom by a passing cosmic ray), it will be accelerated by the electric field existing between the electrodes (whose magnitude is: $E = V/d = 1.5$ kV $/15$ cm $= 100$ V /cm).

The average distance that a free electron will travel in an Ar gas at 133 Pa before colliding with an Ar atom (i.e. the mean free path), is 0.0122 cm (see Chap. 3). Most electron-atom collisions are *elastic*, in which virtually no energy is transferred between the electron and gas atom (because of the very light mass of the electron compared to a gas atom). Thus, the mean distance traveled by an electron before it makes an *inelastic* collision (in which significant energy *is* transferred to the atom, by the excitation of an atomic electron to a higher energy level in the atom, or to escape from the atom) is about ten times the mean free path, or 0.122 cm. If this is the *mean distance traveled by electrons between inelastic collisions,* there must be a significant number of electron path lengths between inelastic collisions in the range of 0.5-1.0 cm. If a free electron travels 1 cm in the 100 V /cm electric field, it will have picked up 100 eV of kinetic energy. This is sufficiently high so that during an inelastic collision, enough energy can be transferred to the orbital electrons of the Ar atoms to cause their excitation or ionization. If the transferred energy, E, is less than the ionization potential (e.g. 11.5 eV$<$ $E<$ 15.7 eV for Ar), the orbital electron will be excited to a higher energy state for about 10^{-8} sec, and then return to the ground state with the emission of visible-light photons. Such excitation is the source of light emission in glow-discharges.

If the energy transferred is greater than the ionization potential (i.e. >15.7 eV for Ar), a second free electron (and positive ion) will be created. Subsequently, both free electrons will become accelerated again, and the opportunity for cascading the number of free electrons and creating a condition known as *gas breakdown* exists. Figure 4 shows the breakdown voltage required to initiate discharge, as function of the product of the pressure, P, and the electrode spacing, *d*. When the condition of gas breakdown is reached, current flows in the external circuit as the collision-generated free electrons are collected by the anode. Each new ionization event, however, will take place closer to the positively charged anode (as the electrons are accelerated in its direction). Therefore, the current will increase to a maximum and quickly decay to zero, unless there is a mechanism available for generating additional free electrons for sustaining the current flow. When a sufficient number of electrons are available to maintain the discharge, the discharge is said to be *self-sustained*. The source of such electrons is discussed later.

Structure of Self-Sustaining Glow Discharges and Their Dark Spaces

A self-sustaining discharge in a system (as shown in Fig. 3) exhibits certain characteristics. First, at equilibrium the voltage drop between the electrodes is reduced (e.g. in our example, from 1.5 kV to about 150 V), and the discharge current builds up to the point that the voltage drop across the current limiting resistor is equal to the difference between the supply voltage and the electrode potential difference. Second, the discharge has a particular structure, as shown in Fig. 3a. The most important region of the discharge is the *Crookes dark space* between the negative glow and the cathode. In Fig. 6b, we see that the positive ions of the discharge are present in higher density in front of the negatively charged cathode, producing a localized *space charge* there. Any electrons near the cathode are rapidly accelerated *away* from it, due to their relatively light

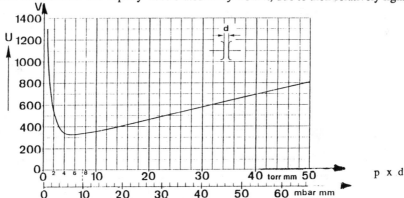

Fig. 4 Breakdown voltage, U, between two parallel plane electrodes in a homogeneous electric field as a function of gas pressure, p, and electrode distance, d, for air (Paschens curve).

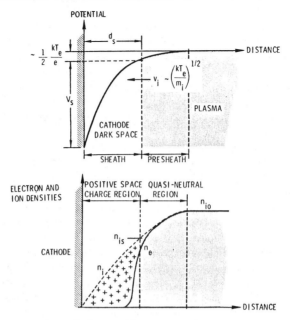

Fig. 5 Schematic representation of the positive space charge region that develops over a cathode[13]. Reprinted with permission of Academic Press.

mass. The much more massive ions are accelerated *toward* the cathode, but much less quickly. Thus, on average, they spend more time than electrons traversing the Crookes dark space, and at any instant their concentration in the dark space is greater than that of electrons.

This has the net effect of greatly increasing the electric field immediately in front of the cathode, and also of screening the remainder of the discharge from the cathode voltage. As a result, the electric field in the rest of the discharge is rather low and uniform, and the glow regions of the discharge are therefore more truly plasma-like (as defined at the outset of this section), than the dark space regions, which do not contain equal concentrations of positve and negatively charged particles. The greatest part of the voltage between the anode and cathode is thus dropped across the Crookes dark space, and therefore charged particles (ions and electrons) experience their largest acceleration in this region.

The reduced luminosity of the dark space is also due to the electric field present across the dark space. That is, since electrons in the dark space are strongly accelerated and traverse through the space quickly, the electron density at any instant in the dark space is drastically reduced (Fig. 5). Although there are fewer electrons, those present can gain high energies from acceleration by the electric field. When they collide with gas atoms in the dark space, they are thus more likely to cause *ionization*, rather than *excitation* events. As a result fewer light-generating electron-atom collisions occur than in the negative and positive glow regions.

When positive ions from the negative glow region enter the Crookes dark-space, they experience the strong local electric field and become accelerated toward the cathode. They also have a high probability of exchanging their charge with neutral atoms in the dark space. In doing so, they retain their momentum, while losing their charge. The formerly neutral recipient atom becomes an ion, and only at that time can begin accelerating towards the cathode. The effect of this *charge transfer mechanism* was studied by Davis and Vanderslice, and a summary of their results is shown in Fig. 6.   The data concludes that *virtually no ions reach the target with the full dark space energy*.   In addition, it implies that the *cathode is bombarded by energetic neutral atoms as well as energetic ions*, and thus both species apparently produce sputtering events.

The source of electrons that sustains the discharge is the cathode, which emits secondary electrons when struck by ions. (The mechanisms of such emission are discussed in more detail in a subsequent section.) Thus, upon entering the dark space from the cathode, the electrons are accelerated by the dark space field toward the anode. Upon colliding with atoms in the dark space, they cause ionization, and are also slowed. As slower electrons, they are less energetic and thus the probabiltiy of colliding with Ar atoms and causing excitation rather than ionization events is increased. The edge of the negative glow region therefore provides an indication of where such slowed electrons start to be created in large numbers (as a result of high energy, inelastic collisions). If the pressure in the tube is reduced, the distance that the emitted electrons travel before colliding and becoming slowed, becomes longer. Consequently, the dark space lengthens. If the pressure becomes too low, the dark space extends the full length between the electrodes, and the glow-discharge extinguishes. For dc diode sputtering systems, this effect limits the minimum practical pressure for sputter deposition to the range of 10-40 mtorr (1.3-5.2 Pa). The long dark spaces of dc diodes are also an indication that the electrons are not efficiently utilized to produce ions. Thus, we shall see later how magnetic fields are employed to increase the electron path length in the course of traversing the region between cathode to anode [magnetron sputtering]. This increases the probability that the electrons undergo an ionizing collision before being collected by the anode. The dark space in such magnetron sputtering systems is typically reduced from 1-2 cm to ~0.5 mm.

Obstructed Glow Discharges and Dark-Space Shielding

In dc diode configurations, there must be sufficient electrons emitted at the cathode to keep the discharge self-sustaining, and these electrons must undergo an adequate number of

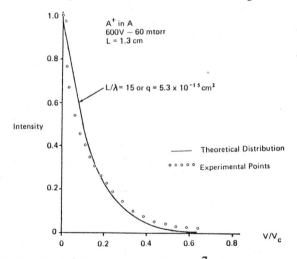

Fig. 6 Energy distribution for Ar^+ from an argon discharge[7]. Reprinted with permission of the American Physical Society.

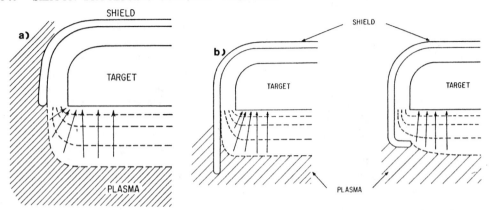

Fig. 7 (a) Potential distribution in vicinity of cathode shield (b) Reducing rim effect by extending cathode shield. (c) Reducing rim effect by wrapping shield around the cathode[15]. From L. Maissel and R. Glang, Eds., *Handbook of Thin Film Technology*, 1970. Reprinted with permission of McGraw-Hill Book Company.

ion-producing collisions with the sputter gas. These electrons are produced by ion bombardment of the cathode. Thus, there only needs to be a sufficient number of ions produced in the dark space (and in the glow regions) to produce enough secondary electrons to sustain the glow-discharge. As a result, as long as enough ions are produced in the dark space and glow regions, the exact location of the anode is not important. In fact, if the anode is gradually moved closer to the cathode, the positive column shown in Fig. 3 first disappears, then the Faraday dark space vanishes. Finally, a substantial fraction of the negative glow may be extinguished, before any significant effect on the electrical characteristics is observed. If the anode gets too close to the dark space, the ion production rate becomes reduced, and the voltage across the electrodes must rise to increase the secondary electron emission. Such a glow is known as an *obstructed glow*. In most practical sputter deposition systems the glow is obstructed. That is, in order to most effectively collect the sputtered material onto the substrate, the anode (on which the wafers are sometimes mounted) is placed as close to the cathode as possible (typically just far enough away to avoid extinguishing the negative glow).

On the other hand, it is typically necessary to insure that sputtering is allowed to occur only at the front side of the target, as the back side contains cooling coils, attachment fixtures, etc., which are definitely not to be sputtered. To guarantee that no sputtering takes place except from desired surfaces, a shield of metal (at a potential equal to that of the anode) is placed at a distance less than the Crookes dark space at all other cathode surfaces (Fig. 7). Since no discharge will take place between two electrode surfaces separated by less than this distance, such shielding (termed *dark-space shielding*) is effective in preventing sputtering from unwanted cathode surfaces.

Physics of Sputtering

When a solid surface is bombarded by atoms, ions, or molecules, many phenomena can occur. The kinetic energy of the impinging particles largely dictates which events are most likely to take place. For low energies (<10 eV), most interactions occur only at the surface of the target material. At very low energies (<5 eV) such events are likely to be limited to

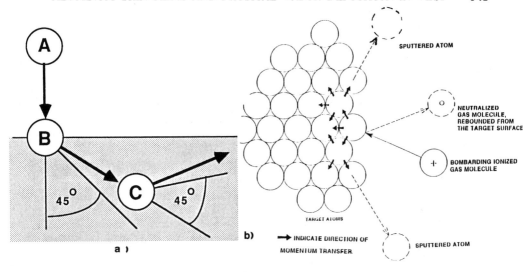

Fig. 8 (a) Binary collision between atom A and B, followed by a binary collision between atom B and C. (b) Collision process responsible for sputtering and fast neutral generation.

reflection or physisorption of the bombarding species. For low energies which exceed the binding energy of the target material (5-10 eV), surface migration and surface damage effects can take place. At much higher energies (>10 keV), the impinging particles are most likely to be embedded in the target, and this mechanism is the basis of ion-implantation. At energies between the two extremes, two other effects also arise: 1) some fraction of the energy of the impinging ions is transferred to the solid in the form of heat, and lattice damage; and 2) another fraction of such energy causes atoms from the surface to be dislodged and ejected into the gas phase (*Sputtering*).

Billiard Ball Model of Sputtering

The exact mechanisms which lead to the ejection of atoms under ion bombardment are not known, and a comprehensive theory of sputtering is not likely to be developed in the near future since many parameters are involved, including the kinetic energy of the ions, lattice structure, and binding energy of lattice atoms. Some of the details, however, are reasonably well understood and can be aptly described with a relatively simple momentum-transfer model. G.K. Wehner, whose theoretical work first established a solid scientific basis for sputtering, often described sputtering as a game of three-dimensional billiards, played with atoms[7]. Using this analogy, it is possible to visualize how atoms may be ejected from a surface as the result of two binary collisions (Fig. 8b & c) when a surface is struck by a particle with a velocity normal to the surface (e.g. atom A in Fig. 8b). Note that a *binary collision* is one in which the primary particle strikes a single object (e.g. atom B in Fig. 8b), and gives up a significant fraction of its energy to the struck atom, while retaining the remaining fraction. As a consequence of the collision, atom B may leave the point of impact at an angle greater than 45°. If atom B then undergoes a secondary collision with atom C, the angle at which atom C leaves the secondary impact point may again be greater than 45°. Thus, it is possible that atom C can have a velocity component greater than 90° (and thus be directed outward from the surface). As a result, there is a finite probability that atom C will be ejected from the surface as a result of the surface being struck by atom A.

When the directions of sputtered atoms from the surface of polycrystalline materials (and most cathode materials in sputter applications are polycrystalline) are measured for the case of *normal incidence*, it is found that the ejected atoms leave the surface in essentially a cosine distribution. A cosine distribution, however, does not describe the sort of small-angle ejections that would be expected from the simple collision processes described above. Evidently in actual sputtering events, more than two collisions are involved, and the energy delivered by impinging ions *during normal incidence* is so randomly distributed that the effect of the incident momentum vector is lost. Note that the energy range of sputtered atoms leaving the target is typically 3-10 eV, and that the bombarding species also recoil from the cathode face with some energy. Thus, the target surface is a source of sputtered atoms as well as energetic backscattered species.

For the case when the surface is bombarded by ions at an oblique angle (i.e. >45°), there is a higher probability that the primary collision between the incident ion and the surface atom will lead to a sputtering event. Furthermore, oblique incidence confines the action closer to the surface, and thus sputtering is enhanced. In cases of oblique bombardment, the incident momentum vector becomes important, and sputtered atoms are ejected strongly in the forward direction. In addition, the *sputtering yield*, defined as the number of atoms ejected per incident ion, may be as much as an order of magnitude larger than that resulting from normal incidence by bombarding ion (Fig. 9). This effect also leads to *faceting*, which is discussed in a later section.

Sputtering Yield

The *sputtering yield* is important because it largely (but not completely) determines the rate of sputter deposition. Sputtering yield depends on a number of factors besides the direction of incidence of the ions, including: a) target material; b) mass of bombarding ions; and c) the energy of the bombarding ions. There is a minimum energy threshold for sputtering that is approximately equal to the heat of sublimation (e.g. 13.5 eV for Si). In the energy range of sputtering (10-5000 eV), the yield increases with ion energy and mass. Figure 10 shows the sputtering yields of copper as a function of energy for various noble gas ions. The sputtering yields of various materials in argon at different energies is given in Table 2[14].

Several matters related to sputtering yield should also be noted. First, although the sputtering yields of various materials are different, as a group they are much closer in value to one another than, for example, the vapor pressure of comparable materials. This makes the

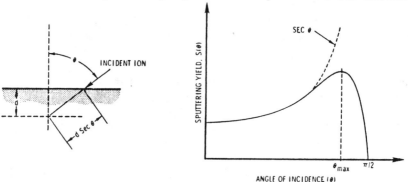

Fig. 9   Schematic diagram showing variation of sputtering yield with ion angle of incidence.

Table 2. SPUTTERING YIELDS FOR METALS IN ARGON IN ATOMS /ION

Target	At.Wt. /Dens.	100 eV	300 eV	600 eV	1000 eV	2000 eV
Al	10	0.11	0.65	1.2	1.9	2.0
Au	10.2	0.32	1.65	2.8	3.6	5.6
Cu	7.09	0.5	1.6	2.3	3.2	4.3
Ni	6.6	0.28	0.95	1.5	2.1	
Pt	9.12	0.2	0.75	1.6		
Si	12.05	0.07	0.31	0.5	0.6	0.9
Ta	10.9	0.1	0.4	0.6	0.9	
Ti	10.62	0.08	0.33	0.41	0.7	
W	14.06	0.12	0.41	0.75		

deposition of multilayer films or multi-component films much more controllable by sputtering. We discuss the details of sputtering from multi-component targets in the section on *Process Considerations in Sputter Deposition.* Second, since the bombarding ions are by no means monoenergetic in glow discharge sputtering, it is not necessarily valid to use the sputtering yield values for pure metals when alloys, compounds, or mixtures are being sputtered. The tabulations of sputter yields, however, are useful for obtaining rough indications of deposition or etch rates of various materials.

Selection Criteria for Process Conditions and Sputter Gas

The information gleaned from sputtering yields and the physics of sputtering can also be usefully applied toward an understanding of how process conditions and materials are selected for sputtering including: a) type of sputtering gas; b) pressure range of operation; and c) electrical conditions for the glow discharge[4,15]. That is, in pure physical (as opposed to reactive

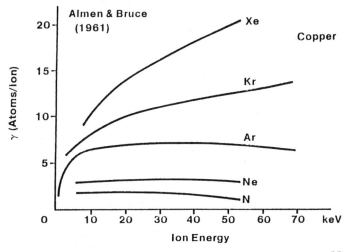

Fig. 10 Sputtering yields of the noble gases on copper, as a function of energy[55].

sputtering) it is important that the ions or atoms of the sputtering gas not react with the growing film. This limits the selection of sputtering species to the noble gases. Furthermore, argon is virtually the unanimous choice, since it is easily available, is less costly than xenon or krypton, and still gives adequate sputtering yields. The *pressure range of operation* is set by the requirements of the glow discharge (lower limit ~2-3 mtorr for magnetron sputtering) and the scattering of sputtered atoms by the sputtering gas (upper limit 100 mtorr). In addition, a desired goal of sputter deposition is to obtain maximum deposition rates. So *electrical conditions* are selected to give a maximum sputter yield per unit energy. That is, as energy is increased, each incremental energy addition gives a progressively smaller incremental sputtering yield increase. This occurs because higher energy ions implant themselves, and thus end up dissipating a greater proportion of their energy via non-sputtering processes. The most efficient ion energies for sputtering are typically obtained for electrode voltages of several hundred volts.

In general, the higher the current at the cathode, the higher is the deposition rate, since more ions are striking the cathode (and thus are causing more sputtering). The product of the cathode current and the electrode voltage gives the input power of the sputtering process. In magnetron sputtering, cathode current densities of 10-100 mA /cm^2 at a few hundred volts are typical.

From the foregoing discussion on the mechanism of sputtering, it is also reasonable to surmise that sputtering is a highly inefficient process. This is true, and in fact, ~70% of the energy consumed during the sputtering process is dissipated as heat in the target, and ~25% by emission of secondary electrons and photons by the target. The target heating can raise target temperatures to levels capable of damaging the target, associated vacuum components, or the bonding of the target and the backing electrode. Cooling of the target with water, or other suitable liquids is typically used to maintain low target temperatures.

Secondary Electron Production for Sustaining the Discharge

Near the outset of the discussion on glow-discharges, it was also observed that the glow-discharge must be continuously provided with free electrons to keep it self-sustaining. In most dc sputtering systems, the source of such electrons is secondary electron emission from the target. An important mechanism for such secondary electron emission is Auger emission, although other emission mechanisms are also responsible for generating secondary electrons. The *Auger emission* process occurs according to the following sequence of events (Fig. 11): when a positive ion comes close to the target surface it appears as a potential-well of 15.67 eV to the free electrons of the target. The target electrons require a minimum of energy, $q\Phi$, to escape from the target (where Φ is the work function of the target material, and $q\Phi$ is typically 3-5 eV). Thus, some of the target electrons can tunnel into the Ar potential-well and escape the target. The energy difference between 15.67 eV and the minimum escape energy is released in the form of a photon. If this photon (e.g. of energy 15.67 eV-5 eV = 10.67 eV), is absorbed by another target electron near the surface, this electron may then posses enough energy to be emitted from the target (i.e. the Auger electron). This sequence of events is however, rather improbable. As a result, several ion bombardment events must occur at the cathode in order for a single secondary Auger electron to be emitted. The secondary electron yield per bombarding ion, γ_i, has been measured to be ~0.1 for metal targets, but is considerably higher for dielectric targets.

Besides providing a source of electrons to sustain the discharge, electron emission is also important to the sputtering process in other ways. First, ions bombarding the cathode are neutralized. Thus, it is highly probable that each ion that closely approaches the target will extract an electron from the target, and return to the discharge as a neutral atom. Second, the

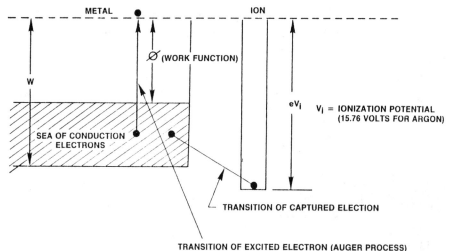

Fig. 11 Potential energy diagram for ion approaching a metal target. Courtesy of MRC.

total target current, I, is the sum of the ion flux striking the target, I_i, and the secondary electron current leaving the electrode, I_e. Since γ_i is larger for dielectric materials, their I_e is also larger. This implies that for the same cathode current, I, dielectrics will sputter more slowly than metals That is, in dielectric sputtering a larger fraction of I is due to electron emission, and thus for equal values of I, a smaller ion flux is striking the dielectric target.

Sputter Deposited Film Growth

Upon being ejected from the target surface, sputtered atoms have velocities of $3\text{-}6\times10^5$ cm /sec and energies of 10-40 eV. It is desirable that as many of these sputtered atoms as possible be deposited upon the substrates and form the specified thin film. To accomplish this goal, the target and wafers are closely spaced, and target-to-wafer spacings of 5-10 cm are typical. The mean free path, λ, of such sputtered atoms at typical sputter pressures is less than 5-10 cm (e.g. at 5 mtorr, $\lambda \cong 1$ cm). Thus, it is likely that the sputtered atoms will suffer collisions with the sputter gas atoms before reaching the substrate (Fig. 12). The sputtered atoms may therefore: a) arrive at the substrate with reduced energy (~1-2 eV); b) be backscattered to the target or the chamber walls; or c) lose enough energy so that they are thereafter transported by diffusion in the same manner as neutral sputter gas atoms. As a result of these events during the transport of sputtered atoms to the substrate, we can see how that sputtering gas pressure can alter various deposition parameters (such as deposition rate).

The formation and growth of the thin film on the substrate proceeds according to the general discussion on thin film formation given in Chap. 4. Therefore, we restrict our discussion to the events that uniquely impact the formation and growth of glow-discharge sputtered films.

The substrate surface on which we desire to deposit a film of the sputtered target material, is also subjected to impingement by many species during this process. The sputtered atoms arrive and condense onto the substrate. For a deposition rate of 200Å /min, a monolayer of deposited film will form approximately every second (assuming the size of a typical atom is ~3Å). For this case, and even for much higher deposition rates of Al (e.g. 6200Å /min) the heat of

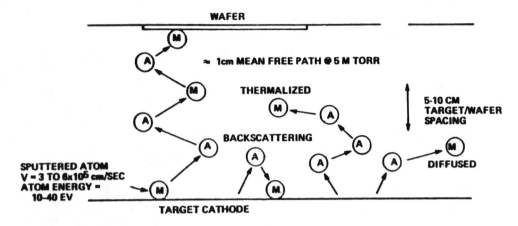

Fig. 12 Gas scattering events.

condensation is not necessarily the most important source of substrate heating.

Species that Strike the Substrate During Film Deposition

In addition to the sputtered atoms, the substrate is also struck by many other species (Fig. 13), the most important being: a) *fast neutral sputter gas atoms*, which retain significant energy after having struck and recoiled from the cathode. As they impinge upon the substrate, some may embed themselves in the growing film; b) *negative ions*, formed near the cathode surface by the reaction of secondary electrons and impurity gas atoms, such as O and N. These can acquire substantial energy from the Crookes dark space electric field, and thus also strike the substrate with appreciable energy; c) *high energy electrons*, which are accelerated across the Crookes dark space, and then impinge on the substrates. (In dc diode and rf diode sputtering

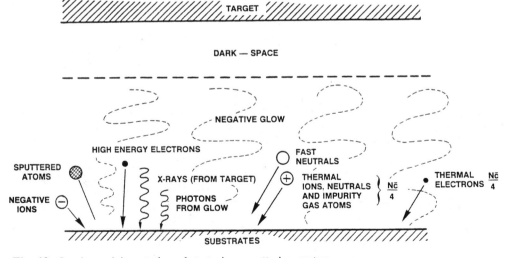

Fig. 13 Species arriving at the substrate in a sputtering system.

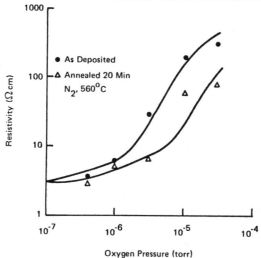

Oxygen Pressure (torr)

Fig. 14   Room temperature resistivity vs oxygen pressure during evaporation of 5000Å Al films at 200°C and 20 Å/s. Reprinted with permission from TRANSACTIONS OF THE METALLURGICAL SOCIETY, Vol. 242, pp. 205, a publication of the Metallurgical Society, Warrendale, Penn.

they can represent a major source of substrate heating); d) *low energy neutral sputter gas atoms* (Ar), which strike the substrate at a high flux (e.g. at 20 mtorr, the Ar flux is several thousand times greater than the arrival rate of sputtered material).   However, the sticking coefficient of low energy neutral Ar atoms is considered to be negligibly small, and thus the only incorporation of Ar atoms into the growing film arises from fast neutral embedment [species (a)]; and e) *contaminants present in the form of residual gases* in the chamber can be incorporated into the growing film, particularly if they are chemically active. For example, if a sputtering system contains oxygen in the partial pressure of 10^{-6} torr, the substrate surface will be struck by $\sim 10^{15}$ oxygen atoms /sec (or 1 monolayer /sec).   In Fig. 14[8] we observe that at O_2 partial pressures of 10^{-6} torr and Al deposition rates of 1,200 Å /min, the resistivity of the Al film has increased significantly due to the oxidation of some of the Al material.

The sources of residual gases can include: a) impure sputtering gas; b) outgassing from wafers /pallette (or wafer holder); c) outgassing from chamber walls; and d) leaks. To minimize the presence of residual gases in sputtering, it is advisable to use highly pure sputter gases, and to continually evacuate and refill the sputter chamber to flush any residual gases being introduced through outgassing or leaks. Additional details on the effects of residual gases on sputter deposited films is given in the section on *Sputter Processing Issues*.

Up to this point in our discussion, the glow-discharge and the sputtering phenomena we have been describing, were produced by the application of a *dc voltage* between the electrodes. Sputtering performed with such systems is known as *dc glow-discharge sputtering*. The diode configuration is the simplest glow-discharge based sputtering system, and consequently it is appropriate to consider its characteristics at the outset of the discussion on sputtering. In general, however, dc diode sputtering processes possess severe limitations that render them unsuitable for most VLSI sputtering applications. As a consequence, two other glow-discharge based sputtering modes are used instead to sputter deposit thin films in semiconductor manufacture: a) radio-frequency (rf) sputtering; and b) magnetron sputtering.

Radio-Frequency (RF) Sputtering

One of the drawbacks of dc diode systems is that they cannot be used to sputter insulators (dielectrics). As we shall explain, this is due to the fact that glow-discharges cannot be maintained with a dc voltage if the electrodes are covered with insulating layers. This represents a significant limitation, as there are several important applications which call for the sputtering of insulators, or the maintenance of a continuous discharge in the face of dielectric-covered electrodes, including: a) sputter-deposition of SiO_2 films, which can be deposited as planarizing inter-metallic dielectric layers; b) the *in situ* sputter removal of thin *native-oxide* layers from silicon and aluminum surfaces prior to the deposition of overlying films (i.e. *in situ* implies that sputter etch and deposition are performed sequentially without breaking vacuum); and c) the condition of glow-discharge must be continuously maintained in some dry-etching process chambers that are equipped with electrodes coated with insulating materials.

The claim that a glow-discharge cannot be sustained if an insulator is sputtered using dc voltages is supported by the following argument: When the negatively charged cathode in dc diode systems is bombarded by ions, an electron is stripped from the cathode surface each time an impinging positive ion is neutralized. If the cathode material is a *conductor*, such electrons can be replaced by electrical conduction, and the cathode surface maintains the negative potential required to sustain the discharge. If the material is an *insulator*, the electrons lost from the cathode surface are not replaced, since electrical conduction from the insulator interior to the sputtered surface, is not possible. Therefore, the front surface accumulates a positive charge that increases with time of bombardment. This causes the potential difference (between the cathode sputter surface and the surface of the anode) to decrease (Fig. 15). As soon as it drops below the value needed to sustain the discharge, the discharge extinguishes. In practical glow-discharge configurations, the time for the insulator surface to acquire this charge is ~1-10 µs (see Prob. 5).

A technique was developed to allow replenishment of such lost electrons to the insulator surface. This involves the application of an ac voltage (rather than a dc voltage) to the electrodes, and sputtering processes based on applying this method are known as *rf sputtering*[4,15]. As may be expected, however, this solution introduces other complications. That is, in order to achieve a practical sputtering process by the application of ac voltages, the following conditions need to be established: a) the electrons lost from the insulator surface, by the mechanism observed in dc glow-discharge sputtering, must be periodically replenished; b) the glow-discharge must be continuously sustained under ac conditions over the full period of the ac waveform; c) an electric field configuration must be created in the process chamber which allows ions of sufficient energy to bombard and sputter the target insulator; d) sputtering in the process chamber should be

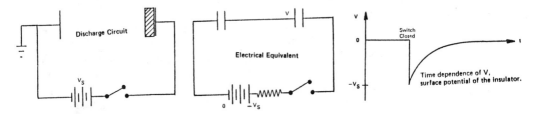

Fig. 15 Surface charging of an insulating cathode. From B. Chapman, *Glow Discharge Processes*, 1980, Copyright © John Wiley & Sons. Reprinted with permission of John Wiley & Sons.

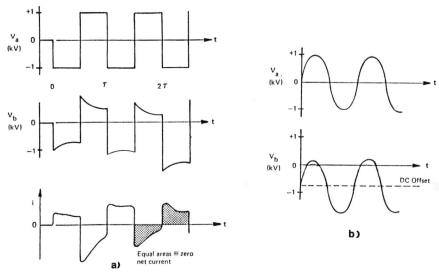

Fig. 16 (a) Voltage and target current waveforms when the circuit of Fig. 15 is square wave excited. (b) Voltage waveforms at generator (V_A) and target (V_B) in a conventionally sinusoidally excited rf discharge. From B. Chapman, *Glow Discharge Processes*, 1980, Copyright © John Wiley & Sons. Reprinted with permission of John Wiley & Sons.

suppressed at all surfaces except the target surface (i.e. sputtering should not occur at the chamber or non-target electrode surfaces); and e) the rf power must be efficiently coupled to the glow-discharge to maximize sputter deposition rates. We will discuss rf sputtering by explaining how each of these requirements is met. The system used to apply an ac signal to the process chamber is shown in Fig. 16. Note that we refer to the two electrodes as *A* and *B* , and no longer as cathode and anode, since the signal polarity applied to both electrodes alternates between positive and negative values.

As point out earlier, Assume that the rf generator in Fig. 16 produces a signal of alternating polarity with an amplitude, V_o, that is large enough to cause breakdown of the process chamber gas. When such a signal is applied to the electrodes, positive ions from the discharge are accelerated toward an electrode when it is subjected to the negative-bias portion of the waveform (and electrons during the positive-bias portion). In *steady-state conditions* (to be defined), the positive charge at the target surface (e.g. electrode *A*) accumulated during the negative part of the cycle, is replaced by impinging electrons during the positive part of the cycle. The net surface charge accumulated by the insulator during each complete cycle of the waveform is zero, and the lost electrons are thus replenished, and condition (a) is satisfied (Fig. 17).

As pointed out earlier, the time required to sufficiently charge the insulator surface during the negative part of the cycle to extinguish the glow-discharge is only 1-10 µs. Therefore, low frequency ac signals to the electrodes (e.g. 60 Hz) are ineffective at maintaining a continuous glow-discharge. That is, the discharge remains ignited for only a very small fraction of the input signal period (e.g. 1-10 µs of the 16 ms period of a 60 Hz signal). Thus, in order to satisfy condition (b) above, the charging time of insulator surfaces implies that an ac signal with a minimum frequency of 100 kHz-1 MHz must be applied to the electrodes. In practice, most ac glow-discharges are operated at 13.56 MHz (i.e. glow-discharge sputtering processes and plasma

/RIE dry etch processes). The applied power has a frequency in the radio-frequency range (hence the name *rf sputtering*), and the 13.56 MHz frequency is utilized because it is one of the frequencies designated by international communications authorities at which electromagnetic energy can be radiated without interfering with other radio-transmitted signals.

At this point it is also important to mention that rf excited discharges can be sustained without relying on the emission of secondary electrons from the target. The mechanism that allows rf excited glow-discharges to be thus sustained is not well understood, although several theories have been advanced. One theory postulates that ionization is produced by free electrons oscillating in the very weak rf field that penetrates the plasma. Such oscillations, combined with properly timed elastic collisions between the electron and gas atoms, permit the electrons to gain sufficient energy to cause ionization (despite the weak field). Two important effects are associated with this phenomenon. First, it is observed that rf excited discharges remain sustained at lower minimum pressures than dc discharges (e.g. 10 mtorr versus 40 mtorr). Second, we see that rf excitation provides a mechanism to create glow-discharges for dry etch processes without needing to be concerned about a source of electrons to sustain the discharge.

To meet condition (c) above, an electric-field in front of the target electrode must exist to accelerate the glow-discharge ions to sufficient energy to produce sputtering. Such an electric-field is produced in rf excited systems by a phenomenon known as *self-bias*. To study this phenomenon, let us examine the events that occur when the ac signal is initially applied to the insulating electrode (Fig. 16). At the time the switch is closed (t = 0), the electrode surfaces contain no charge. The applied negative bias causes positive ions from the discharge to strike electrode *A* (an insulator), and to leave behind a positive charge on the surface as electrons are removed. During the positive portion of the cycle, the electrode attracts and collects electrons. The mass of the electrons is much smaller than that of ions, and the electric field accelerates them more rapidly to the electrode surface. Thus, during the two halves of the first ac cycle, many more electrons are initially collected at the electrode than ions. At the end of the first complete period, the electrode surface therefore has a net negative charge. As more cycles transpire, the negative charge continues to build up. Such negative surface charge, however, also has the

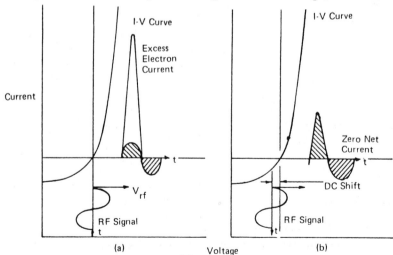

Fig. 17 Self-biasing of a dielectric surface[56]. Reprinted with permission of American Phys. Soc.

effect of repelling some electrons during the positive part of the cycle, and attracting positive ions more strongly during the negative parts. In the steady state, an equal number of positive ions and electrons strike the electrode during each complete signal period. The *average* charge build-up on the electrode which occurred during the first few cycles is negative, and remains constant in value under steady-state conditions. This also results in a negative dc offset voltage between the electrode and the discharge. Such a *self-bias voltage* is almost equal to the peak of the applied voltage (Fig. 16b) of the rf input signal. If the potential difference between the discharge and the self-biased electrode is sufficiently large, ions will be accelerated strongly enough toward the cathode to cause sputtering at the electrode [thus satisfying condition (c)].

Unfortunately, this effect can occur at both electrodes, as well as at any surface in the chamber connected electrically to the rf signal. To restrict the sputtering only to the target surface, the electrode configuration in the system must be altered. First, the non-sputtering electrode is grounded, and this may also be a convenient location for placing the substrate wafers onto which the sputtered film is to be deposited. Next we note that there are voltages V_A and V_B (called *sheath voltages*) occurring between the discharge and electrodes A and B respectively. To restrict sputtering only to the target (e.g. electrode A), it is necessary for sheath voltage V_A to be large, and V_B to be as small as possible. We assume that the electrodes A and B have areas A_A and A_B, respectively.

To see how to establish such sheath voltages, we make use of an expression that describes the relationship between ion flux through the sheath, and the voltage across the sheath (Fig. 18). If a space-charge limited current is assumed, the ion current flux, J, is given by Child-Langmuir equation[10]:

$$J = \frac{K \, V^{3/2}}{\sqrt{m} \; D^2} \tag{1}$$

where V is the voltage drop across the sheath, D is the dark-space thickness, m is ion mass, and K is a constant. Since the current density of positive ions must be equal at both electrodes, Eq. 1 indicates that:

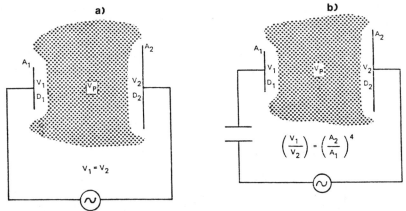

$$\left(\frac{V_1}{V_2}\right) = \left(\frac{A_2}{A_1}\right)^4$$

Fig. 18 (a) Voltage distribution with equal area electrodes and no blocking capacitor. (b) Voltage ditribution with unequal area electrodes and blocking capacitor[10]. Copyright 1970 by International Business Machines Corporation; reprinted with permission.

$$\frac{V_1^{3/2}}{D_1^2} = \frac{V_2^{3/2}}{D_2^2} \tag{2}$$

Since there is a large voltage drop across the dark space, this implies that the dark space is a region of very limited conductivity. Thus, we can model the electrode-dark space-plasma as a capacitor, with capacitance proportional to electrode area and inversely proportional to D:

$$C \propto A/D \tag{3}$$

Now, an rf voltage will divide between two capacitances in series according to:

$$V_A / V_B = C_B / C_A \tag{4}$$

Using Eqs. 3 and 4, we can write:

$$V_A / V_B = (A_B /D_B)(D_A / A_A) \tag{5}$$

and substituting into Eq. 2, we get:

$$V_A /V_B = (A_B / A_A)^4 \tag{6}$$

Thus, the larger dark-space voltage will develop at the electrode having the smaller area. The fourth-power dependence is not observed in real systems (the value of the exponent is more typically determined experimentally to be <2), but the principle of increasing the sheath voltage ratio by controlling the relative electrode areas is valid. In actual sputtering systems, the larger electrode consists of the entire sputtering chamber, including the wafer holder, which has a common electrical ground with the rf power supply. The relatively large area causes a small dark space voltage to exist, and no sputtering takes place from these surfaces. The target material is attached to the smaller electrode, which therefore develops a large dark-space voltage, and can thus experience strong sputtering. In this manner, condition (d) is satisfied.

Since the dark-space voltages V_A and V_B will not be equal if the areas of the electrodes are not equal (and the plasma is a region of equipotential), a dc voltage between the electrodes will exist. If the target is a conductor, a blocking capacitor (Fig. 19) is needed to prevent this self-bias voltage from being grounded through the rf generator. If the target was an ideal

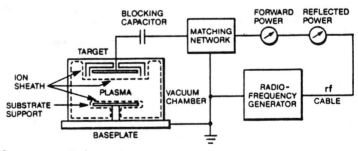

Fig. 19 Radio frequency sputtering system.

insulator, it would not be necessary to use an external *blocking capacitor*. It is common practice however, to design systems with external blocking capacitors to remove any possibility of leakage through or around the edges of the target in the presence of the plasma.

The final requirement (e), that must be satisfied in order to realize an useful rf sputter process, is to be able to couple the maximum power from the rf generator to the discharge. The output impedance Z_0 of rf power supplies is designed to be purely resistive (with the value $Z_0 = 50 \, \Omega$), while the discharge has a larger and partially capacitive impedance. Therefore, an *impedance matching network* is required between the generator and discharge chamber (Fig. 19). Most systems use feedback control to tune the network to the generator by automatically maintaining minimum reflected power. This also to protects the generator and coupling lines.

Rf diode sputtering systems behave in other respects much like dc diode systems. That is, secondary electrons are still emitted by the target, and subsequently accelerated through the dark space. Most still end up being collected by the grounded electrode, and thus such electrons usually represent a source of substantial substrate heating. In addition, the target is also the source of fast neutral sputter gas atoms, negative ions, and electromagnetic radiation. These species are generated in the same manner as in dc diode systems. Furthermore, rf diode configurations suffer the same inefficiency of secondary electron utilization. That is, since most electrons emitted from the target are collected by the anode, and do not cause ionization, ion production (and consequent ion bombardment of the target) is not significantly greater than in dc diode systems. Thus, for processes that require conductive materials to be deposited at high rates (e.g. aluminum or aluminum alloys), rf diode sputtering is not an optimum process.

Magnetron Sputtering

In both dc and rf diode sputtering, most secondary electrons emitted from the target do not cause ionization events with Ar atoms. They end up being collected by the anode, substrates, etc., where they cause unwanted heating. Since most electrons pass through the discharge region without creating ions, the ionic bombardment and sputtering rate of the target (and the deposition rate on the substrates) is much lower than if more of these electrons were involved in ionizing collisions. Thus, dc diode and rf diode sputtering processes normally exhibit inadequately low deposition rates.

A technique know as *magnetron sputtering* increases the percentage of electrons that cause ionizing collisions, by utilizing magnetic fields to help confine the electrons near the target surface. By using magnetron sputtering, current densities at the target can be increased to 10-100 mA /cm^2, from about 1 mA /cm^2 for non-magnetron configurations.

To describe the principle of magnetron configurations, first consider the motion of a charged particle of charge q, mass m, and initial velocity v_0, that is perpendicular to a uniform magnetic field, B, as shown in Fig. 20. The charged particle experiences a magnetic force, $F = q \, v \times B$, which is perpendicular to both its velocity and the direction of **B**. If no other forces are exerted on such a particle, it will continuously be acted upon by the force, F, and this will induce a circular motion, with radius, r_g:

$$r_g = (m \, v) / q \, B \qquad (7)$$

where v is the velocity of the particle.

For electrons and Ar$^+$ ions, the equation for r_g can be expressed in terms of the energy, E, of the particles (in eV), and of the magnetic field, **B** (in gauss), or:

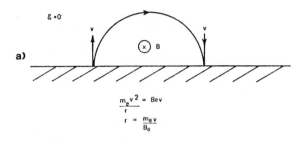

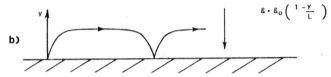

Fig. 20 (a) Motion of an electron in a region of magnetic field B parallel to the surface. (b) Motion of an electron ejected from a surface with velocity v into a region magnetic field B parallel to the surface, with no electric field and , (c) with a linearly decreasing field.   From B. Chapman, *Glow Discharge Processes*, 1980, Copyright © John Wiley & Sons.   Reprinted with permission of John Wiley & Sons.

$$r_g \text{ (electron)} = 3.37 \text{ x } [E^{1/2} \text{ (eV) } /B \text{ (gauss) }] \tag{8}$$

$$r_g \text{ (Ar}^+) = 9.11 \text{ x } 10^2 [E^{1/2} \text{(eV) } /B \text{ (gauss) }] \tag{9}$$

From these equations, we can see that Ar$^+$ ions have motion radii which are ~300 times larger than electrons for equivalent conditions of B and particle energy. Thus, as an ion moves thorough the region containing the magnetic field, its r_g is so large that it essentially moves in a straight line when crossing the dark space. This illustrates that the direction of electron motion in magnetron discharges is strongly influenced by the magnetic field, while the magnetic field does not significantly change the direction of ions as they cross the dark space.

In the next case, examine the situation of an electron which is emitted from a sputtering target surface. Let there be a magnetic field, **B**, parallel to the surface, and an electric-field perpendicular to the surface (due to the dark space). Assume the electron is initially at rest (e.g. having just been emitted from the target), and that the magnitude of the electric-field is given by:

$$E = E_o (1 - y /L) \tag{10}$$

where E_o is the magnitude at the surface of the target, L is the length of the dark space, and y is the vertical distance above the target. Now the electron will be initially accelerated away from the target in a direction perpendicular to the target surface by the electric field, but simultaneously it will experience a force due to the magnetic field, $F = qv \times B$.

The electron velocity will therefore be altered from a direction perpendicular to the surface. If the magnetic field is strong enough to deflect the electron velocity so that it begins to return to the target surface before it leaves the magnetic field, its motion will be roughly cycloidal as shown in Fig. 20c. That is, as it approaches the target surface after having been deflected, the electron will be *decelerated* by the electric field, and eventually will come to rest (momentarily) at the target surface. At that point, the next period of the cycloid motion will be initiated, as it is once again repelled by the electric-field. The maximum excursion of the electron (with charge e and mass m) from the target during such motion, y_{max}, is given by:

$$y_{max} = \frac{1}{B} \left[\frac{2m}{e} (V - V_T) \right]^{1/2} \tag{11}$$

where V_T is the negative target voltage, and V is the potential in the dark space at y_{max}.

The net result is that secondary electrons are trapped near the target surface by the combination of the magnetic and electric fields, and continue to move with cycloid (or *hopping*) motion until they collide with an Ar atom. Even though not every such collision produces an ionization event, many more ions are generated than if the secondary electron escaped the discharge without undergoing any collisions with Ar. The magnitude of the electron velocity *parallel* to the target surface, v_D (also known as the *E* x **B**, or *magnetron drift velocity*), is given by:

$$v_D = \frac{10^8 E_\perp \ (V/cm)}{|B| \ (gauss)} \tag{12}$$

There are two principal magnetron electrode configuration designs used in VLSI sputtering systems: a) the *planar magnetron*; and b) the *circular magnetron*. Figure 21a shows a variety of

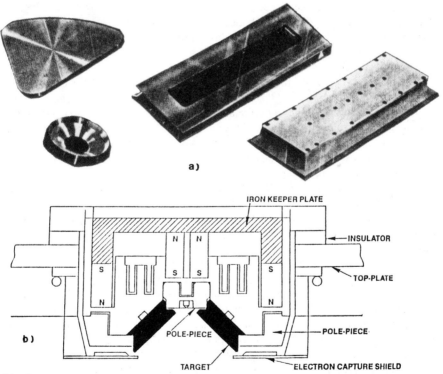

a)

b)

IRON KEEPER PLATE

INSULATOR

TOP-PLATE

POLE-PIECE

POLE-PIECE

POLE-PIECE

TARGET

ELECTRON CAPTURE SHIELD

Fig. 21 (a) Sputtering targets. Reprinted with permission of Semiconductor International. (b) Focest® sputtering targets for MRC sputtering systems. Courtesy of MRC.

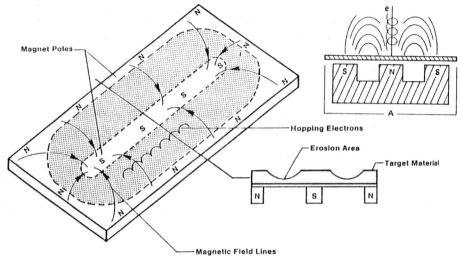

Fig. 22   Schematic drawing of planar magnetron target and magnets.

target shapes used in these systems. Note that the planar magnetron targets can be rectangular or wedge shaped. Some of such rectangular targets (e.g. MRC Focest® targets, Fig. 21b) are still thought of as planar targets, although they no longer look very planar. The shape of the planar targets are dictated by system design, although they all basically share the same principle of operation. Therefore, we limit our discussion to the rectangular planar magnetron for purposes of describing the characteristics of all planar magnetrons.

In *planar magnetrons*, the target surface is planar, and the B-field is created by magnets behind the target[4,5,12,39]. The anode is usually a rod placed along side the cathode. (Apparently the location of the anode relative to the cathode is not critical to effective operation.) During deposition the substrates pass in front of the target. A planar-magnetron target is shown in Fig. 22a, and its cross-section, cut along A-A, to show the magnets and magnetic-field orientation, in Fig. 22b. In examining these illustrations, three things should be noted: a) half-way between the magnet poles, the magnetic field lies parallel to the target face. This is the region where the magnetron drift velocity, v_D occurs; b) the E x B field that formed in this region closes on itself. Thus, a continuous path that the "hopping" electrons can traverse is established; and c) the plasma density is greatest where E x B is maximized. This region is called the *race-track* and sputtering is very rapid in the race-track portion of the target. Since the race-track path is continuous, and rectangular planar magnetron targets are normally long and narrow, they form a *two-line sputter deposition source* for wafers that pass in front of the target during deposition.

Figure 22a also shows the hopping motion of the electrons as it follows the race track. Near the center of the track the maximum excursion, y_{max}, is ~0.5 cm. The dark-space is also significantly reduced (to about 0.5 mm). Since magnetrons are much more effective in causing secondary electrons to undergo ionization collisions, much higher target currents occur for comparable voltage levels. Thus, magnetron devices have impedances of ~50 Ω, as opposed to 40,000 Ω for dc diode devices. Planar magnetron devices can be operated with either dc or rf power supplies.

The particular configuration of *circular magnetrons* used in VLSI sputters systems was invented by Clarke[13], and are known as the Sputter Gun® (registered trademark of Sloan Associates), and the S-Gun® (registered trademark of Varian Associates). Figure 23a shows an example of the target shape used in such magnetrons, and Fig. 23b shows a cross-section of an S-Gun. As in other magnetrons, an intense plasma region is formed near the cathode, due to the E x B field orientation (Fig. 23b). The anode is concentrically located, and the magnetic field is generated by annular permanent magnets as shown in Fig. 23a. In modern high-rate circular magnetron systems, the substrates are placed close to the cathode (Fig. 23d), and only a single wafer is deposited at a time. This maximizes deposition rates and efficiently collects a large percentage of the sputtered species. Figure 23d also qualitatively illustrates the flux of the sputtered species that are produced in such a circular magnetron. Note that in early systems which used circular magnetrons, the substrates were mounted on a planetary or other movable support system, as discussed in the *Evaporation* section. This also provided an inexpensive vehicle for implementing a sputtering capability, because existing evaporation systems could be retrofitted with an S-Gun.

The hopping motion along the target surface that constrains electrons to the region near the cathode surface of the circular magnetron is shown in Fig. 23b. The target erodes as material is sputtered away as shown in Fig. 23c. Efforts to maintain satisfactory film thickness uniformity across large diameter wafers together with a higher percentage of material utilization from the target, have lead to such modifications in circular target designs as: a) multiple concentric cathodes; and b) non-planar target cathode surfaces.

Bias Sputtering

The surfaces of the substrates during glow-discharge sputtering always acquire a negative potential, albeit usually small, with respect to the plasma. Thus, they are subject to some

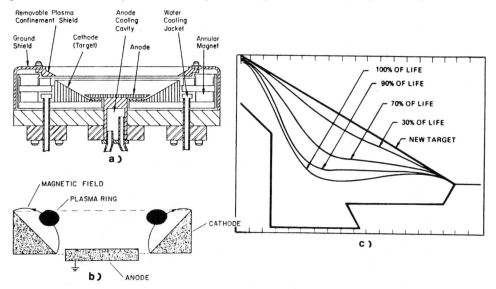

Fig. 23 Conical magnetron target and magnet configuration. Courtesy of Varian Associates.

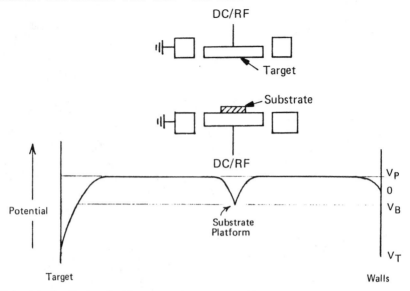

Fig. 24 Potential distribution in bias sputtering systems. From B. Chapman, *Glow Discharge Processes*, 1980 Copyright © John Wiley & Sons. Reprinted by permission of John Wiley & Sons

bombardment by positive ions. This bombardment may cause sputtering of the substrate surface. The value of the potential differenece between the plasma and substrate can be left uncontrolled. If, however, the substrates are deliberately given a *negative bias* with respect to the plasma (usually smaller than the negative bias applied to the target), this *bias sputtering* effect can be harnessed to modify some of the properties of the films deposited on the substrates[4,15]. Most commonly, an rf bias on the substrate is used, with dc or rf power applied to the target.

In Fig. 24a, it is shown how the substrates in such a configuration are electrically isolated from ground, and biased. In Fig. 24b the potential distribution of the configuration is illustrated. Since this arrangement allows the bias voltage on the substrate to be adjusted, this also permits the flux and energy of electrically-charged particles striking the substrates to be varied. In fact, use of bias sputtering enables a large number of film properties to be controlled including gas incorporation, step coverage, stress, reflectivity, grain size, resistivity, surface roughness, hardness, and alloy composition.

As an example, Winters and Kay[16] measured the incorporation of Ar in a growing Ni film as a function of substrate bias (Fig. 25). For low bias voltages less Ar was found in the films, because incoming Ar ions sputtered the previously sorbed Ar, but did not embed themselves in the film. An increase in the bias voltage, however, also raises the kinetic energy of the impinging ions. Thus, they eventually become energetic enough (at ~100 eV) to occasionally be implanted into the growing film. For energies greater than this value, the implantation effect becomes incrasingly dominant, and the percentage of incorporated Ar starts to rise again.

The bias voltage can also impact the film properties in several other ways including: a) the incoming ions add energy to the surface ions and thus increase surface migration or chemical reaction rates at the surface; b) if the ions possess sufficient energy (~100 eV), resputtering of the deposited film atoms can occur; c) the incoming ions can damage the surface of the growing film; and d) the growing film can be heated by the ion bombardment. All of these mechanisms

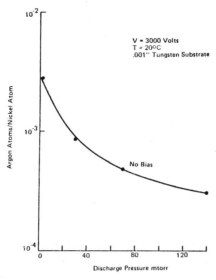

Fig. 25 Argon content in sputtered nickel film as a function of argon discharge pressure[16]. Reprinted with permission of the American Physical Society.

can alter the properties of the growing film, often in unexpected ways. References 4,17,18,31,35 describe how bias sputtering has been used to modify a variety of specific film properties.

Sputter Deposition Equipment

A large and stringent set of material properties must be exhibited by VLSI interconnect films. In order to successfully deposit such films by sputtering, complex and sophisticated sputtering systems must be utilized. These are offered commercially in a variety of configuration designs. Each has certain advantages and limitations, and it is up to the user to determine the best equipment type for specific applications. In this section we discuss the various equipment types, after first describing the generic components of a VLSI fabrication sputtering system.

Before discussing the details of sputtering equipment, it is useful to list the most important requirements of VLSI interconnect films, and the equipment used to deposit them. The necessary film properties, and how they are achieved by various sputtering processes, are the subjects of a later section. The requirements placed on sputter deposition equipment will be explored here.

In general, there are some desired qualities shared by all processing equipment including: a) safe operation; b) adequate process throughput; c) compatibility with the required processes to be performed; and d) compatibility with the VLSI manufacturing environment. Let us see what attributes need to be possessed by sputtering equipment to meet goals (b), (c), and (d), as any machine not capable of safe operation would automatically be disqualified.

The *throughput* of a sputtering system depends on such factors as: maximum deposition rates; single-wafer vs. batch operation; pumpdown times; pre-processing cycle times; reliability (i.e. percentage uptime of equipment); and scheduled maintenance requirements (e.g. time between regeneration of cryopumps, or time between target replacement). In general, magnetron sputtering with power supplies capable of supplying high power and optimally designed targets are utilized to maximize deposition rates.

The *process compatibility* characteristics of a sputter system involve such questions as: a) should the system be flexible or not? (i.e. will it be called upon to deposit a wide variety of materials, multilayers, etc., or will it only be dedicated to a narrowly defined set of sputter depositions?); b) what size wafers can the system handle?; and c) can the system perform all the aspects of the desired processes, such as sputter etching, wafer heating, dc or rf magnetron sputtering, rf diode sputtering, or bias sputtering?

The *compatibility with the VLSI manufacturing environment* addresses the following issues: a) compatibility with clean-rooms (i.e. the systems must be clean, should occupy as small a space as possible in the clean-room, and ought to be serviceable from outside the clean-room); b) capabilty of interfacing with factory automation-computer systems; and c) compatibility with robotic loading and unloading.

The schematic of a sputtering system is shown in Fig. 26. It consists of the following components: a) sputter chamber, in which reside the electrodes and sputtering targets and possibly shutters; b) pre-processing chamber, which may contain wafer heaters, rf sputter etch electrodes, and also function as a load-lock; c) vacuum pumps; d) power supplies (dc and /or rf); e) sputtering targets; f) sputtering gas supply and flow controllers; g) monitoring equipment (pressure gauges, voltmeters, and residual-gas analyzers); h) wafer holders and handling mechanisms; and i) microcomputer controller. We will present information only on items (a), (b), (c), (d), and (e), as these components operate somewhat differently in sputtering systems than in other equipment. The remaining subsystems, on the other hand, play the same role in sputtering systems as they do in many other semiconductor equipment applications.

The *sputter chamber* (and its associated subcomponents), not only make up the heart of a sputtering system, but each design is also the most unique part of the commercial systems. Therefore we will delay the discussion on sputter chambers until the various system configurations are compared.

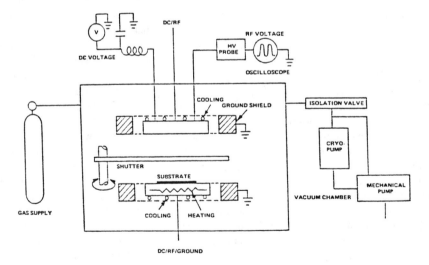

Fig. 26 Schematic drawing showing some of the components of a sputtering system.

The *pre-processing chamber* (or *station*), is normally a chamber into which the wafers are first transferred from the ambient. Several procedures may be conducted in the pre-process chamber (PPC), depending on the system design, including: a) pumpdown of the PPC; b) heating of the wafers; and c) cleaning of the wafer surface by sputter etching or other means (e.g. plasma etching). The purpose of the *pumpdown* is to allow the sputter chamber to be loadlocked (i.e. wafers can be introduced into the sputter chamber without requiring that it be vented to atmosphere). After the deposition is completed, the wafers can be transferred through the same PPC, or another loadlock chamber, and brought out of the system. The sputter chamber itself is normally kept under high vacuum when sputtering is not taking place. *Wafer heating* may be used to accomplish several purposes, including desorption of moisture, etc., from the wafer surfaces, and /or preheating to improve step coverage during deposition. In some systems, however, such step coverage heating may be done in the sputter chamber during deposition.

The *vacuum pumps* are also an important part of the sputtering system. For cleanliness sake, most sputtering equipment manufacturers have opted to use cryopumps to keep the sputtering chamber under high vacuum between times of sputtering[19]. This pump is normally also used during the sputter process to continually pump the chamber (at pressures ~1 Pa). In order to use high vacuum cryopumps to maintain such medium vacuum pressures, the pump must be throttled, and this reduces its pumping speed. Since the purpose of pumping during a sputter process is to remove evolving gas contaminants (especially moisture) on a continuous basis, the reduction of pumping speed is undesirable. Thus, *Meissner traps*[20] (described in Chap. 3) are sometimes placed in sputter chambers to maintain high water vapor pumping speed. Some systems are also offered with *titanium sublimation pumps* for gettering the large volumes of hydrogen evolved in the chamber during sputtering (i.e. from residual moisture on wafers and holders that dissociates into H and O, and from the H_2 absorbed in stainless steel chamber fixtures). This allows the cryopump to be operated for longer periods between regeneration, as saturation with H_2 is usually the condition that mandates pump regeneration[20].

The vacuum in the loadlock PPC may be created by a rough pump or turbomolecular pump. Some reports indicate that the loadlock should be pumped to <1 mtorr in order to prevent excess N_2 from being introduced into the sputter chamber from the PPC[20]. This would require a high vacuum pump to evacuate the loadlock.

The *power supplies* in a sputtering system may be dc or rf. Dc power supplies can be built to supply up to 20 kW of power, whereas rf power supplies are limited to ~3 kW. As a result, dc magnetron sputtering can provide higher deposition rates than rf magnetron sputtering, and consequently is the operational mode of choice for high deposition rate processes. Dc power supplies for sputtering, however, must be designed to be capable of withstanding arcs without damage[12]. Such arcs are thought to arise from the local dielectric breakdown of native oxides on freshly installed targets, or from other surface regions on which contaminant insulator layers have formed. Conversely, rf sputtering removes such regions as a matter of course, and thus rf supplies are not subjected to arcing. The dc supplies, however, must be designed to handle such local arcing without shutting down or damage, as the regions responsible for such arcing are normally cleaned within minutes after the power is supplied to the system, and arcing then ceases. On the other hand, the supply must be able to sense large arcs caused by such events as metal flaking between the target and ground, and to be able to switch off soon enough to prevent damage to the target. Saturable reactor power supplies are used for dc magnetron applications, as they are tolerable of arcs, and yet can maintain a constant current output for a wide range of loads. When bias sputtering is carried out, a separate supply may be provided, or power from a single

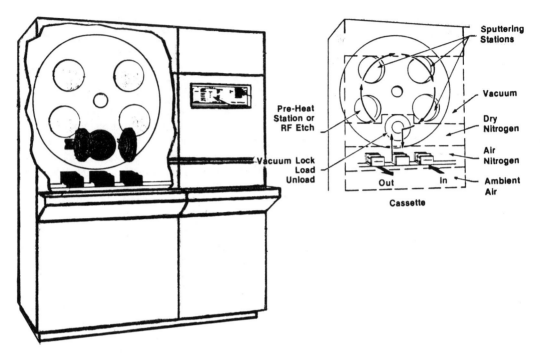

Fig. 27 Static sputtering system, Varian 3180. Courtesy of Varian Associates.

supply may be split and fed to both target and substrate electrodes at the required voltage and phase difference.

The *target* consists of the material that is to be sputter-deposited onto the substrate. Target shapes depend on the material to be deposited and the sputtering system design. Figure 21 shows some of the target shapes in common use[21]. Sizes range from round targets only a few inches in diameter, to rectangular targets several feet long. Targets are mounted to the cathode in such a way that they are in intimate contact with a water-cooled backing plate. Since it is estimated that over 70% of the energy incident on the target goes into heating, targets must be adequately cooled to prevent warpage, deposition rate changes due to thermal restructuring, or even melting. Most metal targets are fabricated from high-purity, vacuum cast materials (e.g. 99.999 - five nines, 99.9995 - five nines five, or 99.9999 - six nines pure, for aluminum targets, as four nines purity is generally unacceptable for semiconductor metallization). Some applications also demand material that is very low in alpha-particle emitting elements (e.g. low thorium and uranium content). Other desirable characteristics of targets include: high percent utilization of the target material; the ability to perform a large number of runs between target replacements; and accurate alloy composition. *Composite targets* consist of two or more metals in unalloyed form with a selected ratio of surface areas mechanically arranged, so that sputtering takes place simultaneously to yield an alloy in the deposited film.

Presputtering of targets is done to clean their surfaces prior to film deposition. Shutters may be used to protect wafers present in the chamber from being deposited with the target surface contaminants, during such target cleaning. Once the target surface is clean, sputter deposition can be initiated. As discussed in the power supply presentation, arcing may occur at the target surface in dc sputtering until the surface is cleaned. Thus, it is common practice to *condition* freshly installed targets by slowly increasing the applied power, until all the material that causes the arcing has been sputtered away.

Commercial Sputtering System Configurations

The systems commercially offered for VLSI sputtering applications[29] differ primarily in the design of the sputter chamber, electrode arrangement, and wafer-target relationship. Note that virtually all systems have load-locked sputter chambers, and offer wafer heating, rf sputter etching, and bias-sputter deposition capabilities. The following are examples which represent most of the variations offered in sputter system design. They are:

a) The *static* sputter configuration (Fig. 27), in which the electrodes are in an S-gun® (or Sputter gun®) configuration, and the wafers (normally smaller in diameter than the S-gun cathode) are placed within 5-10 cm of the target[13]. Only one wafer at a time is sputter deposited from each target, and the wafer and the target are held in fixed positions during the deposition, hence the name, *static deposition*. Note that systems may have more than one sputter station, so that several wafers may be sputtered simultaneously, and multi-layered films can be deposited by moving a wafer between stations. Normally, the wafer is held vertically, and this positioning is known as *side-sputtering*. In such systems, each wafer can be individually heated during deposition. Advanced S-gun, circular magnetron configurations have been designed to allow large diameter wafers (150-200 mm) to be deposited with films possessing ±5% thickness uniformities.

b) The *in-line sputter* configuration (Fig. 28)[5,11], in which the wafers travel in a straight

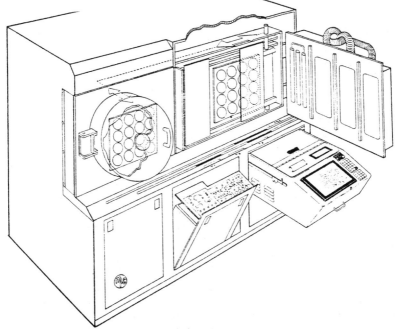

Fig. 28   MRC model 603 in-line side sputtering system. Courtesy of MRC.

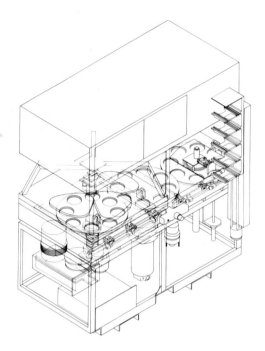

Fig. 29   Planar rotation sputtering system.   Courtesy of Gryphon Products.

path and in front of a rectangular shaped planar magnetron target (Fig. 21b & 22). Normally the deposition occurs as the result of a single pass, but multiple passes under the target can also be performed, if desired. In such systems, the deposition rate is determined, and then the transit speed in front of the target is adjusted to yield the desired film thickness. Such systems are batch systems, and the number of wafers per batch is dependent on the wafer diameter. In-line systems can contain more than one sputtering station. This allows several types of materials to be deposited without having to change targets, or to deposit multilayered structures in one vacuum pumpdown. Wafers are again held vertically, and thus side-sputtering is used.

   c) In *rotating platter* [20,23] sputter systems (Fig. 29 &30a) the wafers make multiple passes in front of the planar magnetron target, as the platter on which they are mounted rotates. Such systems are batch systems, and the size of the batch depends on the wafer diameter, although most systems have a maximum wafer size that can be accommodated. Multiple targets can be used, so that multiple layer films can be easily deposited in a number of various structures (e.g. layered films of Al-Si and Ti, which are reported to prevent hillock formation). The target, typically wedge-shaped as shown in Fig. 21a, can be above or below the wafers. In most rotating platter systems (Fig. 30), the wafers pass below the wafer target (sputter-down mode). For processes in which spalling and flaking result in large particle generation (i.e. Ti:W or silicide process film depositions), this configuration would not be feasible, as the particles would fall onto the wafers. For aluminum depositions, or multi-layer depositions in which underlying

Ti:W layers are immediately covered with Al, the films are highly adherent. Any micrometer size particles that might exist in the sputter environment, are directed by electric fields, and not by gravitational fields. Hence, it is claimed that the spatial orientation of the sputtering process is not relevant, and that sputter-down systems produce no more particles on wafers than do systems which sputter up or sideways. The combination of rotary motion, target design, and heating has been reported to produce enhanced step coverage in Al films with some of these systems.

d) In *rotary drum* sputter systems, wafers are mounted on the surface of a rotating cylinder, and make multiple passes in front of planar magnetron targets as the cylinder or drum rotates. The principle is similar to the in-line systems, except that the motion is rotary rather than linear. If multiple passes before the target are advantageous to a process (compared to a single pass, or vice versa), the rotary versus in-line configurations offer a choice. Side-sputtering is used in most drum type systems.

Process Considerations in Sputter Deposition

The goal of a deposition process for producing an interconnect layer is to create a film having all of the properties required for the VLSI technology that is being fabricated. In this section, we list the desirable properties of an interconnect film and discuss factors that impact the quality of such characteristics. Some of the relevant information was presented earlier in the sputtering discussion, but other practical considerations still need to be addressed.

A deposited interconnect film must satisfy a large number of requirements in order to effectively perform its role in an integrated circuit. (Note that the film properties necessary to meet these requirements must be exhibited by both sputtered and evaporated films.) A "wish list" of some of the most important of these characteristics that depend on the deposition process includes: a) correct nominal thickness; b) thickness uniformity of at least ±5% (across the wafer, and wafer-to-wafer); c) low resistivity; d) uniform resistivity of at least ±5%; e) good adhesion to underlying and overlying layers; f) good step coverage (e.g. usually >50% is desired); g) high electromigration resistance; h) good resistance to hillock formation; i) controllable reflectivity; j) film hardness is controllable to a desired specification, for wire-bonding compatibility; k) the layer can make low-resistance contacts to Si and /or other interconnect layers; l) the films can be

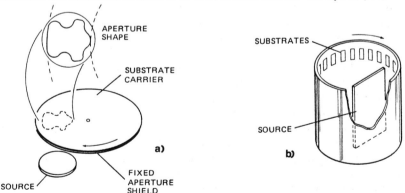

Fig. 30 Planar magnetron deposition with substrate motion: (a) planar rotation with aperture shield. (b) drum rotation[39]. Reprinted with permission of Academic Press.

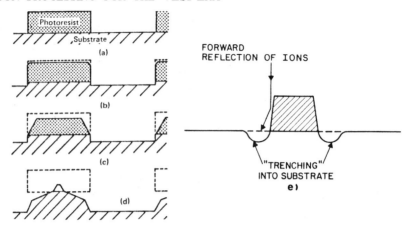

Fig. 31 Faceting results from the dependence of sputter etch rate on the angle of incidence of ions striking the surface. (a) prior to etching, (b) initiation of the facet, (c) facet intersects the substrate surface, (d) substrate is exposed and forms its own facet. (e) Trench formation arises from an excess flux of ions resulting from reflection off the sidewall.   Copyright, 1983, Bell Telephone Laboratories, Incorporated, reprinted by permission.

deposited as an alloy with tightly controlled composition; m) the film may be required to possess a multilayer structure; n) grain sizes in the film can be controlled; o) deposited films are low in alpha-particle emitting elements (uranium and thorium); p) stresses in the deposited film are low; and r) there is a reasonable latitude in the process conditions that effect the desired film properties.

In the following discussion we will focus on the aspects of the sputtering process that impact many of the above properties, especially, alloy composition, step coverage, resistivity, hillocks, electromigration, hardness, stress, and reflectivity. Additional information on stress, adhesion, and resistivity of thin films is found in Chap. 4, and a detailed discussion of electromigration and hillocks is undertaken in Vol. 2 of this text.

Faceting and Trenching

Faceting and trenching are two effects arising from sputtering that can impact the results of sputter etch processes, as well as the the reactive-ion etching processes discussed in Chap. 16.

The effect of *faceting* is illustrated in Fig. 31a-d, in which the sidewall of a feature is seen to develop an increasingly larger facet as the sputtering continues. (The term *facet* originated as a description of the *small faces* that are cut from the surface of a jewel.) The faceting effect arises from the fact that sputtering yield is greater from surfaces which are inclined at a non-90° angle to the incoming ions (Fig. 10). The facet usually starts on corners, which always have some rounding, but is also more pronounced in bias-sputtering processes due to the increased electric field at sharp corners. The facet is inclined in the direction of the incident angle corresponding to the angle of maximum sputtering yield. The faceting effect can *directly* alter substrate step profiles if an unprotected substrate is sputter etched. On the other hand, as shown in Fig. 31a-d, if a protective layer (e.g. photoresist) is used during a sputter etch or RIE process, the facet developed in the masking layer can still be *indirectly* transferred into the etched film, if the etching proceeds long enough that the facet intersects the surface.

The phenomenon of *trenching* is illustrated in Fig. 31e. It stems from the enhanced ion flux at the base of a step due to ion reflection off the side of the step, which causes etching to occur more rapidly there.

Wafer Heating During Sputter Deposition

Heating of the wafers can occur during sputter deposition. This heating may be a natural consequence of the process, or may be deliberately produced by auxiliary heating procedures. In some cases heating produces beneficial effects, especially if it can be controlled, while in other instances the heating can degrade the deposited film properties.

Unintentional heating is primarily caused by the following: a) high-energy electrons striking the substrates; b) the heat-of-condensation of the depositing film (this component is more important in high deposition rate processes); and c) the kinetic energy of the arriving film atoms. In magnetron sputtering, fewer high energy electrons escape from the target region to strike the substrates, and electron capture shields are utilized to suppress this effect even further. Thus, in modern sputtering systems, this heating effect has been significantly reduced. The heat of condensation and kinetic energy of the incoming film atoms will cause only small temperature increases in low-deposition-rate processes, while in high deposition-rate processes (e.g. 1 μm /min Al depositions), temperature increases of ~50°C[24] to 100°C[25] have been reported.

Heating by auxiliary means to improve film properties (e.g. step coverage), is conducted by lamps, backside heating with hot Ar gas, and other techniques. It should be noted that although it is difficult to measure the exact temperature on the wafer surface during deposition, in practice it is possible to reproduce the same heating from run-to-run.

Deposition of Alloy Films

In many sputter deposition applications, it is desired that alloys, rather than pure films be deposited (e.g. Al-alloys, or Ti:W). They may be deposited by co-sputtering from multiple targets, but it is more common that single, multicomponent targets are used. In steady-state sputtering conditions, the elements are sputtered from the target in the same ratio as are present in the alloy. Let us discuss the mechanism associated with such single target depositions that causes the composition of the sputtered film to track that of the target alloy composition[4,15,21].

When the surface of a metal alloy is initially exposed to the impinging ions in a sputter process, it will have its nominal composition (e.g. 2at%Cu: 98at%Al). Since Cu has a slightly higher sputter yield than Al (e.g. ~2.7 vs. ~2, for 1000 keV Ar ions), the ratio of Cu to Al atoms sputtered would be higher than the ratio of Cu to Al in the target, if 1000 keV ions are used. But this situation would not persist for very long. As the surface gets relatively depleted of Cu, the sputtering rate of Al atoms increases (and that of Cu decreases), until in the steady-state, the material is leaving the surface in the ratio of 2Cu : 98Al. It is useful to use a shutter during the period in which the surface concentration is stabilizing, to prevent deposition of material with non-desired alloy composition onto the substrate.

The composition of the deposited film, however, may not exactly mirror the composition of the target because of other factors. For example, if the alloy constituent atoms differ greatly in mass, the lighter atoms will be more strongly scattered by Ar atoms as they move from the cathode to the substrate. In addition, if bias-sputtering is utilized, resputtering of the deposited film may alter the film composition (often in unexpected ways). Thus, the film composition must be measured for each process condition. Again, once a set of process conditions is fixed, it is possible to obtain reproducible results from run-to-run.

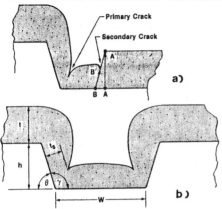
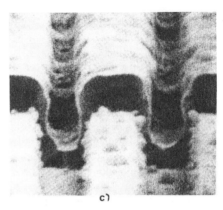

Fig. 32 Computer simulated film growth on a step: (a) Point source planetary system; (b) dc magnetron sputtering system[47]. Reprinted with permission of Solid State Technology, published by Technical Publishing, company of Dun & Bradstreet. (c) Example of poor metal step coverage.

Sputter Deposition Process Control of Step Coverage

In most applications it is desired that thin films maintain a uniform thickness and freedom from cracks or voids. As thin films cross steps that occur on the surface of the underlying substrate, they may suffer unwanted deviations from the ideal, such as thinning or cracking[47]. A measure of how well a film maintains its nominal thickness is expressed by the ratio of the minimum thickness of a film as it crosses a step, t_s, to the nominal thickness of the film on flat regions, t_n (Fig. 32). This film property is referred to as the *step coverage* of the film, and is expressed as the percentage of the nominal thickness that occurs at the step:

$$\text{Step Coverage (\%)} = (t_s / t_n) \times 100\% \qquad (13)$$

Step coverage of 100% is ideal, but each process is normally specified by a lesser minimum value that is acceptable for a given application. It is important, however, that the circuit designers understand the actual step coverage value, and derate the maximum allowable current in an interconnect line to reflect the step coverage limitation. The height of the step and the *aspect-ratio* of the feature being covered, also determine the expected step coverage. That is, the greater the height of the step, or the aspect ratio (i.e. the height-to-spacing ratio of two adjacent steps), the more difficult it is to cover the step without thinning of the film, and hence the worse the expected step coverage. Besides the step height and aspect ratio, the step coverage depends on two other factors which are not controllable by the sputtering process: 1) the *shape*; and 2) the *slope,* of the step. In general, the smoother the contour and the smaller the slope of the step, the better the expected coverage. Assuming that the four above factors have been fixed as a result of previous process steps, there are then several conditions of the sputtering process that can be adjusted to improve step coverage, including: a) sputter etching the underlying substrate prior to film deposition, to reduce the slope of surface steps; b) optimization of the target design; c) substrate heating; and d) substrate-bias sputter deposition.

One purpose of *sputter etching* of the substrate surface is to reduce the severe slopes of the steps. This phenomenon makes use of faceting effect described earlier. It should be noted, however, that much more material from the surface layer must be removed to achieve significant slope-reduction of steps, than is removed during the more common sputter-etch steps that remove thin [~50Å] native oxides. The utilization of this effect to enhance step-coverage in a three-level metallization process is discussed in Ref. 27.

By optimizing the *shape of the target* and by *controlling the target regions where maximum sputtering occurs*, step coverage can also be enhanced. The details of such target and related system design practices are generally unique to each sputtering system, and some designs are apparently more effective at improving step coverage than others. The general idea is that the extended source configurations of the sputtering systems can offer the potential of such improvements. As progress is continuing to be made in this area, it will be up to users to monitor the claims of enhanced performance by equipment suppliers, and decide for themselves which improvements can best benefit their processes.

Heating the substrate has been shown to improve step coverage[20]. This effect occurs because the migration of the surface atoms of deposited films increases with substrate temperature (Fig. 33a & b). As a result, the atoms tend to move toward regions where the least amount of material has been deposited, thereby averaging out the film thickness (i.e. the concentration gradient of material favors diffusion toward less populated locations). For aluminum films, the substrates must be heated to >250°C before significant coverage improvements are observed. It has been postulated that intermittent deposition, or thin ~200Å thick films (e.g. from multiple passes before the sputter target) is especially effective. That is, during the times that the heated film is not being coated, surface atoms are given the opportunity to migrate. If they are sufficiently mobile, they will equalize their distributions across the surfaces. The optimum temperature and deposition parameters will depend on the particular step topography, and must be determined for each process. An understanding of surface migration phenomena is necessary for developing adequate models of this mechanism. It should also be noted excessive heating during deposition can cause hillock formation in Al films. Thus, the step-coverage heat cycle often represents a compromise temperature. This effect also emphasizes the importance of being able to control the temperature.

Bias-sputtering (introduced earlier) is another technique that has been utilized to increase step coverage of deposited films (Fig. 34)[23,26]. That is, an rf bias is applied to the wafers, causing them to be bombarded by energetic sputter gas ions (e.g. Ar+), as the film is being deposited.

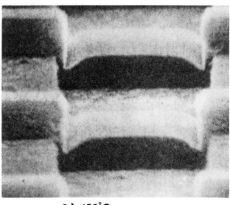
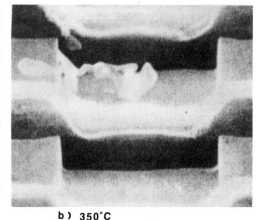

a) 150°C

b) 350°C

Fig. 33 Effect of heat on step coverage. Courtesy of Varian Associates.

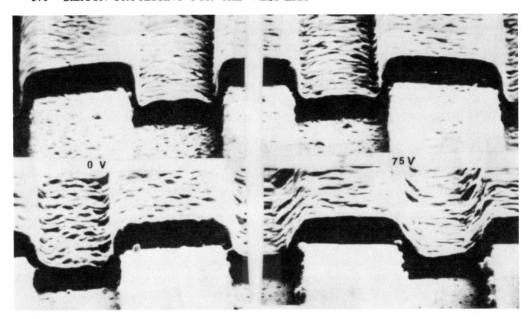

Fig. 34 Effect of different bias voltages on sputter deposited films.   Courtesy of MRC.

The applied rf causes the wafers to acquire a negative self-bias, and thereby accelerates ions from the discharge toward their surfaces.   The impinging ions transfer energy to surface atoms, and cause them to be transported to the sidewalls of steps, where they accumulate and locally increase the film thickness.   Two mechanisms have been suggested as being responsible   for such atomic transport: a) the transferred energy from the bombarding ions increases *surface atom migration*, thereby improving step coverage (much as in substrate heating); and b) the transferred ion energy   causes *re-sputtering from the surface of the depositing film*[35].   Since the   directions of the atoms emitted in such re-sputtering is thought to occur according to an

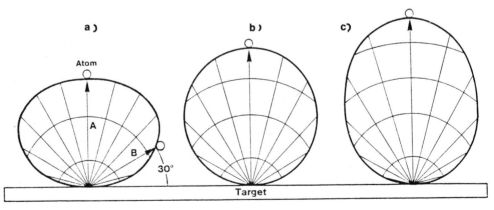

Fig. 35 (a) Under-cosine emission distribution. (b) Cosine emission distribution. (c) Over-cosine emission distribution.

under-cosine distribution (Fig. 35), more atoms will be directed at the sidewalls of steps, where they accumulate. The surface migration model enjoys more support, as it is argued that energies of impinging ions in bias sputtering appear to be too small to cause appreciable resputtering.

Sputter Processes for Various Thin Film Materials

This section discusses process conditions used to sputter deposit a variety of films for VLSI applications, including: a) aluminum and aluminum alloys; b) titanium-tungsten; c) platinum; d) refractory metal silicides; and e) fused silica (also referred to as quartz).

Aluminum and aluminum alloy films of ~3000-12,000Å thickness are deposited by dc magnetron sputtering, since high deposition rates are normally required (3000-10,000Å /min). Substrate heating and/or bias-sputtering are frequently used to enhance step coverage. Film reflectance and alloy composition to some degree are also impacted by bias-sputtering conditions. Residual gases (O, N, and H)[46] can cause deleterious effects in the deposited film. For example, in the discussion on process of film formation during sputter deposition, it was noted that small amounts of O_2 can lead to large resistivity increases in the Al films. In addition, O_2 can produce an increase in the film *hardness* (which can cause difficulty in forming strong wire-bonds to such hardened films). The presence of N_2 has been shown to produce stress in Al films. This leads to cracking and voids, which can contribute to early failures from electromigration[28].

It has also been reported[20] that an excess quantity of H_2 present in the chamber also increases the propensity of the Al films to form hillocks (Fig. 36). In addition, the pumping of large volumes of H_2 will cause the cryopump to saturate more quickly, requiring increased system down-time to regenerate the pumps.

To minimize problems from residual gas incorporation, some or all of the following procedures are utilized: a) the chamber is pumped to a low base pressure before being back-filled with Ar at the start of the deposition process; b) the wafers and wafer holders are pre-heated in the load-lock chamber to desorb surface moisture; c) high purity Ar gas is used; d) in some systems Ti sublimation pumps are offered for pumping H_2; e) residual-gas analyzers are used to monitor the gas composition during the sputter process; f) the chamber is pumped during the process with a high-throughput pump (e.g. cryopump and Meissner trap) so that the residence time of any desorbing contaminant gases is short; and g) high deposition rate processes are used, so that the

Fig. 36 Comparison of hillock-free and hillock containing films.

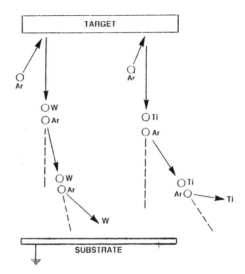

Fig. 37   Titanium-tungsten gas scattering model.   Courtesy of MRC.

growing film is exposed to any impinging contaminant gases for the shortest possible time.

Aluminum alloy films are usually deposited from a single compound target, but successful deposition by co-sputtering has also been reported, especially if the alloy material is to be introduced as a series of thin layers[20,23]. In many cases, an underlying layer (e.g. Ti:W), or overlying layer (e.g. TiSi$_2$) is applied during the same pumpdown, and in such instances the sputtering system must have the capability of depositing several materials in a single process.

Titanium-tungsten is also deposited using magnetron sputtering (dc or rf). The deposited film contains less Ti than the target alloy (~ 50% less), since the Ti atoms are lighter than W, and are strongly scattered by the Ar gas atoms (Fig. 37). The trajectories of the heavier W atoms, on the other hand, are less perturbed by collisions with Ar. As a result, more of the scattered Ti ends up on the chamber walls, and less on the substrate. (Thus, to get a film of 10% Ti, 90% W, by weight, a target containing ~19% Ti must be used.[30]) This effect has two important implications: 1) the resistivity of the deposited film depends on the Ti concentration since ρ_{Ti} = 48 μΩ-cm, and ρ_W = 5.5 μΩ-cm. In some applications the resistivity of the Ti:W film is a critical parameter[31] (e.g. in fuse links of some memory devices); and 2) the back-scattered Ti atoms form dendrites on the target, which eventually flake off[30]. A sputter-sideways technique must therefore be used in systems that are dedicated to Ti-W deposition, so that these flakes fall down in the space between the target and substrates (and not on the substrates). It has also been found that the addition of N$_2$ into the chamber during Ti:W deposition both increases the diffusion-barrier properties of Ti:W films and alters the film resistivity. If N$_2$ is added (e.g. by using pre-mixed N$_2$: Ar gases), and the second film cannot tolerate substantial nitrogen partial pressures, a high-vacuum pumping step must be used between sequential deposition steps (e.g. Al following Ti:W). The intrinsic stress and resistivity in Ti:W sputter deposited films can also be controlled by altering the Ar pressure and bias-sputter voltage levels[31].

Platinum is normally used in VLSI applications to form a PtSi layer in Si /PtSi /Ti:W /Al contact structures (see Vol. 2 for more details on such contact structures), and to form Schottky diodes. The Pt layers required for these uses are very thin (e.g. 500Å). In order to controllably

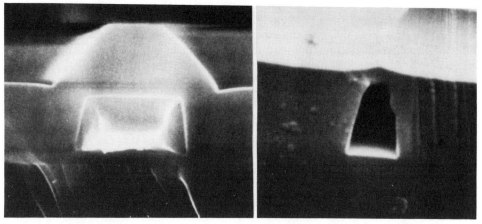

Fig. 38 Partially and fully planarized bias sputtered quartz films. Courtesy of Temescal.

deposit thin layers, a slower deposition method than dc magnetron sputtering is advantageous. Thus, sputtering of Pt is usually carried out using an rf diode or rf magnetron configuration.

Refractory metal silicides (e.g. Me(metal)Si$_x$ is the desired silicide film, where x for the best quality is determined experimentally, and may range from ~2.0-2.6) are sputter deposited both by co-sputtering[32,33,50] and by sputtering from a single source (such as a composite or sintered target). In co-sputtered films, high purity targets are used, and calibration runs must be conducted to measure the deposition rate vs. power for each target. Then x is measured in the deposited films, and the procedure is adjusted until desired values of x are obtained. In compound target processes, several targets of different composition must be initially purchased to home-in on the best value of x. There is still some question, however, as to how reproducibly vendors will supply targets so as to maintain the necessary x in the deposited film from target to target. Bias sputtering is also normally used to promote good step coverage. The as-deposited sputtered silicide films are amorphous and possess high resistivities. A high temperature anneal (in Ar) is used to form the silicide and reduce the resistivity (e.g. a 1000°C anneal reduces resistivity by a factor of ~10). A full discussion on refractory metal silicides in VLSI is undertaken in Chap. 11.

SiO_2 is typically sputter deposited to provide dielectric layers for multilevel interconnect structures. The SiO_2 is rf sputtered from a silica glass target (although it is commonly, but incorrectly, referred to as a quartz target), and an rf bias is applied to the substrates as well, hence the name *bias-sputtered quartz*, or BSQ[34,35,36]. By this procedure, reduced slope steps in the deposited BSQ are achieved. In fact, the surface can be entirely planarized under certain deposition conditions (Fig. 38). The process, however, has several limitations, including: a) the deposition rate is quite slow (e.g. several hours generally are needed to deposit suitably planarized films of adequate thickness); b) the purity of glass targets is still an area of uncertainty; c) the SiO_2 deposits on chamber walls, as well as on the substrate. If these deposits get too thick they will flake off and form particulates. To avoid the problem, the system must be regularly and thoroughly cleaned. This may be a ardous task.

Reactive Sputtering

The introduction of reactive gases into the sputtering chamber during the deposition process allows material sputtered from the target to combine with such gases to obtain chemical compound films[37]. In addition, compound targets of some substances can be rf sputtered, and controlled partial pressures of reactive gases can be added to compensate for any loss of gaseous constituents of the compound if the sputtering target dissociates. Although these advantages appear to offer interesting process possibilities, the technique has not found wide application in the silicon microelectronic industry.   Two examples of its use, however, do include: a) the addition of N_2 into the Ar ambient during Ti:W depositions, as discussed earlier; and b) in the deposition of the barrier film material, TiN[54], by the sputtering of a pure Ti target with a sputter gas of Ar mixed with N_2.   The complexity of the mechanisms that can take place in the chamber when reactive gases are added to the Ar, and the difficulty in obtaining adequately pure compound targets, have otherwise restricted reactive sputtering to VLSI research laboratories.

Future Trends in Sputter Deposition Processes

It appears that sputter deposition will continue to be the most important PVD process in silicon VLSI fabrication for some time. It is also likely that even more demands will be made on sputter deposition equipment as processing technology proceeds to evolve. The following are some of the challenges to equipment design that will arise: a) wafers will grow larger, and uniform films on such wafers will need to be deposited with high throughput; b) the machines will need to operate at higher levels of cleanliness (i.e. less particulate production); c) greater reliability (e.g. >90% up-time) will be expected; d) the machines will become progressively more automated and controllable from a central factory computer; e) the ability to improve step coverage or even planarize deposited films is an important capability being developed, and this feature will be avidly sought in new sputtering equipment designs[51,52]; f) alternative techniques to replace (or enhance) sputter etching for *in situ* cleaning of small area openings will be developed[53]; and g) as multilevel interconnect technologies are developed, more multilayer films (e.g. Ti:W /Al-Cu; or Ti /Al-Si /Ti /Al-Si /Ti) will be utilized. Sputter equipment will be called upon to deposit such complex films with high throughput.

PHYSICAL VAPOR DEPOSITION BY EVAPORATION

Evaporation Basics

Thin films can be deposited by applying heat to the source of film material, thereby causing evaporation[38]. If the heated source resides in a high-vacuum environment, the vaporized atoms /molecules are likely to strike the substrates (or chamber walls) without suffering any intervening collisions with other gas molecules. The rate of mass lost from the source per unit area per unit time, R, as a result of such evaporation, can be estimated from the Langmuir-Knudsen relation:

$$R = 4.43 \times 10^{-4} (M /T)^{1/2} p_e \qquad (14a)$$

$$R = 5.83 \times 10^{-2} (M /T)^{1/2} p_e \qquad (14b)$$

where: M is the gram-molecular mass; T is the temperature in °K; and p_e is the vapor pressure in Pa (Eq. 14a), or torr (Eq. 14b).

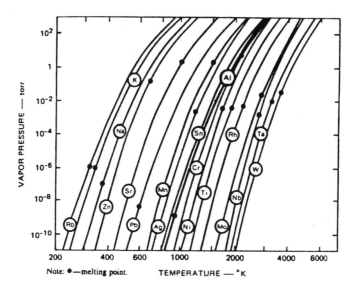

Fig. 39 Vapor pressure of metals commonly deposited by evaporation.

Figure 39 shows the vapor pressure (defined in Chap. 3) of a variety of metals commonly deposited by evaporation. To achieve deposition rates which are high enough for practical manufacturing processes, vapor pressures of >1.5 Pa (>10 mtorr) must be achieved. From Fig. 39 we see that there is a significant variation in the temperatures which must be applied to various materials in order to achieve such vapor pressures (e.g. Al must be heated to 1200°K, while W must be at 3230°K, before their vapor pressures exceed 10 mtorr).

Evaporation has been widely utilized for depositing Al and other metallic films in microelectronic fabrication. Some of the characteristics that were responsible for the widespread use of evaporation include: a) films can be deposited at high rates (e.g. 0.5 μm /min. for Al); b) the low energy of the impinging metal atoms onto the substrate (~0.1 eV) leaves the substrate surface undamaged; c) due to the high vacuums under which evaporation is performed, films can be deposited with very little residual gas incorporation, and thus the deposited film is about as pure as the source material; and d) unintentional substrate heating is only caused by the heat of condensation of the depositing film, and by heat radiation from the source.

For VLSI applications, however, evaporation suffers from the following limitations: a) accurately controlled alloy compositions are more difficult to achieve with evaporation than with sputtering; b) *in situ* cleaning of the substrate surfaces is not possible with evaporation systems, but is an option in sputter systems (i.e. by performing a sputter-etch step prior to deposition); c) use of extended sources in sputter deposition can improve step coverage vis-a-vis evaporation; and d) x-ray damage, caused by electron-beam evaporation processes, is avoided in sputter deposition.

The successful deposition of thin films onto a wafer surface also implies that the following film characteristics can be achieved: a) the nominal film thickness can be controlled; b) adequate film thickness uniformity exists across the wafer; c) uniform thickness is exhibited when the film crosses over steps (i.e. good step coverage). In evaporation processes, a different approach is used to achieve each of these three goals. To obtain the desired nominal thickness the rate of mass loss per unit area of source R, as well as data about the directions of evaporated atoms, is needed. Equation 14 can be used to calculate R. Information about the directions of evaporating atoms is more difficult to determine.   The flux of atoms leaving a small area *plane source* (Fig. 40a) was theoretically postulated to exhibit a cosine distribution, and this predicts that the mass dM_e collected by an area $d\omega$ in Fig. 40a can be calculated from:

$$dM_e = (M_e / \pi) \cos \phi \; d\omega \qquad\qquad (15)$$

where, M_e is the total mass emitted by the source. Experimental information which confirmed this model was first obtained by Knudsen, who measured the evaporation from a small area plane source onto the inner surface of a sphere (Fig. 40b).   In this experiment, $\cos \vartheta = \cos \phi = r / 2r_o$, for each point on the sphere. Thus, the amount of deposit according to Eq. 15 should be uñiform for every point on the sphere, and this was indeed observed. From this data, it would seem that to obtain uniform coating of wafers, it would only be necessary to mount them on an inner surface of a sphere, and evaporate from a source as shown in Fig. 40b.

Unfortunately, actual sources do not have infinitesimally small areas, but are in fact *extended* sources. Thus, the directions of emission do not follow the ideal cosine distribution. In addition, the region directly above high-rate evaporation sources is populated by evaporant atoms with a density large enough to cause them to exhibit *viscous* rather than *molecular* flow. This disturbs the directional distribution even further (Fig. 40c). As a result of these (and other non-ideal emission effects), uniform distribution at every point on a sphere surface does not occur.

The technique developed to achieve maximum uniformity during evaporation, in spite of deviations from the ideal, is to mount the wafers onto a mechanical *planetary* fixture inside the chamber (Fig. 41). During deposition, the entire fixture rotates about the vertical axis of the vacuum chamber, and each wafer is also rotated about a second axis. The combination of rotary motions results in maximum thickness uniformity across a substrate, as well as substrate-to-

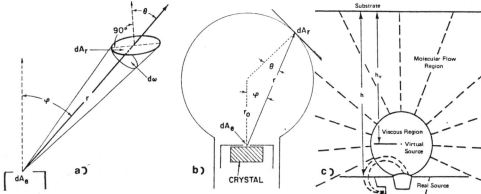

Fig. 40 (a) Surface element dA_r receiving deposit from a small-area source dA_s. (b) Evaporation from a small-area source onto a spherical receiving surface[38].   From L. Maissel and R. Glang, Eds., *Handbook of Thin Film Technology*, 1970.   Reprinted with permission of McGraw-Hill Book Co. (c) Schematic diagram showing regions of viscous flow and molecular flow around an electron beam heated crucible.   Courtesy of Temescal.

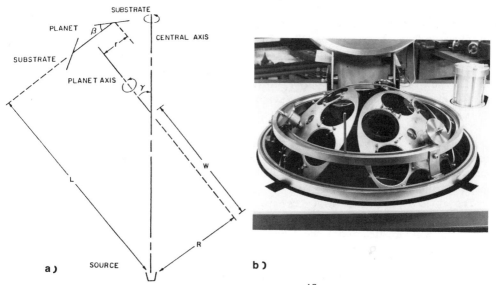

Fig. 41 (a) Schematic of planetary evaporator geometry[48]. Reprinted with permission of the American Physical Society. (b) Photograph of a planetary system.   Courtesy of Temescal.

substrate. As a result, adequate control of film thickness and thickness uniformity is achievable.

The problem of maintaining adequate thickness as the film crosses steps, however, is not solved satisfactorily by the planetary[48]. That is, even when a planetary is used, thinning of the film and microcracks can occur, because of the shadowing effect of the step. Substrate heating (as discussed earlier), is one technique used to reduce the severity of this problem.

Evaporation Methods

Evaporation is carried out under high-vacuum conditions. Typically the chamber is pumped down to ~5×10^{-7} torr before the evaporation is initiated. At such pressures the mean free path, λ is ~100 m. As the source is heated, the pressure rises somewhat as contaminants are desorbed from its surface.

The least complex method of depositing thin films by evaporation is with *resistance-heated* sources. That is, a wire of low vapor pressure metal (e.g. W), is used to support small strips of the metal to be evaporated. The W-wire is then resistively-heated. The metal to be evaporated first melts and wets the heated filament, and evaporation ensues. Although the technique is simple, it has several drawbacks which make other evaporation techniques more useful for microelectronic applications, including: a) evaporation from the filament may contaminate the deposited film; b) refractory metals cannot be deposited because of their high melting points; and c) the small charges limit the ultimate film thickness. As a result, two other techniques have come to be widely adopted for performing evaporation depositions in microelectronic applications: 1) electron-beam evaporation; and 2) inductive-heating evaporation.

Electron-Beam Evaporation

In electron beam evaporation, a stream of electrons is accelerated to high kinetic energy

(5-30 keV)[40]. The beam is directed at the material to be evaporated, and the kinetic energy is transformed to thermal energy upon impact. The electron stream can melt and evaporate any material, provided the beam can supply energy to the evaporant at an equal or greater rate than the rate at which heat is lost, as the material is held at high temperature. Electron beam *guns* (Fig. 42a and b) can be built to supply up to 1200 kW of highly concentrated electron beam power for evaporation applications. Very high film deposition rates can thereby be attained (e.g. 0.5 μm /min), as a result of the high power available.

The beam energy is concentrated on the surface of the target, and thus a molten region can be supported by a cooled structure. In fact, the target material itself provides a solid layer that separates the molten portion of the evaporant material from the wafer-cooled crucible. This eliminates the problem of reaction with, or dissolution of the crucible by the melt, and allows highly pure films to be deposited.

The self-accelerating, 270° beam guns have become the standard gun design. Earlier electron beam guns aimed the electron beam directly at the source, and hence the gun filaments would get coated and short circuited during the deposition cycle. In the 270° guns, a magnetic field simultaneously bends the beams through 270° and focuses them. The electron emission surface is hidden from the evaporating source, and the substrates are also protected from contamination by material evaporating from the heated filament. Movement of the beam (which allows the source to be scanned) is accomplished by electromagnetic deflection. This avoids the problem of nonuniform deposition that would be caused by the formation of a cavity in the molten source if the beam were stationary.

Two problems can arise from use of electron-beam evaporation: 1) At voltages >10 kV, the incident electron beam will cause Al to emit K-shell x-rays. These x-rays cause such damage as the creation of trapped charges in the gate oxides of MOS devices. The silicon devices interconnected by electron-beam Al must therefore be subjected to subsequent annealing to remove such damage, and as a result, restore device characteristics to their pre-metal deposition values; and 2) If excessive power is applied to the source, metal droplets that have been blown out of the source by the expanding metal vapor may be deposited on the substrates.

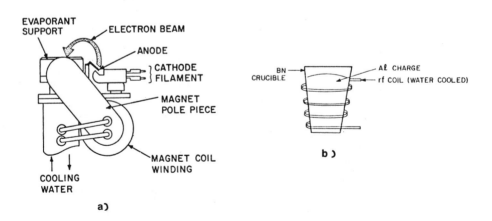

Fig. 42 (a) and (b) Bent beam electron gun. Courtesy of Temescal. (c) Inductively heated evaporation source. The molten Al charge is contained by the BN crucible.

Inductive Heating Evaporation

High-rate evaporation deposition can also be accomplished by heating the source with rf energy. As shown in Fig. 42c, an rf induction heating coil surrounds a crucible containing the evaporation source. The crucible is commonly made of BN, because it is not attacked by molten Al, has excellent heat-shock resistance, and is an insulator that is easy to machine. The main advantage of inductive heating over electron-beam heating is that the wafers are not subject to ionizing radiation. Two of its disadvantages are: a) the mandatory use of a crucible that is in direct contact with the molten evaporant; and b) the increased complexity of rf equipment and process development.

Evaporation Process Considerations

In addition to the information provided on evaporation, some topics related to practical processes need to be addressed including: a) evaporation of alloys; b) evaporation of multilayer films; c) effect of substrate temperature on aluminum film deposition; and) use of shutters.

Evaporation of *alloys* is a complex process, since alloys consist of mixtures of elements that seldom have similar vapor pressures at a given temperature. In addition, *Raoults Law* states that the vapor pressure of a solution is lower than that of a pure solvent by an amount proportional to the concentration of the solute. Alloy films are therefore evaporated using one of two techniques: a) evaporation from a *single source* composed of an alloy, but not necessarily containing the constituent elements in the same percentage ratio as in the deposited film; and b) simultaneous co-evaporation using *two electron guns and two sources*, with each source containing only one of the alloy constituents.

The *single-source method* is only practical if the vapor pressures are within a factor of 100 of each other at a given temperature. The composition of the film, however, is difficult to control, since as evaporation occurs, the compositions of both source and vapor change continuously. That is, the source becomes richer in the less volatile element. Alloy films of Al-0.5wt%Cu are typically evaporated using a single source of Al-2wt%Cu[43].

The *dual-source* method offers the promise of better control, at the price of added system complexity. In addition, the uniformity of deposition across large substrates from two sources is reduced, and the relative power applied to the two sources must be well controlled to insure that the desired evaporation rates are maintained. By using the two-source method, however, it is even possible to co-deposit materials that form neither compounds nor solid solutions. Alloys of Al-Si are normally co-evaporated in two-source systems[41,42]. On the whole, however, it is significantly more difficult to deposit alloys with highly controlled compositions by either evaporation method than by sputter deposition.

Multilayer films can be deposited in a system containing several sources and a single electron-gun, or several guns can be aimed at multiple crucibles, to evaporate film layers in a planned sequence.

The variation of film thickness of Al films as they *cross steps* can be reduced in evaporation processes by applying heat to the substrates during deposition. For example, a significant improvement in Al film step coverage is observed as the wafer temperature is increased to 350°C during evaporation. The temperature increase is also seen to increase the grain size of the deposited film. As described in the sputtering discussion, substrate heating has the effect of creating greater surface mobility of the deposited material. Film atoms are thereby more likely to migrate toward regions which have received less material deposition. This enhances thickness uniformity. Infrared lamps are commonly used to provide heat to the wafers.

Shutters are frequently employed in evaporation systems to prevent contaminants adsorbed on the source surface from being incorporated into the deposited films. That is, if the evaporation chamber and source are exposed to ambient conditions in the loading and unloading of wafers, the source surface may adsorb moisture, or form a native oxide layer. When the source is initially heated, such surface contaminants will vaporize (together with the source material). By interposing a shutter between the source and substrates (and postponing deposition by keeping the shutter closed until the source surface has been cleaned), the purity of the deposited film can remain uncompromised. Normally, the gas pressure rises upon initially heating the source and substrates, as a result of the contaminant desorption, and shutter opening is delayed until the chamber pressure returns to an acceptable level.

METAL FILM THICKNESS MEASUREMENT and MONITORING

Metal thickness is usually monitored during evaporation, but is generally deposited without monitoring in sputter deposition. In both cases, the films are normally measured after deposition.

Post-deposition measurements of metal film thickness are most commonly performed either directly with a surface profiling device (known as a *stylus profilometer*), or indirectly by electrical measurement of sheet reisistivity[45]. The basic aspects of sheet resistivity measurements are covered in Chap. 4, but it also should be mentioned at this point that the technique offers an important benefit for VLSI fabrication applications. That is, the probes can be stepped over a test wafer to provide contour maps of sheet resistance or thickness. This particular sheet resistance measuring method is discussed in more detail in Chap. 9, in the section dealing with

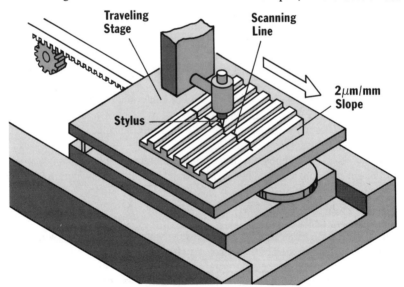

Fig. 43 Schematic drawing of a surface profilometer. Courtesy of Sloan Technology Corporation.

the monitoring of ion implant doses.

In the technique that uses the stylus profilometer, a step in the deposited film is first created, either by masking during deposition, or by a post-deposition etch process. Next, the profiling instrument draws a fine stylus along the surface containing the stepped film (Fig. 43). Whenever the stylus encounters a step, a signal variation (based on differential capacitance or inductance techniques), yields an indication of the step height. The information is displayed on a chart recorder or CRT. Two examples of commercially available instruments are the Sloan Dektak, and the Tencor Alpha-step. Films less than 1000Å thick can be measured with such instruments, but films much thinner than 1000Å are normally measured with sheet resistivity techniques. Vibration, surface roughness, and greater criticality of leveling the profilometer, all make such thin film measurements more difficult with the stylus-type instruments.

Periodic calibration of stylus profilometers is also required, using calibration standards traceable to the National Bureau of Standards. If the instrument is to be used to measure films whose thickness differ greatly from the calibration standard, a reference can be created by etching a step in thermally grown SiO_2, and accurately measuring its height with an ellipsometer.

REFERENCES

1. D. Pramanik and A.N. Saxena, "VLSI Metallization Using Aluminum and Its Alloys, pt I", *Solid State Tech.*, Jan. '83, p. 127, and pt. II, Mar. '83, p. 131.
2. F. Fischer and F. Neppl, "Sputtered Ti-Doped Al-Si for Enhanced Interconnect Reliability", *Proceedings IEEE Rel. Phys. Symposium,* 1984, p. 190.
3. D.S. Gardener, *et al*, "Layered and Homogeneous Films for Al and Al-Si with Ti and W for Multilevel Interconnects", *IEEE Trans. on Electron Devices*, Feb. '85.
4. B. Chapman, *Glow Discharge Processes*, John Wiley & Sons, New York, 1980.
5. *The Book of Basics*, 3rd Ed., Materials Research Corp., no date, Orangeburg, N.Y.
6. W.D. Davis and T.A. VanderSlice, *Phys. Rev.* **131**, 219, (1963).
7. G.K. Wehner and G.S. Anderson, "The Nature of Physical Sputtering", in *Handbook of Thin Films*, Eds. L.I. Maissel and R. Glang, Chap. 3, McGraw-Hill, New York, 1970.
8. F. d'Heurle, L. Berenbaum, and R.A. Rosenburg, *Trans. Met. Soc.* (AIME) **242**, 502, (1968).
9. A.D. MacDonald and S.J. Tetenbaum, in *Gaseous Electronics*, Vol. I, Eds. M.N. Hirsh and H.J. Oskam, Academic Press, New York (1978).
10. H.R. Koenig and L.I. Maissel, *IBM J. Res. Develop.*, **14**, 168 (1970).
11. W.H. Class, "Focest Cathode DC Magnetron Deposition of Conductor Metallization", *Solid State Technology*, June, 1983, p. 103.
12. J.L. Vossen and J.J. Cuomo, "Glow Discharge Sputter Deposition", in *Thin Film Processes*, Eds. J.L. Vossen and W. Kern, Academic Press, New York, 1978, Chap. II-1, p. 11-73.
13. J.A. Thornton and A.S. Penfold, "Cylindrical Magnetron Sputtering", in *Thin Film Processes*, Eds. J.L. Vossen and W. Kern, Academic Press, New York, 1978, p. 75.
14. G.K. Wehner and G.S. Anderson, "The Nature of Physical Sputtering", in L.J. Maissel and R. Glang, Eds., *Handbook of Thin Film Technology*, McGraw-Hill, New York, 1970, Chap. 3.
15. L. Maissel, "Application of Sputtering to Deposition of Films", in L. Maissel and R. Glang, Eds., *Handbook of Thin Film Technology*, McGraw-Hill, New York, 1970, Chap. 4.
16. H.F. Winters and E. Kay, *J. Appl. Phys.*, **38**, 3928 (1967).
17. J.L. Vossen and J.J. O'Neil Jr., *RCA Review* **29**, 566 (1968).
18. J.L. Vossen and J.J. O'Neil Jr., *RCA Review* **31**, 276 (1970).

19. S.R. Wilder and R.G. Johanson, "High Vacuum Systems for Sputtering Applications", *Solid State Technology*, Nov. '85, p. 95.

20. D.R. Denison, "Sputtering System Design for Optimum Deposited Film Quality", *Microelectronics Manufacturing and Testing*, July '85, p. 12.

21. R. Iscoff, "Trends in Metallization Materials", *Semiconductor International*, Oct. '82, p. 57.

22. F.T. Turner and D.J. Harra, "Advances in Cassette-to-Cassette Sputtercoating Systems", *Solid State Technol.*, July '83, p. 115.

23. R. Nowicki, *et al.*, "Dual RF Diode/DC Magnetron Sputtered Aluminum Alloy Films for VLSI", *Semiconductor International*, March '82, p. 105.

24. W. Class and R. Hieronymi, "Measurement and Soruces of Substrate Heat Flux Encountered with Magnetron Sputtering", *Solid State Technol.*, Dec. '82, p. 55.

25. V.E. Hoffman and H.M. Chang, "Individual Wafer Metallization Utilizing Load-Locked, Close-Coupled Conical Magnetron Sputtering", *Solid State Technol.*, Feb. '81.

26. J.F. Smith and Y.H. Park, "Sputtering and Dry Etching Technology for VLSI and ULSI Devices", *Advanced Aluminum Metallization*, MRC Corp., Orangeburg, N.Y., 1985, J-111-1.

27. L.J. Fried, *et al.*, "A VLSI Bipolar Metallization Design with 3-Level Wiring and Array Solder Connections", *IBM J. Res. Devel.*, **26**, May '82, p. 362.

28. J. Klema, R. Pyle and E. Domangue, "Reliability Implications of N_2 Contamination During Deposition of Sputtered Al/Si Films", *Proceedings Intl. Rel. Phys. Symposium*, IEEE, New York, 1984, p. 1.

29. P. Burggraaf, "Magnetron Sputtering Systems", *Semiconductor International*, Oct. '82, p. 37.

30. M. Hill, "Magnetron Sputtered Titanium-Tungsten Films", *Solid State Technol.*, Jan.'80, p. 53.

31. L. Hartsough, "Resistivity of Bias-Sputtered Ti-W Films", *Thin Solid Films*, **64** (1979) p. 17.

32. D. Nichols, "Co-Sputtering of Metal Silicides for Semiconductor Applications", Presented at *Semicon /East*, Sept. '81, Varian Inc., Palo Alto, CA.

33. R.S. Nowicki, "The RF Diode Co-Deposition of Refractory Metal Silicides", *Solid State Technology*, Nov. '80, p. 95.

34. C.Y. Ting, V.J. Vivalda, and H.G. Schaefer, *J. Vac. Sci. Technol.*, 15 (3), (May/June, 1978).

35. J.F. Smith, *Solid State Technology*, 27 (1), 135 (Jan. '84).

36. K. Urbanek, "Magnetron Sputtering of SiO_2, an Alternative to CVD", *Solid State Technol.*, Apr. '77, p. 87.

37. P.S. McLeod, "Reactive Sputtering ", *Solid State Technol.*, Oct. '83, p. 201.

38. R. Glang, "Vacuum Evaporation", in L. Maissel and R. Glang, Eds., *Handbook of Thin Film Technology*, McGraw-Hill, New York, 1970, Ch. 1, p. 1-130.

39. R.K. Waits, "Planar Magnetron Sputtering", in J.L. Vassen and W. Kern, Eds., *Thin Film Processes*, Academic Press, New York, 1978, Chap. II-4, p. 131.

40. R.J. Hill, *Physical Vapor Deposition*, Temescal, Berkeley, CA, 1976.

41. F. Hegner and A. Feuerstein, "Aluminum-Silicon Metallization by Rate Controlled Dual E-Beam Gun Evaporation, *Solid State Technol.*, Nov. '78, p. 49.

42. "Controlling Evaporation of Co-Deposited Al-2% Si", *Solid State Technol.*, Dec. '79, p. 94.

43. L.C. Hecht, "Use of Method of Successive Dilution for Reproducible Control of Al-Cu Alloy Evaporation", *J. Vac. Sci. Technol.*, **14**, Jan./Feb. '77, p. 648.

44. *ibid*, Ref. 38, p. 1-107.

45. S.C.P. Lim and D. Ridley, "An Overview of Thickness Measurement Techniques for Metallic Thin Films", *Solid State Technology*, Feb. '83, p. 99.

46. A.R. Nyaiesh and L. Holland, "Effects of Gas Composition on Discharge & Deposition Char-

acteristics when Magnetron Sputtering Aluminum", *Vacuum*, **31**, No. 8/9, 1981, p. 371.

47. I.A. Blech, "Step Coverage By Vapor Deposited Thin Aluminum Films", *Solid State Technology*, Dec. '83, p. 123.

48. I.A. Blech, D.B. Fraser, and S.E. Haszko, "Optimization of Al Step Coverage Through Computer Simulation and SEM", *J. Vac. Sci. and Technol.*, **15**, 13, (1978), errata *J. Vac. Sci. Technol.*, **15**, 1856, (1978).

49. J.L. Vossen, Ed., "Bibliography on Metallization Materials and Techniques for Silicon Devices", Thin Film Div., *American Vacuum Society*, 1974-82.

50. S.P. Murarka and D.B. Fraser, "Silicide Formation in Thin Cosputtered (Titanium & Silicon) Films on Polysilicon and SiO_2, *J. Appl. Phys.*, **51**, 350 (1980).

51. Y. Homma and S. Tsunekawa, *J. Electrochem. Soc.*, **132** (6), 1466, (1985).

52. I. Wagner, J.F. Smith, and Y.H. Park, "Planarization of Sputtered Aluminum Films", in *Sputtering and Dry Etching Technology for VLSI and ULSI Devices*, MRC Corporation, Orangeburg, N.Y., Aug. '85, p. J-VII-C-1.

53. J.L. Vossen, *et al*, *J. Vac. Sci. & Technol.* A2(2), 1984, p. 212.

54. C.Y. Ting, *J. Vac. Sci. and Techol.*, **21**, 14, 1982.

55. O. Almen and G. Bruce, *Nuclear Instrumentaytion Methods*, **11**, 257, (1961).

56. H. S. Butler and G.S. Kino, *Phys. Fluids*, **6**, 1346, (1963).

PROBLEMS

1. Calculate the mean free path of a sputtered atom in an Ar glow discharge at 10 mtorr pressure. If the cathode-to-substrate spacing is 5 cm, how many gas-phase collisions is a sputtered atom likely to encounter as it traverses this distance?

2. Aluminum is bombarded by 1000 eV Ar ions. (a) What is the maximum energy that can be transferred to the target atom. (b) What is the maximum value that the mean energy of a struck target atom can attain if the threshold sputtering energy for the Ar-Si ion-target combination is 13 eV.   The atomic number and mass of Al and Ar, respectively are Z_{Al} = 13 M_{Al}= 27, Z_{Ar}=18, M_{Ar}= 39.44

3. The sputtering yield of aluminum in Ar is known to be S = 1.2 at an ion energy of 600 eV. The measured rate of deposition in a specific sputtering system is 400 Å / min when the external electron current is I_0 = 100 mA. (a) Calculate the ion current, I, using the Townsend equation:

$$I = (I_0 \exp [\alpha \, d]) / (1 - \gamma [\exp \{\alpha \, d\} - 1]),$$

where, α = 6 ions /cm and γ, the secondary electron yield = 0.08 electrons per ion, and the interelectrode spacing is d = 5 cm. (b) Find the new deposition rate when the ion energy is incresed to 1000 eV, and S = 1.9, α = 3 ions /cm, and γ = 0.1 electrons per ion.

4. Explain how the faceting effect is utilized to produce planarized films of so-called bias-sputtered quartz.

5. If a target is made of fused silica with a thickness of 2.5 mm, it has a capacitance of ~1 pF /cm^2.   If the current density during rf sputtering of this target is 1 mA /cm^2 and the applied voltage is 1000 V, calculate the time required to charge this capacitance. This roughly represents the time interval during which a glow discharge can remain "on".

6. Explain why the higher secondary electron emission from insulators leads to slower sputter deposition of these materials.

11

REFRACTORY METALS

and THEIR SILICIDES

in VLSI FABRICATION

As the features of VLSI circuits continue to shrink to 1 μm (or less), the necessity of decreasing the resistance and capacitance (RC) associated with interconnection paths becomes ever more pressing. This is particularly true for MOS devices, in which the RC delay due to the interconnect paths can exceed the delays due to gate switching. The higher the value of the interconnect R (resistance) x C (capacitance) product, the more likely is the circuit operating speed to be limited by this delay. A simplified approach to the problem shows that:

$$RC = R_s L^2 \varepsilon_{ox} / x_{ox} \qquad (1)$$

where: R_s is the sheet resistance of the connection (given by, $R_s = \rho / x_{int}$); L is the length of the connection; x_{int} is the thickness of the interconnect conductor; x_{ox} is the oxide thickness over which the connection runs; and ε_{ox} is the permittivity of SiO_2. In reducing x_{ox} and x_{int}, RC depends only on L, and not the width of the conductor, W. This simple model however, is inaccurate for W<3 μm, where capacitance fringing-fields become important. Since the R_s decreases with decreasing line size, the RC product also decreases. Figure 1 shows the RC time constant for a 1 cm line as a function of design rule for polysilicon, tantalum silicide ($TaSi_2$), and aluminum (Al). Shown for comparison is the gate delay per stage for a MOS device. For small feature sizes, the interconnect delay could predominate, depending on the maximum length of the interconnect lines in the circuit. *The conclusion is that low resistivity interconnection paths are critical in order to fabricate dense, high performance devices*[1].

There are several potential approaches to reduce the resistivity of the interconnect to less than the 15-30 Ω /sq exhibited by polysilicon. The poly-Si could be replaced by Al, but due to the low melting and eutectic temperatures of Al, all subsequent processes would have to be held to less than 500°C. Since several post-gate formation processes must be carried out at temperatures >500°C (e.g. source-drain implant anneals, oxidation, and glass flow /reflow), Al is not a suitable alternative material. The poly-Si could also be replaced by a refractory metal (e.g. W, Ta, or Mo), a refractory metal silicide (e.g. WSi_2, $TiSi_2$, $MoSi_2$, or $TaSi_2$), or a multilayer structure, consisting of a low resistance material (e.g. a refractory metal silicide) on top of a doped polysilicon layer (such a structure is termed a *polycide*). Figure 1a shows a polycide configuration. The refractory metals have adequately high melting temperatures, but their oxides are typically of poor quality, and in some cases volatile (e.g. Mo and W oxides). In addition, it may be difficult to obtain consistent threshold voltages in MOS transistors due to

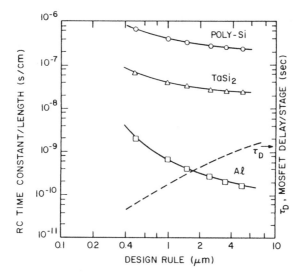

Fig. 1 RC time constant per unit length for three conductive materials as a function of feature size. Also shown is delay per stage of ring oscillators as a function of feature size. (After Sinha[1]) Reprinted with permission of American Physical Society.

impurities in the sources of the refractory metals. Use of refractory metal silicides alone as the gate /interconnect layer, suffers similar problems as those associated with the use of refractory metals alone. The *polycide* structure has therefore become the predominant gate /interconnect film for replacing polysilicon. In this way, the advantages of the known work function of poly-Si and the highly reliable poly-Si /SiO$_2$ interface are preserved, as poly-Si is still directly atop the gate oxide. In addition, the poly-Si provides a source of silicon for oxidation, and thereby allows isolation oxides for subsequent poly-Si or metal layers to be formed on the silicide.

Numerous studies on the preparation and properties of silicides, particularly using the polycide structure, have been published, and are surveyed in extensive detail in an comprehensive book by Murarka[2]. The major emphasis of this chapter is to present a brief overview of the

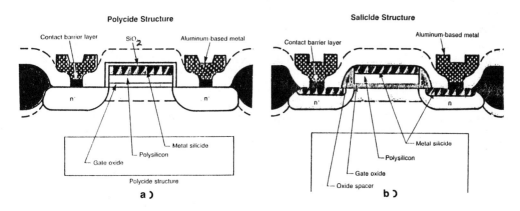

Fig. 2 (a) Polycide, and (b) Salicide structure. Reprinted by permission Semiconductor Internatl.

Table 1. SILICIDE MATERIAL PROPERTY REQUIREMENTS FOR VLSI[2]

- Low Electrical Resistivity
- Ease of Formation
- Ease of Fine Line Pattern Transfer
- Controlled Oxidation Properties and Stability in an Oxidizing Ambient
- High Temperature Stability
- Smooth Surface Features
- Good Corrosion Resistance
- Stable Contact Formation to Aluminum Metallization
- Excellent Adhesion and Low Stress
- Good Electromigration Resistance
- Ohmic and Low Contact Resistance
- Stability throughout Subsequent high-temperature Processing, including Ion Implant and Diffusion

properties and preparation of silicides and refractory metals for VLSI applications, including some new data that has been reported since the time Murarka's text was published.

CANDIDATE SILICIDES FOR VLSI APPLICATIONS

In order to effectively utilize refractory metal silicides in VLSI fabrication (either alone, or in the polycide configuration), very stringent materials requirements must be satisfied. Table 1 lists some of the most important of these requirements.

There are a group of refractory metal silicides (MSi_x) that meet most, or all of these criteria. For example, the refractory metal silicides can withstand much higher temperatures than aluminum, and their eutectic temperatures with Si are in excess of 1300°C. On the other hand, the silicides of Pd (Pd_2Si), Pt (PtSi), and Ni ($NiSi_2$), have eutectic temperatures of 720°C, 830°C, and 966°C, respectively.   Hence these three silicides are not suitable for processes in which they are subjected to temperatures that approach their silicide-Si eutectic temperature.

Resistivity

Resistivity is the key parameter of a silicide that reduces interconnect delay, and Table 2 lists the resistivities of the most commonly used silicides. The resistivity values quoted in the Table 2 are for specific annealing times and temperatures. The resistivity values depend on many factors, including the method of formation, the sintering time and temperature, the stoichiometry of the compound, and its chemical purity.   Table 2 clearly points out the significant differences between the materials that depend upon the method used to form them. It

Table 2. Resistivity of Silicide Films Annealed at ≤1000°C (in $\mu\Omega$-cm).

Material	Metal + Poly-Si	Metal + Si Crystal	Co-Sputter	Co-Evaporation	CVD
$TiSi_2$	13	15	25	21	21
$TaSi_2$	35		50		38
$MoSi_2$	90	15	100	40	120
WSi_2			70	30	40
PtSi	28		35		

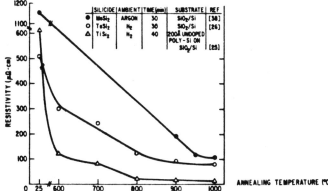

Fig. 3 Sheet resistivity of metal silicide ($MoSi_2$, $TaSi_2$, $TiSi_2$) films versus annealing temperature[3]. (© 1983 IEEE.)

should be noted that the lowest resistivity of the group (~13 $\mu\Omega$-cm) is achieved by $TiSi_2$ formed by direct metallurgical reaction.

Figure 3 shows the resistivity of $MoSi_2$, $TaSi_2$, and $TiSi_2$ as a function of *annealing temperature*[3]. $TiSi_2$ achieves its minimum resistivity at 800°C, while $MoSi_2$ requires temperatures in excess of 900°C[3]. Figure 4 shows the effect of *annealing time* on the sheet resistance for several silicides. In all cases, equilibrium is achieved in less than 30 minutes. The use of rapid thermal processing (RTP), as described in Chap. 2, allows reaction times for WSi_2 formation to be under 60 sec at 1200°C[4].

The resistivity dependence on *stoichiometry* can be illustrated with a few examples. The resistivity of chemically vapor deposited WSi_x is shown in Fig. 5, where x varies from 2.2 to 2.6. The resistivity of the film increases as it becomes progressively richer in Si. Figure 6 shows the resistivity of co-sputtered $TiSi_2$ films as a function of Si /Ti ratio before and after sintering (900°C /30 min). The resistivity of the unsintered material increases with increasing Si content. For sintered films, however, the resistivity goes through a minimum at Si /Ti = 3 (nominal atomic ratio). $TiSi_2$ is formed only where the ratio exceeds 2. The resistivity decreases

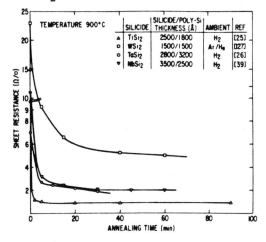

Fig. 4 Sheet resistance of polycide (WSi_2, $NbSi_2$, $TiSi_2$) films versus annealing time. (© 1983 IEEE.)

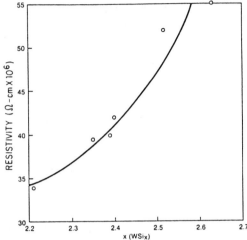

Fig. 5 Resistivity of CVD WSi_x versus x, where x varies from 2.2 to 2.6.Courtesy of Genus Inc.

for $TiSi_2$ with excess Si up to a value of Si /Ti = 5 /2, and then increases at higher Si content as the Si may undergo precipitation[6] (exact composition is determined by RBS, see Chap. 17).

Oxygen contamination in the silicide film also has a significant effect on the film resistivity. Figure 7 shows the effect of oxygen in as-deposited and sintered 2500Å thick $TaSi_2$ films. The sintering conditions were 900°C for 30 min in argon[3].

SILICIDE FORMATION

The silicides of interest can be formed by basically three techniques, each of which involve a deposition followed by a thermal step to form the silicide: 1) deposition of the pure metal on

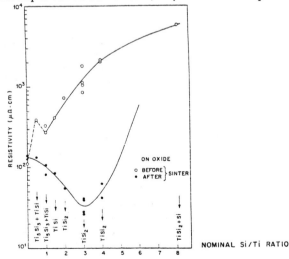

Fig. 6 Resistivity of cosputtered films on SiO_2 before & after sintering at 900°C for 30 min in H_2 as function of nominal Si /Ti ratio in film[6]. Reprinted with permission of American Phys. Soc.

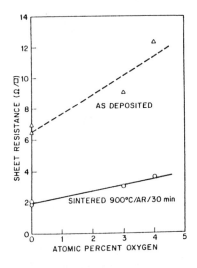

Fig. 7 Sheet resistance of the as-deposited and sintered cosputtered $TaSi_2$ films as a function of the oxygen concentration in the film[2]. Reprinted with permission of American Physical Society.

silicon (i.e. onto single crystal and /or polycrystalline Si); 2) simultaneous evaporation of the silicon and the refractory metal from two sources (co-evaporation); and 3) sputter-depositing the silicide, either from a composite target, or by co-sputtering or layering. Table 3 lists the methods of formation and some of the advantages and disadvantages of each.

Table 3. METHODS OF SILICIDE FORMATION

Method of Formation	Advantages	Disadvantages
Direct Metallurgical Reaction. $M + xSi \Rightarrow MSi_x$ Metal deposited by evaporation, sputter, or CVD.	Both polycide and salicide structure can be formed. Selective etch possible.	[M] /[Si] depends on phase formed. Sensitive to sintering environment. Rough surface.
Co-evaporation from an Independent Si and M Source.	Smooth surface. Sintering environment not as critical	[M] /[Si] control difficult but possible. No selective etch possible. Poor step coverage.
Co-sputtering from Independent Si and M Targets.	Good control of [M] /[Si]. Smooth films. Sintering environment not as critical. Deposition of sandwich possible.	Difficult calibration to achieve [M] /[Si] control.
Sputtering from a Composite MSi_x Target.	Excellent [M] /[Si] control if correct target chosen. Good step coverage.	Contamination from Target.
Chemical Vapor Deposition: atmospheric, low-pressure, or plama-enhanced.	High throughput. Excellent step coverage.	Rough surface. [M] /[Si] control difficult but possible. Possible poor adhesion.

Direct Metallurgical Reaction

When a refractory metal film is deposited directly on a silicon surface, and the wafer is subjected to heating, the metal and the silicon can react to form a silicide. One advantage of this method of silicide formation is that it usually yields a lower resistivity than most other methods. A significant body of work has been published on the formation and kinetics of such intermetallic compounds of thin film metals and silicon, both for single-crystal and polycrystalline silicon. Specialized techniques including RBS, AES, XPS, SIMS, and electrical measurements, have been used to follow the phase formations as the reactions proceed. See Chap. 17 for additional information on the analytical techniques.

For a metal film deposited on silicon, and annealed at relatively low temperatures, *metal-rich* silicides form first, and continue to grow until the metal is consumed. Typically, at that point, the next *silicon-rich* phase begins to grow. For example, Fig. 8 shows the effect of temperature on the formation of TiSi and $TiSi_2$. Above 600°C, no evidence of $TiSi_2$ exists, while TiSi begins to grow. The growth of TiSi peaks at 700°C, and the compound is not stable above 800°C. $TiSi_2$ begins to grow at 600°C, reaching a maximum at 800°C, and no more polysilicon is consumed above this temperature. After the complete conversion to $TiSi_2$, the system is stable. At 700°C it takes approximately 60 minutes until the $TiSi_2$ has completely formed, while no silicide is detected even after 10 hours at 500°C.

The Pt /Si reaction proceeds at 350°C. RBS studies on a sample sintered for 20 min shows that it is converted to Pt_2Si, with small amounts of PtSi present. After 60 minutes, nearly 65% of the Pt_2Si film is converted to PtSi. Canali[9] analyzed the time dependence of the Pt silicide formation, and found that the layer thickness x, is given by:

$$x^2_{Pt_2Si} = D_1 t \qquad (2)$$

$$x^2_{PtSi} = D_2 (t - t_o) \qquad (3)$$

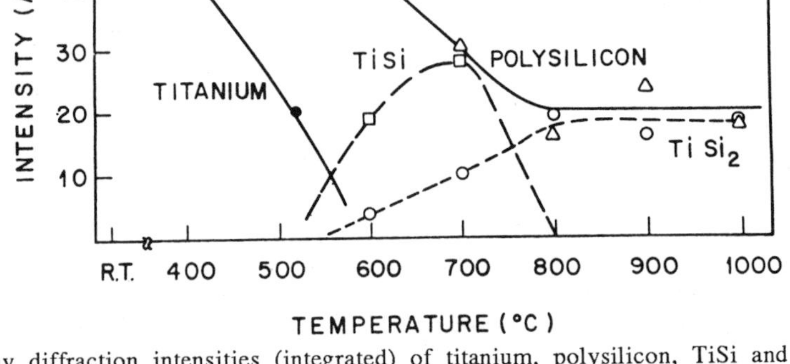

Fig. 8 X-ray diffraction intensities (integrated) of titanium, polysilicon, TiSi and $TiSi_2$ as a function of vacuum sintering temperature[8]. Reprinted with permission of American Physical Society.

where D_1 and D_2 are the interface diffusivities for Pt_2Si and $PtSi$, respectively, and t_o is the time at which $PtSi$ starts forming. Plots of D_1 and D_2 versus reciprocal temperature show an Arrhenius behavior (see Appendix 4), which yields activation energies of 1.3 and 1.5 eV for Pt_2Si and $PtSi$ formation, respectively.

The $t^{1/2}$ dependence, which has been observed for most silicide formation, is indicative of a diffusion-limited process. The linear dependence, as seen during the formation of some other silicides, is indicative of a surface-reaction rate-controlled process. Difficulty in interpreting the reaction data results from the variation in the cleanliness of the interface, which has the effect of preventing the initiation of the reaction.

When a silicide is formed by direct reaction, silicon is consumed. For example in the WSi_2 reaction, for each angstrom of tungsten thickness consumed, 2.53Å of silicon are consumed, resulting in 2.58Å of WSi_2 being formed. Care must be exercised that sufficient silicon is available when using this technique.

Co-Evaporation

The evaporation method utilizes the simultaneous deposition of the metal and the Si under high vacuum conditions. The metal and silicon can be vaporized by electron beam, rf induction, laser, or resistive heating. Since the refractory metals (i.e. Ti, Ta, Mo, W) have very high melting points (1670°C to 2996°C), and silicon has a low vapor pressure, e-beam evaporation is the technique of choice (see Chap. 10). Typically, two guns are employed with separate power supplies. Careful determination of the evaporation rate as a function of power for both metal and Si must be determined. The correct amount of power is supplied to each gun to provide the proper (M /Si) ratio. X-ray damage from e-beam evaporation is generally annealed out during the high temperature sinter operation. In addition, the gate oxide is typically protected by polysilicon during the evaporation, which prevents radiation damage of the gate oxide. Careful control of the evaporation base pressure ($<10^{-6}$ torr), evaporation rates, purity of the elements, and residual gases in the vacuum chamber are critical to insuring that films with reproducible characteristics are deposited. The as-deposited resistivity and thickness should be monitored, and the (metal /Si) ratio measured by analytical techniques, as discussed in Chap. 17.

The very high temperatures needed to evaporate the refractory metals puts stringent requirements on the vacuum systems, and causes much higher susceptiblity to contamination due to outgassing. For example, to obtain a vapor pressure of 10^{-2} torr, Ti, Mo, Ta, and W must be heated to 1700°C, 2500°C, 3100°C, and 3200°C, respectively. Thus it appears that only $TiSi_2$ is likely to be satisfactorily deposited by co-evaporation in a manufacturing environment[10] (although $MoSi_2$, $TaSi_2$, and WSi_2 have all reportedly been successfully deposited by co-evaporation in research laboratories[4,11,12]).

Sputter Deposition: Co-Sputtering and Sputtering from Composite Targets

Sputtering is an excellent method for preparation of silicides. Both rf and magnetron sputtering can be employed (see Chap. 10). As in co-evaporation, the sputtering rates of the metal and the Si must be carefully established. Sputtering can be done in older-style sputtering systems from two "S" type guns (see Chap. 10) onto wafers mounted on a planetary. Care must be exercised to maintain the gun stability during the processing. Sputtering from two targets in more recently designed multi-pass sputtering systems can also achieve an appropriate mixture of Si and metal, resulting in a layered structure. After sintering, complete reaction between metal and Si occurs, forming the silicide.   As described in Chap. 10, the base pressure of the sputtering system should be lower than 10^{-6} torr, to minimize incorporation of residual gases during deposition. In the newer sputtering systems the short distance between the target and substrate allows high deposition rates, and therefore shorter deposition times, resulting in lower concentrations of included residual gases.   The step coverage of co-sputtered films is also superior to that of evaporated films, and can be further improved by applying dc bias to the substrate. The bias also results in changes in surface texture (generally rougher), and increased residual gas incorporation and film stress.

Sputtering from a composite target (MSi_x) would appear to be ideal for compositional control and desired stoichiometry. Composite targets for silicide deposition, however, are usually manufactured by powder metallurgical techniques, which employ a mixture of fine particles of metal and Si. The powders are pressed together, and sintered at high temperatures and pressures, to 70-80% of their theoretical density. This fabrication process may result in sodium contamination of the targets. Improvements have been made to these processes, so that targets are available with low levels of contamination. Resistivities of silicide films comparable to those achieved with co-evaporation have been reported. In general, targets with varying ratios of (metal /Si) must be purchased to achieve the optimum stoichiometry in the deposited film. Care must be exercised to insure the sputtering yields of the metal and Si remain equal, otherwise compositional changes will occur over time. Reference 10 compares the properties of co-sputtered and layered sputter-deposited silicides, and reports that both deposition methods are capable of producing good quality silicide films. Reference 3 is another source that describes the properties of various sputtered silicide films.

Chemical Vapor Deposition (CVD) of Silicides

CVD offers several advantages over other techniques for silicide formation including, improved step coverage, higher purity films (low O_2 content), and higher throughput. In order to use CVD techniques, volatile components of the metal must exist. Considerable effort has been

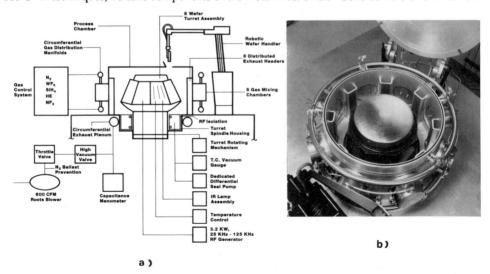

a)
b)

Fig. 9 (a) Schematic of cold-wall reactor for CVD of WSi_2 and W. (b) Photograph of system. Courtesy of Genus Inc.

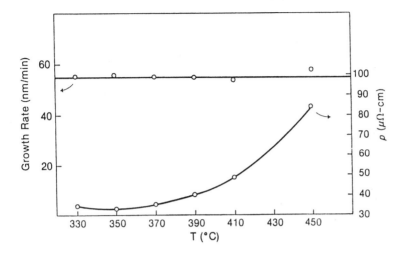

Fig. 10 Growth rate and bulk resistivity of WSi$_2$ vs. deposition temp. Courtesy of Genus Inc.

expended to produce CVD WSi$_2$ and TiSi$_2$ films. TaSi$_2$, MoSi$_2$ have also been prepared by CVD techniques. The reactions used to produce WSi$_2$ and TiSi$_2$ are shown below (note that they are performed at elevated temperatures):

$$WF_6 \text{ (vapor)} + 2SiH_4 \text{ (vapor)} \rightarrow WSi_2 \text{ (solid)} + 6HF + H_2 \qquad (4)$$

$$TiCl_4 \text{ (vapor)} + 2SiH_4 \text{ (vapor)} \rightarrow TiSi_2 \text{ (solid)} + 4 HCl + 2H_2 \qquad (5)$$

The WF$_6$ is a corrosive gas, with a relatively high density, and a moderately-high vapor pressure at room temperature. TiCl$_4$ is a corrosive liquid, with a 300°K vapor pressure of 11 torr.

A commercial cold-wall system has been successfully used for WSi$_2$ deposition[13], and is shown schematically in Fig. 9. Substantial data on WSi$_2$ formation exists for this system. Figure 10 shows the effect of deposition temperature on growth rate and resistivity. The growth rate is relatively insensitive to temperature while the resistivity shows a factor of 3 increase over the range investigated. The resistivity increase is due to the increased silicon in the compound. Figure 11 shows the effect of the WF$_6$ flow on the film growth rate and stoichiometry. The growth rate increases linearly with WF$_6$ flow, while the silicon concentration decreases from x = 2.6 to x = 2.2 over the range investigated. The as-deposited resistivity of the WSi$_x$ is 600 to 900 $\mu\Omega$–cm, which drops to 35-80 $\mu\Omega$–cm, depending on the annealing conditions (i.e. time and temperature). C-V characterization of MOS polycide capacitors fabricated with 2500Å WSi$_2$ /500Å poly Si /1000Å SiO$_2$ /Si, shows a fixed charge Q$_f$ of ~5 x 10^{10} cm^{-2} and a fast interface trap density at midgap of 3x10^{10} cm^{-2} eV^{-1}. These values are comparable to those achieved with polysilicon gate technology. Other systems for CVD deposition of WSi$_2$ have also been developed, including single-wafer, plasma-enhanced deposition machines (see Chap. 6).

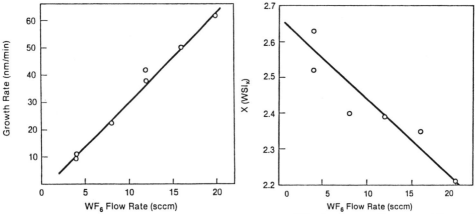

Fig. 11   Growth rate and stoichiometry of WSi_2 films vs. WF_6 flow rate.   Courtesy of Genus Inc.

STRESS IN SILICIDES

During the formation of silicides there is a net volume shrinkage which could possibly result in a large tensile stress in the film.   This stress can result in delamination and other problems during subsequent processing.   As a result of such mechanisms, the stress value in a silicide film is an important parameter.   Stress has been measured in metallurgically formed and co-sputtered samples of $TaSi_2$, $MoSi_2$, $TiSi_2$, $CoSi_2$, $NiSi_2$, and $PtSi$.

Figure 12 shows room temperature measured stress as a function of sinter temperature for $TaSi_2$ and $TiSi_2$.   The stress in these films tends towards compressive after low-temperature sintering, probably as a result of silicon and /or oxygen diffusion in the metal.   At elevated temperature, the stress becomes tensile and independent of temperature.   Stress can be minimized by co-deposition techniques with the ratio of (Si /M) ≥ 2.   The stress in co-sputtered films is sensitive to oxygen contamination of the sputtering gas as well as sputtering pressure.

Stress can result from differences in the thermal expansion coefficient of the film and the substrate, as well as intrinsically from the structure of the film.   Silicides formed by direct reaction result in large volume decreases.   A measure of this shrinkage can be obtained by determining the film thickness of the as-deposited film, and comparing this value to the thickness after sintering.   The volume change can be calculated from[16]:

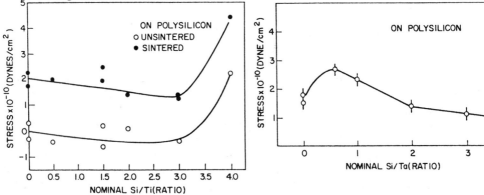

Fig. 12   Room temperature stress in sintered $TiSi_x$ and $TaSi_x$ films on poly-Si as a function of the nominal values of x[14].   Reprinted with permission of American Physical Society.

$$\Delta V (\%) = \frac{[(xV_M + y V_{Si}) - V (M_xSi_y)]}{[xV_M + yV_{Si}]} 100 \qquad (6)$$

where V_M, V_{Si}, and V_{MxSiy} are the molecular volume of the metal, silicon, and silicide, and x and y are the number of metal and silicon atoms in the silicide M_xSi_y. The volume change measured varies from approximately 15% for PtSi to 31% for $ReSi_2$. These large volumetric changes should result in extremely large tensile stresses. A 25% volume shrink corresponds to a linear contraction, ε, of 0.63. Assuming an elastic modulus, E, for the silicide of $1x10^{12}$ dyn /cm^2, the stress ($\sigma = \varepsilon E$) is calculated to be $6.3x10^{11}$ dyn /cm^2. This is 1.5 orders of magnitude larger than generally observed in these materials. Sintered co-deposited films, which show significantly less shrinkage, exhibit stresses similar to a film prepared by direct reaction. It is believed that either the room temperature stresses do not result from the contraction effect, or that the stresses are relieved at the high formation temperature. Retajczyk and Sinha, on the other hand state that the stress is primarily due to differential thermal expansion, and calculated a value of $1x10^{10}$ dyn /cm^2 for a $TaSi_2$ film prepared at 1000°C[17]. This appears to be closer to the measured values of stress in these films.

OXIDATION OF SILICIDES

Often during VLSI processing, an insulating layer must be formed on the top surface of the polycide structure to isolate it from subsequently deposited layers. Vapor deposited SiO_2 and Si_3N_4 can be used for this application, but thermally grown SiO_2 generally possess better material properties. As a result, it is desirable that silicides in a polycide structure be readily oxidizable. The oxide formed must be stable and exhibit adequate electrical and physical properties.

All metal silicide films on polysilicon or single-crystal silicon appear to form SiO_2 by heating in a oxidizing ambient. The silicide film is generally unchanged. Bartur and Nicolet have compiled data on the oxidation kinetics of various silicides from published reports, and Table 4 summarizes their findings[18]. The oxidation process for silicides is presumed to consist of four distinct steps. These are:

1. Diffusion of oxidizing species through the SiO_2 layer;
2. Reaction at the silicide /oxide interface;
3. Transport of Si atoms relative to metal atoms in the silicide; and
4. Reaction at the poly-Si /silicide interface, which releases Si from the poly-Si substrate.

The *transport of the oxidizing species through the oxide* is diffusion controlled. Bartur and Nicolet have plotted the *parabolic rate constants* for wet and dry oxidations (see Chap. 8 for a discussion of these terms) on single crystal, polysilicon, and various silicides. They conclude that the diffusion process through the oxide is essentially the same during the oxidation of all silicon sources. That is, the physical properties of the SiO_2 do not depend on whether it grew from silicon or a silicide. The reaction at the silicide /oxide interface is more difficult to treat. Simple thermodynamic arguments can be used to predict the existence of the oxide. The generalized chemical equilibrium reaction at this interface, assuming limited oxygen supply, and unlimited Si, is written as:

Table 4. DATA ON THE OXIDATION KINETICS OF VARIOUS SILICIDES[18].

Silicide film	Substrate	Kinetics x = thickness t = time	Condition	Rate constants at temperature B/A (cm/s)	B (cm²/s)	(°C)	Activation energy (eV) B/A	B
Si	Si <?>	$x^2 + Ax = B(t + \tau)$	wet	3.5×10^{-8}	8.0×10^{-13}	1000	1.96	0.71
		$x^2 + Ax = B(t + \tau)$	dry	2.0×10^{-9}	3.3×10^{-14}	1000	1.99	1.24
TiSi₂	Si <?>	$x^2 + Ax = B(t + \tau)$	wet	5.6×10^{-8}	6.4×10^{-13}	1000	2.00	1.39
TiSi₂	Si(poly)	$x^2 + Ax = B(t + \tau)$	wet	6.6×10^{-8}	6.4×10^{-13}	1000	2.10	1.51
MoSi₂	SiO₂	$x^2 + Ax = B(t + \tau)$	dry	2.8×10^{-9}	1.9×10^{-14}	1000	1.90	1.60
MoSi₂	Si <?>	$x^2 + Ax = Bt$	wet	1.2×10^{-7}	1.0×10^{-12}	1000	0.84	1.10
TaSi₂	Si(poly)	$x^2 = Bt$	wet		1.4×10^{-13}	1000		1.40
	Si(poly) Si <100> <111>	$x^2 = Bt$	dry		2.8×10^{-14}	1000		1.20
TaSi₂	Si <?>	$x^2 + Ax = Bt$	wet	8.2×10^{-8}	1.0×10^{-12}	1000	0.93	1.10
WSi₂	Si(poly)	$x^{1.82} = B \cdot t$	wet		$B^* = 4.4 \times 10^{-12}$ ª	1000		1.00
	SiO₂	$x^{1.82} = B \cdot t$	wet		$B^* = 6.9 \times 10^{-12}$ ª	1000		0.40
WSi₂	Si <?>	$x^2 + Ax = Bt$	wet	1.3×10^{-7}	6.8×10^{-13}	1000	1.0	1.30

$$[\frac{y}{x} + \frac{1}{2}]\ \text{Si} + \text{M}_y\text{O} \Leftrightarrow \frac{y}{x}\text{M}_x\text{Si}_y + \frac{1}{2}\text{SiO}_2 \qquad (7)$$

where M_xSi is the generalized silicide (e.g. x = 1/2 for WSi₂), and M_yO is the generalized metal oxide. Calculation of the heats of formation of reaction show that all transition metal silicides except, Zr and Hf, should form SiO₂. In addition, Ti, Ni, and Ta are marginal regarding their propensity to form an oxide. TiSi₂ forms an oxide, but only in temperatures in excess of 900°C[19]. Thermodynamics does a reasonable job of predicting which oxides form.

The reaction rates at the SiO₂/silicide interface are factors that affect the *linear rate constant, B/A*. Silicon transport relative to the metal in the silicide may also affect B/A. Bartur and Nicolet discuss several cases which allow preservation of the silicide. Their conclusion is that dissociation during oxidation occurs when the metal is the moving species during silicide formation (NiSi₂, Pd₂Si, PtSi).

Reaction at the Si/silicide interface during oxidation is similar in nature to the reaction at that interface during silicide formation. Silicide formation is usually diffusion (not reaction) controlled at low temperatures. The rate of silicon consumption appears to be roughly the same at high temperatures as for silicide formation a low temperatures. The reaction at the Si/silicide interface is relatively fast at low temperatures, and should be extremely rapid during oxidation at elevated temperatures. This reaction does not control the oxidation rates.

The conclusion is that the variation in oxidation rates of silicides is due to different linear rate constants (B/A)[20]. Silicides in which the *metal* is the moving species oxidize more rapidly than silicon, and only parabolic growth is observed. Silicides in which *silicon* is the moving species are usually formed at high temperature, and the limited Si diffusion may impede the oxidation, thereby reducing the (B/A) term. Oxidation of silicides on SiO₂ (not silicon) results from decomposition of the silicide to form SiO₂ and metal oxides.

A recent study by Hsieh and Nesbit on the oxidation of WSi₂/polysilicon has shown the presence of voids in the polysilicon layer[21]. These voids range in size from several hundred Å to over 1 μm in diameter. They tend to nucleate at the interface and grow into the polysilicon. These voids result from oxide on the polysilicon, preventing diffusion of the Si from the polysilicon to the silicide surface. Localized diffusion occurs at pinholes in the oxide, leading to depletion of the silicon from regions under the pinholes.

The electrical breakdown strength of oxides grown on silicide/polysilicon structures have also been measured for TaSi₂, WSi₂, NiSi₂, and MoSi₂, and are found to range from $2\text{-}5\times10^6$ V/cm (or close to the values measured for oxidized polysilicon)[22,23,24].

PROCESS INTERACTION

In an MOS process the polycide is normally formed after the gate oxide step. Since the polycide approach to reducing interconnect resistance represents an "add on" process to the conventional Si gate process, compatibility with the original process must be maintained. In this section, we discuss some of the properties of silicides that can impact the original process.

First there are some steps that must be added and others that must be modified, in order to *form, pattern,* and *oxidize* the silicide. In addition, the polycide must then remain compatible with the rest of the steps in the process sequence, including: a) the source-drain ion-implant and anneal steps; b) the flow /reflow cycles involving the CVD dielectric (e.g. PSG or BPSG); and c) the Al alloying, passivation, bonding, and package sealing steps. The subjects of silicide *formation* and *oxidation* were covered earlier. The remaining topics will be discussed here.

The *patterning of the polycide* can involve several techniques, depending on how the structure is formed. If the underlying poly-Si is patterned first (e.g. by dry-etching), pure metal is deposited on top, and the silicide is formed by direct metallurgical reaction. The metal that is not deposited on poly-Si remains unreacted, and can be subsequently removed by a selective etch. This is one method for the formation and patterning of a $TiSi_2$ /poly-Si structure. If the silicide is formed on an unpatterned poly-Si layer, the polycide is patterned by dry-etching. The subject of dry-etching polycides is discussed in Chap. 16 and in Refs. 3 and 25.

The highest temperature thermal steps associated with post-polycide formation can occur during the annealing /drive-in of the source-drain implantations (600-900°C, depending on the implant species and conditions), oxidation of the polycide, and flow /reflow of the deposited glass layer (up to 1000°C). The polycide must not exhibit any undesirable properties during these thermal steps. That is, the following characteristics must be demonstrated: a) the silicide must remain chemically stable; b) the poly-Si /silicide interface must not move, and the poly-Si thickness remain unchanged (except during oxidation); c) the silicide should not react with the deposited glass; d) the stress of the silicide film should not increase to unacceptably high levels; and e) the resistivity of the silicide should not degrade. Reports indicate that polycides of $MoSi_2$, WSi_2, $TaSi_2$, and $TiSi_2$ are capable of satisfying the above process compatibility requirements. Note that the oxidation of the silicide proceeds by the diffusion of the silicon from the underlying poly-Si. Therefore, sufficient poly-Si must remain under the silicide (after silicide formation), to supply Si for SiO_2, and still leave an adequate poly-Si underlayer.

At the final steps of VLSI manufacture, one or more layers of aluminum are deposited and patterned, and the chip is passivated, bonded, and packaged. The metal annealing, chip bonding, and sealing temperatures are in the 350-500°C range. Since Al makes contact with the silicide layer, it is important that the Al /silicide interface remain stable at these temperatures. Fortunately, the lowest temperatures at which Al and the refractory metal silicides interact, are above the typical alloy and assembly procedure temperatures [e.g. WSi_2 (~500°C), $TaSi_2$ (550°C), $MoSi_2$ (500°C), and $TiSi_2$ (450-600°C)]. Note that at such temperatures, intermetallics of Al and the metal are formed, as well as free Si, that precipitates into the Al[26]. Such interactions affect the electrical characteristics and the stability of silicide /Si contacts. A thin diffusion barrier (e.g. W) has been suggested to prevent this reaction, if necessary.

SELF-ALIGNED SILICIDE (SALICIDE) TECHNOLOGY

As the contact dimensions of VLSI shrink, the contact resistance increases, and in addition, the sheet resistivity of the shallow-junctions of the source /drain regions also increases. To reduce these resistance values, while simultaneously reducing the interconnect resistance of the polysilicon lines, self-aligned silicide (or *salicide*) technology can be used[27]. That is, metal is deposited over an MOS structure, and reacted with the exposed Si areas of the source and drain, as well as the exposed poly-Si areas on the gate, to form a silicide. Note that side-wall oxidation structures along the gate (known as *oxide spacers*) are used to prevent the gate and source /drain areas from being electrically connected by avoiding silicide formation on this oxide.) Following the silicide formation, a selective etch removes the unreacted metal without attacking the silicide.

Figure 13 shows the key processing steps and final salicide structure[28]. A much lower contact resistance between silicide and Si is achieved than with a conventional contact structure because the area of this interface is much larger than the area of a conventional metal-Si contact structure. (The silicide /Al contact resistance is much lower than metal /Si contact resistance). The salicide structure is also formed after the source-drain implant and anneal steps, and thus must only experience high-temperatures during oxidation and flow /reflow steps.

The most widely used silicide for the salicide process is $TiSi_2$, although PtSi and $MoSi_2$ have also been employed[27,29]. $TiSi_2$ is attractive for this application because it exhibits the lowest resistivity of the refractory metal silicides, and since it can reduce native layers, it is the only known refractory metal that can reliably form a silicide on both poly and single-crystal Si by thermal reaction. However, it also has the following drawbacks: a) the reactivity of Ti with SiO_2 can cause unwanted reaction of Ti and the oxide spacers during the silicide formation process; b) $TiSi_2$ is less stable than WSi_2 or $MoSi_2$; and c) Ti films have a high propensity to oxidize, and hence the silicide reaction must be conducted in ambients that are free of oxygen.

In the salicide structure, the silicide is formed both in the diffusion areas and on the poly-Si gate. The oxide spacers separate these two regions by only about 2000-3000Å. Thus, any lateral formation of silicide can easily bridge this separation and cause the gate to become shorted to the source /drain (and is referred to as *bridging*). It has been observed that if $TiSi_2$ is formed by conventional furnace annealing (i.e. anneal times of ~30 min) in an inert gas (e.g. Ar) atmosphere, such lateral $TiSi_2$ formation rapidly occurs. By annealing in an ambient of N_2, the Ti absorbs a significant amount of N_2 (e.g. >20at%) and at the same time reacts with N_2 and forms a nitride phase at the Ti surface. Once Ti is "stuffed with" (or reacted with) nitrogen, the

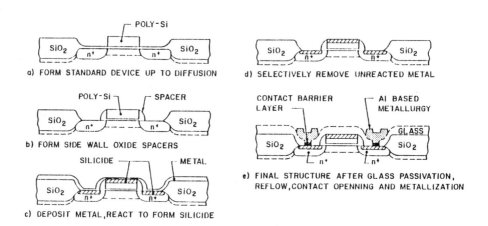

a) FORM STANDARD DEVICE UP TO DIFFUSION

b) FORM SIDE WALL OXIDE SPACERS

c) DEPOSIT METAL, REACT TO FORM SILICIDE

d) SELECTIVELY REMOVE UNREACTED METAL

e) FINAL STRUCTURE AFTER GLASS PASSIVATION, REFLOW, CONTACT OPENNING AND METALLIZATION

Fig. 13   Salicide process flow and final structure. [28]. (© 1985 IEEE.)

lateral silicide reaction is essentially suppressed. Thus, annealing in pure N_2 (i.e. oxygen and moisture content of less than 10 ppma), or pure forming gas (90% N_2 + 10% H_2), results in $TiSi_2$ formation without bridging.

During the $TiSi_2$ formation, the Ti and the spacer SiO_2 can react. Any residues of this reaction can degrade device performance by compromising the oxide integrity, or by producing bridging. To avoid such effects, it is recommended that the $TiSi_2$ formation temperature be held to <700°C, and that a minimum field oxide thickness of 1000Å be utilized. In practice, a two-step formation process has been suggested. During the first step, the temperature is kept at ~650°C. After selectively etching and removing the unreacted Ti in a room temperature mixture of DI H_2O, H_2O_2, and NH_4OH (5:1:1), a second temperature step of ~800°C is used to lower the $TiSi_2$ sheet resistance, and to stabilize the $TiSi_2$ phase[30].

Rapid thermal processing (RTP) has also been used to effect the silicide formation. Wang and Lien report that $TiSi_2$ is formed by RTP at 600-800°C in Ar (reaction time depends on the temperature selected). After selectively removing the unreacted Ti, a stabilization anneal of 1000°C for 30 sec in Ar is conducted to reduce the $TiSi_2$ resistance[31].

Once the $TiSi_2$ is formed and stabilized, it can be subjected to somewhat higher temperatures. Because of instability of the $TiSi_2$ above ~900°C, however, it is recommended that all processing steps after silicide formation be kept below 900°C[28].

REFRACTORY METAL INTERCONNECTS FOR VLSI

As features shrink below 1 µm, and chip sizes increase beyond 1.0 cm^2, polycide sheet resistances of 1-5 Ω /sq can still become the limiting performance factor for VLSI circuits. In these cases it is necessary to use even lower resistance interconnects (such as metal films)[32]. Although aluminum metallization has been widely used for VLSI interconnects, it suffers from its inability to withstand high temperatures processing, which precludes its use in self-aligned MOS processing. This is not the case, however, for the refractory metals (i.e. W, Ti, Mo, and Ta). The applicability of these materials to VLSI interconnect applications has been considered, and the conclusions have been reviewed in several excellent articles. As a result of these findings, extensive efforts have been directed towards developing the chemical vapor deposition

Table 5. PROPERTIES OF SILICIDES AND REFRACTORY METALS

Material	Melting Point (°C)	Resistivity (µΩ-cm)	Thermal Coefficient of Expansion (10^{-6}/°C)
Si	1420	500 (heavily doped poly)	3.0
$TiSi_2$	1540	13-17	10.5
$MoSi_2$	1870	22-100	8.2
$TaSi_2$	2400	8-45	8.8
WSi_2	2050	14-17	6.2
Ti	1690	43-47	8.5
Mo	2620	5	5.0
Ta	2996	13-16	6.5
W	3382	5.3	4.5

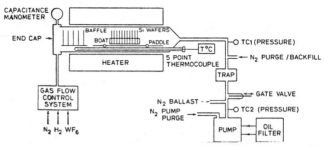

Fig. 14 Schematic of hot-wall low-pressure CVD reactor for tungsten deposition[38]. Reprinted with permission of Electrochemical Society.

(CVD) of tungsten (W) thin films for such applications. Processes for forming CVD-W films both in selective[33] and blanket deposition[34] modes have been successfully pursued. These films have found application as low resistance gate interconnections, ohmic contacts, planarized low resistance vias, and contact barrier materials. This section will be limited to the formation and properties of such CVD-W films. It should be noted that a substantial body of literature also exists on the preparation and properties of other refractory metals. That is, although the subjects are not covered here, the reader should be aware that information is also available on sputtered, evaporated, and CVD films of Ta, Ti, and Mo[35] (as well as W). The refractory metal films typically have a lower resistivity than do their corresponding silicides. Table 5 compares some properties of refractory metals and their silicides.

CVD tungsten appears to be an excellent candidate material for interconnect applications as a result of its low resistance, low stress ($<5 \times 10^9$ dyn /cm^2), excellent conformal step coverage, and because its thermal expansion coefficient closely matches that of silicon. Tungsten also has good electromigration resistance, can form low resistance contacts to silicon, has none of the stoichiometry control problems, that often plague silicides, and it can be selectively deposited. Some of the disadvantages of W include: a) poor adhesion of W films to oxides and nitrides; b) oxides form on W at temperatures > 400°C; c) silicides form at temperatures > 600°C.

Deposition of CVD Tungsten

The chemical vapor deposition (CVD) of tungsten is generally performed in either a hot-wall low-pressure system (see Fig. 14) or a cold-wall, low-temperature system (Fig. 15). Tungsten hexafluoride, WF$_6$, is well-suited as the W source gas, since it can be either reduced by hydrogen or silicon. The silicon reduction is given by[36]:

$$2WF_6 \text{ (vapor)} + 3Si \text{ (solid)} \rightarrow 2W \text{ (solid)} + 3 SiF_4 \text{ (vapor)} \qquad (8)$$

while the hydrogen reduction is given by:

$$WF_6 \text{ (vapor)} + 3H_2 \text{ (vapor)} \rightarrow W \text{ (solid)} + 6HF \text{ (vapor)} \qquad (9)$$

Other source gases such as WCl$_6$ have also been employed in the hydrogen reduction, and the other refractory metals are generally reduced by hydrogen from their respective chlorides.

The hydrogen reduction may result the selective deposition of W. However, the reaction requires good nucleating surfaces. Silicon, metal, and silicide surfaces provide good sites, while SiO$_2$ and Si$_3$N$_4$ (especially at low temperatures), do not. On a silicon surface the deposition

Fig. 15 An SEM photograph of a blanket deposited W film. Courtesy of Genus Inc.

starts by the Si reduction, but after the reaction becomes self-limiting, the H_2 reduction takes over. At the outset of the deposition the carrier gas is Ar. After the Si reduces WF_6, H_2 is added to the gas flow, and the Ar flow is stopped.

Broadbent and Ramiller[36] studied the deposition kinetics of hydrogen reduction of WF_6. Their study spanned the temperature range of 250-500°C, and the pressure range of 0.1-5 torr. Figure 16 shows an Arrhenius plot for the deposition rate of W for several temperatures. An activation energy of 0.71 eV is calculated for all pressures. The pressure dependence of the deposition rate is shown in Fig. 17. The dependence can be written as $R \propto P^{1/2}$. A growth rate dependence was also observed for the H_2 partial pressure at constant WF_6 pressure, namely $R \propto P_{H2}^{1/2}$. The growth rate however was found to be independent of WF_6 partial pressure. These dependencies suggest that *surface adsorbed H_2 dissociation* is the rate limiting reaction mechanism. The activation energy of 0.71 eV is in close agreement with the value of 0.69 eV

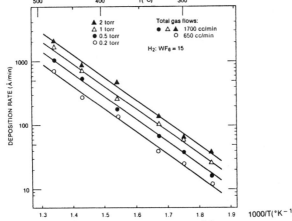

Fig. 16 Arrhenius plot of deposition rate of CVD tungsten[36]. Reprinted with permission of Electrochemical Society.

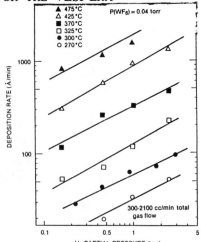

Fig. 17 Deposition rate dependence of CVD tungsten on pressure[36]. Reprinted with permission of Electrochemical Society.

for the H_2 surface diffusion on W^{37}.

The silicon reduction of WF_6 can be used to produce thin films of selectively deposited W. The reaction of Eq. 9 is self-limiting, since once the W layer is deposited, it serves as a diffusion barrier between the Si and the WF_6. Recent studies over the temperature range of 200-400°C show that the deposition proceeds rapidly at first (>1000Å /min) and essentially stops when the film reaches a thickness of 100-150 Å[38,39]. These films have excellent surfaces and sheet resistances of 10-15 Ω /sq. For each 2 atoms of W deposited, 3 atoms of Si are consumed and volatilized as SiF_4. It is found that typically 150-200Å of Si are consumed during the 100-150Å W deposition. No deposit occurs on the SiO_2 during the reaction.

Selective Deposition of Tungsten

It is possible to selectively deposit W on silicon using a two step process. The first step

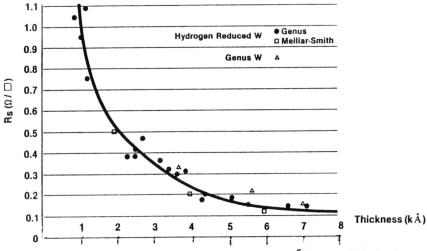

Fig. 18 Sheet resistance of CVD tungsten as function of film thickness[5]. Courtesy of Genus Inc.

involves the reduction by Si, and forms ~100Å of W. The second step utilizes the H_2 reduction, but is carried out under conditions in which the W deposits only on the metal W layer deposited in the first step, and not on the oxide. The deposition rate for this step is ~ 80Å /min. A set of conditions that favor selective deposition by H_2 reduction have been published[38]. They include: a) lower temperature of deposition; b) lower total and /or partial pressures of WF_6; c) reduced number of wafers in the deposition chamber (and use of dummy wafers near the exhaust); d) reduced selective deposition area on a wafer; e) short deposition times (low thickness); f) proper cleaning of the wafers; and g) use of undamaged SiO_2 (i.e. free from damage produced by dry etch or ion implantation). A favorable set of deposition conditions is also reported as being: a) T = 300°C; b) P_{total} = 0.25 torr; c) H_2 flow 75cc /min; and d) WF_6 flow 10cc /min. This results in selectively deposited W films of 1000Å in thickness. Other studies have reported films grown selectively, as thick as 3000Å.   Blewer and Wells[40] reported on selective films in excess of 7000Å at 300°C and 0.75 torr with (H_2 /WF_6) = 220. This is significantly higher pressure and H_2 flow than reported in previous studies and is somewhat contrary to the findings of Ref. 38.

Properties of CVD Tungsten for VLSI Contacts

Figure 18 shows the resistivity of CVD tungsten as a function of film thickness, which increases as the thickness decreases below 8000Å. This probably results from decreased grain size in the thinner films. Gaseous impurities (C, N, O) also tend to increase resistivity.   The contact resistance of CVD-W (400Å thick) to n^+ and p^+ Si has been measured, and is shown in Fig, 19 as a function of surface doping. It was also reported that a 450°C anneal did not affect this contact resistance, indicating that 400Å of W behaves as a good barrier to Si diffusion at 450°C. The contact layer is very clean since ~150Å of Si is etched during the initial process, along with the native oxide on the silicon surface. Schottky barriers fabricated from the selective W on n-type Si have a barrier height of 0.62 eV, and an ideality factor[42] of 1.02.

Two problems associated with selective W deposition in oxide cuts are: a) *encroachment*; and b) *tunnels*. Figure 21 shows a schematic of these effects. The effects of temperatures and WF_6 partial pressure on lateral encroachment have been investigated.   It has been determined that

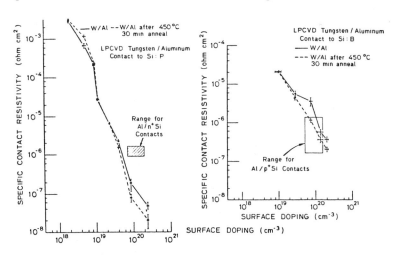

Fig. 19 Specific contact resistivity of CVD tungsten to B and P doped junctions before and after a 450°C anneal in forming gas[38]. Reprinted with permission of Electrochemical Society.

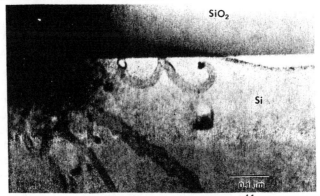

Fig. 20 (a) Detail showing tunnel deflected by SiO_2-Si interface[41]. Reprinted with permission of Electrochemical Society.

below 300°C, and at low WF_6 partial pressures, encroachment is negligible. The interfacial structures of selective W on silicon have also been studied[41]. It has been reported that the W film itself consists of two layers, separated by an interfacial oxide. The first layer results from the Si reduction, while the second from the hydrogen reduction. The surface preparation plays a strong role on the behavior of the W /Si interface. A thin uniform oxide, or no oxide at all, results in a smooth interface with minimum amount of Si consumption. A non-uniform oxide results in irregular interfaces and "tunnels" which extend 1 μm or more into the Si substrate (Fig. 20). One model holds that the tunnels are formed by the etching action of trapped F adsorbed on W particles, but this has not yet been conclusively demonstrated.

Future Trends

By 1986, many of the issues regarding the use of selective and blanket W have been resolved, while others are under continuing investigation. The well-understood phenomena, and resultant processes, include: 1) a demonstration of a suitable selective CVD process for contact protection and via refill; 2) acceptable adhesion of W to underlying substrate films; 3) the causes of encroachment have been identified, and measures which prevent its occurrence have

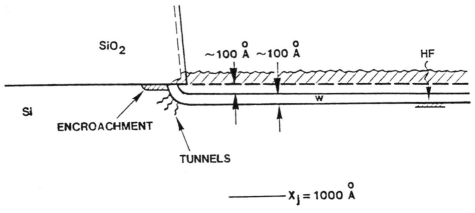

Fig. 21 Schematic drawing of encroachment and tunnels accompanying deposition of CVD W.

been developed; 4) deposition has been successfully accomplished in contact windows as small as 0.5 μm; and 5) selectivity has been achieved on Al and TaSi$_2$. Problems still under investigation include: 1) the origin and elimination of *tunnels* is not yet in hand; 2) complete control over selectivity has still not been accomplished; 3) unwanted silicidation at temperatures above 600°C still occurs; and 4) control over cleaning effects in small contacts has not been solved.

REFERENCES

1. A.K. Sinha, "Refractory Metal Silicides for VLSI Applications", *J. Vac. Sci. and Technol.*, **19**, 778 (1981).
2. S.P. Murarka, *Silicides for VLSI Applications*, Academic Press, Orlando, Fla., 1983.
3. T.P. Chow and A.J. Steckl, "Refractory Metal Silicides: Thin-Film Properties and Processing Technology", *IEEE Trans. on Electron Devices*, Vol. ED-30, No. 11, Nov. '83, p. 1480.
4. T.O. Sedgwick, F.M. d'Huerle, and S.A. Cohen, "Short Time Annealing of Co-evaporated WSi$_2$ Films", *J. Electrochem. Soc.*, **131**, Oct. '84, p. 2446.
5. S. Sachdev and R. Castellano, "CVD Tungsten and Tungsten Silicide for VLSI Applications", *Semiconductor International*, May '85, p. 306.
6. S.P. Murarka and D.B. Fraser, *J. Appl. Phys.*, **51**, 350 (1980).
7. *ibid.*, Ref. 1, p. 115.
8. S.P. Murarka and D.B. Fraser, *J. Appl. Phys.*, **51**, 350 (1980).
9. C. Canali, *et al.*, *Appl. Phys. Letts.*, **31**, 43, (1977).
10. L. Kammerdiner and M. Reeder, "Co-Deposition vs. Layering of Sputtered Silicide Films", *Semi. Intl.*, Aug. '84, p. 122.
11. F. Neppl, G. Menzel, and U. Schwabe, "Properties of Mo-Silicides in Si-Gate Technology", *J. Electrochem. Soc.*, **130**, May '83, p. 1174.
12. F. Neppl and U. Schwabe, *IEEE Trans. Electron Devices*, **ED-29**, 508 (1982).
13. D.L. Brors, *et al.*, "Properties of Low Pressure CVD WSi$_x$ as Related to IC Process Requirements", *Solid State Technol.*, April '83, p. 183.
14. S.P. Murarka, *J. Vac. Sci. Technol.*, **17**, 775 (1980).
15. S.P. Murarka and D.B. Fraser, *J. Appl. Phys.*, **51**, 342 (1980).
16. *ibid.*, Ref. 2, p. 60.
17. T.F. Retajczyk and A.K. Sinha, *Thin Solid Films*, **70**, 241 (1980).
18. M. Bartur and M.A. Nicolet, "Thermal Oxidation of Transition Metal Silicides on Si: A Summary", *J. Electrochem. Soc.*, **131**, Feb. '84, p. 371.
19. F. D'Huerle, E.A. Irene, and C.Y. Ting, *Appl. Phys. Letters*, **42**, 361 (1983).
20. J.M. DeBlasi, R.R. Razouk, and M.E. Thomas, "Characteristics of TaSi$_2$/Poly-Si Films Oxidized in Steam for VLSI Applications", *J. Electrochem. Soc.*, **130**, Dec. '83, p. 2478.
21. N. Hshieh and L. Nesbit, "Oxidation Phenomena of PolysiliconTungsten Silicide Structures", *J. Electrochem. Soc.*, **131** Jan. '84, p. 202.
22. S. P. Murarka, *et al.*, *J. Appl. Phys.*, **51**, 3241 (1980).
23. H. J. Geippl, *et al*, *IEEE Trans. Electron Dev.* **ED-27**, 1417, (1980).
24. T. Moshizuki and M. Kashiwagi, *J. Electrochem. Soc.*, **127**, 1128 (1980).
25. R.W. Light and H.B. Bell, "Patterning of Tantalum Polycide Films", *J. Electrochem Soc.*, **131**, Feb. '84, p. 459.
26. *ibid.*, Ref. 2, p. 156-61.
27. C.M. Osburn, *et al.*, *Proceedings VLSI Science and Technol.* **82-7**, Electrochem. Society, Pennington, N.J., p. 213, 1982.

Conf., 1985 IEEE, New York, p. 307, 1985.

29. H. Okabayashi, *et al.,* "Low Resistance MOS Technology Using Self-Aligned Refractory Silicidation", in *IEDM Tech. Dig.*, p. 555, 1982.

30. C.Y. Ting, *et al*, "Interaction Between Ti and SiO_2", *J. Electrochem Soc.*, **131**, p. 2934,1984.

31. A. Wang and J. Lien, "A Self-Aligned Titanium Polycide Gate and Interconnect Formation Scheme Using Rapid Thermal Annealing", *VLSI Science and Technology, 1985,* Eds. W.M. Bullis and S. Broydo, Electrochem. Soc, Pennington, N.J., 1985, p. 203.

32. P. Burggraaf, "Silicide Technology Spotlight", *Semi. International.*, May '85, p. 293.

33. N.E. Miller and I. Beinglas, *Solid State Technology*, Dec. '82, p. 85.

34. K.C. Saraswat, *et al., IEEE Trans. on Elect. Dev.*, **ED-30**, No. 11, p. 1497, (1983).

35. M. Suzuki and K. Asai, "Characteristics of Sputter-Deposited Mo Films", *J. Electrochem. Soc.*, **131**, Jan. '84, p. 186.

36. E.K. Broadbent and C.L. Ramiller, *J. Electrochem. Soc.*, **131**, No. 6, p. 1427 (1984).

37. G. Ehrlich, in *Metal Surfaces*, p. 236, American Society for Metals, Metals Park, OH, (1963).

38. K.C. Saraswat, S. Swirhun, and J.P. McVittie, "Selective CVD Tungsten for VLSI Technology", *Proc. Second Int'l Symp. on VLSI Science and Technol.*, Electrochem. Society Proceedings, **84-7**, p. 409 (1984).

39. Y. Pauleau and P. Lami, "Kinetics and Mechanism of Selective Tungsten Deposition by LPCVD", *J. Electrochem. Soc.*, **132**, Nov. '85, p. 2779.

40. R.S. Blewer and V.A. Wells, "Thick and Selective Tungsten Films by LPCVD Processes", *Proceedings Multilevel Metallization and Packaging Symposium,* Electrochem. Society, Pennington, N.J., 1984, p. 200.

41. W.T. Stacy, E.K. Broadbent, and M.H. Norcott, "Interfacial Structure of Tungsten Layers Formed by Selective LPCVD", *J. Electrochem. Soc.*, **132**, Feb. '85, p. 444.

PROBLEMS

1. Calculate the resistance of a line that is 2 µm wide, 1500 µm long, and 5000Å thick, when it consists of each of the following materials (a) n^+ polySi (R_s = 30 Ω / sq), (b) aluminum (0.25 Ω /sq), (c) WSi_2, (d) $TaSi_2$, (e) $TiSi_2$, and (f) $MoSi_2$ (see Table 2 for the R_s values of the silicides).

2. Repeat problem 1 for a polycide structure whose silicide layer is 3500 Å thick, and whose polysilicon layer is 1500Å thick and has the same R_s as in problem 1. Assume the silicide is $TiSi_2$ in one case, and $MoSi_2$ in another.

3. What parameters (material property and process fabrication procedures) must be controlled in a polycide fabrication process that are not of concern when polysilicon alone is used as a gate/interconnect material?

4. If a polycide structure is to be fabricated by producing WSi_2 by a direct metallurgical reaction, how thick must the polysilicon layer be in order for the final polycide structure to consist of a 3500 Å layer of WSi_2 on a 1500 Å layer of polysilicon? What would happen if all of the polysilicon were consumed by the reaction that forms the silicide?

5. Discuss the difficulties associated with each of the methods used to form refractory metal silicides.

6. Stresses occur in silicide films as a result of various fabrication steps. Explain how such stresses can have negative impact on the circuits fabricated with such stressed films.

7. Using Table 4, estimate which of the silicides shown would grow the thickest oxide under the same growth conditions. Does polysilicon grow oxide more or less rapidly?

8. Explain how the silicide layers formed in the source /drain regions of a *salicide structure* are kept from becoming shorted to the silicide formed on the gate region of the salicide structure.

12

LITHOGRAPHY I :

OPTICAL PHOTORESIST MATERIALS

and PROCESS TECHNOLOGY

Microcircuit fabrication requires that precisely controlled quantities of impurities be introduced into tiny regions of the silicon substrate, and subsequently these regions must be interconnected to create components and VLSI circuits. The patterns that define such regions are created by lithographic processes. That is, layers of photoresist materials are first spin-coated onto the wafer substrate. Next, the resist layer is selectively exposed to a form of radiation, such as ultraviolet light, electrons, or x-rays. An exposure tool and mask, or data tape in electron beam lithography, are used to effect the desired selective exposure. The patterns in the resist are formed when the wafer undergoes the subsequent "development" step. The areas of resist remaining after development protect the substrate regions which they cover. Locations from which resist has been removed can be subjected to a variety of additive (e.g. lift-off) or subtractive (e.g. etching) processes that transfer the pattern onto the substrate surface.

Three chapters are devoted to the details of lithographic processing for VLSI. The first is concerned with the properties of photoresist materials, and the resist processing technology utilized in VLSI fabrication. The discussion is restricted to resists exposed by optical (e.g. UV) radiation. The subject of the second chapter deals with the tools used to expose the resist. That is, optical aligning equipment and photomasks are described. The third chapter covers advanced lithography, including electron beam, ion beam, and x-ray patterning technology.

BASIC PHOTORESIST TERMINOLOGY

The basic steps of the lithographic process are shown in Fig. 1. The photoresist (PR) is applied as a thin film to the substrate (e.g. SiO_2 on Si), and subsequently exposed through a *mask* (or *reticle* in step-and repeat projection systems). The mask contains clear and opaque features that define the pattern to be created in the PR layer. The areas in the PR exposed to the light are made either soluble or insoluble in a specific solvent known as a *developer*. In the case when the irradiated (exposed) regions are soluble, a positive image of the mask is produced in the resist. Such material is therefore termed a *positive* resist. On the other hand, if the *nonirradiated* regions are dissolved by the developer, a negative image results. Hence the resist is termed a *negative* resist. Following development, the regions of SiO_2 no longer covered by resist,

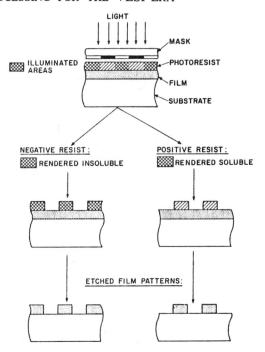

Fig. 1 Exposure and deveiopment of negative and positive photoresists, and the resulting etched film patterns. From L. Maissel and R. Glang. Eds. *Handbook of Thin Film Technology,* 1970. Reprinted with permission of McGraw-Hill Book Company.

are removed by etching, thereby replicating the mask pattern in that oxide layer.

The resist is seen to perform two roles in this process. First, it must respond to exposing radiation in such a way that the *mask image* can be replicated in the resist. Second, the remaining areas of resist must protect the underlying substrate during subsequent processing. In fact, the name *resist* evolved from the ability of these materials to "resist" etchants.

Although both negative and positive optical resists are used to manufacture semiconductor components, the higher resolution capabilities of positive resists have virtually made them the exclusive choice for VLSI applications. As such, almost all of the discussion in this chapter will be limited to positive resists and their processing. Conventional positive optical lithographic processes and resists are capable of producing images on VLSI substrates with dimensions as small as 0.8-1.5 µm. For submicron features, however, diffraction effects during exposure may ultimately cause other higher resolution techniques to replace optical lithography.

In general, users of resists are not overly concerned with the complexities of resist chemistry, but rather how well the resist will function in their process. The majority of the information in this chapter is presented in this vein.

Conventional optical photoresists are three-component materials, consisting of: a) the matrix material (also called *resin*), which serves as a binder, and establishes the mechanical properties of the film; b) the sensitizer (also called the *inhibitor*), which is a photoactive compound (PAC); and c) the solvent (different than the developer solvent),which keeps the resist in the liquid state until it is applied to the substrate being processed. The *matrix material* is usually inert to the incident imaging radiation. That is, it does not undergo chemical change upon irradiation, but provides the resist film with its adhesion and etch resistance. It also determines other film properties of the resist such as thickness, flexibility, and thermal flow stability.

The *sensitizer* is the component of the resist material that reacts in response to the actinic radiation. The term *actinic* relates to the property of radiant energy (especially in the visible and ultraviolet regions) by which photochemical changes are produced. The sensitizer gives the resist its developer resistance and radiation absorption properties.

PHOTORESIST MATERIAL PARAMETERS

As previously indicated, photoresist performs two primary functions: 1) precise pattern formation; and 2) protection of the substrate during etch. A large group of material properties possessed by the resist play a role in how effectively these functions are performed. The material parameters can be grouped into three categories: a) *optical properties*, including, resolution, photosensitivity, and index of refraction; b) *mechanical /chemical properties*, including, solids content, viscosity, adhesion, etch resistance, thermal stability, flow characteristics, and sensitivity to ambient (e.g. oxygen) gases; and c) *processing and safety related properties*, including, cleanliness (particle count), metals content, process latitude, shelf life, flashpoint, and threshold limit value (TLV, a measure of toxicity). In this section we discuss these parameters. By understanding the role that each parameter must fulfill to allow the resist to function effectively, users can better select and match resists to their particular applications[1,2,3]. Methods for measuring the values of most of the parameters are also given.

Resolution

The resolution of a lithographic process is formally defined in Chap. 13 in terms of the *modulation transfer function* of the lithography exposure equipment and how well it is matched to the resist being utilized. The term, however, is also less formally, but still widely used in the industry in a more practical sense, to *specify the consistent ability to print minimum size images under conditions of reasonable manufacturing variation*. Thus, it is fair to say that in order to build devices with submicron features, lithographic processes with submicron resolution

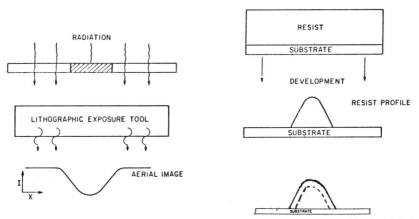

Fig. 2 Idealized photolithographic system. Final drawing shows resist profile (——) before and (- - -) after an anisotropic etch[5]. Reprinted with permission of Academic Press.

capabilities must be available.

It is also useful to define the term *linewidth* at this point. *Linewidth* is the horizontal distance, L, between the resist material /air boundaries in a given cross-section of the line, at a specified height above the resist-substrate interface (Fig. 2). Since different measurement methods (e.g. SEM or optical) may in effect, measure the linewidth of the same line at different heights of the cross-section, it is convenient to identify the technique used to obtain the linewidth (e.g. *SEM* linewidth, or *optical* linewidth). Linewidth measurement techniques are discussed in *Linewidth Measurements*, which is part of the *After Development Inspection* section.

The resolution of a lithographic process can be limited by many aspects of the process, including: a) hardware (e.g. diffraction of light, lens aberrations, and mechanical stability of the system); b) optical properties of the resist material; and c) process characteristics (e.g. softbake step, develop step, postbake step, and etching step)[1,3,4]. The *hardware limitations* to resolution are considered in Chap. 13, and the resolution limitations due to the *resist process* are discussed in the appropriate resist processing sections later in this chapter. In the present section, the *resist material properties* that impact resolution, such as contrast, swelling behavior, and thermal flow are considered. The *contrast* and *swelling* behavior of resist materials are manifested in response to the exposure and development steps of the resist process, while the *thermal flow* impacts the resolution during the post-bake and dry-etching steps. We consider contrast and swelling at this point, and elaborate on thermal flow in the *Resist Processing: Postbake* section.

Resolution: Contrast

Theoretically, the *contrast* of a resist has been shown to directly influence resolution, resist wall angle, and linewidth control over steps. A measure of contrast is given by the value of a quantity called γ. The descriptive explanation of the concept of *contrast*, is best undertaken by considering negative and positive resists separately. For a *negative resist*, the contrast is related to the *rate of crosslinked network formation* at a constant irradiation dose. Therefore, if one negative resist type has a higher crosslinking rate compared to another, it also possesses the higher contrast of the two. For *positive resists*, the contrast is related to the rate of *chain scission* and *change of solubility* at a constant irradiation dose.

Both the contrast and sensitivity of resists are usually measured by exposing resist layers of a given thickness to varying radiation doses, and then determining the film thickness remaining after a fixed development step. (Note that methods of measuring resist dissolution are discussed in the section entitled *Resist Processing: Development*.) This information is plotted in the form of a *response*, or *sensitivity* curve (see Fig. 3).

In the case of *negative resists*, the onset of crosslinking, as evidenced by gel formation, is not observed until a critical dose is received (D_g^i in Fig. 3a). Thus, below this dose no image has yet formed. At higher doses, the image film thickness increases, as a result of increasing gel content, until the thickness of the imaging film equals that of the resist film prior to exposure (i.e. at dose D_g^o in Fig. 3a). The value of the contrast of a negative resist, γ_n, is found from the slope of the response plot as it intersects the log dose axis (Eq. 2a).

In *positive resists*, the resist thickness becomes smaller with increasing irradiation dose, until complete film removal occurs (i.e. when a critical dose D_c is reached). The measurement technique is the same, except that a solvent is used that does not appreciably attack unexposed positive resist. The value of film thickness in the exposed regions is measured and *normalized to the thickness prior to exposure and development*. These residual thickness values are plotted as shown in Fig. 3b. Contrast for positive resist films, γ_p, is found from the absolute value of extrapolated slope of the linear portion of the response curve (Eq. 2b).

Contrast can also be calculated using the following expressions[6]. Since γ_n and γ_p are found

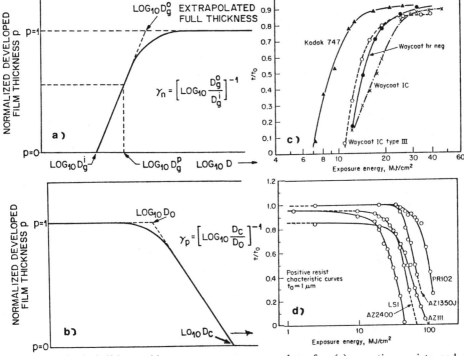

Fig. 3 (a) Typical lithographic-response or contrast plots for (a) negative resists and (b) positive resists in terms of the developed thickness normalized to initial resist thickness (p) as a function of log (dose). Reprinted from Ref. 1 with permission of the American Chemical Society. c) and d) Contrast plots for actual resists. Reproduced with permission, from the Annual Review of Materials Science, Volume 6, © 1976 by Annual Reviews Inc.

from the slope of the linear portions of the response curves, and the slope in general is given by:

$$\text{Slope} = \frac{Y_2 - Y_1}{X_2 - X_1} \qquad (1)$$

where: $Y_2 = 1.0$, $Y_1 = 0$, $X_2 = \log_{10} D_g^o$, and $X_1 = \log_{10} D_g^i$ for negative resists; and $Y_2 = 0$, $Y_1 = 1.0$, $X_2 = \log_{10} D_c$, and $X_1 = \log_{10} D_o$ for positive resists.   Then:

$$\gamma_n = \log_{10} \left[\frac{D_g^o}{D_g^i} \right]^{-1} \qquad (2a)$$

$$\gamma_p = \log_{10} \left[\frac{D_c}{D_o} \right]^{-1} \qquad (2b)$$

Resists with higher contrasts result in better resolution than those with lower contrast. How this arises is illustrated with the aid of Fig. 4.   An ideal exposure process would deliver the desired exposing radiation only to the resist region whose dimensions are equal to the pattern of the mask, and there would be no energy delivered elsewhere.   In real exposures, however, energy is delivered in a more diffuse fashion, due to diffraction and scattering (as shown in Fig. 4). This energy then strikes the resist layer (positive in this example).   In areas of the resist where exposure energy is greater than D_c, the resist will be completely dissolved during development. Since actual resists do not have infinite contrast, there will be other regions of the resist that, upon receiving energy less than D_c, but greater than D_o, will partially dissolve. Thus, the resultant resist profile after development will exhibit some slope.   The higher the resist contrast value, the more vertical would be the resist profile.   Since linewidth is measured at a specified height above the resist /substrate interface, as the resist profile is less vertical, the less accurately will the resist linewidth represent the mask dimension. In addition, the actual linewidth *after etch* is the dimension that must actually be controlled. Since dry-etch processes usually erode resist materials to some degree (Fig. 2), a less vertical resist profile will lead to more linewidth erosion in the etched film, thereby further reducing resolution.

The angle of the resist profile at the point where the resist thickness falls to zero has been derived by King[5], and is shown to be directly proportional to γ and the resist thickness. Ideal resists would therefore exhibit infinite contrast, and would thus have response curves as shown in Fig. 4a.   Typical response curves of actual resists are shown in Figs 4b.   Positive optical organic resists exhibit higher contrasts ($\gamma_p \cong 2.2$) than negative optical resists ($\gamma_n \cong 1.5$). Inorganic resists based on Ge-Se chalcogenide glasses have much higher contrast than either positive or negative organic resists ($\cong 6.8$)[31].

Besides being used to calculate resist wall angles, contrast values are useful in predicting the smallest feature sizes printable by given exposure system.   A term has been defined[7], called the *critical resist modulation transfer function,* $CMTF_{resist}$:

$$CMTF_{resist} = \frac{I_{100} - I_o}{I_{100} + I_o} \qquad (3)$$

where I_{100} is the minimum exposure energy for 100% film loss, and I_o is the maximum energy for zero exposure.   In order for an exposure system to adequately print a given feature, its modulation transfer function MTF (defined in Chap. 13) for that feature size, must be greater than or equal to the $CMTF_{resist}$ for the resist used.   Knowing the contrast of a resist is useful as a result of the relation between it and the $CMTF_{resist}$, as shown in Fig. 5.

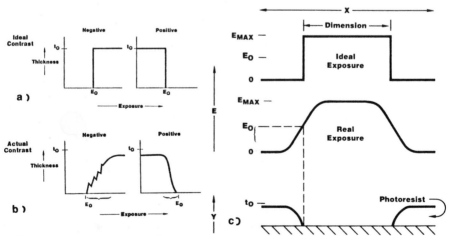

Fig. 4   (a) Ideal contrast curves. (b) Actual contrast curves. (c) The actual dimensions and slope of an exposed and developed resist[4].   Reprinted with permission of SPIE.

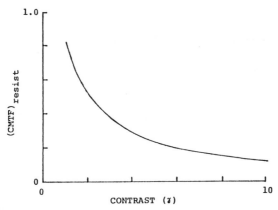

Fig. 5 Relationship between critical resist MTF (CMTF$_{resist}$) and contrast, γ^7.   Reprinted with permission of the American Physical Society.

This relationship is defined by:

$$\text{CMTF}_{resist} \; = \; \frac{10^{1/\gamma} \; - \; 1}{10^{1/\gamma} \; + \; 1} \tag{4}$$

Thus, if the MTF of an exposure system is known for various feature sizes, knowledge of the γ of a resist, and Fig. 5, will allow a prediction of the smallest size features that are effectively printable with that system. Note that an MTF of 1 roughly corresponds to perfect image transfer, and the MTF value at 1 μm for projection printers available in 1986 (see Chap. 13) is ~0.6. Thus, from Fig. 5, positive resists with $\gamma = 2$ can produce satisfactory images of ~1.0 μm with such projection printers. The contrast value also impacts the linewidth variation of resist patterns that cross over steps. This issue is discussed in the section *Resist Processing: Exposure.*

Resolution: Swelling, Proximity Effects, and Resist Thickness

If the resist volume increases during the development step (*swelling*), as a result of penetration of the resist material by developer solution, the feature size of the pattern created in the resist is altered. Such swelling in most negative resists. For features smaller than 3 μm this produces a change in feature size that is unacceptably large compared to the specified dimension. As a result, negative resists are inadequate for controllably producing such features. Positive resists (of low molecular weight) do not exhibit such swelling, due to a different dissolution mechanism during development, as is discussed in the sections on *Positive and Negative Resists.*

The resolution of both positive and negative resists is also impacted by *proximity effects.* That is, an isolated line is most easily resolved in negative resist, while an isolated hole is easier to define in positive working materials. When evaluating the resolution of a particular resist, the feature that is most difficult to resolve should always be selected as the ultimate benchmark[8].

Resolution is also improved by thinning the photoresist. Conversely, since positive resist has higher contrast, and exhibits less swelling than negative resists, it can be used in thicker layers than negative resist, and still provide equal resolution. Thus, if equal resolution can be obtained, thicker resist is advantageous in that it provides better step coverage, more defect protection, and greater dry etch resistance.

Sensitivity

The *sensitivity* of an optical photoresist can be defined as the input energy (measured in terms of the number of photons per unit area) required to cause a specified degree of chemical response in the resist, which results (after development), in the desired resist pattern. For photo-chemical reactions, the response is expressed quantitatively in terms of the photo-efficiency, or quantum yield, Φ, defined as[9]:

$$\Phi \; = \; \frac{\text{Number of photo-induced events}}{\text{Number of photons absorbed}} \qquad (5)$$

A positive resist that operates by the photo-scission of sensitizer molecules, must possess a high Φ in order to exhibit high sensitivity. After resolution, *sensitivity* is the most important resist parameter, with *high sensitivity* generally being a desirable characteristic. This stems from the circumstance that as the resolution requirements increase, the irradiating wavelengths must shift deeper into the UV region. In this range, the brightness of existing light sources is severely reduced, and the efficiency of the optical elements is also reduced (i.e. they absorb more of the energy passing through them). Since the total input energy incident on a resist is the product of light source brightness, time, and absorption efficiency of the aligner optics, a decrease in the first and third of these factors must be compensated by longer exposure times. This translates into a smaller throughput of wafers per hour, thereby requiring an increase in the number of exposure tools required. The availability of resists with increased sensitivity would allow shorter exposure times. It should be noted, however, that in PR processes which utilize advanced steppers, sensitivity alone is not the limiting throughput factor. That is, such procedures as alignment, focus, and wafer loading and unloading, take significant time, and limit throughput.

In discussing the resist sensitivity parameter, however, it turns out that there are several aspects of interest to the process engineer. The first involves the match between the resist and the energy source of the exposure tool. This information is provided in the form of a *spectral-response curve,* which is a measure of the response of the resist to the spectrum of the light emitted by the source that illuminates the resist. The spectral-response curve (Fig. 6a) gives a general indication of the effectiveness of a source in exposing a resist. If the resist absorbs strongly in the ranges where the source has strong emission lines, relatively short exposure times should result (implying high resist sensitivity). Spectral-response curves do not, however, provide quantitative information on exposure times.

It is also important that the absorbance be tuned to the resist system, so that stray light (e.g. yellow and green ambient light) which strikes the resist will cause minimal exposure effects. Furthermore, it is generally desirable to have no more than about 40% absorbance by the unexposed resist at the actinic wavelength. When the percent absorbance exceeds this value, the exposure gradient through the resist depth degrades the image profile, since a large percentage of the light will be absorbed by the upper layers of resist. On the other hand, if the absorbance is too low, exposure times become impractically long.

The *actinic absorbance* is the next aspect of the resist sensitivity parameter that must be considered. This property is defined as the difference between the absorbance of the unexposed and exposed resist. Figure 6b shows the actinic absorbance of S-1400® positive resist vs. wavelength. A sensitive resist must have a high actinic absorbance over the irradiation spectral range. That is, light passing through unexposed resist should be strongly absorbed. Upon exposure and photo-chemical transformation, however, the material should become transparent to

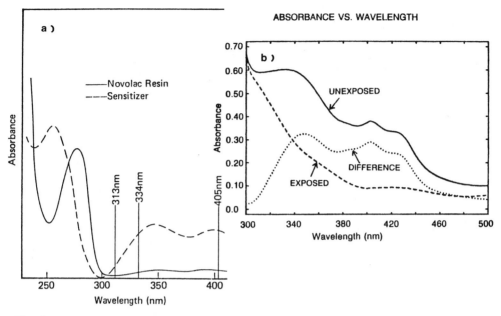

Fig. 6 (a) Absorbance spectrum of a typical diazonaphthoquinone sensitizer (in solution) and a novolac film. The wavelengths of the principle mercury emission lines are labeled. Reprinted from Ref. 3 with permission of the American Chemical Society. (b) Actinic absorbance vs wavelength of S-1400 series resists.   Courtesy of Shipley Company.

further irradiation. If this occurs, incident light can pass efficiently through upper regions of the resist film, which are exposed first, and be absorbed by lower as yet unexposed resist material.

If several candidate resists have been identified as having acceptable spectral response and actinic absorbance properties, a *quantitative* comparison of their sensitivities needs to be conducted. The response plots (Fig. 3) used to determine resist contrast, also yield quantitative resist sensitivity data. That is, for positive resists the dose, $D_o{}^P$, is required to bring about complete solubility of exposed positive resist regions, under conditions where the unexposed regions remain essentially insoluble. If the $D_o{}^P$ values for various resists are known, a quantitative comparison of their sensitivities is also in hand.

It is relevant to mention that there are practical limits to sensitivity. Resists whose sensitivities are too high may undergo thermal reactions at room temperature[3], and consequently may exhibit unacceptably short self-lives. In addition, if the sensitivity is too high, only a small number of photons per pixel would be needed to expose the resist. Since the number of impinging photons per pixel is subject to statistical fluctuations, this could result in exposure uniformity control problems. The lower limits of sensitivity, as discussed earlier, are governed by throughput considerations.

Etch Resistance and Thermal Stability

Etch resistance specifies the ability of a resist to endure the etching procedure during the pattern transfer process. Resist materials typically exhibit excellent resistance to most wet etchants. This is not the case for many dry etch processes. Fortunately, conventional positive optical resists of the novolac resin /diazide quinone family possess a reasonable resistance to a resistance to a variety of dry-etching conditions, a circumstance that contributed significantly to the initial success of dry-etching. The superior dry-etch resistance apparently stems from the strong backbone and side-chain bonds in the material molecules. This advantage in positive resists is obtained at the price of reduced radiation sensitivity. Other approaches, such as using fluorinated polymers to increase resistance to chlorine-based plasmas, or polysiloxanes to increase their resistance to oxygen plasmas, have been investigated to overcome the reduced sensitivity obstacle of conventional optical positive resists[10]. Many of the x-ray and electron beam resists possess much poorer dry-etch resistance than optical positive resists.

Some of the dry-etch processes are also conducted at higher temperatures than wet etch processes. As a result, the resist in such applications must show thermal stability at the process temperatures. Requests for resists able to withstand 200°C (or higher), have lead to development and commercial availability of products designed to be used at such elevated temperatures[11].

Adhesion

Photoresist is deposited on various substrates in the course of semiconductor processing, including metal films (e.g. Al, Au, Ti), insulators (e.g. SiO_2, PSG, and Si_3N_4), and semiconductor surfaces. The resist must adhere to these surfaces through all the resist processing and etch steps. Good adhesion to poly-Si, metals, and highly phosphorus-doped SiO_2 layers have generally posed the biggest problems. Poor adhesion brings about severe undercutting, loss of resolution, or possibly the complete loss of the pattern. Wet etching techniques demand a high level of adhesion of the resist films to the underlying substrates. Various techniques are used to increase the adhesion between resist and substrate including: a) dehydration bakes prior to coating; b) use of adhesion promoters such as hexamethyldisilazane (HMDS) and vapor priming systems (see section on *Resist Processing: Dehydration Baking and Priming*); and c) elevated temperature post-bake cycles. In dry etching applications, adhesion requirements are reduced, but it should be remembered that there must be sufficient adhesion to withstand the developing step. As feature sizes shrink, this is becoming more difficult to achieve.

Solids Content and Viscosity

The *solids content* specifies the percentage of the resist (in liquid form) that will remain as a solid after all the solvent has been evaporated by baking. Solids content is expressed as the percentage of the original liquid mass that is left behind as a dried mass. The procedure for measuring solids content is given in ASTM Std. F 66 84[12]. The solids content is important because it impacts coating thickness and resist flow properties. Photoresists may suffer a change in solids content as a function of time. In positive resists, the sensitizer can decompose with time, causing formation of precipitates, which when filtered from the resist, result in a net solids content change.

Resist *viscosity* depends upon the solids content and temperature, and is one of two key parameters that determines the thickness of a deposited resist film (the other being spin speed). For this reason it is important that tight control be maintained on both solids content and resist temperature during coating and spinning, in order to obtain reproducible film thicknesses (e.g. ±100Å thickness uniformities). The unit of *dynamic viscosity* is the *poise*, which equals one dyne sec /cm. Viscosities of photoresist are given in centipoise, cP, (0.01 poise). Some resist suppliers provide viscosity information in the form of *kinematic viscosity* (i.e. dynamic viscosity divided by the resist density, in g/cm^3), in units of *centistokes* (cSt).

Viscosity can be measured according to ASTM Std. F 66 84[12]. Since solids content and viscosity are related, a reference for checking samples of incoming resist can be created by obtaining a plot of one versus the other.   The effective viscosity during coating is also a function of the molecular weight distribution in the resist, and a technique such as *gel permeation* must be employed to detemine the molecular weight distribution as well. Figure 7 shows a solids versus viscosity curve for AZ 1300   positive   photoresists . Most resist manufacturers can supply resists in custom dilutions at almost any viscosity, and to within ±0.2 cSt of the desired viscosity value.

Particulates and Metals Content

The *purity* of photoresist material is an extremely important criterion. The greatest purity concerns involve the *particulate* and *metals content* of the resist. To satisfy the stringent requirements for particulate-free materials, resist suppliers have established extremely tight filtration and packaging procedures[13]. Ultra-filtered products, which are provided in superclean packaging, yield highly pure resists.

Some users also choose to filter the resist, even when obtained in such pure form. Their concern is that particles may still form as the resist ages, after having been filtered and packaged. For example, air in the coater lines can lead to particle formation. Typically, the higher the degree of filtration that is performed, the higher the cost. A two-filter system using an *absolute filter* downstream from a *depth filter*, has been recommended as best for in-house filtration. The resist is forced through the filter membranes with a dry, inert gas (e.g. nitrogen). Particle elimination with this system can be achieved to a level of 0.1 μm.

The *metals content* of a resist specifies the level of sodium and other metals present in the material.   It is desirable to have very low levels of metal (especially sodium, to less than 0.5 ppma), as their presence in the resist can lead to subsequent contamination and degradation of the devices being fabricated. Such low levels of sodium (and potassium) are detected by atomic-absorption spectrophotometer methods.

Flash Point and TLV Rating

Resists in their liquid form are usually combustible. The *flash point* identifies the temperature at which the vapor concentration becomes high enough to cause the resist vapors to ignite in the presence of an open flame. Its value is provided by resist vendors so that safe procedures for shipping and handling resist can be established. Typical flash points are ≥30°C.

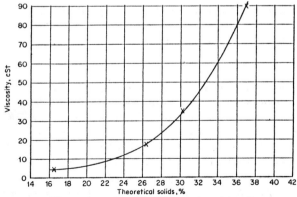

Fig. 7 Viscosity versus solids content for AZ-1300 photoresists. Courtesy of Shipley Company.

The *threshold-limit value* (TLV) is a toxicity rating that specifies the maximum ambient concentration of a substance to which a worker can be safely exposed during a normal work day (in parts per million). A more extensive discussion on TLV ratings is presented in *Dangerous Properties of Industrial Materials* by N. I. Sax[14]. The major concern has focused on the evolution of solvents from the resist. Many resist vendors have been replacing one of the traditional solvent components, *cellosolve acetate*, with less harmful substances.

Process Latitude, Consistency, and Shelf-Life

Process engineers responsible for maintaining a production process within required tolerances set a high store on the *process latitude* of a photoresist. That is, desirable resists should be able to produce critical dimensions within the specification limits in the face of the process variations encountered during production. The resist processing steps that generally call for the most latitude, because their process parameters are most subject to variation, are: a) soft-bake (time and temperature); b) exposure (time, source output, resist thickness); c) focus (wafer flatness); and d) develop (time, soft-bake step, developer concentration, temperature, method, and exposure time). Negative resists typically exhibit more process latitude than do positive resists.

Once a process has been "tweaked in", *it is important that it not be changed.* Therefore, another key feature of a resist is that it performs in an identical manner if subjected to the same process conditions. This requires that the resist be supplied with high *batch-to-batch consistency* (which also has a direct impact on process latitude). Most manufacturers supply resists with ±5% variations in sensitivity, and the ability to tolerate ±20% variations in exposure energy.

Some of the components in resists will undergo changes as a function of time and temperature. Negative resists generally have shorter *shelf lives* as a function of such changes (e.g. autopolymerization, that causes formation of aggregates or gels and changes in resist speed), than positive resists, which exhibit sensitizer decomposition. In ultracritical applications, however, many positive resists should be refiltered after about six weeks to remove decomposed sensitizer precipitates. Use of pre-filtered incoming material, adhering to proper shipping and storage procedures and conditions (~21°C), consumption within the manufacturers shelf-life specification time (date codes should be routinely monitored), and filtration before use, should minimize problems related to aging[15].

OPTICAL PHOTORESIST MATERIAL TYPES

Positive Optical Photoresists

Positive photoresists for microelectronic applications are three-component materials (i.e. matrix, sensitizer, and solvent), whose properties are altered by the photochemical transformation of the photosensitive component, from that of a *dissolution inhibitor* to that of a *dissolution enhancer*. The *matrix* component of positive resists is a low-molecular weight *novolac* resin which forms the resist film properties[16]. The generic term *novolac*, or *new lacquer*, describes the purpose for which these resins were first developed. The vast majority of novolac resins produced are used for such applications as the adhesive of plywood. Only a very small fraction is used for photoresist materials. The novolac resin dissolves or disperses in an aqueous base. For example, if a film consists only of novolac resin (i.e. without the sensitizer that acts to inhibit dissolution of the resin), such a film (assuming it has been adequately

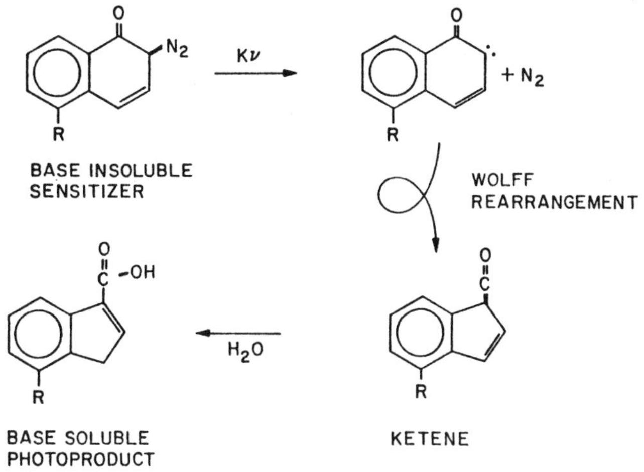

Fig. 8 Sequence of photochemical transformations of the quinonediazide sensitizer (dissolution inhibitor). Reprinted from Ref. 1 with permission of the American Chemical Society.

soft-baked) will rapidly dissolve in developer (for example, let us say, at a rate of ~150 Å /sec).

The sensitizers or *photoactive compounds* (PAC) in positive resists are *diazonaphthaquinones*. These substances are photosensitive, but are insoluble in aqueous developer solution. They also therefore prevent the novolac resin from being dissolved by the developer (i.e. their role is that of the dissolution inhibitor).   Unexposed positive resist films exhibit developer attack rates of ~10-20 Å /sec. Typical ratios of resin to PAC are ~1:1. The *solvent* systems for such resists are typically mixtures of n-butyl acetate, xylene, and cellosolve acetate.

Figure 8 provides a summary of the chemical compositions and reactions that take place in diazoquinone /novolac materials[17]. Upon exposure to light, the diazoquinones, which prior to exposure act as dissolution inhibitors, photochemically decompose. This leads to molecular rearrangement and hydrolosis, with the end product being *carboxylic acid*. The carboxylic acid photoproduct is readily soluble in basic solutions (dissolution enhancer). Hence, in the exposed regions of the film where complete photodecomposition of the inhibitor has been effected, the material dissolves in aqueous developer solutions at rates equal to or greater than materials which consist of novolac resin alone (e.g. ~1000-2000Å /s vs. 150Å /s). On the other hand, the unexposed regions remain much less soluble in developer (~100 times less soluble). This *differential solubility* is the primary mechanism for image formation in positive resists.

Figure 6a shows the ultraviolet absorption spectrum of a typical diazonaphthaquinone and a common novolac resin. The naphthaquinone sensitizer exhibits strong absorbance at the 365 nm, 405 nm (and to a lesser extent the 436 nm) mercury emission lines. Table 1 lists some typical electrical and physical properties of another family of positive photoresists, namely the Shipley S-1400 series.

Positive photoresists have become the dominant resists for VLSI applications because of their higher resolutions (~1 μm), as previously mentioned. This capability arises from the fact that the unexposed film regions are not permeated by the developer. As a result they closely retain the size that they possessed after-exposure, even after being immersed in developer. In negative resists, the developer permeates both the exposed and unexposed regions of the film. In the unexposed regions such penetration leads to film dissolution, but even in the exposed areas where little dissolution occurs, the solvent penetration causes the regions to swell and distort the

Table 1. ELECTRICAL AND PHYSICAL DATA ON SHIPLEY S-1400 PHOTORESIST

	Shipley S-1400		
Electrical Data:		Physical Data:	
Film Thickness (average)	2.8×10^{-4} cm	Vapor pressure (max)	15 mm at 25°C
Volume Resistivity	1.7×10^{16} Ω /cm	Vapor Density	0.88-4.72
Dielectric strength (average)	1.8 kV /mil	Coefficient of Linear	1×10^{-4}cm/cm/°C
Dielectric Constant (average)		Expansion (Average)	
At 100 Hz	1.42	Ash Content (ppm)	< 10
At 1000 Hz	1.39	Type of solution	Solvent base
Power factor (average)		Appearance	Clear, amber red
At 100 Hz	0.031	Film Density	20.65 g /in^3
At 1000Hz	0.016	Toxicity (MAC)	100 ppm
		Gamma (γ)	2.2

resist size. This degrades the negative resist resolution capability. In addition, positive resists exhibit improved dry etch resistance, and better thermal stability, than negative resists. Resist vendors also offer "custom" positive resist formulations to match specific lithographic systems, as was historically done with negative resists. Although positive resists are slower than negative resists, in stepper-based PR processes (in which throughput is dictated by alignment, focus, and wafer handling), such lower sensitivity does not significantly reduce throughput.

The large variety of commercially available positive resists[11], however, provides an indication that some variation in their capabilities exists. In fact, one or more of the following properties are purported to be exhibited by particular resists, although users must determine which resist advantages are most useful for each application: consistency from batch-to-batch of resist; superior resolution; wide process latitude; excellent adhesion to oxides, metals, and other surfaces; available in a variety of viscosities; reduced sensitivity to reflection and topography; unaffected by standing waves; low-metal ion content; utility in stepper systems; ability to withstand high (e.g. 200°C) process temperatures; resistance to aluminum dry-etch processing; low particle-count; high contrast; minimal linewidth change over steps; minimal unexposed film loss during developing; high photospeed; better thickness uniformity; free of cellosolve acetate; and meets OSHA safety and TLV requirements.

Negative Optical Photoresists

Negative photoresist has historically been the workhorse of the microelectronic industry, and has only been replaced when VLSI resolution requirements have approached 2 μm. It is the swelling of negative resists during development that makes them unsuitable for critical dimensions less than about 3 μm. Their long predominance, however, stems from several attractive properties that negative resists possess for lithographic applications. In fact, it is predicted that these advantages will continue to make negative resist a widely used material for many lower resolution lithographic processes, especially on older wafer fab lines, where it would be prohibitively expensive to change masks to accommodate positive resists. The advantages of negative resist over positive resist include the following: 1) better adhesion to some substrate surfaces; 2) faster photospeed, which allows greater exposure throughput, and thus lower fabrication costs; 3) somewhat greater process latitude in terms of developer dilutions and temperatures; 4) less costly (about one-third as expensive as positive resists).

Optical negative resists, like their positive counterparts, are three-component materials, although their chemistry and photoactive behavior is quite different. The film-forming component is a cyclized synthetic rubber resin, which is radiation insensitive but is also extremely soluble in non-polar organic solvents such as toluene and xylene. The photoactive compound (PAC) is a bis-arylazide. A typical chemical structure of the cyclized rubber matrix and a commonly employed PAC are shown in Fig. 9. It should also be noted that prior to being applied to a wafer surface, the negative resist is in liquid form, dissolved in an aromatic solvent.

Most optical negative resists function by becoming less soluble in the regions exposed to light. That is, a photochemical reaction generates a cross-linked three-dimensional molecular network that is insoluble in the developer. The photochemical transformations associated with the generation of these cross-linked molecules is shown in Fig. 9[17]. The most important photoreaction is the evolution of nitrogen from the excited state of the arylazide to form an extremely reactive intermediate compound called *nitrene*. The reactive nitrene can undergo various reactions which result in the formation of more stable molecules, and in several cases, the generation of a polymer-polymer linkage, or *crosslinkage*.

In projection printers, the light that is invariably scattered off the projection optics crosslinks a thin layer in the top surface of the negative resist. This thin layer can become punctured and slide down between the features, resulting in a *scumming* effect. It has been observed that as little as 1% scattered light can induce this unwanted mechanism and ruin the imagery of fine (e.g. ~1μm) patterns. Note that in positive films, such scattered light only results in a slight reduction in resist thickness, and no scumming is produced[18].

In some negative resists, the sensitizer decomposes without crosslinking the polymer if exposure takes place in the presence of oxygen. The reaction remains confined to a thin layer near the free surface, which therefore remains soluble in developer. This effect may reduce the thickness of thin resist layers (e.g. ≤ 0.5 μm) so much, that not enough protection remains for the subsequent etch step. Therefore, keeping the resist surface exposed to a nitrogen ambient is recommended as a method for reducing the effect.

Image Reversal of Positive Photoresist

An interesting and useful technique has been developed which allows the *tone of the image in positive resist to be reversed,* so that it functions as a negative resist. There are several attractive applications for such a process. First, there are some mask layers in a VLSI process that call for a light-field together with a positive resist. Darkfield masks, however, are preferred over light-field masks. This is because they reduce light scattering effects, and because particles

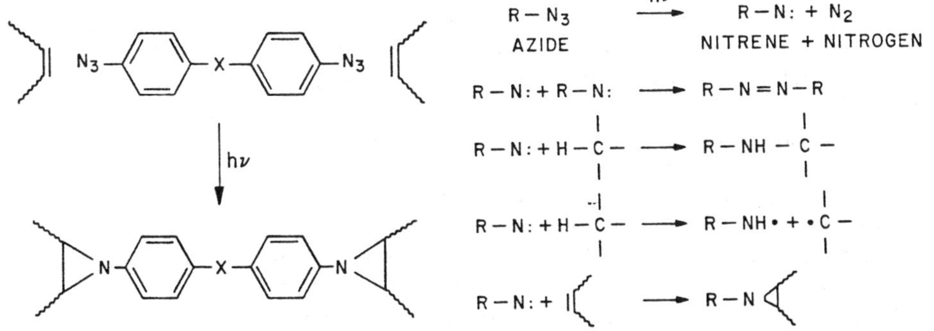

Fig. 9 Photochemical transformations of a bisazide sensitizer in negative photoresists based on cyclized polyisoprene. Reprinted from Ref. 1 with permission of the American Chemical Society.

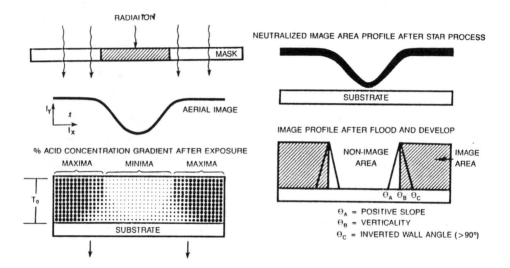

Fig. 10 Sequence of steps in an image reversal process[20]. Reprinted with permission of SPIE.

are more likely to fall onto optically-opaque (dark) regions of the mask, and thus not be printed. Light and dark-field masks are discussed in Chap. 13. The ability to perform image reversal would allow the use of positive resist *and* darkfield masks throughout an entire process. Another possible application of image reversal is the situation where a positive photoresist process has been established to manufacture new products, but because of expensive remasking costs, the existing mask sets must still be used to fabricate older products. The use of an image reversal process would permit the utilization of positive resist with the existing negative masks.

The first reported image reversal process involved the doping of positive resists with a material (monazoline) that would allow the dissolution enhancer (carboxylic acid) to be thermally degraded to form a substituted indene after initial exposure of the resist[19]. The indene is insoluble in aqueous base, and thus functions as a dissolution inhibitor. The resist is then exposed a second time (flood-exposure), causing the previously unexposed regions (in which the photo-active compound, diazoquinone still remains) to become soluble in developer. Thus, during development, the areas which were subjected to the first (image forming) exposure, dissolve at a much slower rate because of the presence of the substituted indene, leaving about 75% of the original resist thickness. This process was not extensively pursued because it was feared that the addition of a doping material to the resist would lead to contamination and consistency problems.

An alternative approach to introducing reversal-causing substances into the resist was subsequently developed, and is being offered commercially (Fig. 10)[20]. That is, after the first exposure the resist layers are subjected to amine vapors. The amine diffuses through the resist, and reacts with the carboxylic acid photoproduct in the exposed resist areas, in the same chemical fashion as the original image reversal process. The process is conducted in a vacuum bake chamber, which allows the amine vapors to be uniformly delivered to the resist and for the temperature to be uniformly controlled to ±1.5°C. Following the vapor treatment, a second (flood) exposure by UV light at high doses is performed, and the resist is developed in a conventional manner.

The process reportedly provides resolution equal to that obtainable for positive resist images, gives good critical dimension control, good contrast, and adequate depth of focus. It also has the potential for allowing the slope of the resultant images to be controlled without the film thickness loss normally present in conventional positive resist processing. Wall angles in excess of 90° (i.e. inverted wall angles) may even be produced, a feature that could be useful in metal lift-off processing.

Multilayer Resist Processes

As resist films cross over steps their local thickness is altered. This arises because the resist that crosses the top of the steps is much thinner than the resist that covers wafer regions which are low-lying. Thus, during exposure either the thin resist becomes overexposed, or the thicker resist underexposed. Upon development, a resist pattern crossing a step will therefore possess a linewidth variation (i.e. narrower on the top of the step). For lines in which step heights approach the size of the linewidth (e.g. for linewidths or spaces of 1 μm or less), such variations in dimension become intolerable. In addition, standing wave effects in thick resist layers reduces their minimum resolution. Finally, reflective substrates also degrade resolution in thick resist films (see section *Linewidth Variation Over Steps*).

The use of thin resist films could relieve the problems from standing waves and reflective substrates, but would not overcome the step coverage limitation.   A thick planarizing layer under a thin-imaging layer would, however, reduce the impact of all of the problems listed above. Approaches using such underlying planarizing layers are known as *multi-level resist* (MLR) processes, since two or more layers are required to implement them[21].

In MLR processes, an organic layer is therefore first spun onto the wafer, thicker than the underlying steps, to provide a surface which is smooth and significantly more planar than the original wafer topography.   After pre-baking this bottom layer, a thin imaging layer is deposited. In some cases a third thin transfer layer, such as SiO_2, is deposited on the thick layer prior to depositing the imaging layer. High resolution patterns are then created in the thin top layer. These are next precisely transferred into the bottom layer using the delineated imaging layer as a blanket exposure mask, or as an etching mask to pattern the planarizing layer. Patterns with resolutions less than 0.5 μm have been delineated with MLR processes.

The advantages gained by MLR processes, however, are obtained at the expense of added complexity. This can result in diminished throughput, possible increase in defects, and increased costs. The degree of additional complexity that can be tolerated, and the expected increase in defect density depend on the maturity of the alignment tools and the overall device process.

Three multilayer resist processes will be described in this section: a) a portable conformable mask process, or PCM; b) an organic tri-layer process; and c) a bilayer process utilizing an inorganic resist (Ag_2Se /Ge_xSe_{1-x}) as the imaging layer.

The *PCM* is the simplest MLR scheme, and is the only one that has reportedly been used by 1985 in a VLSI production mode[22]. In this process, a thick underlying layer of resist (e.g. PMMA, which is sensitive to deep UV), is used to planarize the surface. A thin resist that is sensitive to the near UV, but opaque to deep UV, is then deposited over the PMMA.

The upper layer is conventionally patterned, and subsequently used as a conformal deep UV mask for PMMA. The PMMA is then flood exposed (using deep UV) and developed. The resolution of the PCM process, and that of the tri-layer processes are equal, since they utilize the same imaging resist layer.   The resolution of Ge_xSe_{1-x} layer process is higher because of its edge sharpening effect. The instability of the PMMA under dry etch, however, is a limitation of the PCM process. To improve the durability during dry-etching, the imaging resist is often retained on top of the PMMA. This is often referred to as a *capped structure*. Another attraction of the PCM process is that it can be done on conventional lithographic equipment, together with a source of deep-UV for the flood exposure of the PMMA. On the other hand, film interface layers at the PMMA /optical resist interface can cause exposure and develop problems.

In the *tri-layer resist* process (Fig. 11), a thick organic layer is again used to partially planarize the surface[23]. Materials investigated for this layer have included conventional positive optical photoresist, polyimide, and polysulfone. A thin layer of SiO_2 (spin-on or PECVD) is then deposited to act as an etch mask for the bottom layer. An imaging resist layer is finally applied, and patterned in a conventional manner (e.g. near-UV exposure and wet development). To minimize effects from standing waves and reflections from the substrate, the lower layer is treated to increase its optical density (see the section on *Resist Processing: Exposure*). The SiO_2 and lower layer are dry etched anisotropically to precisely transfer the pattern from the thin imaging layer into the thick lower layer. The main disadvantage of the tri-layer process is the added complexity that it requires.

In Ge-Se based bilayer processes, the high contrast and resolution advantages of Ge-Se chalcogenide glasses are exploited (see section on *Inorganic Resists*). These materials are used for the imaging layer of a bilayer structure. The bottom layer is again a thick polymer, such as AZ2400 resist. This process exhibits the highest resolution limit of the three MLR processes discussed, but is also the most complicated to develop. Since its resolution capability is potentially so small, however, it has been identified as a process that might compete favorably with x-ray or electron beam lithography[21].

Contrast Enhancement Layers

An alternative technique to the multilayer resist processes for increasing the maximum resolution obtainable with projection aligners, uses a photobleachable top layer called a *contrast enhancement layer* (CEL)[24]. The penalties incurred for obtaining the improved resolution are

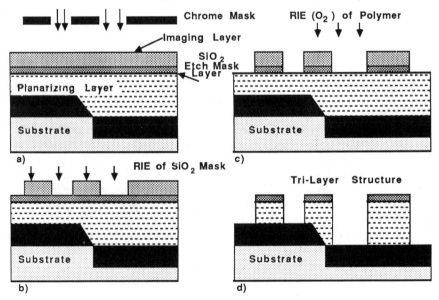

Fig. 11 Schematic of the tri-layer process[21]. Reprinted with permission of Solid State Technology, published by Technical Publishing, a company of Dun & Bradstreet.

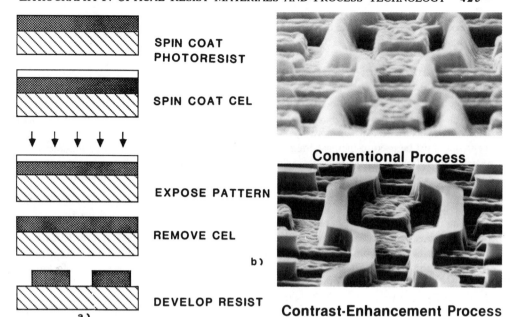

SPIN COAT
PHOTORESIST

SPIN COAT CEL

EXPOSE PATTERN

REMOVE CEL

DEVELOP RESIST

a)

b)

Conventional Process

Contrast-Enhancement Process

Fig. 12 (a) Photoresist process incorporating the CEL concept. Added CEL steps are outlined in boxes. (b) SEMs (8000X) of identical microcircuits produced with and without a CEL. Courtesy of the General Electric Research and Development Center.

increased process complexity, longer exposure times, and longer development times. Note, however, that use of a CEL increases exposure time by ~3x, but stepper time by only ~15%.

The CEL layer consists of a material that is spun onto a resist-coated wafer to a thickness of 1000-3000Å after soft-bake. This layer is normally opaque, but becomes transparent (bleached), when exposed to the light of the aligning tool. When the mask image from a projection printer is focused onto the wafer, regions of the opaque CEL layer that are struck by high-intensity light become transparent. This creates *windows* in the CEL through which light can shine and expose the resist. The unbleached CEL regions continue to absorb the relatively weak diffracted light. Since the CEL is in direct contact with the resist, it acts much like a thin mask used in contact printing. As discussed in Chap. 13, contact printers have higher resolution capabilities than projection printers. In addition, because the CEL is applied as a relatively thin film, it has an inherently high resolution capability. As such, use of a CEL layer allows the advantages of projection printing to be exploited, together with the benefits, but without the limitations of contact printing. In addition, unlike the PCM process, no interface layer problems are reported with CEL processes. The name is derived from the fact that the presence of the CEL layer increases the contrast of illumination that effectively reaches the resist.

Following exposure, the CEL layer is completely stripped prior to developing. This requires that an extra dispense head in a spray developer (see section on *Resist Processing: Development*) be used to spray on the stripping solution. A spin dry step is then performed before the normal development cycle is begun. The CEL stripping operation adds about 30 seconds to the development process.

Linewidths of ~0.5 μm using commercially available optical photoresists and stepper projection systems have been achieved with the aid of a CEL compared to the 1 μm resolution limit without CEL. Figure 12 shows SEM photographs of 2 μm wide lines in photoresist patterned with and without use of a CEL layer, and demonstrates the improvement that use of a CEL may provide.

Inorganic Photoresists

The finite contrast, γ, of organic-based resists limits their resolution capability. In order to increase the resolution capabilty of a photoresist process, without having to increase the numerical aperture NA of the system, or decrease the wavelength, λ, of the system source, a technique must be used to increase the resist contrast. It has been demonstrated that *inorganic resist materials* exhibit extremely high contrasts, and therefore high resolution capability (e.g. $\gamma = 6.8$ for inorganic resists, versus $\gamma \cong 2$ for organic resists). Delineation of 0.4 μm lines and spaces using conventional optical lithography has been demonstrated by utilizing $\lambda = 405$ nm, a lens of NA $= 0.35$, and such an inorganic resist[21].

The first such materials were based on the Ge-Se, that we introduced in the tri-layer MLR process discussion[31]. In such resists, Ge_xSe_{1-x} glasses are deposited as thin films by rf sputtering or evaporation. Next, a 1000Å layer of silver (Ag) is deposited by plating from an $AgNO_3$ solution onto the Ge_xSe_{1-x} surface, by a dip into an aqueous solution of $AgNO_3$. When this composite film is exposed to UV light (or electron radiation), the Ag diffuses into and dopes the Ge-Se matrix (Fig. 13). This renders the Ag-doped Ge-Se almost insoluble to alkaline solutions that would normally dissolve an amorphous Ge-Se film. Thus this process exhibits a negative resist type behavior. Note that although this mechanism implies that wet developing is used to create patterns in Ge-Se films, dry development in CF_4 is also feasible.

The resist also demonstrates a unique *edge-sharpening* behavior. That is, rapid lateral diffusion of the Ag during exposure (photodoping) takes place across the boundary between exposed and unexposed regions. The resulting Ag profile compensates for diffraction at the image edges, which leads to the sharp edges.

Inorganic resist materials also exhibit other attractive features for VLSI lithographic applications, including the following: a) there is no swelling during development and the edges have anisotropic profiles; b) the material is resistant to oxygen plasmas and hence is compatible with dry etched bilayer MLR processes; c) the materials possess a broad-band spectral response to all regions of the UV spectrum, while appearing opaque to light of all wavelengths up to 450 nm (thereby eliminating standing waves on reflective substrates) and is transparent to visible light.

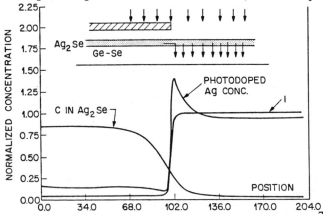

Fig. 13 Mechanism of edge sharpening effect in inorganic resist material[77]. Reprinted with permission of the publisher, the Electrochemical Society.

In spite of their very impressive properties, inorganic resists have not been widely adopted by the industry for several reasons. First, conventional organic resists still posses adequate resolution for current device needs, and all aspects of organic resist processing are well understood. Second, inorganic resists require new deposition techniques, with accompanying equipment costs, etc. Third, process complexity rivals that of MLR. Finally, data on potential defect sources in such vacuum-deposited films has not yet been conclusively amassed and disseminated.

Dry-Developable Resists

The disadvantages of resist swelling during development, the defects and contamination introduced by wet development, and the necessity of disposal of spent developer solutions, could all be overcome if dry development of the resist were possible. Wet deposition (e.g. spin coating of the resist) is still attractive, since it results in planarization by the thick lower resist layer in multilayer processes. Two dry development processes have therefore been explored: 1) the self-dissolution of the resist upon exposure and vacuum heating; and 2) the selective removal of the resist by a plasma after exposure (negative resist action).

In one form of the self-dissolution type, certain poly(olefin sulfones) have been found to undergo scission after exposure[25]. The products are volatile, and thus spontaneous image formation results. Other materials (e.g. plasma deposited methylmethacrylate and tetramethylin) will likewise develop an image when heat is applied during postbake.

In the second types, a differential resistance to plasma etching is produced between exposed and unexposed film regions[26]. Various materials and chemical reactions have been explored to induce such an effect. In one instance, an oxygen plasma removes the unexposed regions more rapidly than those which have been exposed. The major thrust of research is to improve the differential plasma resistance, while maintaining the sensitivity at about 10 mJ /cm^2. Information on development uniformity, yield, and defect density, however, will be needed before a meaningful comparison between these materials and other resist types can be fully made.

Mid-UV and Deep-UV Resists

Mid-ultraviolet (UV) and deep-UV technology offers a significant increase in resolution, because of the shorter, and therefore more energetic, wavelength used), while maintaining the optical projection approach, and the capital equipment with which it is processed. In such approaches, exposure would be achieved using the 313 nm (mid-UV), or the 254 nm (deep-UV) line of Hg. See Chap. 13 for more information on the mercury arc spectrum.

Mid-UV lithography in fact, is already a relatively mature technology, and projection printers with mid-UV capability are commercially available[27]. Filters are used in these printers to allow the 313 and 334 nm lines to be transmitted to the resist surface, but effectively block light with wavelengths shorter than 300 nm, and longer than 350 nm.

Commercially available resists, however, exhibit less sensitivity to mid-UV than to near-UV. This is due to several factors: a) the resist sensitizers are less absorptive to mid-UV; b) the sensitizer photoproducts are bleachable by near-UV but are not bleachable by mid-UV (i.e. they absorb at 313 nm but are transparent to 450 nm radiation); and c) the novolac resins are essentially transparent at 350 nm, but are relatively opaque at 313 nm. The upshot of these characteristics is that the resist is inefficient in absorbing mid-UV light. That is, these resists require greater exposure doses, and patterns exhibit degraded wall profiles. Commercially available resists for use in the mid-UV range are unable to offer comparable throughput to that obtained under near-UV conditions.

Unfortunately, currently available near-UV resists are completely unsuitable for deep-UV (254 nm). Thus, for deep-UV entirely new resist chemistries must be developed. The major drawback of conventional positive resists is that the diazonaphthaquinones are opaque to deep-UV light, and thus short wavelengths will not penetrate such films. In addition, the novolac resin also strongly absorbs at 254 nm.

The exposure hardware also exhibits some limitations for *deep-UV*. Lens materials with acceptable transmission characteristics below 300 nm are limited. Reflecting systems do not suffer from this limitation, and thus fully reflective designs for deep-UV tools show the most promise. In addition, the Hg lamp is an inefficient source at shorter wavelengths. That is, at 254 nm, significant self-absorption by the lamp of such radiation occurs. Furthermore, since each photon possesses more energy, a greater lamp energy is required to obtain a photon flux equal to that at longer wavelengths. Thus, alternate sources, such as excimer lasers (e.g. based on KrCl and KrF), which can provide a power output in excess of 10 W in the deep-UV region, as opposed to tens of milliwatts by Hg lamps, are being investigated. However, implementation of such sources requires extensive redesign of projection optics.

Deep-UV resists based on various chemistries have been formulated[28], including: a) PMMA chemistries; b) conventional positive resist-type chemistries; and c) some totally new systems. *PMMA-type deep-UV resists* have absorption peaks around 215 nm, and from 260 nm to the visible, cease to be sensitive as a photoresist. Unfortunately, PMMA and even some of its copolymers, exhibit poor resistance to dry-etching processes, and research efforts are being pursued to overcome this drawback. Modified *novolac-type systems* have been formulated in attempts to make them more suitable for deep-UV. For example, a photosensitizer based on a diazo-Meldrums acid (which absorbs strongly at 254 nm, and which functions as a dissolution inhibitor) replaces the diazonaphthaquinone. The deep-UV opaque novolac resin is also replaced with substitute novolac-related but transparent resins. Some *new systems* involve modified poly (olefin sulfones), or poly (methyl isopropenyl ketones) for positive resists, and cyclized polyisoprene and appropriate bisazides, or poly (vinylphenol) and a bisarylazide, which avoids the swelling proble, have also shown promise.

Photosensitive Polyimides

Polyimides are polymers that have found application in VLSI processing as organic insulators, primarily in the roles of interlevel dielectrics, passivation overcoats, and alpha-particle barriers. Their properties of relatively high-temperature stability (up to ~450°C), low dielectric constant, and their ability to planarize topography, make them attractive for such functions. The drive to save processing steps has also lead to the concept of a *permanent photoresist*. This is a photosensitive polymer that first acts as a photoresist, but can subsequently be left behind as an organic insulator. Since conventional photoresists cannot tolerate the 400°C temperatures encountered in routine "back-end" semiconductor processing sequences, polyimides have been investigated as candidates for permanent resist films. Photosensitive polyimide precursors (polyamic acids) are spun onto wafers, and upon exposure to UV light, undergo crosslinking. During development, the unexposed regions are dissolved, and final curing by further heat treatment leads to a chemical transformation (known as *imidization*) of the remaining crosslinked material, which yields polyimide. At 275°C more than 99% of the polyimide precursor is converted to polyimide, and most of the photo-crosslinked groups are volatilized. The final cured film is essentially equivalent in its properties compared to non-photosensitive polyimide (dielectric constant $\cong 3.3$, resistivity of ~10^{16} Ω-cm)[29].

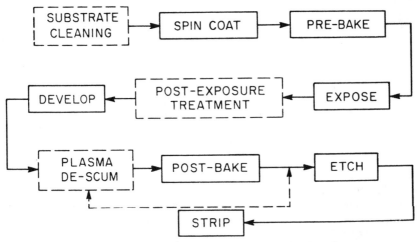

Fig. 14 Flow chart of a typical resist process. Steps in broken lines are not used for all materials. Reprinted from Ref. 8 with permission of the American Chemical Society.

One of the major limitations of photosensitive polyimide material has been their short shelf-life stability. Research to improve this shortcoming, however, is being actively pursued, with apparently successful promise[30].

PHOTORESIST PROCESSING

The basic sequence of steps that comprise a complete photoresist process is shown in Fig. 14. The remaining sections of this chapter describe these steps in detail. There are a few steps that are covered in other chapters, including *Wafer Cleaning* (Chap. 15), *Etching* (Chaps. 15 and 16), and *Photoresist Stripping* (Chap. 15).

It will be seen that the steps of a complete resist process are not independent of one another. That is, in specifying one step, several other steps will be impacted. As a result, the development of a resist process requires substantial effort. Many facets of the process development are largely empirical, and thus, as much (if not more) art as science, is performed in establishing an effective resist process.

Resist Processing: Dehydration Baking and Priming

Many resist processing problems can be attributed to dirty or contaminated surfaces. Thus, it is important that a substrate be clean in order not to adversely affect the lithographic process. Some common problems traced to dirty surfaces include poor adhesion and defects (such as pinholes and opaque spots). Loss of linewidth control, and in extreme cases, the entire loss of pattern elements, can result from poor adhesion. Since most surfaces in VLSI fabrication are formed either by thermal oxidation or by reduced pressure deposition techniques, *in most cases the surface is in its cleanest form immediately after having been formed.* Thus, further cleaning can be avoided by *coating the surface as quickly as possible with resist after deposition.* Wafer cleaning techniques in some cases may be necessary, and are discussed in Chap. 15.

Moisture from the atmosphere can be rapidly absorbed by substrate surfaces, and such hydrated surfaces have been shown to reduce adhesion. Therefore, a *dehydration bake step* is often performed before priming and spin-coating a wafer with resist. When carried out at atmospheric pressure, dehydration baking evolves moisture at three temperature plateaus[32]: a) surface water molecules are liberated at 150-200°C; b) loosely held water of hydration is evolved at about 400°C; and c) total dehydration has been postulated to occur at temperatures in excess of 750°C. It is believed, however, that upon cooling, the water of hydration removed in step c) is reabsorbed from atmospheric moisture. Thus, some reports argue that no significant improvement over baking at 400°C should be expected. What is apparently important is the time delay between the dehydration bake and coating. In fact, to assure uniformity of processing, the recommended procedure is to carry out coating immediately after the dehydration bake. Dehydration bakes of up to 400°C are generally carried out in convection ovens, while 800°C bakes are performed in furnace tubes (both, of course, in dry-gas ambients).

Following the dehydration bake, the wafer is normally *primed* with a pre-resist coating of a material designed to improve adhesion even further[33]. The most widely used priming substance is hexamethyldisilazane (HMDS)[34]. Figure 15 shows how one end of the HMDS molecule reacts with wafer-oxide surfaces to tie up molecular water on a hydroxilated SiO_2 surface. It also shows how the other end of the HMDS molecule forms a bond with the *resist.* Thus, it is seen that the HMDS behaves as surface-linking *adhesion promoter* (e.g. SiO_2 surface-to-resist surface linkage). The wafers should be coated with resist as quickly as possible after priming, and it is recommended that coating be performed no later than 60 min after completing the priming step.

The HMDS is typically applied to the surface in one of two methods: a) spin coating; and b) vapor priming. In *spin coating*, the HMDS is dispensed onto the wafer surface through an additional nozzle present in the resist spin coating system. After the HMDS is dispensed, the wafer is spun at 3000-6000 rpm for 20-30 sec, causing approximately a monolayer of HMDS to remain on the wafer surface. In *vapor priming*, the HMDS is introduced in vapor form into a chamber containing the wafers. An exposure of the wafer surfaces to the HMDS vapor for ~10 min primes the surface so that good adhesion is obtained. The advantages of vapor priming are the following: a) wafers can be batch primed; b) since only vapors come in contact with the wafer surface, potential contamination from particles present in the HMDS solution is avoided; and c) less HMDS is used per wafer, thus providing a cost savings. Equipment that combines a vacuum dehydration bake of the wafers in the same chamber, prior to introducing HMDS vapors, is also commercially available. Such systems offer the opportunity to prime the wafers after vacuum dehydration baking (i.e. at a temperature less than is required for an atmospheric pressure bake), and without subsequently having to expose the wafers to atmospheric moisture.

Resist Processing: Coating

Following cleaning, dehydration baking, and priming, the wafers are ready to be coated with photoresist. The goal of the coating step is to produce a uniform, adherent, defect-free polymeric

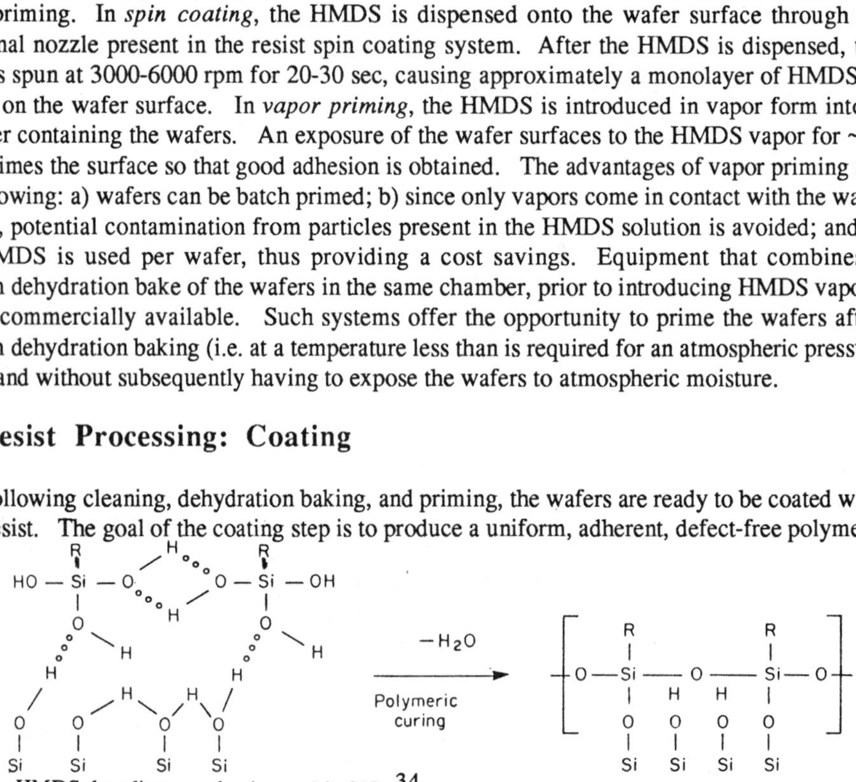

Fig. 15 HMDS bonding mechanism with SiO_2[34].

film of desired thickness over the entire wafer. *Spin coating* is by far the most widely used technique to apply such films[35,36]. This procedure is carried out by dispensing the resist solution onto the wafer surface, and then rapidly spinning the wafer until the resist is essentially dry. In order to maintain reproducible linewidth in VLSI fabrication applications, resist film uniformity across the wafer (and from wafer-to-wafer) should be within at least ±100Å.

The spin coating procedure involves three stages: a) dispensing the resist solution onto the wafer; b) accelerating the wafer to the final rotational speed; and c) spinning at a constant speed to establish the desired thickness (and to dry the film).

The *dispensing stage (a)* can either be accomplished by flooding the entire wafer with resist solution before beginning the spinning, or by dispensing a smaller volume of resist solution at the center of the wafer and spinning at low speeds (e.g. 200 rpm for ~1 sec) to produce a uniform liquid layer across the wafer (*spread cycle*). In the latter method ~3.0 ml of resist is dispensed onto 100 mm wafers, and ~5.0 ml onto 150 mm wafers. It has also been found that *static dispense* (in which the wafer is stationary as the resist is dispensed), provides more uniform coatings than if the wafer is rotating (*dynamic dispense*).

In *Stage (b)* the wafers are normally *accelerated* as quickly as is practical to the final spin speed. High ramping rates have been shown to yield better film uniformities than low ramping rates. This is due to the fact that the solvent in the resist evaporates rapidly from the resist after it has been dispensed onto the wafer. Since film thickness depends on the viscosity of the liquid resist solution, the more time that is allowed for the solvent to evaporate during the spin-ramping to the final speed, the greater are the drying and film setting-up tendencies that contribute to thickness non-uniformity. A spin-ramp of 20,000 rpm /sec has been suggested as an adequate compromise to provide maximum coating uniformity, without causing motor-wear problems or excessive wafer breakage.

During *Stage (c)* the resist layer acquires a relatively uniform, symmetrical flow profile

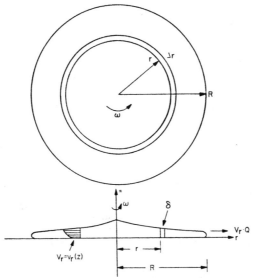

Fig. 16 Symmetrical flow pattern of a homgeneous liquid on a rotating disk showing flow rate, Q, and velocity profile v_r (z). Copyright 1977 by International Business Machines Corporation; reprinted with permission.

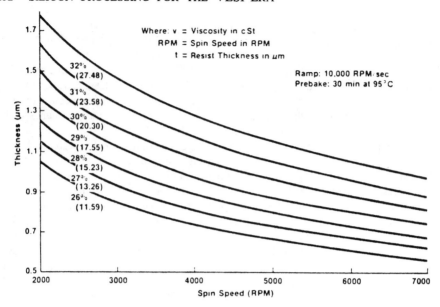

Fig. 17   Thickness vs spin speed of Kodak 820 resist.   Courtesy of Eastman-Kodak Company.

(as shown in Fig. 16). Once this structure has been formed, the remainder of the high speed spinning cycle consists of solvent evaporation to produce the solid film. Most spin coating processes are performed with a final spin speed in the 3000-7000 rpm range for 20-30 sec, as this range provides good film uniformity, as well as limited spin-motor wear and wafer breakage. The highest degree of thickness uniformity ($\pm 100 Å$) is observed at the upper end of the range (6000-7000 rpm).

The spin coating fluid dynamics have been studied in some detail[35]. It has been determined that prior to reaching the "equilibrium" form of Fig. 16[42], the resist layer undergoes some intermediate shapes. At the start of spinning (i.e. within a few hundredths of a second), a wave of resist is created that then moves toward the wafer edge. The *corona stage* is next observed, in which the bulk of the resist that supported the wave runs out to the wafer edge to form a crown-like structure. When most of the resist has been driven off the wafer, the wave and corona disappear. At that point, centrifugal forces drive the remaining resist material off the surface in a fine spray that exhibts a spiral-like appearance.   This *spiral stage* generates thousands of resist droplets as the material reaches the wafer edge and is flung off.   The droplets leave the wafer at high speed and are thus subject to being bounced off of the spinner bowl sides, and redeposited on the wafer surface.   To avoid this possibility, plastic splashguards and special exhausted bowl designs have been developed. An *edge bead* (a remnant of the corona stage, several times as thick as the deposited resist layer), also develops along the wafer edge during the spinning process[39]. Edge-rounding of the wafers significantly reduces the edge-bead, and wafer backside washing with solvent after spin-coating can entirely remove it (as described further in Chap. 15).

The resist film thickness after spin coating, for a resist of constant molecular weight, solution concentration, etc., depends only on the spinning parameters. As an example, Fig. 17 shows the thickness curves of Kodak Micro Positive Resist 820 as a function of spin speed. Various mathematical relationships have been developed to predict the resist thickness as it depends upon such factors as percent solids, spin speed, etc. Most users, however, utilize data from the resist manufacturer to plan a baseline spin-coating process, and then empirically generate spin speed curves for their own specific process applications (i.e. that will depend on such specific parameters as resist formulation, spinner type, wafer size, spin parameters, and ambient temperature during deposition)[40]. A study of thickness variation across a wafer as a function of many such parameters, including volume of resist dispensed, wafer diameter, resist viscosity, wafer spin-speed during dispense, wafer acceleration, and final spin-speed, was published in Refs. 37 and 38.

Resist film thickness can be measured in many ways. *Contacting types* of thickness measuring instruments are *surface profilometers*, which are capable of resolving 200Å steps. These are described in more detail in Chap 10. *Noncontact types* of thickness instruments utilize optical techniques of ellipsometry or interferometry. Automatic interferometer-based instruments have become widely used for resist thickness measurement, since they are easy to use, accurate, and perform the measurement quickly. Such instruments are discussed in Chap. 8.

There are also a variety of other aspects of the coating procedure that must be controlled in order to maintain an effective coating process, including: a) defect generation; b) the environment in which the coating is performed; and c) wafer handling and storage.

Defects can be introduced during the coating process in a number of ways, and some steps to minimize their generation include the following: a) Since the resist film is sticky until being soft-baked, it can easily entrap airborne particles. Thus, spinning should be done in a Class-100, or better, environment; b) The resist itself should be clean and free of all particulate matter above 0.2 μm in diameter. (Resist cleanliness and filtration prior to use is considered in an earlier section of this chapter.); c) The resist solution should be free of all entrapped air, as air bubbles can cause defects in the resist image that have the same effect as those caused by particles. Allowing a resist to sit for several days after filtration, and prior to application, will let dissolved air and gases to escape; d) radial resist striations (i.e. radial streaks of thickness different than the nominal resist thickness) are a form of defect caused by non-uniform solvent evaporation. One cause of striations has been attributed to excessive venting of the spin bowl[35]. This depletes the solvent content from the ambient around a spinning wafer, and apparently leads to striation formation. Spin bowl exhaust flow can be controlled by special flow meters designed to sense the flow of gases in the bowl exhaust lines. When multiple bowls are exhausted by a single pump, the exhaust flow in a single line can vary if it is not controlled. This can cause resist thickness nonuniformities[83]; e) the spin bowl should be designed to prevent splashback by resist droplets being spun off of the wafers; f) the nozzle that dispenses resist onto the wafer should be set close to the wafer surface to prevent splashing effects; g) the pump that controls the resist being dispensed through the nozzle should be designed to provide a *suck-back* action after the resist volume is dispensed onto the wafer. This suck-back function draws excess resist from the nozzle tip back into the supply line, thus preventing unwanted dripping onto the spinning wafer. The suck-back stroke must be adjustable so that air is not inadvertently drawn up into the nozzle tip. This could cause bubbles to become entrapped in the resist present in the nozzle line; and h) the chuck should be properly designed. That is, the chuck diameter should be large enough to consistently hold wafers and prevent breakage, and the vacuum ports should fasten wafers to the chuck in such a fashion that does not cause a resist puddle to form at the center of the wafer (due to *dishing*). Multiple vacuum ports are therefore preferred, as they distribute the vacuum force across the wafer area. (Note that new wafer hold-down mechanisms may need to be developed as wafer diameters and thicknesses continue to increase.)

The *environment in which the spin-coating is carried out*, must also be controlled. Since resist viscosity, and hence film thickness, depends on temperature, the latter should be controlled to ±1°C. In addition, the wafers, resist, and spinner hardware all need to maintained in thermal

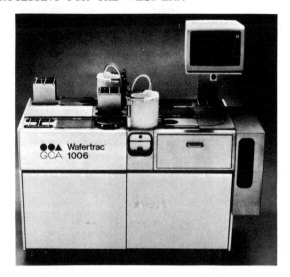

Fig. 18 Photograph of an automated in-line wafer spin coater. Courtesy of the GCA Corp.

equilibrium with one another. During spinning, solvent evaporation can lead to wafer cooling, and this may cause condensation of moisture on the resist surface. To avoid this problem, it has been recommended that the humidity be maintained at less than 50% RH. Since solvent vapors are flammable, exhaust equipment should also be spark-resistant to prevent the possibility of fire.

Upon completing the spin-coating cycle, wafers should be transported to the next step of the process, namely soft-bake, as soon as possible. Unbaked freshly spun films should not be stored in excess of a few hours, since their tacky consistency makes them vulnerable to particulate contamination. If particles contact the resist surface prior to soft-bake they become almost impossible to remove, and they will lead to pinhole or opaque defects after exposure and development. The wafers should be placed in a dry box for short-term storage, and should be soft-baked immediately before exposure.

Resist coating equipment (e.g. Fig. 18) is manufactured by a variety of suppliers[41]. Most suppliers offer equipment with cassette-to-cassette operation. Wafer handling and transport systems vary from one model to another.   Microprocessor control, including programming capability, and a *lock-out* feature, is also typically offered (i.e. in lock-out, once the coating process recipe has been programmed into the machine, it cannot be altered unless the system is "unlocked", presumably by a responsible authority). Equipment process control and diagnostic capabilities are also offered.   Since more than one wafer size may need to be processed, the ability to handle multiple wafer sizes may be important. Equipment that can perform priming, baking, and coating on one track in an automated manner, is also coming into widespread use.

Resist Processing: Soft-Bake

After wafers are coated with resist, they are subjected to a temperature step, called *soft-bake* (or *pre-bake*)[42,43]. This step accomplishes several important purposes, including: a) driving off solvent from the spun-on resist, reducing its level in the film from ~20-30% to ~4-7%; b) improving the adhesion of the resist, so that it is better able to adhere during the development step; and c) annealing the stresses caused by the shear forces encountered in the spinning process.

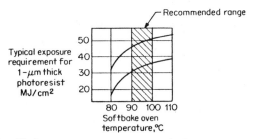

Fig. 19 Example of softbake temperature versus typical exposure requirements for AZ-1300 series and AZ-111 series resists. Courtesy of Shipley Company.

The conditions under which photoresist is soft-baked determine a number of parameters in subsequent steps of the process. Consequently, the soft-bake step must be carefully optimized and controlled. As a result of the solvent loss during soft bake, the thickness of the resist is also reduced, typically by ~10%.

The degree of soft-baking (i.e. due to the temperature, time, and soft-bake equipment used) determines the residual solvent content of the resist. The rate of attack of the resist by a developer is significantly dependent on solvent concentration. In general, the more residual solvent, the higher the dissolution rate in developer. Therefore, under-soft-baked resists are readily attacked by the developer in both the exposed and unexposed regions. This property makes it appear that the resist possesses increased photosensitivity (normally a desirable feature). However, since the unexposed resist is also eroded to a greater degree, this apparent advantage is obtained at the cost of a thinner patterned resist layer. This can lead to higher pinhole concentrations, or decreased protection during etching. On the other hand, there is also an upper limit to soft-bake temperatures. That is, excessive soft-baking causes some or all of the PAC to undergo reaction during the bake, rendering the resist less photosensitive during exposure[44].

Since soft-baking is so intimately linked to the exposure and development, it is necessary to conduct soft-bake characterization tests before finalizing an entire resist processing sequence. First, the *relationship between soft-bake and exposure* needs to be established (an example is shown in Fig. 19a). From these curves, it can be seen that the greatest process latitude (i.e. the smallest variation in required exposure dose to compensate for fluctuations in the soft-bake temperature) occurs over some range (e.g. 90-100°C). In this example, the curves show that required exposure doses are significantly lower at soft-bake temperatures below 80°C, but the process latitude is much lower, and unacceptably large erosion of unexposed resist during development will probably occur as well. The second relation that should be experimentally determined is the *solubility rate of the exposed resist as a function of soft-bake temperature*. An example of a plot of such data is shown in Fig. 19b.

Several types of ovens have been developed to implement soft-bake processes including: a) convection; b) infrared (IR); and c) hot plate ovens[45,46]. Each employs a different heat transfer mechanism (i.e. convection, radiation, and conduction, respectively). Nevertheless, they all seek to provide uniform, controlled temperatures, and cleanliness. The cleanliness requirement is important since the resists are heated above their softening point, and any particulates that contact the surface will become embedded and thereafter very difficult to remove.

Convection ovens were the first types used for production soft-baking (Fig. 20a). Hot air transports the heat to the wafers in the oven cavity, and the duration of a convection oven soft-bake cycle is ~30 minutes. Wafers are processed in a batch mode, and it takes 6-8 minutes until their temperatures approach that of the oven ambient. An important advantage of convection ovens is their stability over time. Once a process is established, and the oven is properly characterized, the soft-bake step is highly repeatable. There are several limitations of convection ovens, however, that have caused them to be replaced with other methods of soft-baking. First, temperature can vary within the oven cavity, and this may lead to non-uniform baking. Second, when wafers are moved into and out of the oven, the oven temperature fluctuates, again leading to bake nonuniformities. Oven temperature profiles and loading and unloading tests under conditions of actual use, can be used to characterize such ovens and produce more uniform results. Third, since good air movement is necessary to exhaust solvent vapors, particles may enter the oven cavity and become embedded in the heated resist. Convection ovens must therefore be fitted with highly filtered air-intakes. Fourth, since hot air transports the heat to the wafer, the film is heated from the top surface downward. This can cause the formation of resist surface crust, bubbles and wrinkles, and also leads to solvent entrapment near the resist-substrate interface. Crust formed on the resist surface retards its dissolution rates, due to the induction period before the onset of development. Incomplete solvent removal at this interface can lead to degraded adhesion, which results in problems during development and etch.

Infrared ovens (IR) have also been widely utilized for soft-baking resist. The advantage to using IR is that soft-bake times are much shorter than convection oven soft-bake times (3-4 min vs. ~30 min), and drying is done from "inside out". That is, long wavelength IR penetrates the resist film and is reflected back by the resist substrate interface. The thermal gradient is such that the coolest part of the resist is the surface. Solvent is most effectively removed nearest the interface, and such removal inhibits entrapment or bubble formation. The wafer is also rapidly heated to the desired temperature by the IR (Fig. 20b).

Infrared soft-baking also has some drawbacks. First, different wafer surfaces (e.g. oxide, silicon, and aluminum) have different reflectivities and absorption characteristics. Therefore, the same level of incident IR energy can cause variations in the temperature between wafer types. Second, solvent vapors coat the IR lamps, changing their energy output. Thus, frequent cleaning of the lamps, and calibration of the ovens, needs to be performed to minimize such "solvent clouding" effects. Third, the gases used to exhaust vaporized solvents can also become heated by the IR. This can lead to convection heating of resist surfaces, and contribute to crust formation.

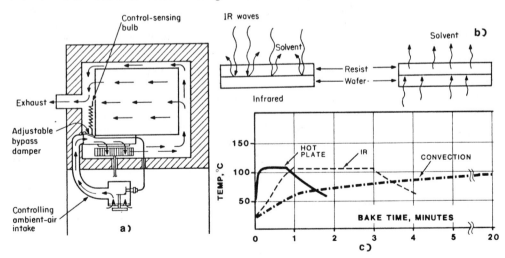

Fig. 20 (a) Convection oven cross section. Courtesy of Blue M, a Unit of General Signal. (b) Infrared and conduction solvent removal mechanisms. (c) Profiles for various bake methods[45]. Reprinted with permission of the Eaton Corporation.

Hot-plate soft-baking has several attractive features compared to the other techniques. The resist is heated quickly to the desired temperature (Fig. 20c), the soft-bake cycle can be short (30-60 sec /wafer), and heating takes place "inside out"(as in IR soft-baking). To promote heat transfer and to insure uniform results, vacuum ports should be used to clamp the wafers. Automatic wafer handling, however, must be utilized to make single-wafer hot-plate heating feasible. Single wafer heating also offers the potential of a compact bake station. Reports in the literature indicate that hot-plate soft-baking cycles of 30-60 sec produce residual solvent levels and thickness reductions comparable to the other bake methods. Such cycles allow throughputs of roughly 50-120 wafers /hour, which are comparable to the throughputs of batch convection oven and batch IR oven soft-baking. To allow wafers to be cooled prior to exposure in an in-line system, use of a cold-plate immediately after the hot-plate bake has been proposed.

Resist Processing: Exposure

After a wafer has been coated with resist and suitably soft-baked, it is ready to be exposed to some form of radiation in order to create a latent image in the resist[47,48]. The degree of exposure is adjusted by controlling the energy impinging on the resist (a product of the intensity of the source and the time of exposure). An *energy integrator* (described in more detail in Chap. 13) is used to detect the total energy striking a unit area of resist, and automatically adjusts the time of exposure to compensate for aging variations in the source.

Exposure of the photoresist is a critical step in the resist processing procedure for several reasons. First, exposure is one of steps in the imaging sequence in which the wafers are individually processed. A lengthy exposure process may thus lead to a throughput restriction relative to the other steps. In addition, attempts at increasing throughput during exposure typically result in decreased resolution. This is due partly to aspects of the resist chemistry. Longer exposure times (and consequent lower throughputs) accompany the use of high-resolution resists. In projection step-and-repeat aligners, however, this factor is not as important, as the time to load and align the wafers consumes a large fraction of the total exposure process time.

As discussed earlier in the chapter, during exposure, photochemical transformations occur within the resist. The goal of an optimized exposure process is to produce the desired photochemical effects in the resist in the shortest period of time, and in a highly reproducible manner. Successfully optimizing such exposure steps is a complex task because it is dependent

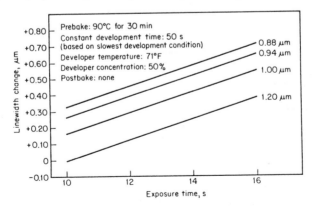

Fig. 21 Resist thickness and exposure-time relationship versus linewidth change[79]. Courtesy of Eastman-Kodak Company.

on other steps of the resist processing sequence. That is, it requires: a) determining an optimum resist thickness, and thereafter insuring that this thickness is uniformly produced across the wafer and from wafer to wafer; b) choosing a soft-bake process that complements the selected exposure conditions, as well as other process steps; c) determining the line-width specifications and dimensional tolerances with respect to the mask and exposure tool; d) establishing the developer concentration, temperature, and agitation requirements based upon the preceding steps; e) selecting post-bake conditions compatible with all earlier steps; and f) specifying the etching process. Let us discuss the first two of these steps in detail, as the others are considered further in subsequent sections. The hardware used to perform the exposure steps (e.g. exposure tools and energy sources) is the subject of Chap. 13.

The resist must possess a specific thickness which is consistent from wafer-to-wafer, and uniform across each wafer. Methods for ensuring resist thickness uniformity are covered in the *Resist Processing: Coating* section. Variations in the resist thickness result in changes in the linewidth. Figure 21 shows an example of how linewidths may vary as a function of resist thickness. In this example, a variation of 20% in resist thickness causes a linewidth variation of a process to undergo a variation of 0.25 μm. For some applications, if this were the only element of a process to undergo variation, the 0.25 μm change might be tolerable. In actual processes, however, other parameters that impact linewidth can also change (e.g. lamp intensity and developer concentration). If these all were to simultaneously undergo worst-case variation, a larger cumulative change in the linewidth would result. As a consequence, a 20% thickness variation would be unacceptable as part of a set of allowable resist process variations.

The degree of solvent removal during soft-bake also impacts the choice of exposure time. Positive resists which are partially soft-baked can be exposed by a dose that will not completely photoreact the film. The resist, however, will appear to be *more* photosensitive than a more fully baked film, because the developer will more quickly dissolve any resist regions having relatively higher solvent content. The unexposed resist, however, will also be subject to more dissolution than if it were baked at a higher temperature. The dissolution versus soft-bake temperature should be established whenever a new resist process is being developed. An acceptable soft-bake range from such curves can be chosen. With this (and other data), resist processes can be optimized for particular combinations of resolution and throughput. For example, high throughput would involve a process with less soft-baking, underexposure, and standard development. A high resolution process would utilize oversoft-baking, standard exposure, and a diluted developer solution.

Standing Waves

Standing waves are caused when the actinic light waves propagate through a resist film down to the substrate, where they are reflected back up through the resist. The reflected waves constructively and destructively interfere with the incident waves and create zones of high and low exposure with a separation of (λ/4n), where n is the index of refraction of the photoresist. This causes unwanted effects in resist layers in two ways. First, the periodic variation of light intensity in the resist causes the resist to receive non-uniform doses of energy throughout the layer thickness (Fig. 22a). The resulting pattern in the resist due to this phenomenon is shown in Fig. 22b. The second effect causes linewidth variations as the resist crosses a step, and is due to the variation of the total energy coupled to the resist by interference effects at different resist thickness. Both effects contribute to resolution loss of resists, and become significant as 1 μm resolutions are approached. The mathematics that explains the basis of such interference phenomena is resists is well covered in a paper by Ilten and Patel[49].

Highly reflective substrates accentuate the standing wave effects, and thus attempts to

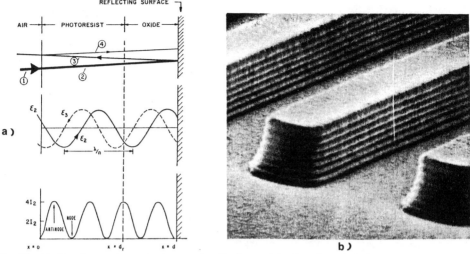

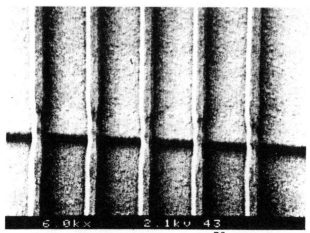

Fig. 22 (a) Standing light waves in a resist film caused by interference between the incident and refelcted film[80]. (b) Photograph of standing wave effects in exposed resists.

suppress such effects have involved the use of dyes and anti-reflective coatings below the resist layer. These two techniques will be discussed in more detail in the following section. Other approaches have also been investigated, including: a) the use of incident UV radiation of multiple wavelengths, to reduce the variation of energy coupling by averaging; and b) post-baking between exposure and development.

Linewidth Variation as Resist Crosses Steps

Photoresist linewidth variations are especially difficult to control in the vicinity of steps, where large resist thickness changes occur. These variations are primarily due to: a) differences in the energy that is coupled into the resist at different resist thicknesses; b) the scattering of light at the steps due to diffraction and reflections; and c) standing wave effects. The minimum

Fig. 23 Notching of photoresist lines over topography[78]. Reprinted with permission of the publisher, the Electrochemical Society.

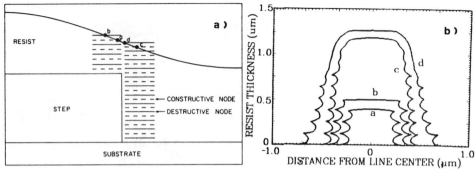

Fig. 24 (a) Schematic diagram of the resist thickness changing through constructive and destructive nodes at a step. (b) Simulated resist profiles corresponding to resist thicknesses of a-d in (a)[50]. Reprinted with permission of SPIE.

linewidth which can be specified for production with projection printers is often dictated by the degree of linewidth narrowing which occurs at the top corners of steps as a result of these effects (also known as *notching*, Fig. 23). The dramatic increase in notching observed as substrate reflectivity increases is the basis for the rule of thumb that *the minimum linewidth is 0.75 λ /NA on silicon, but is 1.1λ /NA for aluminum*, where λ is the exposure wavelength (e.g. 430 nm), and NA is the numerical aperture of the exposure tool[50] (e.g. 0.28).

The fact that the resist thickness changes very significantly in crossing a step is illustrated in Fig. 11. In this example, the thickness of the resist at the bottom of the step is about twice the thickness at the top of the step. This implies that if the linewidth of resist as it crosses a step is to be uniform, the resist must be simultaneously printable over a wide range of resist thicknesses. In practice, it is necessary that the narrowest remaining resist linewidth at the top of steps still be acceptable after the resist is exposed with an energy dose that is sufficiently large to clear out the thick resist below the step.

A model derived by King[5] for linewidth variations in resist, due to differences of coupled energy with thickness, as steps are crossed, predicts that the change in linewidth, ΔL, can be found from:

$$\Delta L = 2 \, \Delta T_o \, / \tan \theta \qquad (6)$$

where ΔT_o is the change in resist thickness between the resist at the top and bottom of the step, and θ is the angle of the resist wall at the resist-substrate interface. Equation 6 emphasizes the need for steep resist wall profiles when crossing over steps (i.e. large values of $\tan \theta$). King also showed that $\tan \theta$ is proportional to the contrast γ, and to the thickness of the resist T_o. Thus high γ values, and thick resist layers, help to maintain linewidth uniformity over steps.

A computer simulation program SAMPLE, has also been used to model the linewidth variation of resist when it crosses steps (including both energy coupling and interference effects). Figure 24b shows SAMPLE simulated resist profiles for various points on top and below the step of shown Fig. 24a[50]. The resist profile at point A is thinnest, narrowest, and is also affected by standing wave interference. The resist profile at the bottom of the step (point D) is the thickest and widest. The extreme linewidth differences occur between constructive nodes above the step (A), and destructive nodes below the step (D). Net linewidth variations of 0.35 μm in 1 μm lines were predicted for this example.

A variety of procedures have been investigated to minimize linewidth variation over steps. They fall into two categories: a) those discussed earlier which involve minimal increase in process complexity (but also effectiveness) including use of high γ resists, thick resist layers, multiple wavelength light sources, and post exposure bakes; and b) those which impact throughput or require increased process complexity, but offer greater improvement in performance, including dyes to make the resist more optically absorptive, and anti-reflective coatings (ARC).

The addition of unbleachable dyes is designed to increase the optical absorbance of resists, and thus reduce the effects of standing wave interference and light scattering, especially over highly reflective substrates[51,52]. Dyes are simpler to implement than ARCs, but require increased exposure times (i.e. reduced throughput). They have been reported to generally reduce standing wave effects and notching over steps, but at the expense of reduced contrast and resist sidewall slope. In addition, high dye concentration has been observed to actually increase the magnitude of the standing wave near the resist-substrate interface. In extreme cases, a large toe in the resist from this effect can lead to resist scumming.

One type of *anti-reflective coating* (ARC) is a polymer film which is highly absorbing and non-bleaching at the exposure wavelength. It is applied directly to the substrate to a thickness of ~0.5 μm, and resist is spun on top of it (Fig. 25). As a result, the ARC absorbs most of the radiation that penetrates the resist (70-85%). Standing wave effects are substantially reduced, as there is much less reflection off of the substrate. Scattering from topographical features is also suppressed. In addition, the ARC partially planarizes the wafer topography, further helping to improve linewidth variation over steps, since the resist thickness is more uniform. Layers of ARC can also be used between the imaging resist and PMMA in mutilayer resist processes[53,53,55]. Thin layers of Ti or Ti:W are also as ARCs on films of aluminum.

The price paid for these benefits is increased process complexity and possible loss of dimensional control. Two extra process steps incurred with polymer film ARCs are: 1) spin-coating the ARC material; and 2) prebaking the ARC before spinning on the resist. In most ARC types the pattern is transferred to the ARC during the development step following exposure, although some experimental ARC materials call for a dry-etch process for this step. Those ARC materials that are patterned during the development step are removed isotropically. In addition, their rate of attack by developer is a strong function of the soft-bake conditions (i.e. temperature and time). If insufficient baking is carried out, the ARC will undercut excessively, and the resist will lift. If too high a bake temperature is used, the ARC will not be completely removed during the development step. Therefore, to insure repeatable results, an optimum prebake cycle for each ARC type must be established, and thereafter be tightly controlled. Some linewidth loss, related to the ARC undercutting, has been reported during SiO_2 and Si etching (note that anisotropic etching still results in some linewidth loss), but not during Al etching.

Resist Processing: Development

Following exposure, the resist film must undergo *development* in order to leave behind the

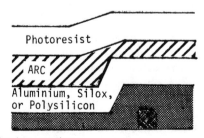

Fig. 25   Anti-reflection coating.

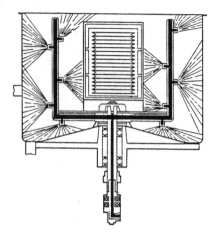

Fig. 26 Batch spray developer: Process chamber action.   Courtesy of FSI Corp.

image which will serve as a mask for etching, ion-implantation, lift-off, etc.   *Development* is another of the critical steps in the photoresist process, and often entails both an extensive process development effort and stringent control measures, after it is put into production[56,57].   The goals of an effective development process include: a) the original film thickness of unexposed (positive) resist should not be measurably reduced during development; b) the development time should be short (e.g. less then one minute); c) the development should cause minimum pattern distortion or swelling; and d) the specified pattern dimensions should be precisely produced.   After development, it is necessary to rinse and dry the substrate, because the developing action continues until the developer is completely removed from the resist surface.

We stated that the first goal of the development process is that the unexposed positive resist film should be minimally attacked by the developer.   Actual developers always erode unexposed resist to some degree, but what makes a resist process viable is the diffferential solubility of exposed and unexposed areas.   In practice, more selective developers also take longer to fully develop the image.   Thus, when choosing a developer, tradeoffs between differential solubility and throughput must be made.   A developer should be selected so that throughput requirements can be satisfied without causing excessive resist thinning.   In addition, the developer should also allow sufficient latitude so that the process can be maintained in a manufacturing environment.

It should be noted that *positive resist developers* are aklaline solutions diluted with water. As a result, they have the advantage of requiring only a water rinse.   *Negative developers* are organic solvents, and must be rinsed in other organic solvents (e.g. n-butlyl acetate).

There are three main methods by which development is carried out: immersion, spray, and puddle developing.   In *immersion developing*, cassette-loaded wafers are batch-immersed and agitated in a bath of developer at a specific temperature and time, often using a mechanical agitating arm.   The advantages of immersion development are: high throughput; good uniformity of development; low-capital equipment costs; and small clean-room footprint by the development equipment. The *spray development* process can be carried out in *batch* or *single-wafer* modes.   In *batch systems,* fresh developing solution is directed across wafer surfaces by fan-type spray devices as cassettes of wafers are rotated by a turntable (Fig. 26)[58].   At the end of each development cycle (~30-45 sec), a controlled fraction of the developer is drained off, and an equivalent amount of fresh developer is added, allowing a relatively uniform bath strength to be maintained. In *single-wafer spray development systems*, the developer is sprayed onto a spinning wafer, and each wafer is treated with a fresh dose of developing solution. The wafer is then spun-dry and transported to the postbake module. In the *puddle technique*, a fixed amount of developer is first dispensed onto a static wafer[59]. After the required develop dwell time, the developing action is stopped by directing a stream of deionized water onto the developed wafer. A spin-dry again follows the rinse step. The puddle technique is reported to be less effective for 125 mm and larger wafers.

Positive resist is developed according to chemical reactions postulated by Sus and Levine[3], whereby the carboxylic acid photoproduct is neutralized by the alkaline developing solution. The reaction products are amines or metallic salts, which are rapidly dissolved into the developer solution. Since there are no such groups formed in the unexposed areas, they remain essentially unaffected by the developer. The dissolution of the salts causes the developer normality to drop, since the active ingredient (e.g. NaOH or KOH) is consumed by the reaction. As a result, the developer bath is typically monitored by pH level, number of lots processed, or time since replenishment. When it is indicated that the developer is depleted, the bath solution is replaced.

Positive optical resists also dissolve layer-by-layer, with a minimum of swelling. That is, since positive resist is relatively impervious to developer, at any instant only the surface of the resist is being dissolved. The entire negative resist film on the other hand is penetrated by the developer, and in unexposed regions, a thick gel layer is first formed and dissolution follows. As discussed earlier, this is an important reason why positive resist systems are able to achieve higher resolution. The required developer strength is established by determining the rate of dissolution of unexposed resist at various developer dilutions.

The dissolution rates of resists can be measured in several ways[60]. The simplest method is the *dip and dry technique*. Pieces of resist-coated wafer are immersed in the developing solution, and removed after a measured development time. The samples are then dried and baked for a short time (to remove residual solvent), and the thickness of the remaining resist is measured (e.g. with a profilometer or interferometer). By removing samples after sequentially longer develop times, a complete thickness vs. develop time curve can be generated. However, while the method is simple and inexpensive, it is also tedious. A second technique involves the use of *laser interferometry* (Fig. 27). That is, a laser beam is reflected from the two surfaces of a

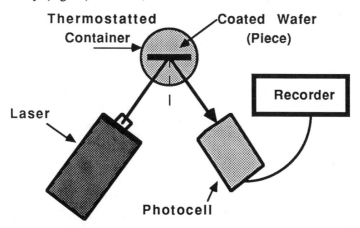

Fig. 27 Basic laser interferometer for monitoring polymer dissolution[60]. Reprinted by permission of Solid State Technology, published by Technical Publishing, company of Dun & Bradstreet.

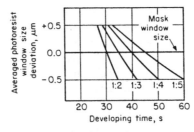

Fig. 28 Example of photoresist window-size deviation versus developing time.   Courtesy of Shipley Company.

dissolving polymer film immersed in a solvent.   The second surface is the solid surface (such as an SiO$_2$-coated wafer), to which the polymer is attached.   Variations in the intensity of the refelected beam provide data about the dissolution process.   The intensity of the reflected beam typically oscillates with a constant period and amplitude as the dissolution process causes the resist thickness to decrease, if the resist dissolves without swelling.   The oscillation period can be used to infer the dissolution rate.   An instrument that carries out this measurement technique has been commercially introduced[71].

In earlier sections it was discussed how various process parameters can impact development. Therefore, they must be optimized, and thereafter carefully controlled.   These include: a) exposure time; b) softbake temperature; c) photoresist thickness; d) developer concentration; e) developer temperature; and f) developer agitation method.   In typical photoresist processes all of these parameters are fixed, except developer concentration, which in batch process can vary with the number of wafers developed in a bath.   The develop time is then adjusted to compensate for concentration variations.

The ability to produce contollable linewidths in the face of developer concentration variations and consequent developer time adjustments, involves some preliminary experimental work.   That is, several developer dilutions are first prepared and then plots of line size versus development time, for a fixed exposure, are generated (Fig. 28).   From this data the range of developer concentrations that gives the most process latitude (i.e. smallest linewidth variation per development time change), and also provides a reasonable development time (30-60 sec) should be selected.   Shorter times than this have the effect of increasing the margin for error during development, while longer times reduce throughput.

Developer temperature is another process parameter that must be tightly controlled (Fig. 29).   Generally, most developers are formulated to perform best at ~21°C.   In order to maintain reproducible development results for 1 μm geometries, this temperature should thus be held constant to within ±0.5°C.   Some fabrication lines elect to set developer temperatures at

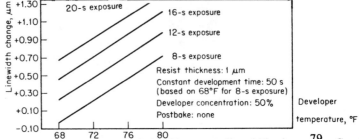

Fig. 29 Developer temperature and exposure relationship versus linewidth change[79].   Courtesy of Eastman-Kodak Company.

23°C, since it is easier to control a bath at temperatures slightly higher than ambient. In addition, at 23°C there is a small increase in photospeed, which aids exposure throughput. In spray development applications, however, a drop in developer temperature of 3-8°C can occur as the developer is dispensed from the spray nozzle. Heated nozzles are frequently employed to compensate for this effect. A recent development is the *vapor jet nozzle*, which utilizes a piezoelectric cell to atomize the developer into a fine mist. This avoids the temperature drop that occurs when developer is sprayed by pressurized N_2 and also consumes 40% less developer[84].

A fresh supply of developer needs to be transported to the wafer surface during the entire development step. This is accomplished by *agitation* of the wafers in immersion development. Several agitation techniques are used, including motion of resist-coated wafers with a mechanical agitator, and bubbling gas through the developer (nitrogen is used to prevent depletion of the active ingredient by the carbon dioxide from air). Pressure dispensing of solution onto a wafer spinning at ~1500 rpm is another method for introducing fresh developer to the wafer surface. Finally, spray developing is utilized. In this method the solution is broken up into small droplets, which produces a scrubbing action. Such scrubbing action effectively carries away dissolving resist, as well as continuously bringing fresh developer to the surface.

Since most conventional developers for positive resists contain metal ions, there has been concern about the possible contamination of the wafers from the developer residues. As a result, *metal-free developers* (with less than 15 ppma of total metals) have been formulated as alternatives[61]. Most metal-free developers, however, attack unexposed resist more than do conventional developers. Typically, metal free developers remove 25-30% of the unexposed resist thickness during development. As devices shrink, the need for metal-free developers will increase, and this should result in stimulating efforts to produce improved low-metal products.

It is important that wafers be developed promptly after exposure. If there is a delay between exposure and development, under some circumstances the resist chemistry will proceed down one or more of the side reaction paths and form basic-insoluble product. This can result in critical dimension variations as a function of the delay times[62].

Techniques to automatically control the development process on an in-line basis are being pursued. That is, in single-wafer spray or puddle development systems, resist thickness during

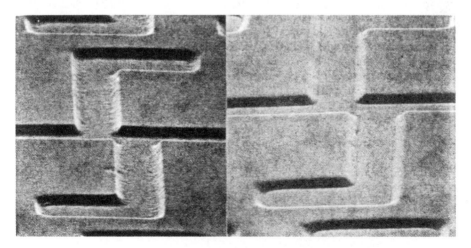

Fig. 30 Effect of plasma descum on pattern quality. Reprinted from Ref. 8 with permission of the American Chemical Society.

development is monitored by measuring the reflectance from the wafer of a non-exposing mono-chromatic light beam. A control point (CPT) can be determined that represents the time to achieve a predetermined critical dimension (CD). This time can be adjusted to compensate for variations in dose, resist thickness, and soft-bake temperatures, to yield more consistent CDs from lot-to-lot.

In some instances it is not possible to remove all the resist during the development step. In such cases, a mild plasma treatment referred to as *plasma descumming*, is typically used to remove, in an almost surgical-like manner, very small quantities of resist in unwanted areas (e.g. a few hundred angstroms)[67]. Figure 30 illustrates plasma descumming on pattern quality.

Resist Processing: After Develop Inspection

Following development, an inspection (sometimes referred to as an *after-develop-inspection*, or ADI) is performed. The purpose is to insure that the steps of the PR process up to this point have been performed correctly and to within the specified tolerance. Mistakes or unacceptable process variations can still be corrected, since the resist process has not yet produced any changes (e.g. through an etch step) to the wafer itself. Thus, any inadequately processed wafers detected by this inspection (known as *rejects*) can have their resist stripped and reworked.

In high production fab lines, individual reject wafers may be collected by device type and mask level, until enough are assembled to be reprocessed as a group. In lower volume production lines (or for special "hot lots", that need quick completion), the rejected wafers may be immediately re-run, while holding the acceptable wafers at the inspection station.

The wafers at ADI are typically inspected with an optical microscope, although SEMs and laser-based systems are being introduced for performing some inspection tasks such as linewidth measurement, as well. The following are some of the aspects of the resist process that are monitored by the inspection procedure: a) correct mask has been used; b) resist film qualities are acceptable (i.e. resist is free from contamination, scratches, bubbles, striations, etc.); c) image quality is adequate (i.e. look for good edge definition, linewidth uniformity or indications of bridging); d) critical dimensions are within the specified tolerances; e) defect types and densities are recorded (and this data is used to correlate the occurrence of defects and product yield): and f) registration is within specified limits (see Chap. 13 for more information on *registration*).

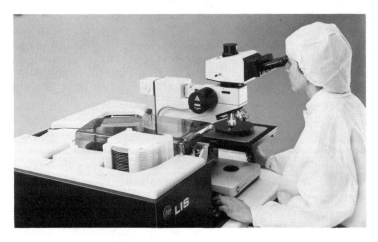

Fig. 31   Photograph of a wafer inspection station.   Courtesy of E. Leitz, Inc.

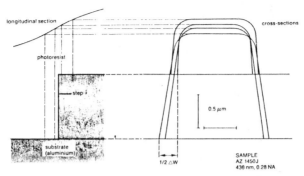

Fig. 32 Longitudinal section and cross sections of a photoresist line running across a one micron aluminum step. The resist profiles are simulated by SAMPLE. The nominal linewidth is 1.8 μm[52]. Reprinted with permission of SPIE.

On some microscopy-based inspection stations, all wafer handling and data processing functions have been automated. Only human vision remains as a non-automated aspect of the inspection procedure. That is, wafers are transported by belts or vacuum shuttle from an input cassette to a pre-aligner, then onto an inspection stage under the microscope (Fig. 31). Automatic handling allows the operator to concentrate on inspection, and to minimize the likelihood of airborne or human handling contamination. Inspection data is entered with a keypad, and many stations include host computer interfacing capabilities for processing and storing the data[63].

In more automated systems, the human operator is completely removed from the *defect inspection task.* That is, *in-process wafer inspection systems,* based on automatic image processing have been introduced. Defect detection is accomplished either by die-to-die or die-to-database comparison. Manufacturers of these systems claim defect detection sensitivities well into the sub-micron range. Such instruments, however, often have difficulty detecting particles on substrates that have surface granularity, or on wafers containing surface topography. In addition, for particles near the minimum-size detection limit, such machines can be prone to miss the presence of some particles, and signal the detection of others that may be non-existent.

The remainder of this section discusses linewidth measurement techniques used to verify that critical dimensions have been produced. In addition, procedures are described for monitoring the variation of linewidths produced in a production environment as a function of time. Such data can serve as a gauge for tracking the performance of a lithographic process line.

Linewidth Variation and Control

There are two aspects of feature sizes that must be controlled in the lithographic /etching process: a) the absolute size of a minimum feature, including linewidth, spacing, or contact dimensions (also referred to as a *critical dimension,* or CD); and b) the variations of the minimum feature sizes as they cross steps on the wafer surface. Linewidth (and spacing) measurements are regularly performed to determine the actual sizes of CDs at each masking level of a process. The variation of linewidths over steps are also monitored, and the causes of the variation were discussed in the section on *Resist Processing: Exposure.* These two aspects are mentioned together because there is also a tradeoff between absolute linewidth size and variation of the size over steps. That is, over-exposure and over-development can improve linewidth control, but at the expense of linewidth size. Figure 32 shows a SAMPLE simulation which calculates linewidth variation, ΔL, across a 0.5 μm step, as the line sizes vary with changing exposure and development. It shows that linewidth variation over steps can be considerably reduced by over-exposure, but at the expense of dimensional accuracy[52].

Another issue involving linewidth control is that correct feature *sizes* must be maintained across an entire wafer, and from one wafer to another. The ability to do this is referred to as *linewidth control*. As feature size is reduced, the tolerable error on feature size control is also decreased. For example, the required tolerance on a nominal 1 μm linewidth of polysilicon features is typically ±10%, or ±0.1 μm. Note that when exposure is performed by a wafer stepper, feature size must be controlled across *every exposure field*, and *field-to-field* variations in feature sizes must also not exceed acceptable limits.

Linewidth control is impacted by a variety of factors, and depends on hardware, processes, and materials. The degree of control is determined by measuring a series of test structures with known feature sizes across a wafer, and then plotting the feature dimension as a function of location on the wafer. The standard deviation at the one and two sigma level then becomes a measure of the linewidth control capability of a particular exposure /resist process. For example, a process that has minimum linewidths of 1.5 μm, and spaces of 1.0 μm, the controllability specification might be a maximum 7% linewidth variation, with 95% (2σ) confidence. Such data are then plotted as a function of time (Fig. 33) and are utilized to monitor the performance of a lithographic line. Note that this subject is covered in more detail in Vol. 2 of this text.

Linewidth Measurements

The width of features produced on wafers are measured in many phases of the fabrication process, including: a) after development; b) after etching; c) during photomask production; and d) to monitor overlay registration. A number of techniques are currently utilized to perform such feature size measurements. For VLSI applications the measurement technique must be repeatable to 0.1 μm in order to verify that the ±10% size tolerance specification (cited as an example in the previous section) is satisfied during fabrication. In fact, it is argued that the measurement uncertainty of the metrology system must even be smaller. That is, in the 1 μm polysilicon line described above, if the uncertainty in the etching process is ±0.03 μm, then it must be necessary to determine that its photoresist linewidth is 1 μm ±0.07 μm.[72] The National Bureau of Standards (NBS) is involved in developing linewidth measurement standards that can be used to

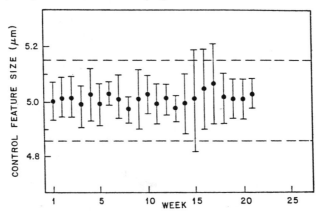

Fig. 33 Linewidth control data for a typical process line. Weekly averages are shown for a 5.0 μm control feature. The dashed lines represent limits. Note weeks 15-17 represent a processing problem that was corrected in week 18. Reprinted from Ref. 8 with permission of the American Chemical Society.

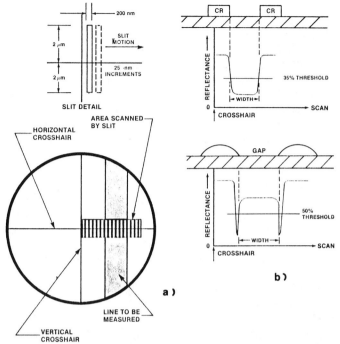

Fig. 34 Slit scan technique. (a) When viewed through the binocular, the crosshair and line to be measured are superimposed. An open field is placed at the vertical crosshair, with the line of interest to the right on the horizontal crosshair. The 2000Å wide by 4 µm wide slit then scans a path 4 µm wide by 10 µm long in 250Å increments[65]. (b) Typical edge sensing profiles, corresponding to available edge-sensing programs, particularly suited for chromium lines on photomasks. Reprinted by permission, Micrelectronics Manufacturing and Testing.

calibrate commonly used linewidth measurement equipment[64]. Calibration standards for photomasks have been available for some time from the NBS and are in widespread use. They were the first to be developed because line edges on masks are well defined, and good contrast exists between light transmitted through the mask and opaque feature edges. Measurements on processed wafers are much more complicated, and the NBS has not yet released such standards. Calibration by semiconductor manufacturers is commonly made to a known "good" wafer, which is thus used as the standard.

Most of the linewidth measurement systems employ either visible-light optics, He-Ne laser light, or electron-beam optical techniques[65]. Current optical techniques based on ordinary microscopes are satisfactory for use with 1.5 µm (or larger) feature sizes. These systems will thus continue to find wide applicability because of their low cost, ease of use, and high throughput. Future development will be directed at improving these attributes, while maintaining the required accuracy and precision. For smaller geometries, scanning electron microscopy, and laser scanning techniques will become more frequently employed.

Measurement of linewidth by optical techniques is accomplished with the following types of systems: a) mechanically scanning an optical slit across the magnified feature image; b) video scanning across the feature of interest; c) image shearing; and d) scanning a laser spot across the feature, and detecting the reflected image. In the *scanning slit technique* (Fig. 34), the light passing through the slit is measured by a photomultiplier tube (PMT) to form a microdensitometer profile, which is then used to perform the linewidth measurements. The operator views the image through a binocular microscope or color crt. The narrow slit (e.g. 2000 Å wide) is moved across the image (e.g. in 250 Å increments), and the intensity profile is acquired via the PMT. The profile is then analyzed by an edge-sensing algorithm to determine the dimension. An auto-focus algorithm can also be used to find the "best" focus, by moving the motorized stage in the z-axis until the maximum slope of the selected edge is located (or, in some systems, by use of an independent laser focus technique). In *video based systems*, a video camera is used to capture and store the profile of the feature of interest (Fig. 35). An operator moves a computer-controlled cursor to identify the specific feature to be measured. The optical profiles are then acquired and processed in a manner similar to that described for the scanning slit. In general, video systems are faster, but slit scans offer greater sensitivity. (The NBS photomask calibration system uses a slit-scan technique.) The third optical technique is *image shearing,* in which an operator visually positions the edges of a pattern that has been sheared into two images. At the start of the measurement, the image edges are butted against each other. The shearing control is the adjusted until the images are rejoined into one. The difference between the initial and final values of the shearing control vernier is used to determine the linewidth. Due to the increased precision of slit scan and video scan techniques, they have largely supplanted image shearing methods.

Linewidth measurement systems based on measuring reflected light from a *scanning laser spot* offer higher resolution than ordinary optical microscopy techniques (Fig. 36). They utilize a He-Ne laser focused to a 1 μm spot, which is scanned across the line to be measured. The reflected light is detected in various ways, including: a) a pair of photodetectors positioned on either side along the scan axis; b) through a confocal scanned microscope[73]; and c) a high

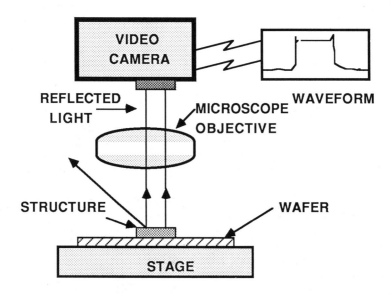

Fig. 35 The video scan technique utilizes an image profile obtained by analyzing the digital data obtained from a video camera. Reprinted with permission of Semiconductor International.

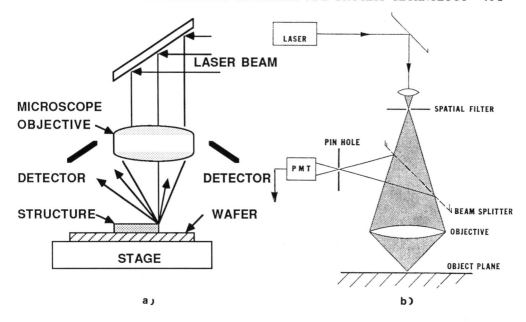

Fig. 36 (a) The principle of the laser scanning measurement technique is shown. The distance between edge detection pulses determines the linewidth[74]. Reprinted with permission of Semiconductor International. (b) One laser scanning system utilizes a laser and a confocal microscope[73]. Courtesy of SiScan Systems.

resolution laser interferometer[74]. Resolution capabilities of 0.25 μm, and repeatability of 0.05 μm are claimed for some of such systems. Their increased measurement limits are also accompanied by significantly higher cost than the optical techniques described earlier.

Linewidth measurement systems based on scanning electron microscopes (SEM) can overcome the accuracy limitations of optical techniques for submicron geometries[72,75]. As the number of applications that demand such accuracy and precision continues to grow, the use of SEMs for linewidth measurement will also increase. SEMs also provide data about the *resist profile* (i.e. information that optical techniques cannot provide). When low-selectivity dry-etch processes (see Chap. 16), or high energy implants are used (see Chap. 9), knowledge about the shape of the resist profile can be quite valuable. To optimize throughput for production-line use, totally dedicated single-purpose IC metrology SEMs are expected to be used in such roles.[76] Readers are also directed to Chap. 17 for additional information about the characteristics of SEMs that are well suited for linewidth measurement in VLSI production applications.

The accuracy and precision of linewidth measurement techniques, however, are not simple parameters to determine. Figure 37 illustrates that there are a host of significant factors inherent in the characteristics of both the linewidth sample, and the components of the measurement system, that impact the linewidth value. In the example of Fig. 37, an optical measurement technique is examined, but similar complexities are associated with SEM-based methods. This is one of the reasons that establishing uniform calibration standards for feature sizes on patterned wafers has been so difficult. Readers are directed to the Proceedings of the International Society of Optical Engineering (SPIE), Vol. 480, *Integrated Circuit Metrology ll*, 1984, for a set of

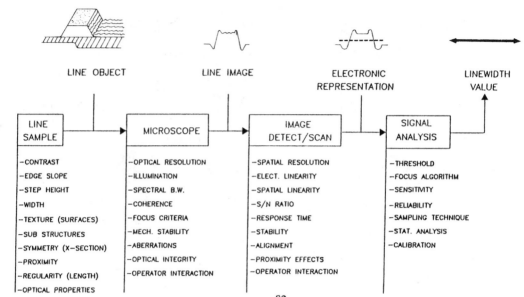

LINE OBJECT	LINE IMAGE	ELECTRONIC REPRESENTATION	LINEWIDTH VALUE

LINE SAMPLE	MICROSCOPE	IMAGE DETECT/SCAN	SIGNAL ANALYSIS
−CONTRAST	−OPTICAL RESOLUTION	−SPATIAL RESOLUTION	−THRESHOLD
−EDGE SLOPE	−ILLUMINATION	−ELECT. LINEARITY	−FOCUS ALGORITHM
−STEP HEIGHT	−SPECTRAL B.W.	−SPATIAL LINEARITY	−SENSITIVITY
−WIDTH	−COHERENCE	−S/N RATIO	−RELIABILITY
−TEXTURE (SURFACES)	−FOCUS CRITERIA	−RESPONSE TIME	−SAMPLING TECHNIQUE
−SUB STRUCTURES	−MECH. STABILITY	−STABILITY	−STAT. ANALYSIS
−SYMMETRY (X−SECTION)	−ABERRATIONS	−ALIGNMENT	−CALIBRATION
−PROXIMITY	−OPTICAL INTEGRITY	−PROXIMITY EFFECTS	
−REGULARITY (LENGTH)	−OPERATOR INTERACTION	−OPERATOR INTERACTION	
−OPTICAL PROPERTIES			

Fig. 37 Factors affecting the linewidth measurement[82]. Reprinted with permission of SPIE.

excellent papers describing progress on many aspects of this formidable measurement task.

Resist Processing: Post-Bake

Post-baking is a process which subjects the resist to an elevated temperature after completion of development, and is usually performed just prior to etching[66,67]. Its chief purposes are to remove residual solvents, to improve the adhesion, and to increase the etch resistance of the resist. In addition, since post-baking often causes the resist to flow, this effect is sometimes utilized to reduce the incidence of pinholes or thin spots in the resist prior to etching, or to modify the edge profile of the resist (Fig. 38). Note that the onset of flow occurs at a temperature which is close to the temperature of the resist softening point, known as the *glass transition temperature*[68].

The residual solvent in a resist after soft-baking is usually less than 3-4% and post-baking reduces this value even further. Additional solvent removal is important if the patterned wafers are to be processed in an ion-implanter, dry-etcher, or any other vacuum environment in which the presence of the solvent in the film could produce solvent-burst effects in the resist.

Resist adhesion may be improved by post-baking, but excessive temperatures may cause the opposite to occur. It has been observed that the adhesion of resists to SiO_2 increases up to post-bake temperatures of 170-180°C, after which it rapidly decreases (presumably due to the rupture of the resist /SiO_2 bonds). This information should be determined for the particular resist being utilized if strong resist adhesion is a critical requirement. It has also been observed that the adhesion during development may also be strengthened by a post-exposure, and pre-development, bake. That is, a second soft-bake of 5-10 minutes at ~90°C is especially useful on highly doped SiO_2 or polysilicon substrates.

The etch resistance of the resist film may also be influenced by the post-bake process. Several other techniques have been investigated to produce the same effect including: a) ion implanting the resist surface with a high-dose, low-energy beam (e.g. using As as the implant

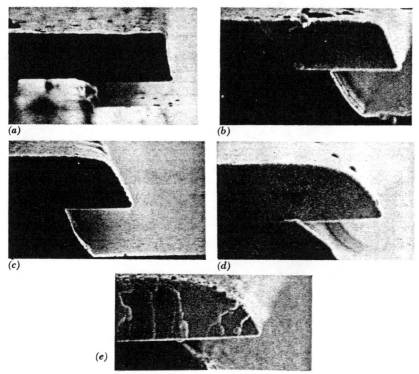

Fig. 38 Postbake temperature versus thermal flow (1350J).   (a) 100°C, (b) 110°C, (c) 120°C, (d) 130°C, (e) 140°C.   Courtesy of the Shipley Company.

species); b) plasma hardening; and c) UV hardening (discussed in the following section).

The thermal flow of resist that can occur during post-bake is sometimes used to modify the edge profiles of the resist.   Higher thermal flow temperatures permit higher post-bake temperatures without modifying wall profiles (and feature sizes).   If a tapered resist profile can be created, followed by an anisotropic dry-etch process, tapered wall profiles in the patterned features are covered more uniformly by subsequently deposited films.   The optimum post-bake cycle is one which will produce the best combination of adhesion, wall-profile, and resist removal after etch (i.e. higher post-bake temperatures also usually mean more difficult post-etch resist removal).   *Stripping* of photoresist is covered in Chap. 15.

Resist Processing: Deep UV Hardening of Resists

It is necessary to maintain tight control of fine feature sizes in photoresist patterns during a variety of processes that involve elevated temperatures.   During such processes as high-current ion implantation, plasma etching, and ion milling, the temperature can rise high enough (125-200°C) to cause the resist to flow.   This reduces the resolution of the original pattern.   It has been demonstrated, however, that if positive photoresist patterns are irradiated with a deep UV light source (e.g. <320 nm) and simultaneously heated with a ramped temperature (ramp endpoints 120-190°C) for 60 sec, they can then withstand bake cycles up to about 200°C before the onset of flow[69,85].   The flood irradiation is believed to increase the molecular weight of the resin at the surface, most likely through crosslinking.   Since deep UV light is strongly absorbed

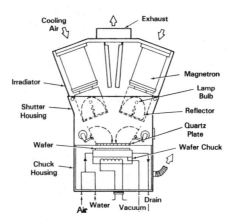

Fig. 39 System for UV-photostabilization[85].   Courtesy of Fusion Semiconductor Systems.

by positive resists, only the upper few thousand angstroms of resist is subjected to deep UV irradiation.   Therefore the mechanism of deep UV hardening involves the formation of a surface skin.   Flood irradiation with near-UV (e.g. >300nm) on the other hand, apparently decreases the thermal flow resistance of the resist[62,70].   Thus, longer wavelength UV components must be filtered from the illuminating beam in order to maximize the effectiveness of this technique.

Deep UV photostabilization, as described above, is also used to condition resists prior to subjecting them to the harsh dry-etching conditions that must be employed to etch Al-Cu films in single-wafer etchers (see Chap. 16)[85].   Special systems, designed to generate intense deep-UV light with magnetron microwave sources, are commercially available for this process (Fig. 39).

PHOTORESIST SELECTION

An important aspect of implementing an effective photolithographic process is the matching of the most appropriate resist with the application.   The variety of resists available makes this a formidable task.   A useful approach to resist selection is to identify the characteristics of the application for which the resist will be utilized, and to compare those with the properties of the candidate resists.   The following is a list of some of the questions about an application that will impact the selection of the resist:

1) Properties of the surface onto which the resist will be applied, including:

   a) What is the surface reflectivity (highly reflective or not) ?

   b) What is the wafer flatness specification for the PR step being performed ?

   c) Are there any irregularities that must the resist must cover (hillocks, trenches, etc.)?

   d) Is the surface material lightly or heavily doped ?

   e) Is special cleaning, scrubbing, or priming required ?

   f) What are the step heights to be covered by the resist ? (The answer to this question will yield one criteria for minimum resist thickness, because sufficient resist must exist on top of steps to protect the surface be patterned during etching.)

2) Etching Procedure to be Used:

   a) What is the minimum resist thickness to provide pinhole protection ?

   b) If wet etching is to be used, have the optimum postbake time and

temperature been determined to yield the desired undercut ?
c) If dry etching is to be used, how susceptible is the resist to erosion?
d) Does the etching procedure make the resist difficult to remove ?
3) Sizes of Patterned Features:
   a) Can the critical dimension (CD) objectives (linewidth and spacing) be
      met by the resist resolution under all normal process variations ? (i.e. is
      there sufficient process latitude in resist system ?)
   b) Have optimization experiments for bake, exposure and development been conducted?
4) Equipment Considerations:
   a) Are the resists and developers compatible with existing processing equipment ?
   b) Is the resist sensitive enough to yield adequate exposure throughput ?
   c) What is the technological life of the equipment with which the resist is used ?
      (i.e. If it is short, will the resist still be compatible with replacement equipment ?)
   d) Have other users of the resist system and equipment combination under consideration
      been identified ? If so, it may be possible to learn from their experience.
5) Ecological and Disposal Considerations:
   a) Do the resist or developer materials contain any toxic or hazardous components
      that cannot be safely handled with existing current processing equipment ?
   b) Have disposal recommendations been obtained form the resist vendor ?
   c) What are the current guidelines for legal disposal ?

With the answers to these questions, the number of candidate resists can probably be reduced to a small enough group to construct a test matrix to allow the final selection to be made.

REFERENCES

1. M.J. Bowden, "A Perspective on Resist Materials for Fine Line Lithography", in *Materials for Microlithography*, Advances in Chemistry Series, No. 266, American Chemical Society, 1984, Washington, D.C., Chap. 3, p. 39-117.

2. J. Bargon, "Lithographic Materials", in *Methods and Materials in Microelectronic Technology*, J. Bargon, Ed., Plenum Publishers, New York, 1984.

3. C.G. Willson, "Organic Resist Materials - Theory and Chemistry", L.F. Thompson, C.G. Willson, M.J. Bowden, Eds., *Introduction to Microlithography*, Advances in Chemistry Series, No. 219, American Chemical Society, Washington, D.C., 1983, Chap. 3, p. 87-157.

4. P.S. Gwozdz, "Positive vs. Negative: A Photoresist Analysis", *SPIE Proceedings, Semiconductor Microlithography VI*, Vol. 275 (1981), p. 156.

5. M.C. King, "Principles of Optical Lithography", in N.G. Einspruch, Ed., *VLSI Electronics, Microstructure Science*, Academic Press, New York, 1981, Chap. 2.

6. R.W. Wake, M.C. Flanigan, "A Review of Contrast in Positive Resists", Proceedings SPIE, Vol. 539, Advances in Resist Technology and Processing, II, 1985, p. 291.

7. K.L. Tai, *et al.*, J. Vac. Sci. Technol., **17**, 1169 (1980).

8. L.F. Thompson and M.J. Bowden, "Resist Processing", in L.F. Thompson, C.G. Willson, M.J. Bowden, Eds., *Introduction to Microlithography*, Advances in Chemistry Series No. 219, American Chemical Society, Washington, D.C., 1983, Chap. 4, 172.

9. *ibid.*, Ref. 1, p. 49.

10. H., Gokan, K. Tanigaki, Y. Ohnishi, "Dry-Etch Resistance of Metal-Free and Halogen-Substituted Resist Materials", *Solid State Technol.*, May '85, p. 163.

11. P.H. Singer, "Trends in Resist Design and Use", *Semiconductor International*, Aug. '85, p. 68.

12. American Society for Testing and Materials, *Published Methods and Procedures*, "Standard

Methods for Testing Photoresists in Microelectronic Fabrication"; Test Std F-66-84, (1984).

13. M.L. Long, "Photoresist Particle Control for VLSI Microlithography", *Solid State Technol.*, Mar. '84, p. 159.

14. N.J. Sax, *Dangerous Properties of Industrial Materials*, Reinhold, New York, 1980.

15. D.J. Elliot, *Integrated Circuit Fabircation Technology*, McGraw-Hill, New York, 1982.

16. T. Pampalone,"Novalac Resins Used in Pos. Resist Systems" *Sol. State Tech.*, June 84,p 115.

17. R., Rubner, E. Kuhn, *ACS Div. Org. Coatings and Plast. Chem.* Preprint 1977, **37**, (2), 118.

18. *ibid.*, Ref. 5, p. 51.

19. S.A. MacDonald, *et al.*, "The Production of a Negative Image in a Positive Photoresist", *Kodak Microelectronics Seminar*, San Diego, 1982.

20. E. Alling and C. Stauffer, "Image Reversal Process of Positive Resist" Proceedings SPIE, Volume 539, *Advances in Resist Technology and Processing II*, Mar. 11-12, 1985, p. 194.

21. E. Ong and E.L. Hu, "Multilayer Resists for Fine Line Optical Lithography," *Solid State Technol.*, June '84.

22. K. Bartlett, *et al.*, *Proceedings SPIE, Vol. 394, Optical Microlithography II*, 1983, p. 49.

23. J. M. Moran and D.J. Magdan, *J. Vac. Sci. Technol.*, **16**, 1620 (1979).

24. B.F. Griffing and P.R. West, "Contrast Enhanced Lithography", *Solid State Technology*, May '85, p. 152.

25. M.J. Bowden and L.F. Thompson, *Polymer Eng. and Sci.*, **14**, 525 (1974).

26. *ibid.*, Ref. 1, p. 103-106.

27. *ibid.*, Ref. 1, p. 56.

28. *ibid.*, Ref. 1, p. 57-65.

29. T.D. Berker and S.E. Bernacki, *IEEE Trans. Electron Dev. Lett.* **EDL-2**, 281, (1981).

29. R. Rubner, H. Ahne, E. Kuhn, and G. Kolodziej, "A Photopolymer - The Direct Way to Polyimide Patterns", *Phot. Sci. Eng.*, **23** (5), 303, (1979).

30. O. Rohde, *et al*, "Recent Advances in Photoimagable Polyimides," *Proceedings SPIE, Vol. 539, Advances in Resist Technology and Processing II*, 1985, p. 175.

31. A. Yoshikawa, *et al*, *Appl. Phys. Letts.*, **31**, 167 (1977).

32. R.D. Lussow, *J. Electrochem. Soc.*, **115**, p. 660-663, (1968).

33. K. Mittal,"Factors Affecting Adhesion of Lithographic Matls" *Sol. State Tech.*, May 79, p.89.

34. R.H. Collins and F.T. Deverse, U.S. Patent No. 3,549,368, Dec. 22, 1970.

35. *ibid.*, Ref. 15, Chap. 6, "Spin Coating", p. 125-44.

36. *ibid.*, Ref. 8, p. 186-94.

37. P.O'Hagan and W.J. Daughton, "An Analysis of the Thickness Variance of Spun-On Photoresist," *Kodak Interface Conference Proceedings*, 1977, p. 95.

38. W. Daughton, P. O'Hagan, and F.L. Givens, "Thickness Variance of Spun-On Photoresist, Revisited", *Kodak Interface Conference Proceedings*, 1979, p. 16.

39. M.W. Chan, "Another Look at Edge Bead", *Kodak Seminar Proceedings*, 1975, p. 16.

40. American Society of Materials and Testing; *Published Methods and Procedures* "Standard Practice for Producing Spin Coating Resist Thickness Curves", Test Standard F-804-83.

41. *ibid.*, Ref. 15, Chap. 7, "Softbake", p. 145-163.

42. B. D Washo, *IBM Journal of Research and Development*, **21** (2), 190 (1977).

43. *ibid*, Ref. 8, p. 195-8.

44. T. Batchelder, J. Piatt, "Bake Effects in Positive Resist" *Solid State Technol.*, Aug.'83, p.211.

45. G. MacBeth, "Prebaking Positive Photoresists", *Proc. Kodak Interface Seminar*, 1982, p. 87.

46. J.A. Irvin and T.J. Weber, "Characterization of Baking Operations in Photolithographic Processes", *Proceedings Kodak Interface Seminar*, 1982, p. 31.

47. *ibid.*, Ref. 15, Chap. 8, "Exposure", p. 165-188.

48. *ibid.*, Ref. 8, p. 199.
49. D.F. Ilten and K.V. Patel, "Standing Wave Effects on Photoresist Exposure", *Image Technology*, Feb.-Mar., 1979, p. 9.
50. A.R. Neureuther, P.K. Jain, W.G. Oldham,"Factors Affecting Linewidth Control Including Multiple Wavelength Exposure and Chromatic Aberation", *SPIE Proceedings, Vol. 275, Semiconductor Microlithography*, 1981, p. 110.
51. I. Bol, *Kodak Interface Microelectronics Seminar*, San Diego, 1984.
52. W. Arden & L. Mader, "Linewidth Control in Optical Projection Printing: Influence of Resist Parameters", *SPIE Proc. Vol. 539, Adv. in Resist Technol. and Processing II* (1985), p. 219.
53. K. Harrison and C. Takemoto, "The Use of Antireflection Coatings for Photoresist Linewidth Control", *Kodak Interface Microelectronics Seminar*, 1983, p. 107.
54. A.T. Jeffries, *et al*, "Two Anti-Reflective Coatings for Use Over Highly Reflective Topography", *SPIE Proc., Vol. 539, Advances in Technology and Processing II*, 1985,p. 342.
55. R.D. Coyne and T. Brewer, "Resist Processes on Highly Reflective Surfaces Using Anti-Reflection Coatings", *Kodak Interface Seminar Proceedings*, 1983, p. 40.
56. *ibid.*, Ref. 15, Chap. 9, "Development", p. 209-31.
57. *ibid.*, Ref. 8, p. 204-210.
58. D. Burkman and A. Johnson, "Centrifugal On-Center, Flood Spray Development of Positive Resist", *Solid State Technol.*, May '83, p. 125.
59. R.F. Leonard and J.A. McFarland, "Puddle Development of Positive Photoresists", *SPIE Proceedings, Vol. 275, Semiconductor Microlithography VI*, (1981).
60. T. Rodriguez, P.D. Krasicky, and R.J. Groele, "Dissolution Rate Measurements", *Solid State Technol.*, May '85, p. 125.
61. J.S. Peterson, A.E. Kozlowski, "Optical Performance and Process Characterizations of Several High Contrast Metal-Ion Free Developer Processes", *SPIE Proceedings, Vol. 469, Advances in Resist Technology* (1984), p. 46.
62. D.W. Frey, J.R. Gould, and E.B. Hryhorenko, "Edge Profile and Dimensional Control for Positive Resist", *Kodak Interface Seminar Proceedings*, 1981, p. 40.
63. P. Burggraaf, "Wafer Inspection for Defects", *Semiconductor International*, July, 1985, p. 57.
64. D. Nyyssonen, "Optical Linewidth Measurement on Patterned Wafers," *SPIE Proceedings, Vol. 480, Integrated Circuit Metrology*, 1984, p. 65.
65. G. Toro-Lira and R. Mellen, "Critical Dimension Control in the Late Eighties", *Microelectronic Manufacturing and Testing*, Part 1, Jan. '85, p. 9, Part II, Feb. '85, p. 19.
66. *ibid.*, Ref. 15, Chap. 10, "Postbaking", p. 233-243.
67. *ibid.*, Ref. 8, p. 211.
68. K. Massau, R.A. Levy, and D.L. Chadwick, "Modified Phosphosilicate Glasses for VLSI Applications", *J. Electro. Chem. Soc.*, Vol. 132, No. 2, Feb. '85, p. 409.
69. P. van Pelt, "Processing Deep UV Resists"*SPIE Proc.Vol. 275, Semicon. Litho.* 1981, p. 150.
70. Y.T. Yen and M. Foster, "Deep UV and Plasma Hardening of Positive Photoresist Patterns", *Proceedings Kodak Interface Seminar*, 1982, p. 125.
71. DRMR Development Rate Monitor PEO-4587, Perkin-Elmer Corp, Garden Grove, CA, Aug '84.
72. D.G. Seiler and D.V. Sulway, "Precision Linewidth Measurement Using a Scanning Electron Microscope", *SPIE Proceedings, Vol. 480, Integrated Circuit Metrology*, 1984, p. 86.
73. J.T. Lindow, S.D. Bennet, and I. Smith, "Scanned Laser Imaging for Integrated Circuit Metrology", *SPIE Proceedings, Micron and Submicron Circuit Metrology*, Vol. 565, 1985, p. 81.
74. P. Singer, "Linewidth Measurement & Process Control", *Semiconductor Int'l.*, Feb.'85, p. 66.
75. M.T. Postek, "Critical Dimension Measurement in the Scanning Electron Microscope", *Proceedings SPIE, Vol. 480, IC Metrology II*, 1984, p. 109.

76. P.E. Russell, *et al*, "Development of SEM-based Dedicated IC Metrology System", *Proceedings SPIE, Vol. 480, IC Metrology II*, 1984, p. 101.

77. K.L. Tai, *Proc. Symp. on Inorganic Resist Systems*, 1982, Electrochemical Soc. 82-9, p.49.

78. T.R. Pampalone and F.A. Kuyan, " Improving Linewidth Control Over Reflective Surfaces Using Heavily Dyed Resists", *J. Electrocem. Soc.*, **133**, 192, (1986).

79. E.B. Hryhorenko, "A Positive Approach to Resist Process Characterization for Linewidth Control", Eastman-Kodak, Rochester, N.Y., 1980.

80. J.D. Cuthbert, "Optical Projection Printing", *Solid State Technology*, p. 59, Aug. 1977.

81. H. Yanzawa, *et al*, "Chemical Characterization of Photoresist to Silicon Adhesionin Integrated Circuit Technology", *Kodak Interface Seminar*, 1977.

82. J. J. Chisholm, "A New Linewidth Measurement on 1 μm Geometry Process Wafers Using Fluorescence", *SPIE Proc. 480, Integrated Circuit Metrology*, 1984, p. 49

83. "Flow-Monitors for Controlling Spin Bowl Exhausts", Sierra Instruments, Carmel, California.

84. D. Ditmer and M.V. Hanson, "Stepper Exposed Critical Dimension Tolerances Using a Vapor Jet Developer Nozzle", *SPIE Proc, Optical Microlithography III*, Vol. 470,1984, p. 203.

85. J.C. Mathews and J. I. Wilmott, "Stabilization of Sinle Layer and Multilayer Resist Patterns to Al Etching Environments", *SPIE Proc., Optical Lithography III*, Vol. 470, 1984, p. 194.

PROBLEMS

1. Explain why a resist with higher contrast can lead to higher resolution of features transferred from a mask to the wafer surface than a resist with a lower contrast.

2. In examining Fig. 3, which of the positive resists exhibits (a) the highest contrast, and (b) the highest sensitivity. Determine this also for the negative resists shown in the figure.

3. Why is it important that a resist exhibits a high *actinic absorbance?*

4. Explain why a multi-layer resist process offers the potential of higher resolution than a single-layer resist process. What prices are paid for this benefit?

5. In a photoresist process it is important that the resist film thickness be highly uniform. List five or more parameters that impact the thickness of the resist film by the time it has been spun-on and soft-baked.

6. Explain how under-softbaking leads to an apparent increase in the photosensitivity of the resist. Why is not generally possible to utilize under-softbaking to exploit this effect?

7. Calculate the number of standing wave maxima that will be produced in a positive resist layer that is 1.0 μm thick, and has an index of refraction of 1.68 at the G-line of exposing radiation.

8. Explain how higher resolution is achieved by the use of (a) contrast enhancement layers, (b) anti-reflective coatings, and (c) incorporation of dyes in the resist. Cite the advantages and disadvantages of each of these methods.

9. Describe four techniques utilized to measure the linewidth of patterned features on a substrate. Why is accurate linewidth measurement more difficult on wafer surfaces than on the mask and reticle substrates?

10. Rounding of the resist profiles by elevated post-bake temperatures (Fig. 38) is sometimes used to produce a tapered profile in the etched feature (by anisotropically dry-etching the resist and the layer to bepatterned at an equal etch rate). Why does this procedure become ineffective when the dimension of the etched feature approaches that of the resist film thickness?

13

LITHOGRAPHY II:

OPTICAL ALIGNERS and

PHOTOMASKS

In the previous chapter, the material properties of photoresists and their processing technology was covered. In this chapter, we describe the remainder of the topics involved in transferring patterns to the silicon wafers, including: a) the equipment used to project the image of the patterns onto the wafer surface (i.e. so-called *aligners* or *printers*), thereby allowing the photoresist at the desired pattern locations to be exposed; and b) the pattern transfer tools that contain the patterns to be printed onto the photoresist-coated wafers (i.e. the photomasks and reticles). As a prelude to the description of aligners or masks, we briefly present some optical theory which is used to design and describe the operation of aligners, and which underlies the specifications of masks and reticles.

Even before beginning the discussion on optics, however, it is useful to identify the key issues of microlithography hardware. That is, the most important characteristics of the machines and masks used to project the image of patterns onto wafer surfaces are the following: a) resolution; b) pattern registration capability (alignment); c) dimensional control; and d) throughput.

In general, the term *resolution* describes the ability of an optical system to distinguish closely spaced objects. Specifically, we will refer to the *minimum resolution* of a microlithographic printing machine as the dimension of minimum linewidth or space that the machine can adequately print (or *resolve*). We will see, however, that such a minimum dimension also depends on the photoresist and etching technology, as well as the resolution of the aligner. The subject of resolution will be dealt with more thoroughly in the section on *Optics of Microlithography*, but it is important to emphasize that *high resolution* is usually the most sought after property of an aligner.

The *pattern registration capability* is a measure of the degree to which the pattern being printed can be "fit" relative to a previously printed pattern. As the feature sizes are so small in microlithographic applications, and the number of layers that must be correctly fit on top of one another is large (e.g. 10-15), a very tight fit indeed is required.

The *dimensional control* requirement in microlithography refers to the ability to produce device feature sizes over the entire wafer surface with high accuracy and precision. In order to accomplish this task, the feature patterns must first be correctly reproduced on the pattern transfer tools (masks and reticles), and then be accurately printed (imaged and etched) onto the wafer surface.

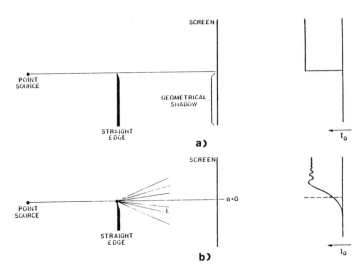

Fig. 1 Intensity distribution produced by a spatially coherent ray of light as it passes by a straight edge: (a) according to geometrical optics; (b) with diffraction.   Reprinted with permission of Ref. 1 Copyright 1983, American Chemical Society.

The *throughput* of the pattern transfer process is an important, but not always overriding, process characteristic. For example, if a device can only be fabricated on a low-throughput, but high resolution printer, it may still be economically sound to carry out the fabrication on such a machine. Nevertheless, in most production applications the throughput of a process or machine *is* very important, and the selection of one machine over another may be decided by this factor.

OPTICS OF MICROLITHOGRAPHY

In VLSI fabrication, the resolution of the optical lithography printing system is of major importance, since it is the main limitation of minimum device size. In modern projection printers the quality of the optical elements is so high that their imaging characteristics are limited by diffraction effects, and not by lens aberrations (i.e. they are said to be *diffraction limited* systems). Since the resolution is determined by diffraction limitations, and it is necessary to have an understanding of some concepts surrounding diffraction limited optics including coherence, diffraction, numerical aperture, modulation, and most importantly the modulation transfer function. It is the purpose of the following sections to present information on these subjects, although in a brief and fundamental manner. Consult Refs. 1 and 2 for more details.

Diffraction, Coherence, Numerical Aperture, and Resolution

All optical lithography systems involve light diffraction imaging, insofar as radiation is directed from an illumination source past an edge or through a slit (e.g. both features occur on photomasks). Due to diffraction effects, the radiation spreads into the region which is not directly exposed to the oncoming waves, and if projected on a screen, appears in patterns dependent on the diffraction effects (see Figs. 1 & 2). The intensity distribution produced on the screen may be a series of alternating light and dark bands, and is dependent on the distance

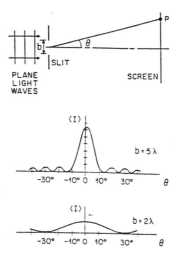

Fig. 2 Fraunhofer diffraction pattern of a single slit illuminated with monochromatic light; the intensity distribution is shown for two different slit widths.   Reprinted with permission of Ref. 1, Copyright 1983, American Chemical Society.

between the slit (mask) and the screen onto which the image is projected (e.g. a wafer surface), the geometrical configuration of the slit, and the chromatic purity of the illumination source. The chromatic purity of the radiation source is usually of little concern in microlithography, since most optical lithography systems use filters to select particular wavelengths, or employ resists that have been sensitized to a particular range of wavelengths.

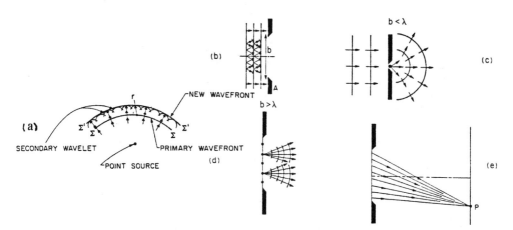

Fig. 3   (a) Schematic of Huygens Principle showing construction of a new wave front Σ' from the preceeding wave front Σ; (b) plane wave front incident on a slit of width b; (c) diffraction for case where $b < \lambda$; (d) $b > \lambda$ showing subdivision of slit into distribution of Huygens sources; (e) superposition of waves at P. Reprinted by permission of Ref. 1, Copyright 1983, American Chemical Society.

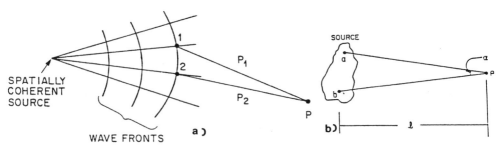

Fig. 4 (a) Schematic showing a perfectly coherent source. (b) An incoherent source of radiation of finite size. Reprinted with permission of Ref. 1, Copyright 1983, American Chemical Society.

It is also important to point out at the outset that all the information contained in the edge of an image results only from the diffracted light.

If the distance between the screen and slit is short (i.e. in *contact* or *proximity* aligners), the form of diffraction which occurs is known as *Fresnel (or near-field) diffraction*. If the distance between slit and screen is large, *Fraunhofer (or far-field) diffraction* takes place. Because most modern optical lithography systems are *projection type systems*, in which the slit-to-screen (e.g. mask-to-wafer) distance is large, our discussion is limited to Fraunhofer diffraction.

Invoking Huygens' principle, each point on a wave front of light may be regarded as a new point source of waves. Each of these Huygens sources are in phase and equal in intensity. As shown in Fig. 3, when the primary wave front approaches a slit, new wavefronts are constructed from point sources at the slit. Representation of the number of sources is dependent upon the width, b, of the slit. The intensity profile that is imaged on a screen (shown in Fig. 2), results from summing all the radiation emitted by the sources at the plane of the screen. Figure 2 illustrates that for values of b larger than the wavelength λ, pronounced diffraction effects arise from the summation of wavefronts, arriving from different distances, at each point on a screen. This geometrical condition is encountered in current optical microlithography applications (i.e. $b \geq 1.0$ μm, $\lambda \leq 0.43$ μm).

Spatial coherency, s, is a measure of the degree to which the light emitted from a source is in phase at all points along the emitted wave fronts. A point source, of infinitely small dimension, represents an ideal coherent source (Fig. 4a). Because all lithography systems have radiation sources of *finite* size, the degree of coherence exhibited by light incident on a plane is dependent on the source size (Fig. 4b), its distance from the plane, and the angular range of lightwaves from the source allowed to pass through an aperture between the source and the plane. We will show later that perfect coherency is not needed, nor even advantageous (i.e. partial coherence is better), for imaging patterns with projection printers. This is fortunate, since the brightness of an ideal point source reduces to zero, and would require an infinite exposure time.

To determine the resolution capability of an optical system, diffraction gratings are typically employed, both experimentally and in mathematical analysis. Diffraction gratings consist of features of equal widths arranged at equidistant intervals (known as the grating period v,). The grating can be treated as a series of diffraction slits which, when irradiated with coherent light, will emit identical diffraction patterns from each slit. (Note that in IC fabrication, one would not print a diffraction grating *per se*, but such features as a set of interconnect lines designed with dimensions equal to the resolution limits of the lithography system have essentially the same structure. Hence, the principles of this discussion are still valid.) The patterns interfere with each other to produce an intensity pattern on the screen, whose principal maxima are given by the equation:

$$d \sin \theta_N = N \lambda \qquad (1)$$

where: $N = 0, \pm 1, \pm 2 \ldots$ is the order of diffraction; d is the distance between the slits; and θ_N is the angle at which the diffraction pattern exists. Note that the width of the principle maxima is determined by the number of slits used, and that the intensities of all the maxima are governed by the diffraction pattern of a single slit of width equal to that of any one slit used.

In any optical projection system the major factor that limits its resolution capabilities is the physical design of the objective lens and its numerical aperture (NA). The NA is a measure of a lens' capability to collect diffracted light from an object (photomask) and project it onto the wafer. The NA is defined by:

$$NA = n \sin \alpha \qquad (2)$$

where n is the refractive index (typically 1.0 for air), and 2α equals the angle of acceptance of the lens (shown in Fig. 5). The NA of lenses in projection aligner ranges between ~0.16-0.40.

The resolution of a lens depends on the wavelength and the degree of coherence of the incident light, and the NA of the lens. The definition of resolution is based on *Rayleighs criterion*. That is, diffraction causes even a geometrically perfect lens to image an ideal infinitesmally small point source into a blurry disc, called the *airy disc*. When two points are so close that the two airy discs look like a single blurred disc, the two points cannot be resolved. Rayleighs criterion defines two images as being *just resolved* when the intensity between them drops to 80% of the image intensity (Fig. 6). Mathematically, this criterion is expressed as[3]:

$$2b = 0.61 \lambda / NA \qquad (3)$$

where: 2b is the separation distance of two images (i.e. as shown in Fig. 8b, it is the sum of the dimensions of a line and a space on a wafer, each of dimension b); and λ is the illuminating wavelength. Figure 7 plots the Rayleigh limit (on the right of the plot) for lenses of various NA (on left of the plot), when the illuminating wavelength is $\lambda = 436$ nm (i.e. the G-line of the mercury-arc spectrum, which is a widely utilized exposure wavelength in many aligners)[4]. For the Zeiss 10-77-82, which has an NA of 0.28, and is a lens used in several projection stepper aligners, the Rayleigh limit of the minimum resolvable dimension, b, is shown to be 0.47 μm.

Because the image is constructed from diffracted light, and the collection of higher orders of

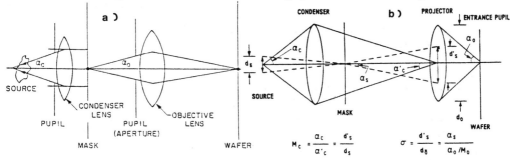

Fig. 5 (a) A refractive lens imaging system using partially coherent light, condenser lens, and objective lens. (b) Kohler illumination system. Reprinted with permission of Ref. 1, Copyright 1983, American Chemical Society.

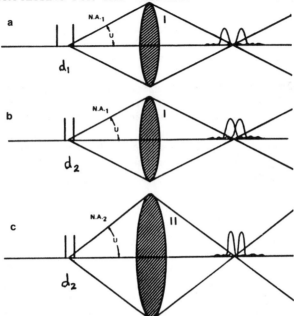

Fig. 6 Example of Rayleigh criterion. (a) Example lens I with NA = 0.1 can resolve objects separated by distance d_1. (b) Example lens I cannot resolve objects separated by distance $d_2 < d_1$. Image intensities are shown on right of lens. (c) Example lens II with larger NA (e.g. NA = 0.2) can resolve two objects at a distance smaller than the resolution limit of lens I (e.g. $d_2 < d_1$).

diffracted light enhances the resolution of the image, a larger NA would allow a larger angle of collection, and therefore better resolution (also seen in Figs. 6 and 7). This benefit, however, also has a price, in that the depth of focus, σ, is inversely proportional to the square of the NA, or:

$$\sigma = \lambda / (NA)^2 \qquad (4)$$

For example, for an NA = 0.3 and a λ = 0.4 nm, σ = 4.4 μm. From this example, it can be readily seen that variations in silicon substrate flatness could easily render an image outside the

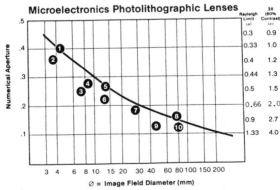

Fig. 7 The relationship between the NA and the effective field size of commercially available (1980) microelectronic photolithographic lenses[13]. Reprinted with permission of SPIE.

focal plane. Therefore, an appropriate compromise between high resolution and long depth of focus must be established in designing aligners. Nevertheless, it is predicted that at some dimension between 1 μm and 0.5 μm, the depth of field becomes intolerably small, and this will force optical lithography to give way to electron or x-ray lithography, as higher resolution is pursued. Others argue that use of planarization techniques, and shorter wavelengths (e.g. I-line), will be employed to combat the depth of focus limitation, allowing optical resolution to be extended.

As also shown in Fig. 7, for refractive-type projection printing systems (defined in a later section), higher NA achieves greater resolution at the expense of another property, that of *image field size*. Since image field size decreases significantly (e.g. the largest image field size of lenses used in 5X reduction steppers is 20 mm x 20 mm), in order to print a pattern on a 100 mm or larger wafer, several consecutive exposures must be conducted, using a *step-and-repeat* type exposure method.

Modulation Transfer Function

The *modulation transfer function*, or MTF, is a parameter which, when appropriately plotted, provides a rapid and convenient indication of the capability of an optical projection system to reproduce a mask image on a wafer surface[1,5]. This system imaging capability depends on several parameters, including the illuminating wavelength, the mask feature size being imaged, the NA of the lens, and spatial coherence of the source. A clear understanding of the MTF concept is essential for appreciating the tradeoffs between various projection aligner systems. To begin our explanation of this parameter, we define the concept of modulation, M, as it applies to incident light energy on a plane:

$$M = \frac{I_{max} - I_{min}}{I_{max} + I_{min}} \tag{5}$$

where I_{max} is the image intensity at the center of a bright area, and I_{min} is the intensity at the center of a dark pattern. As shown in Fig. 8, knowledge of M reveals the degree to which diffraction effects cause incident radiation to fall between the images on a screen of two slits in the mask. If $M = 1$, $I_{min} = 0$, and this is the highest degree of modulation that can be achieved. In Fig. 8 the modulation can be determined both at the object plane (i.e. the plane of the mask), M_{mask}, and at the image plane, M_{im}. The MTF of an exposure system is then defined as the ratio of the modulation in the image plane to that in the object plane, or M_{im}/M_{mask}. Note that having said this, since the intensity of the light in the mask plane at the center of an opaque line on the mask is essentially zero, $M_{mask} = 1$, and we can simply consider the MTF and M_{im} as identical. Making note of this simplification makes the concept of MTF easier to describe.

As stated at the outset, MTF depends on the NA, on λ, on mask feature size, and on the degree of spatial coherency of the illuminating system. Since λ and the NA of the lens are typically fixed by system hardware design considerations, it is most useful to plot MTF versus feature size and parametrically vary the spatial coherency. (Actually as shown in Fig. 8, MTF is plotted versus line pairs per mm, v, which can be converted to feature size, b, by b [μm] = 1 /2v.) Let us study the plot in more detail, with the assumption that $M_{mask} = 1$.

At the extreme limit of the optical system resolution (i.e. the minimum distance that satisfies Rayleigh's criterion), given by point A in Fig. 9a the MTF is ~0.1. As the size of the images to be printed becomes larger, however, the value of MTF increases. The maximum value of the MTF function is 1 (point C in Fig. 9), corresponding to a condition at which some point of the space between two images on the image plane receives no light (Fig. 8). Under such

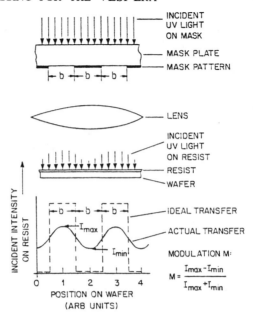

Fig. 8   Schematic representation of image transfer efficiency for a 1:1 projection printer. Reprinted with permission of the publisher, the Electrochemical Society.

conditions, two images incident on photoresist could certainly be printed as two distinct features. To summarize, the MTF curves shown in Fig. 9 describe the degree of modulation produced by a given optical projection system on a wafer surface, for different feature sizes on a mask.

Wafer surfaces are coated with a photoresist designed to respond to the incident radiation, and the relevant response characteristic for this discussion is the *resist contrast*, as defined in Chap. 12. An ideal resist would have an infinite contrast, and in such ideal material image dimensions equal or greater than the Rayleigh limit of the optical system could be printed. *Real resists, however, have finite contrast values, and thus a greater degree of modulation (MTF) than occurs at the Rayleigh limit must exist before an adequate image can be formed in a real resist.* The minimum MTF of an optical system to adequately define an image in a resist depends on the resist contrast value, and is defined as the *critical MTF*, or $CMTF_{resist}$[6]. Figure 5 in Chap. 12 is a plot of $CMTF_{resist}$ versus resist contrast.   For example, the contrast of positive optical resists is ~2, and from Fig. 5, Chap. 12, it is seen that *$CMTF_{resist}$ when using positive resist is ~0.6.* Examining Fig. 9, this implies that in a system with an NA of 0.3, a λ of 436 nm, and a spatial coherency of 0.7, the maximum spatial grating frequency that can be printed on a film of positive optical photoresist is 442 line pairs /mm, corresponding to a minimum feature size of 1.13 μm (point B in Fig. 9). This example demonstrates the *2.5x rule of thumb*, which relates the minimum resolution of a projection aligner given by the Rayleigh limit and the actual working resolution achievable with positive optical resist[4].

The fabrication of increasingly smaller features on VLSI relies on the availability of increasingly higher resolution lithography equipment.   From the discussion on MTF, the approaches used to design such equipment can now be described.   First, referring back to the Rayleigh criterion, the illuminating wavelength can be decreased, or the NA of the system lens can be increased.   Second, the contrast of the resist can be increased, by modifying the resist chemistry, by creating entirely new resists, or by using contrast enhancement layers, which allows a smaller MTF to produce adequate images. Third, the coherence of the optical system can be adjusted. Let us close the discussion on MTF with a brief description on this last subject.

Varying the spatial coherency alters the shape of the MTF curve[5,12]. *Completely coherent radiation* (s = 0), yields an MTF curve as shown in Fig. 9. *Completely incoherent radiation* (s = ∞) produces an MTF curve that has a greater Rayleigh resolution limit, but whose MTF value increases only slowly as the feature size increases. *Partially coherent radiation* yields MTF curves in-between these two extremes, resulting in a higher MTF at a given spatial frequency than if completely incoherent radiation is used. Use of partially coherent radiation is also more advantageous than highly coherent radiation for two other reasons: 1) radiation with coherence values s < 0.4 produces an MTF of higher value at a given spatial frequency, but such coherence is achieved by reducing the effective source size (thereby increasing exposure time); and 2) such highly coherent radiation degrades the edge integrity of isolated edges by a phenomenon known as *ringing* (Fig. 10). Based on various tradeoffs, a coherence value of ~0.7 is typically selected for reduction steppers and ~0.45 is used for 1X steppers.

The configuration used in many refractive projection systems is a Kohler illumination system, in which the exposure source is imaged through a condenser in the entrance pupil of the

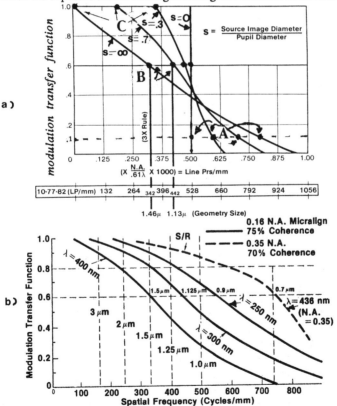

Fig. 9 (a) The modulation transfer function for a projection sytem with NA = 0.28, λ = 436 nm and three coherency factors[4]. Reprinted with permission of SPIE. (b) Modulation transfer functions for scanning projection aligner with NA = 0.16 (exposure wavelengths 250 nm, 300 nm, and 400 nm), and S/R aligner with NA = 0.35 (exposure wavelength 436 nm)[18]. Reprinted by permission of Solid State Technology, published by Technical Publishing, a company of Dun & Bradstreet.

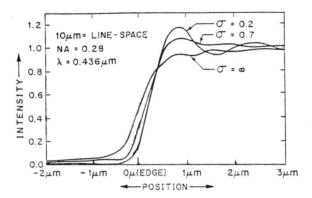

Fig. 10 Image intensity profile at the edge of an opaque line in a grating of 10 μm lines and spaces, showing effect of coherency on ringing. Reprinted with permission of Ref. 1. Copyright 1983, American Chemical Society.

projection lens, Fig. 5b. The degree of spatial coherence, s, in such systems is given by:

$$s = d_s / d_o \qquad (6)$$

where d_s is the effective size of the source imaged by the condenser lens, and d_o is the diameter of the entrance pupil which is determined either by the diameter of the projection lens or a stopped-down aperture. In general, the distances are chosen to keep the source size less than the entrance pupil to produce the desired partial coherence value as described above.

OPTICAL METHODS OF TRANSFERRING PATTERNS TO A WAFER: OPTICAL ALIGNERS

The optical theory presented in earlier sections is applied to the design of the machines used to image the circuit patterns onto the resist-coated wafers. Such machines contain a variety of subsystems that work together to perform the imaging function, including: a) an illumination source that provides the optical energy for transforming the photoresist; b) an optical subsystem that focuses the circuit patterns onto the wafer surface and allows for controlled exposure times; c) a movable stage that holds the wafer to be exposed. The stage position can be finely and accurately adjusted so that the image from the optical pattern transfer tool (i.e. the reticle or mask) can be aligned to previously printed patterns on the wafer. This step, in fact is so critical, that the machines we are describing are most commonly referred to as *aligners*, as well as printers, exposure tools, etc.; d) some form of alignment subsystem (i.e. manual alignment, by a human operator using a microscope and manual mechanical adjustments, has been the traditional method used. Automatic alignment methods have been developed for wafer steppers, which require no human intervention); e) a wafer handling subsystem; and f) an exposure meter.

There are three major methods of optically transferring a pattern on a mask to a photoresist coated wafer[7,8]. These are *contact printing, proximity printing*, and *projection printing* (Fig. 11). Projection printing is used almost exclusively for VLSI fabrication, and it is accomplished using three different types of projection aligning systems: 1) projection scanning systems; 2) reduction (e.g. by 10X or 5X) step-and-repeat projection systems; and 3) 1X step-and-repeat

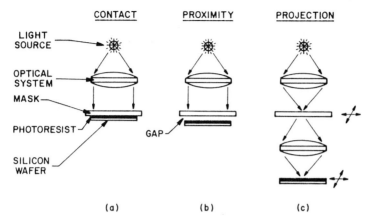

Fig. 11 Schematic of three optical lithographic techniques. (a) Contact. (b) Proximity. (c) Projection[27]. Copyright 1983, Bell Telephone Laboratories, Incorporated, reprinted with permission.

projection systems[9,10]. This section primarily describes projection aligners, and the advantages and limitations each type possesses. Light sources and exposure meters are discussed first.

A few words about the environment that is necessary for operating high performance aligners and for performing the majority of photoresist processing steps is in order. First, the light in the rooms needs to be filtered (i.e. yellow light) so that resist is minimally exposed when subjected to ambient light. Next the temperature in the photoresist processing and aligner rooms must be controlled to within $\pm1°C$, and the entire aligner apparatus itself must be kept at the same temperature. The room should maintained to Class 10 cleanliness standards (see Vol. 2), and the humidity set at ~50% (and controlled to $\pm5\%$ of the set point). The aligners must be highly isolated from vibration by use of vibration isolation tables, as well as through appropriate vibration-isolation building design practices.

Light Sources and Light Meters for Optical Aligners

Virtually all optical aligners used in IC fabrication utilize mercury-vapor lamps as the illumination source. In such lamps a discharge arc of the high-pressure (e.g. 230 atm) mercury vapor emits a characteristic spectrum. The mercury-vapor emission spectrum is shown in Fig. 12. It is not of uniform intensity at all wavelengths, and in fact contains several intense, sharp

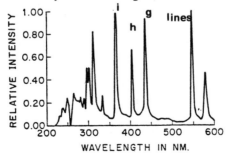

Fig. 12 Typical high pressure mercury-arc spectrum. Reprinted with permission of Ref. 1, Copyright 1983, American Chemical Society.

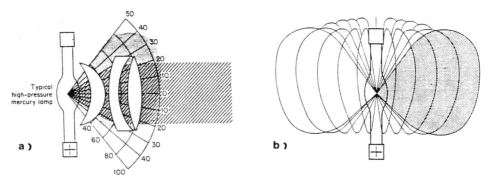

Fig. 13 (a) Schematic showing the relative intensity distribution with respect to the output of a mercury-arc lamp. (b) Three dimensional view of the output of a typical mercury-arc lamp. Reprinted with permission of SPIE[4].

lines. In the UV wavelength range from 350 to 450 nm, there are three strong lines, the I-line (365 nm), the H-line (405 nm), and the G-line (436 nm). Most reduction steppers are designed to operate using the G-line (although development on I-line reduction steppers is being pursued[11]). The 1X steppers utilize a wider portion of the spectrum than just a single line (i.e. the wavelength range between 390-450 nm). The 1:1 projection scanners are available in models that can operate at all three emission lines, as well as a deep-UV model that operates at 240 nm.

To obtain maximum intensity, high power mercury-arc lamps are used (e.g. 200-1000W). In some systems, air jets cool the bulb, and the heated air is removed by an exhaust fan. This is important because heat will not only cause shifts in the spectral output of the lamp, but can also cause dimensional changes in the mask and optical path length of projection aligners. The relative intensity distribution of the lamp in two dimensions is shown in Fig. 13a, and in three dimensions in Fig. 13b[13]. Collecting optics systems, which surround the lamp, are used to direct as much of the emitted light as possible to the surface being exposed. Filters are also used to limit exposure wavelengths to the specified frequencies and bandwidth.

The spectral output of the lamps vary with age, and together with variations in glass transmission and optical element coatings, exposure energy can change with time, or from machine-to-machine (even with the same lamp model). As a result, a spectral exposure meter must be used so that a correct and repeatable total energy dose is provided for each exposure. As

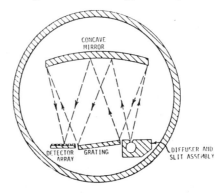

Fig. 14 Schematic of detection of a spectral exposure meter[14]. Reprinted by permission of SPIE.

the lamp output varies, the exposure time can be adjusted to maintain a constant incident energy dose onto each wafer. The meter must also be spectrally matched to the photoresist type being exposed. Typical exposure meters for these applications use filters to set their spectral response, but the limited variety of transmission curves available in filters results in a relatively crude approximation of the response of actual photoresists. More recently, exposure meters which contain a complete UV spectrometer have been developed (Fig. 14)[14]. They can be adjusted to match the response of any photoresist over the wavelength range from 254-436 nm.

Contact Printing

Contact printing was the earliest method used to produce patterns on silicon wafers[8,15]. A schematic representation of a contact printer is shown in Fig. 11a. The mask containing the circuit pattern is first correctly positioned (aligned) relative to existing patterns on the wafer. Next, the mask is clamped to the resist-coated wafer (while maintaining alignment), and exposed with UV light. This form of printing yields the most faithful image transfer and best resolution.

Unfortunately, as a result of the repeated mask-to-wafer contacting process, defects in the mask are generated. These defects print on the next wafer that is exposed through the mask. To minimize this effect, hard surface masks must be inspected and cleaned regularly. If the defects cannot be removed by cleaning, the mask must be replaced. Particles between the mask and wafer also prevent intimate contact which reduces resolution in the local area. This has led to the virtual abandonment of contact printing for VLSI, and the development of techniques in which the wafer and mask do not make contact.

Proximity Printing

The next form of pattern transfer to evolve was *proximity printing*, in which the mask and wafer are placed close to one another during exposure, but do not make contact[8,15]. By introducing such a gap between the mask and wafer, the defect problem of contact printing should be avoided. As the gap size is increased, however, the resolution rapidly degrades . For example, a 10 μm gap with 400 nm exposing radiation results in a minimum resolution of W_{min} = 3 μm.

The use of this technique also requires extremely flat masks and wafers. In spite of the fact that many of todays wafers are flat enough to allow the use of a 10 μm gap, many VLSI circuits require features smaller than 2 μm. As a result, this technique is also not adequate for VLSI.

Projection Printing

In projection printing, lens elements or mirrors are used to focus the mask image on the wafer surface, which is separated from the mask by large distances. Several types of projection printing techniques have been developed, including: a) 1:1 projection scanners (based on reflective optics); b) reduction step-and-repeat projection aligners (based on refractive optics); and c) non-reduction (i.e. 1X) step-and-repeat projection aligners (based on a reflective /refractive system).

1:1 Scanning Projection Aligners

The 1:1 wafer scan system marketed by Perkin-Elmer (Micralign® series aligners), uses a reflective spherical mirror to project the image onto the wafer surface[15,16,17]. Figure 15 shows a schematic diagram of such a system. It is seen that a narrow arc of radiation is imaged from the mask to the wafer, with the light traveling an optical path that reflects it 5 times, most importantly by a spherical mirror of NA = 0.16. The wafer and mask are scanned through this arc of radiation by means of a continuous scanning mechanism. The scanning technique minimizes mirror distortions and aberations by keeping the imaging illumination in the "sweet spot" of maximum optical correction. The image formed is a mirror image of the mask. These systems use Hg lamps with the range of 240 to 440 nm. A resolution of better than 1.50 μm is reportedly attainable when 350-400 nm light is used. At deep UV (240 nm), exposure resolution close to 1 μm can be achieved[18]. The depth of focus of the system is ±6 μm, and overlay accuracy of ± 0.25 μm (1σ) is possible. A throughput of 100, 150 mm wafers /hr can also be accomplished. In most applications, such scanning projection aligners are used to align patterns with minimum feature sizes of ~3.0 μm, while step-and-repeat aligners (steppers) are utilized for smaller sized patterns. In processes that utilize some mask layers with minimum feature sizes of ≤ 3μm, and others of ≥ 3 μm, both aligner types can be used in a *mix-and-match* approach, to take advantage of the high throughput of scanners[19].

Reduction Step-and-Repeat Projection Aligners (Reduction Steppers)

The second type of projection exposure system uses *refractive,* instead of reflective, optics to project the mask image onto the wafer. It is impractically complex and expensive to produce a lens capable of projecting a mask over an entire 100 mm or larger wafer. Therefore, refractive systems are designed to project an image only onto a portion of the wafer. This field is then stepped-and-repeated (S /R) across the wafer, and aligners that operate in this manner are known as *steppers*. Wafer size is therefore no longer a limitation. Current systems employ reticles that have a pattern that is an enlargement of the desired image on the wafer. Thus, the reticle pattern is reduced when it is projected onto the wafer during exposure.

The ultimate advantage of stepper technology over scanner-type aligners is higher image resolution[9,18]. In addition, stepping each die allows correction for wafer distortion. The necessity of conducting a step-and-repeat exposure reduces the throughput of these machines, but this approach also offers the possibility of outstanding overlay accuracy and resolution. Precise control of a mechanical stage is required to ensure accurate step and repeat exposures. Figure 16 shows a schematic of a step-and-repeat (S /R) reduction projection system.

Increasingly more attention is being given to lithography with wafer steppers near and below 1 μm. It was estimated in 1986 that ~80% of the installed steppers were being used to produce high-density memory chips, including 256K and 1M DRAMs, static CMOS RAMs, and EPROMs. The remainder are being used to develop new products, bipolar, and MOS logic[9].

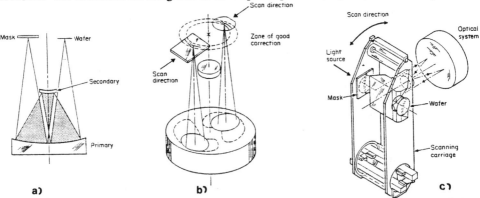

Fig. 15 Scanning projection optical system. (a) basic optical system consisting of two concentric spherical reflecting surfaces. (b) zone of good correction. (c) Micralign® projection and scanning system. Courtesy of Perkin-Elmer.

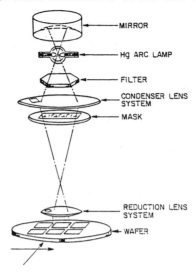

Fig. 16 Schematic of a reduction step-and repeat system. Reprinted with permission of Ref. 1, Copyright 1983, American Chemical Society.

The first reduction steppers developed were 10X, but 5X steppers have largely replaced them. This has occurred primarily because of the greater utility of 5X reduction lithography[9,20,23]. That is, a typical field size of a 5X lens projected onto a wafer is 14 mm x 14 mm. The size of this pattern at 5X is thus 70 mm x 70 mm, which can be accommodated on a standard 5" x 5" reticle substrate. The same field size at 10X becomes impractically large, requiring a 7" x 7" reticle substrate. 5X lenses also have larger field sizes (up to 20 mm x 20 mm, versus 10 mm x 10 mm for a 10X lens), which allows greater throughput.

Non-Reduction Step-and-Repeat Projection Aligners (1X Steppers)

A recently introduced system uses a 1:1 catadioptic optical system (i.e. a system consisting of a mirror, two folding prisms, and a two element achromatic lens), rather than a reduction lens (Ultratech 1000, Fig. 17). The principal optical element is a mirror (NA = 0.315), but apart from the combination mirror /lens, the system operates in a conventional step-and-repeat mode[21,22]. Since the optical effects of defects cannot be decreased by reduction, as in 10X and 5X steppers, 1X reticles must be defect-free. Once a defect-free mask is fabricated, a protective pellicle (described in later sections) must be applied to it. The lens offers a larger field in one dimension than conventional refractive S /R lenses, thereby allowing more than one pattern to be printed at each exposure. When operated in the 390-450 nm wavelength range, the system achieves a working resolution of ~1.0 μm, an overlay accuracy budget of ±0.25 μm, and a depth of field of 5.0 μm. If suitably adapted for deep-UV illumination, it could possibly offer ~0.5 μm resolution. Registration (next section) is performed by an automatic local alignment technique.

PATTERN REGISTRATION

Up to this point, we have emphasized the quality of the lithographic imagery. Of comparable importance is the accuracy with which an image can be positioned on the surface of a wafer. That is, integrated circuits are fabricated by patterning a sequence of masking layers, and

Fig. 17 1X Stepper. Courtesy of Ultratech Stepper.

the features on succesive layers bear a spatial relationship to one another. Thus, as a part of the fabrication process, each level must be aligned to the previous levels. Figure 18 is an example where a device feature on *masking level 2* is designed to be positioned within a feature on *masking level 1*, with the restriction that the edges of the *level 1* feature can never touch the edge of the feature on *level 2*. To insure that such edge overlap will not occur, a minimum *registration* or *overlay* tolerance must be allowed between the edges of level 1 and level 2. This tolerance then becomes one of the design rules used when laying out the circuit patterns.

Three factors contribute to the the magnitude of registration tolerance, and the effects of these factors must be considered together when establishing registration tolerance value:

1) The location of a feature edge on a silicon wafer may not be exactly at the location specified in the original circuit layout. The uncertainty in edge location can be

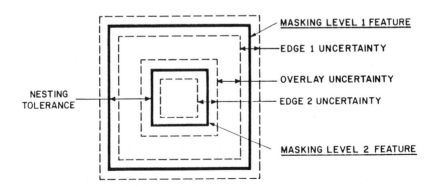

Fig. 18 Components of the nesting tolerance required between two mask levels registered to one another[27]. Copyright 1983, Bell Telephone Laboratories, Incorporated, reprinted by permission.

due to the fact that: a) there is typically a measurable variation in line sizes across a mask, although such variation may be insignificant on reticles used with reduction steppers; b) the resist image on wafers can vary, because of non-uniformities in such parameters as resist thickness, baking temperature, exposure, and development; and c) non-uniformities in etching processes can lead to size variations in the etched features.
2) Dopants in the silicon substrate can diffuse laterally during processing, and the edge of a diffused layer may therefore not be coincident with the drawn edge location.
3) There is an uncertainty in *registering*, or *aligning* the image of level 2 to the previously etched pattern of level 1. This inability to perfectly align two layers may be due to several factors, including: a) the mask-making equipment may not produce a set of masks that perfectly overlay; b) temperature differences between times of exposure may cause relative mask expansion (or contraction), so that even a perfectly manufactured mask set may not produce perfect alignment; and c) the alignment tool has a limited registration capability. *In fact, from the standpoint of aligner fabrication capability, its overlay accuracy and machine-to-machine repeatability, and not linewidth control, commonly becomes its limiting characteristic.*[5]

To determine the magnitude of the registration tolerance that must be allowed between two mask levels, the distributions of the edge locations arising from all of the process parameter variations for *each of the two edges*, σ_{p1} and σ_{p2}, must be estimated, as well as the distributions of the edge locations due to registration variations, σ_r (we ignore the edge shift due to lateral diffusion in this discussion). Assuming that σ_{p1}, σ_{p2}, and σ_r are independent random variables with normal distributions (see Chap. 18), the probability that edges from level 1 and 2 will touch can be estimated for various registration tolerance values. The magnitude of the dimension, T, that will ensure that the probability of the two edges touching is only 0.1%, is given by[27]:

$$T = 3 \left[\left(\frac{\sigma_{p1}}{2} \right)^2 + \left(\frac{\sigma_{p2}}{2} \right)^2 + \sigma_r^2 \right]^{1/2} \qquad (7)$$

Typical values of σ_{p1} and σ_{p2} are ±0.15 μm in a well controlled VLSI process line, and σ_r depends on factors as described above. Assuming a value of σ_r of ±0.15 μm, Eq. 7 predicts that a registration tolerance of $\cong 0.6$ μm must be allowed.

Alignment of one pattern layer to previous layers is done with the assistance of special alignment patterns designed onto each mask level. When these special patterns are aligned, it is assumed that the remainder of the circuit patterns are also correctly aligned. The adjustment of the image of the mask being exposed to the previously produced patterns was originally performed by human operators, who compared the image locations under a microscope and adjusted the position of the mask to bring it into alignment with wafer patterns. Decreasing feature sizes, and the increasing number of alignments per wafer with steppers, have been the impetus for developing automatic alignment systems[24,26] for use with projection aligners.

The principle of one type of automatic alignment procedure is illustrated in Fig. 19[25]. Alignment marks consisting of two rectangular patterns, each set at a 45° angle to the directions of the stage motion are fabricated on the wafer (the black rectangles in Fig. 19). Two corresponding rectangular patterns are located on the reticle, and their image is projected onto the wafer (the larger clear rectangles in the figure). The superimposed alignment target and the reticle image are reflected back into the main optical element of the aligner, and then into an on-axis microscope. The image from the microscope is focused onto the face of a TV camera, and is subsequently digitized into a form that can be analyzed by a computer. When alignment is achieved, a signal as shown in Fig. 19 is obtained. The 45° orientation of the alignment marks makes it possible to obtain both x and y registration information from the horizontal

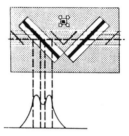

Fig. 19 Target for automatic alignment. Reprinted with permission of Microelectronics Manufacturing and Testing.

scan-lines of the video camera. The relative position of the wafer marks with respect to the reticle windows determines the registration of the two images.

Alignment in wafer steppers can be performed *globally* and *locally*. *Global alignment* performs rotational and translational alignment of the entire wafer. *Local alignment* provides alignment to a target within the particular die which is in position for immediate exposure, and is also referred to as *through the lens, site-by-site, die-by-die, or field-by-field alignment*. Global alignment is usually done at a remote alignment station before a wafer is sent under the projection lens for exposure, and is often used as the sole alignment in cases where the overlay tolerance is > 0.7 μm. Substantial time is saved by performing only a single global alignment, and hence exposure throughput can be significantly increased[24]. In such cases, following global alignment, the wafer is moved to each step-and-repeat site by *dead reckoning*, and then each site is exposed. Such an exposure procedure is also referred to as *blind stepping* . When local alignment is used, a global alignment always precedes the sequence of local alignments.

MASK and RETICLE FABRICATION

The pattern "tools" utilized in IC processing are known as reticles and masks. We define a *reticle* to be a tool containing a pattern image that must be stepped and repeated in order to expose an entire substrate. Usually the pattern size is enlarged from 2X to 20X the size of the image on the substrate, but in some instances is of equal size. A *mask* is defined as a pattern tool which contains patterns that can be transferred to an *entire wafer (or to another mask) in one exposure*. Reticles are used in two applications: 1) for printing images of patterns onto masks; and 2) for printing images directly onto wafers in step-and-repeat aligners. In 1X wafer steppers, the pattern on the reticle is the same size as the image projected on the wafer, while the patterns on reticles of reduction steppers, consist of enlarged versions of the actual device patterns.

Figure 20 shows the various steps that can be utilized to transfer the circuit pattern data, as entered into a computer by the designer, to the wafer. For VLSI, e-beam writing on a 10X or 5X reticle, or direct e-beam generation of 1X masks, is becoming predominant[23], although optical pattern generation of reticles, and reduction-camera step-and-repeat of masks is still widely practiced. In this section the steps necessary to generate high quality masks and reticles are described. Such pattern tools must provide high resolution, low defect levels, and be optically compatible with exposure equipment, resists, and cleaning and etching processes. One pattern tool term worth defining at this point is its *polarity*. That is, a mask or reticle is said to be a *dark-field* (or *negative*) *tool*, if the field (or background) areas are opaque, and to be a *clear-field* (or *positive*)*tool* if the field is transparent (Fig. 21).

Glass Quality and Preparation

The glass used for photomasks and reticles must be free of defects on both surfaces, as well as internally, and should have high optical transmission at the resist exposure wavelength. Several types of glasses have been used for making photomasks, including: a) soda-lime glass; b) borosilicate glass; and c) quartz[28]. *Green soda-lime glass* and *low sodium white soda-lime glass* (~50% more expensive) are easy to draw into large flat sheets and have shown excellent quality. Their high thermal expansion coefficient (93×10^{-7}cm /cm°C), however, has made them largely unsuitable for projection applications. In applications where low expansion is required *borosilicate glasses* and *quartz*, with their lower thermal expansion coefficients

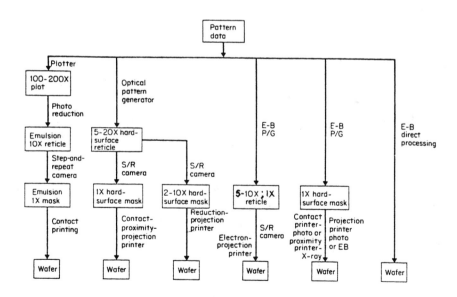

Fig. 20 Mask paths from design to wafer.

clear-field (or positive) tool

dark-field (or negative) tool

Fig. 21   Photomask polarity.

(37 and 5 x10^{-7} cm/cm°C, respectively), have become the materials of choice. This is generally the case for projection printing where ambient temperature changes can result in dimensional and positional errors of the image on the wafer (run-out). *Quartz plates* are classified as ultra low expansion glasses, and result in very small amounts of thermal run-out. Quartz also has very high transmission in the near and deep-UV. Quartz, is more expensive, although it has recently become more affordable with the development of high quality synthetic quartz material[29]. Natural quartz is manufactured by the Verneuil method, in which rock crystal is fused with oxygen and hydrogen gases. Synthetic quartz uses ultrapure silicon tetrachloride ($SiCl_4$. It offers a wider range of light transmission, a lower level of impurities and fewer physical defects. Its use is increasing with the number of applications that call for low run-out or deep UV. Figure 22 shows the transmission characteristics of various glasses used for masks.

The glass is prepared by cutting from large sheets. The inclusion-free plates are cleaned to remove chips, and then graded for flatness. Since non-flat masks can result in run-out, the grading specifies their run-out over the usable area of the mask. For step and repeat applications the flatness is somewhat less critical. Plates can also sag significantly under their own weight. Thus minimum plate thicknesses are recommended for each particular mask size. For example, a plate 5" x 5" should be 90 mils thick, while a 6" x 6" plate must be 120 mils thick.

The plates are then polished, cleaned, and inspected before coating with the mask forming material. Polishing is a multistep process using consecutively finer grade abrasives on both

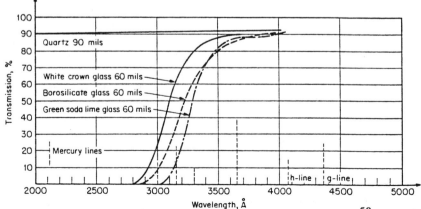

Fig. 22 Transmission versus wavelength of different glass substrate materials[50].

sides of the plate. Plates are cleaned, rinsed, and dried prior to inspection and coating.

Glass Coating (Chrome)

The glass plates are next coated with a material in which the pattern is eventually formed. These materials include emulsion, chrome, and iron oxide. Emulsion is not used in VLSI because of its poor linewidth control, and its inability to stand up to use and cleaning. Chromium (chrome) is the most widely used material, and it is deposited onto the plate by sputtering or evaporation. Generally, the thickness is less than 1000Å. Sputtering has the advantage of high throughput, good adhesion, and excellent uniformity. It is possible to sputter etch the glass plate prior to the chrome deposition, resulting in improved adhesion. Vacuum deposition (e-beam evaporation) has also been used to prepare chrome plates in a vacuum of 10^{-5}-10^{-6} torr. Although sputtered chrome can be more reflective than evaporated chrome (an undesirable property), use of anti-reflective coatings can overcome this drawback. Such coatings consist of a thin (~200 Å) layer of Cr_2O_3 on the surface, which reduces the reflectivity. Figure 23 shows the reflectivity of chrome coatings as a function of wavelength, with and without anti-reflection coatings.

Mask Imaging (Resist Application and Processing)

The application of resist and its subsequent processing to produce images (either optically or by e-beam) is similar to that used for wafers (Chap. 12), except that the resist coating is much thinner, and different types of exposure equipment are used.

Prior to coating, the blanks are cleaned and dried, and then spin-coated with filtered photoresist. Such optical resists as AZ-1370 or Kodak 820, or e-beam resists as PMMA and COP (described in Chap. 14), are used for mask fabrication. Thick coatings (~0.5 μm) result in improved line size control and pinhole protection, while thinner resists (0.2-0.3 μm) result in better resolution.

The resist is soft-baked rapidly because of the thin coatings used. Exposure control is critical since the high reflectance chrome surface results in standing waves. Exposure of the resist can be performed either by an optical or e-beam method.

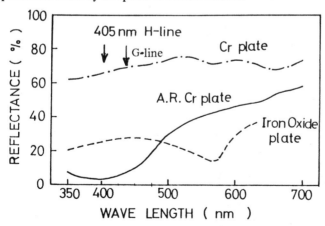

Fig. 23 Reflectance of chrome and anti-reflective chrome films used to produce reticles and masks.

Table 1. SELECTED WET ETCHANTS FOR CHROME FILMS

			Ceric Ammonium Nitrate:	310g Ce$(NH_4)_2$ $(NO_3)_6$ 120 ml HNO_3 (conc.)
Alkaline Etch:	50g NaOH, 100 ml H_2O	1 part		1970 ml H_2O
	100 g K_3 (Fe $(CN)_6$) 300 mL H_2O	3 parts	Ceric Sulfate:	9 parts saturated Ce $(SO_4)_2$ (solution) 1 part HNO_3 (conc)

The resist is developed, typically with spray techniques. Tight control of developer concentration and temperature is critical for maintaining line size control. After this step, critical dimension (CD) and resolution measurements are made to insure they are in specification, and placement error is checked with a mask pattern or on a Nikon 2I®. The resist is then hard-baked (to ~100°C) prior to etching.

The pattern transfer into the chrome has largely been accomplished using wet etching, since the thin chrome films can be effectively wet-etched. Some etchants used are listed in Table 1.

There have been reports in the literature describing the plasma etching of chrome photomasks using CCl_4, Cl_2[30], or CF_4[31], all with O_2. In order to form volatile compounds, a large proportion of oxygen must be used, but this results in resist erosion. Good selectivity between Cr and resist can be achieved at a pressure of 250 mtorr, a power of 35 watts, and a CCl_4 /O_2 ratio of approximately 1.6. The maximum selectivity is about 1.5 at a chrome etch rate of 300 Å /min. Good selectivities can be obtained using e-beam resists such as PBS and COP. The sidewalls of the plasma etched features are sinusoidal in nature, but such structures do not cause a change in linewidth of the mask when viewed in projection.

It has also been reported that a reversal of resist tone (positive to negative) can be achieved by plasma processing[32]. Treatment of the exposed and developed resist pattern in a plasma (O_2, SF_6, or Ar) passivates the exposed chrome areas against wet etching. The resist is then stripped by conventional techniques, and the plate wet-etched. The unpassivated areas are attacked more rapidly by the etchant than the passivated areas. Black chrome (anti-reflective) passivates much better than shiny chrome. It is believed that the passivation results from material (SiO_2) sputtered from the backing plate.

Pattern Generation

Optical Pattern Generation of Reticles
During the previous decade, significant advances were made in the field of optical pattern generation (PG) for reticles[33]. Patterns can be conveniently generated by imaging and exposing a set of accurately positioned rectangles onto the chrome plate. Since the patterns generated by integrated circuit designers are, in general, polygons, they must be decomposed into rectangles. This is termed "fracturing" the data. Figure 24 shows an example of a design shape fractured into its composite rectangles. The challenge of pattern generation is to reduce any image to a set of rectangles, each specified by a height and width, by x and y coordinates, and by an angle relative to the coordinate axes. Each rectangle can then be unambiguously defined with the following five coordinates: H, W, x, y, and θ.

The key components of the optical pattern generator (OPG) are apertures (or shutters), and a movable stage. The *shutters* are mounted on a movable head, which can also be rotated to give

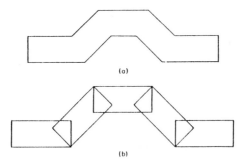

Fig. 24   (a) A polygon before fracturing into rectangles.   (b) The fractured polygon[15].   Reprinted with permission of Academic Press.

desired values of θ, and the shutter size can be controlled to produce patterns with the specific rectangular dimensions, H and W.   The rectangles are imaged through a 10X lens onto a photosensitive film that covers a chrome-coated glass plate. This plate is accurately positioned by a *laser controlled x-y stage*.   The stage sets the center coordinates, x and y, of the rectangle defined by the apertures.

The variable aperture size can be controlled to within $\pm 7.5 \ \mu m$, which results in a ± 0.75 μm size uncertainty on a reticle (after a 10x reduction).   This is the critical dimension (CD) control. If a 10X reticle is produced, a $\pm 0.075 \ \mu m$ error on a wafer results. Note that errors due to processing latitude both on the reticle and on the wafer occur in addition to the aperture size error. Position errors in x and y are quite small since the stage is under laser control. Stage control to the desired position, however, does result in positional errors.   Such error can be traded off for throughput.

The throughput of an OPG depends on the complexity of the mask designs being generated, the optimization of the data fracturing program, and the positional accuracy required. A well tuned machine running an optimized pattern in AZ1370 has a throughput of 10,000 rectangle exposures (flashes) per hour. A complex VLSI circuit level may have in excess of 100,000 flashes, requiring more than 10 hours to produce a reticle. During this period temperature control better than $\pm 0.5°F$ must be maintained so as not to result in additional positional errors. The quality of the reticle cannot be ascertained until after development and etch of the chrome. Flaws detected at that time may mean the scrapping of a 10 hour job. The low throughput along with reduced feature sizes, has made optical pattern generation of reticles much less attractive than electron beam generation for VLSI.

Directly-Stepped Photomasks

Photomasks for projection scanning aligners contain a large number of die patterns with the same dimensions as the patterns on the wafer[33,34]. Such masks can be patterned by step-and -repeating the individual die patterns from a 10X reticle with a high-speed step-and repeat camera. Two examples of such cameras are the GCA /MANN 3696 and the TRE / ELECTROMASK. The stepping motion of these cameras is metered with a He-Ne laser interferometer, and their stepping precision is $3\sigma \leq 0.25 \ \mu m$. The 10X reticle contains fiducial marks that align with the X and Y directions of the pattern on the reticle. When the reticle is mounted on the camera head, these fiducials are aligned to targets fixed to the head. The X and Y motions of the stage, on which the blank photomask plate is placed, are aligned to the camera head targets.   The stage moves in the X and Y directions, precisely positioning the photomask plate prior to each exposure of the die pattern. Step-and-repeat cameras can produce from ~0.5 to 15 plates per hour, depending on the plate size and the number of multiple patterns that must be stepped onto the plate (e.g. besides the circuit pattern, *test drop-in* patterns may be used).

Although the throughput of 10X step-and-repeat cameras can be high, they also have the following drawbacks: a) the maximum die size that can be printed is limited by the working diameter of the camera lens, to typically a 14-mm diagonal; b) the minimum resolution of the mask features is limited by the camera lens to ~1.0 μm; and c) the best array registration is ~0.5 μm, compared to ~0.15 μm for e-beam pattern registration.

Electron Beam Pattern Generation

The electron beam exposure system (EBES) was originally developed at Bell Laboratories during the 1970s. It utilizes raster scan techniques to write the pattern. Commercial versions for mask fabrication are available and find wide use at both commercial and in-house mask shops. They are very well suited for 10X, 5X, and 1X reticle fabrication and also for 1X masters[23,24].

Figure 25 shows a schematic of the e-beam stage of such a system. A sensitized (e-beam resist) chrome plate is placed on the table and moved uniformly in X direction under the beam. The width of the scan is one beam diameter, and the scans are contiguous on the plate. A rectangle of 512 address units (AU) in height and 32,768 AU in length is contained in memory and is called a "stripe". The pattern is constructed of these stripes. How well these strips abut (as well as their size uniformity), both contribute to the precision of the pattern. If the scan is parallel to the pattern edge, the edge will be smooth. If it is perpendicular, however, it will show roughness, that is dependent on the spot size. Nevertheless, such edge roughness is not a problem for 10X and 5X reticle fabrication. Roughness can also result when writing an angled line, and this is called the *raster effect*[35]. The abutment of the stripes, and drift in the electron beam also result in CD errors. The stripe abutment error arises from the inability to exactly align the stage motion to the beam motions and is generally less than 0.125 μm. Electron beam

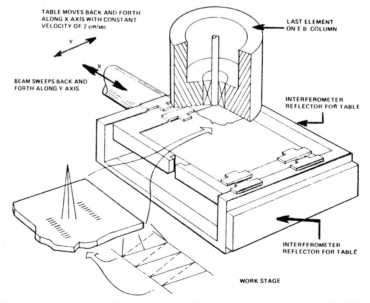

Fig. 25 The work stage of an e-beam system. The electron beam scans in the Y-axis as the stage traverses under the column in the X-axis. Courtesy of Perkin-Elmer - ETEC.

drift error is likewise quite small (< 0.0625 µm) in a five minute period. Variations in resist processing and exposure effects also add to the CD error. Consideration of all errors for a 1 µm spot (10X reticle) has a calculated RMS value of 0.27 µm if angled features are not written, and 1.04 µm if angled features are used. Again, for 10X and 5X reticles this presents no problem, but for 1X masters or reticles, a finer beam size (0.25 µm) must be used to reduce this error size.

Registration precision is the ability to *repeatably* place a feature at its site, while *registration accuracy* is the ability to place it where it is intended to go. In addition to abutment and beam drift, temperature variations during the reticle fabrication affect the placement of the feature. Values of registration accuracy of 0.3 µm have been achieved.

E-beam has an advantage over optical pattern generation in that uniform exposure is given to all rectangles of the pattern. Optical patterns, depending on the fracturing, can have areas that are double exposed (see Fig. 24)[36]. Due to such additional exposure, the feature size can grow. The e-beam, on the other hand, turns on and off as required, thereby avoiding double exposure. As described in Chap. 14, however, for small features in dense patterns, the effect of back-scattering (leading to pattern distortion by proximity effects) must be accounted for in e-beam exposure. This is generally not a problem in 10X or even 5X reticles, but can be for 1X reticles.

A recently introduced potential alternative to e-beam pattern generation of reticles (at least for 5X and 10X reticles), is a laser-based system designed to produce reticles with a minimum feature size of 1.6 µm, a CD uniformity of ±0.12 µm and an overlay accuracy of ±0.12 µm, and a positioning accuracy of ±0.25 µm over a field size of 128 mm^2 (e.g. the CORE 2000 by ATEQ[37]). Such systems cost one-third as much as e-beam systems, are 5-10 times as fast, and use optical resists.

Mask and Reticle Defects and their Detection and Repair

Defects in masks and reticles have always been a source of yield reduction in integrated circuit manufacture. As the minimum pattern sizes approach 1 µm and below, and the circuits are designed with higher device densities, defects that were once tolerable can no longer be accepted. For example, a single defect that in the past killed only an individual die, now becomes a repeating defect in stepper systems, and will kill every die in single die reduction reticles. Such defects can be due to incorrect design of the mask patterns, or flaws introduced into the patterns during the pattern generation process. Even if the design is correct, and the pattern generation process is performed correctly, so that the desired mask patterns are produced, defects in the mask or reticle can be generated by the mask /reticle fabrication process, as well as during subsequent processing and handling.

It is true that reduction steppers do provide some reduction in the severity of such defect effects. That is, many defects are effectively rendered invisible by the simple expedient of a 10X reduction, and 5x steppers also enjoy this benefit, but to lesser extent. 1X systems, of course do not offer this opportunity for defect reduction, and reticles in such systems must therefore be prepared with greater care to avoid repeating defects, and are usually protected with pellicles (as described in a following section). 1X and 5X systems, however, may be able to place more than one die pattern on a reticle, and in such cases a defect will not kill every die.

It is the mask fabrication and wafer processing induced defects (and their repair), that are the subjects of this section. The detection of design or pattern generation flaws is discussed in a subsequent section on mask and reticle inspection. Figure 26 illustrates the types of defects that may be found on a mask as a result of mask fabrication and wafer fabrication problems. Defects which result in inoperative devices or which would cause a die to be rejected at final visual inspection are termed *fatal defects*, while others are nonfatal defects.

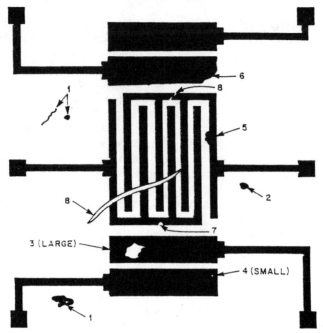

Fig. 26 Types of defects commonly encountered on a process line. (1) contamination, (b) opaque spot, (3) large hole, (4) pin hole, (5) excess material, (6) lack of adhesion, (7) intrusion (mouse nip) and (8) scratch. Reprinted by permission of Ref. 1, Copyright 1983, American Chemical Soc.

The causes of mask fabrication defects are first considered[38]. *Raw glass substrates* can contain bubbles, scratches, pits and fractures. Bubbles are relatively rare in good quality plates and can be readily detected (and plates containing bubbles, rejected). Scratches, pits, and fractures are surface defects, which are harder to detect, and thus more likely to be found on a finished mask. Pits inhibit chrome adhesion, causing pinholes, while scratches produce nonuniform etching in local areas, especially at pattern edges. *Chrome defects* include: a) particulate inclusions in the film; b) pinholes or voids in the chrome surface; and c) invisible chemical anomalies (such as nitrides or carbides), which lead to erratic local etching that produces undesired patterns. *Resist defects* include voids, which produce pinholes (that lead to chrome spots), and resist gels that may locally affect resist solubility. Masks and reticles made with e-beam lithography require use of e-beam resists. Until recently, e-beam processing has been more prone to resist defects than conventional optical resist processing. Recent advances in e-beam resist and material technology have made them about equal. Of course *dirt particles* can also be introduced onto the mask plate during any of the pattern transfer steps, again leading to mask defects.

Upon completion of the mask or reticle fabrication sequence, a series of inspections to qualify the product are performed. The inspection procedure examines several characteristics of the mask or reticle, including: a) linewidth measurement; b) measurement of the registration among the arrayed die patterns (if more than a single die per mask or reticle is present); c) determining that all the features present in the design database have been transferred to the mask; and d) determining if any mask fabrication defects have been produced. Different inspection tools are utilized for each of the above inspections. Inspection procedures and tools for detecting fabrication defects are discussed in this section, while the other inspections are discussed later.

Mask fabrication defects are located by using transmitted light. In the past such inspections were carried out by a human operator working with a microscope. As masks have become more complex, this task has been relegated to automatic defect detection systems that perform the task much more rapidly, and with fewer errors[39,40]. These systems are also able to plot the distribution and size of defects over an entire mask. Defects (e.g. pinholes) as small as 0.35 μm are detectable with ~95% probability on optically transparent substrates with the most recently designed systems. Since the detection of small defects by automatic systems is a probabilistic process, mask makers commonly run a reticle through an automatic system several times, to guarantee that all defects have been located.

Several automatic mask /reticle defect detection systems are being sold, including the KLA KLARIS[39], and the Cambridge Instrument's Chipcheck[41]. The most advanced models have the capability of performing both die-to-die and die-to-database inspection (Fig. 27). The *die-to-die* inspection allows defects unique to an individual die to be identified. *Die-to-database* inspection allows repeating defects to be found (which would not be detected by a die-to-die comparison), as well as errors made in converting data from the computer memory format to the mask /reticle pattern. With systems that can perform both such inspections, only one die of a mask array must be inspected against the database, while die-to-die inspection checks each die for random defects. A defect detection system based on holographic principles is also available[40]. It is specified to detect particles $\geq 0.5\mu m$, and to be able to inspect masks with pellicles attached.

As a final check before going from reticle or mask to wafer, a glass wafer can be used[42]. With only a few die exposures on a glass wafer, an automatic reticle inspection system can qualify a reticle within minutes. Such glass wafers are dimensional replicas of standard silicon wafers offered in 76, 100, 125, and 150 mm sizes, and coated with a thin film of aluminum.

Repairing Defects in Masks and Reticles
Since fatal defects in a mask or reticle are obviously highly undesirable, or in cases of one die per reticle, entirely unacceptable, it would be useful if such defects could be repaired, thereby rendering the mask free of fatal defects. Mask repair methods for accomplishing this purpose

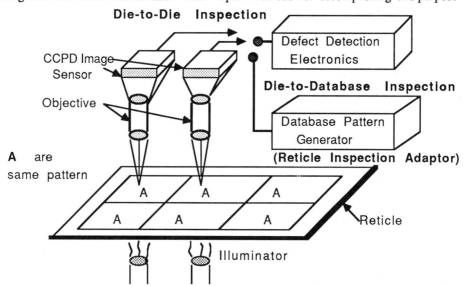

Fig. 27 Automatic defect detection instrument for masks and reticles. Courtesy of KLA Instruments Corp.

have been developed. The repair of *opaque defects* (i.e. removal of chromium spots from areas in which they do not belong) has been performed with lasers for many years (i.e. *laser zappers*). A focused laser beam merely evaporates unwanted material. One concern with laser evaporation is potential damage of the glass substrate. Large chrome spots may require several laser pulses to remove them, and if damage (*laser burn*) occurs, it can become another printable defect.

The repair of *clear defects* (i.e. the deposition of chromium in areas from which it should be missing) has proven more difficult. The most widely used process is a chromium lift-off procedure, which requires hours of processing time, and may introduce additional defects (Fig. 28a). As a result, several alternative methods have been developed. One uses a local pyrolytic decomposition of a chromium bearing-gas at the spot where the clear defect exists. The gas is delivered by a process-gas delivery needle into the space between a 100X microscope objective and the reticle surface. An argon laser beam is delivered through the microscope optics, and this beam increases the temperature of the desired spot (~1 µm in size) on the substrate surface to 150°C. This causes decomposition of the gas, and consequent local deposition of the chromium at the heated spot. Such deposited patches can be controlled in the 1-2 µm size range, and since they are composed of chrome, can be trimmed by a YAG laser to submicron dimensions for *mouse-nip* filling (Fig. 28b). Another method involves the application of epoxies (dispensed by pipette, while observing with a high-magnification, wide-field microscope). A third technique utilizes a focused ion beam to deposit chrome onto local regions of the substrate, or to locally etch a fine diffraction grating pattern onto the glass surface, so that it appears effectively opaque.

Pellicles

Even though masks and reticles used in projection printing can be fabricated without defects (i.e. by utilizing repair techniques if necessary), and no damage-creating contact between mask and wafer occurs, mask defects due to handling or airborne contamination can still be generated. In the regime of ≤ 1 µm resolution, particulates that are large enough to cause defects are also harder to detect and remove.   Thus, even in projection printing processes, a method for

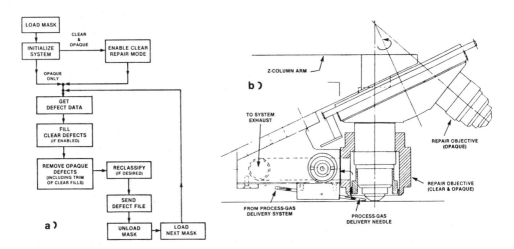

Fig. 28 (a) Typical photomask lift-off repair process cycle. (b) Opaque mask defect repair process by pyrolytic decomposition of chromium-bearing gas.   Courtesy of Quantronix Corp.

protecting masks and reticles against such defects is highly desirable. Such a method is available, and it involves the use of pellicles[44,45].

A *pellicle* is a membrane that seals off the mask or reticle surface from airborne particulates and other forms of contamination.   The membrane is mounted on a metal frame which is securely attached to the chrome side of the mask, as shown in Fig. 29a. Particles (typically <100 μm) which adhere to the pellicle surface are kept at a distance far enough from the wafer surface that they are not imaged onto the wafer surface (Fig. 29b). The membrane is thin enough to be optically transparent, strong enough to be stretched across a support ring that covers the entire printable area of a mask or reticle, and is sufficiently durable to withstand cleaning and handling.   The membrane material must also be stable enough to retain its shape over long periods of time and exposure to 100,000 flashes of UV radiation, and be inexpensive enough to be cost effective. The pellicle material is usually nitrocellulose, with a thickness of 1-15 μm (typically ~2.9 μm).

Although it is easy to understand how a pellicle can prevent particles from directly contaminating the chrome surface of a mask, it is useful to elaborate upon the reason why particles which adhere to the pellicle surface may not be imaged onto the wafer surface. From Fig. 29 it can be seen that light directed at the wafer surface will be obscured by particles present both on the back side of the mask and the surface of the pellicle.   To insure that a particle on either of these surfaces causes no more than a 10% reduction in the light intensity on the wafer surface, a minimum standoff distance from the mask for a given size of particle can be calculated. This standoff distance, D (in mm), for a particle size (in μm) is calculated from[44]:

$$D = n F P / 280 \qquad\qquad (8)$$

where n is the refractive index (1.0 for air, 1.5 for glass), and F is the condenser F-number at the mask or reticle (F = 1 /NA). Figure 29 gives some standoff distances versus particle size, for several values of F-number. For example, the F /4 curve is valid for the optical systems in the Perkin-Elmer 1:1 scanning aligner and the Ultratech Model 900 1:1 stepper. It shows that a mask plate thickness of 2.2 mm offers immunity against particles up to 50-100 μm, while a pellicle standoff distance of 2.5 mm provides excellent immunity against 100-200 μm particles. For 10:1 steppers, which have a greater depth of focus, a 2.2 mm plate only provides protection against 10 μm particles on the reticle backside. A second pellicle frequently attached to the reticle backside to provide double-sided protection.   A study by Flamholz[46] indicates that because of their increased depth of focus, reduction-type steppers have their images impacted by particles at somewhat greater standoff distances than predicted by the simple model of Eq. 8. Furthermore, only limited protection is afforded against extended defects, such as hairs and fibers. Flamholz concluded that in order for pellicles to adequately provide protection in such systems, all pellicle parameters must be conservatively chosen (e.g. standoff distance, focus position).

The optical transmission losses of the pellicle material (due to reflection, absorption, and scattering) affects the exposure time, and thus throughput. (In the 350-450 nm range a 2.9 μm thick nitrocellulose pellicle, with an anti-reflection coating, has an average transmission of ~99%.) Nitrocellulose and mylar (another pellicle material), however, both exhibit strong absorption near 300 nm, which limits their usefulness for deep UV applications. Optical interference effects can also cause significant variations in transmission through a pellicle if the pellicle thickness exhibits even submicron nonuniformity.   Fortunately, pellicles can be manufactured with thickness uniformities of better than 1%, minimizing such problems.

Masks or reticles with pellicles attached are cleaned with a DI water rinse to remove most

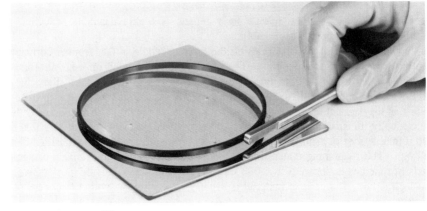

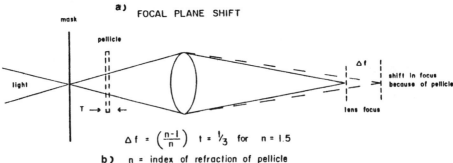

a) FOCAL PLANE SHIFT

$$\Delta f = \left(\frac{n-1}{n}\right) \ t = \frac{1}{3} \quad \text{for} \quad n = 1.5$$

b) n = index of refraction of pellicle

Fig. 29 (a) Photograph of a pellicle. Courtesy of EKC Technology. (b) Focal plane shift of a pellicle.

particulates. Fingerprints can be cleaned by using a mild surfactant and manual scrub. The glass (back) side of the mask is usually cleaned at the same time. Damaged pellicles can be successfully replaced if careful mask cleaning and inspection is done prior to attaching the new unit. With the exception of a few IC manufacturers, use of pellicles has become an industry standard.

Inspecting Masks and Reticles with Pellicles Attached

The process of attaching a pellicle to a defect-free mask or reticle should itself not create defects on the mask. For example, a zero defect reticle may become useless due to contamination under the pellicle. To insure that the assembly procedure has not introduced any defects, a post-pellicle inspection procedure should be performed. Van Asselt and Brooks[47] suggest a method for performing such an inspection. The mask-pellicle assembly is mounted in a projection aligner and is used to expose round, thin glass plates (glass wafers), coated with chromium and resist. The wafers are exposed, developed, etched, and stripped to produce an image suitable for inspection by transmitted light. The processed glass wafers are then inspected on an automatic mask inspection system with a special mask holder. Defects located by the inspection system that coincide in location on two or more glass wafers are attributed to defects that occurred as a result of the pellicle attachment procedure, while defects occurring in only one location can be associated with the processing of that particular wafer. The procedure can be extended to multiple-die reticles used in 1X and 5X stepping systems.

Critical Dimension and Registration Inspection of Masks and Reticles

Upon completing the fabrication of a reticle or mask, it must be inspected to verify that the fabrication was correctly carried out[48]. The detection of fabrication process-induced defects was described earlier. In this section inspection procedures of the following mask and reticle characterisitics are discussed: a) physical characteristics (e.g. title correctness, polarity of the field of the primary and test patterns, substrate [glass] type, chrome type [i.e. shiny, low reflective, or medium reflective]; b) critical dimensions (CD); and c) registration. The ability to effectively perform the full gamut of inspections on VLSI masks or reticles involves the use of a set of complex and expensive inspection tools.

Checking the *physical characteristics* requires the least complex inspection equipment (e.g. an optical microscope and video measurement device), but detected flaws can still force the rejection of a finished plate. Since the inspection can be relatively easily performed, some users request a *check plate* prior to initiating the full tooling fabrication sequence. A check plate can be made on a low-cost blank, allowing for early detection and correction of physical flaws.

The subject of linewidth measurement was covered in Chap. 12, and the same types of linewidth measurement techniques used to measure CDs in developed resist patterns can be used to measure mask feature dimensions. As pointed out in Chap. 12, the task is considerably easier on a mask or reticle, because the chrome pattern layer is relatively thin (making accurate edge detection easier), and the topography under the chrome is flat, unlike the topography on a wafer near the end of the fabrication cycle. Linewidth standards are available from the NBS for mask and reticle feature dimension measurements.

The ability to perform both inspection of *critical dimensions* and *registration* on VLSI masks or reticles with a single machine, however, requires a sophisticated metrology instrument, such as the Nikon 2I® system. This is an interferometry tool containing two instruments: 1) a Nikon Lampas® head which uses a laser beam to detect pattern edges; and 2) an x-y stage for precisely determining position. Together, they allow the tool to measure both CDs and registration accuracy. While a 10X reticle commonly contains only a single pattern, 5X reticles and 1X masks typically have an array of patterns. In addition to the repeated primary die patterns, the finished plate may contain test strips, array alignment marks, and global alignment marks. It must be determined that each of these patterns is located at the specified coordinates.

If a specific metrology instrument is utilized for both the registration and CD inspection, the alignment marks and CD test patterns may need to be placed on specific locations of the plate. For example, a popular method used with the Nikon 2I is to combine the CD and registration marks into one feature and to place these at the corner of each die of the array. The dimensions of CD marks are compared to a known standard, and registration marks to other levels within the device. Presently, inspection for process-induced defects and CD /registration inspection is done on different systems. It is postulated that future metrology requirements will dictate that a single machine which combines both capabilities be available (i.e. one that is able to detect, measure, and classify defects, and also to perform a die-to-database inspection and CD /x-y coordinate measurements). The Wafervision® system introduced by Contrex is designed as such a "universal" inspection system[49].

REFERENCES

1. L.F. Thompson and M.J. Bowden, "The Lithographic Process: The Physics" in *Introduction to Microlithography*, American Chem. Society, Advances in Chem. Ser., Vol. 219, 1983, p. 15.

2. B.T. Lin, "Optical Methods for Fine Line Lithography" in R. Newman, (ed.), *Fine Line Lithography*, North Holland, Amsterdam, 1980.

3. E.B. Brown, *Modern Optics* (Van Nostrand-Reinhold, New York, 1965).

4. J.E. Roussel, "Submicron Optical Lithography?", *SPIE Proc. Vol. 275, Semiconductor Microlithography*, 1981, p. 9.

5. M.C. King, "Principles of Optical Lithography", in N.G. Einspruch (Ed.), *VLSI Electronics Micro Structure Science*, Vol. 1, Academic, New York, 1981.

6. K.L. Tai, *et al.,J. Vac. Sci. Technol.*, **17**, 1169, 1980.

7. J.G. Skinner, "Some Relative Merits of Contact, Near-Contact, and Projection Printing", *Proc. Kodak Interface*, **73**, 53 (1973).

8. D.J. Elliot, *Integrated Circuit Fabrication Technology*, McGraw-Hill, New York, 1982, Chap. 8, Exposure, p. 165.

9. P. Burggraaf,"Wafer Steppers and Lens Options", *Semiconductor International*, Mar '86, p. 56.

10. P. Burggraaf, "Advances Keep Optical Aligners Ahead of Industry Needs", *Semiconductor International*, Feb. 84, p. 88.

11. V. Miller and H. Stover, "Sub-Micron Optical Lithography, I-Line Wafer Stepper and Photoresist Technology", *Solid State Technol.*, Jan. 85, p. 127.

12. M.M. O'Toole and A.R. Neureuther, "Influence of Partial Coherence on Projection Printing", *SPIE Proc. Vol. 174, Developments in Semiconductor Microlithography*, 1979, p. 22.

13. J. Roussel, "Step and Repeat Wafer Imaging", *SPIE Proc. Vol. 135, Developments in Semiconductor Microlithography III*, 1978, p. 30.

14. N. Gold, "Spectral Exposure Meter", *SPIE Proc. Vol. 334, Optical Microlithography Technology for the Mid-'80s*, (1982), p. 70.

15. R.C. Bracken and S.A. Rizvi, "Microlithography in Semiconductor Device Processing", in N.G. Einspruch and G.B. Larabee, (Eds.), *VLSI Electronics-Microstructure Science*, Vol. 6, Academic, Orlando, Fla., 1983, Chap. 5, p. 256.

16. D.A. Markle, *Solid State Technology*, p. 50 (June '79).

17. M.C. King, "New Generation of 1:1 Optical Projection Aligners", *Proc. SPIE Vol. 174, Developments in Semiconductor Microlithography IV*, 1979, p. 70.

18. A.N. Broers, "The Submicron Lithography Labyrinth", *Solid State Technol.*, June '85, p. 119.

19. W. Carpenter, "Mixing 1X Scanning and 5X Step-and-Repeat Exposure Systems in a Semi-conductor Processing Line", *SPIE Proc. Vol. 538, Optical Microlithography IV*, 1985, p. 38.

20. H.L. Stover, "Near Term Case of 5X versus 10X Steppers", *SPIE Proc. Vol. 334, Optical Microlithography-Technology for the Mid-80's*, 1982, p. 60.

21. R. Hershel, "Characterization for the Ultratech Wafer Stepper", *SPIE Proc. Vol. 334, Optical Microlithography-Technology for the Mid-80's*, 1982, p. 44.

22. M.S. Chang, "One Micron 1X Stepper Photolithography: An Introductory View", *SPIE Proc. Vol. 538, Optical Microlithography IV*, 1985, p. 32.

23. P. Burggraaf, "Reduction Reticle Trends:Emphasizing 5X",*Semiconductor Intl.*, Aug.'84, p. 59.

24. B. Heflinger, "Comparison of Autoalign Techniques", *Proc. SPIE, Vol. 334, Optical Microlithography*, 1982, p. 70.

25. P. Swanson, G. Alonzo, J.Dey, "Automatic Self-Testing of Machine-to-Machine Matching of Wafer Steppers", *SPIE Proc. Vol. 470, Optical Microlithography III*, 1984, p.185.

26. A variety of articles dealing with automatic alignment are published in the *Proc. SPIE, Optical Microlithography III and IV*, 1984, and 1985.

27. D.A. McGillis, "Lithography", in *VLSI Technology* [S.M. Sze, Ed.], McGraw-Hill Book Co., New York, 1983, Chap. 7, pp. 267-301.

28. *ibid.*, Ref. 8, Chap. 14, "Mask Fabrication", p. 339.

29. R. Iscoff, "Photomask and Reticle Materials Review",*Semiconductor Internatl*, Mar '86, p. 82.

30. T. Yamazaki, *et al., Japanese Jnl of Appl. Phys.*, **19**, 1371, (1980).

31. B. Hasler, German Pat. 2,738,839 (Siemens AG).

32. B.J. Curtis, H.R. Brunner, and M. Ebnoether,"Plasma Processing of Thin Chromium Films for Photomasks", *J. Electrochemical Soc., 130*, 2242 (Nov. 1983).

33. *ibid.,* Ref. 15, p. 260-292.

34. P. Singer, "Merchant Mask-Making in State of Change", *Semiconductor Intl.*, Mar '86, p. 45.

35. *ibid.,* Ref. 15, p. 266-272.

36. S.N. Gupta, *et al.,* "Multiple Flash Overlap Defect in Photomasks", *SPIE Proc. Vol. 538, Optical Microlithography IV,* 1985, p. 153.

37. CORE 2000, *Semiconductor International.,* Oct. '85, p. 29-30.

38. A.C.Titus, "Photomask Defects: Causes and Solutions", *Semiconductor Intl.* Oct. '84, p. 94.

39. K.L. Harris, *et al.,* "Automated Wafer Inspector Characterization", *SPIE Proc. Vol. 538, Optical Microlithography IV,* 1985, p. 138.

40. L.H. Lin, *et al,* "A Holographic Photomask Defect Inspection System", *SPIE Proc. Vol. 538, Optical Microlithography,* 1985, p. 110.

41. R.V. Fraser and B.A. Wallman, "Database Inspection of Wafer Resist Images", *SPIE Proc. Vol. 538, Optical Microlithography IV,* 1985, p. 122.

42. R.T. Hilton, *et al.,* "Glass Wafer Processing and Inspection for Qualification of Reticles in a Fineline Wafer Stepper Production Facility", *SPIE Proc. Vol. 538, Optical Microlithography,* 1985, p. 117.

43. G. Hearn. "Repair of Both Clear and Opaque Mask Defects", *Microelectronics Manufacturing and Testing,* Oct. 1985, p. 19.

44. R. Herschel, "Pellicle Protection of Integrated Circuit Masks", *SPIE Proc. Vol. 275, Semiconductor Microlithography VI,* 1981, p. 23.

45. J. Lent and S. Swayne, "Implementation of Pellicles Into An Established Production Area", *Proc. Kodak Interface,* 1982, p. 93.

46. A. Flamholz, "An Analysis of Pellicle Parameters for Step-and-Repeat Projection", *Proc. SPIE, Vol. 470, Optical Microlithography III,* 1984, p.138.

47. R. VanAsselt and G. Brooks, "Technique for Inspecting Photomasks with Pellicles Attached", *Proc. Kodak Interface,* 1982, p. 158.

48. P. Chipman, "Qualifying Reduction Reticles", *Semiconductor International,* Aug. '84, p. 68.

49. M.L. Baird, *et al.,* "Extending the Limits of Pattern Inspection Using Machine Vision", *SPIE Proc. Vol. 538, Optical Microlithography IV,* 1985, p. 130.

50. G. Zinsmeister, "Hard Surface Mask Technology", Technical Handout from IGC Conference on Microlithography, Amsterdam, Holland, September, 1979.

PROBLEMS

1. Explain why information about the edge of features on a mask is only contained by diffracted light and not by the undiffracted light that reaches the wafer surface.

2. Explain why an extended source appears as an incoherent source of radiation, and how the use of an entrance pupil (Fig. 5b) is used to control the degree of partial coherence in a source.

3. Calculate the minimum distance that can be just resolved (Rayleigh limit) by a 0.16 NA lens when (a) G-line radiation and (b) I-line radiation from a mercury-arc spectrum is used.   Repeat this exercise for an 0.28 NA and an 0.35 NA lens.

4. As the NA of a lens is increased, the image field diameter decreases. For a 150 mm diameter wafer, how many exposures would be necessary to completely expose a resist layer, if a

step-and-repeat refractive projection aligner with a lens of (a) 0.28 NA, and (b) 0.35 NA was used.

5. Determine the minimum sized line that can be printed by a projection aligner that has a lens with an NA = 0.28, uses a G-line radiation source, and has a coherence of (a) s = 0, (b) s = 0.7, and (c) s = ∞. Assume the resist being used has a γ = 2. Express your answer in μm. Repeat the problem for the case when the NA of the lens is increased to 0.35. Repeat the initial problem again if a resist is used which has a γ = 7.

6. Discuss the problems of (a) *ringing*, as the spatial coherence of the source is increased, and (b) *decrease of the depth of focus* as the NA of the lens is increased. How do these limitations effect the quality of the transferred image on a wafer?

7. Discuss the differences in the principles of operation of (a) scanning projection aligners, (b) reduction refractive step-and-repeat aligners, and (c) 1X step-and-repeat aligners.

8. Assume that the values of σ_{p1}, and σ_{p2} in a process line are ±0.1 μm and the value of σ_r = ±0.15 μm. Find the registration tolerance that must be incorporated by the circuit designers into the mask dimensions to insure that there will be a probabilty of < 0.1% that the edges of features patterned by two different mask layers will touch.

9. Show how the intensity pattern obtained from the automatic alignment marks shown in Fig. 19 can yield information about how well the mask is aligned to the patterns already present on a wafer surface.

10. Explain the difference between the *die-todie* and *die-to-database* inspection methods of masks (and reticles). Where does each of these inspection techniques find application?

11. Explain why *clear defects* in a mask or reticle are more difficult to repair than *opaque defects*. Why has the traditional *lift-off* technique for opaque mask repairs been found less than suitable for the repair of VLSI masks and reticles?

12. Describe the two types of inspection systems that must be used to completely inspect a mask or retcicle after it has been fabricated. Explain the use of *glass wafers* in reticle qualification.

14

ADVANCED LITHOGRAPHY

The resolution and registration (3σ) of optical lithography is limited to somewhat less than 1 μm, and ±0.3 μm, respectively. To extend the capability of the lithographic pattern transfer process beyond these limits, alternatives to optical lithography have been developed, including: a) electron beam lithography, which offers resolution of 0.375 μm, with ±0.05 μm (3σ) registration; b) x-ray lithography, 0.25 μm resolution with ±0.3 μm (3σ) registration; and c) ion beam lithography (~0.2 μm resolution). These advanced lithographic processes are the subject of this chapter. In Chap. 13, the use of electron beam lithography to produce photomasks was described. The present discussion will therefore emphasize the direct wafer exposure applications of electron beam lithography, although the equipment used to perform both tasks will be covered.

ELECTRON BEAM LITHOGRAPHY

Electron beam (or e-beam) lithography (EBL) is the process of forming circuit patterns by using a focused electron beam. Such beams can be readily scanned and accurately positioned on the substrate to expose radiation-sensitive material (e-beam resist). The early e-beam machines were modifications of commercial scanning electron microscopes.

Early workers quickly realized the potential advantages of EBL over optical lithography, including: a) EBL provides the ability to produce features less than 1 μm directly on a wafer, without the use of a mask, and under the control of a computer; and b) an e-beam can be used to accurately detect features on a substrate. This ability can be used to provide extremely accurate layer-to-layer registration. Features as small as 0.1 μm have been made with this technique. This is possible because, although electrons do possess wave-like properties, for the energies used in EBL systems, their wavelengths are on the order of 0.2-0.5Å. As a result, diffraction effects which limit resolution in optical lithography are avoided.

There are, however, limitations which have kept EBL from being widely incorporated into VLSI fabrication environment, including: a) resolution is limited by the fact that the electrons are forward scattered in the resist and backscattered from the substrate. This results in exposure of the resist outside the desired areas; b) resolution is also limited by a resist processing phenomenon termed *swelling*, which occurs upon development of negative e-beam resists; c) EBL is extremely slow when compared to optical projection and step and repeat systems; and d) EBL systems are 3 to 5 times more expensive than optical steppers, and generally require more expensive facilities.

Table 1. E-BEAM SYSTEM ACCURACY SPECIFICATIONS, for an AEBLE™150[1]

Minimum Feature Size	0.5 μm	Overlay Accuracy	≤ 0.15 μm, 3σ
Line Edge Roughness	≤ 0.08 μm, 3σ	Dimensional Accuracy	≤ 0.08 μm, 3σ
Field Butting Error	≤ 0.15 μm, 3σ	Address Structure	0.0156 μm (1 /64 μm)

The potential advantages of EBL, however, have prompted the commitment of significant development efforts towards overcoming its limitations. These efforts are beginning to pay off as machines with improved throughput times (e.g. 4-6 100 mm wafers per hour) are coming to market (e.g. the AEBLE™-150 from Perkin-Elmer)[1]. Table 1 lists the specifications for a state-of-the art EBL system used for direct-write on wafers.

Electron Beam Systems

EBL systems that have gained in popularity use scanned, focused e-beams which expose the resist in a serial manner. Figure 1 shows the main features in a simplified form of an EBL system[2]. There are five main sub-systems: 1) the electron source (or gun); 2) the electron column; 3) the mechanical stage; 4) the wafer handling system; and 5) the control computer.

To write submicron features, the e-beam must be generated, and then focused into a small diameter spot. To obtain short resist exposure times, the current density in the focused spot must also be high. The electron beam is produced by an *electron source*. There are two types of electron sources commonly used in EBL systems: a) thermionic sources; and b) field-emission sources. Note that electron sources are also discussed in Chap. 17, in the discussison on scanning electron microscopes. *Thermionic sources* utilize a material which is heated to the elevated temperatures at which electrons are emitted. Candidate materials for this application are tungsten (W), thoriated tungsten, and lanthanum hexaboride (LaB_6). The *W sources* have been the most widely used as they provide a stable current and exhibit excellent tolerance to vacuum conditions. The maximum brightness of such sources is 1-3 x 10^5 (A /cm^2) /steradian. The

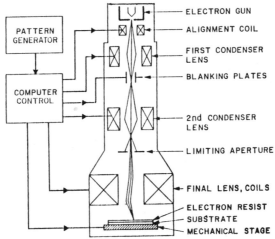

Fig. 1 Schematic of an electron beam machine. From Kern, *et al, Solid State Technol.*, **27**, 1984. p. 127. Reprinted with permission of Solid State Technology, published by Technical Publishing, a Company of Dun & Bradstreet.

thoriated tungsten sources have a lower work function, thereby providing currents comparable to standard W sources, but at lower temperatures. This leads to longer life and better beam stability. The fabrication of thoriated W sources, however, is difficult and their use requires very stable vacuum conditions. The LaB_6 *source* has even a lower work function than the previous two sources and can provide a brightness of 10^6 (A /cm^2) /steradian. LaB_6 sources, however, exhibit less stability and require better vacuum conditions than the other two sources. Nevertheless, they can emit over an extended area, which allows them to be used in shaped beam systems, and because of their advantages they are becoming the standard emitter for EBL systems. For example, the MEBES III system, which is a Gaussian beam type EBL system, uses a LaB_6 source, as do the variable shaped beam systems.

In *field emission sources*, a very high electric field at the sharp tip of an emitter extracts electrons, which form a fine Gaussian spot. These sources are brighter than the thermal emitters, having a brightness of 10^8 to 10^9 (A /cm^2) /steradian. They are difficult to fabricate, however, and require high vacuum (10^{-8}-10^{-10} torr) for effective operation. The Electron Beam Corporation direct-write system uses a thermal field emission source to achieve high brightness[3].

The *electron column* consists of (at a minimum): a) the electron source; b) various lenses used to focus and magnify or demagnify the beam; c) a beam blanker to turn the beam on and off; d) various apertures to limit and shape the beam; and e) a beam deflection system to position the beam on the substrate. Figure 2 shows a schematic of the electron column used in the Perkin-Elmer AEBLE-150 direct-write system. E-beam machines used in research applications must provide the smallest possible spot to achieve the high resolution (e.g. some machines are capable of producing beam spot sizes of ~500Å). On the other hand, e-beam systems dedicated to the production of photomasks with feature sizes of 1.0-4 μm can operate effectively with a relatively large beam diameter (e.g. 0.25-1.0 μm). Since most thermionic electron sources produce electrons from a cathode that is 10-100 μm in diameter, in order to achieve submicron spots, electron-optical demagnifying lenses are required to reduce the beam diameter from 100-1000 times. The lenses of the system are also designed to minimize aberration, which is one of the parameters that limits the resolution of the system. The price paid for reducing aberration is that the area over which the beam can be scanned is limited.

In order to write over an entire wafer (e.g. 150 mm diameter), a *mechanical stage* is used to position the substrate under the e-beam. The position of the stage must be accurately known at all times, and this position is usually controlled by a laser interferometer to better than 1 /64 μm. The velocity of the stage is important in determining the throughput of the system, because patterns can be written after the stage position has been determined. It is also possible to perform a registration at this time. On the other hand, it is possible to write the pattern "on the fly". The stage position must be accurately known at any instant so corrections can be made by the deflection system. The AEBLE-150 has a stage that achieves a velocity of 10 cm /sec, a 50 ms stepping time for 2 mm step, a positional accuracy of 0.5 μm, and no vibration >200 Hz.

An important part of an effective EBL system is its ability to automatically handle wafers, and then feed them into the beam as required. In this way a batch of wafers can be run unattended. Such a handling system must be able to pre-align wafers, and to introduce them into the high-vacuum environment of the stage without degrading the vacuum. The AEBLE-150 wafer handling system is capable of loading a 25-wafer cassette of 150 mm wafers. The wafer is taken from the cassette and transported to the wafer aligner, where is is aligned with respect to its primary flat, and then oriented along its x, y, and z axes. The wafer is next placed on a chuck, which is then loaded on a pallet on rollers that can be pushed through the airlock and loading chambers to the work chamber. Substrate exchange occurs in less than 30 sec. The wafer is held by electrostatic forces within the work chamber. When the exposure is complete, the pallet is

Column schematic

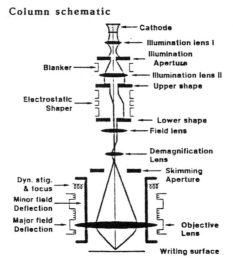

Fig. 2 Electron column used in the Perkin-Elmer AEBLE-150 system. Courtesy of Perkin-Elmer.

removed back through load and airlock chambers to the output cassette. A new wafer is then loaded on the pallet and chuck.

All of these sophisticated sub-systems must operate correctly, and in unison. This requires control by a computer system. The system operates in real time for control functions of the stage position and column. In addition, the computer controls the electron beam size, exposure time, stage position, and data processing for alignment and calibration. It can also monitor the column, beam quality (focus, current density, and alignment), and spot shape and size. The pattern data must also be rapidly transferred to the deflection and blanking systems. The data transfer rate is generally high enough not to limit the throughput of the machine. Higher performance systems are now available as a result of fast, high density computer memory that has become available at low prices.

Writing Strategies

The writing strategy of an EBL system includes: a) the shape of the beam; and b) the method of scanning the beam over the wafer. The substrate to be exposed is subdivided into a grid of addressable locations. Each of these locations represents a *pixel*. A pixel defines the minimum resolution element that can be generated by the specified electron dose. The pixels are joined to create pattern shapes, and the minimum observable pattern is one pixel exposed and one not exposed. If the pixel size is 0.5 μm x 0.5 μm, a 150 mm wafer would contain ~7.5 x 10^{10} pixels. To form an acceptable image in the resist, a minimum number of electrons, N_m, must impinge upon each exposed pixel (Fig. 3). For a given resist sensitivity, S (defined in units of C /cm^2), this minimum is given by:

$$N_m = \frac{S \, l_p^2}{q} \tag{1}$$

where l_p is the minimum pixel dimension, and q is the electron charge.

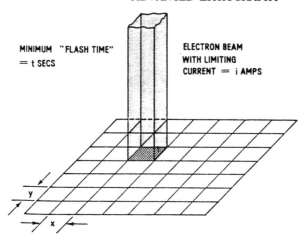

Fig. 3 Schematic of exposure of a unit address of resist surface. Reprinted with permission from Ref. 7, 1983, American Chemical Society.

Beam Shape

Two alternative types of electron beams are used for EBL: a) the Gaussian shaped round beam[2,4] (Fig. 4a); and b) the variable shaped beam[2,5] (Fig. 4b). The *Gaussian round beam* is typically four times smaller than the smallest pattern dimension. The *variable shaped beam* (VSB) is varied to match the feature size (i.e. larger pattern areas can be exposed by increasing

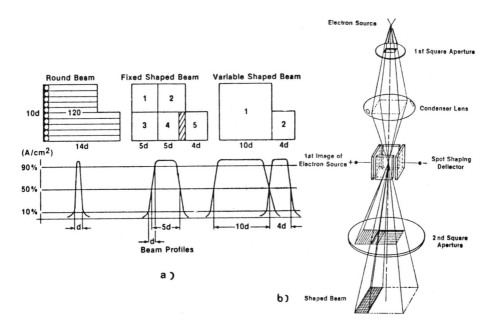

Fig. 4 (a) Pattern generation and beam profiles[28]. (b) Variable shaped beam forming principle[29]. Reprinted with permission of Solid State Technology, published by Technical Publishing, a Company of Dun & Bradstreet.

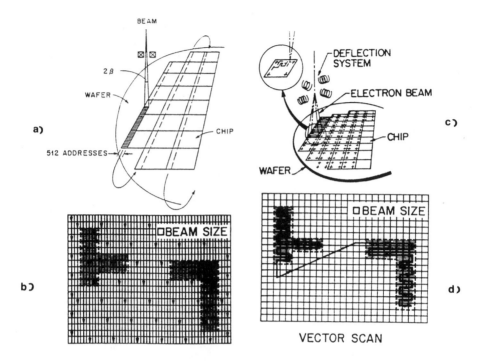

Fig. 5   (a) and (b) Raster scan exposure scheme. (c) and (d) Vector scan exposure scheme. Reprinted with permission of Ref. 7., 1983, American Chemical Society.

the size of the beam). On the average, the shaped beam simultaneously exposes about 30 pixels.

Throughput is higher with the VSB, but not by as much as is indicated by the ratio of the beam areas. This results from electron-electron interactions which impose a maximum limit on the current of a shaped beam. For beam currents greater than a few microamperes, the edge definition becomes limited by aberrations from such electron-electron interactions, rather than by a combination of beam brightness and deflection aberrations. The full brightness of the electron gun, therefore, cannot be used for shaped beams, but can be for the round beams.

The VSB, which is becoming the choice for high throughput EBL systems, is formed by illuminating a square aperture with a uniform flood of electrons. Its image is then projected onto a second square aperture. This produces a rectangular spot. The dimensions of the spot can be set by deflecting the image of the first aperture to impinge on the second aperture in the appropriate position (Fig. 4b). The AEBLE-150 system utilizes a VSB which can produce rectangles or 45° triangles of any size from 0.3-2.0 μm.

Scanning Systems

After the beam is focused and shaped it requires scanning (deflection) over the substrate. Deflection is generally accomplished electromagnetically, although electrostatic systems also exist. The AEBLE-150 utilizes magnetic deflection over a 2 mm x 2 mm field, while an electrostatic deflector provides high speed scanning over a 64 μm minor field.

The *raster technique* scans the beam over the entire substrate and is turned on and off according to the desired pattern[2,6,13]. The chip is broken down into stripes of typically 256 address units (AU), and deflected over that range. A laser controlled table moves the substrate along the AU stripe in the x-direction, while the beam is sweeping out one scan line in the y-direction. Figure 5a shows a schematic representation of this technique.

The *vector scan system* aligns the individual die or chip, then exposes it with a two-dimensional electronic scan that covers the entire area[2,6,13]. Since the scan area is smaller than a chip area, it is required to mechanically move the sample. This can be accomplished by step and repeat movement, or by continuous mechanical scanning. Figure 5b shows a schematic of the vector scan. The AEBLE-150 uses the vector scan mode where a major field is 2 mm x 2 mm, and nested. Within that field is a minor scan field of 64 μm x 64 μm.

Since a single chip is typically larger than the area scanned by the major field, continuity across field boundaries must be considered. The degree of misalignment across a field boundary is called the *butting error*. The AEBLE-150 boasts a butting error of <0.15 μm (3σ) for both the major and minor fields.

The registration of a layer can be accomplished in three ways: a) on a global basis; b) on a die-by-die basis; or c) on a field-by-field basis. In the *global system* the wafer is first accurately aligned, and then accurately positioned by the laser controlled table. This assumes extreme stability of the system, and flat, non-distorted wafers. In the *field-by-field* (and *die-by-die*) methods, special alignment marks are present which allow alignment at each major field (or die). These marks need to be compatible with the process, so they can be recognized by the detector system throughout the entire processing sequence.

Alignment marks may consist of either a high Z (atomic number) metal pedestal, or (more likely) a feature etched in the silicon or SiO_2. The alignment is achieved by detecting backscattered electrons from such marks. Along the edge of the mark, the electron signal is reduced because of scattering at various angles.

Electron Scattering and Proximity Effects in Resists

During e-beam exposure the electrons entering the resist are scattered (forward and back) by interaction with the atoms of the resist and the substrate. Most of the electrons are forward scattered through small angles (< 90°) from their original direction. This, in effect, broadens the beam. Some of the electrons experience large angle backscattering (e.g. >180°), causing them to return toward the surface. Figure 6 shows the results of a Monte Carlo calculation of trajectories for 100 electrons projected onto the x-z plane for 10 and 20 keV point sources[7]. The degree of forward and backscattering predicted by the calculation is illustrated in the pictorial representation. Figure 7 shows the effect of this scattering on a beam that strikes a resist-coated surface[8].

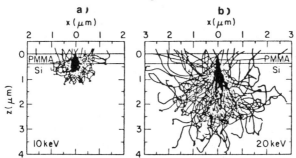

Fig. 6 Simulated trajectories of 100 electrons in PMMA. (a) Simulation for a 10 keV beam. (b) Simulation for a 20-keV beam[31]. Reprinted with permission of the Electrochemical Society.

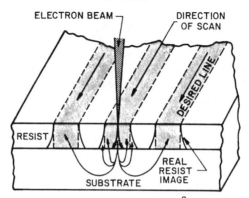

Fig. 7   Electron scattering effects in resist-coated substrate[8].   Reprinted with permission of Semiconductor International.

First, it is seen that the scattering broadens the incident electron beam.   Second, the electrons backscattered from the substrate return through the resist, contributing to additional resist exposure. This results in a developed image wider than anticipated from the beam diameter alone. Such scattering is thus one of the mechanisms that limits the minimum linewidth.   Since the scattering can extend over several microns, closely spaced patterns will receive electrons from the exposure of their neighbors.   This is termed the *proximity effect*.

The absorbed energy density in the resist is given by[8]:

$$E(r) = k \left[\exp^{(-r^2/\beta_f^2)} + n \frac{\beta_f^2}{\beta_b^2} \exp^{(-r^2/\beta_b^2)} \right] \qquad (2)$$

where k is a proportionality constant, n is the ratio of total energy deposited by the backscattered electrons to that deposited by the forward scattered electrons, β_f and β_b are the characteristic widths of the forward scattered and backward scattered Gaussian distributions, and r is the radial distance from the center of the spot. The parameters β_f, β_b, and n vary with energy, while β_b and n also vary with substrate material. The adjustment of these three parameters with exposure and substrate conditions is used for proximity effect corrections in device writing.

Figure 8 illustrates some other consequences of the proximity effect on a developed image. At point A (the center of a large exposed area), we see that there are energy contributions from

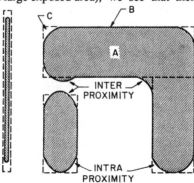

Fig. 8 *Inter* and *intra* proximity effects in e-beam exposure caused by electron scattering[8]. Reprinted with permission of Semiconductor International.

surrounding electrons. At the corners and edge of this large exposed pattern, however, the same total dose is not received. For example, at the edge (point B), only half the dose of point A is received, and at the corner (point C), only one fourth the dose of point A. As a result, upon development, which is optimized for the main pattern, the corners are not developed to their full design size. This phenomenon is referred to as an *intra-proximity effect*, and causes large and small feature sizes to print differently.

For the case of a long narrow line (left of Fig. 8), if the exposure and develop cycle is chosen to obtain the proper edge at B, the narrow lines will be too narrow. On the other hand, since backscattered electrons travel over large distances, adjacent patterns may exhibit cooperative exposure effects if they are closely spaced. This is known as an *inter-proximity effect*, and results in the bulging of patterns towards one another (Fig. 8). Under extreme conditions, bridging of the features can occur.

An understanding of these effects allows the dose to be properly compensated if the circuit pattern and the degree of scattering are identified. That is, the circuit can be broken into primitive shapes (rectangles) and the local dose altered for each shape. Thus, the developed image is produced with the proper dimensions. Programs are available which use Eq. 2 to control the dwell time of the beam in the local area, to adjust for these effects. A throughput penalty may be incurred as a result, since increased computer time will be necessary to partition and expose the subdivided resist features.

The use of multi-layer resist techniques has overcome many of the problems of proximity effects and linewidth control. The thick planarizing layer absorbs backscattering electrons, minimizing their effect on neighboring resist. Multi-level resist systems do, however, add processing complexity and create problems with registration mark recognition[9].

Electron Resists and Resist Development

Electron beam resists are materials in which chemical or physical changes are induced upon exposure to the high energy electron beam, which allows the resist to be patterned. Exposure to such electrons typically causes two types of changes in e-beam resist material: a) radiation-induced polymer cross-linking (in negative resists); and b) chemical bond breaking (in positive resists). Table 2 lists the various resists and compares their sensitivity, resolution, and etch resistance[10]. The sensitivity, S, for negative resists is defined as the electron dose $/cm^2$ required to retain 50% of the resist in the exposed areas, while for positive resists, it is the dose required to generate a solubility ratio of 50.

Negative electron resists are polymer chains which chemically cross-link upon absorption of energy from the e-beam to form a complex molecular network. The average molecular weight (MW) increases as a result of the cross-linking, until the material is no longer soluble in the solvents used for the starting polymer. Unfortunately, the solvents used to remove the unexposed resist result in swelling of the exposed area during development (as in the negative optical resists described in Chap. 12), and limits negative e-beam resist resolution to ~1.0 μm.

The most widely used commercial negative resists are poly(glycidyl methacrylate) monopolymers (OEBR 100, manufactured by Tokyo Okha), and copolymers (COP, manufactured by Mead), which both exhibit sensitivities of 0.7 μC $/cm^2$, and resolutions of 1.5 μm. Recent activity has focused on sensitive poly (styrene) derivatives which have high contrast, require no post-curing, and exhibit high dry etch resistance (e.g. poly (chlorostyrene) and poly (iodostyrene) [RE 400 DN, manufactured by Hitachi]). The formulation of a sensitive (0.5 μC $/cm^2$) resist with 1 μm resolution capabilities has been reported by Hewlett-Packard [poly-(chloromethylstyrene)][10].

Table 2. THE PERFORMANCE OF ELECTRON BEAM RESISTS[10]

	Image Type	Trade Name	Manufacturer	Sensitivity (μCoul/cm^2)	Resolution (μm)
Poly(glycidyl methacrylate)	-	OEBR100	TO	0.7	1.5
PGMA co poly(ethyl acrylate)	-	COP	MD	0.7	1.5
Poly(chloromethylstyrene)	-		HP	0.5	1.0
Poly(iodostyrene)	-	RE4000N	H	1.8	1.0
Poly(methylmethacrylate)	+	PMMA	N	80	0.05
PMMA co poly(acrylonitrile)	+	OEBR1030	TO	30	< 0.5
Poly(trifluro-α-chloroacrylate)	+	EBR 9	T	15	< 0.5
Poly(fluroalkylmethacrylate)	+	FBM	DK	??	< 0.5
PMMA Xlinked	+	PM	M	40	< 0.5
Poly(butene sulphone)	+	PBS	MD	2.0	0.5
Novolac - diazoquinione resist	+	AZ 1350	AZ	15/50	0.5
Novolac-poly(methylpentene sulphone)	+	RE 5000P	H	5.0	< 0.5
Poly(styrene)-tetrathiofulvalene	±		IBM	8.0	< 0.5
Inorganic GeSe /AgSe	-		N	500	< 0.5
Poly(siloxane)	-		IBM	1.5	< 0.5

Company codes: TO = Tokyo Okha, MD = Mead, HP = Hewlett-Packard, TS = Toyo Soda,
N =Numerous, DK = Daikin Kogyo, M = Microimage, AZ = AZ, H = Hitachi.

Positive electron resists are polymer chains in which chemical scission occurs upon exposure to the e-beam. The fractured chain has an average MW lower than the unexposed material, and may then be removed by developer without significantly attacking the unexposed areas. Successful implementation of this process, however, requires careful selection of solvents. The process engineer must find the best combination of solvents which results in the highest differential solubility ratio between the exposed and unexposed regions.

As with positive optical resists, positive electron beam resist development can be treated as an etching phenomenon. Figure 9a shows the resist profile of the most commonly used positive e-beam resist, poly(methylmethacrylate), or PMMA, for various develop times. To obtain a good profile, the etch rate of the exposed area should be maximized with respect to that of the unexposed areas. The resist solubility is characterized by the fragmented molecular weight of the exposed molecules, M_f, which in turn depends on the absorbed energy density in the resist, E. The dissolution rate, R, for exposed resist given by:

$$R = R_o + \beta M_f^{-\gamma} \qquad (3)$$

where R_o is the removal rate of very high molecular weight materials (e.g. material in unexposed regions). M_f is found from:

$$M_f = (g E) / (\rho N_o) \qquad (4)$$

and where: ρ is the resist density; N_o is Avogadros number; and g is the radiation yield (i.e. the number of scissions per eV of absorbed energy). Table 3 lists values of the relevant parameters (R_o, α, and β) for some developers of PMMA resist. The parameter R_o describes the removal

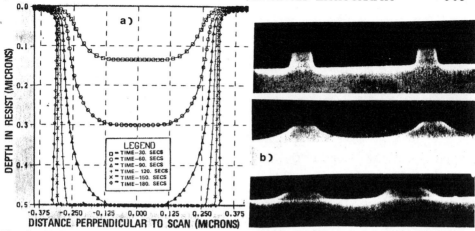

Fig. 9   (a) Developed contours at 30 s intervals at a nominal half micron exposure at 20 keV. Nominal development occurs at 120 s[8]. Reprinted with permission of Semiconductor International. (b) Effect of developer strength on line cross-sections and resulting resolution in a typical negative resist[32]. Reprinted with permission of the American Physical Society.

rate for resist of very high molecular weight (i.e. in the unexposed areas), and this rate should be as small as possible. The parameter γ represents the contrast of the resist. Note that its value should be as large as possible. As discussed in Chap. 12, higher contrast resists result in better resolution and critical dimension control.   There is, however, a limit on increasing γ, since β also increases, and therefore the overall sensitivity of the resist decreases. The actual developed resist profile depends on: 1) the three-dimensional energy density distribution; 2) the amount of electron flux or exposure, in C /cm^2; and 3) the solubility rate of the resist /developer system. Significant effort is continuing to improve e-beam resists in terms of their sensitivity, resolution, adhesion to the substrate, and dry etch compatibility. Three review articles on recent developments in e-beam resist technology are given in the reference list[10,11,12].

   The most widely used positive e-beam resist is PMMA, which has a sensitivity of 80 μC /cm^2. The resolution of this resist is better than 500 Å, since it does not swell. On the other hand, since it is ~100 times less sensitive, it requires exposures ~100 times longer than the negative resist, COP. Commercial products based on PMMA have improved the sensitivity of the original formulation, without losing its favorable dissolution properties (from 80 μC /cm^2 to 15 - 30 μC /cm^2).   An alternative material, poly (butene-1-sulfone) has a high sensitivity (~2 μC /cm^2 at 20 kV), but exhibits poor dry etch resistance and poorer resolution (~0.5 μm).

   Inorganic resists, consisting of GeSe (onto which AgSe is evaporated), can also be used for EBL. Upon irradiation, the Ag migrates into the GeSe layer to form a material that may be wet

Table 3. SOLUBILITY RATE PARAMETERS for PMMA at ROOM TEMPERATURE[8]

Developer	R_0 (Å / min)	β	α
MIBK	84	3.14×10^8	1.5
1:1 MIBK:IPA	0	6.70×10^9	2.0
2:3 MIBK:IPA	0	9.37×10^{12}	2.75
1:3 MIBK:IPA	0	9.33×10^{19}	3.86

or dry developed. The resultant pattern in the resist forms a negative image. High resolution at a sensitivity of 100 μC /cm^2 has reportedly been achieved.

Optical resists such as AZ 1350 can act as a positive or negative e-beam resist depending on the developer. Work also continues on dry developable resists, plasma developed resists, and self-developable (ablative) resists[11,12].

In order to predict post-development resist profiles, a *development model* must be coupled with the *exposure model* previously presented. For example, Fig. 9b shows the effect of developer strength on the developed image in COP resist. The 3:1 developer mixture develops the unexposed resist, but is not strong enough to remove the tail of the resist that was cross-linked by scattered electrons. The 4:1 developer mixture produces good images, while the 5:1 and 7:1 mixtures result in swelling of the resist (which in effect broadens the line). The broadening reduces resolution, and may result in bridging of adjacent features. This implies that tests must be conducted to determine the correct exposure /developer combination for a given e-beam resist to establish a reproducible process.

X-RAY LITHOGRAPHY

X-ray lithography is a process for imaging a full-field mask onto a wafer by using x-rays of very short wavelength (e.g. 4-50Å). At shorter wavelengths, tranmission through the "opaque" mask regions is too great, while at longer wavelengths, transmssion through the clear portions of the mask is too low. The ultra-short wavelengths eliminate the diffraction effects that limit the resolution of optical proximity systems. Although x-ray lithography was restricted to the research environment for a lengthy period, commercial x-ray systems have finally begun to become available[14]. A schematic of an x-ray lithography system is shown in Fig. 10.

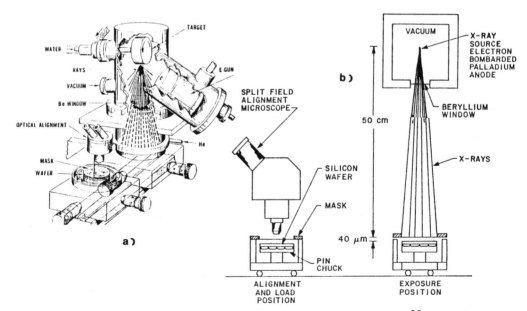

Fig. 10 (a) Line drawing of complete x-ray exposure system with alignment[30]. (b) Schematic of an x-ray lithographic system[33](©1983 IEEE). Reprinted by permission of American Phys. Soc.

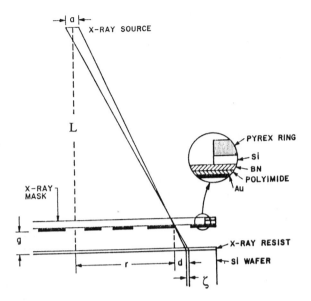

Fig. 11 Geometric effects in x-ray lithography. Insert shows x-ray mask[33]. (© 1983 IEEE).

The major promise of x-ray lithography is its potential for attaining high resolution (~0.5 μm) together with high throughput (e.g. 40-50 150 mm wafers /hr). The technology also offers the advantages of eliminating both the depth of focus problems, and the reflection and scattering effects that are associated with optical and e-beam lithography processes. X-rays also provide high aspect ratios in resist, and are immune to defects caused by organic particulates present on the mask (because x-rays in the 15-40 Å wavelength region are not appreciably absorbed by materials formed of low atomic number elements).

As x-rays are not capable of being focused into fine spots, the lithography process involves a form of shadow printing. As such, there are some geometrical effects associated with shadow printing that limit the resolution of x-ray lithography. For example, consider an x-ray source of diameter, a, at a distance, L, from the mask, and which is separated a distance, g, from the coated wafer as shown in Fig. 11. A *penumbral blur*, ζ, occurs on the resist image given by :

$$\zeta \;\; = \;\; (g\,/\,L)\,a \tag{5}$$

For a typical x-ray system, where a is 3 mm, g is 20 μm, and L is 26 cm, a ζ of 0.23 μm results. This effect can be tolerated if the intensity of the exposure in the blur is not decreased below the threshold intensity required for adequate exposure. In addition, the effect can in fact be advantageous, since, due to blurring, images of x-ray absorbing particles on the mask of ~0.45 μm or smaller, will not be printed.

Another limiting geometrical effect is *lateral magnification error*, which results from x-ray divergence from a point source and the finite mask-wafer separation. The image on the wafer is shifted by an amount, d, given by:

$$d \;\; = \;\; (g\,/\,L)\,r \tag{6}$$

where r is the radial distance measured from the center of the wafer (Fig. 11). The error is zero at the center of the wafer, but increases linearly with distance toward the edge. For large wafers this can result in *run-out* effects as large 5 μm near the edge of the wafer. Compensation for these errors can be achieved by adjusting image positions on the mask. The real problem is created as a result of non-flatness of the mask and the wafers. To avoid such problems, flatness for both masks and wafers must be held to better than 1 μm. A way around maintaining such tight requirements is to reset the mask /substrate gap prior to each exposure. However, local distortions of the Si resulting from processing, do not allow the gap to be set equally over the entire wafer surface. For wafer diameters larger than 75 mm, step-and-repeat on the wafer would improve this condition. The practical use of step-and-repeat systems, however, requires that resists of with less than 1 mJ /cm^2 sensitivity, and much brighter x-ray sources become available. The higher sensitivity resists and brighter sources would allow full wafer exposure systems to operate at a longer L, while still maintaining reasonable exposure times, thereby reducing both the penumbral blur and lateral magnification error. A last major technolgical hurdle that must be overcome before x-ray lithography can be considered for production process applications, is the *development of an effective mask fabrication technology*.

X-Ray Sources

X-ray tubes are the simplest x-ray sources. In such tubes, high energy electrons (e.g. 25 keV) are focused onto a metal (e.g. a palladium) target. X-rays (characteristic and background) are emitted with an efficiency of less than 1%, while the remainder of the energy goes into heating the target. The x-ray power produced is limited by the abilty of the target to dissipate this heat. Such heating requires that the target be water-cooled. The most widely used x-ray tube design has an the inverted cone geometry[15]. This provides a large surface area with a minimum and symmetric projected spot size (Fig. 12). Sources of this type range from 4-7 kW in a 3 mm spot. By utilizing high velocity coolants, the heat transfer can be improved, and this allows targets to operate with power densities of 60 kW /cm^2. Further enhancements to this technology may result in producing sources with 6 kW in a 1.5 mm spot. Such power densities are

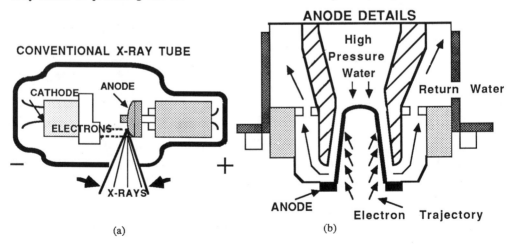

Fig. 12 (a) Conventional x-ray tube. (b) Stationary-anode x-ray tube uses a cone shaped target with bombarding electrons directed into the conical cavity[34]. Reprinted with permission of Semiconductor International.

adequate for full wafer exposure, but not for step-and-repeat use.

The utilization of rotating anodes allows delivery of considerably higher power, but gradual decreases in output are observed as carbonaceous material deposits on the anode. This problem has been eliminated through the utilization of ferrofluidic seals. The complexity, cost, and spot size, however, are some of the remaining drawbacks of this technology.

An alternative approach to achieving high output x-ray sources is *synchrotron radiation*. Synchrotron radiation is emitted by electrons in response to their radial acceleration, while they are maintained in circular orbits within a storage ring or synchrotron. The x-rays from this device are nearly collimated (i.e. no geometrical effects) with flux densities in excess of 10-20 mW /cm^2 at the wafer plane. The radiation beam uniformity from this source is in the 10-50Å range, which is strongly absorbed by the thin absorber patterns of the x-ray masks (Au). The major disadvantage of the synchrotron source is cost, which may exceed $10 million for a multiport system. Furthermore, if system problems caused all ports to be shut down, excessive downtime could make the operation of multiport systems extremely expensive.

A third approach is the use of *laser plasma sources*[16]. In these sources, either a pulsed IR laser or UV excimer laser is used, with pulse widths varying from 50 psec to 10 nsec. The beam is focused on the target, where it creates a plasma of sufficiently high temperature to produce continuous and characteristic x-ray radiation. A 12% conversion from laser power to useful x-rays has been obtained for $Cu\text{-}L_\alpha$ radiation, but the efficiency decreases with increasing Z of the target. The emitted energy density can be as high as 6.3 mW /cm^2, which is some 50 times brighter than electron impact sources. Another major advantage of this approach is that the power supply can reside in a remote location relative to the aligner, thus saving valuable clean room space and avoiding problems due to electromagnetic interference. Presently, work is being performed to increase the pulse repetition rate while reducing peak power, and to decrease the extent of debris deposited on the x-ray window by firing the laser in synchronism with a shutter. The debris is generated from target ablation as it is struck by the laser.

X-Ray Masks

The fabrication of suitable masks has always been, and continues to be, the most difficult challenge of x-ray lithography. The *mask blank* consists of a thin layer of x-ray absorbent material (usually gold), supported on a substrate. Note that in optical masks the absorber is a layer of chrome, and the subtrate is glass. A variety of materials have been investigated to serve as the substrate for the absorber layer. The leading contender to emerge from the group is the multilayer structure consisting of boron nitride (BN) and polyimide. This structure was developed at Bell Laboratories[17], and x-ray blanks using the technology are being commercially manufactured[18]. Such substrates provide good x-ray transmission, strength, inertness, and long term stability. Figure 13 outlines a set of processing steps involved in making such blanks for x-ray masks.

A silicon wafer is used as the starting material. The wafer is coated with BN by LPCVD techniques. The central area of the wafer backside is next etched back to silicon by plasma etching. The wafer is then bonded to a pyrex ring, and a polyimide coating is spun on the front side and cured. The polyimide surface is then sputter deposited with a multilayer metal (absorber) structure [e.g. Ta (300Å) /Au (6000Å) /Ta (800Å)]. The exposed silicon is then wet-chemically etched away, leaving the absorber and polyimide film, which constitute the mask blank.

A *subtractive process* can be used for pattern formation on the mask. A pattern is written using e-beam lithography into a photosensitive coating. The first Ta layer is etched by reactive ion etching, using a CF_4 plasma. The Au layer serves as an etch stop. The resist film

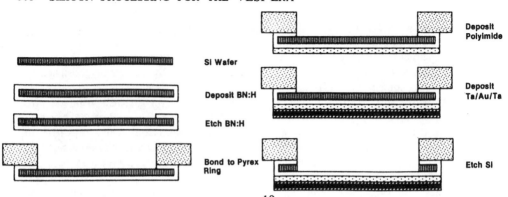

Fig. 13 X-ray mask blank processing sequence[19]. Reprinted with permission of Solid State Technology, published by Technical Publishing, a company of Dun & Bradstreet.

is then stripped, and the 800 Å Ta film serves as the mask for subsequent pattern transfer. The 6000Å Au layer is sputter etched or ion milled in an Ar-O_2 atmosphere. The O_2 is used to increase the selectivity of the etch, so that the Au etches more quickly than the Ta. The bottom 300Å Ta film serves as an etch stop to the Au etching, and also as an adhesion layer. The final step is to remove the Ta by reactive ion etch in a CF_4 plasma. The Au film serves to absorb the x-ray radiation in regions where it has not been etched away. A PMMA film then coats the mask to suppress photoelectron and Auger emission during exposure. The sputter etch process results in angles of the Au film of 65-73° because of *faceting* (see Chap. 10). As a result of this effect, the minimum line spacing in 0.6 μm Au films is limited to ~0.5 μm. This thereby limits the resolution of such masks. Reactive ion milling, or Ar /O_2 ion milling, can result in vertical sidewalls in the Au film, potentially allowing higher resolution masks.

Additive patterning (or *liftoff*) can also be used to produce Au lines with more vertical side-walls, thereby improving the resolution (Fig.14). This is achieved by the electroplating of Au through a vertically walled stencil. (See Chap. 15 for additional information on *Liftoff*.)

PMMA films 1.0 μm thick have been patterned with direct e-beam writing to produce 0.5 μm lines and spaces. Bilevel and tri-level processing techniques have been used to create high

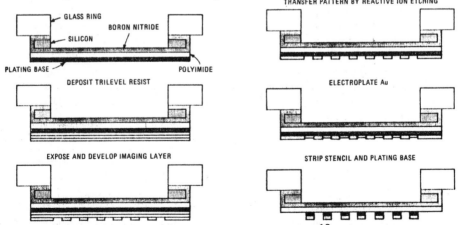

Fig. 14 Additive process for pattern formation on an x-ray mask[19]. Reprinted by permission of Solid State Technology, published by Technical Publishing, a company of Dun & Bradstreet.

resolution stencils. Tri-level processing has also met with considerable success in forming such stencils. A planarizing layer 1.2-1.5 μm thick, and a 750 Å Ta transfer layer (followed by a thin imaging resist) is now commercially available for making additive x-ray masks[18]. (See Chap. 12 for additional information on multilevel resist processing.) The plating base on which the electrodeposited Au grows typically consists of a 350 Å Ta top layer, which serves as a clean and flat Au base during the tri-level resist patterning. An initial 200 Å layer of Au is sputter deposited to provide the necessary conductive path during plating. The Ta layer serves as an adhesion layer between the Au base and the polyimide /BN membrane, since Au adheres very poorly to the polyimide /BN material. The Au is typically plated from commercially available solutions such as BBT 200 (from Sel-Rex Plating Systems). Low stress films are generally achieved during plating of the Au. After the stencil is removed by CF_4 etching, the mask is ready for use. Repair of x-ray mask defects is difficult. Excess material can be removed by focused ion beams, while pinholes require the use of laser activated CVD for selective deposition.

The Au x-ray absorber must possess adequate thickness to block the incident radiation in the desired locations. The thickness of the absorber layer causes the equal lines and spaces to be projected at oblique angles, and become asymmetric as the features are removed from the center of the mask. Thus, absorber layers should be restricted to their minimum effective thickness.

The alignment of subsequent layers in x-ray lithography generally utilizes optical techniques. In addition to x, y, and θ alignment, the mask-wafer gap must also be set correctly. Alignment techniques are discussed in Ref. 19. Values of 2σ misalignments of 0.3 μm over 100 mm wafers have been obtained. Since e-beam lithography is required for fabricating x-ray masks, improvements in resolution, image size control, and placement accuracy in e-beam technology will be required to realize the full potential of x-ray lithography.

X-Ray Resists

X-ray resists must satisfy contradictory requirements. That is, on one hand a high sensitivity to ionizing radiation is necessary for exposure speed, while on the other, good etch resistance in plasma environments is required after exposure[20]. For conventional and pulsed plasma sources, very low x-ray fluxes (10-100 μW /cm^2) mandate the use extremely sensitive resists. With synchrotron sources, conventional electron, or photoresists can be utilized.

In order to formulate x-ray resists with higher sensitivity four approaches have been taken. They are: 1) incorporation or more highly absorbing groups; 2) use of more reactive groups; 3) use of higher molecular weight polymers; and 4) new formulations which exhibit amplification in the exposure or development steps. An attempt at incorporating all four of the above approaches has resulted in the development of a polymer named poly (2,3-dichloro-1-propyl-acrylate), which operates as a negative resist. It is mixed with COP for adhesion, and is termed DCOPA. This resist material is generally used with tri-level processing, and has a sensitivity of 14 mJ /cm^2 and a resolution of ~0.8 μm. The limit of resolution is due to resist swelling during development. The resist must also be exposed in a 0.3% oxygen in nitrogen ambient for optimum resolution. A new negative resist, which shows reduced swelling (called PSTTF)[21], boasts a sensitivity of 50 mJ /cm^2 and 0.5 μm resolution at the Al K_α wavelength. It would be suitable for the x-rays output of plasma sources, but is still too slow for electron-impact sources.

In the positive resist realm, long exposure times are required for adequate exposure. Work by Wilson and Ito[22] incorporated chemical amplification to increase sensitivity. During exposure, a catalytic group is formed which enhances the polymer cleavage. Resolution down to

Source type & Schematic / General Source Characteristics	Brightness (β) (A cm^{-2} sr^{-1})	Energy spread ($\frac{\Delta E}{E}$) (eV)	Angular current density	Virtual Size	Typical Ions
(a) Gas-field ion source — 4.2 keV, H$_2$, Extractor electrode, Ions	10^6	1	10-50 µAsr^{-1}	≈10Å	H$_2^+$
(b) Liquid metal ion source — Liquid metal, Flowing liquid metal, Taylor cone, Extractor electrode, Ions	10^6	5-10	10-50 µAsr^{-1}	100-500 Å	Li Ga Au
(c) Micro plasma ion source — Gas source, Plasma, Extractor electrode, Ions	10^3	5	3 mAsr^{-1}	50 µm	H He Ar

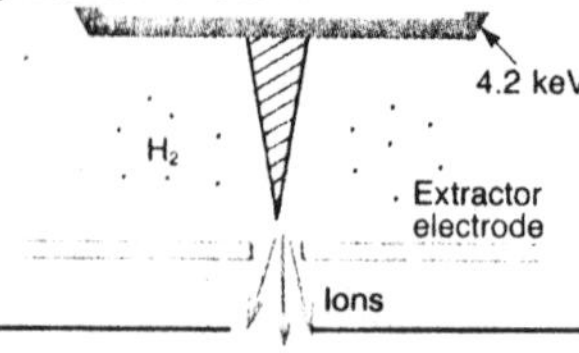

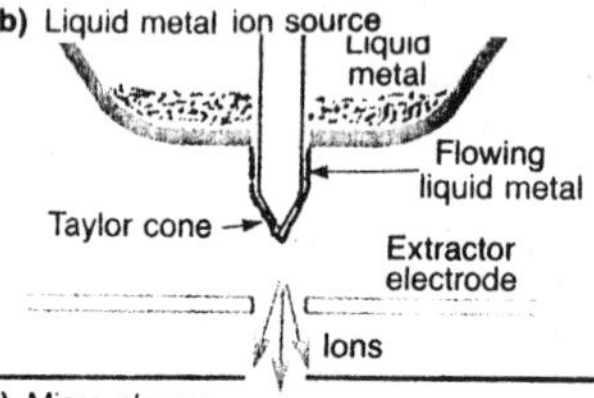

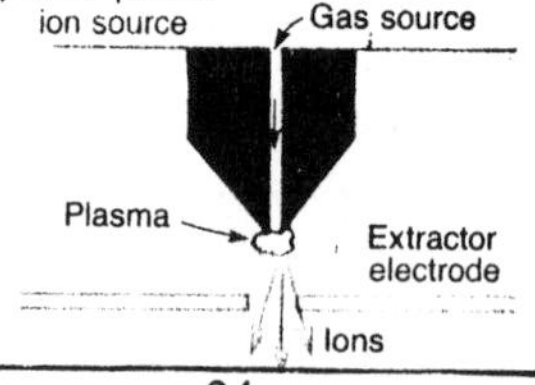

Fig. 15 Basic ion sources[24]. Reprinted with permission of Semiconductor International.

0.5 µm has been achieved at x-ray doses approaching 10 mJ/cm^2.

Work continues on developing new resists to support conventional electron bombardment sources. High sensitivity negative resists have been prepared but resolution is limited (>0.75 µm), and multi-level processing is required. The positive resists have low sensitivity, but outstanding resolution can be obtained with plasma or synchrotron sources.

ION BEAM LITHOGRAPHY

Ion beams can be used in the focused ion beam (FIB) and showered ion beam (SIB) modes[23]. Etching, sputter deposition and ion implantation are all examples of processes that use SIB technology. Ion lithography and direct ion implantation are examples of FIB technology. The application of FIB to silicon processing is in its infancy, but significant research activity is continuing to be conducted at major corporate laboratories and universities in the U.S., Japan, and Europe in an attempt to capitalize on the potential advantages of this technique.

A typical FIB system (Fig. 15) consists of an ion source, a blanking system, deflectors to guide the beam over the wafer, and an electrostatic lens for imaging the beam-defining aperture over the wafer[24]. The advantages of ion beams compared to electron beams are: a) since ions are more massive they are less prone to backscattering, and attendent resolution limiting proximity effects; and b) ion beam resists promise more sensitivity than e-beam resists. These advantages suggest that higher resolution limits are possible with ion beam lithography. On the other hand, since ion beam penetration in resist is only 300-5000Å, tri-level processing must be used.

A FIB system is characterized by its spot size, spot current, deflection frequency, field size, and writing speed. The larger mass of the ions limits focusing with an electrostatic lens, and also produces lower deflection speeds than electron beam systems.

A major hurdle in this technology is the availability of reliable ion sources. Three types of sources are currently used: a) gas field sources; b) liquid metal sources; and c) plasma sources. *Gas field sources* show promise for ion lithography, *liquid metal sources* allow almost any ion to be used, and *plasma sources* can operate at high efficiency[24]. Resist materials exhibit a higher sensitivity to ions than to either electrons or x-rays[25,26]. For example organic polymers have from 10 to 100 times more sensitivity to ions than to electrons. The resolution of these resists, however, is limited by swelling, contrast, and low dose tails. Secondary electrons that are backscattered by ions have very low energies (~15eV) and ranges of only a few angstroms.

Submicron features have been exposed and developed in 0.25 µm thick substituted polystyrene films with focused 140-160 keV Si^+ beams, and the pattern transferred to the underlying layer. Inorganic resists such as $GeSe_2$ with a thin AgSe (~100 Å) surface layer can be exposed by low energy He^+, N^+, Ar^+, Xe^+ and Ga^+ ions. A self developing (vapor development) resist (nitro-cellulose), has been described. Films of this material completely decompose when exposed to an ion beam above a certain dose threshold.

A potentially significant application of FIB technology is that of direct ion implantation, or doping without the use of masks or resists. This could eliminate many of the defect causing steps related to the doping process. Finer geometries are possible because the shadowing problems characteristic of broad-beam implantation through a mask can be eliminated. Other advantages could include: a) the production of lateral doping profiles; b) mixing of different types of devices on the same wafer; and c) low temperature annealing characteristics resulting from the high current densities associated with the beam. The main drawback of the technology is the low throughput (~1% of broad beam ion implantation), and low reliability of liquid metal sources.

Another useful application of ion beam technology is the repair of optical and x-ray masks. Opaque defects can be removed by sputtering away excess metal, while clear defects can be eliminated by etching a fine diffraction pattern into the mask (thereby rendering that region opaque).

The use of showered ion beam (SIB) lithography has taken two approaches. In the first, a broad ion beam illuminates a metal foil mask. The beam is then demagnified by a factor of 10, and projected onto and stepped across the wafer. In the second approach, a 3-6 µm thick <110> silicon membrane, with 1000 Å patterns, is used as the mask. The resist-coated substrate is exposed through the silicon foil using a collimated He^+ beam. The technique utilizes the dechanneling effect of the thin Au layers, as well as the difference in energy loss for the random and channeling directions.

REFERENCES

1. A. Carroll, J. Freyer, "Measuring the Performance of the AEBLE-150", *Proc. SPIE, Electron-Beam X-Ray & Ion Beam Techniques for Submicron Lithography IV*, Vol. 537, 1985, p. 25.

2. D.R. Herriot, "Electron-Beam Lithography Machines", in G.R. Brewer, (ed.), *Electron-Beam Technology in Microelectronic Fabrication*, Academic Press, New York, 1980.

3. W.R. Livesay, *et al*, "A Process-Compatible Electron Beam Direct Write System", *Solid State Technology*, Sept. '83, p. 137.

4. A.N. Broers, "The Submicron Lithography Labyrinth", *Solid State Technol.*, June '85, p. 119.

5. H.C. Pfeiffer, "Direct Write Electron Beam Lithography - A Production Line Reality", *Solid State Technology*, Sept. '84, p. 223.

6. A.D. Weiss, "E-beam Lithography: A Story of Dual Identities", *Semi. Intl.*, Feb. '84, p. 54.

7. L.F. Thompson and M. Bowden, "Lithographic Process: The Physics", in L.F Thompson, C.G. Wilson, and M.J. Bowden, Eds., *Introduction to Microlithography*, ACS Symposium

Series No. 219, American Chemical Society, Washington, D.C., 1983, Chap. 1.

8. J.S. Greenreich, "Process Hurdles in Electron-Beam Exposure of Sub-Micron Patterns", *Semiconductor International*, April '81, p. 159.

9. J.S. Greenreich, *Proc. 9th Intl. Conf. Electron and Ion Beam Sci. and Technol.*; ed. Bakish, Electrochem. Soc., Pennington, N.J. (1980), p. 282.

10. M.P.C. Watts, "Electron Beam Resist Systems ", *Solid-State Technol.*, Feb. '84, III.

11. E. Roberts, "Recent Developments in Electron Resists",*Solid-State Technol.*,June '84, p. 135.

12. Y.Takahashi,"Evaluation of Commercial E-Beam Resists in Japan",*Semicon. Int'l.* Dec 84 p91

13. R.K. Watts, "Electron Beam Lithography", in N.G. Einspruch, Ed., *VLSI Handbook*, Academic Press, Orlando, Fla., 1985, Chap. 20, p. 351.

14. P. Burggraaf, "X-Ray Lithography & Mask Technology", *Semiconductor Int'l*, Apr. '85, p. 92.

15. J.R. Maldonado, *et al., J. Vac. Sci. Technol.*, 16, 1942 (1979).

16. R. Byer, *et al., Proc. SPIE, Vol. 448*, p. 2 (1983).

17. D. Maydan, *et al, J. Vac. Sci. Technol.*, 16, 6 (1979).

18. A.P. Neukermans, "Current Status of X-Ray Lithography" *Solid State Technology*, Part 1, Sept. '84, p. 185, Part II, Nov. '84, p. 213.

19. A.R. Shimkunas "Advances in X-Ray Mask Technology",*Solid State Technol.*, Sept'84, p 192.

20. G.N. Taylor, "X-Ray Resist Trends", *Solid State Technology*, June '84, p. 124.

21. F.B. Kaufman, *et al*, Preprints, *ACS Div. Org. Coatings and Plastic Chem.*, 43, p.375 (1980).

22. H. Ito and C.G. Wilson, *Polymer Engrg. and Science*, Vol. 23.

23. I. Brodie and J.J. Muray, *The Physics of Microfabrication*, Plenum Publ. Co., New York 1983.

24. J.J. Muray, "Physics of Ion Beam Wafer Processing", *Semiconductor Int'l.*, April '84, p. 130.

25. J.E. Jensen, "Ion Beam Resists", *Solid State Technol.*, June '84, p. 145.

26. M.K. Shearer and G. Cogswell, "Focused Ion Beam Technology: Reviewing Applications Research", *Semi. Intl.*, April '84, p. 145.

27. "A New Mask Could Finally Open Up X-Ray Lithography", *Electronics*, Sept 16, 1985, p. 48.

28. R.D. Moore, "EL Systems", *Solid State Technol.*, Sept. '83, p. 127.

29. B.P. Pywczyk and A.E. Williams, "High Throughput Variable Shaped Electron Beam Lithography", *Solid State Technol.*, Sept. '83, p. 145.

30. D. Maydan, "X-Ray Lithography for Microfabrication", *J.Vac.Sci. Technol.* 17, 1164 (1980).

31. D. Kyser and K. Murata, "Monte Carlo Simulation of E-Beam Scattering and Energy Loss", *Proc. 6th Intl. Conf. on Electron and Ion Beam Science,* 1974, Electrochem. Soc.

32. E.D Feit, *et al, J. Vac. Sci. and Technol.*, 16, 391, (1979).

33. M.P. Lepselter, *et al.*, "A System Approach to 1-μm NMOS", *Proc. IEEE,* 71, 640, (1983).

34. S. Harrell, "X-Ray Source Technology for Microlithography", *Semi. Int'l.*, Sept.' 83, p. 74.

PROBLEMS

1. Calculate the number of electrons that must impinge upon each pixel if the minimum pixel dimension is 0.5 μm x 0.5 μm, and the electron resist being used is (a) COP, and (b) PMMA (see Table 2).

2. Explain why a shaped beam promises higher throughput than a Gaussian beam in e-beam lithography.

3. Expain how a *butting error* arises in electron beam lithography.

4. Explain why point C in Fig. 8 is not completely exposed, thereby causing the pattern in the beam resist to be smaller than desired at that point.   Explain how intra-proximity effects can cause the space between metal lines patterned with a positive e-beam resist to be smaller than the

dimension specified in the design.

5.   Explain how the penumbral blur in x-ray lithography causes even x-ray absorbing particles present on the mask, with a size of <2X the penumbral blur value, not to be printed.

6.   Elaborate on the problem of *run out* that can occur in x-ray lithography. That is, expalin why it arises, and give techniques that could reduce its effects.

7.   What are the main factors that limit the resolution in x-ray lithography? In addition to resolution, explain why alignment in x-ray lithography is difficult. (i.e. How is alignment performed in x-ray lithography?)

8.   Advances in e-beam, x-ray, and ion beam lithography will improve the capabilities of these processes beyond those reported in this text, as time goes on. Thus, it is valid to ask the reader to determine the state-of-the-art performance specifications of these techniques at the time this problem is assigned.

15

WET PROCESSING:

CLEANING, ETCHING, and LIFTOFF

In this chapter we describe a variety of wet processes in which wafers are immersed in liquid reagents to achieve some beneficial fabrication effect. These include: a) wafer cleaning procedures; b) wet etching techniques; and c) *lift-off*, a patterning process wherein material is *additively deposited* on the desired locations of a wafer surface. We also introduce terminology associated with patterning technology, information applicable to the subjects covered in this chapter as well as to Chap. 16, which deals with *Dry Etching*.

WAFER CLEANING

Scrupulously clean wafers are critical for obtaining high yields in VLSI fabrication. Wafer cleaning is a complex subject as there are many possible kinds of contamination. In this section, the contamination types are listed, and then methods and equipment to remove them are described.

It should also be emphasized that if a source of contamination exists it may be more effective to eliminate the source of contamination, rather to remove the contaminant after it contacts the wafer. While cleaning procedures can remove an immediate problem, they represent additional processing steps which can possibly lead to even more contamination. In many instances eliminating contamination merely involves cleaning of process equipment.

Sources of Contamination and Their Detection on Wafer Surfaces

The two general categories of wafer contamination are particulates and films[1]. *Particulates* are any bits of material present on a wafer surface that have readily definable boundaries. As feature sizes shrink, the sizes of particulates that can cause defects also increase. Particulate sources include silicon dust, quartz dust, atmospheric dust, and particles originating from clean room personnel and processing equipment, lint (from street clothing that escapes from around protective clean room garments), photoresist "chunks" (e.g. resulting from tweezers gripping wafer edges), and bacteria (which can grow in DI water supplies).

Layers of foreign material on wafer surfaces are sources of *film contamination*. Portions of films may, however, break loose and become particles, as often happens with photoresist scums. Examples of films that contaminate wafers include *solvent residues* , such as acetone, isopropyl alcohol, methyl alcohol, xylene, *photoresist developer residues* from dissolved photoresist in the developer, or from inadequate post-development rinsing, *oil films* introduced through improperly filtered air or gas lines, and *metallic films* deposited during immersion of wafers in etchant or resist stripper baths, both of which may contain metal ions and free metal in solution).

The chemical cleaning and photoresist stripping operations used to remove film contamination have also been identified as significant sources of particle contamination[2]. Thus, one stage of a cleaning process may in fact reduce the effectiveness of other cleaning procedures. To prevent particle contamination during chemical cleaning, the chemical process must be carefully monitored and particulate deposition effectively controlled. It is recommended, for example, that ultra-pure chemicals be utilized together with in-line point-of-use microfiltration for both chemical cleaning and resist stripping procedures[2,36].

In order to minimize wafer surface contamination and particulates, techniques to detect their presence must be available. For example, the concentration levels of particulates needs to be quantifiably measured after each process step. Only with such data in hand, can the effectiveness of the steps taken to reduce particulate counts be evaluated, and improved wafer cleaning procedures be implemented. Optical microscopy is used to detect particulates down to 1-2 μm in diameter, scratches, and solvent residues. Light-field, dark-field, Nomarski, and fluorescent illumination modes (see Chap. 17), are all useful for detecting such contamination and defects. Scanning electron microscopes are another tool used to analyze wafer surface contaminants.

Automatic laser scanners are also available for measuring surface defects and contamination (Fig. 1). Such instruments utilize a sharply focused laser beam and integrating light collector, to scan the entire wafer surface[3,4]. On a clean smooth surface, laser light is reflected at predictable angles, while any defects or contaminants on a surface cause light to be scattered. The light scattered by a defect is collected through a high efficiency light collection system. The scattered light is totally integrated by this optical system and then amplified by a photomultiplier tube. The output is a graphic display showing the number and type of defects. Such scanners can detect particles, pits, epi spikes, protrusions, cracks, scratches, and fingerprint residues of sizes as small as 1 μm in diameter. Total scanning time is on the order of a few seconds for a 100 mm

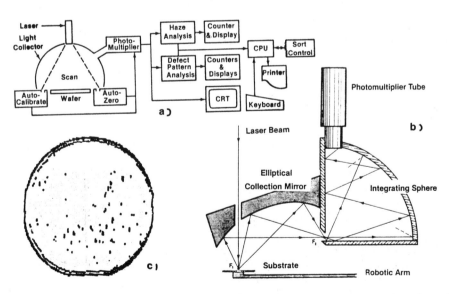

Fig. 1 (a) Functional block diagram of automatic defect detection system, utilizing a laser scanning detector. (b) Optical unit of the system (c) Typical surface patterns form an automatic laser scanning system. Courtesy of Tencor Instruments.

wafer. The major limitation of such systems is they are only effective on pattern-free substrates.

Surface analysis techniques such as Auger emission spectroscopy (AES) and secondary ion mass spectroscopy (SIMS) are capable of compositional analysis, typically elemental rather than molecular, of surface contaminant films, and in some cases of particulates (see Chap. 17). Such techniques are particularly useful for identifying the elemental composition of contamination which may have first been detected by other means.

Wafer Cleaning Procedures

As there are two classes of contamination, so there are separate cleaning procedures to remove each of them. That is, both chemical cleaning procedures and particulate cleaning techniques must be employed to produce a completely clean surface. As noted earlier, when one technique follows another, the latter steps must not recontaminate the surface and degrade the effectiveness of former cleaning procedures.

Chemical Removal of Film Contaminants

Chemical cleaning is used to remove chemically bonded films from wafer surfaces. Conventional chemical cleaning is performed with a series of acid and rinse baths. Many chemical cleaning processes are considered proprietary by device manufacturers, especially for some of the surfaces being cleaned (metal silicides, refractory metals, aluminum, and sputtered or deposited glasses). The chemical cleaning of silicon and silicon dioxide prior to high temperature operations, such as oxidation, diffusion, epitaxy, or anneal, however, is widely practiced according to the techniques presented here.

When bare silicon, or a silicon wafer with only thermally grown oxide, is chemically cleaned prior to a furnace step, the following procedure formulated by Kern and Puotinen at RCA[5], and hence often referred to as the *RCA method*, is widely used. The technique first removes organic film contamination and then applies an inorganic ion and heavy metal cleaning step. Note that the procedure outlined is adapted from Ref. 6, in which it is noted that a wide ranges of solutions in Steps 2 and 4 have also been successfully used.

1. *Preliminary Cleaning* - If photoresist is present on the wafers, it is removed by plasma oxidation stripping and /or immersion in an inorganic resist stripper (e.g. H_2SO_4-H_2O_2). In many processes, even if the resist has previously been stripped, the first step of the cleaning is a second immersion into a sulfuric acid-oxidant mixture. Upon removal from this solution, the wafers are rinsed in 18-23°C deionized and filtered water, with a resistivity of 10-18 MΩ-cm. Such water is also used for all other rinse steps of this cleaning procedure.

2. *Removal of Residual Organic Contaminants and Certain Metals* - A fresh mixture of H_2O-NH_4OH-H_2O_2 (5:1:1 by volume) is prepared and heated to 75-80°C. The wafers and their holder are submerged in the solution for 10-15 minutes, with the temperature being maintained at 80°C. The wafers are then rinsed in DI water for one minute.

3. *Stripping of the Hydrous Oxide Film Formed During Step 2* (Note that this step is not necessary if a thermally grown SiO_2 film completely covers the wafer.) - Wafers are submerged for 15 seconds into a mixture of 1 volume HF (49% electronic grade) and 10 volumes H_2O directly from rinse tank of Step 2. Exposed silicon (but not SiO_2) should repel H_2O as the wafers are pulled out of this solution. Following immersion in HF, the wafers are transferred to a rinser, but are rinsed for only 20-30 seconds. The short rinse minimizes regrowth of the oxide. The holder and wafers are then transferred, without drying, into the solution of Step 4.

4. *Desorption of Remaining Atomic and Ionic Contaminants* - A fresh mixture of

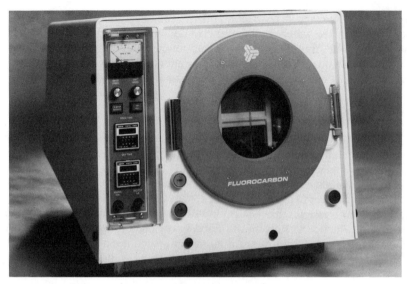

Fig. 2 Single carrier rinser /dryer.   Courtesy of Verteq Inc.

H_2O: HCl : H_2O_2 (6:1:1 by volume) is prepared and heated to 75-80°C. The still-wet wafers from Step 3 (or Step 2, if Step 3 is not necessary) are submerged into the solution for 10-15 minutes. The wafers are next rinsed in DI water to resistivity.

5. *Drying of Wafers* - Wafers are dried in a rinser-drier that uses DI water to rinse, and heated N_2 to dry. Remove wafers by dump transfer to a high-temperature boat.

6. *Storage* - Avoid storage of cleaned wafers, preferably by immediately continuing processing. If storage must occur, store in closed glass containers in a *nitrogen dry-box*.

Several other comments regarding this process are also pertinent. First, vapors of NH_3 and HCl form a particulate smoke of NH_4Cl when intermixed. It is therefore recommended that the solutions of Step 2 and 4 be separated by using two separate exhaust hoods, to avoid wafer contamination from colloidal NH_4Cl particles .

Second, reasonable care must be taken to prevent the cleaning solution of Step 2 from being depleted of H_2O_2, as NH_4OH in the absence of H_2O_2 will etch silicon. Such depletion may occur if the solution temperature is allowed to rise above 80°C, at which point rapid decomposition of the H_2O_2 ensues. In addition, if impurities are allowed to accumulate in the solution, they can also accelerate the decomposition of H_2O_2. Thus, use of fresh solutions is advised.

Third, the use of centrifugal spray cleaning instead of immersion in cleaning solutions is also favorably mentioned. In this method, the wafers are enclosed in a chamber purged with N_2. A sequence of fine sprays of cleaning solutions and high-purity water wets the wafers. The chief advantages of spray cleaning include: 1) smaller volumes of chemicals and DI water are consumed (about 2 /3 less); 2) wafer surfaces are continually exposed to fresh reagent solutions; 3) the environment of the process is carefully controlled; and 4) the process sequence is automated.

Finally, *dump rinsers* are used for DI water rinsing. In such rinsers, water is sprayed onto wafers and the cassette, filling the rinse tank at the same time. When the water reaches a predetermined level, doors at the bottom of the tank are opened and the water is quickly dumped. Several rinse /dump cycles are used, the exact number determined by the specific process, during a single rinse step. *Rinser-driers* are used to dry the wafers after the DI rinse step (Fig. 2). Modern models typically hold one cassette of wafers. Following an initial spray rinse with DI water after the cassette is loaded, the wafers are spun dry while being sprayed with heated N_2. Static electricity may build up during the dry cycle and rinser-dryers can be equipped with static eliminators. In addition, loading and unloading of cassettes has been found to abrade the cassettes and generate particulates, and alternative drying techniques are therefore being explored.

Photoresist Removal

Photoresist must be removed following a wide variety of processing steps, including etching (wet and dry), ion implantation, lift-off processes, high temperature postbake (for improving resist adhesion or etch resistance), or merely simple removal of misaligned resist patterns for reimaging after development and inspection ("rework"). In addition, wafer surface patterns of several different materials may be present under the resist (e.g. SiO_2, aluminum, polysilicon, silicides, deposited SiO_2 or Si_3N_4 or polyimide). The main objective in resist stripping is to insure that all the photoresist is removed as quickly as possible without attacking any underlying surface materials[1]. In fact no single resist stripping chemical or technique is suitable for all applications. Resist stripping techniques are thus divided into three classes: 1) organic strippers; 2) oxidizing-type (inorganic) strippers; and 3) dry type stripping techniques.

Organic strippers perform resist removal by breaking down the structure of the resist layer. Commercially sold phenol-based strippers (such as *J-100*® by Indus-R-Chem, and *A20*® by Allied Chemical) were once quite popular, but their use is becoming more limited due to their relatively short pot life and the problem of phenol disposal[7]. In addition, some phenol-based strippers will attack metals. A class of low-phenol and phenol-free organic strippers (such as *Burmar 712*® [EKC Chemical], *Ecostrip*® [Allied Chemical], or *Remover 1112A*® [Shipley]) have been formulated to overcome these drawbacks. That is, they are safer to use and easier to dispose, although some may still attack aluminum to a limited degree.

Oxidizing-type strippers (wet inorganic) are solutions of H_2SO_4 and an oxidant, heated to ~125°C. They are used to remove resist from non-metallized wafers. The oxidant originally used was H_2O_2. It oxidizes the carbon in the resist to CO_2, which leaves the bath as a gas. H_2O_2, however, decomposes into water, making it difficult to maintain baths of consistent composition. As a result, *ammonium persulfate* has been suggested as an alternative to H_2O_2. Ammonium persulfate also liberates the carbon in the bath, but does not decompose into H_2O to dilute the H_2SO_4. This substantially increases bath life compared to H_2O_2. Oxidizing-type strippers are also commonly used to provide residue free removal of postbaked and other difficult-to-remove resists (e.g. those exposed to heavy ion implant doses or harsh dry etch environments).

Dry etching of resist is done using oxygen plasmas in plasma etching equipment. It offers several advantages over wet resist strippers including safer operating conditions, no metal ion contamination, reduced pollution problems, and no attack of most underlying substrate materials. Dry etching of organic films, including resist, is discussed in further detail in Chap. 16.

Wafers with defective photoresist imaging are frequently "reworked". In this process the photoresist is typically stripped with a commercially available stripper and then subjected to either: 1) a short dip in a 10:1 H_2O : HF solution and DI water rinse and dry; or 2) a 20 minute immersion in an 80°C solution of 7:3:3 (by volume) H_2O : H_2O_2 (30%) : NH_4OH (29%) and DI water rinse and dry[8]. These two post-strip steps are used to remove the monolayer of resist that usually remains after stripping, and which may lead to subsequent resist adhesion failure.

Photoresist may also remain attached to the *edge* of a wafer even after a conventional resist stripping process. It has been reported that failure to remove such resist clinging to the wafer edge, can cause significant defects. That is, such resist may later flake off and thus become a source of particulates. Reference 9 discusses the problem, and suggests a process of flowing a solvent off the back of slowly rotating wafers, immediately afer spin-coating with resist, to thereby dissolve resist on the wafer edge (and remove the resist edge bead).

Cleaning Techniques for Removal of Particulates

The removal of insoluble particulate contamination is commonly carried out by ultrasonic scrubbing, or by techniques combining high pressure spraying and mechanical scrubbing. In *ultrasonic scrubbing*, wafers are immersed in a suitable liquid medium to which sonic energy (at 20,000-50,000 Hz) is applied[33]. Microscopic bubbles in the liquid medium are rapidly formed and collapsed under the pressure of the sonic agitation, producing shock waves which impinge on wafer surfaces. The bubble collapsing is known as *cavitation*. These shock waves displace or loosen particulate matter. To prevent shock waves from carrying particles from the liquid back to wafer surfaces and redepositing them, the particles must be removed from the liquid through overflow or filtration after they are initially detached. Another problem also observed is mechanical failure of the substrate film as a result of the ultrasonic energy imparted during the cleaning cycle. This frequently results in film loss in certain regions, and in the extreme, the entire film may be removed.

Cleaning systems using higher frequency sonic waves (~850kHz) are also commercially available (Fig. 3a). They can be operated with solutions used in the RCA chemical film removal process. With such systems chemical cleaning and contaminant desorption can be accomplished while simultaneously removing particulates.

Particulate removal by the combination of high pressure spraying and brush scrubbing is also commonly carried out *after* a variety of process steps (e.g. sawing, lapping, and polishing), and *before* others (e.g. metallization, CVD, and epitaxy [double-sided scrubbing]).

The scrubbing process operates by rotating a brush across the surface of wafers (Fig. 3b). It cleans by imparting motion to appropriate solvents, and the moving solvent dislodges particulates. In fact, the brush hydroplanes over, and does not actually contact the wafer surface. Two types of brushes are in common use: bristle and PVC sponge material. High pressure jet spraying is also almost always used with brush scrubbing. The high pressure, 13.8-20.7 MPa

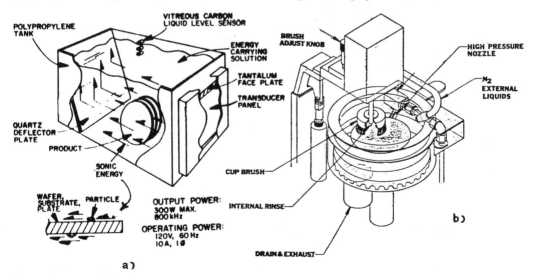

Fig. 3 (a) Megasonic wafer cleaning system. Courtesy of Verteq Inc. (b) Diagram of a cup-type brush cleaning station. Courtesy of Solitec Inc.

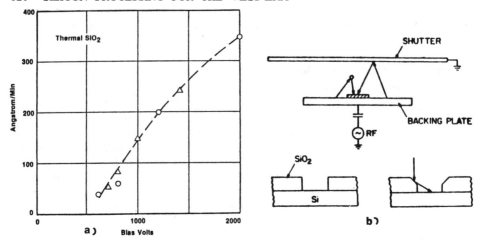

Fig. 4 (a) Example of sputter-etch rate of thermal SiO_2. Courtesy of Varian Associates. Contamination of contacts during sputter etching by (b) backscattering events, and (c) faceting. Material sputtered from the facet deposits in the contact at a rate that can exceed removal of material from the bottom of the opening[10]. Reprinted by permission of the American Physical Soc.

(2000-3000 lb /in^2) DI water jet sweeps across the wafer surface and removes the microscopic debris dislodged by the brush, as well as any residual particles generated by the brush.

In-Situ Sputter Etch Removal of Native Oxide Films

A thin oxide (5-50 Å) grows on silicon (SiO_2) or aluminum (Al_2O_3) when these materials are exposed to air. This thin oxide (known as *native oxide*) can adversely effect subsequent processing steps, for example by causing high contact resistance, or impeding interfacial reactions of films deposited on the substrate materials. Thus, it is important to remove this oxide layer and keep it from reforming before depositing the overlying film.

Concerns have arisen about whether chemical cleaning techniques will be adequate for removing native oxide films, especially in contact holes or via regions smaller than 2 μm. As a result removal of such films is also being conducted in the same vacuum environment in which the overlying film will be deposited (*in situ*). Sputter etching is used to remove up to several hundred angstroms of the wafer surface including, it is surmised, the unwanted native oxide at the bottom of the contacts or vias. Some questions have also been raised about the effectiveness of this technique for small (e.g. < 2 μm) contacts[10]. It is argued that such sputter etching (Fig. 4) will cause more contamination of the contacts through redeposition by backscattering of material sputtered from wafers and chamber surfaces, and by sputtering of contact sidewall material into the contact bottom. *In situ* removal of the native oxide by plasma chemical reactions, instead of physical sputtering mechanisms, has been proposed to circumvent this problem. Several sputter equipment suppliers offer such alternative *in situ* cleaning capabilities.

TERMINOLOGY OF ETCHING

Etching in microelectronic fabrication is a process by which material is removed from the silicon substrate or from thin films on the substrate surface. When a *mask layer* is used to protect specific regions of the wafer surface, the goal of etching is to precisely remove the material which is not covered by the mask (Fig. 5). In this section we will discuss the terms used to describe the basic aspects of etch processes.

Bias, Tolerance, Etch Rate, and Anisotropy

In general an ideal etch process is not completely attainable. That is, the etching processes are not capable of precisely transferring the pattern established by the protective mask into the underlying material. The degree to which the process fails to satisfy the ideal is specified by two parameters: *bias* and *tolerance*. As shown in Fig. 6d, *bias* is the difference in lateral dimension between the etched image and the mask image. *Tolerance* is a measure of the statistical distribution of bias values that characterizes the uniformity of etching. The tolerance parameter can be specified for a single wafer (bias distribution across a wafer), for an entire lot (bias distribution throughout the lot) or from run-to-run (bias distribution across a group of runs).

The rate at which material is removed from the film by etching is known as the *etch rate*. The units of etch rate are typically expressed in Å /sec, μm /min, etc. Generally, high etch rates are desirable as they allow higher production throughputs, but in some cases high etch rates make control of lateral etching a problem. That is, since material removal can occur in both the horizontal and vertical directions, the *horizontal etch rate* as well as the *vertical etch rate* may need to be established in order to characterize an etch process. Normally the uniformity of these etch rates is also of interest, and is expressed for three conditions (across a wafer, from wafer-to-wafer, and from run-to-run), as *etch rate % uniformity*, according to:

$$\text{Etch Rate Percent Uniformity} = \frac{(\text{Etch Rate}_{high} - \text{Etch Rate}_{low})}{(\text{Etch Rate}_{high} + \text{Etch Rate}_{low})} \times 100\% \qquad (1)$$

Highly uniform etch rates are almost always desirable in an etch process.

The lateral etch ratio, L_R, is defined as the ratio of the etch rate in a horizontal direction to

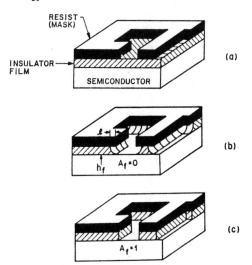

Fig. 5 Comparison of (b) isotropic, and (c) completely anisotropic etching. From E.C. Douglas, *Solid State Tehnol.*, **24**, 65, (1981). Reprinted with permission of Solid State Technology, published by Technical Publishing, a company of Dun & Bradstreet.

that in the vertical direction. Thus:

$$L_R = \frac{\text{Horizontal\ \ Etch\ Rate\ of\ Material}}{\text{Vertical\ \ Etch\ Rate\ of\ Material}} \qquad (2)$$

In the case of an ideal etch process the mask pattern would be transferred to the underlying layer with zero bias. This would then create a vertical edge profile in the etched layer coincident with original edge of the mask. Therefore the lateral etch rate would also have to have been zero. For non-zero L_R, the film material is etched to some degree under the mask and this effect is called *undercut* (Fig. 5d).

When the etching can proceed in all directions at the same rate, it is said to be *isotropic* (Fig. 5b). By definition, however, any etching that is not isotropic is *anisotropic*. If etching proceeds exclusively in one direction (e.g. only vertically), the etching process is said to be *completely anisotropic*. Since many etch processes fall between the extremes of being isotropic and completely anisotropic, it is useful to define a degree of anisotropy, A, as:

$$A = 1 - L_R \qquad (3)$$

Thus, when $L_R = 0$, $A = 1$, and this condition corresponds to completely anisotropic etching. When $L_R = 1$, the vertical and horizontal etch rates are equal, and the degree of anisotropy is $A = 0$. This corresponds to an isotropic etching condition. Most wet etching processes and some dry-etching processes exhibit uniform etch rates in all directions, and hence are isotropic.

An example of an etch profile in the film being removed versus time is shown for an isotropic etch, $L_R = 1$ (Fig. 6a) and for a process in which $L_R = 0.1$ (Fig. 6b). If the films are etched just to completion, the profile for $L_R = 1$ has the shape of a quarter circle, whereas the profile of $L_R = 0.1$ is vertical except near the bottom (where it is rounded). If this etch is allowed to continue, however, even the profile with $L_R = 1$ becomes more vertical, though

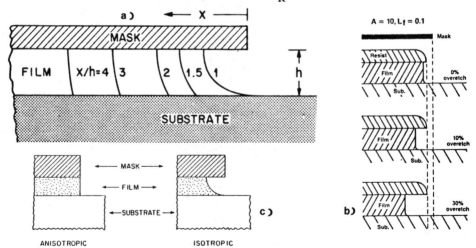

Fig. 6 (a) Isotropic etching of a film vs time ($L_R = 1$). Overetching results in profiles are more vertical. (b) Etching of film versus time when $L_R = 0.1$. (c) Etch bias is a measure of the amount by which the etched film undercuts the mask at the mask film interface. Fig. (c) Copyright, 1983, Bell Telephone Laboratories, Incorporated, reprinted by permission.

lateral etching proceeds more rapidly than the process in which $L_R = 0.1$, and thus leads to more severe undercutting. As a result, we see that L_R is one of the variables which impacts feature size, as well as edge profile. We shall also see that other parameters play a role in controlling such characteristics of the feature attributes as size and edge profile.

In fabrication technologies that are performed using isotropic etching processes, the problem of etch bias is handled by specifying an appropriate amount of compensation in the mask dimensions. For example, if the bias of an etch process is 1 μm, a 6 μm feature on the *mask* can be used to produce a desired 5 μm feature on the *wafer*. Unfortunately, for VLSI technologies in which the pattern dimensions approach the thicknesses of the films being patterned, the margin for compensation diminishes, and a higher degree of anisotropy is required. For practical purposes this situation arises when pattern features become smaller than ~3 μm. Under these circumstances, isotropic etching processes become inadequate, and processes that provide higher degrees of anisotropy need to be employed.

Selectivity, Over-Etch, and Feature Size Control

In earlier sections, only the etching characteristics of the film were considered when examining the relationship between bias and edge profile, and degree of etch anisotropy. We assumed that the mask was not attacked by the etchant, and did not consider that the layers under the etched film can also be attacked by the etchant. In fact, both mask material and underlying layer materials are generally etchable, and these effects may play a significant role in specifying etch processes. Note that the underlying material subject to etchant attack may be either the silicon wafer itself, or a film grown or deposited during a previous fabrication step. The ratio of etch rates of different materials is known as the *selectivity of an etched process*[12]. Thus both: 1) the selectivity with respect to the mask material; and 2) the selectivity with respect to the substrate materials are important characteristics of an etch process.

The *selectivity with respect to the mask material*, S_{fm}, plays a role in determining the etched feature sizes. The *selectivity with respect to substrate*, S_{fs}, can impact performance and yield. Film thickness and etch rate non-uniformities increase the required values of S_{fm} and S_{fs} because the etch processes need to be continued beyond the point at which the mean film thickness is completely etched (cleared). Such additional etching is referred to as *overetch*. For example, due to overetch requirements, when contact holes are to be etched in SiO_2 it is desirable that the etch rate decrease when the silicon substrate is reached. In this case, a process with high selectivity with respect to substrate is necessary.

For many wet-etch processes, both S_{fm} and S_{fs} are very high, and thus neither the mask or substrate materials are affected very much during such well-controlled wet-etch procedures. However for dry-etch processes, these desirable circumstances are rarely encountered. Thus, it is necessary to calculate the selectivities that an etching application will require, so that dry-etch processes which are able to meet such specifications can be selected or developed.

Determining the Required Selectivity With Respect to Mask Materials, S_{fm}

The required selectivity with respect to the mask, S_{fm} is dependent on several factors including: a) film thickness uniformity; b) film etch rate uniformity; c) mask etch rate uniformity; d) the edge profile of the mask; e) the anisotropic etch rate of the mask; and f) the maximum acceptable loss of line width of the patterns being etched[11]. These factors can be quantified with the assistance of the information given in Fig. 7.

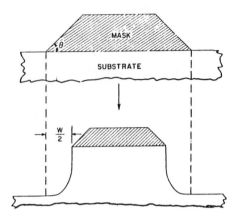

Fig. 7 The evolution of an etched feature when the mask has a finite etch rate. The difference between the intended pattern width and the actual linewidth is W. Copyright, 1983, Bell Telephone Laboratories, Incorporated, reprinted by permission.

For example, assume we wish to etch a film that has some degree of thickness non-uniformity. This film has a mean thickness, h_f, a maximum thickness, $h_f(1 + \delta)$, and a minimum thickness, $h_f(1 - \delta)$, where δ is a dimensionless parameter with a value $0 \le \delta < 1$. Assume also that the film etch rate is somewhat nonuniform. That is, a mean etch rate, v_f, of the film exists over the wafer area, and that v_f varies between $v_f(1 + \phi_f)$ and $v_f(1 - \phi_f)$, where ϕ_f is again a dimensionless parameter of value $0 \le \phi_f < 1$. In order to insure that the maximum loss of linewidth is less than or equal to the allowable linewidth loss, this effect must be determined for the worst-case etching condition on the wafer. Maximum linewidth loss occurs where the film is thickest, $[h_f(1 + \delta)]$, and at wafer locations where the film etch rate is slowest, $v_f(1 - \phi_f)$. Thus, the time required to etch the film at such locations is the longest, and is given by:

$$t_c = \frac{h_f(1 + \delta)}{v_f(1 - \phi_f)} \qquad (4)$$

If it is independently determined that a fractional overetch time, Δ, is also required, then the total etch time, t_t, is increased to:

$$t_t = \frac{h_f(1 + \delta)(1 + \Delta)}{v_f(1 - \phi_f)} \qquad (5)$$

During the etch time, the mask will be eroded as shown in Fig. 7. If the mask material is removed with maximum vertical and horizontal etch rates, v_v and v_1, respectively, then the edge of the mask will retreat from its original locations by a distance, $W/2$, given by:

$$\frac{W}{2} = [v_v \cot \theta + v_1] t_t \qquad (6)$$

where θ is defined in Fig. 7, and the total loss of linewidth is $2(W/2)$. The loss of linewidth dimension of the mask, W due to erosion during t_t, is then found by substituting Eq. 5 into Eq. 6 to give:

$$W = 2 \frac{v_v}{v_f} h_f \frac{(1 + \delta)(1 + \Delta)}{(1 - \phi_f)} \left[\cot \theta + \frac{v_1}{v_v} \right] \quad (7)$$

The mask etch rates are generally also nonuniform, and the nonuniformity is expressed as $v_v = v_m (1 \pm \phi_m)$, where v_m is the mean mask etch rate and ϕ_m is a dimensionless parameter. For the worst case again, we select the condition where the mask etch is fastest. Thus, $v_m (1 + \phi_m)$ is substituted into Eq. 7 for v_v. In addition, selectivity with respect to the mask is defined as $S_{fm} = v_f / v_m$. From Eq. 3 the mask lateral etch rate ratio is $v_v / v_L = L_R = 1 - A_m$. Using these definitions and rearranging Eq. 7 we get the required S_{fm} as:

$$S_{fm} = \frac{h_f}{W} U_{fm} [\cot \theta + (1 - A_m)] \quad (8)$$

where;

$$U_{fm} = \frac{[(1 + \delta)(1 + \Delta)(1 + \phi_m)]}{(1 - \phi_f)} \quad (9)$$

and U_{fm} is the uniformity factor that accounts for the simultaneous occurrence of worst-case conditions that leads to the greatest mask erosion, and thus maximum linewidth loss. Figure 8 shows a set of curves of required S_{fm} which apply to the specific case of $\phi_m = \phi_f = 0.1$, $\delta = 0.05$, and $\Delta = 0.2$. For these conditions $U_{fm} = 1.54$, and the curves are plotted for various h_f / W ratios and various degrees of mask edge angles, θ, and etch anisotropy, A.

Example: Assume that a 1 μm film is to be patterned with a completely anisotropic etch process and thus the only loss of feature linewidth is due to mask erosion. In addition, let the mask and film etch rate uniformities both be 10%, and the film thickness uniformity be 5%. In addition, a 20% overetch is required. Find the required selectivity to the mask for a maximum 0.2 μm linewidth loss if the resist profile angle is, a) 60%, and b) 90% for, 1) isotropic mask etching, and 2) completely anisotropic mask etching.

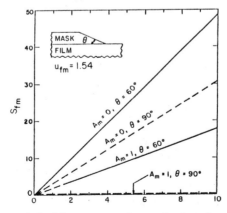

Fig. 8. Selectivity, S_{fm}, needed with respect to the mask, plotted as a function of the ratio of film thickness to loss of linewidth for various mask profiles, and the extremes of isotropic & anisotropic mask etching. Copyright 1983 Bell Telephone Laboratories, Inc, reprinted by permission.

Solution: Figure 8 can be used to solve this problem since the curves were generated for the given worst case nonuniformity values from Eq. 8. Note that for other etch rate and thickness non-uniformities, different curves need to be derived. For this film $h_f = 1.0$ µm, and $W = 0.2$ µm, so $h_f/W = 5$. From the curves of Fig. 7 we see that the following S_{fm} are required:

a) S_{fm} ($\theta = 60°$, $A = 0$) = 23
b) S_{fm} ($\theta = 60°$, $A = 1$) = 9
c) S_{fm} ($\theta = 90°$, $A = 0$) = 15
d) S_{fm} ($\theta = 90°$, $A = 1$) = 0

It can be seen from this example that anisotropic etch processes, combined with vertical walled etch mask profiles, result in the most precise pattern transfers. However, there is also another phenomenon associated with sputtering and dry-etching, referred to as *faceting* which can also contribute to mask erosion. Faceting is discussed in more detail Chap. 10.

Determining Required Selectivity With Respect to Substrate, S_{fs}

The necessary selectivity with respect to substrate, S_{fs}, is also calculated by considering the worst-case condition[13]. That is we assume the thinnest part of the film to be etched lies over the region of the substrate that experiences the highest etch rate. We use this assumption to calculate a uniformity factor, U_{fs}. We then multiply U_{fs} by the ratio h_f/h_s (where h_f is the mean film thickness, and h_s is the maximum allowable penetration depth of the substrate layer) to arrive at the required S_{fs}, or:

$$S_{fs} = \frac{h_f}{h_s} U_{fs} \qquad (10)$$

and

$$U_{fs} = \left[\frac{\phi_f(2 + \Delta + \Delta\delta) + \delta(2 + \Delta) + \Delta}{(1 - \phi_f^2)} \right] \qquad (11)$$

where ϕ_f, Δ, and δ are defined as in the last section that derived S_{fm}. We see from Eq. 10 that if the film is perfectly uniform ($\delta = \phi_f = 0$) and if no overetching is required ($\Delta = 0$), selectivity with respect to the substrate would not be an issue of concern since U_{fs} would equal zero. However, since these conditions are not representative of actual conditions, Eq. 10 is useful in determining realistic S_{fs} values.

There is another factor, even more important than film and etch rate non-uniformities, which dictates the degree of overetching when anisotropic etch processes are employed. That is, as shown in Fig. 9a, when material is cleared from a planar region on a wafer, residual material at steps has still not been removed. Thus, additional etching beyond the point at which the planar regions have cleared must be used to remove such residual material (sometimes referred to as *stringers* or *picket fences*). As shown in Fig. 9b, failure to remove this material can lead to unwanted electrical shorting paths between adjacent lines. From Fig. 9a and b, it can be seen that for completely anisotropic etch processes ($A = 1$), the fractional overetch, Δ, required to clear such residual material, is h_1/h_2.

Example: Given a 2500 Å polysilicon layer that passes over both 5000 Å field-oxide regions and 250 Å gate oxide layers. Assume that a completely anisotropic

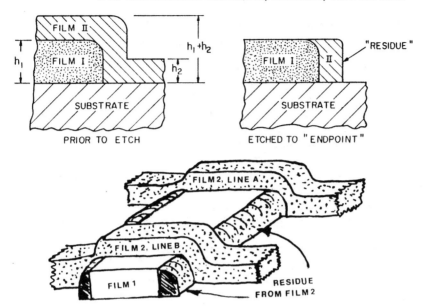

Fig. 9 If etching is anisotropic, overetching is needed to remove residual materials at steps. The degree of anisotropy, A = 1 in the example shown.   Copyright, 1983, Bell Telephone Laboratories, Incorporated, reprinted by permission.

etch process will be used to etch the polysilicon, and that the polysilicon film thickness uniformity is 5%, and the uniformity of etching is 10%. Find the required S_{fs} for this process, if the etching is to be stopped immediately after the polysilicon is completely etched.

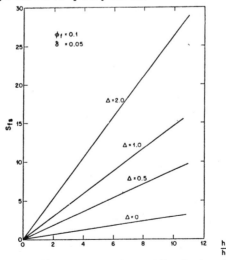

Fig. 10 The selectivity needed with respect to substrate, S_{fs}, is plotted as a function of the ratio of film thickness to the amount of substrate removed for various amounts of overetching. Copyright, 1983, Bell Telephone Laboratories, Incorporated, reprinted by permission.

Solution: Since the polysilicon must pass over 5000 Å steps (e.g.where field and gate oxide meet), the thickness of the poly at the step will be (from Fig. 8):

$$h_1 + h_2 = 5000 \text{ Å} + 2500 \text{ Å} = 7500 \text{ Å}.$$

This is the thickness of polysilicon that must be removed by an anisotropic etch procedure. Thus, $\Delta = h_1/h_2 = 2$. The maximum allowable penetration of the fractional overetch required to completely remove the residual material at the steps underlying gate oxide layer is $h_s = 250$ Å. Any more penetration will remove all of this oxide and expose the silicon substrate material to the etchant. Thus, $h_f/h_s = 2500/250 = 10$. From Fig. 10, which can be applied to this problem, we find that:

$$S_{fs} = 25.$$

Note the curves of Fig. 8 and Fig. 10 were derived for specific mask, film, and etch rate nonuniformities. Each etching process, however has its own set of characteristics and appropriate curves for a specific process need to be derived using Eq. 8 and Eq. 10 with those values.

Combined Impact of the Requirements of Anisotropy and Selectivity

A high degree of anisotropy is a desirable feature of an etch process for fine feature patterning since in such applications very little etch bias can be permitted. A highly selective etch rate with respect to the mask is needed to maintain feature size control. An adequate degree of selectivity with respect to underlying materials is also necessary in order to prevent removal of previously processed portions of the circuit. However, when anisotropic etching is performed in the presence of stepped topography, we have shown that the remaining residual material at the steps requires additional overetching beyond even that required by etch rate and "normal" film thickness nonuniformities. Thus, the mandate that calls for anisotropic etch processes ends up also driving the selectivity requirements even higher.

Loading Effects

When the etch rate is dependent upon the amount of etchable surface exposed to the etchant, the phenomenon is called a *loading effect*. The effect arises when there are a limited number of etchant species available to etch a material. When these are depleted, etching cannot continue until new etching species arrive at the surface. Loading effects are most commonly encountered in dry-etch processes where they can occur in a variety of different conditions. These include: a)

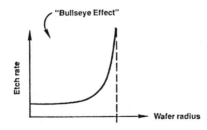

Fig. 11 Example of "bullseye" etch nonuniformirty.

etch rate is dependent upon the number of wafers present in the chamber. In such cases, as the number of wafers in the system is increased, the rate of consumption of etchant species also increases. This may result in a loading effect that will lead to a slower overall etch rate; b) the etch rate depends on the amount of area on the wafers, or on a single-wafer in single-wafer etching systems, that has etchable material exposed. Note that this exposed etchable area can even change during an etch process. That is, upon the clearing of an etching film from the planar regions of a surface, much less residual material remains. Thus, more etchant species may be available to attack any residues, but also the exposed sidewalls of the etchable film. As a result, lateral etch rate changes during overetch time may be a side effect of local loading effects.

Gas flow effects may also combine with loading effects to cause etch rate nonuniformities. That is, the location of a wafer in a chamber, as well as the number present in the chamber, may impact the etch rate. The *bullseye effect* sometimes observed in dry etching, in which the film at edges of a wafer is etched more quickly than at the center (Fig. 11), is due to depletion of etchant species at the wafer center. More etchant species can arrive at the wafer near its edge than from just above the wafer surface. Techniques to minimize loading effects are described in Chap. 16.

WET ETCHING TECHNOLOGY

Wet etching processes are generally isotropic. As such we have pointed out that they are inadequate for defining features less than about 3 μm wide. Nevertheless for those process that involve patterning of linewidths greater than 3 μm, wet etching continues to be a viable technology. Since it turns out that a significant fraction of semiconductor products are still being fabricated with such large geometries, wet etching should not be ignored. In this section we present some of the more important aspects of wet etching technology for current processing needs. In addition more detailed information can be found in references listed at the end of the chapter[1,14,17,18,10].

The reason wet etching has found widespread acceptance in microelectronic fabrication is that it is a low cost, reliable, high throughput process with excellent selectivity for most wet etch processes, with respect to both mask and substrate materials. Some recent refinements to wet-etching equipment have in fact increased these advantages, including: a) the automation of wet stations; b) the placing of wet etch process steps under microprocessor control, to improve reproducibility of etching conditions from run-to-run, and to give the process engineers better control of the equipment functions by preventing unauthorized process changes; c) point-of-use filtration of etchants to prolong their use by removing etch process generated defects; and d) the development of spray etching. Such advances make it likely that wet etching will continue to find extensive use in semiconductor fabrication for the foreseeable future[15,16].

On the other hand, besides the 3 μm limitation, wet etching is subject to the following disadvantages: a) higher cost of etchants and DI water compared to dry etch gas expenses; b) increased personnel safety hazards from chemical handling; c) exhaust fumes and the potential of explosions; d) resist adhesion problems; and e) bubble formation and incomplete wetting of wafer surfaces by the chemical etchants leading to incomplete etching and etching non-uniformities.

In general a wet etch process can be broken down into three steps: 1) diffusion of the reactant to the reacting surface; 2) reaction; and 3) diffusion of the reaction products from the surface[20]. The second step can obviously be further differentiated into adsorption prior to, and desorption subsequent to, the actual reaction step. The slowest of the steps will be *rate-controlling*. That is, the rate of that step will be the rate of the overall reaction.

Chemical etching can occur by several processes. The simplest involves dissolution of the

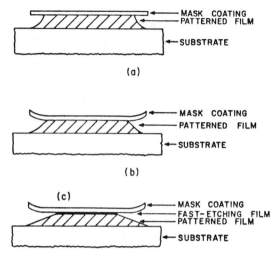

(a)

(b)

(c)

Fig. 12 Different etch profiles produced from various degrees of undercutting during wet etch. (a) Good mask-to-film adhesion. (b) Undercutting has occurred at mask-film interface. (c) Use of fast-etching film to achieve controlled undercutting[14]. Reprinted with permission of Academic Press.

material in a liquid solvent without any change in the chemical nature of the dissolved species. Most etching processes, however, involve one or more chemical reactions (Step 2 above). Various types of reactions may take place, although one commonly encountered in semiconductor fabrication is oxidation-reduction (redox). That is, a layer of oxide is formed, then the oxide is dissolved and the next layer of oxide is formed, etc (e.g. in wet etching of Si and Al).

In semiconductor applications, wet etching is used to produce patterns on the silicon substrate or in thin films. A mask is typically used to protect desired surface regions from the etchant and this mask is stripped after the etching has been performed. Thus, when choosing a wet etch process, in addition to selecting an etchant, a suitable masking material must be picked to have good adhesion to the underlying films, good coating integrity and the ability to withstand attack by the etchant. Photoresist is the most commonly encountered masking layer, but sometimes it falls short in this role. Problems encountered with photoresist as a mask layer in wet etching applications include loss of adhesion at the edge of the mask-film interface due to etchant attack (Fig. 12), and large area failure of the resist. Edge attack is combatted by use of adhesion promoters such as hexamethyldisilazane (HMDS). Large area failures of resist are usually due to differential stress buildups in the substrate and mask layers. Such problems are minimized by using mask materials with enough elasticity to conform to stresses between mask and substrate (that build up from differential coefficients of expansion).

Etching processes which produce bubbles can also lead to poor pattern definition due to the clinging of such bubbles to the substrate, particularly along the pattern edges. Bubbles deny the etchant local access to the film being etched, and at those locations etching temporarily slows down or ceases, until the bubble is dislodged. Therefore, use of wetting agents in the etchant together with agitation during etching are measures taken to assist in dislodging such bubbles.

Developer residue or *scum* is also a common cause of etch blocking in which non-etching or poor wetting of the etchant is observed. The two most common causes are: 1) developer baths that have had their active ingredients depleted; and 2) underexposed resist. When the developer is no longer at full strength, it may not completely remove the exposed positive resist, and a thin resist layer is left behind. Such residual layers are very difficult to detect prior to etching.

Therefore, to avoid such problems it is necessary to use fresh developer after a specific number of wafers have been processed in the bath, in batch immersion development processes. In addition, a descum step, using an oxygen plasma, is used to remove such residues (see Chap. 12).

We will now discuss wet-etching aspects of the most commonly encountered materials which are etched in the microelectronic fabrication environment: silicon; silicon dioxide; silicon nitride; and aluminum.

Wet Etching Silicon

Both single crystal and polycrystalline silicon are typically wet etched in mixtures of nitric acid (HNO_3) and hydrofluoric acid (HF). The reaction is initiated by the HNO_3 which forms a layer of silicon dioxide on the silicon, and the HF dissolves the oxide away. The overall reaction is:

$$Si + HNO_3 + 6\,HF \rightarrow H_2SiF_6 + HNO_2 + H_2 + H_2O \qquad (12)$$

Water can be used to dilute the etchant, but acetic acid (CH_3COOH) is preferred as a buffering agent, since it causes less dissociation of HNO_3, and thus yields a higher concentration of the undissociated species.

The mixture compositions can be varied to yield different etch rates. Figure 13 shows the isoetch curves of such mixtures for various constituents by weight[21]. We see that at high HF and low HNO_3 concentrations (the region near the upper corner of the triangle), the etch rate is controlled by the HNO_3 concentration, because in such mixtures there is an excess of HF to dissolve the SiO_2 created during the reaction. On the other hand at low HF and high HNO_3 concentrations, the etch rate is limited by the ability of the HF to remove the SiO_2 as it is created. In such etchants the etching is isotropic, and they are often used as polishing agents.

In some applications, it is useful to etch silicon more rapidly along some crystal planes than others[22]. This allows the etch to significantly slow down or to etch specific shapes or structures in the silicon. In the diamond lattice we observed that the (111)-plane is more densely packed than the (100)-plane, and thus the etch rates of (111)-oriented surfaces are expected to be lower than those with (100)-orientations. An etchant that exhibits such orientation-dependent

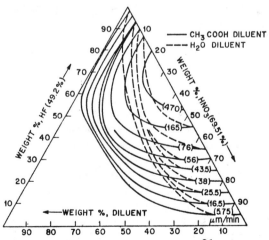

Fig. 13 Isoetch curves for silicon (HF: HNO_3 diluent system)[21]. Reprinted with the permission of the publisher, the Electrochemical Society.

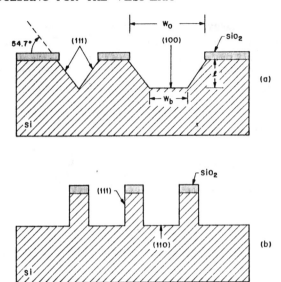

Fig. 14 Orientation-dependent Si etching (a) etch pattern profiles on <100>-Si, (b) etch pattern profiles on <110>-Si[22]. (© 1978 IEEE).

etching properties in silicon consists of a mixture of KOH and isopropyl alcohol (e.g. 23.4 wt % KOH, 13.3 wt % isopropyl alcohol, and 63 wt % H_2O). The etch rate of this etchant is about 100 times faster along (100)-planes than along (1$\bar{1}$1)-planes (e.g. at 80°C, 0.6 μm /min vs. 0.006 μm /min).

If features in a (100)-plane of silicon are patterned with SiO_2, this orientation-etchant will create precise V-shaped grooves in the silicon, and the edges of the grooves will be (111)-planes at an angle of 54.7° from the (100)-surface (Fig. 14a). If (110)-surfaces are used, straight-walled grooves with sides having (111)-planes are formed (Fig. 14b).

Various etchant mixtures have also been devised to allow the delineation of crystalline defects in silicon. Most defect analysis using such wet chemical etchants utilizes one of three formulations which are listed in Table 1. The Sirtl etch[23]is fast and effective for (111)-surfaces, but tends to cause clouding and thereby produces confusing results on (100)-surfaces[30]. The Secco[24] and Wright[25,29] etches provide better results for these applications.

Wet Etching Silicon Dioxide

Wet etching of SiO_2 films in microelectronic applications is usually accomplished with various hydrofluoric acid (HF) solutions[17,20]. This is because SiO_2 is readily attacked by room temperature HF, while Si is not. Etching takes place according to the overall equation:

$$SiO_2 + 6HF \rightarrow H_2 + SiF_6 + 2\,H_2O \tag{13}$$

The concentration of HF supplied by chemical manufacturers is 49% in water. Such concentrated HF, however, etches SiO_2 too quickly for good process control (e.g. thermally grown SiO_2 is etched at approximately 300 Å /s at 25°C). Thus diluted HF is generally used instead. A common etchant formulation contains buffering agents such as ammonium fluoride (NH_4F), which help prevent depletion of the fluoride ions, and thus maintain stable etching characteristics.

Table 1. ETCHES FOR DEFECT ANALYSIS ON SILICON WAFERS

Etch	Formulation	Comments
1. Sirtl [4]	50g CrO$_3$ to 100 ml H$_2$O.	Fast, good defect analysis on (111) material.
	Mix 1:1 with 48% HF prior to use.	Develops stains and cloudy surfaces on (100) making interpretation difficult.
2. Secco [5]	44g K$_2$Cr$_2$O$_7$ to 1000 ml H$_2$O.	Good delineation of defects on (100) surfaces.
	Mix 1:2 with 48% HF.	Slower etch rates.
		Requires ultrasonic agitation to avoid bubble formation.
3. Wright [6]	2g Cu (NO$_3$) $\times$ 3H$_2$O in 60 ml H$_2$O.	Good for both (100) and (111) surfaces.
	Add:	
	60 ml 48% HF	Moderate Etch rate.
	30 ml 69% HNO$_3$	Minimum of etch anomalies.
	30 ml 5 MCrO$_3$	
	(1g CrO$_3$/2 ml H$_2$O)	Useful for wafer cross-sections.
	60 ml acetic acid	Formulation is complex.

Unbuffered HF also causes both excessive undercutting at the resist oxide interface and lifting of the resist. A typical buffered HF mixture (BHF) contains 6:1 (by volume) NH$_4$F:HF (40%:49%) which etches thermally grown SiO$_2$ at ~20 Å /sec (or ~1000 Å /min) at 25°C.

The etch rate of SiO$_2$ for a given etchant and temperature, also depends on several other factors. For example, SiO$_2$ thermally grown in steam etches slightly faster than SiO$_2$ grown in dry O$_2$. The presence of impurities in the oxide can also strongly affect the etch rate. A high concentration of boron results in a reduced etch rate, while a high concentration of phosphorus rapidly increases it. Ion implantation can produce damage that will increase the SiO$_2$ etch rate.

CVD SiO$_2$ (Chap. 6), generally etches much more rapidly than thermally grown SiO$_2$, but this rate also depends on many other factors[26], including deposition conditions, impurity concentration, and densifying heat treatments after deposition. As a general rule, SiO$_2$ films deposited at low temperatures exhibit higher etch rates than films annealed or deposited at higher

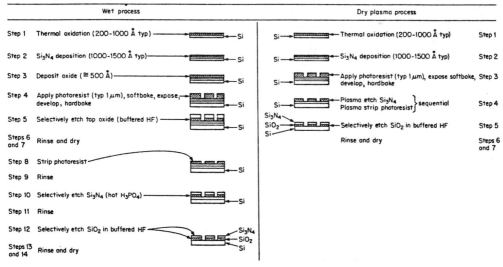

Fig. 15 Wet vs. dry-etch process for Si$_3$N$_4$. Courtesy of L.F.E. Corp., Plasma Systems Group.

temperatures. CVD SiO_2 is more commonly etched in diluted HF etches, since slower and more controllable etch rates can be achieved (e.g. 10:1-100:1 H_2O:HF).

Wet Etching Silicon Nitride

Silicon nitride (Si_3N_4) can be etched by reflux boiling 85% phosphoric acid at $180°C^{17}$. However photoresist is lifted during such etching and does not make a good etch mask for this application. Most wet silicon nitride etching thus utilizes a thin SiO_2 layer (either thermally grown or deposited), to mask the nitride. The SiO_2 layer is first etched using a resist mask, then the resist is stripped, and the patterned oxide serves as the etch mask for the nitride in the phosphoric acid etch. The Si_3N_4 etch rate is about 100 Å /min, but only 0-25Å /min for CVD SiO_2. Films of plasma-enhanced CVD Si_3N_4 have much higher etch rates than high temperature CVD Si_3N_4. The rates depend strongly on the film composition, which may be expressed as $Si_xN_yH_z$.

The added complexity of using an SiO_2 etch mask makes dry etching of Si_3N_4 an attractive alternative. In fact, the first widely used dry etch process for microelectronic applications was developed for etching Si_3N_4 for this reason (Fig. 15).

Wet Etching Aluminum

Wet etching of aluminum and aluminum alloy films is generally done in heated solutions (35-45°C) of phosphoric acid, nitric acid, acetic acid, and water[27]. A typical etch composition may be 80% phosphoric acid, 5 % nitric, 5% acetic and 10% water. The etch rate is in the range of 1000-3000 Å /min and depends on several factors including etchant composition and temperature, type of resist used, agitation of wafers during etch, and impurities or alloys in the predominantly aluminum film.

The chemical mechanism of wet etching aluminum proceeds as follows: The nitric acid forms aluminum oxide, and the phosphoric acid and water dissolve this material. Conversion to Al_2O_3 takes place simultaneously with the dissolution process.

One of the difficulties encountered in wet etching of aluminum is H_2 gas bubble evolution. These bubbles tend to adhere firmly, locally inhibiting etching. Mechanical agitation during etching, and the addition of agents which lower the interfacial tension, are used to minimize this problem. Periodic removal of the wafers from the etching solution also breaks the bubbles.

The H_2 bubble formation and other problems (e.g. local contamination of the developed

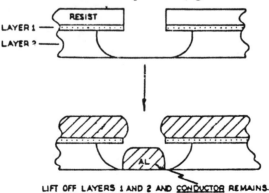

LIFT OFF LAYERS 1 AND 2 AND CONDUCTOR REMAINS.

Fig. 16   Lift-off process for metallization.

metal, local oxidation, and delayed etching in certain locations [particularly in narrow spaces due to incomplete removal of resist residue]) delay the start of etching or prolong the time to clearly etch all the areas on the wafer. Thus, once the minimum etching time for a given pattern is established, 10-50% overetch time is usually added to assure the complete isolation of features.

LIFT-OFF TECHNOLOGY FOR PATTERNING

Lift-off is a technique for forming patterns on a wafer surface by an *additive process*, as opposed to the removal (or *subtractive*) process utilized in etching[28]. That is, in lift-off an inverse pattern is first formed in a so-called *stencil layer* present on the wafer, thereby exposing the substrate in specific locations (Fig. 16). Next the film to be patterned is deposited over the inverse-patterned stencil layer and the exposed substrate. Those portions which are deposited on the stencil layer are removed when the wafer is immersed in a liquid capable of dissolving the stencil layer. In other words, the depositied film on the stencil layer is *lifted-off* during the dissolution of the stencil. The film material that was deposited on the exposed substrate regions remains behind as the required pattern.

The key to a successful lift-off process is to insure that a distinct break exists between the film material on top of the stencil and that deposited on the exposed substrate. This separation allows the desolving liquid to reach and attack the stencil layer and also insures that the film atop the stencil is free to be lifted off. Cold evaporation over steep steps achieves such breaks. The major applications of lift-off techniques involve patterning of metals for interconnections.

The advantages of lift-off include the following: a) composite layers (e.g. Al / Ti / Al) can be sequentially deposited and then patterned with a *single* lift-off, as opposed to removal by multiple etch steps in an etching process; b) hard to remove residues, such as copper residues that are left when dry etching Al-Cu alloy films, are avoided since no etching of the patterned film is necessary; and c) the patterned film features can have sloped side walls, which makes them more easily covered by subsequent films (good step coverage).

On the other hand, although lift-off has received considerable attention because of the above advantages, it has several severe disadvantages which have made etching a far more popular patterning technology. These primarily involve difficulties in the creation of stencil layers that are compatible with the deposition, photolithographic, and dissolution processes that must also be successfully performed in order to create patterns by lift-off. We will list some of the lift-off processes as reported in the literature and describe in detail the process outlined by Fried, *et al*[32], for use in bipolar VLSI technology by IBM. The latter provides an example of the complexity

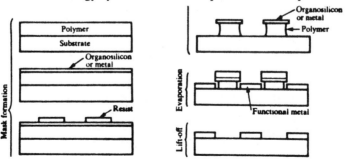

Fig. 17 Example of a lift-off sequence for bipolar VLSI circuits[32]. Copyright 1982 International Business Machines Corporation, reprinted with permission.

involved in fabricating suitable stencil layers.

Proposed stencil layers for lift-off have included: a) photoresist; b) two photoresist layers[29]; c) a photoresist / aluminum /photoresist layer[30]; d) polyimide /molybdenum layer; e) polyimide / polysulfone / SiO layer; f) inorganic dielectric layer /photoresist, as well as many others[31].

The lift-off process discussed by Fried[32] (Fig. 17) begins with the application of a thin polysulfone release layer onto a wafer, followed by a layer of positive photoresist (with a thickness exceeding that of the eventual metal layer). This composite structure is baked at a temperature of above 200°C in order to make it thermally stable at the elevated temperatures that will occur during subsequent metal deposition. Next a polysiloxane film (2000-3000Å thick) is spun on. Following another 200°C bake, the imaging photoresist is then applied. The imaging resist layer is now exposed and developed. A reactive-ion-etch step is used to reproduce the photoresist pattern into the polysiloxane, underlying resist and polysulfone layers. This step requires two sequential etch processes: one to etch the polysiloxane and the other for the remaining layers. The second etch also undercuts the polysiloxane and leaves an overhang that ensures that the deposited metal on the exposed substrate will be cleanly broken away from the metal deposited on top of the stencil. Upon completion of etching the stencil, any remaining imaging resist is stripped and the metal is evaporated. Lift-off is accomplished by immersing the wafers in N-methylpyrrolidone, which releases the polysulfone, and thereby also the resist, polysiloxane and the metal deposited thereon.

REFERENCES

1. D.J. Elliot, *Integrated Circuit Fabrication Technology*, McGraw-Hill, New York, 1982.
2. I. Bansal, "Particle Contamination during Chemical Cleaning and Photoresist Stripping of Silicon Wafers", *Microcontamination*, Aug./Sept. '84, p. 35.
3. Automated Defect Detection System, *Solid State Technol.*, Aug. '85, p. 89.
4. C. Logan, "Analyzing Semiconductor Wafer Contamination", *Microelectronic Manufacturing and Testing*, March '85, p. 1.
5. W. Kern and D.A. Puotinen, "Cleaning Solution Based on Hydrogen Peroxide for Use in Semiconductor Technology", *RCA Review*, June '70, p. 187.
6. W. Kern, "Purifying Si and SiO_2 Surfaces with Hydrogen Peroxide", *Semiconductor International*, April, 1984, p. 94.
7. P. Van Zant, "Handling and Disposal of Hazardous Chemicals and Photoresist Strippers", *Microelectronic Manufacturing and Testing*, July '84, p. 39.
8. *ibid.*, Ref. 1, p. 113.
9. N. Durrant and P. Jenkins, "Defect Density Reduction Utilizing Wafer Edge Resist Removal", *Microcontamination*, Apr. '85, p. 45.
10. J.L. Vossen, *et al, J. Vacuum Science and Technol.*, A2 (2), 1984, p. 212.
11. C.J. Mogab, "Dry Etching" in *VLSI Technology*, S.M. Sze, Ed., McGraw-Hill, New York, 1983, p. 307.
12. M. Hutt and W. Class, "Optimization and Specification of Dry Etching Processes", *Solid State Technol.*, Mar. '80, p. 92.
13. *ibid.*, Ref. 11, p. 309.
14. W.A. Kern and C.A. Deckert, "Chemical Etching", in *Thin Film Processes*, Ed. by J.L. Vossen and W. Kern, Academic Press, New York, 1978, p. 401-481.
15. P. Burggraaf, "Wet Etching Today", *Semiconductor International*, Feb. '83, p. 48.
16. "Wafer Cleaning Equipment, a Special Report", *Microeltrnc Mfg.&Testing*, May '85, p. 43.
17. W. Kern, "Chemical Etching of Dielectrics", in *Etching for Pattern Definition*, Electrochem.

Society, 1976, p. 1.

18. L.T. Romankiw, "Pattern Generation in Metal Films Using Wet Chemical Etching, A Review", in *Etching for Pattern Definition*, Electrochem. Soc., 1976, p. 161.

19. D. MacArthur, "Chemical Etching of Metals", in *Etching for Pattern Definition*, Electrochem. Soc., 1976, p. 76.

20. J.S. Judge, "Etching Thin Film Dielectric Materials", in *Etching for Pattern Definition*, Electrochem. Soc., 1976, p. 19.

21. H. Robbins and B. Schwartz, "Chemical Etching fo Silicon II, the System HF, HNO_3, H_2O, and $HC_2H_3O_2$", *J. Electrochem. Soc.*, **107**, 108 (1960).

22. K.E. Bean, "Anisotropic Etching of Si", *IEEE Trans. Electron Devices*, **ED-25**, 1185 (1978).

23. E. Sirtl and A. Adler, *Z. Metallki*, **52**, 529 (1961).

24. F. Secco d'Aragona, *J. Electrochem. Soc.*, **119**, 948, (1972).

25. M.W. Jenkins, *Electrochem. Soc. Ext. Abstracts*, May, 1976, **76-1**, No. 118, 317 (1976).

26. W.A. Plisken and R.P. Esch, "Etches for the PSG-SiO_2 System", in *Etching for Pattern Definition*, Electrochem. Soc., 1976, p. 37.

27. *ibid*, Ref. 1, p. 257.

28. *ibid*, Ref. 1, p. 27.

29. B.C. Feng and G.C. Feng, U.S. Patent No. 4, 204, 009, May 20, 1980.

30. N.L. Dunkelberger, U.S. Patent No. 4,218,532, Aug. 19, 1980.

31. L.B. Rothman, P.M. Schaible, and G.C. Schwartz, "Lift-Off Process to Form Planar Metal, Sputtered SiO_2 Structures", *Proceedings VLSI Multilevel Interconnection Conference*, IEEE, New York, June, 1985, p. 131.

32. L.J. Fried, *et al*, "A VLSI Bipolar Metallization Design with 3-Level Wiring and Array Solder Connections", *IBM J. Res. Devel.*, **26**, May '82, p. 362.

33. M. O'Donaghue, "Ultrasonic Cleaning Process", *Microcontamination*, Oct./Nov., 1984, p. 63.

34. M.W. Jenkins, "Characterization of Etching Defects in Silicon Surfaces Using Wright Etch", in *Etching for Pattern Definition*, Electrochem. Soc., 1976, p. 63.

35. D.G. Schimmel, "A Comparison of Chemical Etches for Revealing <100> Silicon Crystal Defects", *J. Electrochem. Soc.*, **123**, 734, (1976).

36. Dataquest Report, "Pushing Impurities to 1 ppb: Japanese Chemical Industry Leads the Way", *Solid State Technology*, Nov. '85, p. 69.

37. P. Gise, "Principles of Laser Scanning for Defect and Contamination Detection in Microfabrication", *Solid State Technol.*, Nov. '83, p. 163.

PROBLEMS

1.   If a layer is etched isotropically in a process in which neither the mask nor the substrate material is attacked by the etchant, sketch the profile of the etched feature at the instant when the layer is etched, and again after an overetch of 100% and 200% have been performed.   What shape does the etched profile begin to assume as the overetch time is increased?   From this data, if an etched feature is inspected after stripping the resist and vertical walls are observed, can it be concluded that the etch process has been achieved in an anisotropic manner?

2.   Why has ammonium persulfate been recommended as a replacement for sulfuric acid in inorganic resist-stripping solutions?

3. Explain how faceting, sputtering from the backing plate, gas phase collisions, and collisions with shutters or cathode surfaces can cause contamination during *in situ* sputter etch cleaning steps.

4.   Show that Eq. 11 is the valid expression for the uniformity factor when the worst-case condition of *the thinnest and fastest etching portion of a film is assumed to lie over the fastest*

etching region of the substrate.

5. A window in SiO_2 which is 5000Å thick is to be etched over a region of Si substrate containing a 2500Å deep pn junction. It is decided that no more than 1500Å of this junction thickness can be removed during the etch. Determine the S_{fs} if the oxide thickness uniformity is ±5%, the etch rate uniformity is ±10%, and a 20% overetch is specified.

6. Polysilicon features 4000Å thick cross over a field oxide step of 7000Å in height and over a gate SiO_2 layer that is 300Å thick. Determine the S_{fm} and S_{fs} if the polysilicon etch rate uniformity is ±5 %, and A = 1 and A_m = 0.4. In this problem the polysilicon thickness uniformity is ±3%, the mask edge profile is 80°, it is desired to control the linewidth to 0.1 μm, and the mask etch rate uniformity is ±3%.

7. Explain how the isoetch curves of Fig. 13 are interpretted. If silicon is to be etched in a solution of HF: HNO_3: CH_3COOH = 1:x:1, show how the etch rate changes as x is varied from 1 to 20.

8. Describe some of the problems associated with wet etching of aluminum.

9. Summarize the advantages and prolems of lift-off as a patterning technique.

10. Suggest some problems that may be encountered when resist is stripped by dry-etching.

16

DRY ETCHING for VLSI

FABRICATION

In Chapter 15 the procedure of transferring patterns onto regions of silicon wafers by wet etching was described. Wet etching was the standard pattern transfer technique in the process sequences used to fabricate early generations of integrated circuits. Its widespread use stemmed from the facts that the technology of wet etching was well established, and that liquid etchant systems are available with very high *selectivity* to both substrate and the masking layer (defined in Chap. 15). As noted in Chap. 15, however, wet etching processes are typically isotropic. Therefore, if the thickness of the film being etched is comparable to the minimum pattern dimension, undercutting due to isotropic etching, becomes intolerable (Fig. 1a & b). Since many films used in VLSI fabrication are 0.5-1.0 μm thick, reproducible and controllable transfer of patterns in the 1-2 μm range becomes difficult if not impossible with wet etching. Alternative pattern transfer processes must therefore be employed to fabricate devices with such dimensions.

One alternative pattern transfer method that offers the capability of non-isotropic (or *anisotropic*) etching is "dry" etching. As a result, considerable effort has been expended to develop dry etch processes as replacements for wet etch processes. Dry etching also offers the important manufacturing advantage of eliminating the handling, consumption, and disposal of the relatively large quantities of dangerous acids and solvents used in wet etching and resist stripping processes. Dry etching and resist stripping operations utilize comparatively small amounts of chemicals (although, as will be discussed later, some of these may still be quite toxic or corrosive). This chapter deals with the technology of dry etch processes for VLSI fabrication, although many of the terms generic to both wet and dry etching are defined in Chap. 15.

Before launching into a description of the details of dry etching, it is worthwhile to identify the characteristics that a useful etching process should exhibit. This approach helps to define the problems that must be overcome when developing adequate dry etch processes, and shows why some types of dry etch processes may not be suitable for all VLSI applications.

The *overall goal of an etch process for VLSI fabrication is to be able to reproduce the features on the mask with fidelity.* This should be achievable together with control of the following aspects of etched features: a) the slope of the feature sidewalls (e.g. the slope of the sidewalls of the etched feature should have the desired specific angle, in some cases vertical, Fig. 1c - f); and b) the degree of undercutting (i.e. usually the less undercutting the better). In addition to this capability, a useful etch process should have the following characteristics:

1) It should be highly selective against etching the mask layer material;
2) It should be highly selective against etching the material under the film being etched;
3) The etch rate should be rapid, or the throughput of a machine performing the etch should be suitably high;

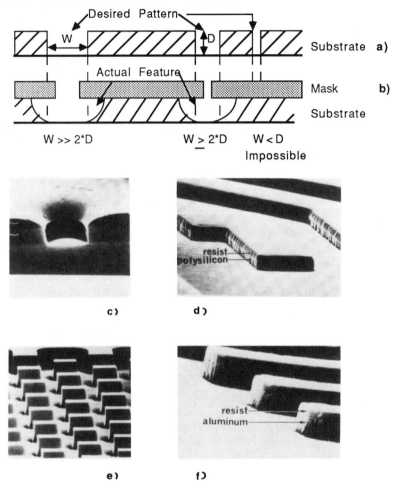

Fig. 1 (a) and (b) Isotropic etching of narrow and deep grooves. (a) shows the desired pattern and (b) shows how the mask must be dimensioned in order to obtain a pattern which resembles the desired pattern. (c) - (f) SEM micrographs show the results of highly anisotropic etching for several materials. (c) Anisotropically etched contact hole in an SiO_2 layer over Si. (d) Anisotropically etched poly-Si film over SiO_2. (e) 1 μm-wide features in single-crystal Si. (f) Anisotropically etched 1.5 μm thick Al-0.7% Cu film, with an SiO_2 substrate[33]. Copyright 1983, Bell Telephone Laboratories, reprinted with permission.

4) The etching should be uniform across the entire wafer, from wafer-to-wafer, and from run-to-run;

5) The process should be safe;

6) The etch process should cause minimal damage to substrates;

7) The etch mask material should be easily removable after the etching is completed;

8) The process should be clean (i.e. low incidence of particulate and film contamination);

9) The process should be conducive to full automation.

As shown in Fig. 2 there is a variety of dry etch process types. This figure also indicates that the mechanism of etching in each type of process can have a physical basis (e.g.

glow-discharge sputtering [Chap. 10], or ion milling), a chemical basis (e.g. plasma etching), or a combination of the two (e.g. reactive ion etching, RIE, and reactive ion beam etching, RIBE).

In processes that rely predominantly on the physical mechanism of sputtering (including RIBE), the strongly directional nature of the incident energetic ions allows substrate material to be removed in a highly anisotropic manner (i.e. essentially vertical etch profiles are produced). Unfortunately, such material removal mechanisms are also quite non-selective against both masking material and materials underlying the layers being etched. That is, the selectivity depends largely on sputter yield differences between materials. Since the sputter yields for most materials are within a factor of three of each other, selectivities are typically not adequate. Furthermore, since the ejected species are not inherently volatile, redeposition and trenching (see Chap. 10) can occur. Another major problem of pattern transfer by physical sputtering involves the redeposition of nonvolatile species on the sidewalls of the etched feature[7]. As a result of these drawbacks, dry etch processes for pattern transfer based on physical removal mechanisms have not found wide use in VLSI fabrication applications.

On the other hand, dry processes relying strictly on chemical mechanisms for etching can exhibit very high selectivities against both mask and underlying substrate layers. Such purely chemical etching mechanisms, however, typically etch in an isotropic fashion. Although some applications in VLSI fabrication (e.g. photoresist stripping in oxygen plasmas) utilize such processes, the problem of undercutting associated with isotropic etching is not solved by them.

By adding a physical component to a purely chemical etching mechanism, however, the shortcomings of both sputter-based and purely-chemical dry etching processes can be surmounted. Dry etch processes based on a combination of physical and chemical mechanisms offer the

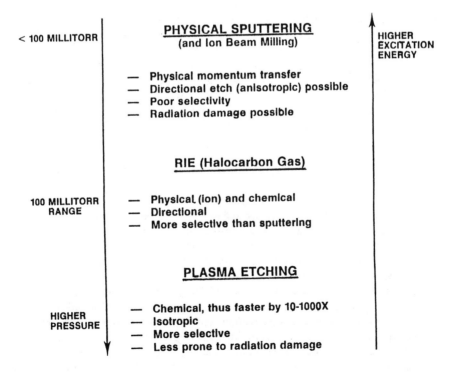

Fig. 2 The dry-etching spectrum.

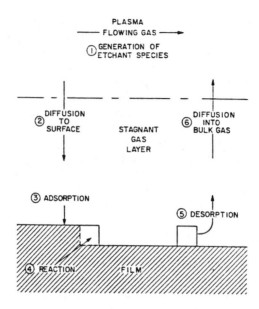

Fig. 3   Primary processes occurring in a plasma etch process.   Reprinted with permission of Ref. 1, Copyright, 1983, the American Chemical Society.

potential of controlled anisotropic etching, together with adequate selectivity.

In this chapter we primarily focus on *plasma etching processes* (i.e. purely chemical) and *reactive ion etching (RIE) processes* (i.e. physical /chemical processes, which are more aptly described as *ion-assisted etching processes*).   Even though purely chemical processes find less application than RIE for VLSI pattern transfer, an understanding of RIE processes is facilitated if the mechanisms which underlie plasma etching processes are understood first.   Sputter-based etching processes receive little attention here, as their use for pattern transfer in VLSI production is limited, and substantial information on sputtering processes is presented in Chap. 10.

BASIC PHYSICS and CHEMISTRY of PLASMA ETCHING

The basic concept of plasma etching is rather direct. A glow discharge is utilized to produce chemically reactive species (atoms, radicals, and ions) from a relatively inert molecular gas. The etching gas is selected so as to generate species which react chemically with the material to be etched, and whose reaction product with the etched material is volatile. An ideal dry etch process based solely on chemical mechanisms for material removal, can thus be broken down into six steps, as shown in Fig. 3.   These steps are listed as: 1) reactive species are generated in a plasma; 2) these species diffuse to the surface of the material being etched; 3) the species are adsorbed on the surface; 4) a chemical reaction occurs, with the formation of a volatile by-product; 5) the by-product is desorbed from the surface; and 6) the desorbed species diffuse into the bulk of the gas[1].   If any of these steps fails to occur, the overall etch cycle ceases. Product desorption is the most important step. Many reactive species can react rapidly with a solid surface, but unless the product has a reasonable vapor pressure so that desorption occurs, no etching takes place. Steps 1, 2, and 6 involve events occurring in the gas phase and plasma, while steps 3, 4, and 5 are steps that take place at the surface of the solid layer being etched.

Hence, it is useful to briefly consider the physics and chemistry of events that involve the etching process that occur in a) the plasma, and b) the etched surface.

The Reactive Gas Glow Discharge

In Chap. 10, the methods for producing a glow-discharge using a dc diode and an rf diode configuration are described. In plasma-etching processes an rf diode configuration is normally used to establish the glow discharge, for the reasons listed in Chap. 10. The glow discharge in Chap. 10 was treated primarily as a source of energetic ions which are used to bombard target surfaces and cause sputtering. In plasma etching applications the glow discharge can be used to produce energetic ionic bombardment of the etched surface, but it also has another even more important role, that of producing reactive species for chemically etching the surfaces of interest. Thus, it is necessary to examine the properties of glow discharges related to this function.

Since plasmas consisting of fluorine-containing gases are extensively used for etching Si, SiO_2, and other materials used in VLSI fabrication, it is appropriate to study the glow discharge of CF_4 gas as an example. Before a glow discharge is established, the only species present are CF_4 molecules. Over the pressure range at which an rf glow discharge can be maintained, 1 Pa (7.5 mtorr) - 750 Pa (5.6 torr), the gas density ranges from 2.7×10^{14} - 2×10^{17} molecules /cm^3. When a glow-discharge exists, some fraction of the CF_4 molecules are dissociated into other species. As will be described in a later section dealing with the *utilization factor*, it is better to have a CF_4 flow rate that is much larger than the dissociation rate. For moderately sized etching systems this represents a flow rate of ≥ 10 sccm^2.

A plasma is defined to be a partially ionized gas composed of ions, electrons, and a variety of neutral species. A glow discharge is a plasma that exists in the pressure range given above, containing approximately equal concentrations of positive particles (positive ions) and negative particles (electrons and negative ions). The density of these charged particles in glow discharges ranges from 10^9-10^{11} /cm^3, and the fraction of ions-to-neutral species is thus $\sim 10^{-4}$-10^{-6}. The average energies of electrons in glow discharges is between 1-10 eV. The reactions that occur in the gas phase (plasma) are called *homogeneous* reactions, while those that occur at the surface are termed *heterogeneous* reactions. Table 1 lists the general types of homogeneous electron-impact

Table 1. HOMOGENEOUS REACTIONS (ELECTRON-IMPACT) and HETEROGENEOUS REACTIONS THAT OCCUR IN PLASMAS

Homogeneous Reactions - Electron Impact Reactions	Heterogeneous Reactions
Excitation (rotational, vibrational, electronic): $e + A_2 \Rightarrow A_2 + e$ $(e + F \Rightarrow F* + e)$ $(F* \Rightarrow F + h\nu_F)$	Atom Recombination: $S - A + A \Rightarrow S + A_2$ Metastable deexcitation:
Dissociation: $e + A_2 \Rightarrow A + A + e$ $(e + O_2 \Rightarrow O + O + e)$	$S + M^* \Rightarrow S + M$ Atom abstraction (etching):
Ionization: $e + A_2 \Rightarrow A_2^+ + 2e$ $(e + O_2 \Rightarrow O_2^+ + 2e)$	$S - B + A \Rightarrow S^+ + AB$ Sputtering (etching):
Dissociative Ionization: $e + A_2 \Rightarrow A^+ + A + 2e$ $(e + O_2 \Rightarrow O^+ + O + 2e)$	$S - B + M^+ \Rightarrow S^+ + B + M$
Dissociative Attachment: $e + A_2 \Rightarrow A^+ + A^- + e$	

reactions and heterogeneous surface-plasma reactions that can take place.

The properties listed above, impart glow discharges with unique and useful capabilities. The first ionization potential of most gas atoms and molecules is ≥ 8 eV. Since the energies of the plasma electrons have a distribution whose *average* is between 1-10 eV, some of these electrons will be energetic enough to cause such ionization. Collisions of these energetic electrons with neutral etch gas molecules (Table 1), in fact, are primarily responsible for the production of the reactive species in a plasma (*electron-impact reactions*). These reactive species, however, can also react with themselves in the plasma (*inelastic collisions among heavy particles,* Table 1), and alter the overall plasma chemistry.

The most abundant *ionic specie* found in CF_4 plasmas is CF_3^+, and such ions are formed by the electron-impact reaction[3]: $e + CF_4 \Rightarrow CF_3^+ + F + 2e$. Other ionization reactions also occur, but the products of these reactions are found in less abundance than CF_3^+ ions, because the probability that such reactions occur is smaller, and their products also react with CF_4, while CF_3^+ ions do not. In addition to CF_4 molecules, ionic species, and electrons, there are a large number of radicals that are formed. A ra*dical* is an atom, or collection of atoms, which is electrically neutral, but which also exists in a state of incomplete chemical bonding, making it very reactive. Some examples of radicals include F, Cl, O, H, and CF_x, where x = 1, 2, or 3. In CF_4 plasmas, the most abundant radicals are CF_3 and F, formed by the reaction: $e + CF_4 \Rightarrow CF_3 + F + e$. In general, radicals are thought to exist in plasmas in much higher concentrations than ions, because they are generated at a faster rate, and they survive longer than ions in the plasma. This view is substantiated by measurements of radical concentrations in plasmas, but in fact, only a few such measurements have been reported. One measurement determined that the F atom pressure was 20 percent of the total gas pressure in the system[4]. To summarize, the gas in an etch chamber when plasma etching is underway, generally consists of the following species[5] (in order of decreasing concentration, and estimated concentration ranges): a) etch gas molecules (70-98% of the total species in the chamber); b) etch-product molecules (2-20%); c) radicals 0.1-20%); charged species, including positive ions, electrons, and negative ions (0.001-0.01%).

The radicals, in fact, are responsible for most of the actual chemical etching phenomena that occur at the surface of the material being etched (except for the etching of Al, which is apparently etched by molecular Cl_2). As will be described later, the ionic species are believed primarily to enhance the etching that occurs, by causing events that are not in them- selves chemical reactions. Thus, the term *reactive ion etching*, that is commonly used to denote processes in which plasma etching is accompanied by ionic bombardment, is actually somewhat of a misnomer. Since the etching by the reactive radicals is principally *enhanced* by the ionic bombardment, these processes would more aptly be described as *ion-assisted etching processes*.

Electrical Aspects of Glow Discharges

It is important to have some information about the electrical potential distribution in systems containing glow discharges. In Chap. 10, details about this subject were given for both dc and rf diode glow discharges, and the information was used to explain how glow discharge sputtering occurs. In plasma etching systems, high frequency (13.5 MHz) rf diode configurations are primarily used, and readers are directed to Chap. 10 for more general information on rf glow discharges. In plasma etching systems, knowledge about the potential distribution is useful because the energy with which particles impinge on the etched surface depends on the potential distribution. In addition, the plasma potential determines the energy with which ions strike other surfaces in the chamber, and high energy bombardment of these surfaces can cause sputtering and consequent redeposition of the sputtered material (as contamination on the wafers).

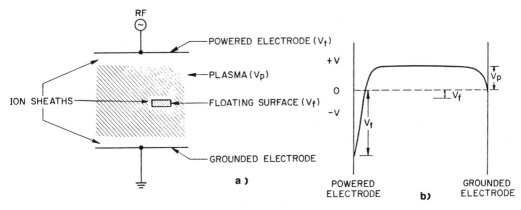

Fig. 4 (a) Schematic view of an rf glow discharge. (b) Potential distribution in a parallel plate plasma etcher with a grounded surface area larger than the powered electrode area.

As shown in Fig. 4 the potential of the plasma is positive relative to that of the grounded electrode (which is usually connected electrically to the chamber walls, grounding them as well), and the powered electrode develops a negative dc self-bias voltage relative to ground, as described in Chap. 10[6]. The magnitude of the self-bias voltage depends on the amplitude of the rf signal applied to the electrodes. As also noted in Chap. 10, if the electrodes of the rf plasma system are of comparable area, the potential difference across the dark space of both electrodes will be equal. Since the powered electrode develops a negative dc self-bias voltage, in order for a potential difference of equal magnitude to exist across the dark space of the grounded electrode, the plasma must assume a positive potential of comparable magnitude. Thus, even if wafers are placed on the grounded electrode of such systems, they will be subjected to substantial energetic ion bombardment. In systems in which the area of the powered electrode is much smaller than that of the grounded electrode, smaller potential differences exist between the plasma and the grounded electrode, and thus grounded surfaces are subjected to less energetic bombardment.

If a 13.56 MHz frequency is used for the applied rf power, this frequency is high enough so that the ions require several rf cycles to traverse the dark space between the bulk plasma region and the wafer surface. Some investigation of systems using lower frequency power has also been conducted (100-450 kHz). Under such circumstances, the ions can cross the dark space in a relatively small fraction of an rf cycle. This can enable ions to strike the surface with greater energies than in the high frequency case, a condition that can be useful in some applications.

Heterogeneous (Surface) Reaction Considerations

The reactive etch species that undergo chemical reactions at the surface that result in etching, and the ionic species that bombard the surface to enhance such etching are produced in the plasma. The events that take place at the surface are interactions between the gas phase species and the solid material to produce etching. The issues related to the mechanisms that occur on the surface include[32]: a) the sticking probabilities of radicals and ions; b) chemical recombination processes that form films, cause species to be adsorbed, or lead to other gas phase species; c) reaction paths which are followed, from adsorption to the eventual formation of volatile products; d) desorption of species from the surface; e) effect of ion and electron fluxes on the surface; and f) the synergistic effects on the surface of of multiple species bombardment (i.e. by ions, electrons, and photons). As shown in Fig. 5, some of the parameters that impact heterogeneous reactions[8], include the surface temperature, the surface electrical potential, the nature of the surface, and

Table 2   **EXAMPLES OF SOLID-GAS SYSTEMS USED IN PLASMA ETCHING**

SOLID	ETCH GAS	ETCH PRODUCT
Si, SiO_2, Si_3N_4	CF_4, SF_6, NF_3	SiF_4
Si	Cl_2, CCl_2F_2	$SiCl_2$, $SiCl_4$
Al	BCl_3, CCl_4, $SiCl_4$, Cl_2	$AlCl_3$, Al_2Cl_6
Organic Solids	O_2	CO, CO_2, H_2O
	O_2 + CF_4	CO, CO_2, HF
Refractory Metals (W, Ta, Mo...)	CF_4	WF_6, ...

geometrical aspects of the surface (e.g. the angle of incidence of impinging ions depends upon whether they are striking the bottom, or the sidewall of an etched feature).

The gases adopted for plasma etching processes have been selected on the basis of their ability to form reactive species in a plasma, which then react with the surface materials being etched and lead to volatile products. Table 2 lists the gas-solid systems for various solids to be etched in VLSI fabrication, together with their resultant etch products.

Parameter Control in Plasma Processes

One of the more challenging aspects of implementing a useful and reproducible etch process

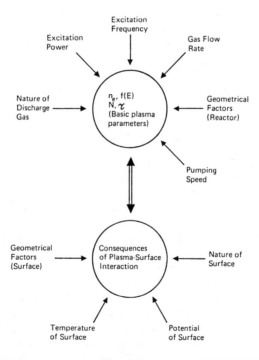

Fig. 5 Representation of the parameter problem in plasma etching systems (n_e is the electron density, f (E) is the electron energy distribution function, N is the gas density, and τ is the residence time[8]. Reprinted with permission of Springer-Verlag Publishing Company.

involves the control of the large number of parameters that affect the process. Figure 5 illustrates some of the parameters that impact the gas-phase interactions, as well as the surface-plasma interactions. Although many macroscopic parameters can be controlled , such as the type of feed gas, power, and pressure, the precise effect of making any changes in these parameters is usually not well understood. In fact, a change in a single macroscopic parameter typically alters two or more basic *plasma* parameters, and possibly one or more of the surface parameters, such as temperature or electrical potential. This makes process development in plasma systems a challenge, and the use of factorial experimental design techniques for such tasks very useful[9] (see Chap. 18). In the introduction to the section on *Dry-Etch System Configurations*, a discussion is presented on how gas flow, pumping speed, and pressure are interrelated, and how this interrelationship is used to control pressure.

ETCHING SILICON and SILICON DIOXIDE in FLUOROCARBON-CONTAINING   PLASMAS

The etching of silicon and SiO_2 in fluorocarbon plasmas is described in this section in substantial detail. This is done because these etching processes are very important in silicon VLSI fabrication. In addition, when the mechanisms of plasma etching were being first studied, the etching of silicon and SiO_2 in plasmas containing CF_4, mixtures of $CF_4 + O_2$, and mixtures of $CF_4 + H_2$, yielded important data about many of the fundamental mechanisms that are operative in plasma etching, as well as information about the specific materials system under investigation. The conclusions from these studies led to the development of two models for organizing chemical and physical information on plasma etching.   These models are the *fluorine-to-carbon ratio model (or F /C model)*[10], and the *etchant-unsaturate model*[11]. Since the models are conceptually similar, although they emphasize different aspects of plasma etching, we describe only the F /C model. Details of the etchant-unsaturate model are given in Ref. 25.
We begin the discussion by considering several basic phenomena related to plasma etching processes. First, it is known that in the absence of a glow discharge, the gases commonly used in plasma etching do not react with the surfaces to be etched. For example, CF_4 does not etch silicon without a discharge. This is due to the fact that CF_4 does not chemisorb on Si, and thus

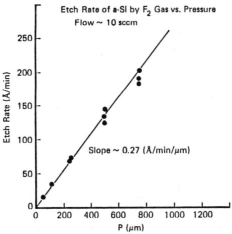

Fig. 6 The fluorine pressure dependence of the etch rate of amorphous silicon at room temperature[12]. Reprinted with permission of the publisher, the Electrochemical Society.

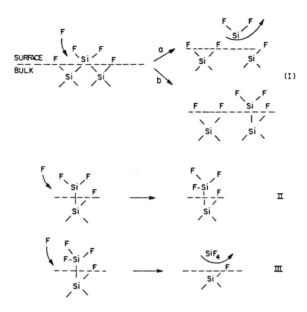

Fig. 7 Proposed mechanism for F-atom reaction with a silicon film leading to the products SiF_2 and SiF_4[13]. Reprinted with permission of the American Physical Society.

the step 3 of the dry-etching process described earlier does not occur. On the other hand, fluorine has been found to spontaneously etch Si, even without the presence of a discharge (Fig. 6)[12]. Thus, when a discharge of CF_4 is created, it is not the CF_4 molecules themselves that participate in the etching reaction. Instead, the etching is accomplished by the radical species which are created by the dissociation of CF_4 molecules; namely fluorine atoms. The products of the Si-etching reaction are SiF_4 and SiF_2. A mechanism for the F-atom reaction with a Si film leading to gasification products has been proposed[13], and is summarized in Fig. 7. The

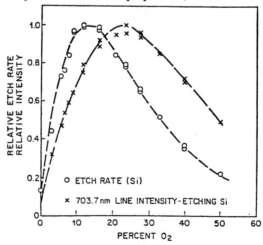

Fig. 8 The normalized etch rate for Si and the normalized intensity of the emission from electronically excited F atoms (703.7 nm line) versus the O_2 concentration in the CF_4-O_2 etch gas[4]. Reprinted with permission of the American Physical Society.

steady-state surface seems to be a stable "SiF$_2$-like" that must be penetrated by impinging F atoms in order for the SiF$_4$ to be formed.   The etch rate of silicon (and SiO$_2$) in pure CF$_4$, however, is relatively low.

If small concentrations of O$_2$ are added to the CF$_4$ feed gas, however, the etch rates of both Si and SiO$_2$ are observed to dramatically increase (Fig. 8)[14]. The addition of the O$_2$ is also accompanied by an increase in the density of F-atoms in the discharge. Although several reasons have been advanced for this effect, it is certain that reactions between the oxygen atoms (or molecules) and the CF$_4$ molecules are responsible for the increased F-concentration. One of these reactions might be the gas phase oxidation of CF$_3$, to first form COF$_2$ + F, which then dissociates into CO + F$_2$. Another suggested reaction could involve CF$_3$ radicals that reach the silicon surface, and upon adsorption, contribute one C atom and 3 F-atoms to the surface. If an oxygen atom reacts with the adsorbed C atom, the 3 remaining F atoms are available to etch the Si.   In any case, the etch rate of Si continues to increase until ~12% O$_2$ (by volume) is added. The etch rate of SiO$_2$ reaches its maximum value when ~20% O$_2$ is added. At greater concentrations, the additional O$_2$ dilutes the F concentration, and causes the etch rate to decrease. Figure 8 also shows that Si is etched much more rapidly than SiO$_2$ in CF$_4$-O$_2$ plasmas, and thus high selectivity of Si over SiO$_2$ is in such plasmas is easy to obtain.

If H$_2$ is added to the CF$_4$ feed gas, the etch rate of silicon decreases monotonically to almost zero for H$_2$ additions $\geq$ 40%. The etch rate of SiO$_2$, however, remains nearly constant for H$_2$ additions of up to 40% (Fig. 9)[15].   The silicon etch rate decrease occurs because the molecular hydrogen reacts with fluorine to form HF, and this drastically reduces the F atom concentration in the plasma. (It is said that the hydrogen *scavenges* F atoms).   Although the effect of the Si etch rate decrease by itself may not be useful, the fact that the SiO$_2$ etch rate does not substantially decrease at the same time is valuable, because the *SiO$_2$-to-Si etch rate ratio* increases. As a result, this allows a higher *selectivity* with respect to the substrate to be achieved when etching SiO$_2$ over Si.   This selectivity is necessary when SiO$_2$ films must be etched down to an underlying Si layer, without significantly etching the Si.

The mechanism responsible for the high SiO$_2$-to-Si etch rate selectivity involves the

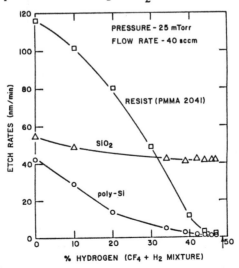

Fig. 9 The etch rate of Si, resist, and SiO$_2$ (measured in a reactive ion etching configuration) as a function of the concentrations of H$_2$ in the CF$_4$-H$_2$ etch gas[15]. Reprinted with permission of the publisher, the Electrochemical Society.

combination of two phenomena: 1) the deposition of a nonvolatile residue; and b) the role of oxygen in the etching of SiO_2. If an nonvolatile layer (e.g. carbon residue) deposits on a surface during etching, and it is not removed, etching will cease. While such carbon residues are found to deposit on all surfaces inside an etch chamber containing CF_4-H_2 plasmas, less accumulation is observed to occur on oxide surfaces than on non-oxide surfaces. There are several ways in which carbon can be deposited on a surface in fluorocarbon discharges. One way involves the dissociation of CF_3, or other fluorocarbon radicals, upon being chemisorbed on a surface. Less residue accumulates on SiO_2 surfaces because some of the carbon combines with the oxygen in the SiO_2 to form CO and CO_2, which are volatile. This in turn allows the SiO_2 layer to continue to be etched under conditions when etching of the Si has ceased. Nevertheless, if the deposition rate of the carbon residue becomes too great, etching eventually stops on *all* surfaces in the chamber, including SiO_2 surfaces. Other gases which also consume F atoms have been found to produce high SiO_2-to-Si selectivities[16] (even without the use of H_2), including CHF_3, C_3F_8, and C_2F_6. The reason for this effect is described as a part of the discussion on the *F /C ratio model*.

In practice, the exact process conditions that produce selective etching of SiO_2 over Si are generally empirically derived for each reactor. This is due to the fact that high selectivity requires that the process be operated very close to the demarcation between etching and polymerization (where etching abruptly ceases). Although the adjustment of plasma conditions to achieve high SiO_2 /Si selectivity remains an art, high selectivity is achievable. For example, selectivities of >20:1 at oxide etch rates of 600-1000Å / min have been reported[17].

Fluorine-to-Carbon Ratio Model

The *fluorine-to-carbon ratio (F /C) model*[10] is one of the two models which have been evolved to assist in assimilating the large amount of information on chemical and physical mechanisms observed in plasma etching. That is, the model represents an attempt to organize such information into a framework that allows processes to be developed more efficiently, by providing some basis for predicting the effects of various parametric variations. The F /C ratio is the ratio of the fluorine-to-carbon species, which are the two "active species" involved in the etching of Si and SiO_2 (as well as other materials etchable in fluorocarbon plasmas, including Si_3N_4, Ti, and W). The F /C ratio model does not attempt to account for the specific chemistry taking place in the glow discharge, but instead treats the plasma as a ratio of F to C species which can interact with the Si or SiO_2 surface. The generation or elimination of these "active species" by various mechanisms or gas additions then alters the initial F /C ratio of the inlet gas. Increasing the F /C ratio increases Si etch rates, and decreasing the F /C ratio lowers them.

For example, a pure CF_4 feed gas has an F /C ratio equal to four. If the plasma environment causes Si etching, however, this phenomenon consumes F atoms without consuming any carbon, and thus the F /C ratio is reduced. If more Si surface is added to the etching environment, the F /C ratio is further decreased, and the etch rate is also reduced. The addition of H_2 to the CF_4 feed gas causes the formation of HF, but does not consume any carbon, thereby the F /C ratio and the etch rate are again reduced. Finally, the utilization of gases in which the F /C ratio is <4, such as CHF_3 or C_3F_8, also has the effect of producing an F /C ratio smaller than that present in a plasma of pure CF_4. Their use is found to produce very similar effects in the etching chemistry as the other two procedures. Plasmas in which the F /C ratio is decreased to less than 4, are termed *fluorine-deficient plasmas*.

Conversely, the addition of O_2 has the effect of *increasing* the F /C ratio, because the oxygen consumes more carbon (by forming CO or CO_2), than F atoms (by the formation of COF_2). Other feed gases that can be added to increase the F /C ratio include CO_2, F_2, and NO_2. The cause of the high selectivity of SiO_2-to-Si in $CF_4 + H_2$ plasmas can be elucidated

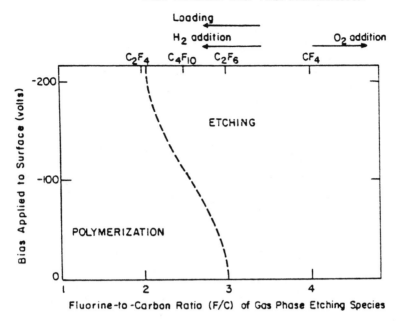

Fig. 10   Illustrative plot of the boundary between polymerizing and etching conditions as influenced by the fluorine-to-carbon ratio of the chemically reactive species and the bias applied to a surface in the discharge[10]. Reprinted with permission of the American Physical Society.

with the aid of the F /C ratio model.   That is, even though the F /C ratio in such plasmas is less than 4, the SiO_2 contributes oxygen (which consumes carbon), thus locally compensating for the decreased F /C ratio at the SiO_2 surface[19].   As long as the F /C ratio at the oxide surface has a value close to that which exists in a pure CF_4 plasma, SiO_2 etching continues to proceed at the same rate.   Meanwhile, since the F /C ratio at the Si surface is less than 4, the local etch rate decreases. To summarize, the F /C ratio model is useful for linking together the effects of a diverse set of phenomena, including the effects of many feed gases and SiO_2 /Si selectivity.

The F /C ratio can also be used to qualitatively portray the demarcation between etching and polymerization as it varies according to changes in some process condition (Fig. 10).   In this figure the boundary between polymerization (x-axis) is shown to vary as a bias voltage (y-axis) is applied to the substrate. The bias voltage has the effect of causing increased bombardment of the surface by energetic ions, which removes the nonvolatile residue layer by sputtering. This allows etching to occur at lower F /C ratios than if the carbonaceous polymer film were not simultaneously removed by sputtering and etching.   This effect, as will be described in the following section, may be utilized as a technique to control sidewall profiles of etched layers.

ANISOTROPIC ETCHING and CONTROL OF EDGE PROFILE

Up to this point in our discussion we have considered the etching of Si and SiO_2 in fluorocarbon plasmas largely as a mechanism that proceeds by chemical action (i.e. the reaction of Si by F-atoms generated by the plasma to form SiF_4).   If etching action is purely chemical, however, the removal of material is isotropic, and no advantage in dimensional control is gained over wet etching (Chap. 15).   In such processes, the plasma plays no role other than to produce the etchant. The attractiveness of dry-etching for VLSI patterning, however, is based on its potential to etch in an anisotropic manner (defined in Chap. 15), and the mechanisms that are believed to produce anisotropic etching need to be considered.

It should be mentioned before discussing details of the directional etching mechanisms, that the desired degree of directionality varies with the specific application in the final device. For example, if the lines etched from a deposited film are designed to carry current, steep-walled profiles are preferred, so that the cross-sectional area of the conductor is maximized. On the other hand, if an etched feature must be subsequently covered by another film, tapered profiles on the walls of the etched feature are more desirable, since highly anisotropic profiles in underlying topography may result in poor step coverage by the overlying film.   As also mentioned in Chap. 15, a highly anisotropic etch will cause "stringers" in the overlying film to be left behind at the base of steep underlying steps (Fig. 9, Chap. 15). Since such stringers represent regions of incomplete etching between adjacent lines, the etch process must remove them. This can be done in several ways including: a) overetching the overlying film; b) introducing a finite lateral etching component into the etch process; and c) insuring that the underlying feature steps have tapered wall profiles. Achieving wall profiles with the desired degree of slope may require the use of a multi-step etch process, in which each of the sub-processes would employ of an etch mechanism with its own degree of anisotropy.   A multi-step etch process for producing arbitrarily shaped wall profiles by such piece-wise anisotropic etching is described in Ref. 19.

The ability to achieve anisotropic etching is thought to depend in some way or another on the bombardment of the etched surface with energetic ions. Other parameters, such as the chemical nature of the plasma, may influence the degree of anisotropy, but unless energetic particles strike a surface, only isotropic etching can be expected. The directional etching effects in ion-assisted etching processes, however, cannot be due to sputtering alone, as product yields of over several hundred substrate atoms per incident ion have been reported in ion-assisted etch processes. Such product yields are much greater than those of typical sputtering yields (e.g. <2

ION - ASSISTED ETCHING OF Si

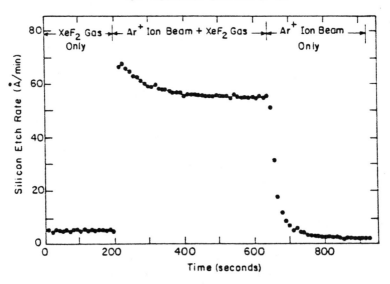

Fig. 11 An example of ion-assisted gas-surface chemistry in the etching of silicon with XeF_2. The XeF_2 flow is $2x10^{15}$ molecules /sec and the Ar energy and current are 450 eV and 2.5 μA, respectively[20]. Reprinted with permission of the American Physical Society.

SURFACE DAMAGE
MECHANISM

SURFACE INHIBITOR
MECHANISM

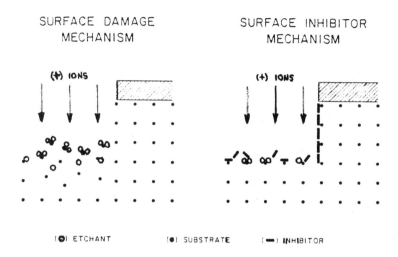

(●) ETCHANT   (●) SUBSTRATE   (━) INHIBITOR

Fig. 12 (a) Surface damage, and (b) surface inhibitor mechanisms, for ion-assisted anisotropic etching[13]. Reprinted with permission of the American Physical Society.

for 400 eV Ar ions, see Chap. 10).   This is fortunate, for as was discussed at the outset of the chapter, purely sputter etch mechanisms result in processes with inadequate selectivities.

The fact that sputtering alone is not operative in such processes, was elegantly demonstrated in an experimental manner by Coburn and Winters[20].   In one experiment, they first exposed a Si surface to a gas of XeF_2 (not a plasma of XeF_2), and observed a low etch rate (Fig. 11). Next, while continuing to expose the surface to XeF_2, an Ar^+ ion beam with an energy of 450 eV was directed at the Si. The observed etch rate was ~10 times as great as with the XeF_2 alone. Finally, when the Ar^+ beam alone was directed at the surface, the smallest etch rate of the three conditions was produced.   The results of this experiment demonstrate that a strong cooperative effect can be result if the etching surface is simultaneously exposed to a reactive gas and bombardment by energetic particles.   The microscopic details of exactly how the ion bombardment enhances the reaction between a reactive gas and a surface, is the subject of substantial research efforts. Evidence indicates that different mechanisms exist for specific chemical systems.

Two principal mechanisms by which energetic ions assist in enhancing the etch rate produced by reactive gases, however, have been postulated to be operative in directional etching processes (Fig. 12)[21]. They are: a) Relatively high energy impinging ions (> 50 eV) produce lattice damage at the surface being etched, extending several monolayers beneath the surface. Reaction at these damaged sites is enhanced compared to reaction at surfaces at which no damage has occurred; and b) lower energy ions (≤ 50 eV) provide enough energy to desorb *nonvolatile polymer layers* (also referred to as *surface inhibiting*, or *blocking* layers) that deposit on the surfaces being etched. In processes in which such polymer deposition occurs, surfaces not struck by the ions do not have the blocking layer removed, and hence are protected against etching by the reactive gas. In features being etched on a wafer, the incident energetic particles generally arrive in a direction perpendicular to the wafer surface, and hence they strike the bottom surfaces of the etched features. The sidewalls of the etched features, meanwhile, are subjected to little or no bombardment. As a result, the bottom of the features exhibit enhanced etching. Ion bombardment effects are enhanced by decreasing the pressure in a high frequency (> 5 MHz) plasma, or by decreasing the frequency of the discharge.

Examples of specific data that illustrates anisotropic etching are given in the discussions on the etching of the various types films, but two idealized examples will be given here to illustrate the kind of approaches that have been suggested to achieve directional etching.

In the first case, shown in Fig. 13, hypothetical Si and SiO_2 surfaces are subjected to positive ion bombardment as a result of a negative bias voltage of -150 V being applied to the wafers[22]. Since the etch rate of SiO_2 is zero without ionic bombardment, but finite rate under ion bombardment (Fig. 13), the SiO_2 film etches anisotropically. The silicon etch rate is finite but smaller on surfaces that receive no ionic bombardment, and thus the lateral etch rate is slower than the vertical etch rate. The etched feature ends up having a profile in which the sidewalls are not vertical.

In the second example, the chemistry of the discharge together with ionic bombardment is used to control the directionality of etching[22]. As was shown in the first hypothetical example, Si is etched more rapidly under energetic ion bombardment (e.g. from 150 eV ions) than when no bombardment occurs. If H_2 is added to the CF_4 feed gas, the Si etch rate decreases (Fig. 14). At some value of H_2 concentration, the non-bombarded surface etch rate decreases to zero, but the bombarded surfaces continue to be etched. Thus, under those conditions, the bottom of the feature is etched, while the non-bombarded sidewalls are not etched, and the resulting etched profile is vertical.

One other technique suggested for introducing a non-vertical shape to a feature being etched by a completely anisotropic process, involves the introduction of some rounding to the sidewall angle of the resist mask[23,24]. The selectivity of the etch process is then adjusted, so that the resist and the layer to be patterned are both etched anisotropically at the same rate. In this manner the slope of the resist wall is replicated in the walls of the etched layer (Fig. 15). In practice, this technique is difficult to successfully implement, because stringent control of all

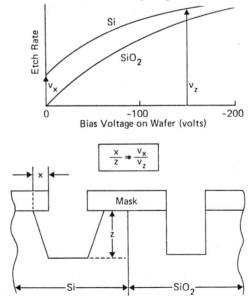

Fig. 13 Illustrative figure which shows the relationship between the shape of the etched wall profile and the dependence of the etch rate on the wafer potential[10]. Reprinted with permission of the American Physical Society.

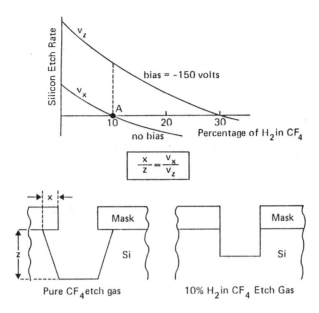

Fig. 14   Illustrative figure which shows the way in which the shape of a wall profile can be influenced by decreasing the fluorine-to-carbon ratio (in this example by H_2 addition)[10]. Reprinted with permission of the American Physical Society.

plasma parameters must be maintained (temperature, ion energy distributions, etc.).   It is a particularly formidable process to control when wafers are etched in a batch mode.

DRY-ETCHING VARIOUS TYPES of THIN FILMS

Silicon Dioxide (SiO_2)

It was described earlier how fluorocarbon-containing plasma can be used to etch SiO_2, and how selectivity with respect to silicon can be obtained by using fluorine-deficient plasmas. An example approach for producing anisotropic etching of SiO_2 was also illustrated. When etching contact holes in SiO_2, however, it is advantageous to have sloped contact sidewalls for easier filling by the following conductor layer. One method for producing such a slope is by controlled resist erosion, as described earlier[24]. By adjusting the resist : oxide etch ratio, the slope can be transferred to the oxide.   A technique for adjusting the *resist : oxide* etch rate ratio, is to add O_2 or SF_6 to the fluorocarbon feed gas. A second method for achieving sloped walls is the use of a reflow step, but this is not an etching procedure (see Chap. 6).   Controllably producing tapered sidewalls on small SiO_2 contacts by an etching process remains a difficult task.

The etch rate of the oxide films depends on several factors including pressure, power, feed gas composition, and film characteristics. For a given set of etching conditions, the film characteristics impact the etch rate. For example, thermal SiO_2 generally etches more slowly than CVD SiO_2 films, and the etch rate may depend on the dopant concentration in the SiO_2 layer.

Fig. 15 SEM photograph of a contact hole in a 1 μm thick phosphorus-doped CVD SiO_2 with tapered profile achieved by controlled resist erosion during etching[23]. Reprinted with permission of the Applied Materials.

Silicon Nitride

Fluorine atoms isotropically etch silicon nitride with a selectivity of silicon nitride : Si ~1:8 in the temperature range of 30-100°C[25]. As described in Chap. 6, two more or less distinct types of silicon nitride are used in VLSI fabrication. The first type is usually deposited by low pressure CVD (LPCVD) at 700-800°C, and results in a stoichiometric compound Si_3N_4 film. The second type is deposited by plasma-enhanced CVD techniques at ≤ 350°C, and such plasma deposited nitrides are really polymer-like Si-N-H materials. Although both nitrides are usually etched in CF_4-O_2 plasmas, the plasma nitride films normally etch more rapidly than the LPCVD films[23,25].

In most applications, nitride films are patterned with relatively coarse lateral dimensions (e.g. the opening of bonding pads in nitride passivation layers, or the patterning of a thin nitride layer [~1000 Å thick] in the LOCOS process), and thus a high degree of anisotropy is not usually necessary for patterning the nitride. In the LOCOS application, however, sufficient selectivity with respect to SiO_2 must exist so that the thin oxide layer under the nitride (~250 Å thick) is not etched away. This would expose the silicon under the SiO_2 to a fast-etching plasma, which as noted above, would etch the silicon ~8 times as rapidly as the nitride. The selectivity of a CF_4 isotropic Si_3N_4 etching process with respect to SiO_2 is ~2-3, but more favorable selectivity (e.g. 9-10) can be obtained with the use of NF_3 plasmas[26].

Polysilicon

In MOS applications, the gate length is a critical, fine line dimension that determines the channel length of the devices. Thus, when polysilicon serves as the gate material, it is paramount that the etched linewidth dimension faithfully reproduces the dimension on the mask (e.g. to within ±5%). A polysilicon etch process must therefore exhibit excellent linewidth control, and high uniformity of etching. In addition, a high degree of anisotropy is also generally required, as the doping of the source-drain and the polysilicon itself is typically performed by ion implantation. If the etch process produced sloped sidewalls in the polysilicon, then portions of the gate would not be thick enough to effectively mask the substrate against the implanation. This would produce devices whose channel length depended on the degree of sidewall taper, and unless the taper could be accurately controlled, would cause a manufacturing control problem.

The degree of anisotropy, however, is dictated by other considerations as well, including the extent of overetching required to remove stringers at the base of steep steps in the inderlying topography (Fig. 9, Chap. 15), and the coverage of the etched polysilicon features by subsequently deposited layers. In the first of these cases, completely anisotropic etching will require extensve overetching to remove the stringers, while in the second, it will produce features that may be difficult for overlying films to cover. Thus, in general, an important characteristic of a process is its ability to produce a profile with the desired degree of slope.

Finally, the polysilicon layer is usually deposited over thin SiO_2 (e.g. gate oxides, 250-500 Å thick). Thus, the etch process must be selective over SiO_2 etching, since if this oxide layer was removed, the shallow source-drain junction regions in the underlying Si substrate would be rapidly etched by the reactants that cause polysilicon etching. In some cases, where buried contacts between polysilicon and the single crystal substrate are made, high selectivity over single crystal Si must also be exhibited (a subject discussed in Ref. 30).

As described in the earlier section on etching of Si with fluorine-based plasmas, fluorine atoms etch Si isotropically, and hence controlled anisotropic etching of Si with fluorine-based chemistries, such as CF_4 or SF_6, is difficult. In addition, such processes exhibit a large loading effect, also an undesirable characteristic in etching processes (see later section on *Loading Effects*).   As a result, several other methods have been investigated to overcome these limitations. First, chlorine plasmas were found to etch polysilicon very anisotropically and exhibited excellent selectivty over SiO_2, but they etch Si more slowly than fluorine containing gases. Thus, etch gases containing both chlorine and fluorine have come to be preferred for polysilicon etching[27]. For example, a detailed description of an etching process utilizing SF_6 and Cl_2 is given in Ref. 28.   In that study, undercutting was controlled to less than 0.3 μm, while maintaining high etch rates and good selectivity over SiO_2, by adding small amounts of chlorine to SF_6 plasmas.   It was concluded that the chlorine together with the resist mask material reacted, to form a thin polymer that deposits on the sidewalls of the etching polysilicon, to produce less etching at the sidewalls.   Other proposed process for etching polysilicon with all of the required etching parameters, involve multi-step etching processes[27, 29].

Refractory Metal Silicides and Polycides

Refractory metal silicides are deposited onto polysilicon to form a low resistance *polycide*

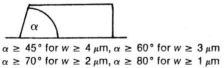

$\alpha \geq 45°$ for $w \geq 4\,\mu m$, $\alpha \geq 60°$ for $w \geq 3\,\mu m$
$\alpha \geq 70°$ for $w \geq 2\,\mu m$, $\alpha \geq 80°$ for $w \geq 1\,\mu m$
Sloped wall improves interconnection step coverage, but limited by ion implantation self-alignment technique

Fig. 16 Polysilicon etching[29]. Reprinted with permission of Solid State Technology, published by Technical Publishing, a company of Dun & Bradstreet.

structure that can serve as both a gate and an interconnect layer (Chap. 11). In many applications, etching is used to pattern such polycide structures, but this is a difficult etching task[25,33]. As in polysilicon etching, the process must provide a vertical profile on the etched polycide, good selectivity over oxide (i.e. >10), and minimal resist erosion. Etching of refractory metal silicides with both fluorine and chlorine based plasmas has been investigated, as both the fluorides and the chlorides of the refractory metals are relatively volatile (Fig. 17a). Etch gases that result in high concentrations of F atoms, however, are not suitable, as they tend to undercut either the n^+ polysilicon or the silicide (or both). On the other hand, fluorine-deficient plasmas can produce anisoropic etching of both the polysilicon and the silicides, but the etch ratio with respect to SiO_2 is less than one.

Chlorine plasmas are known to etch SiO_2 quite slowly, and anisotropic etching of n^+ polysilicon and silicides is easier to obtain. Unfortunately, the vapor pressures of the refractory metal chlorides are much lower than those of the fluorides, and so to assist the etching process, the use of mixtures of gases that consist of both fluorine and chlorine containing gases have been studied (e.g. $SF_6 : Cl_2$). Reference 31 describes the results of etching tantalum silicides and polycides with such mixtures (Fig. 17). It is reported that etch profiles vary widely with process parameters, especially the gas composition. As the silicide etch rates also vary with silicide composition, the successful implementation of a polycide etching process requires stringent process controls. In addition, it has been suggested that in order to achieve a polycide etch process that produces an appropriately-shaped polycide profile, together with high selectivity over SiO_2, a multi-step etch process may need to be employed.

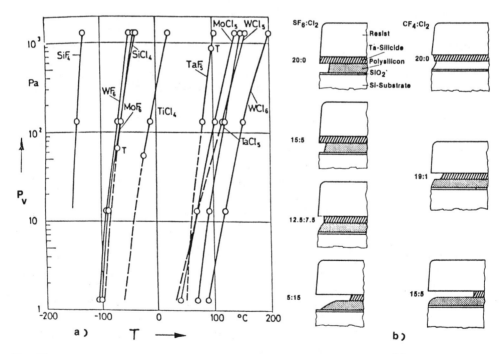

Fig. 17 (a) Vapor pressures of transition metal halides. (b) Polycide etching[31]. Reprinted with permission of Solid State Technology, published by Technical Publishing, a company of Dun & Bradstreet.

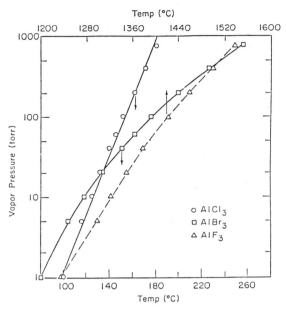

Fig. 18 Vapor pressure of AlF$_3$, AlCl$_3$, and AlBr$_3$ as a function of temperature[34].

Aluminum and Aluminum Alloys

The etching of aluminum and aluminum alloy films is a very important step in the fabrication of integrated circuits. The device density, on many of the most advanced circuits, is limited by the area occupied by the interconnect paths. Anisotropic etching of the metal layers permits the use of smaller minimum metal pitches (i.e. the *pitch* is the sum of the dimensions of a metal line and the space between lines), which increases the interconnect capability. Thus, the isotropic nature of aluminum wet etching processes renders them inadequate for VLSI applications, and there is a need for a directional dry etching process.

The fluorine-containing gases used to etch Si and SiO$_2$, however, are not suitable for etching aluminum since the etch product, AlF$_3$, has a very low vapor pressure (Fig. 18)[34]. On the other hand, other halides of Al (e.g. AlCl$_3$, Fig. 18) have sufficiently high vapor pressures to allow plasma etching of Al, and thus chlorine-containing gases have been exploited to develop dry-etch processes for aluminum films. The chlorine containing gases used to etch aluminum, BCl$_3$, CCl$_4$, SiCl$_4$, and Cl$_2$, however, are all either carcinogenic or highly toxic. They also have a propensity to dissolve and become concentrated in pump oils. Consequently, special care must be taken during system servicing to avoid skin contact or inhalation of vapors from cold traps, oil filters, pumps or pump oils of aluminum etching systems. The reactive compounds, AlCl$_3$ and BCl$_3$, also degrade silicone and hydrocarbon pump fluids, and to avoid pump fluid decomposition, perfluoropolyethers are often used (see Chap. 3). In addition, oxidation of AlCl$_3$, BCl$_3$, or SiCl$_4$ results in particulate formation (e.g. Al$_2$O$_3$, SiO$_2$). Such particles, together with polymeric residues in the oil, can produce vacuum pump bearing failure by plugging lubrication ducts[35]. Hence, it is recommended that oil filters be utilized to prolong pump life and allow longer intervals between oil changes.

It has been determined that a freshly exposed aluminum surface, uncovered by aluminum oxide (Al$_2$O$_3$) will react spontaneously with Cl or molecular Cl$_2$ to form AlCl$_3$, even in the absence of a plasma[36]. If, however, the surface of the aluminum is covered with a thin layer of

Al_2O_3 (i.e. a native oxide of ~30 Å), it will not react with Cl or Cl_2. Thus, etching of Al films is two-step process, involving removal of the native oxide layer, and etching of the Al film.

The successful removal of the native oxide is one of the most important steps in achieving an effective aluminum etching process. This is because removal of Al_2O_3 is far more difficult than the etching of pure aluminum, and the thickness of this oxide can vary from run-to-run, depending on several factors. Thus, an aluminum etch cycle is observed to begin with an *initiation period*, during which the native oxide and the moisture from the chamber is slowly removed. The removal can be accomplished by sputtering with energetic ions, a condition that can be established in reactive ion etching systems, or by chemical reduction. The chemical reduction of the Al_2O_3 requires the availability of oxide reducing species. The dissociation of BCl_3 or CCl_4, for example, produces fragments capable of reducing Al_2O_3. That is, CCl_x (where x <4) can reduce the oxide according to $Al_2O_3 + CCl_x \rightarrow AlCl_3 + CO$, but ion bombardment is still apparently necessary to assist these reducing reactions. If there is water vapor present in the etch chamber, however, it will scavenge the oxide-reducing species and react with the exposed aluminum to form new Al_2O_3. This represents one of the factors which contribute to non-reproducible initiation times. BCl_3 has become the preferred source gas for the role of native oxide removal in many processes[37]. BCl_3 is a much better getter of oxygen or water vapor than is CCl_4, and the etch products of CCl_4 have been identified as carcinogenic (see section on *Plasma Etching Safety Considerations*). Figure 19a shows the results of one report in which the aluminum initiation period was reduced to a relatively short interval, by increasing the flow rate of BCl_3. It should be noted, however, that BCl_3 is highly reactive and

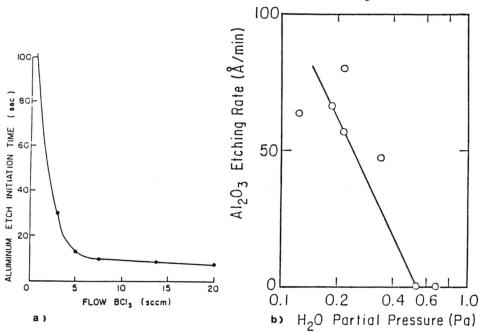

a)   b)

Fig. 19 a) Time necessary for initiation of Al etching vs. BCl_3 flow. Other parameters include: 20 sccm Cl_2, 200 sccm He, 200 Pa total pressure, 1.75 W /cm². Reprinted with permission of the publisher, the Electrochemical Society[38]. b) Effect of small concentrations of H_2O on the initiation period (designated as the Al_2O_3 etching rate)[39]. This figure was originally presented at the Fall, 1981 Meeting of the Electrochemical Society, held in Denver, Colorado.

forms nonvolatile residues upon contact with oxygen or water, for example in the pump exhaust line.

Water vapor, nevertheless, must be excluded from the etch chamber in order to achieve reproducible Al etch processes. As shown in Fig. 19b, the etch rate of Al_2O_3 decreases rapidly with increasing partial pressure of water vapor[39]. If an etch chamber is exposed to ambient after an etch run, moisture can be adsorbed on the chamber walls. This condition becomes more severe if the etch product of aluminum, $AlCl_3$, is allowed to deposit on the chamber walls[40]. Such deposition occurs on surfaces maintained at room temperatures. $AlCl_3$ is very hygroscopic, and thus absorbs considerable moisture on exposure to atmosphere. This moisture may be desorbed after the plasma is struck. The deposition of $AlCl_3$ also has other deleterious effects on the etch process, which are described later, and so minimizing its deposition is important. Techniques used to manage $AlCl_3$ in plasma etching applications include maintaining the temperature in the etch chamber above 35°C (as the $AlCl_3$ evaporation rate at such temperatures is high enough to assist in removing it from the chamber), and using large gas flow rates to keep its partial pressure low. Water vapor in the etch chamber is most effectively reduced by load lock chamber designs. In non-load locked designs, water vapor is decreased by use of extended pumpdowns, heating the chamber walls when the chamber is open (e.g. to 40°C), surrounding the chamber with a nitrogen purge box, use of BCl_3, and minimizing $AlCl_3$ buildup[40].

It should also be noted at this point that the oxygen content of the aluminum film can also play a significant role in determining the etch rate[41], as shown in Fig. 20. This oxygen can be incorporated during the aluminum sputter deposition step, as a result of the partial pressure of oxygen in the sputter chamber during deposition. Since oxygen concentrations of <5% in

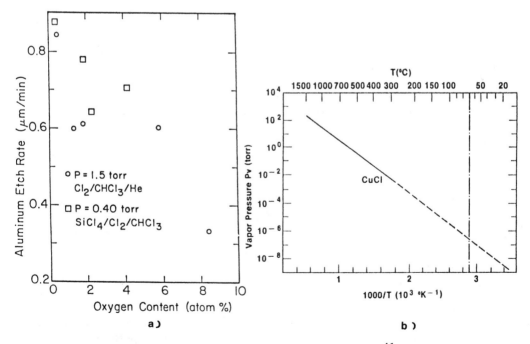

a)   b)

Fig. 20 (a) Effect of oxygen content in Al films on plasma etch rate[41]. (b) Vapor pressure of copper chloride as function of temperature[44]. Reprinted by permission   of Solid State Technology, published by Technical Publishing, a company of Dun & Bradstreet.

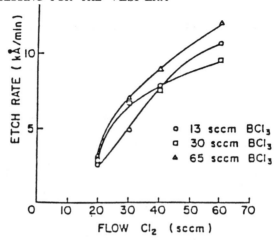

Fig. 21 Etch rate of Al vs. Cl_2 flow for different BCl_3 flows.   Other parameters include: 250 sccm He, 9.4 sccm $CHCl_3$, 160 Pa total pressure[42].   This figure was originally presented at the Fall, 1981 Meeting of the Electrochemical Society, held in Denver, Colorado.

aluminum do not alter the resistivity of the film, this contamination may go unnoticed. In addition, since the oxygen tends to segregate to the grain boundaries, the aluminum in these regions will etch more slowly, leaving a "stencil" of the grains. If a large concentration of oxygen is present, complete etching of grain boundary regions may require undue overetching.

After the native oxide is removed, the aluminum etching proceeds at a rate and with a resultant profile, that is affected by the gases used (e.g. Fig. 21[38]). In general, the demands of high throughput require the highest etch rate consistent with good results. Processes with high concentrations of Cl_2 in the feed gas exhibit isotropic etching, and the etch rate in such processes is not enhanced by energetic ionic bombardment.   Thus, it is postulated that anisotropic etching of aluminum occurs as a result of the formation of an inhibiting layer on the aluminum surface[42], which is removed on surfaces struck by energetic ions, allowing etching to occur there (Fig. 12b).   Such inhibiting layers are believed to arise from the formation of chlorocarbon polymers (e.g. CCl_x), originating from the carbon in CCl_4 or resist etch products. In fact, CCl_4, $CHCl_3$, and other chlorocarbon gases are added to Cl_2 plasmas to reduce undercutting.   It has also been observed that if aluminum is etched with an SiO_2 etch mask, with gases containing large concentrations of Cl_2 and no carbon containing species, isotropic etching occurs[43]. A partial list of the gases and gas mixtures (neglecting rare gas diluents) which have been reported to successfully etch aluminum include: CCl_4, BCl_3, Cl_2, CCl_4 + BCl_3, CCl_4 + Cl_2, BCl_3 + Cl_2, BCl_3 + Cl_2 + O_2, CCl_4 + Cl_2 + O_2, $SiCl_4$, and BBr_3.

Selectivity with respect to SiO_2 is sufficient for VLSI applications.   Selectivity with respect to silicon (or polysilicon), however, is typically inadequate with chlorine-containing gases (e.g. aluminum /silicon selectivities are typically only 2-5:1). As a result, Al patterns must overlap contact windows, and this serves to restrict the density of Al conductor lines.

Small quantities of other materials are added to aluminum to improve some of its properties. That is, 1-2 at% silicon is often added to prevent the aluminum from spiking through shallow junctions (see Chap. 10), and 2-4 at% copper, or 0.1-0.5 at% Ti (often together with Si), are added to enhance the electromigration resistance.   Since $SiCl_4$ is volatile at room temperature, Al-Si films are readily etchable in chlorine-containing gases. Titanium also forms volatile etch products ($TiCl_4$, Fig. 17a), and thus also does not pose a problem, provided the titanium is not oxidized. That is, titanium readily oxidizes, and TiO_2 is difficult to etch in chlorine plasmas.

Copper, on the other hand, forms an etch product with chlorine, CuCl, that is relatively non-volatile below temperatures of 175°C (Fig. 20b)[44]. Thus, copper containing residues often remain after these alloy films have been dry-etched. This makes Al-Cu more difficult to etch in chlorine plasmas. The degree of difficulty increases with increasing Cu concentration, and 4% Cu-containing films pose quite a formidable dry etching challenge. Two methods are used to promote CuCl desorption: increasing the substrate temperature, commensurate with the maximum temperature allowed with the resist material being used; and enhancing the ionic bombardment of the surface, so that significant sputtering occurs. In batch etchers, low pressure operation together with a slower etch rate, promotes high energy bombardment while allowing more time for the sputtering to remove CuCl[45]. In single wafer etchers, higher power is needed to produce sufficient ion bombardment under the conditions of greater operating pressure. This causes the unwanted effect of eroding the resist during etch. In addition, the etch product, $AlCl_3$, is highly reactive, and it can also attack and degrade the resist[40,46]. Thus, it is necessary to employ special UV-thermal resist stabilization steps to enable the resist to withstand such harsh etching environments. See Chap. 12 for more information on UV-stabilization procedures.

Another problem with etching Al is that of post-etch corrosion. The effect arises from the hydrolysis of the chlorine or chlorine-containing residues (mostly $AlCl_3$) which remain on the film sidewalls, substrate, or resist after etch. Upon absorbing moisture, these residues form HCl which corrodes the aluminum. The reaction of HCl and Al produces more $AlCl_3$, and thus as long as moisture is available, corrosion will continue (Fig. 22)[47]. The problem is even more severe in Al-Cu alloys[48], since Al-Cu compounds formed in the film (primarily $CuAl_2$) create a galvanic couple with the Al (Fig. 23), and this drives the corrosion even more rapidly than in pure aluminum films. Consequently, the residual chlorine and ambient moisture levels necessary to induce corrosion are much lower when Cu is present in the aluminum films.

Various techniques have been suggested to deal with the post-etch corrosion problem, all involving the removal of the chlorinated species . These include: a) removing the wafers from the chamber and rinsing in cold deionized water. It has been suggested, however, that this method is not likely to be effective for alleviating corrosion in Al-Cu alloy films[40]; b) plasma ashing of the resist in an O_2 plasma before removal from the etch chamber, which removes the chlorine present in the resist, and restores the passivating Al_2O_3 layer; and c) exposing the aluminum to a fluorine-containing plasma before removal from the chamber (e.g. CF_4 or

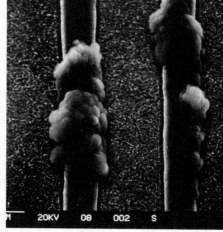

| 20µM | 20KV | 08 | 011 | S |

| 20KV | 08 | 002 | S |

Fig. 22 Corrosion of etched Al lines[47]. Reprinted by permission of Semiconductor Internat'l.

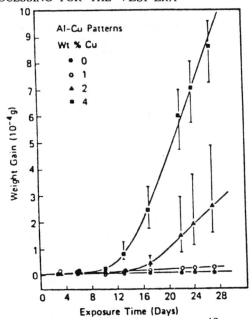

Fig. 23 Corrosion of Al-Cu films in Cl-containing plasmas[48]. Reprinted with permission of the American Physical Society.

CHF_3)[46]. The CF_4 exposure is believed to cause the chemisorbed chlorine to be replaced with fluorine, thereby passivating the Al by the formation of nonhydroscopic AlF_3. The CHF_3 exposure is thought to deposit a polymer film over the aluminum, thus "sealing" the surface and preventing moisture from penetrating to the chlorine residues. Long term reliability studies on the effectiveness of these treatments, especially on circuits in plastic packages, have yet to be reported in detail.

Organic Films

Organic films are exposed to plasma etching environments in many applications during VLSI fabrication.   Photoresist is most commonly used as an etch mask, and in such applications it is usually desired that the resist not be etched by the plasma.   In some cases, however, the resist is deliberately etched as part of a technique used to produce directional etching effects in underlying films (e.g. sloped contact sidewalls), or as a method for producing planarization of layers under the resist. In some of these instances, the etch rate of the resist must be accurately known and controlled. At the conclusion of the pattern etching step the resist must be removed, and this can be achieved by a plasma etch process as well. Organic film etching is also performed in dry-developable resist and tri-layer resist processes (see Chap. 12), and in etching polyimide films.

Plasmas containing pure oxygen at moderate pressures produce species that attack organic materials to form CO, CO_2, and H_2O as end products[1,25,49]. Such oxygen plasmas provide a highly selective method for removing organic materials, since the O_2 plasmas do not etch Si, SiO_2, or Al. The addition of fluorine-containing gases to the O_2 causes the etch rate of organic materials to significantly increase. This occurs because the F atoms extract hydrogen from the organic films to form HF, producing sites that react more rapidly with molecular oxygen.

Ion bombardment can also be used to accelerate organic film etching. Under conditions in which the concentrations of oxygen atoms are small (e.g. low pressure and loading), the etched profiles are completely anisotropic[25].

The resistance of organic materials (e.g. resists) to plasma conditions depends on the chemical composition of the organic layer and the type of plasma. There are some plasma conditions, however, in which the durability of commonly used positive resists is quite high, espcially those used for the selective etching of SiO_2 and Si_3N_4.

PROCESS MONITORING and END POINT DETECTION

Dry etch equipment used in a VLSI production environment requires the availability of effective diagnostic and etch end point detection tools. Extremely tight control of all process parameters must be maintained to ensure wafer-to-wafer and run-to-run reproducibility. In typical production facilities, some of these parameters can be controlled, while others cannot. For example, as will be described, reactor wall conditions, which contribute to the heterogeneous destruction of active reactants, become a *bona fide* variable, if the walls are exposed to atmosphere after every run. Similarly outgassing, virtual leaks, and backstreaming from pumps can sufficiently change the chemistry, so that a calibrated etch-time approach to reproducibility generally proves to be inadequate. Thus, techniques for determining the end point of a cycle become highly valuable as procedures for reducing the degree for overetching, and for increasing throughput and run-to-run reproducibility. In this section we describe four common methods for determining the end point of dry etch processes: 1) laser interferometry and reflectivity; 2) optical emission spectroscopy; c) direct observation of the etched surface through a viewing port on the chamber, by a human operator; and d) mass spectroscopy, which of the four, is least widely used[1,50,51].

Laser Interferometry and Laser Reflectance
Laser interferometry monitors the thickness of *optically transparent* films on reflective substrates by making use of interference effects[49]. The *laser reflectance method* exploits the difference in the reflectivity between a *nontransparent* material being etched and an underlying layer.

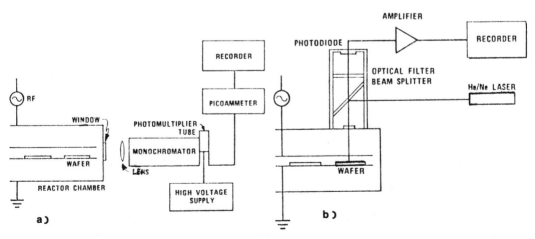

Fig. 24 (a) Experimental apparatus for using emission spectroscopy as an end point detector;   (b) Typical apparatus for the optical reflection method of end point detection[50]. Reprinted by permission of Solid State Technology, published by Technical Publishing, company of Dun & Bradstreet.

The same apparatus can be utilized to carry-out both techniques, and is shown in Fig. 24b. The system is designed to measure the intensity of light reflected from films being monitored.

In the case where a transparent film is being etched (e.g. SiO_2), the amplitude of the intensity of the reflected light varies, in approximately a sinusoidal manner, as interference conditions change with decreasing film thickness. If the incident light is normal to the surface, the film thickness change, Δd, between any two adjacent maxima or minima is given by $\Delta d = \lambda/2n$, where λ is the wavelength of the incident light, and n is the index of refraction of the etched layer. As a result, if the etch time between two adjacent maxima is known, *in situ* etch rates can be inferred. Laser interferometry can also provide end point detection. That is, the interface between two dielectrics is identifiable as a change in slope caused by the different refractive indices, and by a change in the frequency of the reflectance variations, due to the etch rate variations of the two materials.

Opaque /transparent interfaces (e.g. metal /dielectric) are distinguished by a variation from an approximately constant reflectivity to an oscillating one. In the case when *two nontransparent films* are etched there is a change in the reflected signal when the end point is reached, if the reflectivity of the underlying layer differs significantly from the film being etched. This change is proportional to the ratio of the reflectivity of the layer being etched to the underlying layer. Of course, the *laser reflectance* method does not provide any information on the *in situ* etch rate, and therefore does not provide as much information as *laser interferometry*.

These techniques, however, do have several limitations. First, the laser must be focused on a flat region of the wafer on which the film being etched is exposed. Thus, in many etching applications, where the area being etched is too small for good reflectivity measurements (e.g. etching of contacts in an SiO_2 film), a larger test site (>0.5 mm) must be added to the wafer patterning to facilitate this measurement. This requirement can be costly, as the open space must be located in a prime area of the wafer. Even when such a test area is present, each wafer must then be accurately aligned, so that the laser is incident on this area during the etch process. Second, this method provides etching information only on that specific area on a single wafer. If a large batch of wafers is being processed, non-uniformities in the batch etching process cannot be compensated for by this technique. Finally, in some cases, enhanced etching occurs at locations upon which the laser is focused, which yields inaccurate etch rate data.

**Table 3.   SPECIES and EMISSION WAVELENGTH
for OPTICAL EMISSION END POINT DETECTION[50]**

FILM	SPECIES MONITORED	WAVELENGTH (NM)
Resist	CO	297.7, 483.5, 519.8
	OH	308.9
	H	656.3
Silicon, Polysilicon	F	704
	SiF	777
Silicon Nitride	F	704
	CN	387
	N	674
Aluminum	AlCl	261.4
	Al	396

Optical Emission Spectroscopy

Optical emission spectrocopy is the most widely used method for end point detection, because it is easy to implement, can offer high sensitivity, and provides useful information about both etching species and etch products. The technique relies on the change in the emission intensity of characteristic optical radiation, from either a reactant or product in a plasma. Light is emitted by excited atoms or molecules in a plasma when electrons relax from one energy state to another. Atoms and molecules emit a series of spectral lines that is unique to each species. The emission intensity is a function of the relative concentration of a species in the plasma, and emitted light is observed through a viewing port on the etch chamber. A typical apparatus utilized for end point detection is shown in Fig. 24a. It operates by recording the emission spectrum during the etch process in the presence and absence of the material that is to be etched. The comparison of these two spectra indicates the emission lines that are sensitive to the etching process. To detect the end point, the emission intensity of the process-sensitive line (or band) is monitored at a fixed wavelength. When the end point is reached, the emission intensity changes. The change in emission intensity at end point depends on the species being monitored; the intensity due to reactive species increases, while the intensity due to etch products decreases. It is useful to monitor emission from both reactive species and product species simultaneously (Table 3), as in some etching applications one or the other of these measurements may yield a stronger signal[1]. Optical emission spectroscopy is widely used for determining the end point of SiO_2, polysilicon, and aluminum layers. In batch etch processes, the end point signal is derived from the average of etch conditions in the process. As a result, a degree of overetching is still required to insure that all wafers have been completely etched.

Optical emission spectroscopy also some drawbacks, one of the most important being that its sensitivity is determined by the etch rate and the total area being etched. Thus, for slow etch processes, the end point may be difficult to detect. The fact that the sensitivity is also dependent on the total area being etched, in some instances requires that a special test site be established to provide sufficient exposed area to cause a detectable end point signal (e.g. ~1 cm^2 of exposed Si^{50}). Separate test sites are most necessary when small contacts are being etched (i.e. the total area of etched surface is small), or when the etch depths become comparable to the separation between features. In the latter case, the total area (side wall + bottom) of material being etched can remain almost constant, even after the bottom of the film has been reached and only undercutting is occurring.

Mass Spectroscopy

Like emission spectroscopy, mass spectroscopy is a gas phase measurement method that offers the ability to detect and identify individual species in a plasma discharge. It is also able to provide information about the presence of species in the chamber prior to igniting the plasma, or in the effluent extracted from the etching chamber. A diagram of a mass spectrometer apparatus used for monitoring chamber effluents is shown in Fig. 25. Although mass spectroscopy is a useful technique, it is more difficult to implement than emission spectroscopy, and is limited to the sampling of species removed from the plasma, whereas emission spectroscopy obtains its data from the bulk plasma.

As shown in Fig. 25, a small orifice is utilized to sample the process effluent, and electron impact is used to create ions. The species present are analyzed in much the same manner as in a residual gas analyzer (described in Chap. 3), or in a secondary ion mass spectroscopy instrument (described in Chap. 17). Mass spectroscopy is very useful for gaining insights into plasma reactions. For example, the development of the etchant-unsaturate model, that elucidates the mechanism of radical production and quantifies the role of added oxidants in halocarbon plasmas, was evolved with the assistance of experimental data obtained from this method.

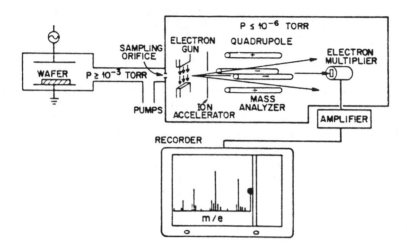

Fig. 25 Diagram for mass spectrometric monitoring of effluents from a plasma etching reactor. Reprinted with permission of Ref. 1, copyright, 1983, the American Chemical Soc.

DRY-ETCH EQUIPMENT CONFIGURATIONS

Plasma etching systems consist of several components: a) an etching chamber, that is evacuated to reduced pressures; b) a pumping system for establishing and maintaining the reduced pressure; c) pressure gauges to monitor pressure in the chamber; d) a variable conductance between the pump and etching chamber so that the pressure and flow rate in the chamber can be controlled independently; e) an rf power supply to create the glow discharge; f) a gas handling capability to meter and control the flow of reactant gases; and g) electrodes. Detailed assembly of such systems from these components has evolved a variety of configurations, depending upon which parameters of a process need to be controlled, as well as the specific applications of the system[52]. In this section we describe several of the most important commercially available plasma etch /RIE etch system configurations, together with some of their applications, advantages, and limitations. They include: 1) barrel etchers; 2) "downstream" etchers 3) parallel-electrode (planar) reactor etchers; 4) stacked parallel-electrode etchers; 5) hexode batch etchers; 6) magnetron ion etchers. After describing each of these configurations, a brief comparison of the characteristics single wafer and batch etchers is given.

Some general comments on two equipment related topic are presented before starting a more detailed look at the various systems. These are: a) the impact of electrode materials on dry etch processes; and b) available procedures for adjusting pressure in the etch chamber.

The type of electrode material can have an impact on the etch process in a number of ways. First, in some systems sputtering of the electrodes can occur, and the sputtered material can redeposit on the wafer surfaces. If these products are nonvolatile, they become a source of contamination. Thus, use of electrodes containing materials that will result in nonvolatile sputter etch products, should be avoided. For example, an aluminum electrode in a system utilized to etch Si and SiO_2 with fluorocarbon gases may be prone to causing such problems. Second, some materials can suppress loading effects (see later section on this topic) which can be advantageous, but may also influence the discharge chemistry. For example, carbon and silicon electrodes may suppress loading effects in fluorine-based plasmas, but they may also decrease the F /C ratio. Third, the etching of some electrode materials may cause etch products that impact or even obscure end point information. Finally, the electrode material may contribute to etch rate non-uniformity effects across the wafer, as is described in a later section.

The pressure in an etch chamber may be adjusted in two ways[2]. The pressure dependence of the etch rates can depend on which of these pressure variation procedures is followed. To adjust the pressure, either the flow rate can be kept constant and the the pumping speed can be varied, or the pumping speed kept constant and the flow rate adjusted. In the former method, the residence time in the chamber changes, while in the latter it remains constant. Possibly the best method for operating a dry-etch system, from the point of view of controlling the gas variables, is to set a fixed flow rate and then maintain a constant pressure by adjusting the pumping speed with a variable conductance and an automatic pressure controller (see Chap. 3). In this manner an adequately high flow rate can be chosen, thereby avoiding the unwanted situation of operating under conditions of a high *utilization factor* (see subsequent section on this topic).

Commercial Dry-Etch System Configurations

Barrel Etchers
The first, and simplest, plasma etchers to be developed[53] were barrel etchers (Fig. 26a). This configuration consists of a cylindrical reaction vessel, usually made of quartz, with rf power supplied by metal electrodes placed above and below the cylinder, as shown in Fig. 26a. A perforated metal cylindrical *etch tunnel* is placed within the etch chamber. This serves to confine the glow discharge to the annulur region between the etch tunnel and the chamber wall. Wafers are placed in a holder or *boat* at the center of the cylinder, and usually no electrical connection is made to them. The reactive species created by the discharge diffuse to the region within the etch tunnel, but the energetic ions and electrons of the plasma do not enter this region. The reactive

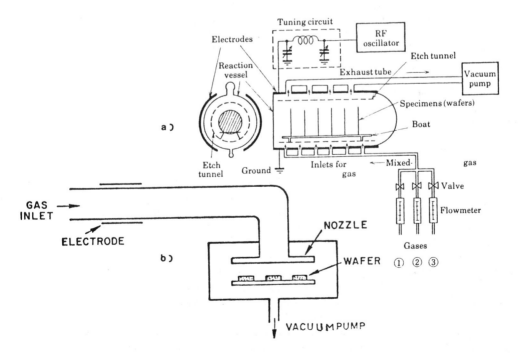

Fig. 26 Schematic of: (a) a barrel-type plasma etching system. (b) a downstream plasma etcher.

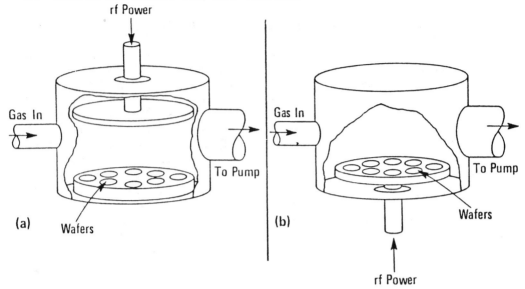

Fig. 27 Parallel-electrode (planar) type dry etcher. (a) When wafers are placed on the grounded electrode, the system is configured in the plasma etch mode. (b) When wafers are placed on the powered electrode, the system is operated in the reactive ion etch, or RIE, mode.

species from the plasma diffuse to the surfaces to be etched, and since there is no ionic bombardment, the etching is almost purely chemical. As a result, the etching tends to be isotropic, and it is possible to obtain good selectivity, with little or no radiation damage. Most barrel etchers are operated in the high pressure range of dry etching (0.5-2.0 torr). The isotropic nature of the etch, however, limits barrel etchers to such applications as resist stripping and "non-critical" etch steps. Nevertheless, these applications are well served by such systems. In fact, for processes that do not require ion bombardment for directional etching, barrel etchers should be considered. Some materials, such as Al, however, cannot be successfully etched in barrel etchers.

Downstream Etchers

Downstream etchers[54] derive their name from the fact that the reactive species are created in a plasma, and are then transported to the etching chamber downstream of the plasma (Fig. 26b). Microwave sources have been used to create the long lived chemical species necessary for this configuration. Since the reactive species are created outside of the etching region, as in barrel reactors, temperature control and radiation damage problems can be minimized or avoided[55]. These systems are mostly operated in the high pressure range (0.5-2.0 torr). In order to obtain directional etching, however, some form of bias must still be applied to wafers in downstream etchers, otherwise etching again proceeds in a purely chemical manner.

Parallel Electrode (Planar) Reactors

As described earlier, wafers exposed to energetic ions of a plasma can be subjected to ion-assisted etching processes. Etcher configurations that utilize parallel electrodes can direct energetic ions at the surfaces being etched, by causing them to be accelerated across the potential difference that exists between the plasma and the electrode surfaces (Fig. 4b). As a result, both a physical and a chemical component can impart directionality to the etch process.

In parallel-electrode systems, the electrodes have a planar, circular shape, and are of approximately the same size[56]. One of the two electrodes of the planar reactor configuration is connected to the rf supply, and the other to ground. Wafers can be placed on either of the electrodes (Fig. 27). When wafers are etched in such systems by placing them on the grounded electrode, the system is said to be operated in the *plasma etch mode*. When wafers are placed directly on the rf-powered electrode, these systems are said to configured in an *reactive ion etch mode*. As discussed in Chap. 10, however, a potential difference between the plasma and the grounded electrode can still exist, since the plasma potential is always above ground potential. Thus, even in the plasma etch mode, wafers are subject to energetic ionic bombardment, although usually to a lesser degree than in the RIE mode. For example, energies of bombarding ions are 1-100 eV in the plasma etch mode, and 100-1000 eV in the RIE mode. Since the chamber walls are also grounded, in the plasma etch mode they may be subject to significant ionic bombardment, causing sputtering and redeposition to occur within the etch chamber (i.e. on wafer surfaces). In the RIE mode, only the powered electrode is subject to energetic positive ion bombardment, and there is less likelihood of sputtering nonvolatile products onto the wafers. Etching in both modes in such systems, nevertheless, is affected by the fact that both physical and chemical mechanisms are operative. Typically operation in the RIE mode is conducted at low pressures (<100 mtorr), while plasma etching is carried out at pressures >100 torr.

Commercial systems built in the parallel-electrode configuration can be batch systems or single wafer systems. Batch systems are typically manually loaded, and usually have the capability of being operated in either the plasma etch or RIE mode. This provides such systems with a useful flexibility. Some such systems, have been offreed with a low frequency (450 kHz) power supply. Single wafer parallel electrode etchers are described later.

Stacked Parallel-Electrode Etchers

The stacked parallel-electrode etcher is a small batch machine capable of handling 6 wafers at a time. Its unique design provides an individual pair of electrodes for each wafer (Fig. 28),

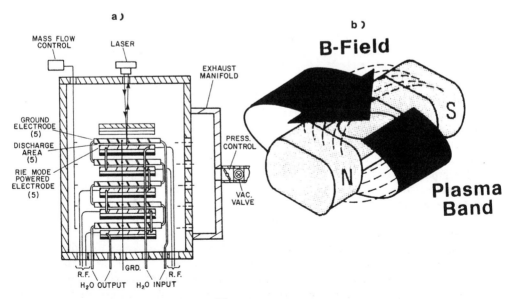

Fig. 28 (a) Stacked electrode etcher[23]. Reprinted with permission of Semiconductor International. (b) Magnetron ion etching system[60]. Courtesy of MRC, Inc.

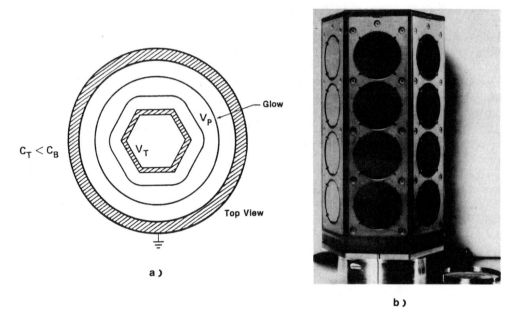

Fig. 29 (a) Schematic of the electrode configuration of the hexode batch etcher. (b) Photograph of a hexode batch etcher. Courtesy of Applied Materials.

thereby combining some of the advantages of single-wafer and batch etchers[57]. (Although, others may hold that it offers less benefits than larger batch systems, while maintaining the same disadvantage of end point detection from only a single wafer). Operating chamber pressures and rf power densities can be kept in the ranges between those of low-pressure, low- power-density hexode batch etchers, and high pressure, high power-density single wafer RIE machines. The system operates cassette-to-cassette, and does not offer load-locks. The major applications of this machine are for etching SiO_2, Si_3N_4, and polysilicon.

Cylindrical Batch Etch Reactors (Hexode Etchers)

Cylindrical batch etchers are designed with a hexagonal inner electrode[45,58,59] (hexode), and the chamber walls serve as the opposite electrode (Fig. 29). Up to 24-150 mm wafers per run are mounted on the hexode, which is the powered electrode, while the other electrode (chamber walls) is grounded. The result is a system in which the area of the hexode is much smaller than the area of the grounded electrode (typically one-half the area). This is just the kind of *highly asymmetrical electrode configuration* needed to cause ionic bombardment of the powered electrode, while minimizing ion bombardment of the grounded electrode and other chamber surfaces (see Chap. 10, *RF Sputtering,* which explains why). Thus, hexode etchers are well designed to operate in the RIE mode and to provide good directional etching capability, but not for operation in the plasma etch mode. Most hexode models are not load-locked and must be manually loaded. Some of the newer models have been designed with load-locks and robotic autoloading. In those non-loadlocked models, more elaborate pumping systems must be provided to remove water vapor from the chamber after exposure to atmosphere (although the load-locked models are also equipped with comparably elaborate pumping systems). Etching is conducted at low pressures (20-100 mtorr) and low power densities.

Comparison of Single-Wafer and Batch Dry Etchers

Dry etchers capable of anisotropic etching are available as systems configured for single wafer or batch processing (the latter being primarily hexode-type designs). Each type of system has its advantages and limitations. In this section, we compare their capabilities, which are summarized in Table 4.

Hexode batch etchers can accommodate a group of wafers (batch) in each run. Because they can process multiple wafers in a single run, their operating power and pressure can be kept low, while maintaining high throughput. The low pressure allows more energetic bombardment of the surface at lower powers. Thus, non-volatile species that must be removed by the physical component of the etching, such as CuCl, are effectively etched without having to resort to the higher power density levels that promote resist erosion, reduced selectivity, and radiation damage. In addition, multi-step etch processes can be performed in the reaction chamber, as sufficient time is available to change the gas composition in the chamber.

The disadvantages of hexode batch etchers are the following: a) an entire batch of wafers can be lost if the etch process is somehow incorrectly performed, while in single wafer etchers the loss due to a flaw in any cycle can be limited to one wafer; b) end point detection is determined from the average of the etching conditions over the entire batch (emission spectroscopy), or from a single wafer in the batch, while a single wafer etcher can monitor the end point of each wafer; c) more laboratory space is required for batch etchers; d) automation and load-locking of hexode etchers is more complex and expensive than in single-wafer etchers; e) the etching characteristics of the batch etcher may depend on the load of wafers in the chamber, and thus several process recipes might have to be developed for the machine, depending on batch size; and f) the batch uniformity represents an extra variable.

Single wafer etchers have the advantage that wafer-to-wafer uniformity problems of large batch systems is avoided. In addition, since only one wafer is at risk at any time, if a flaw in the process is detected, theoretically it can be corrected before allowing the next wafer to be etched. Likewise, end point detection, load-locking, and automation are more easily achieved. New etch processes can usually be more more rapidly developed on single wafer etchers, and the establishment of multiple recipes dependent on batch size is unnecessary. Finally, capital equipment costs are lower, and equipment space requirements smaller.

On the other hand, single wafer etchers must perform their etching at high rates in order to achieve adequate throughputs. There may be unwanted side effects of such rapid etching requirements, such as: a) excessive resist erosion (see *Dry Etching of Aluminum Thin Films*); b) reduced selectivity; and c) multi-step etches may require multi-chamber etching systems, as the time to evacuate a chamber after each sub-step may be prohibitively long. It is therefore likely that single wafer etchers will be utilized in processes that can tolerate the harsher etching environ-

Table 4. SINGLE WAFER versus BATCH ETCHING

	SINGLE	BATCH
Pressure	Low - High (often high)	Low
Throughput	Depends on Pressure	High
Etch Uniformity	Good	Adequate?
End Point Detection	Good Control	Average over batch
Automation	Relatively Easy	More Difficult
Laboratory Space	Smaller	Larger

ments of high-rate etching, but that other, more critical steps, will continue to be performed by the less damaging batch etch processes.

Magnetron Ion Etchers

Magnetron ion etchers (MIE) are single wafer machines in which the magnetron principle from sputter deposition is adapted to make the primary electron excitation process in plasma etchers more efficient at lower gas pressures[60]. As a result, both higher reactant species production and ion bombardment rates of the surfaces being etched can be achieved at lower powers and pressures, than are required in non-MIE single wafer etchers. Thus, some of the drawbacks of single wafer etchers associated with having to operate at high power densities and pressure (such as bulk radiation damage and resist erosion), can be reduced, while still maintaining the high etch rates necessary for adequate throughputs in single wafer etchers.

As shown in Fig. 28b, a band magnetron cathode is utilized to create a uniform magnetic field above the wafer. This causes the electrons emitted by the cathode to be constrained from traversing the discharge region to the collecting surfaces. These electrons are thus more likely to collide with, and ionize, gas particles in a low pressure gas. Etch rates of SiO_2 of 6,000 Å /min have been reported using various fluorocarbon-based gas mixtures, with a 5% uniformity over a 125 mm wafer, and at a low voltage. Polysilicon etch rates of 1 μm /min have also been reported, using an SF_6 /CF_4 gas mixture.

PROCESSING ISSUES RELATED TO DRY-ETCHING

Plasma Etching Safety Considerations

Many dry etching processes require the use of potentially hazardous and /or toxic gases. It is important that these gases be used safely in the course of operating and maintaining dry-etching systems. This section provides a brief, but far from complete introduction to this subject and other dry-etching safety issues. Readers are directed to references 61 and 62, as well as to the Semiconductor Safety Association, and to local safety representatives for further information on these topics. The following safety aspects of dry etching are presented in this section: a) working with compressed gases; b) working with toxic and /or corrosive gases; and c) some safety design considerations in plasma etching equipment.

Gases contained in cylinders at high pressures (>225 psig @ 70°F) should be treated with extreme respect. The total force that could be exerted during the release of the gas in a cylinder 48 inches high with a diameter of 9 inches and a pressure of 2000 psig, is in excess of 24 million pounds. The correct regulator should be used to make connection to the cylinder. Compressed-gas cylinders should be securely fastened in properly designed cabinets. The exhaust capacity of a gas cabinet should be able to rapidly handle the entire contents of the cylinder to limit the exposure area. Cylinders containing flammable materials should be electrically grounded to avoid static charge.

One of the most probable times when a life or health threatening accident can occur is when gas cylinders are being changed, and thus safe procedures must be carefully followed when performing this operation. In some plants, *only* engineers responsible for a particular operation can change cylinders, and the procedure is performed using the "buddy" system.

The constituents of the feed stock gases used in common plasma processes are listed in Refs. 61 and 62. Many of these constituents are toxic. Some of them (e.g. ammonia) cause immediate noticeable symptoms at concentrations below the *immediately dangerous to life and health levels* (IDLH), but many others (e.g. phosgene [$COCl_2$] an impurity constituent gas found in BCl_3 supplies) are colorless and odorless at concentrations well above the IDHL level. This demands that suitable gas detectors be used to monitor and warn of gas leaks, as a means of protecting the work environment from hazardous materials.   All personnel that work with hazardous gases should attend safety courses dealing with such subjects as respirator use, spill clean-up procedures, and fire procedures.

If a toxic gas is accidentally inhaled, both *emergency* and *extended* action needs to be taken. For example, the inhalation of chlorine produces only slight irritation in the first 6 hours. During the following period of up to 8 days, however, *edema* (a swelling of the mucous membrane) may result and acute reaction set in. Thus, *close observation* during this entire period is mandatory to insure the well-being of anyone who has come into contact with chlorine (or any other toxic substance).

Carbon tetrachloride (CCl_4) is an etch gas used in some aluminum dry-etch processes. This etch products from this gas have been identified as being carcinogenic, and its use in *non-load-locked* systems has been pronounced unacceptable[61]. On the other hand, others have suggested that utilization of CCl_4 in *load-locked* systems can be safe[62], provided that NIOSH (part of OSHA) approved respirators with organic vapor and acid gas cartridge, and gloves (made of nitrile rubber, polyvinyl alcohol, or viton) are worn by personnel whenever performing maintenance or cleaning procedures on such systems.

Plasma systems should also be designed for no-fail operation. For example, the shut off valve of gas cylinders should be normally off, and only driven on by high pressure nitrogen that is interlocked to the gas monitor and electrical alarm systems. Thus, if either N_2 pressure or electrical power fails, these cylinders will be isolated. The gas line between the cylinder valve and the last valve in a line should be purged by cycling between over pressure and under pressure conditions, to rapidly dilute the gas trapped in the line to low values. This procedure is much more effective in reducing the concentration of the gas in the line than merely flowing a purge gas through the line.

Plasma system pumps should be oversized and use N_2 *ballast* (see Chap. 3) to eject corrosive gases quickly, and to prevent condensation of corrosive vapors. Pumps should be well marked so that maintenance personnel will follow proper protective procedures when performing oil changes or repairs. Use of synthetic pump oils is recommended, because they require much less frequent oil changes. The spent oil and filter cartridges must be treated as hazardous waste.

The rf emissions from plasma etchers are normally below the U.S. recommended limit for rf exposure (4.9 mW /cm^2 of body area over a 6 min period). However, damage to an etching unit door or housing misalignment can result in excessive rf leakage. Thus, whenever an etcher installed, moved, or modified, its rf emissions should be monitored to ensure the units integrity.

Uniformity and Reproducibility Considerations

It is important that dry etch processes are uniform across a wafer (and from wafer-to-wafer in batch etch systems), and are reproducible from run-to-run. The following factors influence the uniformity and reproducibility of dry etch processes: a) exposure of the etching chamber to room ambient; b) gas distribution within the system; c) gas flow rates; d) end point detection; e) loading effects; and f) electrode-related efffects.

Exposure of Etching Chamber to Room Ambient

If the etching chamber is periodically exposed to room ambient, the reproducibility of the etch conditions from run-to-run may be impacted. The exposed chamber surfaces adsorb moisture which may affect the next etch cycle in a nonreproducible way. This problem is avoided in systems that keep the etch chamber isolated from exposure to the ambient by use of load locks,

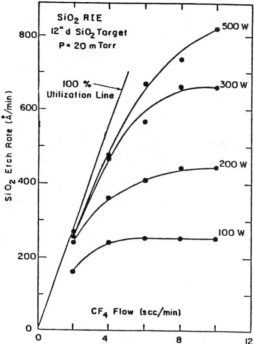

Fig. 30 Etch rate of sputtered quartz, demonstrating the utilization factor concept[63].   Reprinted with permission of the American Physical Society.

but often batch systems, such as some hexode reactor models, are not load-locked, but are opened to atmosphere after every run.   Several techniques are employed to reduce the above mentioned problems in such systems[45], including using a cryopump to more completely remove moisture from the chamber before beginning the etch cycle, employing etch gases that are effective moisture getters (e.g. BCl_3), and surrounding the etch chamber with a nitrogen purged box.

Gas Distribution

A uniform supply of reactive species to the wafer surfaces is necessary to bring about uniform etch rates across a wafer, and from wafer-to-wafer in batch etchers.   Thus, reactors must be designed with gas feed systems that produce such uniform reactant distributions.   This is a challenging task, as viscous flow prevails over the range of working pressures and reactor dimensions.   In viscous flow, the local molecular interactions are strong, and gas mixing is poor. Various gas distributions designs have been evolved, depending on the reactor configuration. Batch-type planar etchers usually use *gas rings* to distribute the gas uniformly around the perphery of the circular electrodes. Some single wafer etchers, that have smaller chambers utilize a *shower head* electrode, that emits gas from the top electrode. Hexode etchers employ a *gas tree* distribution design along the axis of the hexagonal electrode.

Gas Flow Rates (Utilization Factor)

Not only must a *uniform* quantity of reactants be provided to all wafer surfaces in order to achieve uniform and reproducible etch rates, but an *adequate quantity* of reactant must also be supplied.   That is, if the etch rate is limited by the supply of reactant, small variations in flow rate or gas distribution uniformity may lead to etch rate nonuniformities.   As a result, there is a minimum flow rate of reactant gas that must be supplied to prevent the process from being limited by the reactant supply. A concept known as the *utilization factor* has been developed to quantify this situation. For example, in a CF_4 based etching process, the utilization factor is defined to be the ratio of the rate at which fluorine is consumed in the etching process, to the rate at which the fluorine enters the system as CF_4. If the etch rate is limited by the supply of reactants, then all the fluorine is consumed, and the utilization factor = 1 (Fig. 30)[63]. It has been determined that etching processes exhibit optimum behavior if the utilization factor is of the order of 0.1 or less. It is therefore recommended that the utilization factor be determined whenever an etching process parameter is changed[4]. Relatively simple techniques for measuring the utilization factor are given in Reference 63.

Loading Effects

In plasma etching systems, etch gas is fed into the reaction chamber. There, a plasma creates reactive species. The etch rate is generally proportional to their concentration, and the concentration in turn is reflected by the partial pressure of the species, P_x. The reactive species are removed from the chamber by one of the following events: a) consumption, by the etching reactions, taking place at a rate S_{etch}; b) recombination, an event most likely to occur when the reactive species come in contact with the non-etchable surfaces within the chamber, and such recombination removes reactants at a rate, S_{recomb}; and c) pumping of the species from the chamber before one of the other two events occurs, and this removal rate is given by the pumping speed, S_{pump}.

In the steady-state, reactive species are being created and removed at equal rates, and the partial pressure of the reactive species, $P_x = G_x/S$, where G_x is the generation rate of these species, and S is the sum of the three removal rates, $S = S_{etch} + S_{recomb} + S_{pump}$. In processes in which the etch reaction is the dominant removal mechanism, the partial pressure, and hence etch rate, is controlled by S_{etch}. Since S_{etch} is directly proportional to the area of exposed etch surface, the etch rate decreases as more etched surface is added to the chamber. This effect is known as the *loading effect*, and an example of the loading effect is shown in Fig. 31b. In this figure, the etch rate of polysilicon decreases with exposed area when the temperature is increased to 140°C. Note that an increase in the gas flow will not alleviate the loading effect. The concentration of reactive species will be increased by a higher flow rate, and for a constant area of etched surface, the etch rate will be faster. But if S_{etch} remains dominant, when more wafers are added the etch rate will still decrease.

In most etch processes the loading effect is undesirable for the followng reasons: 1) since the etch rate drops, some of the throughput benefits of batch processing are lost; 2) since the etch rate depends upon the amount of etchable area in the chamber, this becomes another parameter that leads to process non-uniformities from run-to-run; and 3) when the end point of an etch step is reached, the etchable area undergoes a precipitous decrease, as the material in the field is removed, and only the sidewalls of the etched features remain exposed. The loading effect at that time causes the etch rate to increase. As a result, during any overetch time, the sidewalls are etched at a higher rate than the nominal process etch rate, and this aggravates the undercutting problem. There are several methods that can be utilized to alleviate loading effects, mostly based on increasing the rate of removal of reactive species by other mechanisms than the etching reactions on wafers.

To quantify the procedures utilized to reduce loading effects, a mathematical model that relates the various parameters involved in with the loading effect has been constructed[25,65]. The model provides a good description of silicon etching in CF_4/O_2 discharges, and expresses the etch rate of m wafers, R_m, in terms of the etch rate of an empty reactor, R_0 (which represents the etch rate when the etchable area approaches zero), by the following expression:

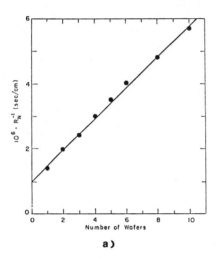

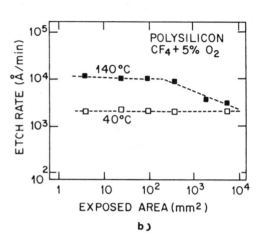

a)

b)

Fig. 31 (a) Reciprocal of Si etch rate vs. number of 3" wafers, demonstrating the loading effect[65]. Reprinted with permission of the publisher, the Electrochemical Society. (b)   The etch rate of Si is increased sufficiently and /or the reactivity is increased sufficiently by increasing temperature[66]. Reprinted with permission of the Japanese Journal of Applied Physics.

$$R_o/R_m = 1 + m\Phi = 1 + m[A_w k_{etch} /(A_{ch} k_{recomb} + V k_v)] \qquad (1)$$

where: Φ is the ratio of reactant species consumed by etching a wafer to the rate of removal of reactant species by all other mechanisms; k_{etch} is the reactant species consumption rate per unit area of etched surface; A_w is the area of exposed etched surface per wafer; k_{recomb} is the removal rate per unit area due to recombination at reactor wall surfaces; A_{ch} is the area of the chamber wall surfaces; k_v is the volumetric removal rate per unit volume due to pumping; and V is the volume of the chamber. When Φ is small, the etch rate approaches that of an empty chamber.

   In plasma etching processes for which the above model is applicable, the sensitivity of etch rate to loading conditions can be reduced by several procedures, including: a) conducting the etching in a chamber with a large volume and high surface area; b) utilizing large gas flows, (large k_v) which will cause the removal rate due to pumping to dominate; c) using an electrode constructed of the same material as that being etched, thereby maintaining a fully-loaded condition in the chamber, regardless of the number of wafers present (in the latter method, however, end pont detection by optical emission spectroscopy becomes difficult); and d) in processes in which the etch rate is controlled by ionic bombardment, it may be possible to reduce Φ by maintaining a sufficiently slow etch rate.

   Not all plasma etching processes are well modeled by Eq. 1, and some of those that are not do not exhibit loading effects. For example, the concentration of chlorine atoms, a reactive species in some chlorine-containing plasmas, is limited by recombination, rather than by the atom-substrate reaction that causes etching. In these plasmas   the etching is independent of etchable area[25].

Electrode-Related Effects

Etching rate nonuniformity effects across a wafer, arising from aspects of the electrode material and configuration, can also occur. Two examples of such effects are: 1) the degree to which the electrode material is etched can play a role. That is, if the electrode is inert, no consumption of reactive species occurs above regions where the electrode surfaces are uncovered. These species then represent an additional supply of reactants near the wafer edge that is not available to central wafer regions. Hence, etching occurs more rapidly near the edges, and the wafer clears from the outside to the center (*bullseye pattern*). Conversely, if the electrode material etches more rapidly than the wafer surface, reactant at the wafer edges may become depleted, causing the center of the wafer to etch more rapidly; 2) With the wafer located on the electrode, the electric field lines of the dark space are disturbed at the wafer edge. This leads to an increase in the ion bombardment near the edge. If the etch rate increases with increasing ionic bombardment flux, etching again becomes greater near the edge than at the center. Techniques suggested for reducing this effect include recessing of the wafers into the electrode surface, or surrounding the wafer with conducting or insulating rings.

End Point Detection

After an etch cycle is completed, it is necessary to terminate the etching action. There are usually too many variables in an etch process performed in a production environment to merely use the etch time as a measure of when the cycle is complete. Instead, the process must be monitored with an end point detection technique, as described earlier, and information obtained from such monitoring must be utilized to decide when the etching action should be terminated. (It should be noted, however, that in processes in which end point detection techniques are not effective, timed etches are used.) Effectively determining the end point of an etch cycle prevents insufficient etching as well as overetching, and can also be used to compensate for film thickness and etch rate variability. End point detection techniques are described in an earlier section.

Contamination and Damage of Etched Surfaces

Contamination

Wafers etched by dry processes can be subjected to contamination from a number of sources including: a) polymeric residues from the etch process; b) deposition of nonvolatile contaminants from sputtering events associated with the etch process; and c) particulate contamination.

Polymeric residue contamination is caused by halogen deficiency in halocarbon plasmas, as described earlier in the section dealing with the etching model of Si and SiO_2 in CF_4, O_2, and H_2. Such contamination can produce rough surfaces on the films being etched (or on underlayers), can lead to high contact resistance, and in the case of Al etching, serve as a "reservoir" for corrosion causing halogens. A technique for avoiding such residues is to somehow increase the F /C ratio near the end of a dry-etching cycle, thereby allowing any residue film to be removed, and not reformed.

Surfaces being etched can be *contaminated by deposits of nonvolatile solids sputtered from surfaces in the etching chamber*. Sputtering events occur in etching chambers as a result of the potential difference that exits between the plasma, and the surfaces being etched and chamber walls. If the chamber surfaces consist of materials containing heavy metals (e.g. Fe, Cr, Ni), they can be sputtered, and then end up being deposited on the surfaces being etched. To reduce this type of contamination: 1) the chamber walls and electrode surfaces should be made of (or coated with) materials compatible with the surfaces being etched (e.g. aluminum, or SiO_2); and 2) the system should be designed so that sputtering events are minimized from surfaces of nonvolatile solids. The latter is accomplished by electrically connecting together the chamber walls and non-wafer holding electrode, and making their combined area large relative to that of the wafer-holding electrode.

Four main sources of *particulate contamination* have been identified in dry-etching systems[67]: 1) wafer transport materials and operation; 2) processing by-products; 3) particulates from paper products utilized nearby the etcher (e.g. operations manuals and strip chart recorders); and 4) the environment. *Wafer transport related particulates* are kept to a minimum by effective system design. *Processing by-product particulate* densities depend on the specific process. In those processes that are prone to particulate production from polymer build-up and subsequent separation from the reactor walls, however, more frequent system cleaning to maintain low particulate densities is required. *Paper product related particulates* can be reduced by using low particulate stationary, or by keeping paper products entirely out of the clean room in which the etchers are located. *Clean room environment particulates* are minimized by a host of elaborate procedures, and these are described in detail in Vol. 2 of this text.

Radiation Damage

As described earlier, in reactive ion etching (RIE), surfaces being etched are subject to bombardment by energetic ions, electrons, and photons. These incident species can produce various forms of radiation damage in the materials being etched including: a) electron traps in gate oxides, which if not annealed out, cause shifts in the threshold voltages of MOS devices; b) displaced atoms and implanted atoms in the surface due to ion bombardment; and c) under some etching conditions, destruction of the gate oxide.

The type of *trapping sites generated by RIE* is reported to depend on the specific process[68]. For example, CF_4 plasmas have been shown to induce bulk trapping sites with a centroid of approximately one half the thickness of the oxide, while exposure to O_2 plasmas leads to surface

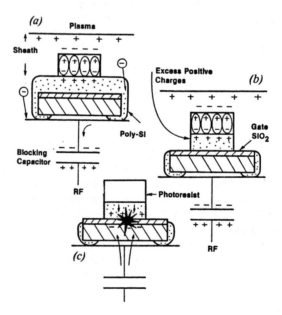

Fig. 32 Gate oxide breakdown[70]. Reprinted with permission of Solid State Technology, published by Technical Publishing, a company of Dun & Bradstreet.

Table 5   ETCH GASES USED FOR VARIOUS INTEGRATED CIRCUIT MATERIALS

MATERIAL	GASES
Silicon	CF_4, CF_4/O_2, CF_3Cl, SF_6/Cl, $Cl_2 + H_2$, C_2ClF_5/O_2
(including polysilicon)	SF_6/O_2, SiF_4/O_2, NF_3, ClF_3, CCl_3F_5, C_2ClF_5/SF_6
SiO_2	CF_4/H_2, C_2F_6, C_3F_8, CHF_3
Si_3N_4	CF_4/O_2, CF_4/H_2, C_2F_6, C_3F_8
Organic Solids	O_2, $O_2 + CF_4$, $O_2 + SF_6$
Aluminum	BCl_3, CCl_4, $SiCl_4$, BCl_3/Cl_2, CCl_4/Cl_2, $SiCl_4/Cl_2$
W, WSi_2, Mo	CF_4, CF_4/O_2, C_2F_6, SF_6
$TaSi_2$	SF_6/Cl_2, CF_4/Cl_2
Au	$C_2Cl_2F_4$, Cl_2

layer trapping sites. The bulk traps are removed by a 600°C anneal, and so are of concern only if an RIE step is performed after an Al thin film is in place. Polysilicon and Al gate materials also effectively shield the oxide from such radiation damage. Such radiation damage is minimized by low power or low dc voltages, or by placing the wafers on the grounded electrodes[69]. Batch etch systems tend to use lower power than conventional single wafer etchers, but magnetron ion etching single wafer systems employ low power as well.

The *gate oxide can also be destroyed by dielectric breakdown during RIE of polysilicon gate layers under some conditions.* Watanabe and Yoshida[70] propose that as polysilicon is being etched (Fig. 32), it builds up a positive charge in response to the electrons from the discharge, which accumulates on the wafer and then become stored in the resist and blocking capacitor (see Chap. 10 for an explanation as to how electrons accumulate ion the surfaces in contact with an rf plasma). As long as the poly-Si layer is continuous, it contacts the backside of the wafer, and there is no potential difference across the gate oxide. Once the poly-Si is etched, the gate and the substrate are no longer in contact and a sufficient voltage may develop across the gate oxide to result in dielectric breakdown. In addition, whenever the power is turned off, positive ions in the dark space recombine with the electrons in the resist, and negative charges stored on the blocking capacitor recombine with the darkspace positive ions through the conductive material of the cathode. If, however, the negative charges on the blocking capacitor are attracted by the positive charges on the isolated gate material, before enough recombination takes place, the transient surge current may also produce a sufficiently large potential difference across the SiO_2 gate to cause breakdown. It is noted, that in conventional RIE processes the cause of such breakdown cannot be removed, and that a solution to this problem requires a more complete understanding of the details of this failure mechanism.

Process Gases for Dry Etching

Many different gases are used in dry etching processes for VLSI fabrication, either individually or as the components of mixtures of several gases[73]. Table 5 lists most of the important gases used in dry etching applications. Chlorine-based gas mixtures (e.g. containing Cl_2, BCl_3, CCl_4, and $SiCl_4$- usually in some combination of Cl_2 and one of the other species) are primarily used to etch aluminum and aluminum alloys. Earlier we discussed the use of fluorocarbon-based plasmas for etching Si, SiO_2, and Si_3N_4, and also described the use of various flourochlorocarbon gas mixtures for etching polysilicon. In dry etching processes for Si and SiO_2 that utilize CF_4, it was found that the active etching species is atomic fluorine.

More recently SF_6 and NF_3[71] have been exploited as sources of atomic fluorine for dry etching Si and SiO_2. These gases exhibit the advantages of higher Si and SiO_2 etch rates than CF_4 (up to 5000 Å /min vs. ~1000 Å /min), excellent silicon to silicon dioxide selectivities (e.g. 50:1), and the fact that they exhibit a smaller loading effect than CF_4 etching processes. The NF_3 etching processes reportedly suffer from more pronounced loading effects than SF_6-based etching processes, but the NF_3 etching by-products formed from side reactions consist of only volatile compounds[72]. Therefore, polymer deposition on the silicon surface is minimal when etching with NF_3. Etching polysilicon with either SF_6 or NF_3, alone, is typically isotropic, but anisotropic etching has been successfully performed by: a) adding small amounts of chlorine-containing gases, such as $CFCl_3$ or $CHCl_3$, to the SF_6; or b) by decreasing the pressure of the NF_3 plasma to ~100 mtorr, causing higher ion bombardment energies. Furthermore, highly anisotropic etching of $MoSi_2$ in such low-pressure NF_3 plasmas has been reported[72].

REFERENCES

1. J. A. Mucha and D.W. Hess, "Plasma Etching", in *Introduction to Microlithography*, Advances in Chemistry Series, Vol. 219, American Chemical Society, 1983, p. 215.
2. J.W. Coburn, "Plasma Etching and Reactive Ion Etching", American Vacuum Society, New York, 1982, p. 15.
3. *ibid.*, Ref. 2, p. 15.
4. C.J. Mogab, A.C.Adams, and D.L.Flamm, *J. Appl. Phys.*, **49**, 3796 (1978).
5. J.W. Coburn, from foils presented in the American Vacuum Society Education Course,"Plasma Etching and Reactive Ion Etching", 1986, foil 3-16, not published.
6. J.L. Vossen, *J. Electrochem Soc.*, **126**, 319 (1979).
7. J.W. Coburn, "Pattern Transfer", *Solid State Technol.*, April, 1986, p. 117.
8. E. Kay, J.W. Coburn and A. Dilks, *Topics in Current Chemistry*, Vol. 94, Springer-Verlag Publishing Co., Heidelburg, W. Germany, 1980, p. 1.
9. M.W. Jenkins, *et al.*, "The Modelling of Plasma Etching Processes Using Response Surface Methodology", *Solid State Technol.*, April, 1986, p. 175.
10. J.W. Coburn and H.F. Winters, *J. Vac. Sci. Technol.*, **16**, 391 (1979).
11. D.L. Flamm, *Plasma Chem., Plasma Process,* **1**, 37 (1981).
12. M. Chen, V.J. Minkiewicz, and K. Lee, *J. Electrochem. Soc.*, **126**, 1946, (1979).
13. D.L. Flamm and V.M. Donnelly, *Plasma Chem. Plasma Process,* **1**, 317, (1981).
14. *ibid.*, Ref. 4.
15. L.M. Ephrath and E.J. Petrillo, *J. Electrochem. Soc.*, **129**, 2282, (1982).
16. R.A. Heineke, *Solid State Electronics*, **18**, 1146, (1975).
17. J.S Chang, "Selective Reactive Ion Etching of SiO_2", *Solid State Technol*, April, '84, p. 214.
18. G. Francis, *Handbuch der Physik*, **22**, 53 (1956).
19. A.S. Bergendahl, D.L. Harmon, and N.T. Pascoe, *Solid State Technol.*, Nov. '84, p. 107.
20. J.W. Coburn and H.F. Winters, *J. Appl. Phys.*, **50**, 3189 (1979).
21. *ibid.*, Ref. 13.
22. *ibid.*, Ref. 10.
23. A. Weiss, "Plasma Etching of Oxides and Nitrides" *Semiconductor Internat'l*, Feb., '83, p. 56.
24. J. Poisson, *et al.*, "Via Contact Dry Etching Using a Plasma Low Resistance Photoresist", *Proc. Third VLSI Multilevel Interconnection Conf.*, Santa Clara, CA., June, 1986, IEEE.
25. D.L. Flamm, V.M. Donnelly, and D.E. Ibbotson, "Basic Principles of Plasma Etching", in *VLSI Electronics, Microstructure Science,* Vol. 8, Academic Press, Chap. 8, p. 190.
26. *ibid.*, Ref. 25, p. 227.
27. A. Weiss, "Plasma Etching of Poly-Si: Overview", *Semiconductor Internat'l*, May, '84, p. 215.
28. M. Mieth, *et al.*, "Plasma Etching Using SF_6 and Chlorine Gases", *Semiconductor International*, May, 1984, p. 222.

29. U. Winkler, "VLSI Polysilicon Etching: A Comparison of Different Techniques", *Solid State Technol.*, April, 1983, p. 169.

30. P. Chang and S. Hsia, "Selective Plasma Etching of Polysilicon", *Solid State Technology*, Aug., 1984, p. 225.

31. W. Beinvogel and B. Hasler, "Reactive Ion Etching of Polysilicon and Tantalum Silicide", *Solid State Technology*, April, 1983, p. 125.

32. H.F. Winters, J.W. Coburn, and J.T. Chang, "Surface Processes in Plasma Assisted Etching Environments", *J. Vac. Sci. Technol.*, B1, (2), 469, 1983.

33. C.J. Mogab, "Dry Etching", in *VLSI Technology*, Ed. S.M. Sze, McGraw-Hill, NewYork, 1983, Chap. 8, p. 303.

34. D.R. Stull, *Ind. Engr. Chem.*, 39, 517, (1947).

35. J.F. O'Hanlon, *Solid State Technol.*, October, 1981, p. 86.

36. D.L. Smith and R.H. Bruce, *Ext. Abs. of the Electrochem. Soc.*, Abs. No. 258, Fall Meeting, 1981, Denver, CO., Electrochem. Soc., Pennington, N.J.

37. K. Tokunaga and D.W. Hess, *J. Electrochem. Soc.*, 127, 928 (1980).

38. R.H. Bruce and G.P. Malafsky, *J. Electrochem. Soc.*, 130, 1369 (1983).

39. T. Tsukuda, *et al..*, in *Plasma Processing*, Eds. J. Dieleman, R.G. Frieser, and G.S. Mathad, The Electrochemical Society,Inc., Pennington, N.J.

40. J.E. Spencer, "The Management of $AlCl_3$ in Plasma Etching Aluminum and Its Alloys", *Solid State Technol.*, April, 1984, p. 203.

41. M. Pender and P.C. Lindsey, 1982, Zylin Corp., unpublished results.

42. R.H. Bruce and G. Malafsky, *Ext. Abs. of the Electrochem Soc. Meeting*, Fall, 1981, Denver, CO., Oct. 1981, Abs. No 288, Electrochemical Society, Pennington, N.J.

43. M. Oda and K. Hirata, *Jap. Jnl. Appl. Phys.*, 19, 405 (1980).

44. F. Daniels and R. Alberty, *Physical Chemistry*, John Wiley & Sons, New York, 1981, p. 126.

45. S. Broydo, "Important Considerations in Selecting Anisotropic Plasma Etching Equipment", *Solid State Technol.*, April, 1983, p. 159.

46. D.W. Hess and R.H. Bruce, "Plasma Assisted Etching of Al and Al Alloys", in *Dry Etching for Microelectronics*, Ed. R.A. Powell, North Holland Publishing Co., 1984.

47. B. Chapman and M. Nowak, "Troublesome Aspects of Aluminum Plasma Etching", *Semiconductor International*, November, 1980, p. 139.

48. W.Y. Lee, *et al.*, *J. Appl. Phys.*, 52, 2994 (1981).

49. M. Sternheim, W. van Gelder, and A.W. Hartman, "A Laser Interferometer System to Monitor Dry Etching of Patterned Si", *J. Electrochem. Soc.*, 130, 655, 1983.

50. P.J. Marcoux and P.D. Foo, "Methods of End Point Detection for Plasma Etching", *Solid State Technol.*, April, 1981, p. 115.

51. A.D. Weiss, "End Point Monitors", *Semiconductor International*, Sept. 1983, p. 98.

52. S.J. Fonash, "Advances in Dry Etching Processes- A Review", *Solid State Technol.*, January, 1985, p. 150.

53. D.L. Tolliver, "Plasma Etching in Microelectronics-Past, Present, and Future", *Solid State Technol.*, Nov., 1980, p. 99.

54. D.L. Smith, "High Pressure Etching", in *VLSI Electronics*, Vol. 8, N.G. Einspruch, Ed., Academic Press, Orlando, FL., 1984, Chap. 9, p. 253.

55. J. Dieleman and F.H.M. Sanders, "Plasma Efluent Etching: Selective and Non-Damaging", *Solid State Technol.*, April, 1984, p. 191.

56. A.R. Reinberg, "Dry Processing for Fabrication of VLSI Devices", in N.G. Einspruch, Ed., *VLSI Electronics*, Vol. 2, Academic Press, New York, 1981, Chap. 1, p. 1.

57. J. Maher, "Selective Etching of SiO_2 Films", *Semiconductor Int'l.*, May, 1983, p. 110.

58. A. Weiss, "Plasma Etching of Aluminum: Review of Process Equipment and Technology",

Semiconductor International, Oct., 1982.

59. A. Weiss, "Etching Systems", *Semiconductor International,* Oct., 1982, p. 69.

60. D.C. Hinson, *et al.,* "Magnetron-Enhanced Plasma Etching of Silicon and Silicon Dioxide", *Semiconductor International,* Oct., 1983.

61. G.K. Herb, *et al.,* "Plasma Processing: Some Safety, Health, and Engineering Considerations", *Solid State Technology,* Aug., 1983, p. 185.

62. G. Corn and D.G. Baldwin, "Safety Considerations for Plasma Aluminum Etching", *J. Vac. Sci. Technol.,* **83** (3), 909, (1985).

63. B.N. Chapman, T.A. Hansen, and V.J. Minkiewicz, *J. Appl. Phys.,* **51**, 3608 (1980).

64. *ibid.,* Ref. 2, p. 2.

65. C.J. Mogab, "Loading Effect in Plasma Etching", *J. Electrochem. Soc.,* **124**, 1262 (1977).

66. T. Enomoto, *et al.,* "Loading Effect and Temperature Dependence of Etch Rate in CF_4 Plasma", *Jpn. J. Appl. Phys.,* **18**, 155 (1979).

67. R. Lachenbruch, T. Wicker, and J. Peavey, "Contamination Study of Plasma Etching", *Semiconductor International,* May, 1985, p. 164.

68. L.M. Ephrath and D.J. DiMaria, "Review of RIE Induced Radiation Damage in SiO_2", *Solid State Technology,* April, 1981,

69. S.W. Pang, "Dry Etching Induced Damage in Si and GaAs", *Solid State Technology,* April, 1984, p. 249.

70. T. Watanabe and Y. Yoshida, "Dielectric Breakdown of Gate Insulator Due to RIE", *Solid State Technol.,* April, 1984, p. 213.

71. A.J. Woytek, *et al.,* "Nitrogen Trifluoride-A New Dry Etchant Gas", *Solid State Technol.,* March, 1984, p. 172.

72. C.S. Korman, *et al.,* "Etching Characteristics of Polysilicon, Silicon Dioxide, and $MoSi_2$, in NF_3 and SF_6 Plasmas, *Solid State Technol.,* Jan., 1983, p. 115.

73. J. Webber, "Choosing Gases for Plasma Dry Etching", *Microel. Mfg. & Test.,* Jan. 85, p.40.

74. R.F. Reichelderfer, *et al., J. Electrochem. Soc.,* **124**, 1926 (1977).

PROBLEMS

1. Cite five advantages that dry-etching possesses when compared to wet etching. Also list three of the most important disadvantages associated with dry-etching when compared to wet etching.

2. Why have *ion-milling* and *reactive ion beam etching* found little application in the fabrication of silicon VLSI circuits.

3. Explain the difference between the terms *glow discharge* and *plasma.*

4. Explain why the term *ion-assisted etching* is a more apt description of dry-etching processes which rely on both chemical and physical etching effects, than is the term *reactive-ion etching.*

5. Explain why the potential of a glow discharge used in dry-etching or sputter deposition, is positive relative to ground.

6. When etching Si over SiO_2 in a CF_4 plasma, the problem of obtaining an adequately high S_{fs} does not usually exist. Explain why this is the case.

7. Explain why a short-lived reactive species, such as atomic chlorine, does not exhibit loading effects when used to perform dry-etching.

8. If the selectivity of SiO_2 over Si can be improved by adding H_2 to the CF_4 feed gas, why have processes that use gases that contain no H_2, but still offer improved selectivity compared to pure CF_4 (e.g. C_2F_6), been investigated as alternatives?

9. The addition of H_2 to a CF_4 plasma is accompanied by the deposition of polymers *(polymerization)* on surfaces within the reaction chamber. Discuss the chemistry of such polymer formation. Note that it may be necessary to consult a chemistry text or the technical literature to

completely answer this problem.

10.   Cite an etch process that is apparently controlled by the *surface damage mechanism,* and another that is apparently controlled by the *surface inhibitor mechanism.*

11.   Explain why the formation of sloped sidewall in etched features by controlled resist erosion is such a difficult process to implement in a batch-etch production manufacturing environment.

12.   Discuss some of the reasons why the development of an adequate dry-etch process for polycide layers is such a challenging task.

13.   If a dry-etch process for an Al-4at% Cu alloy is conducted at 70°C, use the vapor pressure of CuCl versus temperature curve, given in Fig. 20b, to calculate the maximum etch rate of such a film which would still allow the CuCl etch product to completely evaporate as the film is being etched.

14.   In practice, the evaporating CuCl molecules of problem 13 would suffer collisions with the argon sputter gas to a signifiant degree, and therefore be returned to the surface from which they evaporated in relatively large numbers, in effect making the apparent evaporation rate substantially lower.   Calculate the mean free path of a CuCl molecule in an Ar plasma at a pressure of 2 torr.

15.   In some cases a loading effect is necessary in order to be able to monitor an etch process. Cite at least one important aspect of a dry-etch process that takes advantage of loading effects.

16.   Explain the difference between the *utilization factor* and *loading effects.*

17

MATERIAL CHARACTERIZATION

TECHNIQUES FOR

VLSI FABRICATION

The fabrication of complex integrated circuits involves the correct performance of a large number of sequential steps, as well as the use and interaction of many materials. The monitoring, evaluation, and characterization of each step during process development and production requires the use of a variety of measurement techniques. Many of them are discussed in the chapters that describe the processes that they commonly characterize (e.g. measurement of thickness and resistivity of films and layers prepared by sputtering, CVD, or implantation).

The remaining diagnostic techniques that find wide use in semiconductor applications for identifying or characterizing the morphology, chemical composition, and crystallographic structure of materials are discussed in this chapter. Since we can only offer a brief introduction to each of these techniques, the approach will be to state the principles upon which each is based, and then specify the capabilities and limitations that are of most importance to VLSI applications. The purpose is to acquaint readers with each of the techniques, so that they can select the most appropriate of them for particular analytical applications. With this in mind, in the last section we summarize and compare the capabilities of diagnostic techniques. A selection approach and method for interfacing with facilities that provide characterization services, including sample preparation considerations, is also given. Finally, a list of references which treats each of the techniques in significantly more detail is given at the end of the chapter.

WHAT ARE WE TRYING TO DETECT and HOW IS IT DONE?

There are three classes of material properties commonly characterized by the diagnostic techniques discussed in this chapter:
   a) Morphology (form and structural properties, such as thickness or surface roughness);
   b) Elemental /Chemical Composition;
   c) Crystallographic Structure and Defects.
The morphology of the VLSI features is primarily observed by the various microscopy techniques - optical, scanning electron (SEM), and transmission electron (TEM). Crystal structure and defects are evaluated mainly by x-ray diffraction methods, transmission electron microscopy (TEM), and Rutherford backscattering spectroscopy (RBS). The principles which underlie morphology and crystal structure analysis will be discussed in further detail in the sections relating to the specific techniques used for such analysis.

586

The elemental and /or chemical compositional analysis of materials involves a variety of different techniques, and as such deserves some introductory discussion.   Most often such analysis is concerned with the presence of elements (and /or molecules) in the material being evaluated. Sometimes only *qualitative* information is required (i.e. What is present?), but other applications may require *quantitative* data (or *How much* of what elements is present?).   If quantitative data is needed, usually a *standard* is also required. That is, a sample containing a known quantity of the substance whose presence is being quantifiably identified (determined with confidence by an alternate diagnostic technique), must be analyzed to establish reference signal magnitudes for the characterization technique being used. In fact, all compositional analysis techniques except RBS require such standards as a prerequisite to obtaining accurate quantitative data.

Information may also be sought about composition of substances in differing kinds of locations on a substrate.   That is, data might be needed on the composition of localized patterns (e.g. residue in contact holes) or on regions of large area films (e.g. composition of deposited films). Furthermore, data concerning the surface, or the bulk, might be of interest, or information on the variation of the composition as a function of depth might be needed.   Finally, there might be a need to know how a substance is distributed laterally in or on a film, in some cases requiring an instrument with fine lateral resolution, but in others tolerating coarser lateral resolution.

Next, consider the problem of determining which materials are present on the sample of interest.   In order to identify if a substance is present we must be able to measure some quality of an element that distinguishes it from others.   There are two such qualities unique to each element for this purpose, the *atomic mass*, and the *electronic structure*. When an atom /molecule is ionized, the mass-to-charge ratio can be used to (magnetically) selectively filter only the desired mass constituents, and thereby determine the presence of the species.   The uniqueness of the electronic structure of each element or molecule is used to identify the source of emitted electrons and photons from materials under analysis, and thereby identify the constituent species.

In spite of the fact that all the elemental /chemical analysis techniques are based on the mass and electronic structure of the elements, there are a large number of characterization techniques available.   The reason for their variety is that *no single technique is best, or even suitable,* for all applications.   In fact, often more than one must be used to accomplish an unambiguous analysis.   In addition, the task of interpreting data from most of the techniques is quite subtle, even though the data is gathered by sophisticated, expensive instruments.   Thus, interpretation of signal data from analytical instruments is usually best accomplished by an experienced characterization analyst.   Some procedures, however, can be employed to help the user to achieve a successful analysis.   A list of these will be presented.

It also needs to be emphasized that advances in VLSI fabrication have made measurements of material properties increasingly difficult. The form of the material has evolved from *bulk* to *film* to *surface*; the physical size of the area or volume to be analyzed has decreased; and the elemental concentrations have grown smaller.   In addition, decreasing feature sizes make *physical defects* of small dimensions play a greater role in the degradation of both materials performance and device reliability. Thus, techniques for characterizing such small defects must also be available.   It seems that in order for material characterization to continue to be effectively performed, it will become even more necessary to utilize a variety of complementary techniques. In such cases, the role of each technique, and its strengths and weaknesses will need to be well understood.

Energy Regimes and Energy Levels in Material Characterization

The energies used to describe the source beams and emitted species in material characterization techniques are usually expressed in electron volts (eV) - which we note are

Table 1.   ENERGY REGIMES IN THE MICROELECTRONIC MILLIEU

1 eV	- energy possessed by evaporated atoms arriving at a substrate
5 eV	- energy possessed by sputtered atoms arriving at a substrate
10-20 eV	- energy required to ionize neutral atoms (Ar $\approx$15eV)
20 eV -1 keV	- energy possessed by emitted Auger electrons
1- 20 keV	- energy possessed by primary beam species in SEM, AES, SIMS, XES
100 keV	- energy possessed by primary beam of electrons in TEM, and ions in ion implantation.
1 - 3 MeV	- energy possessed by primary ion beam in RBS, and primary beam of neutrons in NAA.

not units of potential (or potential difference), but units of *energy*.

> 1 electron volt (eV) = the amount of energy required to move a unit of electronic charge through a potential difference of 1 volt.

Since 1 electronic charge = 1.6×10^{-19} C, then

$$1 \text{ eV } = (1.6 \times 10^{-19} \text{ C x 1V}) = 1.6 \times 10^{-19} \text{ J.} \tag{1}$$

The energy regimes encountered in the materials characterization millieu are given in Table 1.

As mentioned earlier, one of the methods of determining the presence of elements in the materials being evaluated is through the detection of electrons or photons, emitted by the material under bombardment by a primary electron or photon (e.g. x-ray beam). The notation for describing the lower electronic energy levels in an atomic structure are shown in Fig. 1.

There is a unique energy associated with each electronic level in an atom, relative to free space. For example, if an electron is present in the L_1 level, there is a specific minimum energy that must be absorbed by the electron in order to be ejected into free space.   The detection of emitted electrons or photons in response to specific incident energy is one technique used to identify the presence of an element in the material being analyzed.

Some Definitions of Material Characterization Terminology

Primary beam (electrons, ions, photons) - constituents of the beam which emanates from the instrument energy source and is then directed at the test sample.

Secondary beam - objects (electrons, x-rays, ions) that are emitted by the material of the

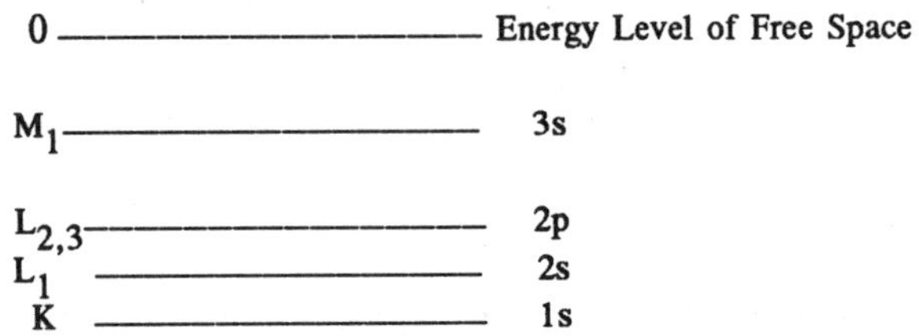

```
0 _____  Energy Level of Free Space


M₁ _____   3s



L₂,₃ _____   2p
L₁  _____   2s
K  _____   1s
```

Fig. 1   Notation for describing energy levels in atomic structure.

test sample as a result of being struck by the primary beam.

Backscattered (ions, electrons) - primary beam objects that "bounce back" after striking the material of the test sample .

Auger electrons - secondary electrons emitted by indirect energy transfer mechanisms, which are discussed in the section on Auger emission spectroscopy, AES.

Matrix - that from which something takes form. For example, oxygen found in plasma silicon nitride takes the *form* of an impurity (or additive) in the matrix of the Si_3N_4; or P, B, and Sb can be impurities in a Si matrix.

Morphology - form and structure of (e.g. the structural attributes of a feature on a substrate, such as the thickness, width, shape, surface structure, grain structure, and interface structure).

Detectability - specifies the elements which can be detected by a given analytical technique.

Sensitivity - defines the minimum concentration level of an element that an analytical technique can detect.

Specificity - the ability of a compositional analysis technique to distinguish different elements from one another. The specificity can be limited either by: a) intrinsic aspects of the analytical technique itself (such as insufficient energy resolution capability of an instrument detector); or b) by interferences in the detected spectrum (such as two elements emitting species with nearly identical energies, and which thus cannot be resolved by the measuring instrument).

Vacuum Requirements of Compositional Analysis

Compositional analysis measurements are performed in vacuum because the primary beam objects must arrive at the sample without interacting with the gas molecules in the intervening distance between the primary beam source and target.   Such collisions could disturb the focus, energy, or ionic charge of the primary beam.   There are three vacuum ranges encountered in the use of compositional analysis instruments: 1) Those techniques that monitor (but do not dislodge elemental material from) the surface of the samples (AES, XPS), place the most stringent requirements on the cleanliness of the residual vacuum in the sample region.   Surface contaminants will attenuate the analytical signal emanating from the true sample, and add their own characteristic spectrum to the output signal. To avoid this, the vacuum in AES and XPS instruments must be in the $1-3 \times 10^{-10}$ torr range; 2) In SIMS, the surface is continually being sputtered, and consequently contaminants adsorbed from the residual gases are continually removed. Thus, less severe requirements are placed on the SIMS vacuum system ($1 \times 10^{-8}-1 \times 10^{-9}$ torr); and 3) RBS uses an energetic beam that penetrates the surface of the sample. Thus the technique is not sensitive to the physical surface, nor is its analytical signal affected by a monolayer surface contaminant.   As a result, a pressure of $\sim 1 \times 10^{-6}$ torr is adequate for RBS instruments.

MICROSCOPY FOR VLSI MORPHOLOGY

The morphology of VLSI structures is generally determined by three microscopic techniques: a) optical microscopy; b) scanning electron microscopy (SEM); and c) transmission electron microscopy (TEM). Since the maximum magnification values of these three methods are 1000x, 100,000x, and 500,000x, respectively, and since the magnification ranges overlap, we have the ability to conduct a complete morphological analysis on any feature of interest.

Optical Microscopes

Optical microscopes are one of the most important analytical tools available for monitoring VLSI fabrication processes[1]. Fabrication personnel must inspect the wafers at all steps of the process to understand and monitor what is occurring during the manufacturing sequence. In semiconductor wafer processing applications, the most important capabilities of the optical microscope are the following: a) resolution; b) magnification; c) mechanical stability; and d) wide-field-of-view. Other useful features of optical microscopes include brightfield, darkfield, and Nomarski interference contrast capabilities[2], fluorescence microscopy capability, and television option.

Resolution, Magnification, and Numerical Aperture

In general the term *resolution*, d, is defined as a measure of the ability to separate closely spaced features. That is, it corresponds to the distance between two points in an image when those two points are recognized as being separated.   The resolution is calculated from:

$$d \ = \ \frac{\lambda}{2 \, (NA)} \tag{2}$$

where λ is the wavelength of illumination, and (NA) is the numerical aperture of the objective lens of the microscope.

The resolution limit of an optical microscope with an NA = 0.95 is approximately 0.25 μm if a λ of 0.5 μm is used (i.e. the λ of green light at the center of the white light spectrum). In routine diagnostic work in a development environment optical microscopes are typically expected to resolve ~0.5 μm. The numerical aperture (NA) is determined by the angle of the cone of light accepted by the objective of a microscope (angle θ in Fig. 2). It is defined by:

$$NA \ = \ n \ \sin \ \left(\frac{\theta}{2}\right) \tag{3}$$

where n is the index of refraction of the medium between the objective and the specimen.   Since (θ /2) cannot exceed 90°, the theoretical limit of sin(θ /2) is 1.0.   The highest NA of manufactured lenses for dry systems (in which air is the medium, and n = 1) is 0.95.

The *magnification*, M, describes the ability to enlarge a pattern (e.g. the ratio of the size of an image to the size of the corresponding object), and the approximate maximum magnification required for the eye to see all the possible detail that a microcope can reveal is:

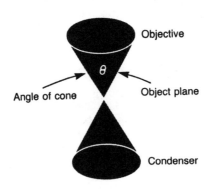

Fig. 2 Numerical aperture is determined by cone of light accepted by the microscope objective.

$$M_{max} \quad = \quad 1000 \; (NA) \qquad\qquad\qquad (4)$$

For an NA = 0.95 the maximum magnification needed for optical microscopes is ~1000x. This upper magnification limit arises from the fact that the resolution of the unaided human eye is ~0.25 mm. At about 1000x, a 0.25 μm separation appears as 0.25 mm. Since the microscope cannot resolve any finer features, at 1000x the eye sees all the microscope resolvable detail.

The magnification range for the inspection of wafers and masks is 20x to about 1000x. Most microscopes sold for semiconductor use can be provided with combinations of 10x to 20x *eyepieces*, and *objectives* in the range of 2x to 100x, to cover this entire magnification range. On high magnification microscopes, the mechanical *stability* is as important as the quality of the objective lenses. Accordingly such instruments are rather massive, and must be mounted on vibration isolated tables.

In some applications (e.g. examination of large areas for the presence of particulates or other surface defects), optical microscopes may be more useful if they have a relatively low magnification, but are equipped with lenses that have the same resolution as the high magnification lenses (but with a *field-of-view* that is 10 times as great).

Brightfield and Darkfield Illumination Modes

In the *brightfield mode* of microscopy, the wafer under observation is seen by the light that it reflects. Light travels along the optical axis of the microscope, through the objective lens, to the sample being examined. The image is formed by the *reflection* of the light received by the sample, which then travels back through the same optical elements. This is the most commonly used mode, and for the majority of applications gives the best overall image and information.

In the *darkfield mode*, light is directed at the wafer from angles *outside* of the cone that the objective encompasses, so that this light strikes the wafer surface obliquely. Only light that is reflected or refracted by *features* on the wafer surface is collected by the objective. Thus the sample appears as a black background, and the features that reflect or refract the light appear bright. Darkfield illumination enhances the visibility of details that might be washed out by brightfield illumination. Even small structural details below the resolution limit are often visible in the darkfield mode, making the technique useful for quickly scanning a wide viewing field to find particles, scratches, and chemical residues.

Nomarski Interference Contrast Microscopy

Under illumination in the *Nomarski interference contrast mode*[2], features on a substrate which are at different relative elevations can appear as different colors or different shades of grey. Such *contrast* is achieved by splitting the illuminating beam into two beams. The two beams strike the surface of the wafer a short distance apart, and are then reflected back into the microscope and reconstituted (Fig. 3a). The optical path length of the two beams will be different if one of the beams encounters a step or change in the index of refraction (e.g. caused by a phase boundary) not encountered by the other. These differences in optical path length produce a microscope image that contains contrast effects. Figures 3b-d show how the intensity variations observable when the Nomarski contrast mode is used, are caused when monochromatic light illuminates a substrate. The polarized light (Fig. 3a), enters the *Wollaston prism* and is split into two mutually perpendicular polarized components. The recombined components will have altered intensities if optical path length differences are encountered, and these are observed by using an analyzer. The polarizer, analyzer, and prism can all have their settings adjusted, and their adjustments will affect the emerging wavefronts and resultant intensity profiles.

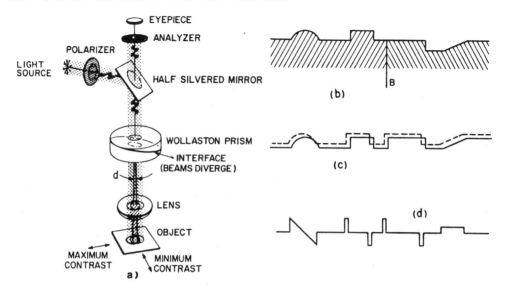

Fig. 3 (a) Features of a Nomarski interference contrast microscope operating in the reflectance mode. (b) Representation of a crosssection of a sample at a surface. (c) The wave fronts of the reflected beams after emerging from the prism, and (d) an intensity distribution in the image plane in a Nomarski interference microscope[40]. Copyright, 1983, Bell Telephone Laboratories, Incorporated, reprinted with permission.

Interference contrast is maximized when the elevation difference on the substrate is parallel to the direction of maximum displacement of the beams, and is essentially zero in the orthogonal directions. The adjustments of the polarizer, analyzer, and prism can be set to produce such maximum interference contrast. Differences in edge elevations as small as 30-50Å can be resolved with the Nomarski mode, although results do depend upon the quality of the edge. The technique is therefore used to view details on the surface of wafers, such as pits, fissures, and stacking faults that are often invisible in ordinary reflected light illumination.

Fluorescence Microscopy

In fluorescence microscopy, the wafer surface is illuminated with ultraviolet light, and in response, the illuminated materials emit characteristic radiation in the visible light regime. Organic substances fluoresce more brilliantly than the typical inorganic constituents of the wafer. The technique is therefore useful for detecting residues of photoresist and other organic films.

Television System Interface Capability

The ability to interface a microscope with a television is becoming a useful option. For example, in the wafer production environment, remote consultation (e.g. out of the clean room) can be performed, and video-tape documentation of the patterns being observed can be generated.

Scanning Electron Microscopy (SEM)

Scanning electron microscopy (SEM) has become an important tool for VLSI analysis because it has the capability of providing much higher magnification, resolution, and depth of field than optical microscopy[3,4,16]. That is, the resolution of SEM can be up to 10Å (100Å is

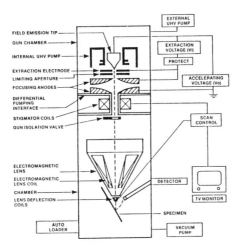

Fig. 4 Block diagram of an electron optical column of an SEM.   Courtesy of Nanometrics, Inc.

routinely obtained), the magnification from 10x-100,000x (a few instruments up to 300,000x), and the depth of fields of 2-4 μm at 10,000x and 0.2-0.4 mm at 100x. (For optical microscopes the maximum resolution and magnification limits are about 1 μm [10,000 Å] and 1000x, and the depths of field are much shallower.) Note that the high depth of field makes SEM especially useful for high magnification (i.e. >2000x) examination of VLSI device surfaces, where film thicknesses rarely exceed 1 μm.   SEM analysis yields information on linewidth, film thickness, step coverage, edge profiles after etch, and other morphology data.

A schematic drawing of a SEM is shown in Fig. 4.   A source is used to create a beam of electrons that is accelerated to energies of 500 eV-40 keV, focused to a small diameter, and directed at the surface of a sample in a raster-scan pattern.   The electrons striking the surface cause a number of physical phenomena to occur, the most important for SEM applications being

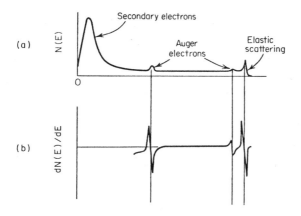

Fig. 5 (a) Energy spectrum of electrons emitted from a surface bombarded by an electron beam. (b) The derivative of the number of emitted electrons with respect to the energy.

the emission of electrons and x-rays. The emitted x-ray signal is useful for chemical analysis, and this is discussed in the section on XES.

Figure 5a shows the energy spectrum of electrons emitted from a surface bombarded by an electron beam. We that see there are three predominant types of emitted electrons: *lower energy electrons* of 0-50 eV, peaking at about 5 eV; *higher energy electrons*, with energies close to those of the primary beam electrons; and *Auger electrons*. The lower energy electrons are called *secondary electrons*, and they are produced by inelastic collisions of the primary beam and the inner shell electrons of the sample atoms. Because they possess such low energies, only secondary electrons created close to the surface actually escape and are detected. *Note that it is these low energy secondary electrons which are generally the most useful for morphology studies of VLSI.* The higher energy emitted electrons are those that have suffered *elastic* collisions with target atoms and thus still possess most of their incident energy. These are referred to as *backscattered electrons*. The mechanism by which *Auger electrons* are produced is described in a later section.

As we shall also see in later sections, the incident electrons undergo multiple collisions as they penetrate the sample, and those that are not backscattered finally come to rest after traversing a range, R (Fig. 6). The electron trajectory also changes with each collision, causing the narrow beam to spread as it penetrates the sample. Because of the short escape depth of secondary electrons, compared to the penetration depth at which beam broadening becomes influential, secondary electrons exhibit better point-to-point resolution than do backscattered electrons.

The detected electron current, which can be due to secondary or backscattered electrons, but as noted is typically the former, is used to intensity-modulate the z-axis of a CRT. An image of the sample surface is produced on the CRT screen by synchronously raster scanning the CRT screen and the electron beam of the SEM.

The *contrast* of the image depends on variations in the flux of electrons arriving at the detector, and is thus related to the *yield* of emitted electrons per incident electron[1]. For secondary electrons the yield depends on the work function of the material, and is significantly higher for oxides and other wide band gap materials than for silicon. This is another important factor that makes the use of secondary electron SEM imaging so valuable for VLSI studies. That is, the effect makes metals, oxide, and silicon patterns readily distinguishable from one another when the SEM is used to produce images in this mode. The second source of contrast in secondary electron images is the dependence of secondary electron yield on surface curvature. Therefore, surfaces that differ significantly in slope can be easily distinguished. Finally, surface

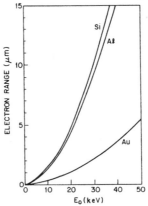

Fig. 6 The electron range in Si, Al, and Au as a function of incident beam energy[41]. Reprinted with permission of the American Physical Society.

regions that face the detector appear brighter than other surface regions[7].

The *resolution* of the SEM depends on several factors, including the type of sample under inspection and the incident beam diameter (which is dependent on the electron source, focusing optics, and accelerating voltage of the primary beam). In early generation SEMs, high voltages were required to achieve the small scanning spot sizes needed to obtain sufficient resolution for operation at high magnification. Unfortunately this leads to a problem when examining insulating surfaces. That is, when beam energies exceed the *secondary electron crossover point* (i.e. the energy at which the number of secondary electrons emitted is less than the number of incident electrons) the surface acquires excess negative charge. This then disturbs the trajectory of the incident beam, and the image is degraded. By applying a thin metallic coating to the surface of the sample (e.g. 100 Å of gold), and attaching a ground wire to the coating, an electrical path to ground is provided. This helps to reduce such charging effects. Unfortunately an Au coating makes the sample unsuitable for further processing, and this procedure is not acceptable for applications where wafers from the fab line must be inspected and returned to production (e.g. for linewidth measurement with a SEM). Another disadvantage of high accelerating voltages is that the electrons may cause damage to the circuit under inspection. Therefore SEM manufacturers have developed techniques which use accelerating voltages below the secondary electron crossover point (i.e. 800-2000 eV), and yet maintain high resolution[5].

The original tungsten hairpin electron beam sources produced beam diameters that were too large to give adequate resolution at beam energies of 1-2 keV. Thus several new sources, including the lanthanum hexaboride (LaB_6), field emission, and Schottky emitter sources have been developed to maintain high resolution at low operating voltages. Table 2 gives some characteristics of these sources[6].

The LaB_6 is very reactive and must therefore be operated at vacuum levels no greater than 10^{-7} torr. Most manufacturers of SEMs with LaB_6 sources equip the source with its own pumping system, generally using an ion pump. If properly operated, however, the LaB_6 source will last 10-50 times longer than tungsten hairpin sources.

The tungsten field emission source offers the highest brightness of these sources and the spot size can be focused to 20-50Å quite easily. The source, however, must be operated at vacuum pressures of 10^{-10} torr, which is more difficult to attain, and the maximum specimen current is approximately three orders of magnitude lower than that achieved by the other sources.

The relatively new extended Schottky field emitter source[6] provides comparable brightness to the field emission source at 1 keV, but is able to functional somewhat lower vacuums ($\sim 10^{-8}$ torr). In addition, it has a longer tip life and greater emission current stability than other electron sources. This allows Schottky emitter sources to be operated and maintained more easily than field emission sources.

SEM Voltage Contrast Microscopy

The energy distribution of secondary electrons in SEMs shows a peak at about 5 eV. These secondary electrons are collected by the detector, where their flux determines the intensity of the image. Local electric fields at the sample surface can significantly restrict the number of secondary electrons that are emitted, and such fields can be generated by applying small voltages (1-5 V) to the wafer surface. For example, if a potential difference of +5V is applied to a conductor on the wafer surface, and an electrical discontinuity exists in the conductor, the portion to which the voltage is applied will appear much darker than the section which is connected to ground (Fig. 7). Thus electrical discontinuities in long conductor lines and reverse-biased diodes can be easily located with voltage contrast microscopy techniques.

Fig. 7 Example of using the SEM voltage contrast mode to locate an open circuit in a long metal stripe. Courtesy of TRW.

Electron Beam Induced Current Microscopy (EBIC)

The difference in the amount of absorbed current by the wafer surface can modulate the intensity of the CRT image and produce enhanced contrast. Thus, this mode is useful for high-lighting electrical paths, such as defects that are leakage sites in capacitors or p-n junction[10,11].

Backscattered Electron Detection

Although the detection of secondary electrons is the primary mode in which the SEM is operated, the detection of backscattered electrons has at least one useful application. Since the yield of backscattered electrons increases with increasing atomic number, Z, contrast is produced in the backscattered electron image between regions of different atomic numbers, such as aluminum particles on a silicon background.

Production SEMs and Failure Analysis SEMs

The first major application of SEMs in microlectronics was for failure analysis. Since the samples inspected in such analysis would ordinarily not be subject to further processing, they could be destructively gold coated to improve SEM image quality. Accelerating voltages and subsequent sample damage were therefore also less important. In addition, failure analysis SEMs can be designed to be equipped with auxiliary analytical capabilities such as x-ray

Table 2. CHARACTERISTICS OF ELECTRON BEAM SOURCES

	Tungsten Hair Pin	Lanthanum Hexaboride	Field Emission	Schottky /Extended
Type of Emission	Thermionic (Heated)	Thermionic (Heated)	(Room Temperature)	Field (Heated)
Brightness (A /cm^2/ster)	10^4	10^5	10^8	10^8
Effective Source Size (Å)	1,000,000	200,000	100	100
Energy Spread (eV	3	3	0.2-0.3	0.28-0.38
Operating Life (hrs)	30-10	100-500	300-1000	2000-10000
Vacuum Required (torr)	10^{-3}-10^{-5}	10^{-5}-10^{-6}	10^{-9}-10^{-11}	$< 10^{-8}$

emission spectroscopy (XES). This increases their flexibility to perform a wider variety of failure analysis tasks. Finally, the time required to get samples into and out of such instruments is relatively long, making the rate of sample inspection (throughput) quite low.

As features decrease to the micron and submicron regimes, optical microscopes become less suitable for providing the resolution needed to give accurate and repeatable measurements of critical dimensions. Therefore the use of SEMs as production tools becomes more attractive. In order to make SEMs compatible with the production environment, they have to be configured differently than the traditional failure analysis instruments. Production compatible SEMs must utilize sources that can operate at low enough voltages to give high resolution on insulating surfaces (resist, oxides, etc.) without charging or damaging the wafer structures. They must also possess the ability to change samples quickly and easily (cassette-to-cassette operation) and to be able to automatically and accurately position wafers to preselected locations. Finally, since the highly trained specialist of the failure analysis staff is not likely to be available to operate such production tools, they need to be simple enough to be run by wafer fabrication personnel[8,9]. Several SEM manufacturers have produced SEMs that satisfy these requirements to some degree, but alternate linewidth measurement techniques (see Chap. 12) still predominate.

Transmission Electron Microscopy (TEM)

Just as shrinking linewidths and vertical feature sizes lead to the displacement of optical microscopes by the SEM, other VLSI applications for which the SEM had been adequate, now require an even higher resolution technique - namely TEM. Whereas maximum SEM resolutions are in the 20-30Å range, TEM offers 2Å resolution. The image in TEM is produced by the differential loss of electrons from an incident beam (60-350 keV, electron wavelength ~0.04Å) as it passes through very thin film samples. The sample must be thin enough to transmit the beam, so that essential information caused by differences in sample thickness, phase composition, crystal structure, and orientation is preserved. The limiting thickness for TEM imaging of a Si sample as a function of accelerating voltage is shown in Fig. 8. In fact, for practical VLSI analysis, the thickness is even smaller than indicated (e.g. 0.8 μm at 200 keV).

In a conventional TEM, the electron beam is focused by a condenser lens, then passes through the sample and is imaged onto a photographic plate or fluorescent screen. The *contrast* in a TEM image arises for different reasons in samples of crystalline and amorphous materials. In crystalline layers, the incident electron beam is diffracted by the materials. Abrupt changes in thickness, phase structure, or crystallographic orientation cause corresponding changes in

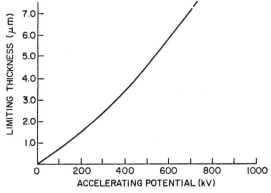

Fig. 8 Limiting thickness of a silicon sample as a function of accelerating voltage[13].

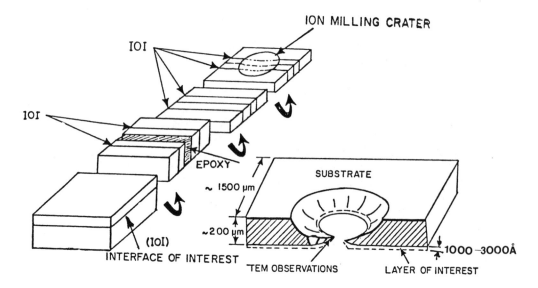

Fig. 9 Method for preparing a cross sectional sample for TEM study.

contrast, and these crystallographic features can be easily imaged at high resolution.   In amorphous regions, contrast is obtained from differences in sample thickness or from differences in chemical or phase composition. TEM images from amorphous materials (e.g. oxides, nitrides) are thus somewhat easier to interpret than images from crystalline layers. Nevertheless, TEM is capable of imaging the grains of polycrystalline films, and is thus a very useful tool for measuring grain sizes and structural anomalies in thin films[12,13,14].

Sample Preparation

The two major factors that have prevented TEM from being more widely used, in spite of its excellent resolution and analytical capabilities, are: 1) the difficulties involved with preparing the required very thin samples; and 2) correctly interpreting TEM images. Related to the sample

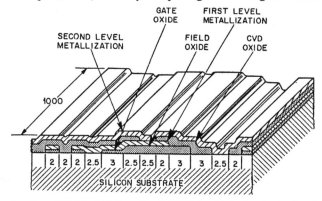

Fig. 10 Schematic of a TEM test pattern for NMOS device technology showing a 29-μm repeat unit all dimensions are in microns[40].   Copyright, 1983, Bell Telephone Laboratories Incorporated, reprinted with permission.

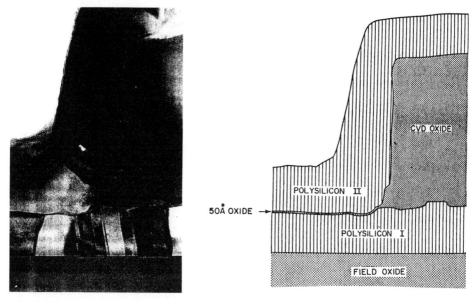

Fig. 11 TEM image and drawing of a highly resistive contact of Poly-Si II to Poly-Si I caused by the presence of a 50Å interfacial oxide. From R.B. Marcus and T.T. Sheng, *Transmission Electron Microscopy of Silicon Devices and Structures,*   Copyright ©, 1983 John Wiley & Sons. Reprinted with permission of John Wiley & Sons, Inc.

preparation problem, is the need to insure that the feature of interest is present within the sample region that has been thinned and prepared for TEM analysis. TEM sample sections of most interest for VLSI studies are vertical cross-sections. Such samples are prepared as shown in Fig. 9. Several hours are required to ion mill samples to the necessary thickness, making their preparation an arduous task. Details on sample preparation procedures are given in Ref. 12.

In order to help insure that the feature of interest will be present in the thinned section of the sample, a TEM test structure must be designed (as shown in Fig. 10), that will contain every morphological feature that will exist in the particular process to be fabricated. Each feature should appear within 1-3 µm in one direction, and should extend about 2 mm in the orthogonal dimension. Thus when 2 mm samples are cut from the wafer, the test structure will be present along the entire 2 mm dimension. Finally all features should be contained in ~23 µm, and this "repeat unit" should be replicated a number of times over a distance of about 2mm. Several TEM photographs appear throughout the text illustrating various applications for which TEM imaging and analysis is used (see also Fig. 11).

ELECTRON /X-RAY COMPOSITIONAL ANALYSIS TECHNIQUES

When the surface of a solid is struck by electrons or x-rays, both x-rays and electrons are emitted from the solid in response. Some of these emitted species contain energy information about elements present at the surface and in the bulk of the bombarded sample. The four analytical techniques based on the detection of such emitted x-rays and electrons are:

  a) Auger Electron Spectroscopy (AES) - electron primary beam /Auger
    electrons are the detected species;

b) X-ray Emission Spectroscopy (XES) - electron primary beam /x-rays
   are the detected species;
c) X-ray Photoelectron Spectroscopy (XPS or ESCA) - x-ray primary beam
   /photoelectrons are the detected species;
d) X-ray Fluorescence Spectroscopy (XRF) - x-ray primary beam /x-rays
   are the detected species.

Auger Emission Spectroscopy (AES)

The technique of Auger emission spectroscopy[15,17,18,38] (AES) involves the
bombardment of a solid by an energetic (up to 10 keV) electron beam. A certain class of
electrons, called *Auger electrons* are generated as shown in Fig. 12b. Auger electron emission
was also discussed in Chap. 10 in connection with the incidence of ions on solid surfaces. In
Fig. 12b we see that a primary electron ejects an electron from the K-shell of a silicon atom. An
electron from the L_2-level of the same atom relaxes into the K-shell, emitting a photon in the
process. In some cases this photon escapes the material, but in others it interacts with and
causes the *ejection* of a lower energy electron (from the L_3-level in this case). Electrons ejected
according to such mechanisms are called *Auger electrons*. Note that 3 electrons must be involved
to create one Auger electron. In this example, one was from the K-level, one from the L_2-level
and another from the L_3-level. Accordingly, this Auger electron is identified as a KLL Auger
electron. The energy of the ejected Auger electrons are thus characteristic of the type of atom
that emitted them. There are several such characteristic Auger transitions for most elements
(Fig. 13). Note that since three electrons must be involved, elements with fewer than three
electrons (i.e. H and He) cannot emit Auger electrons, and hence cannot be detected with AES.

Most Auger electron energies are between 20 and 2000 eV. The depth from which they can
escape from the solid without losing a significant percentage of energy (i.e. the specific energy
value serves to identify their origin) is quite shallow, generally less than 50Å. Note that escape
depth is defined as the distance at which the number of monoenergetic electrons drops to 1 /e
(36.8%) of their original value as a result of inelastic scattering processes. Thus, AES is a
technique that can only provide compositional data about the surface layers of samples.

Many diagnostic problems require information about the composition of the material at
depths greater than the escape depth of Auger electrons. To obtain this information, AES data is
taken from the bottom of a crater, ion sputtered into the surface. The milling process is stopped
at regular intervals, during which an Auger spectrum is taken (Fig. 14a). The Auger peak
heights can be plotted as a function of milling time and a depth profile can be obtained.

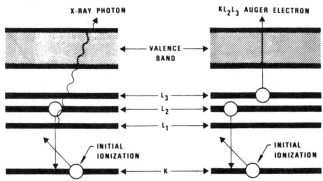

Fig. 12 X-ray and Auger electron emission during de-excitation of atom after initial ionization.

In modern instruments, a differentially-pumped ion-source is used to produce the sputtering ion beam. That is, the ionization compartment is supplied with its own separate pumping station. This is designed to remove all gas except that supplied to the ion beam itself, which is finely focused through a tiny aperture onto the sample. With differential pumping the sample chamber pressures can be maintained in the range between 10^{-7}-10^{-9} torr. This allows lower sputter etch rates to be utilized, or to employ extended scan periods for multiple-point analysis.

The chief advantage of AES is its ability to provide excellent lateral resolution, and consequent analysis of very small areas of 1 μm or less, together with acceptable sensitivity (~1% atomic). This can be coupled with raster scanning and ion-sputtering to yield three dimensional maps of elemental distributions. When AES is performed in such a raster scanned mode the analysis technique is referred to as *scanning auger microprobe* (SAM)[37]. The resolution of both AES and SAM is ~500Å, which corresponds to the minimum spot size of the electron beam in modern AES instruments (~500Å). This size represents a tradeoff among the brightness of the electron source (e.g. LaB_6), probe damage, and the ability of spectrometers to detect the emitted electrons[39]. That is, the LaB_6 brightness [~10^6 (A /cm^2) /steradian] allows conventional spectrometers to detect Auger electrons from a 500Å spot, while surface damage is minimized. If the same number of electrons were focused into a smaller spot, the increased electron density could rupture the bonding of the outer monolayers of some surfaces.

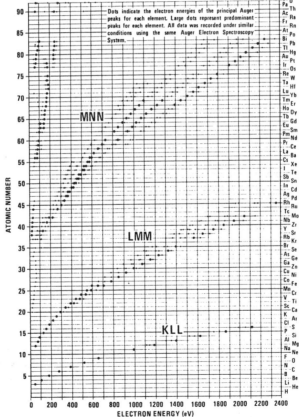

Fig. 13 Most prominent Auger transitions observed in AES. Courtesy of Perkin-Elmer.

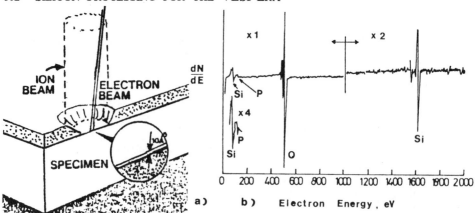

dN/dE

x 1   x 2

Si P

x 4

P

Si   O   Si

0   200   400   600   800   1000   1200   1400   1600   1800   2000

a)   b)   Electron   Energy , eV

Fig. 14 a) Ion milling a crater and performing AES on the exposed surface, thus allowing depth profiling with AES. (b) Auger spectrum of a P-doped SiO_2 film.

In the discussion of SEM, we noted that electrons emitted from solids struck by high energy primary electron beams have a wide energy spectrum, and include secondary electrons and back-scattered electrons as well as Auger electrons. The presence of Auger electrons is manifested as small peaks in the total energy distribution function (Fig. 5). All of these electrons can be collected, but only the Auger electrons are of interest. The remaining emitted electrons constitute strong noise, and to extract valid signal information about emitted Auger electrons from raw data, signal processing methods (e.g. differentiation) must be performed to elucidate the Auger electron peaks. This differentiation is performed by electronic means using lock-in amplifiers. Nevertheless the presence of the background electron signal places a lower bound on the sensitivity limits of AES. Typically AES can detect elements if they are present in concentrations exceeding 0.1-1% at the sample surface. Quantitative information about elemental composition requires the use of standards.

As an example of how an Auger spectrum is interpreted, consider Fig. 14b, which shows an Auger spectrum of a P-doped SiO_2 film. The large peak in the dN/dE spectrum at ~1600 eV is attributed to the KLL Auger transitions of Si (see Fig. 13). Of the two peaks near 100 eV, the larger corresponds to the LMM Auger transitions of Si, while the smaller peak (of slightly higher energy) is due to the LMM Auger transitions of phosphorus. The 500 eV peak can be associated with the KLL Auger transitions of oxygen.

The emission of Auger electrons and x-rays from a solid are competitive processes, with Auger emission being dominant for low-Z elements, and x-ray emission dominating in high-Z elements (Fig. 15). At Z equal to 33 (arsenic) the probability is equal for both processes. As a result AES is somewhat more suitable for detecting low-Z elements.

AES suffers from the following limitations: a) some sample surfaces, especially insulators, are prone to charging when struck by an electron beam; b) some surfaces can be damaged from the high energy primary electron beam, especially organic materials; c) if depth profiling is necessary, the ion sputtering will destroy the sample in the local area of the sample being examined; d) matrix effects can occur. These are signal alterations which are observed when some elements are present in particular matrices. Some of these matrix signal shifts are well understood and can supply information about the chemical state of the atoms present. There is an ongoing effort by ASTM Committee E-42 to study and catalogue the chemical information found in Auger spectra); e) the detection limits depend on the area being examined, the time

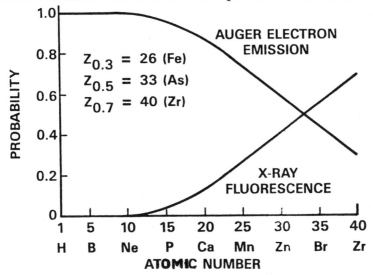

Fig. 15 Probability of X-ray emission and Auger electron emission versus atomic number.

duration of the acquiring scan, and energy range being scanned. (For example, the larger the area the better the detectability. That is, 30 minute scans are more effective than 5 minute scans, and narrow energy range scans can detect lower concentrations than wide energy range scans); and f) as explained earlier, AES is most effective when carried out at vacuums higher than 10^{-10} torr.

Examples of applications in which AES finds use in silicon processing include: 1) surface contamination analysis; 2) detection of very thin SiO_2 layers (native oxides); 3) contaminant concentrations in barrier metals; 4) analysis of corrosion failures in packaged integrated circuits; and 5) P, B, As concentrations in SiO_2 films.

When the SAM mode is used, elemental line scans of two-dimensional maps of surface composition can be obtained. First, a raster scan (about 250 points /line, and 250 lines) in a wide energy range mode, is made to detect all of the elements present on the surface. Then for each of the elements present in which there is interest, a slow raster scan is taken with the detector set only to the energy peak of that element. A sensitive two-dimensional map which locates the element within the area is produced. As mentioned earlier, if the surface is sputtered and the SAM map repeated, a three-dimensional elemental map can be produced[19].

X-Ray Emission Spectroscopy (XES)

XES is an analytical technique in which an electron beam is used to strike the sample, and the x-rays which are emitted in response are analyzed[19,20,21,22,25]. The emitted x-ray spectrum contains peaks characteristic of the target, as well as a background continuum. As x-rays are not readily absorbed by most materials, the x-rays are emitted from the material up to the depth to which the primary electrons penetrate (Fig. 16). Hence XES is not strictly a surface analysis technique, and the depth of the material that emits x-rays cannot be determined within the same resolution as AES or SIMs. Note that some materials, such as lead, strongly absorb x-rays, making it possible to produce effective x-ray shielding. As discussed in the AES section, the electron beam also spreads as it penetrates the sample, hence lateral resolution is also much lower than AES, even if the primary beam has the same diameter.

In spite of the fact that XES is generally not as useful as AES or SIMS, it finds widespread use because SEMs can be easily and inexpensively equipped with a suitable detector to allow XES to be performed. Thus, since SEMs are commonly available at VLSI fab facilities, XES is also likely to be found as an "in-house" characterization capability.

Two types of detectors are used for XES studies (Fig. 17): energy dispersive x-ray detectors (EDX), and wavelength dispersive x-ray detectors (WDX). In EDX, lithium-drifted diodes are the detectors (which convert the x-ray photon to a voltage pulse), and a thin film of berrylium is usually used to provide vacuum isolation from the SEM. Since this window also absorbs low energy x-rays, however, it effectively prevents x-ray peaks from low-Z elements (fluorine and below) from reaching the detector. A second drawback of the EDX is its limited ability to distinguish between two adjacent x-ray peaks. For example, since the resolution of EDX is about 150 eV, the K_α x-ray peak from Si (at 1.74 keV) and the M_α x-ray peak from Ta (at 1.71 keV) would not be distinguishable, since the separation is only 30 eV. Thus, it would be very difficult to analyze the stoichiometry of a $TaSi_x$ film with EDX. On the other hand, WDX detectors can be used to detect one x-ray peak at a time. They are crystal analyzers that are sensitive to the Bragg angle of the emitted x-radiation. This makes the WDX more tedious to operate than EDX. The resolution of WDX, however, is 5-10 eV, and since it does not need to be isolated from vacuum, it can be used to identify low-Z elements (from boron upwards).

SEMs can be equipped with both EDX /WDX and the availability of both detectors increases the characterization capability of the instruments. The detection limits for EDX /WDX are 0.3-1% (about the same as for AES and ESCA). For quantitative data, XES requires the use of standards. XES finds use in the identification of the constituents in thin films and in determining the composition of particles on a wafer surface. The rather poor spatial resolution may not be an insurmountable problem for small particle analysis, if the "background" (e.g. substrate below or region adjacent to) is known to be of different composition. That is, a small (0.5 μm) aluminum particle could easily be identified if it were situated alone on an SiO_2 substrate. But it would be very difficult with XES to positively determine that the particle contained aluminum if it existed in the 1 μm space between two aluminum lines. XES is routinely used in such applications as determining the P concentration in PSG, or the Cu content of Al alloy films.

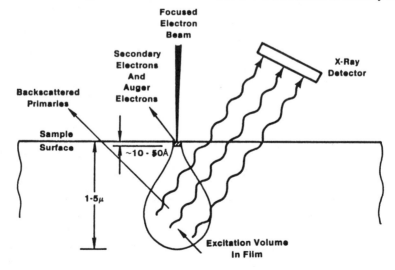

Fig. 16 Spreading of the primary electron beam below the specimen surface, showing the large volume from which backscattered electrons can emanate.

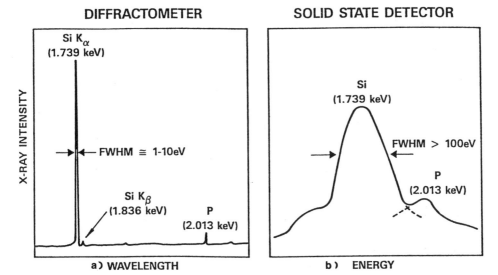

Fig. 17 Comparing energy resolution achievable with energy vs wavelength dispersive analysis.

X-Ray Photoelectron Spectroscopy (XPS, ESCA)

Just as electron bombardment of materials can produce emitted electrons and x-rays, striking the material with x-rays can do the same[18,20,21,22,25]. The technique of XPS (also known as *electron spectroscopy for chemical analysis*, or ESCA), employs low energy x-rays, such as the K-alpha line of aluminum (which has an energy of 1.487 keV), to cause photoelectron emission (Fig. 12a). The emission of photoelectrons in XPS differs from Auger emission in that it is the electron that is "knocked out" of the atom that is analyzed in XPS. Therefore the incident x-ray energy must be both monochromatic and have an accurately known magnitude. It is for these reasons that the K-alpha x-rays of Al and Mg are used in commercial XPS systems.

As in AES, only those electrons from the top 1-10 monolayers are emitted without significant energy loss from collisions. Thus, even though the incident x-ray beam penetrates deep into the sample material, XPS is a surface analysis technique. The electrons are detected in the same manner as in AES, and the collected data (in the form of broad or narrow ESCA scans), provides much the same information as AES. In fact neither technique can detect H, and their detection limits are comparable. Lateral spatial resolution is relatively poor since the smallest x-ray beam diameters are about 150 μm. Such 150 μm x-ray beams do allow a rapid XPS depth profile to be obtained, since ion beams can easily sputter areas of that magnitude.

XPS is often used as a valuable complement to AES, over which it has the following advantages: a) some materials that are subject to dissociation and desorption when bombarded by electrons, are often undisturbed by x-rays (i.e. such materials can be non-destructively studied by XPS); b) insulators that suffer charging problems when irradiated by electron beams are more easily characterized by the neutral x-ray beam; and c) the energy resolution of the XPS peaks due to the emitted photoelectrons is better than for AES (typically 0.5 eV). Since different chemical bonds in a molecular structure cause shifts in the binding energy of the atomic electrons greater than 0.5 eV, these shifts can be detected with XPS and the bond identified. Thus, XPS can be used to obtain information about chemical bonding as well as elemental composition.

X-Ray Fluorescence (XRF)

In x-ray fluorescence (XRF), x-rays constitute both the primary beam and the detected entities. Methods for detecting and analyzing the emitted x-rays are similar to those used in XES[18,20,21,22,25].

XRF, however, has some advantages and limitations relative to XES. First, XRF can be used to analyze layers that would either charge or decompose during the electron bombardment of XES (e.g. oxides or polymers). The large XRF beam (150 μm-1 mm) prevents it from being used to analyze the small features of VLSI circuits, and the penetration of the primary x-ray beam causes emission from material deeper into the substrate than XES. These effects must be considered when XRF is used to analyze substrates that have more than a single layer of material.

ION-BEAM EXCITED COMPOSITIONAL ANALYSIS

Ion beam bombardment of solid surfaces produces substantially different interactions than bombardment with electrons. The resultant methods of materials characterization also tend also to be different. In contrast to the electron /x-ray based techniques, which rely only on the unique electronic structures of each element for identification, ion beam techniques identify the constitutional elements through their atomic mass values.

When ion beams in the 1-30 keV energy range are used to strike a sample surface, and the sputtered positively or negatively charged secondary ions are analyzed using a mass spectrometer, the technique is known as *secondary ion mass spectroscopy* (SIMS). If the exciting beam is kept small (e.g. 2.5-10 μm in diameter), and rastered over the sample, the technique is sometimes referred to as *ion microprobe mass analysis* (IMMA).

In the ion beam energy range of 30-300 keV, there is much less sputtering, and the predominant ion-solid interactions result in ion implantation. As a result, this energy range is typically not used for the characterization of materials. At higher energies (300 keV-4 MeV) the penetration of the ions becomes large (>0.1-1 μm). In this range, backscattering of the incident ions can occur, and this effect is used to analyze the composition of some materials (RBS). Laser ion mass spectroscopy (LIMS), although not ion beam excited, is a related technique which will also be covered.

Secondary Ion Mass Spectroscopy (SIMS)

In the SIMS technique[20,23,24], the energetic bombardment (1- 20 keV) of a material causes *billiard ball-like* collisions with atoms of the surface, leading to their ejection from the material (sputtering). During the energy transfer process, a small fraction of the ejected atoms leave as either positively or negatively charged ions. Over 90% of these secondary ions are emitted from the outer two atomic layers of the sample. The sputtered ions are collected by a mass spectrometer for mass-to-charge separation and detection (Fig. 18a). Typically *quadrupole* type mass spectrometers are used. (See Chap. 3 for more details.) The number of ions collected can also be digitally counted to produce quantitative data on the sample composition. Thus SIMS only analyzes the material removed by sputtering from a sample surface, in contrast to AES, which analyzes the atomic layers closest to the surface without substantial layer removal.

In order to maximize sensitivity, oxygen atoms are used for sputtering and exciting the electropositive elements (i.e. those with low ionization potentials, Na, B, Al, etc.), while beams of cesium (Cs) atoms are employed for producing negative ions from electronegative elements

Table 3. PARAMETERS OF SIMS RELEVANT TO VLSI MATERIAL ANALYSIS

Element (in Si matrix)	Primary Beam	Detected Element	Minimum Detectable Conc. (atoms/cm^3)
Arsenic	Cs^+	$^{75}As^-$	5×10^{14}
Phosphorus	Cs^+	$^{31}P^\pm$	5×10^{15}
Boron	O_2^+, O^-	$^{11}B^+$	1×10^{13}
Oxygen	Cs^+	$^{16}O^-$	1×10^{17}
Hydrogen	Cs^+	$^1H^-$	5×10^{18}

(i.e. those with high electron affinities, C, O, As). The beam choice is important. For example, the detection limit for gold in Si with an O beam is ~100 ppma, but with Cs is ~0.1 ppma[27].

The sputtering process continuously removes surface atoms. Therefore, the shallow analytical zone defined by the secondary escape depth, is advanced into the sample as a function of sputtering time. By monitoring the secondary ion signals with time, a depth profile can be produced. Sputter rates of 2-5 Å /sec, at data acquisition times of 10 sec, produce typical depth increments (vertical resolution) in the 20-50Å range. Layers of up to 10,000 Å thick can be depth profiled with SIMS (Fig 18b). Usually the incident beam is rastered over a small area of the surface to create a crater with a nearly flat bottom. Mass analysis is only performed on the ionic fraction of sputtered material from the center of the crater.

SIMS has several unique capabilities that make it highly useful for VLSI characterization applications. First, it is capable of detecting *all* the elements (whereas AES, ESCA, and RBS cannot detect H and He). Second, it can identify elements present in very low concentration levels. In fact it is the only *surface analysis* technique with the ability to measure doping level concentrations in electronic materials. It is especially sensitive to elements with low ionization potentials (e.g. Na, K), as well elements with favorable electron affinity (group V and VI elements). For applications involving the detection of dopants and contaminants in silicon, Table 3 lists the SIMS sensitivity limits. These show that SIMS is an excellent tool for generating concentration profiles of dopants in Si at levels down to the 1×10^{15} cm^{-3} range.

Like all other analytical methods, however, SIMS also has some drawbacks. First, the range of beam diameters of SIMS is 1.0-200 µm, but maximum sensitivity is achieved when the wider (e.g. 100 µm) beams are used. As the beam is focused to a smaller spot, the sensitivity is correspondingly reduced because fewer total atoms are sputtered from the surface by the smaller beam. Indeed, the technique is much more limited in sensitivity when analyzing very small regions on VLSI. Next, SIMS suffers from the problem of secondary mass interference. As an example, it is difficult to detect the presence of phosphorus in silicon, if water vapor is present in the measuring apparatus. This is because Si reacts with H to form $^{31}SiH^+$, which has a mass peak very close to $^{31}P^+$. Higher system vacuums, higher resolution detectors, and alternate ion sources are techniques utilized to overcome this limitation. Finally, SIMS is locally destructive, is subject to charging, especially when analyzing dielectric layers, and also requires standards to obtain quantitative data about the detected constituents.

Laser Ionization Mass Spectroscopy (LIMS)

LIMS is a relatively new analytical technique[26] in which a microfocused laser beam is used to ablate (remove) material from a sample in one of two modes: laser desorption (LD); and, laser ionization (LI). In both modes, the ions produced are analyzed by a mass spectrometer of the

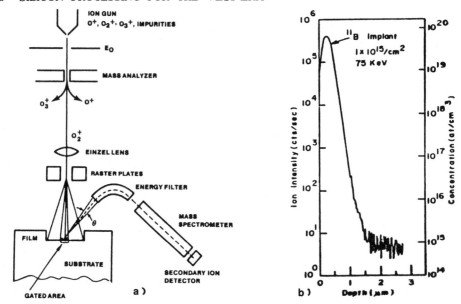

Fig. 18 (a) Schematic diagram of a secondary ion mass spectrometer. (b) Typical SIMS depth profile for boron into boron-doped silicon.

time-of-flight design, in which the transit time of the vaporized ions accelerated across a high vacuum region by a known electric field is used to determine their mass.

Laser desorption produces ions only from those substances that are are adsorbed on the surface, and hence uses low laser power density. In *laser ionization*, the ions are produced during vaporization of the sample surface. The ability to distinguish between adsorbed and bulk materials, and materials incorporated in the matrix, is a valuable characteristic of LIMS. Furthermore, another utility of LIMS is its ability to provide a mass spectrum from a volume of material as small as 1-2 μm in diameter and several monolayers thick (using LD). See Table 4 for additional details about the capabilities of LIMS.

Rutherford Backscattering Spectroscopy (RBS)

RBS employs high energy helium ions (typically with energies from 1-4 MeV) to determine concentration profiles in the outer 0.5-3.0 μm of the material[28,36]. The technique is essentially nondestructive and offers depth resolutions in the 200Å range. Elastic collisions occur between the high energy incident ions and the outer surface and subsurface atoms of the sample. Upon colliding, some of the incident ions are backscattered and experience an energy loss that is characteristic of the atom with which they collide, and also of the distance into the sample at which the collision occurred. Thus, from the energy of the backscattered ions, both elemental analysis and distribution with depth can be determined simultaneously. For example, if a buried film is present as a very thin layer of atoms, then the RBS spectrum will show one discrete energy value which contains two pieces of information: 1) the identity of the buried film atoms; and 2) the depth in the sample at which the atoms are located (Fig. 19a). As a second example, assume the buried film has a finite thickness. In this case, the backscattered ions will have a *range of energies* because of the added energies lost in the depth represented by the thickness of the buried layer (Fig. 19b). Thus, we see that from an ion backscatter energy spectrum, it is possible to obtain the mass, the number of target atoms per unit area causing backscatter, and their depth within the sample. It is also easy to understand from these simple examples that correctly deciphering most "raw" RBS spectrums can be a complex task. When RBS was being developed as an analytical technique, van de Graaf generators were required to supply the high energy necessary for the primary beam, and few facilities had such generators at their disposal. More recently, complete RBS systems have been developed and are being offered commercially. These should make RBS characterization services more widely available.

The instrumentation of an RBS system involves an ion source and ion accelerator. The sample is loaded into a scattering chamber that also contains the energy dispersive silicon barrier detector (which energy-analyzes the backscattered ions). The signal is fed to a multichannel analyzer, and computer automation of both data collection and data reduction is required. The forte of RBS is its ability, unique among compositional analysis tools, to provide quantitative information concerning the composition and thickness of multicomponent thin films *without the use of standards*, and without the problem of sample charging. The technique can detect all the elements except H and He, but its sensitivity to light elements is also poor, especially when they are present in a matrix of heavy elements. On the other hand   RBS is best suited for characterization of heavy masses in light mass matrices. This is exactly the situation that is encountered in refractory metal silicides and alloys of aluminum, both systems which find wide use in VLSI.

The minimum detection limit of RBS is in the 5×10^{18} atoms /cm^3 range, but the technique finds its widest application in the measurement of concentrations in the 10^{20}-10^{22} atoms /cm^3 range. The incident beam diameter is also relatively large (100 μm-1mm), and this unfortunately precludes the use of RBS for the analysis of actual features on VLSI. In addition, although the technique is relatively nondestructive, some structural modifications to the sample films may be caused by radiation damage. Finally, the *specificity* (i.e. the ability to distinguish one element from another) of RBS varies. That is, it is good for low-Z elements, except for the detection of phosphorus in Si, because of the close proximity of the phosphorus and Si peaks, but is poor

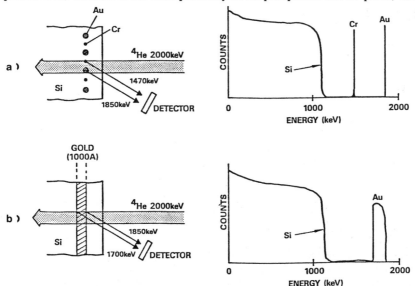

Fig. 19 Schematic illustrating the principles of Rutherford backscattering spectroscopy. (a) RBS signal from a very thin buried layer of Au-Cr. (b) RBS signal from a thicker layer of Au.

for high-Z elements, as a result of the kinematics of the scattering process, combined with the detector energy resolution limitations.

RBS also finds use in determining the crystalline perfection[29] of recrystallized semiconductor films (e.g. after rapid thermal annealing or laser annealing). This application of RBS is referred to as *channeling*. It involves crystallographically aligning the ion beam and the crystal, so that the ions channel down the interstices of the crystal lattice. The ions are only backscattered if they encounter off-lattice-site atoms. Disturbances to crystalline perfection, such as amorphous or polycrystalline regions, or interstitial atoms, can be detected by this technique, especially following ion implantation and thermal processing. In the extreme case of an amorphous layer, the yield is identical to that from a random orientation of a perfect lattice.

CRYSTALLOGRAPHIC STRUCTURE ANALYSIS

In developing and troubleshooting VLSI fabrication processes, many crystallographic structural properties of both the silicon substrate and the deposited films must be determined. These include: a) substrate crystallographic orientation; b) size and crystallographic orientation of the crystal grains in polysilicon, silicides, and metal thin films; c) the identification of crystalline phases; d) the location and identification of amorphous regions; and e) the characterization of crystalline defects, such as dislocations, stacking faults, and precipitates.

In earlier sections we have discussed how RBS can be used to find amorphous regions. We have also pointed out how TEM may be utilized to determine grain sizes and the degree of preferred orientation in polycrystalline deposited films, and to image crystalline defects. In this section, we will discuss the use of *x-ray diffraction* for determining substrate orientation, for identifying crystalline phases and preferred orientations in polycrystalline layers, and for identifying amorphous regions. We will also explore the technique of *x-ray Lang topography*, valuable for characterizing defects in crystalline substrates. Finally, we include a section on *nuclear activation analysis* (NAA).

X-Ray Diffraction

The analysis of the angular position and intensity of x-rays diffracted by crystalline material can reveal information on the crystal structure and crystalline phases present in the sample[30]. X-ray diffraction occurs when the Bragg requirement is satisfied:

$$n \lambda = 2 d \sin \theta \qquad (5)$$

where λ is the x-ray wavelength, d is the interplanar spacing, θ is the Bragg diffraction angle, and n is the integer giving the order of the diffraction. There are various x-ray diffraction methods, but all are based on establishing conditions that allow the Bragg criterion to be met.

The *Laue technique* is typically used to determine substrate orientation. A collimated x-ray beam of *white* radiation (i.e. the continuous unfiltered output of an x-ray tube) is used to strike a wafer placed parallel to a photographic plate. The beam passes through a hole in the film, and diffraction of the beam in the wafer occurs simultaneously from a large number of crystal planes. Each diffracted component beam leaves a spot on the film, and the distribution and symmetry of the spots on the image is directly related to the substrate crystallographic orientation. A pattern from a wafer of unknown orientation can be quickly compared with standard patterns [e.g. from Laue patterns of (100), (111), and (110) silicon wafer surfaces] to determine its orientation.

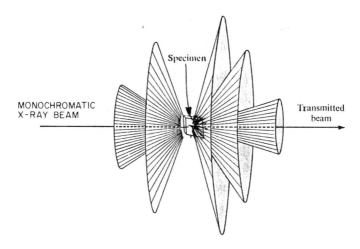

Fig. 20 Schematic diagram of x-ray diffraction analysis using Read camera.

The *Read camera* technique (Fig. 20) is used to determine the preferred orientation of the crystal grains and the phases present in polycrystalline films, and can also be used to identify the presence of amorphous regions. The film in the Read camera is mounted on the inside of a cylinder, at the center of which the sample is placed. The sample, typically a polycrystalline film, is irradiated by a *monochromatic* x-ray beam. The beam strikes the sample at a shallow angle, permitting greater beam penetration into thin films than if the beam were to strike normal to the surface), and a cone of diffraction is created for each set of diffracting planes. The images from these cones are used to identify the material by referring to standard x-ray crystallographic tables. Information by the Read camera on the preferred orientation of the sample film can also be obtained because preferred orientation produces non uniform cone intensities (i.e. elongated spots and arcs instead of constant intensity curved lines).

X-Ray Lang Topography

X-ray Lang topography[31] is a method used to obtain a photographic image of defects in a crystal. One important application is the examination of wafers to detect the presence of oxygen precipitates that are used to getter fast diffusing chemical species such as Cu and Fe (see Chaps. 1 and 2 for more information on *Gettering*). The single crystal wafer is aligned exactly on a specific Bragg angle, and irradiated with a collimated, monochromatic x-ray beam using a Lang camera. After passing through the crystal, only the x-rays diffracted by the lattice planes of the crystal are allowed to strike the photographic emulsion plate. These then form an x-ray transmission topograph of the entire volume of the crystal. If defects or strain centers exist (e.g. due to the presence of oxygen precipitates) in the crystal, the x-rays interact with the lattice distortions to create intensity variations, which are observed as contrast on the photographic plate. The technique, in fact, can also create images due to lattice distortion from slip, stacking faults, dislocations, and metal precipitates. It is applicable to non destructive whole wafer imaging at any point in the IC process, and as mentioned, should be especially useful for understanding and characterizing the role of oxygen nucleation and precipitation. The resolution of the technique is 2-20 µm. It should be noted, however, that before performing x-ray Lang topography, all metallization, oxide, and nitride must be removed from the sample surface.

Neutron Activation Analysis (NAA)

Neutron activation analysis is a valuable diagnostic tool because it can detect the lowest concentration levels for many elements. For example such elements as arsenic, copper, gold, sodium, and tungsten, are detectable down to concentrations of 10^{11}-10^{12} atoms /cm^3. However, the light elements such as boron, oxygen, nitrogen, and carbon do not produce radioactive isotopes that are suitable for NAA. The method is therefore most useful for applications such as: a) evaluating the effectiveness of gettering procedures; b) detecting the presence of trace elements in bulk silicon; or c) measuring the contamination caused by processing furnaces. For example, it has been used to show that the phosphous gettering layer on the backside of a wafer can reduce the bulk gold impurity concentration by a factor of 50[32].

The technique utilizes a flux of thermal neutrons to irradiate silicon samples, which also contain the impurities of interest. This causes radioisotopes of all the elements in the sample to be created. The silicon radioisotopes have a short half-life (2.6 h) and within 24-48 hours the radiation from the silicon falls below the levels from many other elements. The energy of the detected x-rays and the half-life are used to identify the longer lived isotopes which emit the remaining radiation. This radiation is detected by a lithium-drifted germanium detector and analyzed by a multichannel analyser.

SUMMARY OF CHARACTERIZATION TECHNIQUE CAPABILITIES

The most widely used of the diagnostic techniques for VLSI applications that we have discussed in this chapter are (in roughly the following order): optical microscopy, SEM, AES,

Table 4. Comparison of Various Material Characterization Techniques

Analytical Parameter	AES	XPS (ESCA)	SIMS	SEM /EDX	RBS	LIMS
Primary Beam	Electrons	X-Rays	Ions	Electrons	Ions	Photons (Laser)
Detected Entity	Electrons	Electrons	Ions	Electrons/X-rays	Ions	Ions
Detectable Elements	Li ⇒ U	Li ⇒ U	All (H ⇒U)	Be ⇒ U	Li ⇒ U	All (H ⇒ U)
Beam Diameter For Routine Small Area Analysis (µm)	Submicron	150-300	Submicron to several µm.	0.01 (SEM) 1-3 (EDX)	~1000	~1
Surface Sensitivity (Å)	10-40	~40	3-10	>1000	Depth resolution is 2.5-20 µm.	
Routine Detection Limits (atomic %)	0.3 - 1	0.1 - 1	10^{-7}- 10^{-1}	1 EDX	0.01-10	10^{-4}-10^{-2}
Routine Survey Time (Data acquisition time - min)	5	5	5	2 EDX	15	< sec
Specificity	Good	Good	Very Good		Good for low Z	

Table 5. ANALYTICAL TECHNIQUES FOR CHARACTERIZING WSi$_x$/POLYSI FILM

	Thickness	Conformality	Resistivity	Stress	Composition
Best Technique	SEM	SEM	4-Pt Probe	Radius of Curvature	RBS
Accuracy:	95%		99%	90%	97%
Sensitivity:	100Å	100Å	$10^{-6}\Omega$-cm	10^7 Pa	50 ppma
Advantages:	Direct Observation		Rapid, ND	Rapid, ND	ND, Dens.
Sources of Error or Disadvantges	Surface Glass	Voids below resolution	Thickness	Thickness	Thickness, 1 mm beam
Alternatives:	Profilometer	TEM		X-Ray Diffract.	Wet chem., AES, TEM

	Impurity Content	Dopant Redistribution	Surface Glass Properties	Interface Morphology	Crystal Structure
Best Technique:	AES	SIMS	Ellipsometry	TEM	X-Ray Diffraction
Accuracy:	50%	25%	?	?	95%
Sensitivity:	1 at%	10^{16} cm^3	50 Å	5 Å	-
Advantages:	Sensitive to most species	Extreme sensitivity	Rapid, ND	Direct Observation of grains	ND
Sources of Error or Disadvantages:	Ion etch matrix	Ion etch matrix	Surface irregularities	Sample preparation	
Alternatives:	SIMS	-	RBS	-	TEM

XPS (ESCA), EDX and SIMS. AES, scanning Auger microprobe (SAM), and XPS are the most popular compositional analysis instruments, their world-wide installations outnumbering by three times the SIMS facilities. Indeed, the unique combination of a 500 Å spot and acceptable detectability (near 1% atomic) has made AES and SAM the most widely used methods for compositional analysis in VLSI circuits. Although it should be noted that AES and SIMS are often used together to provide a compositional analysis. AES is employed to analyze small features, while SIMS is utilized to detect low-level concentrations. In recent years RBS and TEM have also become important tools for analyzing refractory metal silicides and metal alloy systems. While the other techniques we have covered may not be utilized as widely, occasionally they may still provide the best analytical method or provide an alternative approach for resolving ambiguous or even apparently *contradictory* data obtained using other methods. Therefore the reader should be aware of their capabilities if need to use them should arise. The key parameters of the most important chemical analysis methods (AES, ESCA, SIMS, EDX, RBS) are given in Table 4, along with some comments related to their common applications.

In addition, in Table 5, we illustrate the various materials characterization techniques that are needed to fully evaluate a completely processed sample silicide layer [that is, a tungsten-silicide (WSi$_x$) film that has been deposited, annealed, and oxidized]. This example film was selected because it illustrates the strengths of the tools for each of the characterization requirements that must be met. Let us discuss the evaluation in further detail.

The *thickness* of the WSi$_x$ can be determined using SEM or a surface profilometer. The *coverage of underlying steps* is also best done with SEM. The *film resistivity* is monitored with a four-point probe, and the *film stress* with a radius-of-curvature measurement. The silicide *composition* may ideally be WSi_2, but if during deposition, this stoichiometry is not exactly achieved (or if during a subsequent anneal cycle some W or Si was added /removed from the film), the composition might be different (e.g. $WSi_{2.4}$). Thus, to determine the value of x in WSi_x, RBS can be used (AES can be used as an alternative but with less quantitative accuracy). The *presence of contaminants* in the film (e.g. O, C) can be estimated using SIMS (or AES if such contaminants are present in concentrations >1%). The *impurity concentrations* (i.e. due to possible out-diffusion of P or As from the substrate into the WSi_x) are best monitored using SIMS (since it possesses the required sensitivity capability). The *surface oxide* of the film is evaluated for thickness and index of refraction using an ellipsometer (or with RBS) for thickness and composition. The *interface morphology* (e.g. roughness of the interface between WSi_x and the substrate or the oxide film), is best observed using TEM. Finally, the *crystal structure* (e.g. phases /crystal orientaton of various crystal grains in the film) is observed with x-ray diffraction.

SUGGESTIONS FOR HOW TO ACCOMPLISH AN EFFECTIVE ANALYSIS

In most cases, users of materials characterization services are required to interface with an organization that is providing such services to a large group of clients, either "in-house" or as an independent commercial laboratory. Thus, the services are paid on a fee-for-service basis. The user generally wishes to accomplish the characterization task as quickly and effectively as possible, and at lowest cost[33,34]. Here are some guidelines that should help achieve such goals:

1. Understand as much about the possible techniques that are to be used to perform your analysis[35]. Having a sound awareness of the techniques that are possible for your analysis is often the first step in finding a laboratory which will be capable of performing the analysis. In addition, once one (or more) characterization technique is selected, such information will provide a measure of how effectively the techniques can provide the data that will unravel the problem. Therefore one should know the sensitivity, accuracy, lateral resolution, and depth resolution limits of the candidate techniques. One should also be aware of any possible interference effects that could prevent the technique from getting the data needed for the application (e.g. matrix effects, secondary mass interference).

2. Consider the selection of the most appropriate technique(s) from the point of view of both the problem and the sample. Answer the following questions about what will be needed from the analysis, and what background information you will be able to provide to the analyst:

    a) Is any information known about the elements to be encountered? Is there a need for coverage of the entire periodic table or just a portion?

    b) Is a depth profile required or merely surface analysis? If a profile is required, what layers and thicknesses will be required for this profile? What is the expected contamination depth

    c) How much lateral sensitivty will be necesary? What feature sizes will need to be examined ?

    d) How many samples are available? Can a destructive test be used? Is it required that the sample be returned after the test?

    e) What type of sample is it? Will it be vacuum compatible? Will it dissipate charge when bombarded by a charged beam?

    f) If more than one analytical technique is to be applied, the technique that is least damaging should be performed first (e.g. if AES, ESCA, and SIMS are chosen,

ESCA should be done first, then AES, and finally SIMS).

g) Are any quantitative data required? If so, are standards readily available?

h) How much time is available for analysis? What kind of turn around time can be expected from the laboratory performing the analysis?

i) Are there budgetary /priority considerations that must be followed (such as an open /limited budget, a critical or noncritical need for the information) ?

3. Documentation and Sample Control Measures:

a) If possible, determine from the analyst, and then provide, the maximum sample size that the instrument will hold, so that sample will not need to broken by the analyst.

b) Carefully mark the **back** of each sample (laser ID marking is preferable to scribing) and document the relevant sample history in writing. Do not mark the sample with pencil or ink.

c) Provide the analyst with unambiguous instructions as to where on the sample the analysis is to be performed. A photograph or drawing accompanying the sample might be an appropriate way to provide such instructions.

d) Whenever possible supply reference samples (e.g. known good vs. known bad samples, acceptable vs. unacceptable, before vs. after processing).

e) Provide the analyst with sample histories, including information about how the sample might have been altered (e.g. chemically etched, mishandled) in such a way that the contaminant has been either removed or masked by additional contamination. State, if known, type and composition of the sample matrix. If not known, so state.

4. User /Analyst Interface:

a) Conduct an interview (telephone if necessary) with the analyst to establish the protocol that will be followed for the analysis and to outline your problem. This should include such matters as: 1) Will your presence be required /desired during the analysis?; 2) In what manner will you communicate during the analysis?; 3) Fees and payment schedule; and 4) Written reports and data that will be delivered upon completion of the analysis. You may also wish to solicit the analysts' opinion of the best approach to solving your analysis problem.

b) Trust the analyst. Provide as much background information as possible. Reputable commercial laboratories exercise absolute discretion in handling a customers data and sample information.

c) Interact with analyst at several checkpoints (determined during the initial definition of the analysis protocol). Discussion of the first information available often points to new directions and increases the value of subsqent steps.

Material characterization techniques can be performed at in-house analytical facilities, but may also be obtained through commercial laboratories which provide such services. Many companies may not be able to justify the expense of a full range of in-house characterization techniques. The cost of ESCA, AES, or SIMS instrumentation alone is $300,000-$750,000, and personnel and facility expenses must be added to these figures. When no in-house capabilities exist, or when those that do are inadequate, the commercial laboratories mentioned above are available. One listing of suppliers who provide materials characterization services can be found in the *Semiconductor International Master Buying Guide*, which is published annually.

REFERENCES

1. P. Burggraaf, "Guidelines for Optical Microscopy", *Semiconductor Internatl*, Feb. '85, p. 56.

2. G. Nomarski and A.R. Weill, "Application a la Mettalographie des Methodes interferentielle a Deux Ondes Polarisees", *Rev. Metall*, 52, 121 (1955).

3. J.I. Goldstein, *et al.*, *Scanning Electron Microscopy and X-Ray Microanalysis*, Plenum, New York, 1981.

4. P.R. Thornton, *Scanning Electron Microscopy*, Chapman and Hall, London, 1968.

5. A.D. Weiss, "Scanning Electron Microscopes", *Semiconductor International*, Oct. '83, p. 90.

6. G. Toro-Lira, "Critical Dimension Control in the Late Eighties", *Microelectronic Mfg. and Testing*, Jan. '85.

7. T.E. Everhart, O.C. Wells, and C.W. Oatley, "Factors Affecting Contrast and Resolution in the SEM", *J. Electron. Control*, 7, 97, (1959).

8. R. Carlson, "Manufacturing One Micron", *Solid State Technol.*, Jan. '85, p. 141.

9. P.S. Singer, "Linewidth Measurement Aids Process Control", *Semiconductor International*, Feb. '85, p. 66.

10. M. Kittler and W. Seifert, "On the Characterization of Individual Defects in Si by EBIC", *Crystal Res. Technol.*, 16, 157, (1981).

11. P.E. Russell, "SEM Based Characterization Techniques for Semiconductor Technology", *Proc. SPIE, Vol. 452*, (1983), p. 183.

12. R.B. Marcus and T.T. Sheng, *Transmission Electron Microscopy of Si VLSI Devices and Structures*, Wiley, New York, 1983.

13. G. Thomas and M. Goringe, *Transmission Electron Microscopy*, Wiley, New York, 1979.

14. T.Y. Tan, "Characterization of Semiconductor Si by TEM", *SPIE, Vol. 452*, (1983), p. 170.

15. A. Joshi, I.E. Davis, and P.W. Palmberg, "Auger Electron Spectroscopy", in A.W. Czanderna Ed., *Methods of Surface Analysis*, Elsevier, New York, 1975, p. 164.

16. L. Murr, *Electron and Ion Microscopy and Analysis*, Marcel Dekker Inc., New York, 1982.

17. T.A. Carlson, *Photoelectron and Auger Spectroscopy*, Plenum Press, New York, 1976.

18. K.F.J. Heinrich, "Electron-Beam X-Ray Microanalysis", *Van Nostrand Reinhold Co.*, New York, 1981.

19. D.W. Harris, "Surfaces, Science... Fact, not Fiction", *Solid State Technol.*, Apr. '84, p. 278.

20. G.B. Larrabee, "Materials Characterization for VLSI", in *VLSI Electronics*, Vol. 2, N.G. Einspruch, Ed., Academic Press, Orlando, Fla., (1981), p. 37.

21. R. Woldseth, *X-Ray Energy Spectrometry*, Kevex Corp., Burlingame, CA, 1973.

22. A.W. Casper and C.J. Powell, Eds., *Industrial Applications of Surface Analysis*, ACS 199, American Chemical Society, Wash. D.C., 1982.

23. J.A. McHugh, "Secondary Ion Mass Spectrometry" in A.W. Czanderna, Ed. *Methods of Surface Analysis*, Elsevier, New York, 1975, p. 223.

24. A. Benninghoven, *et al.*, Eds. *Secondary Ion Mass Spectroscopy SIMS II*, Springer-Verlag, Berlin, (1979).

25. E.P. Bertin, *Principles & Practice of X-Ray Spectrometric Analysis*, Plenum, New York, 1970.

26. W.W. Duley, *Laser Processing and Analysis of Materials*, Plenum Press, New York, (1983).

27. P. Williams, *et al.*, "Evaluation of a Cesium Primary Ion Source on an Ion Microprobe Mass Spectrometer", *Anal. Chem.*, 49, 1399, (1977).

28. W.K. Chu, J. Mayer, and M. Nicolet, *Backscattering Spectroscopy*, Academic, New York 1978.

29. L.C. Feldman, "Rutherford Scattering Channeling Analysis of Semiconductor Structures", *Proc. SPIE, Vol. 452*, (1983), p. 192.

30. B.D. Cullity, *Elements of X-Ray Diffraction*, Addison-Wesley, Reading, Mass., 1978.

31. *ibid.*, Ref. 20, p. 55.

32. S.P. Murarka, "A Study of the Phosphorus Gettering of Gold in Silicon by Use of Neutron Activation Analysis", *J. Electrochem. Soc.*, 123, 765 (1976).

33. D.S. Koellen and D.I. Saxon, "Applications of Surface Analysis: Detection and Identification of Contamination on Semiconductor Devices", *Microcontamination*, July '85, p. 47.
34. K. Bakale and C.E. Bryson, "Surface Analysis", *Microcontamination*, Oct./Nov. '85.
35. J.R. Devaney, *et al.., Failure Analysis: Mechanisms, Techniques, and Photo Atlas*, Failure Recognition and Training Services, Inc., Monrovia, CA, 1983.
36. C.A. Evans and M. Strathman, "RBS Technique Exposes Surface Properties of Materials", *Industrial Research and Development*, Dec. '83.
37. R.J. Blattner, "An Introduction to SAM and its Applications", in *Microstructural Science*, Vol. 8, Eds. Stevens, Van derVoort and McCall, (1980), p. 65.
38. *Handbook of Auger Spectroscopy*, JEOL Ltd., Tokyo, 1982.
39. J.J. Shaffner, "Surface Characterization for VLSI", in N.G. Einspruch and G.B. Larabee, Eds., *VLSI Electronics-Microstructure Science*, Vol. 6, Academic, New York, 1983, Chap. 8, p. 497.
40. R.B Marcus, "Diagnostic Techniques", in S.M. Sze, Ed., *VLSI Technology*, McGraw-Hill, 1983, Chap. 12, p. 507.
41. T.E. Everhart and P.H. Hoff, "Determination of Kilovolt Electron Energy Dissipation vs. Penetration Distance in Solid Materials", *J. Appl. Phys.*, **42**, 5837, (1971).

PROBLEMS

1. What is a *standard* and why are standards required for obtaining quantitative data about the composition of materials being analyzed? How are such standards prepared for AES?

2. Why must surface analysis compositional tools be equipped with the capability of pumping the sample analysis chamber to *very high* or even *ultra high* vacuum conditions? If every oxygen atom that struck a surface were to stick to it, how long would it take to form a monolayer of oxygen if it is present in the sample analysis chamber at a pressure of $3x10^{-8}$ torr?

3. Why are optical microscopes designed with a maximum magnification of 1000X?

4. Explain the difference between the principles of how contrast arises in the images obtained in SEM and a TEM. Why is TEM able to achieve higher resolution than SEM?

5. Explain why the type of structure shown in Fig. 10 is important for the success of TEM analysis of a device structure. Describe the difference between *cross-sectional* and *areal* TEM, and list two applications in which might find use.

6. In a SIMS profile of an ion implantation, the peak intensity is 10^5 counts /sec and it occurs at 100 sec. The ion implantation energy was 100 keV and the dose was $1x10^{15}$ cm^{-2}. If an ion implantation with a dose of $4x10^{16}$ ions /cm^2 is performed at an energy of 180 keV, estimate the peak intensity and the time at which this intensity is reached.

7. How are depth profiles calibrated in AES compositional analysis?

8. If AES is being performed, and peak overlaps occur in the observed spectra. Suggest some alternatives that could allow the information desired to be obtained.

9. Describe the difference between the species that are detected in AES and XPS (ESCA). Can one of these techniques yield information about the materials being anlyzed that the other cannot? If so, which one, and what is this information?

10. Give some explanation as to why AES is better suited to detecting lighter elements. Why can't AES or XES detect the presence of Hydrogen or helium?

11. Why does RBS not require the use of standards to obtain quantitiative compositional analysis data?

18

STRUCTURED APPROACH to

DESIGN of EXPERIMENTS

for PROCESS OPTIMIZATION

The process technologies used to fabricate VLSI consist of a sequence of sub-processes. Each of the sub-processes is specified by a recipe that is developed to produce a desired set of outcomes for each sub-process step. The recipes are typically derived from experiments designed especially to identify an optimum set of sub-process conditions. It is the ability of experiments to obtain information about unknown or poorly understood phenomena which is exploited to obtain such information. That is, experiments are used to characterize the relationship between the independent (or controllable) variables and the dependent (response) variables of the process. From this knowledge, the values of the controllable variables which give the best results are derived. The effective utilization of experiments to optimize a process is based upon maximizing the amount of information obtained, while minimizing the cost of experimentation.

A *structured approach*, based on *statistical design and analysis of experiments*, provides investigators with a powerful tool toward maximizing the benefits of experimental projects. The approach to be outlined works best if the process or system being investigated exhibits the following behavior: a) the response functions (i.e. the functions of the process system variables that respond to the controlling variables) are smooth over the range of experimental conditions (but may possibly exhibit an upward or downward curvature); b) the slope of such functions may undergo substantial change when the values of other independent variables are changed; and c) a significant component of the variations in the measured data is due to experimental error.

The statistical basis which underlies the strategy of the structured design of experiments allows an effective experimental plan to be formally specified under the constraints of the system behavior listed above. Fortunately, many if not most, of the sub-processes that are encountered in fabricating VLSI follow such behavior, and statistically designed experiments prove to be very effective for developing and optimizing such sub-processes. The statistical approach to design of experiments is also useful in cases when no definite theory exists about the relationship between independent variables and their responses, and an empirical approach must be followed.

Traditional experimental design has often been conducted haphazardly, with "change one variable at a time, and keep all the others constant", as the single unifying principle. In fact this approach is very inefficient and potentially incomplete in terms of the information obtained. On the other hand, a formal experimental plan, consisting of a sequence of structured experiments (based on statistical principles) is far more likely to produce a maximum of data, with a minimum expenditure of effort and resources.

The purpose of this chapter is to introduce the principles of designing experiments which will yield maximum information, as well as a high confidence that the information is valid, with the least effort. Since these principles are based on statistical concepts, a short discussion on pertinent statistical theory is first presented. Following this, an overall strategy of experimental design is considered, starting from the identification of candidate controllable variables (*factors*). Then the design of *screening experiments* whose role is to identify the most important of these factors, is discussed. Next the definition of *full-factorial experiments,* which are used to more fully evaluate the effects of the most important factors, is presented. Finally, the creation of *response surfaces* is illustrated. These are used to identify detailed response variations and optimum response points within the ranges of the factors (controllable variables) under which the process can be carried out. The ultimate product of a sequence of such experiments should be a model that can be used to predict the response of the process or system that was investigated, if a set of factors within the experimental limits is applied to the system or process. The model may be in the form of an equation or graph.

A detailed discussion on the subjects of *screening tests* and *full-factorial experiments* will also be presented. Some relatively simple examples will be solved to demonstrate the procedures for properly carrying out and analyzing both screening and full-factorial experiments. It will also be described how full-factorial experiments can yield a mathematical model that allows simple prediction of first-order process responses to be estimated with good confidence. The generation of response surfaces, however, involves procedures that are too complex to be adequately treated by this text. Furthermore, in many applications, a sequence of screening and full-factorial experiments can provide enough information to optimize a process. For this reason only a cursory description of response surfaces, and how they are derived, will be given.

In fact, to be able to fully apply the structured approach to experimental design, a deeper understanding of the techniques should be acquired by the reader than we are able to present. Therefore, a statistician would be helpful for reviewing experimental plans, especially if they are complex. In addition a more complete mastery of experimental design can be gained by self-study of the references listed at the end of the chapter, or through enrollment in a formal course in experimental strategy. Such courses are offered both through universities and independent organizations (e.g. the Applied Technology Division of Dupont de Nemours & Co).

Nevertheless significant benefit can be gained in many experimental projects by applying these techniques, even at the level covered in this chapter. Furthermore, such an introduction provides readers with information about the benefits of using a structured design approach to experimental undertakings, as well as a good basis from which to pursue a more complete understanding of the subject. References 1-4 list some texts that offer a more detailed treatment of the general subject of statistical design of experiments. An overview of the methods of applying statistical techniques to developing VLSI fabrication processes is given in Ref. 5.

FUNDAMENTALS OF STATISTICS

Experimental design based on statistics offers the advantage of a structured approach to solving problems with experiments. This approach also reduces the number of experiments that would be performed if an *ad hoc* set of experiments were carried out instead. The underlying statistical basis of such an approach makes it useful for the user to have some acquaintance with statistics. In fact, statistical analysis of data from factorial design experiments can be performed with relatively little mathematics. However, a familiarity with the definitions and approach of statistics will assist users in setting up such analyses. We therefore present an introduction to statistics that will be extensive enough (but still very brief) to enable users to carry out

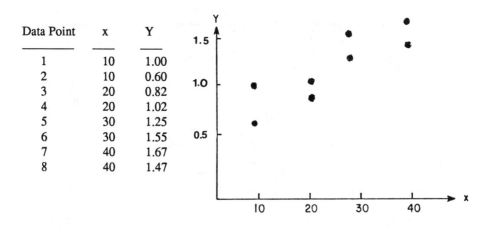

Data Point	x	Y
1	10	1.00
2	10	0.60
3	20	0.82
4	20	1.02
5	30	1.25
6	30	1.55
7	40	1.67
8	40	1.47

Fig. 1 Example of a set of experimental runs involving a single independent variable, x, and a dependent variable, Y. The results are tabulated in (a) and graphed in (b).

experimental design of the level to be presented in the text. The information on statistics of this section is also provided to serve as an reference for understanding other facets of the VLSI fabrication environment that make use of statistics.

Samples, Populations, Mean, Variance and Standard Deviation

When carrying out a series of experiments, it is often of interest to quantitatively describe the changes in the response (controlled or dependent variable) which occur from deliberately induced changes in the factor (independent or controllable variable). Typically, such an investigation is carried out by performing a series of experimental runs, each at different values of the independent variables.

Let us consider a hypothetical example of a group of experimental runs which involve a single independent variable, x, and a dependent variable, Y. The results of the experiments can then be displayed in a number of ways, two which are shown in Fig. 1 (i.e. a tabular form and graphical form). In Fig. 1 there are eight data points - two each at the four values of x.

A statistical analysis produces information about the relationship between x and Y. The statistical information is based on the assumption that actual measurements fluctuate, that is, they are not precisely reproducible. Let us assume that the x-values can be set precisely and that the experimental error is due to the measurements of Y. In our example, we would identify four populations of Y values (one for each setting of x that we use). It is obvious that the number of measurements of Y for a given value of x can be without limit, and thus each population consists of a potentially infinite number of members. The numerical values of such member populations will have a unique central value (the population average or *mean*, μ), and *dispersion* (that is, the population variance, σ^2). These are parameters of the *entire possible population* and are identified by Greek letters. Since it is not possible to conduct measurements to find every member of a population, a *sample population* is determined instead, and μ and σ of the complete population are *estimated* from the measured sample values. The resulting *estimated mean* and *estimated variance* are designated by Y and s, respectively (to distinguish them from the actual mean and variance of the population).

For any specific level of x, say x_o, *the estimate* of the mean (written as Y), is the average of the Y_i observations taken at $x = x_o$:

$$Y_{x = x_0} = \frac{(Y_1 + Y_2 + ... + Y_n)}{n} = \frac{\sum_{i=1}^{n} Y_i}{n} \tag{1}$$

The *estimate* of the dispersion (the variance) is most accurately found from:

$$s^2_{x = x_0} = \frac{(Y_1 - Y)^2 + (Y_2 - Y)^2 + ... + (Y_n - Y)^2}{n - 1} \tag{2}$$

and the square root of the variance is known as the *standard deviation* ($s = [\,s^2\,]^{1/2}$).

Pooled Variance and Degrees of Freedom

Typically, an experiment cannot be run an arbitrarily large number of times to obtain precise estimates of μ and σ. Thus a valid question arises: How many measurements must be taken to insure that Y and s are meaningfully close to the actual population parameters? (i.e. What is the minimum required sample size?)   If just the *mean* needs to be estimated, in some instances, only 2 or 3 measurements can provide an adequate estimate. In order to estimate σ^2 with any precision, however, a larger number of measurements needs to be made. Nevertheless, this information may be obtained without actually performing additional measurements at the specific value of x in question.   That is, if it can be correctly assumed that the population variance σ^2 has about the same value for all levels of x in the experiment, the estimates for σ^2 at each value of $x = x_0$ can be pooled into a single number, s^2_{pooled}, which is defined by:

$$s^2_{pooled} = \frac{(n_1 - 1)\,s_1^2 + (n_2 - 1)\,s_2^2 + ... + (n_k - 1)\,s_k^2}{(n_1 - 1) + (n_2 - 1) + ... + (n_k - 1)} \tag{3}$$

$$s_{pooled} = \sqrt{s^2_{pooled}} \tag{4}$$

and s_j^2 = the estimated variance of s for the jth x level (j = 1, ... , k), and n_j = the number of repeat measurements at the jth x level.

The *pooled variance* gives a weighted average of the individual variance estimates.   The advantage of pooling is that a more precise estimator of the population variance is obtained than from any single value of the variance from which the pooled value was calculated.   The denominators of Eqs. 2 and 3 are called the *degrees of freedom* of the estimates (that is, the s^2 in Eq. 2 has n - 1 degrees of freedom).   The number of degrees of freedom of the pooled estimate is the sum of the degrees of freedom of the individual variance estimates, and thereby gains precision over them.

An example of the concept of degrees of freedom is as follows: The sum of the deviations $(Y_k - Y)$ of n observations from any sample average *must* sum to zero.   This requirement [that $\Sigma\,(Y_k - Y) = 0$], is a *linear constraint* on the deviations that are used in Eq. 2 to calculate the s^2. The constraint implies that if any (n - 1) of the deviations are known, the remaining one can be determined from the knowledge of the others.   The n deviations are thus said to have (n - 1) *degrees of freedom*.   Since the sample variance s^2 is calculated using the deviations, it can only have the same number degrees of freedom, and hence the denominator (n - 1) must be used in Eq. 2.   Thus, if a measurement of Y for a single level of x was replicated three times, its variance would have two degrees of freedom. In later applications dealing with factorial experiments, it will be necessary to estimate mean factor effect values from average effect values calculated from a limited number of measurements. In such cases the determination of each average factor effect also involves the loss of one degree of freedom, since an average value again must be determined in each case. Therefore, the number of degrees of freedom that will be available to calculate an estimate of the variance of the error of the experiment will be [n - 1 - k], where k is the number of degrees of freedom lost in determining factor effects.

In the special case where only two observations were made at each of k-levels of x (in our example, there are 2 observations at 4 levels of x), Eqs. 2 and 3 are reduced to:

$$s^2_{pooled} = \frac{\sum\limits_{j=1}^{k} (Y_{j1} - Y_{j2})^2}{2k} \tag{5}$$

For example, for the data given in Fig. 1:

$$s^2_{pooled} = \frac{(0.4)^2 + (-0.2)^2 + (-0.3)^2 + (0.2)^2}{2\,(4)}$$

$$= 0.041$$

$$s_{pooled} = 0.2$$

A useful observation concerning the standard deviation is as follows: If the measurements in a population are approximately *normally distributed* (to be defined in the following section), about 2/3 of the measurements will be within one standard deviation of the mean (i.e. within [μ - σ] and [μ + σ]).

Normal Distributions

The basis of statistical analysis relies on the concept that random variables exist in the form of populations that have predictable distributions. When measurements of the values of a

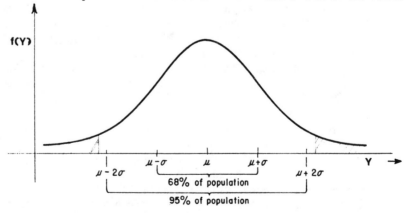

Fig. 2 Curve of the normal distribution function.

population exhibit a *normally distributed* behavior (Fig. 2), it means that the relative frequency of the population values can be expressed according the following mathematical relationship:

$$f(x) = \frac{1}{\sigma\sqrt{2\pi}} \exp^{\frac{-1}{2}[\frac{x-\mu}{\sigma}]^2}$$

(6)

In practice, the normal distribution turns out to describe the distribution of random variables in many sets of experimental observations, and this justifies the study of the normal distribution function. However, in order for any mathematical relationship to satisfy the requirements of being a statistical frequency function, the area under the curve described by the mathematical relation must equal one, and be positive for all values of the variable. In addition, the area under any portions of the density function will give the probability that the random variable will fall in the corresponding intervals of the abscissa.

Distributions of Averages, t-Distributions, and Confidence Intervals

In the course of analyzing experimental data, it is not   necessary to show how to calculate probability information concerning *individual measurements*. What is important, however, is the ability to determine probability information about *average values*. The *normal distribution* introduced above is important because it is possible to show that: *even if the errors from individual measurements from a set of experimental runs do not follow a normal distribution, the average of groups of such errors are distributed in a manner which approaches the normal distribution for very large numbers of groups*. (This property is known as the *Central-Limit Theorem*).

The above property leads to the relationship which states that: the *variance* of the sampling distribution of the mean of a set of n measurements σ^2_Y is just the population variance of the individual σ^2, divided by the number of measurements used to calculate the mean:

$$\sigma^2_Y = \sigma^2/n$$

(7)

That is, if we had taken a set of measurements using n observations then and calculated σ, the *uncertainty (or variance) of the mean* also calculated from this set of measurements would have depended not only on σ^2, but on the number of measurements taken.

At this point we can formulate an answer to the following question: If a set of experiments yields an estimate, **Y**, of the actual population mean (μ) for a given setting of x, and if we do not know the population variance, σ, but only an *estimate*, s, of the population variance, how close is **Y** to μ (i.e. how much confidence do we have about the actual value of μ)? Using an estimate of σ introduces an additional element of uncertainty into the experiment, which can be taken into account by using the so-called *t- distribution* in place of the normal distribution.

Let us say we have observed a number of responses, Y_1, ..., Y_n, from a normal population that has population parameters μ and σ. Then a quantity, t, can be defined as

$$t = (Y - \mu)/(s/\sqrt{n})$$

(8)

This quantity, t, obeys a t-distribution. *T-distributions* have the same shapes as normal distributions, but are wider (Fig. 3). The degree of increased width depends on the number of degrees of freedom associated with the *estimated standard deviation*, s. In the limit, for an infinite number of degrees of freedom (in practice >60 is enough), the t-distribution is identical to

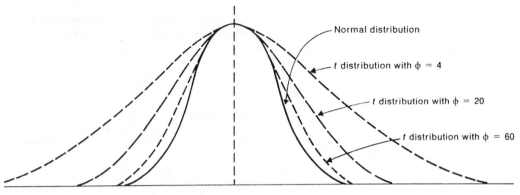

Fig. 3 Comparison of the *t* distribution with the normal distribution.

the normal distribution, and an experimenter could use normal-distribution tables to calculate
probability values in place of t-distribution tables.

The t-distribution can be used to determine the probability that the population mean of a
group of observations will fall between two values *provided that the experiment was performed
with a randomized run sequence.*   (See Ref. 1 for the justification of this requirement and for
more detailed information on the t-distribution.) This is done by first calculating s of a sample
from an experiment with a given number of runs and a known number of degrees of freedom.
Then the t-distribution is used to find the values between which the population mean will fall for
desired probability levels. This is accomplished using the following relation:

$$\text{Values between which } \mu \text{ will fall at some probability level} = Y \pm [\rho_{(1-\alpha,\,d.f.)}][s/\sqrt{n}] \qquad (9)$$

where: $(1-\alpha)$ = the probability level; n = the number of observations used to calculate Y; and
$\rho_{(1-\alpha,\,d.f.)}$ = the value of the t-distribution for the number of degrees of freedom (d.f.) and the
probability level desired. These calculated ranges for the population mean at the desired
probability levels are known as the *confidence intervals* for the mean.

Table 1 gives some values of ρ for varying probability levels and degrees of freedom, d.f.,

Table 1.   Probability Points of t-Distribution: Double-sided; σ^2 unknown.

d.f.	0.005	0.01	0.02	0.05	0.10	0.20	0.30
1	127.00	63.70	31.82	12.71	6.31	3.08	1.96
2	14.10	9.93	6.97	4.30	2.92	1.89	1.39
3	7.45	5.84	4.54	3.18	2.35	1.64	1.25
4	5.60	4.60	3.75	2.78	2.13	1.53	1.19
5	4.77	4.03	3.37	2.57	2.02	1.48	1.16
10	3.58	3.17	2.76	2.23	1.81	1.37	1.09
15	3.29	2.95	2.60	2.13	1.75	1.34	1.07
20	3.15	2.85	2.53	2.09	1.73	1.33	1.06
25	3.08	2.79	2.49	2.06	1.71	1.32	1.06
30	3.03	2.75	2.46	2.04	1.70	1.31	1.05
60	2.91	2.66	2.39	2.00	1.67	1.30	1.05
120	2.86	2.62	2.36	1.98	1.66	1.29	1.05
∞	2.81	2.58	2.33	1.96	1.65	1.28	1.04

Table header: ρ

for the *double-sided* t-tests (i.e. that are valid for experiments designed to detect the occurrence of *significant deviations* [to be defined later] with both positive and negative values].

Example: Using the data from Fig. 1, calculate the values between which the population mean, **Y**, will be found for the value of Y when x = 40, with a confidence level of: a) 95%, and b) 99%.

Solution: The sample mean, **Y**, from Fig. 1 when x = 40 is **Y** = 1.57. The sample variance is calculated from the expression for the pooled variance, Eq. 3, and is found to be:

$$s_{pooled} = 0.2$$

Using Eq. 9 and Table 1 for values of ρ, and the fact that the number of measurements used to calculate **Y** was n = 2, and the number of degrees of freedom used to calculate s_{pooled} is 4, we calculate the ranges within which we can expect to find the population mean, μ, at the two probability levels. From Table 1:

$$\rho \text{ (95 percent probability, 4 degrees of freedom)} = 2.78$$

$$\rho \text{ (99 percent probability, 4 degrees of freedom)} = 4.60$$

Then:

Range within which μ will be found with a 95% confidence level	$= 1.57 \pm [(2.78 \times 0.2) / \sqrt{2}\,]$
	$= 1.17 < \mu < 1.97$

and;

Range within which μ will be found with a 99% confidence level	$= 0.90 < \mu < 2.23$

Thus we see that the t-distribution is useful for letting *confidence bands* to be established on the estimates of population parameters derived from experimental measurements (i.e. when μ and σ are not known).

THE DESIGN OF EXPERIMENTS: BASIC DEFINITIONS

In any experiment there is an *objective* which the experiment is designed to achieve, such as optimizing (e.g. possibly maximizing, minimizing, or stabilizing) some characteristic of a process or product. The variables that are independently controllable are referred to as *factors*, whereas those variables that are measured are the *dependent* variables, and will be called the *responses* of the experiment.

The measured responses contain errors that are an unavoidable aspect of the measurement environment. An awareness of this fact leads to the conclusion that it is necessary to develop an experimental strategy that will give satisfactory results, even in the presence of experimental error. For our discussion errors fall into two groups: a) bias errors; and b) random errors. *Bias errors* are characterized by a numerical value that tends to remain constant, or which follows a consistent pattern over a number of experimental runs. A *constant* bias error might, for example, be present in a set of experiments involving: 1) a single batch of raw material; 2) a particular machine; 3) a specific individual. A bias error that followed a *consistent pattern* might have been dependent on such factors as: 1) time of day; 2) season of the year; or 3) age of the machine. A specific example of a bias effect in wafer fabrication environment might be the seasonal change in humidity that would impact a non-humidity controlled photoresist room.

The statistical tools for minimizing the effects of bias error are *randomization* and *correct use of blocking*. *Randomization* is the running of experiments in a random order. By such randomizing, the *bias effects* do not persistently confuse the *measured effects* when varying values of the factors are applied during the experiment. In *blocking*, an attempt is made to divide up a single experiment with many runs, into several smaller experiments (blocks), each one conducted and statistically analyzed as if the runs in a block constituted an entire experiment. Bias error is reduced because those experimental conditions that are expected to slant (bias) the responses in an ordered manner, are segregated into individual blocks.

For example, consider the suspected effect of humidity on a photoresist process. An experiment is proposed with n runs, but it is considered likely that humidity is a source of bias error. On any day in July, the humidity is expected to be higher than on one, say in September. The experiment could be conducted as a single randomized experiment, with half the runs on a humid July day, and the other half on a non-humid September day. The results in such an unblocked approach, however, would be analyzed as coming from a single experiment. The two halves of the run might have responses offset from one another due to the humidity effect, but the statistical analysis would have no way of separating this effect from other error. On the other hand, it would be more advantageous to run a pair of experiments, each with n /2 runs (and each separately randomized), one on a July day and the other in September. Thus, the effect of humidity would be blocked. By statistically analyzing the responses for each block separately, any bias error from humidity is separated, and does not contribute to the standard deviation of the error from each block. In addition, such blocking provides a way for determining if humidity plays a measurable role in effecting the process. Blocking can also be used to sort experimental materials into categories, each of which is more homogeneous than the aggregate. Without such blocking, unsorted material could produce bias error.

Random error possesses a numerical value that changes from one run to the next. The average value of a set of random errors is zero, and as such, each error can be positive or negative in value. Typically, the magnitudes of random errors are normally distributed (Fig. 2). Furthermore, small values of the error will occur more frequently than larger ones. The chief technique for minimizing random error is replication. That is, since the average value of random error is zero, the average value of a number of measurements is more precise (i.e. a better estimate of the true value of the response being measured), than a single measurement value. For example, if the standard error for a single measurement of response, Y, is σ, and n independent measurements of Y are made at identical levels of x in the experiment, then the standard error of the average, Y, is given by Eq. 7. From this equation we calculate that σ_Y equals σ for one measurement (n = 1), then decreases rapidly for a small number of measurements (e.g. $\sigma_Y = 0.5$ for n = 4), but only decreases slowly as n gets large ($\sigma_Y = 0.25$ for n = 16, and $\sigma_Y = 0.1$ for n = 100). This example demonstrates the idea that a small amount of replication is useful for reducing the error of the average measured value, but excessive replication is wasteful of experimental resources. Thus, an effective experimental strategy should call for sufficient replication, but not too much. The experimental design techniques we present follow this principle.

The design of an effective experimental plan also depends on several aspects of the experimental environment, the most important being the characteristics of the factors (e.g. discrete or continuous), and their number. For example, if it is known that the total number of significant factors is small (e.g. 2 or 3), then a relatively complete investigation of the effects of these variables can probably be performed in a small number of experimental runs. On the other hand, to determine the effects of 30 factors on an experiment with the same thoroughness, a much greater experimental effort would have to be undertaken.

Discrete factors have the property of being able to assume only a limited number of values (and often only qualitative ones). For example, the objective of an experiment may be to optimize a specific photoresist process. Part of the investigation might compare two types of photoresist (e.g. A and B), and thus the *type of photoresist* would represent a *discrete* factor in the experiment. *Continuous* factors on the other hand, are capable of exhibiting any value over some numerical range. For a specific experiment, there will be upper and lower limits for each factor. However within the limits, it is possible to set the factor to any value. Examples of continuous factors that could be varied continuously include pressure, power, flow rate, or temperature.

In the examples used in this chapter, we will examine experiments in which only two levels or settings for each factor are applied. These two levels can be for either discrete or variable factors. The two levels are coded low (-) and high (+). When working with continuous variables these two levels should be sufficiently separate so that a significant response to a variation in factor level can be clearly distinguished from any experimental error in the measured response. Widely spaced (+) and (-) levels improve the chances of identifying significant effects with an experiment, or to conclude that no effect other than experimental error occurs for that process. On the other hand, experiments with more that two levels per factor require many more runs than corresponding experiments with only two levels. Therefore, multilevel experiments (e.g. more than 2 factor levels) are normally only conducted when the number of significant factors has been clearly identified as being very small, or when a significant non-linearity is expected in the relationships between the response and the factors.

CHARACTERISTICS OF FACTORIAL EXPERIMENTS

Since the chief vehicle for statistically based experimental designs is the *factorial experiment*, its basic characteristics will be discussed in this section, and also contrasted with *one-at-a-time* type experiments. In one-at-a-time experiments, only one factor in each run is changed, while the other factors are held constant. As will be illustrated, this method leads to incomplete coverage of the experimental space, missing relationships between factors, and the possibility of erroneous assumptions.

In *factorial experiments*, the use of factorial arrangements for the factors in an experiment makes each run yield information concerning every factor. Thus, not only are factorial experiments highly efficient in the number of runs utilized, but the efficiency is increased by multiple use of each response. The term *factorial* is used to indicate that a factorial experiment permits the simultaneous estimation of several factors, as opposed to "one-at-a-time" experiments in which the measured effects are due to the change of only independent variables. In other words, as the factorial (n!) of an integer is its product with all of the positive integers smaller than itself, in factorial experiments, each factor effect is based on all the individual measured responses. The term *factor* is chosen (in lieu of *independent variable*), to emphasize the fact that in factorial experiments, every run will produce information concerning the effect of every independent variable. In full (or complete) factorial experiments, a set of runs in which all possible combinations of factors at the levels specified by the experiment are conducted. Thus,

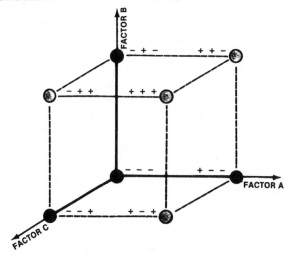

Fig. 4 Experiment space for a three-factor experiment. Only half of the experimental space is examined (black points). No information is obtained using the other half (shaded points). Reprinted with permission of Microelectronics Manufacturing and Testing.

for a 2-level full factorial experiment, with k factors, 2^k runs are required. Since the (+) and (-) levels of each factor can represent the extremes of the ranges over which the factors of the experiment will be set, the complete experimental space of the experiment is covered by a full factorial experiment. For multilevel (m) full factorial experiments m^k runs are required. Thus for a 2-level, 5 factor experiment, $2^5 = 32$ runs are required, while for a 4 level, 5 factor experiment $4^5 = 256$ runs would be required! Table 4 shows the complete set of runs (and the factor levels for each run) that must be performed for full 2-level factorial experiments with up to 5 factors. *Full-factorial experiments*, as opposed to *one-at-a-time experiments*, not only give a complete exploration of the experiment space, but they also offer the additional advantages of *hidden replication*, and information about *interaction effects* among factors. Before discussing these concepts let us examine a graphical tool that helps to visualize the characteristics of factorial experiments.

Graphical aids can be used to help visualize the factor space of full-factorial experiments with 2 or 3 factors. For example, Fig. 4 shows a cube with each cube axis representing one factor in the experiment. Thus, if an experiment has 3 factors, such a cube allows the complete factor space to be easily visualized. If the experiment contains more than 3 factors, such pictures cannot be constructed, but the mathematical relationships used to construct a 3-factor cube are easily extended to experiments with larger numbers of factors.

Each corner of a 3-factor cube represents a set of factor values at which one or more experimental runs will be conducted. The required number of runs at each point, or *direct replicates*, is determined with the aid of statistics, and we will show how this is done in the process of designing a full-factorial experiment. Thus, as can be visualized in Fig. 4, factor A has its low level (-) at the left face of the cube, and its high level on the right face of the cube. The left rear lower corner represents the set of variable values where A, B, and C are all at their low values. Thus, this cube helps to demonstrate that the full-factorial experiment does indeed cover the entire factor space, which is one of the necessary properties of well designed experiments.

The terms factor effect, main effect, and interaction effect need to be defined at this point. A *factor effect* is the *difference in the two measured responses* when two different levels of a factor are applied in an experiment.   An example of a factor effect is $(Y_1 - Y_2)$, where Y_1 is the response due to a factor set to level 1, and Y_2 is the response due to the same factor set at level 2. In the ideal case, only the difference in the factor levels plays a role in determining the value of the difference in the response value, but in actual experiments, the value of the factor effect also depends on the experimental error. As noted, to reduce the role of the random experimental error, replication of the measurement must be performed. In the *one-at-a-time* approach, such replication must be *direct*. That is, identical experimental runs need to be repeated in order to get average values for Y_1 and Y_2.

In factorial experiments, on the other hand, *every* run produces information concerning the effect of *each* factor. Therefore, an average value for each response at a given factor level can be calculated from the data of a full-factorial experiment, without having to replicate the entire factorial experiment (or even any of its individual runs). This ability of factorial experiments to intrinsically produce average values of factor effects without having to directly replicate experimental runs is referred to as a "hidden replication" capability. Information about two types of factor effects are also available from factorial experiments; a) main effects, and b) interaction effects. On the other hand, in one-at-a-time experiments, no information about interaction effects can be extracted. *Main effects* for each individual factor are defined to be the difference between two *average responses* Y_+ and Y_-, or:

$$\text{Main Effect} \ = \ Y_+ \ - \ Y_- \qquad\qquad (10)$$

where Y_+ is the average response for the (+) value of a factor, and Y_- is the average response at the (-) level.

Main effects can be most easily visualized with the aid of the cube that represents a 2-level 3 factor factorial experiment. The main effect of factor A is based on a comparison of the response values at the left and right faces of the cube. Let us study the lower-rear edge of the cube. Along this edge are two corners, each representing a set of factor values. The only difference between the sets is in the values of A. By taking the difference of the responses for the sets of values of these two corners we get an estimate of the effect of factor A (one-at-a-time estimate). However, there are also three other edges along which B and C will remain constant as A is varied. In summary, there are four edges that allow us to estimate the effect of A. The *main effect* of A is the *average* of the *difference in the responses* measured from these four edges, and is a measure of the effect that the varying of A alone will have on the response.

On the other hand, two or more factors are said to interact if the effect of one, say A, is different at different levels of other factors. Such interactions show up as *interaction effects* and they can be estimated from full-factorial experimental data just as well as the main effects. To illustrate an interaction effect, consider the simple case of a 2-level, 2 factor experiment with factors A and B (Fig. 5a).   The interaction effect is defined as :

$$\text{Interaction Effect} \ = \ [(Y_4 - Y_3) - (Y_2 - Y_1)] \ /2$$

$$= \ [(Y_1 + Y_4) - (Y_3 + Y_2)] \ /2 \qquad\qquad (11)$$

and we see that the result involves the comparison of responses at diagonally opposite corners.

The presence of an interaction can also be illustrated by examining the responses of a 2-level, 2-factor experiment (with the factors labeled A and B).   The graph of Y versus A is plotted with B being high and again for B being low.   If the graphs have equal slope (Fig. 5b), there is no interaction effect between A and B.   If the graphs have different slopes (Fig. 5c & d), this indicates the presence of an interaction effect.

One of the roles of statistical analysis is to determine if the measured effects in an

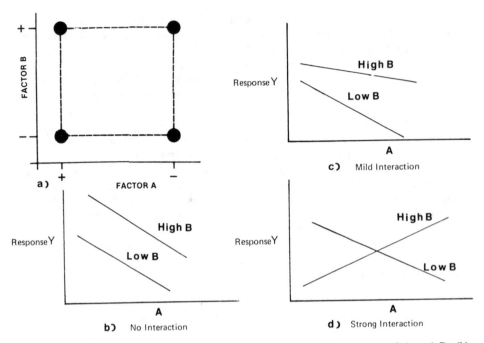

Fig. 5 (a) Experiment space for a two-factor experiment. (b - d) Interaction of A and B: (b) no interaction; (c) mild interaction; (d) strong interaction.

experiment (including both main and interaction effects), are primarily due to the applied factors or are just due to fluctuations (errors) of the measured environment. Those effects which are determined in the analysis of experimental data as being greater than the expected error effects are referred to as *significant* (or *statistically significant*).

We conclude this section by going through a simple hypothetical experimental task that involves two factors. This example shows how a full-factorial experiment can yield significantly more information about the process being studied than a one-at-a-time experimental evaluation.

Let us suppose that it is desired to study the removal rate of photoresist in a dry etch system, in order to develop a resist stripping process of maximum throughput. Let the two factors which will be investigated be the oxygen flow rate and the system pumping speed. (We will assume that a maximum number of wafers is always loaded into the chamber and maximum power is applied, and hence these experimental conditions are not variables). Three replicate runs are to be processed at each combination of factor values. The processing is carried out in random order and the experimental error estimated from the results. The values of the average of the erosion rates over the 3 runs at each experimental point are shown in Fig. 6.

Let us examine the consequences of conducting a one-at-a-time type experiment (Table 2) versus those from a full-factorial one (Table 3). Four trials (of 3 runs each) would have been carried out as shown in Table 2. The number 1 and number 2 trials of the one-at-a-time experiment were conducted first, and it was observed that the high (+) flow rate yielded the highest average erosion rates. Next, the one-at-a-time trials numbered 3 and 4 were conducted with the "best" levels of flow rate (high in this example) fixed, and the pumping speed changed to find its "best" value, at the high value of flow rate. In this experiment, a high pumping speed level was observed to produce faster erosion rates than at the lower pumping speed (for a high flow rate). It appears from these comparisons that the fastest erosion rate is found at both

Table 2. ONE-AT-A-TIME TWO-LEVEL EXPERIMENT

RUN	FACTOR LEVELS A	B	RESPONSE	OBJECTIVE
1	+	+	z_1	Determine "best" level of A at
2	-	+	z_2	high level of B.
3	+	+	z_1	Determine "best" level of B at
4	+	-	z_3	"best" level of A.

high levels of pumping speed and flow rate. It is assumed from the available experimental data that the response when both factors are at their low levels (the missing trial in the one-at-a-time experiment) would exhibit a lower erosion rate than at any of the other factor combinations.

On the other hand, if the experiment was conducted in full-factorial fashion (Table 3), the actual erosion rate at the (-)(-) condition would have been measured. The reason that this condition in fact yields a higher erosion rate is due to an *interaction effect* between A and B. The implicit assumption from the one-at- a-time experiment, that the (-)(-) condition would yield a lower erosion rate, stemmed from the lack of knowledge about the existence of interactions among the factors. The one-at-a-time experiment cannot provide such interaction information. Therefore, if the one-at-a-time experiment had been conducted as described above, the missing experimental point would not have been subjected to a trial, and both the maximum erosion rate condition and the interaction effect of the two factors would have gone undetected. On the other hand, the full-factorial experiment would have found them both!

The reader may object that it would have been easy to add the missing (-)(-) run to the one-at-a-time experiment, and thus obtain the same information. In the full-factorial experiment, however, no "missing runs" are possible, since all possible factor combinations are automatically covered, whereas in one-at-a-time, mistakes can lead to omitting some combinations. Furthermore, as Box and Hunter[1] point out, for k factors a complete one-at-a-time design would require $(k+1)/2$ times as many runs as the full-factorial. Finally, no interaction information data can be directly calculated from the one-at-a-time responses.

As a final note, we have observed that even 2-level full-factorial experiments can easily require a very large number of runs, as the number of factors increase. Under some circumstances, by conducting a properly selected smaller set of runs, useful information can still be obtained. If an experiment is designed such that only a specific fraction of all the runs of a full-factorial experiment are to be conducted, it is designated as a *fractional-factorial* experiment. Such reduced-run experiments find wide applicability in the early stages of an experimental

Table 3. FULL-FACTORIAL TWO-LEVEL EXPERIMENT

RUN	FACTOR LEVELS A	B	RESPONSE	OBJECTIVE
1	-	-	Y_1	Determine "best" combinations of
2	+	-	Y_2	levels A and B.
3	-	+	Y_3	
4	+	+	Y_4	

Factor Level			Resist Erosion	
(A) Oxygen Flow Rate	(B) Pumping Speed		Rate	(Å /min)
+	+		980	
−	+		770	
+	−		840	
−	−		1100	

Fig. 6 Average values of resist erosion rates over 3 runs at each experimental point.

project, where a large number of factors are cataloged, but knowledge about whether they have much effect on the process being optimized is not known. At that point, *screening experiments* designed to identify the significant factors of the group, are utilized. Since the other factors are determined to have no significant effects on the process being investigated, no further experimental effort needs to be expended. Such screening experiments make effective use of fractional-factorial experiments.

STRATEGY OF DESIGNING EXPERIMENTS: AN OVERVIEW

When setting out to conduct an experimental project (for example, to develop and optimize a new VLSI fabrication process), an overall plan to achieve the desired goal with the minimum expenditure of resources and time should be designed. A complete flow diagram of a general sequence of steps that such a plan could encompass is shown in Fig. 7. Note that there are several stages of the experimental procedure, and a particular application may or may not require all of them to be carried out. Part of the strategy of selecting an effective experimental approach is to decide how much information is actually required, and then to utilize as many of the steps shown in Fig. 7 to derive the information. In addition, as each stage is completed, some information is derived that influences subsequent experimental decisions. Hence, even though an overall plan is specified at the outset, all of the experimental details cannot be included yet, since some will depend on the outcome of earlier experimental results.

At the beginning of the project, the *objective* of the experiment needs to be specified.

	Design Type	Purpose	Number of Factors
1.	Screening	Identify important variables; crude predictions of effects.	6 or more
2.	Complete & fractional Factorials	Linear effects and interactions: used for interpolation.	3 to 8
3.	Response surface	Linear effects, interactions, & curvature; used for interpolation.	2 to 6

Fig. 7 Steps conducted in an overall experimental sequence used for optimizing a process.

Sufficient detail about how to determine that the objective has been successfully accomplished should be set down. For example, if the objective involves the characterization and /or optimization of a new process, a specification of acceptable limits (e.g. thickness, uniformity, dielectric strength) that describe the successful outcome of the process, should be included as part of the objective.

Next all the factors (independent variables) that could possibly influence the response of the experiment should be cataloged. Gathering this list of potentially significant factors is an

Table 4. TWO-LEVEL FULL-FACTORIAL DESIGN PATTERNS

RUN	x_1	x_2	x_3	x_4	x_5
1	-	-	-	-	-
2	+	-	-	-	-
3	-	+	-	-	-
$4 = 2^2$	+	+	-	-	-
5	-	-	+	-	-
6	+	-	+	-	-
7	-	+	+	-	-
$8 = 2^3$	+	+	+	-	-
9	-	-	-	+	-
10	+	-	-	+	-
11	-	+	-	+	-
12	+	+	-	+	-
13	-	-	+	+	-
14	+	-	+	+	-
15	-	+	+	+	-
$16 = 2^4$	+	+	+	+	+
17	-	-	-	-	+
18	+	-	-	-	+
19	-	+	-	-	+
20	+	+	-	-	+
21	-	-	+	-	+
22	+	-	+	-	+
23	-	+	+	-	+
24	+	+	+	-	+
25	-	-	-	+	+
26	+	-	-	+	+
27	-	+	-	+	+
28	+	+	-	+	+
29	-	-	+	+	+
30	+	-	+	+	+
31	-	+	+	+	+
$32 = 2^5$	+	+	+	+	+

important task, because at this point there may not be much information as to which factors can significantly effect the response. If the investigator does not successfully identify the key factors that could potentially control the process, the investigation could merely involve characterizing second-order effects. Assembling this list may involve discussions with other workers in the same field, reading of reports and published papers, etc. The final list of candidate factors can be quite long, and may often include 10 to 30 factors. A reasonable range for each factor of interest should also be specified.

The responses that will be measured during the experiment then need to be listed. All of the responses that we suspect will be effected by the factors, and that in some way will measure the performance or quality, should be included. An estimate of the anticipated variation of each response would also be useful.

If at this point the number of candidate factors is small (approximately 2-4), and if they are all known to have a significant effect on the responses, the best strategy might be to immediately undertake a 2-level full-factorial experiment (possibly with a center point, to check for curvature) designed to derive a limited response surface. If desired, a complete set of response surfaces may then be determined using Box-Behnken experiments. If 5 variables are involved, a *half-factorial* experiment (as discussed in the section on screening designs), should be considered.

However, as is quite common, the number of factors may turn out to be rather large (e.g. 6-30), and there may be little knowledge about which of the various factors will be most important in causing response effects. Therefore an effective strategy at this point would be to eliminate (screen out) those factors which cause little or no impact, and to perform such screening with the minimum amount of effort. Special screening experiments have been developed for just this application, with the number of runs required to carry them out depending on the number of candidate factors, the degree of precision desired (i.e. the degree of certainty that a significant effect has been identified) and the type of screening experiment being used. The two types of screening experiments which are most commonly employed are: 1) Plackett-Burman; and 2) fractional-factorial types. The design and analysis of screening experiments using these two screening types will be outlined in a subsequent section. The results of the screening experiment usually identify a small group of the original candidate factors as having significant effects on the process responses. At that point, the determination of limited response surfaces using full-factorial experiments, or full response surfaces using Box-Behnken experiments, can be pursued. A derivation of a theoretical model which would allow the estimation of responses to factor values outside of the tested factor space from response surface information may be an ultimate goal of the experimental project. The complex task of deriving such theoretical models, however, is beyond the scope of this chapter.

Part of the strategy of creating an overall experimental plan also involves deciding on the number of runs and replication that will be utilized in each of the stages of the experimental sequence. Guidelines for making these decisions are presented in the sections dealing with the specific experiment designs.

DESIGNING AND ANALYZING TWO-LEVEL FULL-FACTORIAL EXPERIMENTS

As described in a section that introduced them, full-factorial experiments allow the complete factor space of an experiment to be covered, and also to produce factor effects (main effects and interaction effects) that are based on response information from every run of the experiment. The main effect and interaction effect information can be used to generate limited response surfaces, or a linear polynomial equation that will allow response predictions to be interpolated within the ranges of the factor space. Typically two-level, full-factorial experiments can be effectively

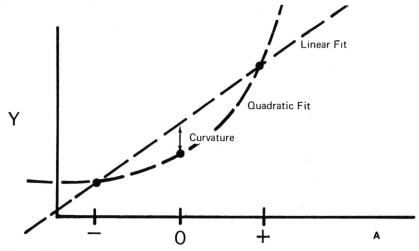

Fig. 8 Determining curvature in a two level factorial experiment.

utilized for providing reasonably detailed data about responses of processes in which 2-4 factors have been identified as being significant. If a larger number of candidate factors needs to be investigated, a screening experiment is usually performed first. This selects the most significant factors for the more extensive full-factorial study (which is subsequently performed. In this section we will discuss the principles of how to design and analyze full-factorial experiments.

Before the method of design is outlined, it is useful to discuss a method for estimating any *curvature* in the response that might exist within the factor space being investigated. That is, since only factor values at the extremes of their ranges are investigated with 2-level factorial experiments, no information about response values in the interior of the experimental space is provided. Thus, only linear (straight-line) variations can be used to predict the values of response within the limits of the factor space. To perform an economical check on the validity of such straight-line assumptions, a small number of additional runs with the factor values all set at their middle points, should also be conducted. It is recommended that 4 replicates of this run be conducted, so that an average value is obtained. The severity of any curvature present is estimated by calculating the difference between the average value of the design points and the average value of the center points (Fig. 8). Unfortunately, if the curvature is found to be severe, the prediction model derived from the full-factorial data is not likely to give accurate results except near the experimental points.

Two case studies that describe the use of full-factorial experiments for optimizing a process are given in Ref. 6 (i.e. development of a deposition process for PECVD silicon nitride), and Ref. 7 (i.e. adapting a conventional positive optical photoresist to a mid-UV process).

Method for Designing 2-Level Full-Factorial Experiments

In this section, a step-by-step approach to the design of 2-level full-factorial experiments with up to 5 factors (and a test for curvature) is presented. Any number of responses may be measured for each set of factor values. The steps of the method are as follows (note that it is assumed that the factors to be investigated have already been selected):

1. The values of the limits of the factors to be applied in the experiment should first be determined. These limits will identify the two levels of each factor that will be applied

in the experiment. The upper level is encoded as (+), and the lower level as (-). As noted earlier, these levels should be sufficiently separated so that a significant response to the variation in factor levels can clearly be distinguished from any experimental error.

2. Any number of responses can be measured at each experimental point. As discussed in the introductory section, the two-level factorial method will yield best results when the responses are continuous and have *uniform* independent errors. Uniform errors are those which have approximately the same magnitude at all experimental points.

3. The design pattern, which specifies the level of each factor for every trial is then obtained from Table 4 (which gives this pattern for up to five factors). Such a Coded Design Table, as well as a Decoded version of the Design Table (in which the (+)'s and (-)'s in the Table are replaced with their actual values) should be prepared for the experiment file.

Table 5. RANDOM ORDERS FOR TEST TRIALS

	16 TRIALS					32 TRIALS				
1st Rep.	2nd Rep.	3rd Rep.	4th Rep.	5th Rep.		1st Rep.	2nd Rep.	3rdRep.	4th Rep.	5th Rep.
1	9	11	10	3		21	12	22	12	28
3	1	15	7	5		18	20	5	24	10
11	16	12	11	10		3	2	29	13	31
5	12	4	5	1		11	16	18	4	1
8	10	14	8	14		26	7	11	15	3
4	5	3	4	16		17	32	17	25	5
12	13	15	15	8		31	3	8	14	8
15	1	10	14	13		1	5	23	28	15
10	11	13	3	9		13	15	15	26	11
2	4	6	2	11		23	22	14	22	26
7	6	8	13	12		7	25	10	16	23
9	8	7	1	6		20	9	16	27	19
6	14	5	9	7		12	24	13	20	12
14	3	2	12	4		16	31	30	5	7
16	7	9	6	15		25	30	32	6	17
						6	10	25	23	2
						14	17	7	21	7
						4	13	4	11	13
						19	18	27	17	29
						32	4	9	18	9
						5	8	12	10	32
						30	29	28	2	16
						15	19	19	31	18
						10	1	20	3	21
						24	6	31	29	24
						27	14	6	1	6
						2	11	1	19	22
						29	21	21	32	20
						9	28	26	8	4
						28	23	2	9	25
						22	27	24	30	30

Table 6. Computing Table for Analyzing Factorial Designs with 2 to 4 Factors

RUN	Mean	x_1	x_2	x_1x_2	x_3	x_1x_3	x_2x_3	$x_1x_2x_3$	x_4	x_1x_4	x_2x_4	$x_1x_2x_4$	x_3x_4	$x_1x_3x_4$	$x_2x_3x_4$	$x_1x_2x_3x_4$
1	+	-	-	+	-	+	+	-	-	+	+	-	+	-	-	+
2	+	+	-	-	-	-	+	+	-	+	-	+	+	-	+	-
3	+	-	+	-	-	+	-	+	-	+	-	+	+	-	+	-
4	+	+	+	+	-	-	-	-	-	-	-	-	+	+	+	+
5	+	-	-	+	+	-	-	+	-	+	+	-	-	+	+	+
6	+	+	-	-	+	+	-	-	-	-	+	+	-	-	+	+
7	+	-	+	-	+	-	+	-	-	+	-	+	-	-	+	+
8	+	+	+	+	+	+	+	+	-	-	-	-	-	-	-	-
9	+	-	-	+	-	+	+	-	+	-	-	+	-	+	+	-
10	+	+	-	-	-	-	+	+	+	+	-	-	-	-	+	+
11	+	-	+	-	-	+	-	+	+	-	+	-	-	+	-	+
12	+	+	+	+	-	-	-	-	+	+	+	+	-	-	-	-
13	+	-	-	+	+	-	-	+	+	-	-	+	+	-	-	+
14	+	+	-	-	+	+	-	-	+	+	-	-	+	+	-	-
15	+	-	+	-	+	-	+	-	+	-	+	-	+	-	+	-
16	+	+	+	+	+	+	+	+	+	+	+	+	+	+	+	+

4. The order of running the trials should be randomized. Randomization reduces bias error. Table 5 gives several random orders for 16 and 32 runs. A new order should be selected each time a complete set of runs is replicated. Persons performing the experiment need to understand that runs must be conducted in the randomized order specified.

5. The number of necessary replicates should next be determined. The *hidden replication* of full-factorial experiments assists in reducing the variance of the average measured values (Eq. 7), by effectively increasing n. However, it must be determined if any additional *direct replication* is needed to estimate the variance of the random error, and to thus satisfy the confidence level requirements when deciding to identify a factor effect as being significant or not. A rule of thumb for selecting the number of required direct replicates is given as follows: *In order to be able to: 1) conclude with a confidence level (e.g. $\geq 95\%$) that the data will yield statistically significant factor effects; and 2) conclude with a confidence level (e.g. $\geq 90\%$) that a factor has an effect whose value is $\geq 2\sigma$ (i.e. twice the value of the standard deviation of a single measurement), the required number of runs is about 12 to 16. To conclude that a factor has an effect greater than σ, the number of required runs is 48 - 64.* Thus, if the experiment in question has 3 factors (8 runs for a full-factorial), and detection of significant factors greater than 2σ is acceptable, two full-factorial groups carried out as a single experiment would have to be conducted (16 runs).

6. In order to use the full-factorial experiments to obtain information about the response within the experimental space, without significantly expanding the number of runs in an experiments, an additional group of runs can be added at a middle value for all factors. Four replicated runs with factor values set at their mid points are suggested to increase the precision of middle point measurement. The degree of curvature can be estimated by the difference between the average of the measured response at the middle points and the average of all the other experimental points. If this turns out to show that the curvature is severe, the prediction model generated from the full-factorial is likely to be quite inaccurate except near the design points (Fig. 8).

7. The experiment is then conducted and the responses for each run are recorded.

Analysis of the Measured Data

1. The factor effects are first computed from response data in the following manner:
   a) Create a Computation Table (CT). To assist in tabulating the analysis, copy the design pattern from Table 6 (see Table 10 as an example for 3 factors).
   b) List the averages of the replicate runs in a column at the right of the CT. (Follow the run order listed at the left of the CT, and not the random run order.),
   c) Total the values of the replicate run averages for the lines that have a "+" sign for each column. Put this total under the correct column of the line SUM +,
   d) Repeat the process for the lines with " - " signs. Enter their totals under the appropriate columns in the line called SUM -,
   e) Add the values of SUM + and SUM -. Enter the total on the line OVERALL SUM. This step is a check. The value of the OVERALL SUM should be identical for all columns,
   f) Subtract SUM + from SUM -. Place the difference on the line called DIFFERENCE,
   g) Finally, divide the difference by the number of + signs in that column, and enter on the line EFFECT. This number is a main effect, or interaction effect (depending on the particular column). For column one, the result on the EFFECT line is the average of the responses of all the experimental points.
2. The curvature is then calculated. This is done by:
   a) Computing the average value of responses of the runs carried out with all the factors set at their middle values (average value of the center points),
   b) Then computing the difference between (1) the average of the responses of all the experimental points and (2) the average value of the center points. This difference is entered on the CURVATURE line of the CT.
3. The significance of the main effects, interaction effects, and curvature are then determined. That is if any of the computed main factors or interaction effects is found to be larger than a *minimum significant factor effect*, MSF (which we will demonstrate how to calculate), it can be concluded that the effect is indeed non-zero. On the other hand, if the value is smaller, it is likely that no significant effect exists, and the variation in experimental response for that effect is just due to experimental error fluctuations. By the same token, if the curvature is greater than the minimum significant curvature, MSC, it can be concluded that at least one response has non-zero curvature associated with it. The MSF and MSC are computed from:

$$MSF = \rho \ s_{pooled} \ (2/rk)^{1/2} \qquad (12)$$

and

$$MSC = \rho \ s_{pooled} \ [(1/rk)+(1/c)]^{1/2} \qquad (13)$$

where: ρ is the value of the t-distribution at the desired probability level $(1 - \alpha)$ for the number of degrees of freedom used to calculate s; s_{pooled} is the pooled standard deviation of a single response; r is the number of + signs in the column (note that the main and interaction effect columns have $2^{p-1}+$ signs, where p is the number of factors in the experiment); k is the number of replicates of each trial; and c is the number of trials conducted with factors set at middle value.

Note that the number of degrees of freedom, d.f., is determined by:

$$\text{d.f.} = n - (\text{number of main effects} + \text{number of interaction effects})$$

$$- 1 \text{ (for calculating the average)} - 1 \text{ (for calculating curvature)} \qquad (14)$$

Thus, if a 3-factor experiment was designed to be replicated twice (16 runs) and to have an additional four center point runs, a total of 20 runs would constitute the experiment. There would be 11 degrees of freedom available for calculating s, since 3 would be used to calculate the main effects, 4 for the interaction effects, and one each for the average of all the responses and the average of the center point responses.

4. The main and interaction effects are compared with MSF, and those that are larger are concluded to be significant. The curvature effect is compared with MSC, and if it is larger, the response is acknowledged as having significant curvature within the factor space being studied.

5. A linear model, in the form of an equation, that will predict the value of the response for any value of the factors within the factor space, from the data of the factorial experiment, can then be formulated. The equation is of the form:

$$Y_{PR} = a_0 + a_1 x_1 + a_2 x_2 + \dots + a_n x_n + a_{12} x_1 x_2 + \dots$$

$$\dots + a_{p-1,p} x_{p-1} x_p + \text{ higher order interactions} \qquad (15)$$

where: Y_{PR} is the predicted response; p is the number of factors; $a_0 = Y$; $a_j = 1/2$ (factor effect for x_j); $a_{jj} = 1/2$ (factor effect for $x_j x_{j'}$); and:

$$x_j = \frac{[(\text{factor level at the experimental point}) - 1/2 (\text{high} + \text{low factor level})]}{1/2 \; (\text{high} - \text{low factor levels})}$$

Table 7. FACTOR VALUE RANGES FOR THE CVD EXAMPLE EXPERIMENT

		RANGE	
x_1	Flow Rate of Gas A (sccm)	10	$\Rightarrow$ 50
x_2	Flow Rate of Gas B (sccm)	60	$\Rightarrow$ 120
x_3	Chamber Pressure (torr)	10	$\Rightarrow$ 80

Example: It is desired to develop a process for depositing a CVD thin film. It has been determined that variations in only three parameters (e.g. flow rate of gas A, flow rate of gas B, and pressure) need to be investigated. Table 7 lists these three parameters and the ranges over which they will be studied. The response that is measured is the film growth rate. Design a 2-level full-factorial experiment that is appropriate for this task. Let there be enough runs so that significant factor effects (both main effects and interaction effects) can be detected with high confidence ($\geq 90\%$) if they are greater than 2σ. Let there also be a test for curvature that will allow significant curvature to be detected with confidence.

Solution: A 2-level full factorial experiment for 3 factors requires 8 trials, plus an additional 2 trials for curvature. Since the rule of thumb requires 16 run to be able to detect factor effects greater than 2σ, the experiment will be performed using two replications (to yield a total of 20 runs). The four middle runs are spread to every 5 runs of the experiment, and the remaining 16 runs are randomized according to Table 5. Let the three factors be identified as x_1, x_2, and x_3, and the response by Y. Table 8 shows the sequence of the coded runs.

Table 8.   TWO-LEVEL FACTORIAL EXAMPLE EXPERIMENT-RUN DESIGN

RUN	CODED			ACTUAL		
	x_1	x_2	x_3	x_1	x_2	x_3
1	-	-	-	10	60	10
2	+	-	-	50	60	10
3	-	+	+	10	120	80
4	+	+	-	50	120	10
5	-	-	+	10	60	80
6	+	-	+	50	60	80
7	-	+	+	10	120	80
8	+	+	+	50	120	80
9	0	0	0	30	90	45

Example: If the average responses (film growth rates) from the 2-level, 3 factor full-factorial experiment (replicated twice and with four center points) are given in Table 9. Calculate : a) number of degrees of freedom of this experiment; b) the value of ρ specified by the problem; c) the pooled standard deviation of a single response measurement; d) the factor effects for each column of the computation table; e) the minimum significant factor effect; f) minimum significant curvature; g) Identify the probable significant factor effects; h) Is curvature significant ?

Table 9. EXAMPLE FACTORIAL EXPERIMENT - SUMMARY OF DATA

RUN	Y (Replicate 1)	Y (Replicate 2)	Y	Range	Variances
1	7	9	8	2	2
2	24	30	27	6	18
3	14	20	17	6	18
4	28	38	33	10	50
5	61	71	66	10	50
6	121	125	123	4	8
7	175	175	175	0	0
8	155	157	156	2	2
9	40	48	44	8	11
	44	45			

Solution: a) The number of degrees of freedom (d.f.) are calulated using Eq. 14. Since n = 20, number of main + interaction effects = 7 (i. e. x_1, x_2, x_3, x_1x_2, x_1x_3, x_2x_3, and $x_1x_2x_3$), d.f. = 20 - 7 - 1 - 1 = 11.
b) The number of runs (20) meets the Rule of Thumb requirements given in Step 5 of the *Method for Designing Full-Factorial Experiments*. Thus, the

Table 10 EXAMPLE COMPUTATION of FACTOR EFFECTS from 2-LEVEL DESIGN

RUN	Mean	x_1	x_2	x_1x_2	x_3	x_1x_3	x_2x_3	$x_1x_2x_3$	Y
1	+	-	-	+	-	+	+	-	8
2	+	+	-	-	-	-	+	+	27
3	+	-	+	-	-	+	-	+	17
4	+	+	+	+	-	-	-	-	33
5	+	-	-	+	+	-	-	+	66
6	+	+	-	-	+	+	-	-	123
7	+	-	+	-	+	-	+	-	175
8	+	+	+	+	+	+	+	+	156
SUM+		339	381	263	520	304	366	266	
SUM-		266	224	342	85	301	239	339	
OVERALL SUM		605	605	605	605	605	605	605	
DIFFERENCE		73	157	-79	435	3	127	-73	
EFFECT		18.25	39.2	19.75	108.75	0.75	31.75	18.25	

Center Point Average = 44.25   /   Curvature = 31.75

data will yield significant factor effects with a confidence level of 0.95. Since there are 11 d.f., we find from Table 1 that $\rho = 2.2$.

c) The pooled standard deviation is found using Eq. 5, $s_{pooled} = 8.6$

d) The factor effects from each of the columns are calculated as follows:

A Computation Table is set up as shown in Table 10. After following the steps of the *Analysis Method* , the factor effects are calculated and listed in the EFFECT row of this table.

e) The minimum significant factor, MSF, effect is found using Eq. 12, where, $\rho = 2.2$, $s_{pooled} = 8.6$, $r = 2^{p-1} = 4$, and $k = 2$. Then:

$$MSF = 2.2 \quad (2/4x2)^{1/2} = 9.0$$

f) The minimum significant curvature is found from Eq. 13, where in this case, $r = 2^p = 8$, and $c = 4$. Thus:

$$MSC = 2.2 (8.6) [(1/8x2) + (1/4)]^{1/2} = 10.6$$

g) The probable significant factor effects are: x_1, x_2, x_1x_2, x_2, x_2x_3, $x_1x_2x_3$

h) The curvature is found from the difference between the center point average ($[40 + 48 + 44 + 45]/4 = 44.25$) and the average of the Y from Table 10, $Y = 75.62$, or the curvature is $75.62 - 44.25 = 31.375$. Hence we conclude that curvature is significant.

DESIGN AND ANALYSIS OF SCREENING EXPERIMENTS

In many processes it is possible to identify a variety of factors that may impact the process responses. The number of runs necessary to perform a 2-level full-factorial experiment with k factors is 2^k, and for more than about 4 factors, this number may become excessive.   In such cases, a more effective strategy than conducting a full-factorial experiment would be to carry out a less extensive experiment designed to identify the most important factors, and then perform the full-factorial experiment using only these.   Such preliminary experiments that eliminate unimportant factors are called *screening experiments*, and can substantially reduce the number of runs that must be used to separate out significant factors.   Such screening experiments are still based on the principles which underlie full-factorial experiments (e.g. hidden-replication and full coverage of the experimental space), but use a smaller number of runs for a given number of factors.

The price that is paid for utilizing a reduced number of runs is that when the test data from a column of the Computation Table (i.e. the values obtained in the EFFECT row of the Table 9) is obtained, it no longer necessarily contains information about a single main or interaction effect.   Instead, it may contain response information from two or more effects.   This is known as *confounding*.   For example, if a screening experiment was performed using three factors and only four runs were conducted (instead of the 8 needed for a full-factorial experiment), the value calculated in each of the EFFECT rows would no longer result from a single main effect or interaction effect, but would have contained response information from a main effect *and* an interaction effect.   Such confounding is due to the fact that the same factor level combination is used to produce the factor effect of more than a single factor or interaction of factors.

As the fraction of the total number of runs compared to the number required for a full-factorial experiment gets smaller, the larger the number of effects are confounded.   That is, if *half* the number of runs is conducted (and in which the experiment is known as a *half-factorial*), each EFFECT value confounds only two effects, while each EFFECT value in an experiment with only one-fourth the number of runs, confounds four effects.   The reason that such confounding is generally tolerable is that interactions involving three or more factors (*higher order interactions*), are almost always less important than main or two-factor interactions.   Thus, by properly designing a screening experiment such that each main effect (and if possible, each two-factor effect) is confounded only with effects due to higher order interactions (i.e. by being careful that two main effects do not confound each other), it is still highly probable that the most important main and two-factor interaction effects will be found.

The following is a rough guideline for obtaining the most information from the least number of screening experiment runs, while paying the least penalty (in the form of losing key information from confounding effects): a) For screening experiments containing 5 factors, use a 16 run experiment (a half-factorial of a 5-factor full-factorial experiment), and take the product of the coded values (+ and -) of the first four factors (for the first 16 runs of a 5-factor full-factorial table), and assign the resulting product level to the fifth factor (i.e. this product value will be either "+" or "-").   Using this design, the main effect of the fifth factor would be the same as the four-way interaction, and thus the main effects are confounded with the four-way interaction.   In addition, the two-way interactions will be confounded with three-way interactions; b) For screening experiments of 6 to 8 factors, a 16 run experiment can still be used in which the main effects are all confounded with third and higher order interactions, and the two-factor interactions are likewise confounded only with such higher order interactions; and c) for screening experiments with 9-12 factors, a 20 run Plackett-Burman design is recommended.   In such experiments, it can only be guaranteed that the main effects are not confounded with each other in a single EFFECT value, but they may be confounded with two-factor interactions.   Nevertheless, for an experiment with such a large number of factors, this is generally the best approach for identifying the most significant factors.   Table 11 gives a 20-run Plackett-Burman table for setting up such experiments.   Consult such references as 1, 2, or 8 for more information on the design of Plackett-Burman type screening experiments.   Results of screening experiments are analyzed with computational methods akin to those used for full-factorial experiments.

Table 11.   TWENTY-RUN PLACKETT-BURMAN DESIGN[8]

Run	Mean	x_1	x_2	x_3	x_4	x_5	x_6	x_7	x_8	x_9	x_{10}	x_{11}	x_{12}	x_{13}	x_{14}	x_{15}	x_{16}	x_{17}	x_{18}	x_{19}
1	+	+	+	-	-	+	+	+	+	-	+	-	+	-	-	-	-	+	+	-
2	+	+	-	-	+	+	+	+	-	+	-	+	-	-	-	-	+	+	-	+
3	+	-	-	+	+	+	+	-	+	-	+	-	-	-	-	+	+	-	+	+
4	+	-	+	+	+	+	-	+	-	+	-	-	-	-	+	+	-	+	+	-
5	+	+	+	+	+	-	+	-	+	-	-	-	-	+	+	-	+	+	-	-
6	+	+	+	+	-	+	-	+	-	-	-	-	+	+	-	+	+	-	-	+
7	+	+	+	-	+	-	+	-	-	-	-	+	+	-	+	+	-	-	+	+
8	+	+	-	+	-	+	-	-	-	-	+	+	-	+	+	-	-	+	+	+
9	+	-	+	-	+	+	-	-	-	-	+	+	-	+	+	-	-	+	+	+
10	+	+	-	+	-	-	-	-	+	+	-	+	+	-	-	+	+	+	+	-
11	+	-	+	-	-	-	-	+	+	-	+	+	-	-	+	+	+	+	-	+
12	+	+	-	+	-	-	+	+	-	+	+	-	-	+	+	+	+	-	+	-
13	+	-	-	-	-	+	+	-	+	+	-	-	+	+	+	+	-	+	-	+
14	+	-	-	-	+	+	-	+	+	-	-	+	+	+	+	-	+	-	+	-
15	+	-	-	+	+	-	+	+	-	-	+	+	+	+	-	+	-	+	-	-
16	+	-	+	+	-	+	+	-	-	+	+	+	+	-	+	-	+	-	-	-
17	+	+	+	-	+	+	-	-	+	+	+	+	-	+	-	+	-	-	-	-
18	+	+	-	+	+	-	-	+	+	+	+	-	+	-	+	-	+	-	-	-
19	+	-	+	+	-	-	+	+	+	+	-	+	-	+	-	-	-	-	+	+
20	+	-	-	-	-	-	-	-	-	-	-	-	-	-	-	-	-	-	-	-

RESPONSE SURFACES

If two factors can be considered as defining a two-dimensional (planar) horizontal surface, then the response can be plotted as a vertical distance above this plane. As various combinations

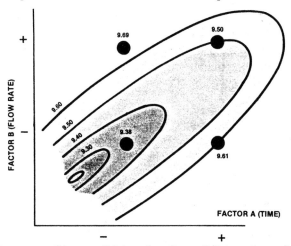

Fig. 9 Contour plot representing possible explanation of interaction of variables in an experiment designed to determine conditions of minimum linewidth variation in a lithography process[10]. Reprinted with permission of Microelectronics Manufacturing and Testing.

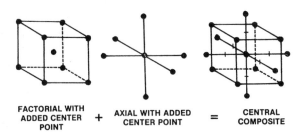

FACTORIAL WITH ADDED CENTER POINT + AXIAL WITH ADDED CENTER POINT = CENTRAL COMPOSITE

Fig. 10 Central composite design in three factors[10]. Reprinted with permission of Microelectronics Manufacturing and Testing.

of factors are tried, their response points would form a surface above the plane. Such a hypothetical surface for an example (in which a lithographic process that yields minimum linewidth variation is being developed) is shown in Fig. 9.

The response surface of Fig. 9 is a descending trough. If the surface is an accurate representation of the linewidth variation response, it shows a region of operation that would yield even lower linewidth variations than the factorial experiment discovered, but this minimum is outside of the factor ranges investigated by the experiment. The objective of the response surface approach is to look for such optimum points.

If such response surfaces were constructed solely on the basis of 2-levels for each factor, only simple planar surfaces would result, and no information on local optimum points would be found. Therefore at least three levels for each factor must be used for response surface construction. From the results of the experimental data, the surface is approximated by a quadratic function, the nature of the fitted surface is examined, and either an optimum point is located, or information about the direction that would move toward the point, is gathered.

Use of full-factorial experiments with p factors at 3 levels requires 3^p runs, and thus even if only a small number of factors are known to be significant, an excessively large number of runs may be necessary.

Two classes of designs, however, have been developed that allow a reduced number of runs

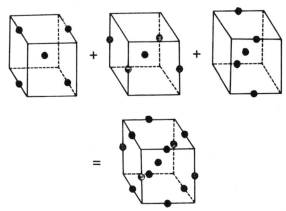

Fig. 11 Three-factor Box-Bhenken design that runs complete two-factorial using all possible combinations of two factors, while holding remaining factors at their center points[10]. Reprinted with permission of Microelectronics Manufacturing and Testing.

to be conducted while exploring response surface curvatures: 1) central composite designs; and 2) Box-Behnken designs.

In *central composite designs*, 2-level factorial experiments (including a center point for curvature determination) are run first.   If significant curvature is found, axial design points are run.   As shown in Fig. 10, *axial points* are outside of the factor space of the cube (but lie along axes that penetrate the center of the cube faces and the cube itself), together with an additional center point run.   Thus, for a 3-factor central composite design, 16 runs would be needed (9 runs for the 2-level full factorial /initial curvature stage, followed by six-axial point runs, and one replicate center run).   Note that a full-factorial, 3-level 3 factor experiment would require 27 runs.

The *Box-Behnken designs* consist of taking all possible two-factor combinations and running them as individual full-factorial experiments, while holding all the other factors at their center points.   As shown in Fig. 11, for a 3 factor problem only 12 runs plus the center runs need to be conducted.   Box-Behnken designs are simpler, but less flexible than central composite designs.   That is, central composite designs can be halted after the first stage if no curvature is found.   Box-Behnken designs must carry out the full number initially specified.

The results of the response surface experiments can be displayed on contour plots as shown in Fig. 9.   Such plots are valuable because they not only locate optimum factor settings, but in addition the relative proximity of the contour lines makes it easy to predict the rate at which the response will change when individual factors are varied.   Thus a measure of how sensitive a process may be to factor variations, can be obtained.

REFERENCES

1. G.E.P. Box, W.B.Hunter, and J. Hunter, *Statistics for Experimenters*, Wiley, New York, 1978.
2. O.L. Davies, Ed., *Design and Analysis of Industrial Experiments*, Hafner, New York, 1967.
3. C. Daniels, *Applications of Statistics to Industrial Experimentation*, Wiley-Interscience, New York, 1976.
4. W.J. Diamond, *Practical Experimental Design*, Lifetime Learning Publ., Belmont, CA (1981).
5. G.R. Bryce and D. Collette, "Optimizing Wafer Processing via Statistical Design", *Semiconductor International*, Mar., '84, p. 71.
6. P.W. Bohn and R.C. Manz, "A Multiresponse Factorial Study of Reactor Parameters in PECVD Growth of Amorphous Silicon Nitride", *J. Electrochem. Soc.*, Aug. '85, p. 1981.
7. J.P. Robic, S.E. Knight, and W.A. Straub, "Mid-UV Lithography with Positive Resist - A Multifactorial Tool and Process Evaluation", *Solid State Technol.*, May '83, p. 147.
8. R.L. Plackett and J.P. Burman, "The Design of Optimum Multifactorial Experiments", *Biometrika* 33, (1946), p. 305-325.
9. R.H. Meyers, *Response Surface Methodology*, Edwards Bros., Inc., Ann Arbor, MI (1976).
10. G.R. Bryce and D. Collette, "Statistically Designed Experiments For Process Optmization", *Microelectronics Manufacturing and Testing*, Oct. '84, p.25.

PROBLEMS

1.   List three of the drawbacks of performing experiments by the "change one variable at a time, and keep all others constant" method.

2.   Describe the kind of information obtained from the following types of experiments: a) screening experiments; b) 2 -level full-factorial experiments; and c) response surface generating experiments.

3.   Given a set of responses **Y** (2 each) from 6 data points as follows:

x	Y	x	Y
2	4.3	8	7.9
2	4.6	8	8.4
4	6.1	10	9.3
4	5.8	10	9.0
6	7.3	12	9.9
6	7.2	12	10.3

Calculate the pooled variance for these measured data, and the values between which the population mean will be found for the value of **Y** when x = 8, with a confidence level of: a) 95%; and b) 99%.

4. Describe the difference between the *t-distribution* and the *normal distribution*.

5. Explain the meaning of therm *confounding* and explain why this effect makes the results obtained from fractional factorial experiments more ambiguous than that from full-factorial experiments.

6. Give an example from one of the processes described earlier in the book of an interaction effect among factors.

7. Why is it recommended that a test for *curvature* be included in 2-level full-factorial experiments?

8. A 2-level full factorial experiment with three factors, A, B, and C is run by conducting 20 runs (4 for curvature).

EXAMPLE FACTORIAL EXPERIMENT - SUMMARY OF DATA

RUN	Y (Replicate 1)	Y (Replicate 2)
1	37.3	33.0
2	0.4	14.6
3	8.1	5.1
4	21.0	14.6
5	33.4	39.2
6	12.1	10.5
7	1.2	2.7
8	18.7	19.3
9	6.4	3.2
	11.6	2.8

Calculate : a) number of degrees of freedom of this experiment; b) the value of ρ specified by the problem; c) the pooled standard deviation of a single response measurement; d) the factor effects for each column of the computation table; e) the minimum significant factor effect; f) minimum significant curvature; g) Identify the probable significant factor effects; h) Is curvature significant?

APPENDIX   1

PROPERTIES OF SILICON at 300° K

Quantity	Silicon
Atomic weight	28.09
Atoms, total (cm^{-3})	4.995×10^{22}
Crystal structure	Diamond
Density (g /cm^3)	2.33 (solid)
	2.53 (liquid, 1421°C)
Density of Surface atoms (cm^{-2})	
(100)	6.78×10^{14}
(110)	9.59×10^{14}
(111)	7.83×10^{14}
Dielectric constant	11.8
Effective density of states	
in conduction band, N_C (cm^{-3})	3.22×10^{19}
Effective density of states	
in valence band, N_v, (cm^{-3})	1.83×10^{19}
Energy gap (eV)	1.12
Index of refraction	3.42
Intrinsic carrier concentration (cm^{-3})	1.38×10^{10}
Intrinsic DeBye length (μm)	24
Intrinsic resistivity (Ω-cm)	2.3×10^5
Lattice constant (Å)	5.43
Linear coefficient of thermal	
expansion, $\Delta L /L \, \Delta T$ (°C^{-1})	2.6×10^{-6}
Melting point (°C)	1421
Mobility (drift) (cm^2 /V s)	
μ_n (electrons)	1500
μ_p (holes)	475
Specific heat (J /g-°C)	0.7
Thermal conductivity (W /cm-°C)	
Solid	1.5
Liquid	4.3
Youngs modulus (kg / mm^2)	10,890

APPENDIX 2

FUNDAMENTAL PHYSICAL CONSTANTS

CONSTANT	SYMBOL	SI - UNIT	DEFINITION
Angstrom unit	Å		$1\text{Å} = 10^{-1}\text{ nm} = 10^{-4}\,\mu\text{m} = 10^{-10}\text{m}$
Atomic mass	$m_u = 1\text{ u}$	kg	$m_u = 1.6605 \times 10^{-27}\text{ kg}$
Avogadro's constant	N_A	mol^{-1}	$N_A = 6.022 \times 10^{23}\text{ mol}^{-1}$
Bohr radius	a_B		0.529 Å
Boltzmann constant	k	$\text{J }^{\circ}\text{K}^{-1}$	$k = 1.38066 \times 10^{-23}\text{ J}/^{\circ}\text{K}$
Electron rest mass	m_e	kg	$m_e = 9.11 \times 10^{-31}\text{ kg}$
Elementary charge	q	C	$1.60218 \times 10^{-19}\text{ C}$
Electron volt	eV		$1\text{ eV} = 1.60218 \times 10^{-19}\text{ J}$
			$= 23.053\text{ kcal /mole}$
Molar gas constant	R	$\text{J mol}^{-1}{}^{\circ}\text{K}^{-1}$	$R = 1.987\text{ cal /mole - }^{\circ}\text{K}$
Permeability in vacuum	ε_o		$\varepsilon_o = 8.854 \times 10^{-14}\text{ F /cm}$
Planck constant	h	J s	$h = 6.626 \times 10^{-34}\text{ J-s.}$
Proton rest mass	M_p	kg	$1.672 \times 10^{-27}\text{ kg}$
Speed of light in vacuum	c	m s^{-1}	$2.9979 \times 10^8\text{ m /s}$
Standard (atmospheric) pressure		$\text{N m}^{-2}, \text{Pa}$	$1.013 \times 10^5\text{ Pa}$
Thermal voltage at 300° K	kT /q	V	0.0259 V

APPENDIX 3

ARRHENIUS BEHAVIOR

Some of the reactions that take place during the course of carrying out VLSI fabrication processes, occur at rates which depend on temperature.   In many cases it is found that the reaction rate dependence is well predicted by a relationship known as the *Arrhenius equation*. This equation was formulated by the Englishman J. Hood, and is named after the Swedish chemist Svante Arrhenius.   The equation is written as:

$$R(T) = R_0 e^{(E_a / k T)} \qquad (1)$$

In this equation, R (T) is the *rate constant* at temperature T (°K), R_0 is the *frequency factor* (also referred to in some cases as the pre-exponential factor), k is Boltzmanns constant $(8.6 \times 10^{-5}$ eV /°K), and E_a is the *activation energy* (expressed in units of eV).   The activation energy, E_a, and the frequency factor, R_0, are also termed *Arrhenius parameters*.   The values of R_0 and E_a can often be used to gain insight into the physical or chemical mechanisms that impact the reaction kinetics of the process being investigated.   Several examples of Arrhenius behavior are cited in the text (see Index).

If a reaction rate depends on temperature, it is necessary to determine if, in fact, it also obeys the Arrhenius relationship.   If the reaction rate does indeed exhibit Arrhenius behavior, it is valid to extract values of E_a and R_0 from measured reaction rate data.   To verify that the reaction rates follow the Arrhenius relationship, the rate constant for a particular reaction is measured at several different temperatures.   The measured values are plotted on graph paper on which one axis plots the log of the rate constant, and the other axis plots the reciprocal of the temperature [1 /T (°K)].   If the measured rate constant values fall along a straight line on this

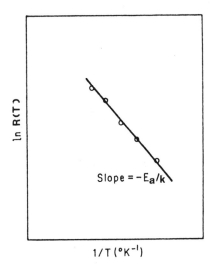

Fig. 1   Arrhenius plot.

plot, it can be concluded that the reaction obeys the Arrhenius relationship. The plot of the rate constants on this type of graph paper is termed an *Arrhenius plot* (Fig. 1). If the plot does not yield a straight line, this is evidence that a more complex reaction is taking place than that described by the Arrhenius equation.

The values of E_a and R_o can be determined in one of two ways: a) graphically, or b) analytically. In the *graphical technique*, the slope of the plotted data, S, represents $S = -E_a/k$. The value of R_o is then found by using this value of E_a, together with a value of R (T) in the Arrhenius equation.

In the *analytical technique*, the rate constants at two temperatures are used to create two simultaneous equations:

$$R_1 (T) = R_o \, e^{(E_a/kT_1)} \qquad (2)$$

and

$$R_2 (T) = R_o \, e^{(E_a/kT_2)} \qquad (3)$$

From the above equations, E_a is found according to:

$$E_a = \frac{k \ln\left(\dfrac{R_2 (T)}{R_1(T)}\right)}{(\dfrac{1}{T_1}) - (\dfrac{1}{T_2})} \qquad (4)$$

and R_o is computed in the same manner as above, once the value of E_a has been determined.

INDEX

NOTE: FOR MAJOR SUBJECT HEADINGS, ALSO CHECK THE TABLE OF CONTENTS. THE LISTINGS ARE NOT ALWAYS DUPLICATED IN THE INDEX.